# Lentiviruses and Macrophages

## Molecular and Cellular Interactions

Edited by

Moira Desport

School of Veterinary and Biomedical Sciences
Murdoch University
Department of Microbiology & Immunobiology
Perth, WA
Australia

Caister Academic Press

Caister Academic Press
Norfolk, UK

www.caister.com

British Library Cataloguing-in-Publication Data
A catalogue record for this book is available from the British Library

ISBN: 978-1-904455-60-8



Cover image taken from Fig. 11.2

Printed and bound in Great Britain

# Contents

# Contributors

**Jenny L Anderson**
Centre For Virology
Macfarlane Burnet Institute for Medical Research and Public Health
Melbourne, VIC
Australia

jla@burnet.edu.au

**Cristian Apetrei**
Center for Vaccine Research
Department of Microbiology and Molecular Genetics
University of Pittsburgh School of Medicine
Pittsburgh, PA
USA

apetreic@pitt.edu

**Giuseppe Bertoni**
Institut für Veterinär-Virologie
Universität Bern
Bern
Switzerland

bertoni@ivv.unibe.ch

**Barbara Blacklaws**
University of Cambridge
Department of Veterinary Medicine
Madingley Road
Cambridge
CB3 0ES
United Kingdom

bab2@cam.ac.uk

**Kelly Cheney**
Centre for Immunology and Infectious Disease
Blizard Institute for Cell and Molecular Science
Barts and the London School of Medicine and Dentistry
London
UK

k.m.cheney@qmul.ac.uk

**Melissa J. Churchill**
Centre for Virology
Macfarlane Burnet Institute for Medical Research and Public Health
Melbourne, VIC
Australia

churchil@burnet.edu.au

**Jodi K. Craigo**
Center for Vaccine Research
Department of Microbiology and Molecular Genetics
University of Pittsburgh School of Medicine
Pittsburgh, PA
USA

craigoj@pitt.edu

**Moira Desport**
School of Veterinary and Biomedical Sciences
Murdoch University
Department of Microbiology and Immunobiology
Perth, WA
Australia

M.Desport@murdoch.edu.au

**Thaidra Gaufin**
Division of Microbiology
Tulane National Primate Research Center
Covington, LA
USA

tgaufin@tulane.edu

**Howard E. Gendelman**
Department of Pharmacology and Experimental Neurosciences
University of Nebraska
Nebraska Medical Center
Omaha, NE
USA

hegendel@unmc.edu

**Paul R. Gorry**
Centre for Virology
Macfarlane Burnet Institute for Medical Research and Public Health
Melbourne, VIC
Australia

gorry@burnet.edu.au

**Margaret J. Hosie**
Retrovirus Research Laboratory
Institute of Comparative Medicine
Faculty of Veterinary Medicine
University of Glasgow
Glasgow
UK

m.hosie@vet.gla.ac.uk

**Tatsuhiko Igarashi**
Laboratory of Primate Model
Experimental Research Center for Infectious Diseases
Institute for Virus Research
Kyoto University
Kyoto
Japan

tigarash@virus.kyoto-u.ac.jp

**Áine McKnight**
Centre for Immunology and Infectious Disease
Blizard Institute for Cell and Molecular Science
Barts and the London School of Medicine and Dentistry
London
UK

a.mcknight@qmul.ac.uk

**Sandy McLachlan**
Dept of Veterinary and Biomedical Science
Murdoch University
Perth, WA
Australia

sandy@nzvp.co.nz

**Wendy Maury**
Department of Microbiology
College of Medicine
University of Iowa
Iowa City, IA
USA

wendy-maury@uiowa.edu

**Ronald C. Montelaro**
Center for Vaccine Research
Department of Microbiology and Molecular Genetics
University of Pittsburgh School of Medicine
Pittsburgh, PA
USA

rmont@cvr.pitt.edu

**J. Lindsay Oaks**
Department of Veterinary Microbiology and Pathology
College of Veterinary Medicine
Washington State University
Pullman, WA
USA

loaks@vetmed.wsu.edu

**Ivona Pandrea**
Center for Vaccine Research
Department of Pathology
University of Pittsburgh School of Medicine
Pittsburgh, PA
USA

pandrea@pitt.edu

**Mary Poss**
Center for Infectious Disease Dynamics
The Pennsylvania State University
University Park, PA
USA

mposs@bx.psu.edu

**Ian Ross**
Institute for Molecular Bioscience
The University of Queensland
Brisbane, QLD
Australia

i.ross@imb.uq.edu.au

**Jasminka Sterjovski**
Centre for Virology
Macfarlane Burnet Institute for Medical Research and Public Health
Melbourne, VIC
Australia

jasminka@burnet.edu.au

**Gilda Tachedjian**
Centre For Virology
Macfarlane Burnet Institute for Medical Research and Public Health
Melbourne, VIC
Australia

gildat@burnet.edu.au

**Sue VandeWoude**
Department of Microbiology
Immunology and Pathology
Retrovirus and Prion Research Lab
College of Veterinary Medicine and Biomedical Sciences
Colorado State University
Fort Collins, CO
USA

suev@lamar.colostate.edu

**Brian J. Willett**
Retrovirus Research Laboratory
Institute of Comparative Medicine
Faculty of Veterinary Medicine
University of Glasgow
Glasgow
UK

b.willett@vet.gla.ac.uk

# Other Books of Interest

# Preface

This unique offering provides both the novice and experienced reader with a 'state of the art' in-depth insight into the biology and pathogenesis of lentiviral induced disease that occurs following infection of mononuclear phagocytes (MP; blood-borne macrophages, tissue histiocytes, dendritic cells, and microglia). This book, as no other, is the sole and most comprehensive scholarly writing to date covering the pivotal role played by these MP as reservoirs and vehicles for pathogen dissemination and in disease induction. For chronic lentiviral infections MP remain an enigma. On the one hand they are among the first cells contacted by virus and, despite a virtual armada of immunological tools, still serve as means to both spread and contain infection. Virus particles can simultaneously assemble and hide in intracellular compartments, largely free from immune attack. Interestingly, the MP are not destroyed by the virus and throughout infection they still contribute to host immunity while at the same time perpetuating lentiviral dissemination. Infected MP are readily observed in lymph nodes and organs such as the lung and brain, where they produce cytotoxic mediators that contribute to the development of disease. For human immunodeficiency virus (HIV) infection functional impairment of infected MP likely accelerates immune deficiency. Thus, the questions most asked are how can a cell that possesses so many intrinsic defence mechanisms harbour such viral pathogens for prolonged time periods? How can MP serve both as sentries and vehicles for disseminating infection and inducing disease? Indeed, in regards to biology, for both ontogeny and phylogeny, MP are the most primitive sensors of tissue injury and as such serve to clear debris and simultaneously protect the host's homeostatic environment. Their roles serve the host in non-specific defence (innate immunity) and in initiating cell-specific protection (adaptive immunity). A consistent evolutionary role resides in the ability of MP to engulf, digest and destroy cell and tissue debris through phagolysosomal fusion. Interestingly, MP can affect neighbouring lymphocytes and other immunocytes to perform similar functions, albeit by divergent mechanisms. Control of viral growth occurs together with the MP's notable possession of a vast repertoire of immune secretory factors that include pro-inflammatory cytokines, chemokines, arachidonic acid and its metabolites, platelet activating factor, nitric oxide, quinolinic acid, amongst others. These serve to regulate immune defence, cell mobility, antigen presentation, immune activation, and cell differentiation. During disease such secretions are induced by infection and affect inflammatory processes that speed cell and tissue injury leading to clinical symptoms and morbidities. This, in the case of common lentiviral tissue injuries, leads to substantive lung, brain, blood, and joint diseases. Clearly, the role played by MP in the pathogenesis of lentiviral infections is seemingly complex and quite multifaceted. Viral spread from MP to MP and from MP to T cells and across cell and tissue barriers is equally vast and complicated.

Historically, how this primary mover for innate immunity serves as a source for restricted and productive lentiviral replication seems almost illogical. First, with an armada of microbial clearance activities that include phagocytosis, intracellular killing, secretion of bioactive antiviral factors such as interferons and biodegradable enzymes, it is of great interest that lentiviruses can enter these cells and replicate over prolonged time periods measured in months or even years. How virus evades innate immune responses, nevertheless, remains poorly understood, even following nearly 30 years of study. Second is the feature of transcriptional control for viral persistence. A host of factors control the viral life cycle and are regulated through the MP external environment which allows virus to continue to replicate in the face of often robust humoral and cellular immune responses and more recently during potent antiretroviral therapies for HIV. In this scenario, virus buds into endocytic compartments with limited cell surface expression of viral proteins and in this context parallels the ancient Trojan horse mythology of being protected inside a sheath of secrecy. It was such a stratagem that Odysseus employed which allowed the Greeks to enter the city of Troy and end a conflict which had lasted 10 years. The Greeks built a huge wooden figure of a horse and left it as a parting gift for Athena as they pretended to sail away from Troy. The jubilant Trojans pulled the wheeled horse into the city unaware of the armed Greek soldiers hiding in the Trojan horse's belly. The Greek soldiers poured out of a trap door in the horse's belly and quickly killed the Trojans, setting fire to the city and thereby winning the war. For lentiviruses, it is analogous to their strategy of entering the brain and overcoming the restrictions imposed by a seemingly impermeable blood- brain barrier through the establishment of a chemokine gradient. Certainly, lentiviruses use MP as a ploy to enter the host and evade immune surveillance thereby escaping detection in a similar manner to the Greek soldiers in the Trojan Horse. Third, there is a curious balance of host defence and dissemination of viral infection. Dendritic cells and tissue macrophages are readily infected in body fluids and are the likely cellular source of viral spread both from animal to animal or person to person through seminal and blood macrophages and/or monocytes and dendritic cells and carriage of virus throughout the body. Fourth are the severe and lasting functional consequences of viral infection and immune activation as it occurs during disease. This is typified by secretion of inflammatory toxins manufactured and released by MP that cause tissue injuries commonly in the lung, the joints, and the central nervous system. The latter perhaps has received the most intense study in regards to relationships between MP function, neuroinflammation, and neurodegeneration. The field spawned biomarker discovery in proteomics and metabolomics and adjunctive therapeutic developments aimed at better understanding and combating disordered MP function during lentiviral disease. Fifth, intra- and inter-cellular MP mechanisms govern the viral life cycle. How lentiviruses hijack subcellular organelles and the cytoskeletal machinery to complete their life cycle is a fascinating area for research activities. Added to such complexities are the relationships between innate and adaptive immunity including spread of virus. Even more importantly, perhaps, is how regulatory and effector T cells amongst other cells that come in contact with macrophages affect macrophage function in disease and either exacerbate or control the tempo of disease.

This book of macrophages and lentiviruses goes a very long way in directly addressing all of the issues that confront an understanding of the interactions between virus and MP. Indeed, lentiviral tissue injuries are linked to sites of active MP infection as described by Ian Ross in his Chapter on a Bird's eye view of macrophage biology. Common features for lentiviral pathogenesis reside in the tropism for MPs and the ability of lentiviruses to evade immune surveillance are discussed by Jodi Craigo and Ronald Montelaro in their chapter on lentivirus tropism and disease. Diseases associated with lentiviral infections are often benign but can also lead to severe debilitation and death. Paul R. Gorry and colleagues cover the important topic of macrophages in HIV-1 infection. This chapter examines the relationship between viral coreceptor usage, cell tropism, and pathogenicity. Intracellular defence systems commonly termed 'restriction factors' have followed vthe discovery of APOBEC3 and TRIM5 cellular proteins as intracellular mechanisms for lentiviral restriction and are developed by Jenny Anderson and Gilda Tachedjian. Understanding how these and other intracellular restriction factors block retroviruses is providing new insights into the barriers of retroviral replication and the cross-species transmission of lentiviruses. Kelly Cheney and Áine McKnight discuss HIV-2 tropism and disease in regards to how the virus can achieve limited pathogenicity with variants that show facilitated transmission. This is thought to represent a successful survival strategy through the virus' inability to kill its host. The adoption of 'accessory genes' by HIV-2 and its more promiscuous pattern of coreceptor usage may allow avoidance of innate restriction. The simian-HIV or SHIV and simian immunodeficiency virus models of HIV disease are developed by Tatsuhiko Igarashi and Thaidra Gaufin, Ivona Pandrea and Cristian Apetrei. These authors discuss macrophage tropism as it relates to disease pathogenesis and the molecular components of the viral *env* gene responsible for tropism. Host factors and disease progression are linked to the molecular properties of the virus and are discussed. Most macrophage-tropic viruses are attenuated when inoculated into immune-competent monkeys. Brian Willett and Margaret Hosie provide an update on the host and viral determinants of feline immunodeficiency virus cell tropism and pathogenicity and the future prospects for a vaccine. Wendy Maury and J. Lindsay Oaks discuss equine infectious anaemia virus (EIAV) and its pathogenesis. Much of the understanding of lentiviral infection of macrophages comes from EIAV studies that have provided insights into molecular regulation of all lentiviruses. As a consequence of this targeted tropism, EIAV causes an acute and sometimes fulminant disease associated with high-titered viraemia with no associated immunodeficiency. Guiseppe Bertoni and Barbara Blacklaws write the important chapter on visna-maedi and the caprine arthritis encephalitis viruses. MPs are the main target cells of both these small ruminant lentiviruses. Monocytes carrying the lentiviral provirus in their genome show little or no viral transcription. These latently infected cells are capable of spreading the virus to different organs, while eluding the host immune response. These viruses manipulate host antiviral mechanisms directly influencing the efficacy of vaccination strategies as documented by the paradoxical effects induced by experimental vaccines on viral load and pathological manifestations in vaccinated and challenged animals. The works of Moira Desport and Sandy McLachlan cover the bovine lentivirus infections and herein the similarities and differences between genetically and antigenically closely related viruses and between other lentiviruses are reviewed in regards to cell tropism, pathogenesis and genetic composition. Finally, lentivirus coinfections and superinfections are reviewed by Sue VandeWoude and Mary Poss. Evaluation of the incidence and biological behaviour of experimental superinfections, either with host-adapted or non-host-adapted strains, provides insights into potential pitfalls and additional strategies to consider during lentiviral vaccine development. Analysis of lentiviral genome recombination events as a sequel of superinfection enhances understanding of mechanisms of lentiviral evolution and virus–host adaptations. This chapter provides a review and comparison of lentiviral superinfections and resultant recombination outcomes in different hosts, including primate, felidae, and ruminant species, highlighting what these studies elucidate in terms of lentiviral pathogenesis, host immune and viral genomic responses to superinfections.

The book, edited by Dr Moira Desport, presents a complete and comprehensive assessment of the basic concepts, current research, and cutting-edge investigations of a broad number of experts. It is an in-depth resource that is clear and well organized and includes all pertinent concepts related to lentiviral pathogenesis in the context of MP infections. It will certainly prove to be a vital resource now and for the future.

Howard E. Gendelman, M.D.
Larson Distinguished Professor of Medicine and Infectious Diseases
Chairman, Department of Pharmacology and Experimental Neurosciences
University of Nebraska Medical Centre

# Lentivirus Tropism and Disease

1

Jodi K. Craigo and Ronald C. Montelaro


## Abstract

Lentiviruses are among the most intensely and extensively studied groups of viruses. They are found worldwide and infect a broad array of animal species. Historically, lentiviruses have been investigated longer than any other virus group. The first viral aetiology ascribed to an animal disease was a lentivirus. The diseases associated with lentiviral infections range from benign and subclinical to severely debilitating and lethal. The diverse group of viruses that compose the lentiviruses have many common and distinctive features. Among the common features is tropism for cells from the monocyte/macrophage lineage. Infection of macrophage affords this assorted group of viruses many evolutionary advantages, including a potential hiding place from the infected host's immune system.


## Introduction to lentiviruses

The terms AIDS (acquired immunodeficiency syndrome) and HIV (human immunodeficiency virus) are acronyms that have become quite familiar in popular society. Conversely, the group of viruses to which HIV belongs, the lentiviruses, are less well known. Lentiviruses are a unique group of widely disseminated viruses from the larger family of viruses known commonly as the retroviruses, formally designated Retroviridae. All retroviruses share structural and genetic properties (Fig. 1.1) including the requirement of synthesizing DNA from their RNA genome during their replication cycle. Specifically, the members of the lentivirus family are non-oncogenic, enveloped viruses, 100–130 nm in size, with a cone-shaped core that contains the bicistronic viral RNA genome. While much has been widely reported on HIV, the most recognized member of the lentivirus family, the animal viruses of this family are not as widely identified. This chapter will present an overview of the unique members and portray in detail the fundamentals of this family of dynamic viruses.

## History

While the nomenclature did not exist at the turn of the 20th century, lentiviruses themselves have been widely studied since the early 1900's (summarized in Table 1.1). Initially important internationally in the agricultural field of study, in the mid 1980s the identification of HIV as the causative pathogen of AIDS and the revelation that HIV was a lentivirus propelled this group of RNA viruses onto the leading edge of infectious disease research. Particularly, HIV-1 was first identified in 1983 as the responsible agent for an immunosuppressive syndrome that led to opportunistic infections and eventual death in homosexual men (Barre-Sinoussi *et al.*, 1983; Gallo *et al.*, 1983). When the disease first emerged in San Francisco (California) it was referred to as the 'gay-related immune deficiency syndrome' (GRID) before being re-named AIDS in 1981 (Barker *et al.*, 1995). After the discovery of the AIDS-associated retrovirus (ARV), it became rapidly apparent that HIV-1 was morphologically and serologically related to ungulate lentiviruses that had been described much earlier (Casey *et al.*, 1985; Gonda *et al.*, 1985, 1986; Montelaro *et al.*, 1988). During the past two decades since the

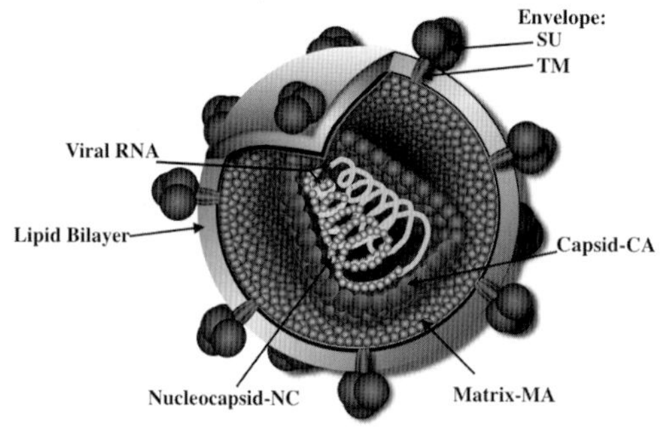


**Figure 1.1** Lentivirus particle. Schematic representation of a generic lentiviral virion.

discovery of HIV-1, research on the animal lentiviruses has increased markedly in an attempt to model the viral replication dynamics and disease processes associated with HIV-1/AIDS.

The original classification of these unique viruses in the family *Retroviridae* as 'slow' or lentiviruses (from Latin, *lenti*, meaning slow) arose in the mid-1900s from classifications of visna–maedi viruses (VMVs). In contrast to many viruses characterized at that time, it takes months or years before VMV result in disease (Marsh, 1923; Sigurdsson, 1954; Palsson, 1976). Icelandic scientists identified VMV (*visna*, Icelandic for wasting; *maedi*, Icelandic for dyspnoea) following an epidemic in Iceland in 1933 after the introduction of German Karakuhl sheep to the region (Sigurdsson, 1954; Sigurdsson *et al.*, 1957). VMV was subsequently demonstrated to exist worldwide and replicate predominantly in monocyte/macrophage cells producing two separate clinical manifestations in either the central nervous or respiratory systems (Gorrell *et al.*, 1992; Brodie *et al.*, 1995; Zink *et al.*, 2006). Later, caprine arthritis encephalitis virus (CAEV) was isolated in 1974 as the cause of paralysis in juvenile goats and identified as a member of the lentivirus family (Cork *et al.*, 1974; Clements *et al.*, 1980b; Crawford *et al.*, 1980; Narayan *et al.*, 1980; Adams *et al.*, 1984).

Historically, CAEV and VMV infections were believed to be restricted to goats and sheep, respectively. However, more recent studies of the genomic sequences of lentiviral strains naturally infecting sheep and goats revealed a number of viral genomic quasispecies common to both animal species. These data indicate that the ovine and caprine lentiviruses actually comprise a highly related family of lentiviruses with a spectrum of tropisms relative to a preference for infection of sheep and/or goats (Narayan *et al.*, 1993; Leroux *et al.*, 1995, 1997a; Shah *et al.*, 2004b; Grego *et al.*, 2005; Pisoni *et al.*, 2005, 2006). As a consequence of these genomic studies, CAEV and VMV have been re-assigned to a single lentivirus species, designated small ruminant lentiviruses (SRLV). The SRLV species is separated into groups A–E, with groups A and B being further divided into subgroups (Zanoni, 1998; Shah *et al.*, 2004b; Reina *et al.*, 2006; Grego *et al.*, 2007). For the purposes of this chapter, the term SRLV will be used when discussing general features of the ovine and caprine lentiviruses; when discussing (in a historical context) the pioneering studies with these lentiviruses, the particular virus strain (CAEV or VMV) used in the cited studies will be indicated for easy reference to published studies. However, it can be assumed that the mechanisms of persistence and pathogenesis described for VMV or CAEV in particular can be extrapolated to the SRLV family in general.

Equine infectious anaemia was actually the first animal disease determined (in 1843) to be caused by an infectious agent, and the first infectious disease of animals to be assigned (in 1904) a viral aetiology, equine infectious anaemia virus (EIAV) (Lignee, 1843; Vallee, 1904). Owing to the difficulties of culturing EIAV *in vitro*, characterizations of the viral pathogenesis, immunology, and molecular biology lagged behind the other lentiviruses (Montelaro *et al.*, 1993; Craigo and Montelaro, 2008). EIAV was not characterized as a retrovirus and a lentivirus replicating in monocyte/macrophage cells until the late 1970s (Henson and McGuire, 1974; Charman *et al.*, 1976; McConnel *et al.*, 1977; Weiland *et al.*, 1977; Gonda *et al.*, 1978; McGuire and Crawford, 1979). The next virus to be revealed in the genealogy of the lentivirus family was bovine immunodeficiency virus (BIV). BIV was first isolated from a dairy cow in Louisiana in 1969 (Malmquist *et al.*, 1969) and identified as 'bovine visna-like virus' in 1972 (Van der Maaten *et al.*, 1972; Gonda *et al.*, 1987). BIV has a worldwide prevalence in cattle causing an array of symptoms including lymphadenopathy and wasting (McNab *et al.*, 1994; Walder *et al.*, 1995). However, for the

**Table 1.1** Lentivirus fundamentals

| Virus | Identification date | Natural host | Target cells | Receptor | Coreceptor | Clinical disease |
|---|---|---|---|---|---|---|
| EIAV | 1904 | Equine (horses) | Monocyte/macrophage<br>Dendritic cells | ELR1 | ND | Anaemia/encephalitis/wasting |
| *Small ruminant lentiviruses (SRLV)* | | | | | | |
| VMV | 1954 | Ovine/caprine (sheep/goat) | Monocyte/macrophage<br>Dendritic cells | ND | ND | Pneumonia/mastitis<br>Encephalitis/wasting |
| CAEV | 1974 | Caprine/ovine (goat/sheep) | Monocyte/macrophage | ND | ND | Arthritis/mastitis/encephalitis/wasting |
| BIV | 1972 | Bovine (cattle) | T-lymphocytes ($CD4^+$ and $CD8^+$)<br>B-lymphocytes<br>Monocyte/macrophage | ND | ND | Lymphocytosis/mastitis/wasting |
| HIV-1 | 1983 | Human | T-lymphocytes ($CD4^+$)<br>Monocyte/macrophage<br>Dendritic cells | CD4 | CCR5, CXCR4 | Immunodeficiency/wasting |
| SIV | 1985 | Simian (monkey) | T-lymphocytes ($CD4^+$)<br>Monocyte/macrophage<br>Dendritic cells | CD4 | CCR5 | Immunodeficiency/wasting |
| HIV-2 | 1986 | Human | T-lymphocytes ($CD4^+$)<br>Monocyte/macrophage<br>Dendritic cells | CD4 | CCR5, CXCR4 | Immunodeficiency/wasting |
| FIV | 1987 | Feline (cats) | T-lymphocytes ($CD4^+$ and $CD8^+$)<br>B-lymphocytes<br>Monocyte/macrophage | CD134 | CXCR4 | Immunodeficiency/wasting |
| JDV | 1993 | Bovine (cattle – *Bos javanicus*) | *Monocyte/macrophage, *B-lymphocytes | ND | ND | Lymphadenopathy/lymphoproliferative disorder/wasting |

ND, not determined.

* Unconfirmed.

most part to date, BIV has not been associated with clinically relevant disease (Van der Maaten *et al.*, 1990; Whetstone *et al.*, 1990; St-Louis *et al.*, 2004).

The classification of HIV-1 in the mid-1980s as a lentivirus accelerated the field of lentiviral research, leading to the expedited discovery of additional human and animal lentiviruses. The first report of a non-human primate lentivirus, simian immunodeficiency virus (SIV) was in 1985 (Daniel *et al.*, 1985; Kanki *et al.*, 1985). Multiple primate research centres discovered SIV in Asian macaques that were potentially exposed to the lentivirus through contact with African monkeys by means of co-housing of primates or blood and tissue inoculations (Daniel *et al.*, 1985; Kanki *et al.*, 1985; Murphey-Corb *et al.*, 1986). SIV infects a multitude of non-human primate species, but in its natural host species, African monkeys, clinical manifestations of disease are not observed (Cichutek and Norley, 1993; Chakrabarti, 2004; Hirsch, 2004; Sharp *et al.*, 2005; Silvestri, 2005; VandeWoude and Apetrei, 2006; Silvestri *et al.*, 2007). A second human immunodeficiency virus (HIV-2) was discovered in West Africa in 1986. This strain of HIV was initially isolated from two Senegalese sex workers presenting with AIDS (Clavel *et al.*, 1986; Marlink *et al.*, 1994). HIV-2 is biologically similar to its predecessor HIV-1 but harbours a somewhat different antigenic and genomic composition. It has since been determined that HIV-1 and HIV-2 arose after separate cross-species transmission events from $SIV_{cpz}$ (from chimpanzee) and $SIV_{smm}$ (from sooty mangabeys), respectively (Hahn *et al.*, 2000).

The final two currently identified members of the lentivirus family are feline immunodeficiency virus (FIV) and Jembrana disease virus (JDV). FIV was originally isolated in 1987 from a colony of cats in Petaluma, California (Pedersen *et al.*, 1987). Initial news reports of the virus caused a great deal of concern amongst cat owners worried about transmission of 'feline AIDS' to humans. FIV is found in the saliva of cats and can be transmitted through biting, making male cat infections twice as likely due to territorial fighting (Pedersen *et al.*, 1989). The disease progression of FIV follows a pattern very similar to the primate lentiviruses. Since its identification, FIV has been found in domesticated and wild cats worldwide (Hosie *et al.*, 1989; Ishida *et al.*, 1989; Brown *et al.*, 1994; Bendinelli *et al.*, 1995; Carpenter *et al.*, 1996; Hartmann, 1998; Troyer *et al.*, 2005; Elder *et al.*, 2008). The newest lentivirus of the family, JDV, was first documented in the Jembrana district of Bali in Indonesia in 1964 in the domesticated form of the wild banteng (I.T. Budiarso, 1976; Wilcox, 1997). JDV was identified as a retrovirus and lentivirus in 1992 and 1993, respectively (Wilcox *et al.*, 1992; Kertayadnya *et al.*, 1993; Chadwick *et al.*, 1995). JDV is an atypical lentivirus related to BIV and appears to only cause disease in Bali cattle (*Bos javanicus*) (Chadwick *et al.*, 1995; Wilcox *et al.*, 1995; Desport *et al.*, 2007). The initial epidemic in 1964 spread over the entire island of Bali within 12 months and killed approximately 26,000 Bali cattle (Wilcox, 1997). The source of the outbreak was never determined. Interestingly, the acute stage of the disease resolves within 2–3 weeks and cattle that recover do not appear to suffer relapses (Hartaningsih *et al.*, 1993; Chadwick *et al.*, 1995; Wilcox *et al.*, 1995).

## Fundamental common features

While there are clear identifiable differences among the assorted viruses that compose the family, lentiviruses in general share a number of distinguishing features that include: infection of cells of the monocyte/macrophage lineage, extended incubation periods, viral persistence despite vigorous host immune responses, and a variable disease course that is usually progressively degenerative resulting in symptoms of wasting and commonly death. Review of the associated pathogenesis and disease inflicted on the hosts of the various lentiviruses separates the viruses into essentially two groups: those that cause immunodeficiency (HIV-1, HIV-2, SIV, FIV and potentially BIV) and those that do not (SRLV, EIAV, BIV, and JDV). Table 1.1 summarizes the target cells as well as the different clinical manifestations of disease for each member of the lentivirus family. The target cells of infection also provide the lentiviruses with a most powerful survival mechanism: latency. Lentiviruses utilize a latent state to disseminate throughout the body of the host as well as to avoid discovery and elimination of the virus population by the immune system. This chapter will review the fundamentals of the different members of the lentivirus family,

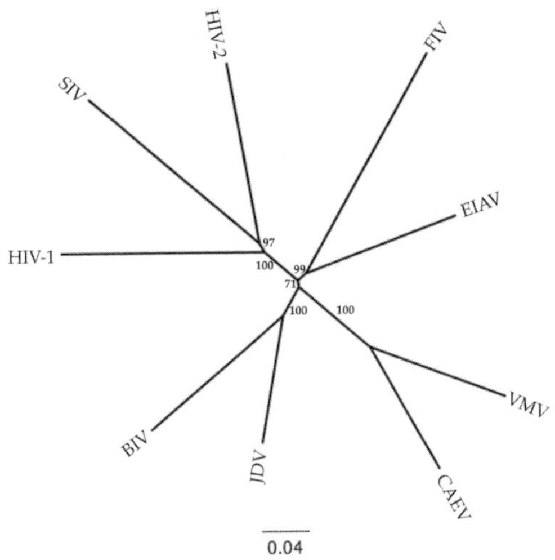


**Figure 1.2** Genetic relationship of members of the lentiviral family. A neighbour-joining, bootstrapped phylogenetic tree of the complete genomes of each identified member of the lentivirus family was constructed from aligned nucleotide sequences.

their associated pathogenesis, and the basic types of latency that have been described for each lentiviral group.

## Lentivirus genomes and proteins

Retroviruses in general have been defined as RNA viruses that reverse transcribe their genome into DNA to enable viral genome translation and ultimately virion production utilizing the host cell machinery. All lentiviruses share a common basic genomic structure that is distinct from other retroviruses. Fig. 1.2 displays the global (phylo) genetic relationship within the lentiviral family. Homology of genomic sequences between the ungulate lentiviruses and HIV-1 confirmed the original identification of HIV-1 as a lentivirus. Fig. 1.3 depicts the general genomic structure of all the lentiviral genomes while Table 1.2 summarizes the proteins encoded by these various genomic templates.

All lentiviral genomes contain the three canonical retroviral genes *gag*, *pol*, and *env* that encode virion core, polymerase, and envelope proteins, respectively. The *gag*-encoded core structural proteins of the virus are designated matrix (MA), capsid (CA) and nucleocapsid (NC). These proteins provide the architecture of the virion (Fig. 1.1). The fourth *gag* gene encoded by all lentiviruses except the SRLV is simply referred to by its size, ranging from p2 in the bovine and feline viruses to p6 and p9 in the primate and equine lentiviruses, respectively (Table 1.2). It remains to be determined where these core proteins reside within the virion. Functionally, data for the feline, primate, and equine proteins demonstrate that they play a very specific role in budding by possessing the late binding domains that interact with the host cell machinery to assist in release of the viruses (Freed, 2002; Luttge *et al.*, 2008). The polymerase proteins include reverse transcriptase (RT), RNase-H (RH), integrase (IN) and, in several of the viruses, dUTPase (DU). These groups of enzymatic proteins perform necessary genomic replication, transcription and integration events required for virus survival. A few of the members of the lentivirus family also bear a nondescript open reading frame within the *pol* coding region that is simply labelled p15 and has yet to be assigned a function (Table 1.2). The *env* gene encoded proteins, surface unit (SU) and transmembrane (TM) proteins are the viral membrane glycoproteins that lentiviruses utilize for recognition and entry into target cells.

In addition to those three common retroviral genes, lentiviruses also encode a variety of accessory genes. HIV-1, HIV-2, SIV, and BIV genomes all contain open reading frames encoding six accessory proteins. The SRLV, EIAV, FIV and JDV genomes have fewer encoded accessory proteins than their primate virus family counterparts (Fig. 1.3 and Table 1.2). A single accessory protein, Rev, is found in the genome of all lentiviruses. All lentiviruses except FIV and potentially SRLV (Villet *et al.*, 2003a,b) contain a *tat* gene, and all except EIAV contain a *vif* gene. The *tat* and *rev* genes encode two of the major viral regulatory accessory proteins. Rev proteins regulate viral mRNA processing through the transport of unspliced and singly spliced genomic transcripts from the host cell nucleus to the cytoplasm (Clements and Zink, 1996; St-Louis *et al.*, 2004; Zink *et al.*, 2006; Craigo and Montelaro, 2008). Tat proteins serve to activate transcription of the lentiviral genome. The primate, bovine and equine Tat proteins enhance transcriptional elongation through a stem loop recognition sequence

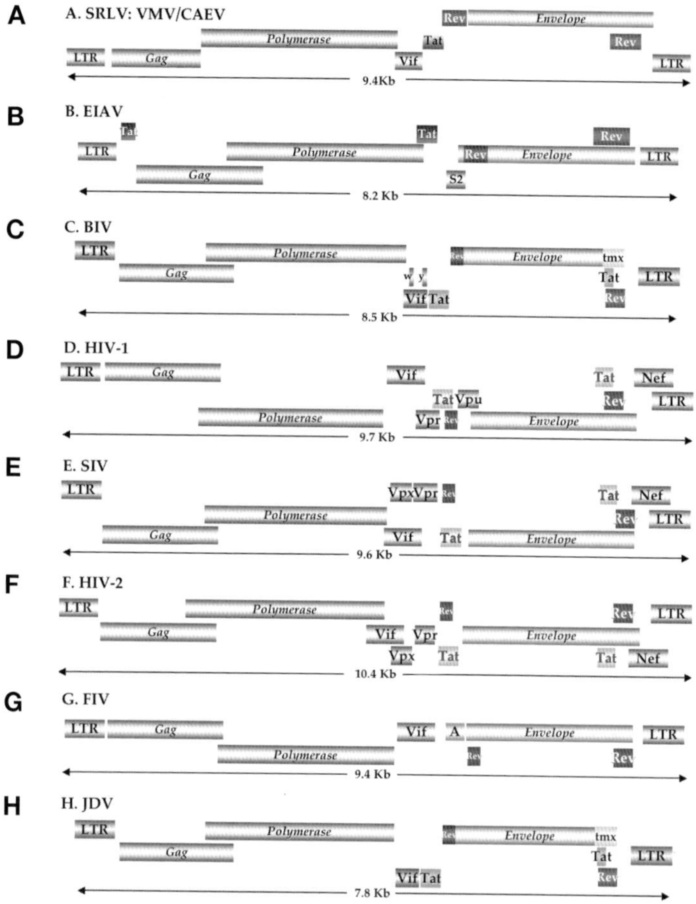


**Figure 1.3** Lentiviral genomes. Schematic representation of all members of the lentivirus family. (A) SRLV. (B) EIAV. (C) BIV. (D) HIV-1. (E) SIV. (F) HIV-2. (G) FIV. (H) JDV.

in the viral LTR designated the Tat activation region (TAR) (Gunnery *et al.*, 1992; Southgate and Green, 1995; Willbold *et al.*, 1998; Barboric *et al.*, 2000). The SRLV Tat weakly enhances transcription in a TAR-independent manner (Harmache *et al.*, 1995) and its functions have recently been reported to potentially correspond to those ascribed for the *vpr* genes of other lentiviruses (Villet *et al.*, 2003a,b). While FIV does not encode a *tat* gene, it does encode what is termed open reading frame 2 or open reading frame A (Orf 2, OrfA or A), which acts in a similar manner to transactivate transcription. However, it also does not work through a TAR sequence element in the viral genome (Sparger *et al.*, 1992; Waters *et al.*, 1996; de Parseval and Elder, 1999; Elder *et al.*, 2008). The Vif protein is believed to increase virus infectivity and be important for virion assembly, although its precise role in assembly remains to be defined. BIV encodes 2 small accessory proteins within the Vif open reading frame, Vpy (Y) and Vpw (W). The function of these two proteins is currently undefined but it is assumed that based upon their genomic location they could have a similar function as Vpr and Vpu (Garvey *et al.*, 1990; St-Louis *et al.*, 2004). Interestingly, EIAV is the only lentivirus that does not encode an obvious *vif* gene. Rather, EIAV contains a distinct accessory gene, *S2*, which to date has no identified function, although it is known to affect *in vivo* replication and virulence (Li *et al.*, 1998; Li *et al.*, 2000; Fagerness *et al.*, 2006).

In addition to the above-described structural and regulatory accessory proteins, some members

**Table 1.2** Lentiviral genomic composition

| | Lentivirus family | | | | | | | |
|---|---|---|---|---|---|---|---|---|
| | | SRLV | | | | | | |
| Viral protein | EIAV | VMV/ CAEV | BIV | HIV-1 | SIV | HIV-2 | FIV | JDV |
| Gag | | | | | | | | |
| MA | p15 | p16 | p16 | p17 | p17 | p17 | p15 | p16 |
| CA | p26 | p17 | p26 | p24 | p24 | p24 | p24 | p26 |
| NC | p11 | p14 | p13 | p7 | p7 | p7 | p13 | p13 |
| Other | p9 | – | p2L, p3, p2 | p6, p2, p1 | p6, p2, p1 | p6, p2, p1 | p2 | p2 |
| Polymerase | | | | | | | | |
| IN | p30 | p35 | p28 | p31 | p31 | p31 | p28 | p28 |
| RT | p66/p51 | p68/p61 | p66/p61 | p66/p51 | p66/p51 | p66/p51 | p66/p51 | p66/p61 |
| PR | p12 | p10 | p10 | p15 | p15 | p15 | p18 | p10 |
| DU | p15 | p15 | – | – | – | – | p15 | – |
| Other | – | – | – | p15 | p15 | p15 | – | – |
| Envelope | | | | | | | | |
| SU | gp90 | gp135 | gp100 | gp120 | gp120 | gp120 | gp100 | gp100 |
| TM | gp45 | gp44 | gp45 | gp41 | gp41 | gp41 | gp36 | gp45 |
| Accessory | | | | | | | | |
| Tat | p8 | p10 | p10 | p16 | p16 | p16 | – | p10 |
| Rev | p18 | p18 | p23 | p19 | p19 | p19 | p22 | p23 |
| Vif | – | p23 | p23 | p23 | p23 | p23 | p23 | p23 |
| Nef | – | – | – | p27 | p27 | p27 | – | – |
| Vpr | – | – | – | p14 | p14 | p14 | – | – |
| Vpu | – | – | – | p16 | – | – | – | – |
| Vpx | – | – | – | – | p12 | p12 | – | – |
| Tmx | – | – | p19 | – | – | – | – | p19 |
| Other | S2/p7 | – | Vpy (Y)/p10 Vpw(W)/p7 | – | – | – | A/p9 | – |

of the lentiviral family encode small accessory proteins involved in integration and assembly or release of new virions (Subbramanian and Cohen, 1994). The Nef protein of HIV-1, HIV-2 and SIV is necessary for the progression and preservation of infection. Nef plays a distinct role in assembly and release by down-regulating expression of the CD4 receptor at the cell surface (Guy *et al.*, 1987; Garcia and Miller, 1991; Anderson *et al.*, 1993; Foster *et al.*, 1994; Aldrovandi *et al.*, 1998). The bovine accessory protein Tmx found in BIV and JDV is located in a similar region to the HIV-1, HIV2, and SIV Nef proteins within their respective genomes; however, the possibility that Tmx exerts Nef-like functions has yet to be determined. The HIV-1 and SIV Vpr and the HIV-2 and SIV Vpx proteins support the transport of the DNA pre-integration complex into the nuclei of non-dividing cells (Lu *et al.*, 1993; Lavallee *et al.*, 1994; Pancio *et al.*, 2000). Interestingly, Vpu is found only in HIV-1, enhances the release of virus from cells and decreases the syncytial formation due to the degradation of newly synthesized CD4 molecules (Willey *et al.*, 1992; Geleziunas *et al.*, 1994; Chen *et al.*, 1996; Piguet and Trono, 1999).

An interesting feature of the equine, small ruminant and feline lentivirus genomes that distinguishes them from the primate and bovine genomes is the presence of a *dUTPase* (*DU*) gene within the *polymerase* region (Table 1.2). The DU protein has been shown *in vitro* to be critical for replication in macrophage, but not in fibroblastic cells (Elder *et al.*, 1992; Threadgill *et al.*, 1993; Lichtenstein *et al.*, 1995) (Turelli *et al.*, 1996; Payne and Elder). The DU protein is believed to minimize the misincorporation of uracil into proviral DNA by maintaining a low dUTP/TTP ratio in macrophage, which, as quiescent cells, naturally contain a lower level of cellular dUTPases (Richards *et al.*, 1984; Vassylyev and Morikawa, 1996). Interestingly, dUTP misincorporation into primate lentiviral DNA is also a potential problem during proviral transcription. In these viruses, uracil misincorportation is controlled by two complementary mechanisms. First, cellular cytosine deamination of viral cDNA is controlled by the viral Vif protein preventing the incorporation of host-derived APOBEC3G cytidine deaminase into viral particles (Sheehy *et al.*, 2002; Harris *et al.*, 2003; Mangeat *et al.*, 2003; Mariani *et al.*, 2003). Second, uracil misincorporation into HIV-1/HIV-2/SIV proviral DNA *via* the direct addition of dUTP during reverse transcription appears to be regulated through the packaging of host cell uracil DNA glycosylase enzyme (UNG2) into viral particles via a specific association with the viral Vpr (Bouhamdan *et al.*, 1996; Chen *et al.*, 2002).

Another distinguishing characteristic of lentiviral genome is the elevated incidence of gene mutations that occur during reverse transcription due to the lack of polymerase editing functions in the viral RT enzyme (Preston *et al.*, 1988; Roberts *et al.*, 1988; Bakhanashvili and Hizi). In this regard, *in vitro* experiments have indicated a $10^6$-fold higher mutational frequency for RT mediated gene replication compared to DNA polymerase mediated gene replication (Gojobori and Yokoyama, 1985). The end result of the error-prone RT is that lentiviruses constantly vary their genomic sequences at random, two to three bases per genome duplication, providing opportunities to select for genetic alterations which have the potential to enhance virus replication (Montelaro *et al.*, 1993). Viruses subsequently rapidly alter protein sequences and produce antigenic variants that can evade established host immune responses. Genomic and antigenic variation during persistent lentiviral infections was first discovered in VMV and EIAV (Scott *et al.*, 1979; Clements *et al.*, 1980a, 1988; Montelaro *et al.*, 1984; Payne *et al.*, 1987a,b; Kono; Rwambo *et al.*, 1990b; Chebloune *et al.*, 1996; Lichtenstein *et al.*, 1996), preceding similar observations of genomic and antigenic variations of HIV-1 and other immunodeficiency lentiviruses. The viral gene that displays the highest levels of variation and rapid evolution is the *env* gene (Salinovich *et al.*, 1986; Clements *et al.*, 1988; Payne *et al.*, 1989; Knowles *et al.*, 1991; Lichtenstein *et al.*, 1996; Leroux *et al.*, 1997a,b), although all viral genes are subject to variation.

## Lentiviral replication

The lentiviral replication cycle, as summarized in Fig. 1.4, is characteristic of retrovirus replication in general (Montelaro *et al.*, 1993; Pedersen, 1993; Barker *et al.*, 1995). Viral recognition, attachment and penetration of target cells are believed to be mediated by specific interaction of the viral envelope glycoproteins with cellular receptor proteins contained in the plasma membrane of the target cells. The currently defined receptor proteins for the members of the lentivirus family are summarized in Table 1.1. Some lentiviruses recognize single receptors (EIAV), while others utilize a coreceptor mechanism of cellular entry (HIV, SIV, FIV). A few of the lentiviruses' (SRLV, BIV, and JDV) receptors are still undefined. The majority of receptor proteins are T-cell receptors (CD4) or are chemokine receptors (CCR5/CXCR4) (Simon *et al.*, 2006; Zink *et al.*, 2006). However, ELRI, a cellular receptor for EIAV, was identified as a member of the tumour necrosis family of receptors (TNFR) (Zhang *et al.*, 2005; Zhang *et al.*, 2008) similar to the identified receptor, CD134, for FIV (de Parseval *et al.*, 2004; Shimojima *et al.*, 2004). Subsequent

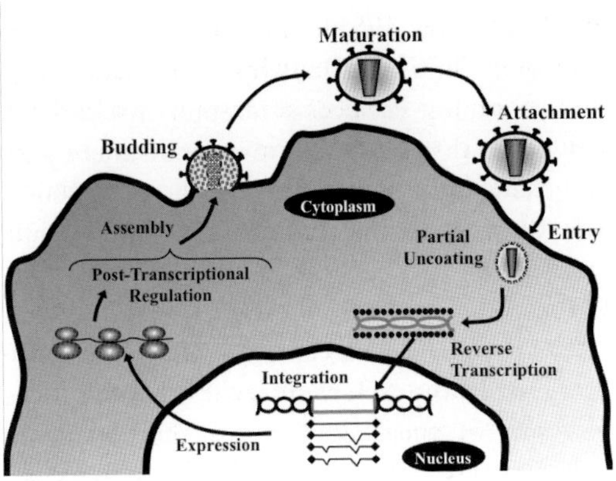


**Figure 1.4** Lentiviral replication. Schematic representation of the lentiviral life cycle.

to receptor binding, a fusion event, mediated by the TM protein, takes place prior to pore formation and the deposition of the viral core into the host cell. Once inside the target cell cytoplasm and following the disassembly of the viral core, the virally encoded reverse transcriptase copies the single-stranded RNA genome into double stranded DNA provirus that is then transported to the nucleus of the cell. There is no evidence for translation of the incoming lentiviral genome at this stage of infection. Once inside the cell nucleus, the viral integrase protein, in conjunction with host DNA repair enzymes, mediates apparently random but limited incorporation of the proviral DNA into cellular chromosomes, although some extrachromosomal viral DNA is always present in productively infected cells. There are typically a total of about 10 copies of proviral DNA per infected cell.

Transcription of the proviral DNA by cellular polymerases produces a complex pattern of viral messenger RNA species whose relative proportions may differ depending on the virus strain/species and target cell(s). In all instances, the predominant lentiviral transcripts are between 8–10 kb transcripts (depending on the virus group) representing the full-length genomic RNA and translated to produce the *gag* and *pol* gene products, and approximately a 3.5 kb mRNA that is a singly spliced transcript translated to produce the viral envelope proteins. Lentivirus-infected cells usually contain in addition to these major viral transcripts, a heterogeneous population of small multiply spliced RNA that are used to produce the various regulatory proteins. In general however, infected cells reveal only minute quantities of these small, multiply spliced RNA species.

The viral Gag and Env proteins are initially produced as polyproteins that are cleaved into the mature virion components by a combination of viral and cellular proteases. The only modification documented in the core proteins is a fatty acylation at the N-terminus of the matrix protein. Both envelope proteins are modified by N-linked glycosylation as they progress through the endoplasmic reticulum of the host cell and the TM proteins are palmitoylated. Although the SU and TM polypeptide contains roughly 400 amino acid residues, the SU contains an average of 10–17 potential N-linked glycosylation sites, while the TM contains only about 2–5 potential glycosylation sites, depending on the lentiviral strain/species. Current data indicates that all potential glycosylation sites are occupied by complex oligosaccharides.

The precise mechanisms of lentiviral assembly have not been completely dissected, although it is assumed that it follows the general model for retrovirus assembly. It appears to be a highly concerted process that is mediated by interactions between the viral Gag polyprotein and host cofactors. Classically, assembly and budding are thought to occur at the plasma membrane of infected cells. However, it has also been demonstrated that budding can occur from intracellular sites within infected macrophage. Current discussions on this controversial topic indicate that the target cell of infection directs the location of assembly and budding, i.e., macrophage vs. lymphocyte (Levy, 2006; Zink *et al.*, 2006; Welsch *et al.*, 2007; Carter and Ehrlich, 2008; Marsh *et al.*, 2009). At the cellular site of budding, viral envelope glycoproteins are initially inserted into the plasma membrane to create distinct sites of virus assembly at which the Gag polyprotein is accumulated beneath the membrane lipid. Subsequently, Gag proteins recruit host cellular components that also regulate the endocytic pathway for assembly and budding. To date, early (for example AP-2) and late endocytic proteins (such as TSG101) as well as the actin cytoskeleton have been associated with lentiviral virion production. The intermediate viral Gag proteins are cleaved into the mature virion proteins as the particle buds from the cell surface and is released to produce progeny virions.

## Lentivirus cellular tropism and associated pathogenesis

The ovine, caprine, and equine lentiviruses all display exclusive monocyte/macrophage tropism; these lentiviruses do not evidently infect lymphocytes or other cell types *in vivo* (Clements and Zink, 1996; Zink *et al.*, 2006). The lack of lymphocyte infection by these lentiviruses appears to correlate with the lack of immune suppression in these viral infections, in contrast to the severe immune suppression observed with the primate and feline immunodeficiency lentiviruses. However, as described below, infection of macrophages by equine and small ruminant lentiviruses produces a number of disease signs that are also observed in HIV-1 infections, including encephalitis, wasting, diarrhoea, anaemia, and thrombocytopenia.

### SRLV tropism and pathogenesis

CAEV and VMV are very closely related, but genetically distinct lentiviruses, which have recently been reclassified as a single family (SRLV) due to the cross-species infectious nature of the two viruses (Zanoni; Shah *et al.*, 2004a,b). SRLV infections of sheep and goats result in a spectrum of disease that is contingent upon both the particular infecting virus strain and the breed of the host sheep or goat. VMV persistently infects monocyte/macrophage and dendritic cells with most data derived from sheep although infection of goats is possible and CAEV is known to infect monocyte/macrophages of goats and sheep (Narayan *et al.*, 1982; Gendelman *et al.*, 1986; Ryan *et al.*, 2000). The primary routes of SRLV transmission are milk and colostrum (Peterhans *et al.*, 2004). Initial reports of the cellular receptor utilized by VMV indicated that it was a MHCII-related protein (Dalziel *et al.*, 1991). More recent evidence indicates the involvement of a 56-kDa chondroitin sulfate proteoglycan membrane protein in VMV-cell surface interactions and infection (Bruett *et al.*, 2000). The chondroitin sulfate proteoglycan was further characterized as a membrane-associated serine/threonine kinase that is constitutively phosphorylated (Barber *et al.*, 2000). The specific receptor for CAEV has not been identified. However, VMV and CAEV do not exhibit reciprocal receptor interference, suggesting that their receptors are distinct (Jolly and Narayan, 1989; Hötzel and Cheevers, 2002). The cross-species infectious nature of the recently reclassified SRLV suggests that some strains of the viruses (CAEV) probably share the chondroitin sulfate proteoglycan receptor identified for VMV; however, in the subgroups, which appear to be goat specific, there feasibly exists a yet unidentified receptor. SRLV causes immune-mediated lesions through the dissemination and infiltration of lymphocytes in infected organs (Mornex *et al.*, 1994). Disease arises three to five years after naturally occurring infections and adheres to a prolonged course wherein a vigorous inflammatory response causes tissue damage (Narayan *et al.*, 1993). Virus-induced disease presents as arthritis, mastitis, chronic interstitial pneumonia, resulting in wasting or encephalitis including paralysis and eventual death. These disease signs are brought on through organ-specific changes, such as smooth muscle hyperplasia in the lung and demyelination in the central nervous system (Narayan *et al.*, 1982; Zink *et al.*, 2000). CAEV-specific immunity, especially antibodies to the viral envelope, has been correlated with the severity of arthritis observed in infected goats (Bertoni *et al.*, 1994, 2000).

### EIAV tropism and pathogenesis

EIAV causes a persistent infection in horses and a chronic disseminated disease of worldwide importance in veterinary medicine (reviewed in Montelaro *et al.*, 1993; Leroux *et al.*, 2004; Craigo and Montelaro, 2008). EIAV is also a monocyte/macrophage-tropic virus. Virus infection of target cells is mediated by the ELR-1 receptor (Zhang *et al.*, 2005). Virus is transmitted via blood-feeding insects or iatrogenic sources such as contaminated syringe needles. Disease occurs in three stages: acute, chronic and inapparent. EIA is characterized during its acute and chronic stages by defined episodes of disease triggered by waves of viraemia and characterized by fever, anaemia, thrombocytopenia, oedema, and various wasting signs. By 8–12 months post infection horses typically progress to life-long inapparent carriers, but continue to harbour various steady state levels of viral replication in monocyte-rich tissue reservoirs (Rwambo *et al.*, 1990a; Montelaro *et al.*, 1993; Hammond *et al.*, 2000; Harrold *et al.*, 2000). Stress or immune suppression of EIAV inapparent carriers can induce an increase in viral replication and potentially a recrudescence of disease (Kono *et al.*, 1976; Montelaro *et al.*, 1993; Craigo *et al.*, 2002). Among virulent lentiviruses, however, EIAV is unique in that despite aggressive virus replication and associated rapid antigenic variation, greater than 90% of infected horses progress from a chronic disease state to an inapparent carrier state. This progression to an inapparent stage of disease is achieved by a strict immunological control over virus replication (Montelaro *et al.*, 1993; Craigo and Montelaro, 2008).

### BIV tropism and pathogenesis

An internationally distributed lentivirus, BIV has been serologically detected in Asia, Europe, Australia, New Zealand and North America (St-Louis *et al.*, 2004; Zink *et al.*, 2006). Regardless of the 'immunodeficiency' in its designation, pathogenically, BIV is more closely related to SRLV and EIAV than HIV and FIV. BIV like all lentiviruses induces a lifelong persistent infection; however, the occurrence of clinical disease and/or immunodeficiency is still somewhat controversial. The majority of BIV infections occur in cattle with no visible evidence of disease. Consequently, BIV is routinely categorized as a subclinical disease syndrome; however, it still can pose a significant impact on the cattle industry. The majority of reports demonstrate that BIV has a monocyte/macrophage tropism; however, there is evidence that it also replicates in B and T-lymphocytes as well (Carpenter *et al.*, 1992; St-Louis *et al.*, 2004). The virus can be spread both vertically through *in utero* transmission or horizontally by the exchange of body fluids such as blood and colostrum (Gonda, 1992; Meas *et al.*, 2002; St-Louis *et al.*, 2004). The BIV receptor for the most part remains unidentified although it has been suggested that it might utilize the chemokine receptor CCR5 (Wright *et al.*, 2002). The major clinical signs associated with BIV infections are lymphadenopathy, progressive weakness, and emaciation (Gonda *et al.*, 1987; Carpenter *et al.*, 1992; Gonda, 1992). While not substantiated there has been the suggestion of the development of bovine paraplegic syndrome after BIV infection (Walder *et al.*, 1995). Potential immune dysfunction due to BIV infections leads to secondary conditions such as mastitis and bacterial infections (Gonda, 1992; St-Louis *et al.*, 2004).

### HIV tropism and pathogenesis

HIV infections are found worldwide with the number of new infections increasing everyday. The geographic distribution of the pandemic reaches every corner of the world. HIV-1 and HIV-2 are biologically related although they are distinguished genomically and antigenically. Through phylogenetic analyses HIV-1 has been subclassified into three groups, M, N, and O with the predominant group M being further categorized into 11 clades, A-K. HIV-2 has also been classified based on phylogenetic relationships into six subtypes A-F (Kuiken *et al.*, 1999). Furthermore, phylogenetic analyses of HIV-1, HIV-2 and SIV led to the determination of a simian origin of AIDS (Hahn *et al.*, 2000; Kuiken *et al.*, 2000). HIV-1 and HIV-2 were introduced into the human population several times (Hahn *et al.*, 2000) yet the M group of HIV-1 viruses appears to have arisen from a single cross-species transmission event (Gao *et al.*, 1999). HIV is spread by sexual contact and by exposure to contaminated blood or blood products via transfusions or sharing needles. Sexual transmission is the predominant mode of transmission worldwide (Levy, 2006; Simon *et al.*, 2006; Lackner and Veazey, 2007). Vertical transmission occurs by the exposure of infants to maternal fluids either in the birth canal or in maternal milk (Thorne and Newell, 2000). HIV enters host cells through binding to CD4 proteins (receptor) and either CCR5 or CXCR4 (co-receptors). HIV is hence inherently tropic for $CD4^+$ T-lymphocytes, and monocyte/macrophage lineage cells. Infection of monocyte/macrophage is particularly imperative for the development of tissue viral reservoirs and latency. Acute disease occurs within the first month of infection and is characterized by high plasma viral loads. Clinical symptoms are flu-like in nature. The initial viraemia drops as the cellular and humoral protective immune responses develop and chronic infection ensues with the development of asymptomatic disease. Chronic, asymptomatic disease lasts for several years. Viral diversity increases throughout all stages of disease. After years of infection a gradual destruction of the naive and memory CD4$^+$ T-lymphocyte populations begins which is the hallmark of HIV infection. This marks the onset of opportunistic infections and AIDS beginning the end stage of the disease. The risk of transmission is highest in the first weeks when viraemia peaks (Barker *et al.*, 1995; Simon *et al.*, 2006). Contrary to this typical disease progression observed with HIV-1, infections with HIV-2 have a very different clinical outcome. Overall HIV-2 has a reduced virulence compared to HIV-1 (Marlink *et al.*, 1988). The rate of CD4$^+$ T-cell loss in HIV-2 infected patients is significantly less than in its counterpart HIV-1 (Marlink *et al.*, 1994).

## SIV tropism and pathogenesis

SIV infects numerous non-human primate species in Africa including African green monkeys (*Chlorocebus*), Sooty mangabeys (*Cercocebus atys*), Mandrills (*Mandrillus*), chimpanzees (*Pan*), sun-tailed monkeys (*Cercopithecus solatus*), L'hoest monkeys (*Cercopithecus lhoesti*), Sykes monkeys (*Cercopithecus albogularis*), baboons (*Papio*), and several others (Hahn *et al.*, 2000; VandeWoude and Apetrei, 2006). SIV infection of these natural hosts lacks observable clinical signs of disease with infected animals living a typical non-human primate lifespan (Cichutek and Norley, 1993; Chakrabarti, 2004; Hirsch, 2004; Sharp *et al.*, 2005; Silvestri, 2005). SIV infection of Sooty mangabeys in the wild appears to be acquired at the time of sexual maturity (2–4 years) (Fultz *et al.*, 1990). Infected natural hosts maintain normal CD4$^+$ T-cell counts and do not develop the signs of disease associated with the development of AIDS. Furthermore, natural SIV infections are characterized by high plasma viraemia which reaches levels that are high or higher than those observed in HIV infections (Silvestri *et al.*, 2003; Chakrabarti, 2004). In contrast, pathogenic SIV has been isolated from non-natural host Asian monkeys such as rhesus, cynomologous, nemestria, and stump-tailed macaques. Unlike African monkeys, feral macaques are not endemically infected with SIV (Gardner, 1996). Experimental infection in these non-natural host Asian monkey species results in an illness similar to AIDS (simian AIDS, SAIDS). SAIDS has a more rapid course compared to its human counterpart, AIDS. Overt signs of clinical disease in SAIDS develop within 1–4 years post infection (Gardner, 1996). As observed with HIV, SIV also infects host cells utilizing CD4 as its receptor. SIV is tropic for both CD4$^+$ T-lymphocytes and monocyte/macrophage lineage cells. However, unlike HIV, both lymphocyte and monocyte tropic viruses use CCR5 as a co-receptor (Edinger *et al.*, 1997). Infection of monocyte/macrophage cells plays an important role in chronic infection and the development of viral reservoirs. Monocyte/macrophage infections are also clearly important for neurological infections as well as infections of non-lymphoid tissues such as the heart and kidneys (Shannon, 2001; Ross and Klotman, 2004).

## FIV tropism and pathogenesis

FIV causes a terminal disease globally in domesticated and wild cats. The immunosuppressive syndrome is comparable to AIDS in humans (Pedersen, 1993). Transmission of FIV occurs through both mucosal exposure as well as via the blood of another animal. Vertical transmission also occurs by both prenatal and postnatal routes (Callanan *et al.*, 1991; O'Neil *et al.*, 1995, 1996). Unlike the other immunodeficiency lentiviruses, FIV has a broader lymphocyte tropism. In addition to CD4$^+$ T-cells, there is also *in vitro* and *in vivo* evidence of FIV infection in at least a subset of CD8$^+$ T-cells as well as B-cells (Brown *et al.*, 1991; English *et al.*, 1993; Dean *et al.*, 1996). FIV utilizes the activation marker, CD134 as a cellular receptor and CXCR4 as a co-receptor (Willett *et al.*, 1997). Disease initiates with a short acute phase within one-month post infection that is accompanied by increasing viral loads, fever, weight loss, and lymphadenopathy. This period of illness is followed by years of asymptomatic infection lacking any clinical signs of disease and is denoted by a strong antiviral immune response with concurrent low viral titres. Eventually there is a gradual decline in CD4$^+$ T-cells (Ackley *et al.*, 1990). Ultimately the animals reach the terminal phase of disease that is demarcated by immunological decompensation with concomitant exacerbation of viral load resulting in immunodeficiency and the onset of opportunistic infections. Neurological manifestations have been demonstrated in cats with feline AIDS (Phillips *et al.*, 1994; Phillips *et al.*, 1996). In general, animals may develop faucitis and chronic gingivostomatitis, rhinitis, cachexia, renal failure due to immune-mediated glomerulonephritis and neoplasia.

### JDV tropism and pathogenesis

JDV, the most recently indentified member in the lentivirus group, causes an acute, sometimes lethal disease syndrome in Bali cattle. Experimentally infected animals demonstrate a fatality rate of approximately 20% (Hartaningsih *et al.*, 1993; Wilcox, 1997). JDV has been primarily documented in natural infections of *Bos javanicus* (Bali cattle) and hence it is widely assumed that JDV will only infect Bali cattle. However, it has been established experimentally that JDV also infects *Bos taurus, Bos indicus,* and cross-bred Bali cattle and buffalo (*Bos javanicus* × *Bos indicus*) (Soeharsono *et al.*, 1990; Soeharsono *et al.*, 1995b). JDV can be detected in cattle bodily secretions (Wilcox *et al.*, 1995; Wilcox, 1997). Dissemination of the virus probably occurs via direct transmission in the animals' secretions through close contact or through blood by haematophagus insects (Soeharsono *et al.*, 1995a). JDV RNA has been detected in macrophage rich tissues and thus is thought to have a monocyte/macrophage tropism although replication in lymphocytes has also been suggested (Chadwick *et al.*, 1998). Acute disease occurs after a short incubation period of 5–12 days. Typical clinical signs during acute disease include lymphadenopathy concomitant with a fever as well as leucopenia. The febrile episode on average lasts 1–9 days and is accompanied by a virus titre of approximately $10^8$/ml and a viral load of up to $10^{12}$ RNA genomes/ml (Stewart *et al.*, 2005; Stewart *et al.*, 2008). Throughout the period of recovery the animal's viral load reduces to approximately 10 $ID_{50}$/ml of blood by 47 days post infection (Soeharsono *et al.*, 1995a). Animals that recover from the acute disease syndrome appear to be resistant to reinfection and also retain infectious virus for at least 25 months post infection (Soeharsono *et al.*, 1995b; Wilcox *et al.*, 1995).

## Macrophage and lentiviral latency

Latency is frequently described as a reversible, non-productive state of infection involving the integration of the viral genome into the host chromosome in a manner that restricts viral gene expression or replication. Classic latency, like that observed in human herpes virus (HSV) infection, necessitates suppression of normal transcription of the viral genes that lead to viral genomic replication and active virion production (Roizman and Sears, 1987, 1995). Suppression of transcription of the viral genome in the case of HSV is in part due to specific viral gene products (for instance the HSV LAT transcripts) that prevent transcription of other viral genes (Roizman and Sears, 1995; Kent *et al.*, 2003). Latency, nonetheless, can also refer to an overall state of genomically active, but clinically inapparent infections. Clinical latency consequently indicates a lack of the evident clinical signs of viral infection and/or dynamic replication. It is widely accepted that all lentiviruses undergo a long-term clinical latency as part of their characteristic 'slow' disease progression. However, does this clinical latency result from an actual viral latency? What role do the host cells play in this latency?

Lentiviruses in general appear to experience both clinical and viral latency, although the mode of the viral latency to date is not well defined. HIV in particular seems to undergo both forms of latency. HIV replicates in many cell types, however, there are two major cellular reservoirs considered in the context of viral latency: latently infected resting T-cells and infected monocyte/macrophage (Carter and Ehrlich, 2008). It is important to note that the concept of a true genetic latency in HIV-1 infections is still controversial. HIV-1 has been demonstrated to persist in reservoirs consisting of small pools of latently infected resting memory CD4$^+$ T-cells that carry the integrated viral genome but do not express viral proteins unless the cells are activated to proliferate (Chun *et al.*, 1997; Finzi *et al.*, 1997; Wong *et al.*, 1997; Pierson *et al.*, 2000; Blankson *et al.*, 2002; Shen *et al.*, 2003; Han *et al.*, 2004; Lassen *et al.*, 2004; Swiggard *et al.*, 2005; Williams and Greene, 2005; Zhou *et al.*, 2005), Macrophage cells, unlike T-cells, do not divide and are incapable of establishing a 'resting state' as observed with these HIV-1-infected CD4$^+$ T-cells. All lentiviruses infect cells of the monocyte/macrophage lineage but some do not infect cells of the lymphocyte lineage. Since lentiviral infection of macrophage cells typically results in a productive cytolytic infection, macrophage do not appear to offer a reservoir for viral or 'true genetic' latency. Therefore, is it possible for the lentiviruses that infect only monocyte/macrophage to experience viral and clinical latency?

### Clinical latency

All lentivirus disease progression is characterized by periods of clinical latency. In SRLV infections of sheep or goats, clinical latency is observed during the first months to years post infection, with disease signs apparently correlating with a gradual increase in virus replication and the accumulation of lymphocytes in affected tissues (Clements and Zink, 1996; Perk, 1999). In contrast, EIAV infection of horses is characterized by rapid and dynamic disease during the first year post infection, followed by progression to a long-term state of clinical latency in inapparent carriers that is associated with an establishment of effective and enduring immune control of EIAV replication (Craigo and Montelaro, 2008). JDV has a very short-lived period of disease that typically resolves within 2–3 months post infection followed by a sustained period of clinical latency while BIV infection of cattle displays clinical latency throughout the period of the infection. Whether the lack of clinical disease is due to a stringent immunological control of BIV or adaptation of the virus to the host environment is not completely clear (Wilcox *et al.*, 1995; Snider *et al.*, 1997; Wilcox, 1997; St-Louis *et al.*, 2004). The primate and feline lentiviruses experience a temporary clinical latency brought about by the transitory immunological control of virus replication that has been well characterized in the time period between acute disease and the onset of immunodeficiency (Levy, 2006; Simon *et al.*, 2006; Assogba *et al.*, 2007; Elder *et al.*, 2008). Among the primate lentiviruses SIV infections of natural hosts, African monkey species, in particular are clinically latent from the point of initial infection. The reasons behind this clinical latency have also not been well defined to date (Pandrea *et al.*, 2003; Silvestri *et al.*, 2003, 2007; Silvestri, 2005; VandeWoude and Apetrei, 2006).

The manner in which the individual lentivirus interacts with the infected host during the disease process and accomplishes clinical latency is not always the same. The inherent differences between EIAV and SRLV infections demonstrate this concept well. Comparison of the disease processes observed in persistent infections by EIAV and SRLV indicates a contradictory role of host immune responses in the establishment of clinical latency by lentiviruses. In the case of EIAV, the development of virus-specific immune responses during the first year post infection produces enduring control of virus replication and disease (clinical latency) that is unique among lentiviruses. In contrast, the development of virus-specific immune responses in SRLV infections is associated with the development of inflammatory disease and a loss of clinical latency. These observations highlight the intricate interplay between lentivirus replication, host immunity, and clinical latency.

### Viral latency

Viral latency in lentiviruses occurs in infected resting $CD4^{+}$ T-cells and monocyte/macrophage. The latency associated with monocyte/macrophage not only extends to perhaps all lentiviruses, but also is almost certainly responsible for the dissemination of the virus to tissue reservoir sites and thus provides the lentiviruses with their best survival mechanism. A seminal discovery in comprehending lentiviral persistence regardless of robust host immune responses was the pioneering studies of SRLV infections of monocytes that led to the 'Trojan horse' model of dissemination of lentivirus infection (Gendelman *et al.*, 1985; Peluso *et al.*, 1985). While equine and small ruminant lentiviral infection of monocytes has been demonstrated *in vitro*, active replication in monocytes has never been observed. Infection initially results in a non-productive infection, and differentiation or maturation of virus-infected monocytes to macrophage is required to activate virus replication (Narayan *et al.*, 1983; Gendelman *et al.*, 1985; Gendelman *et al.*, 1986; Maury; Clements and Zink; Sellon *et al.*, 1996; Harrold *et al.*, 2000). The infected monocytes can become widely disseminated throughout various organs of the infected host undetected by the host immune system, as the infected cells are not presenting viral antigens due to the lack of active viral replication. After seeding in organ tissues, differentiation of latent monocytes activates viral replication and thus establishes reservoir sites in spite of the host immune response.

The mechanisms of latent infections of monocytes were first described in VMV through measurements of viral RNA in infected monocytes and macrophage (Gendelman *et al.*, 1986). The relevance of these initial observations with VMV to lentivirus persistence was further supported by subsequent studies of *in vitro* and *in vivo* infections of equine monocytes by EIAV (Maury; Sellon *et al.*, 1996). More recently, preliminary studies of EIAV infection of monocytes *in vitro* have indicated that the lack of virus replication may be due to a lack of proviral integration in monocytes, leading to a transcriptionally inactive unintegrated provirus harboured in the nucleus of the cell (Montelaro *et al.*, 1993; Craigo and Montelaro, 2008). Differentiation of the infected monocyte to macrophage is required to mediate proviral integration and expression by induced macrophage-specific transcription factors. A similar mechanism for HIV-1 latent infection of human monocytes *in vitro* has also been described (Pierson *et al.*, 2002; Bisgrove *et al.*, 2005; Lewinski *et al.*, 2005; Alexaki and Wigdahl, 2008). Hence, viral latency may occur through suppression of proviral integration or suppression of transcription from integrated provirus DNA in the infected monocyte. The invasive distribution of the infected cells that can result from latently infected monocytes allows pervasive dissemination of lentiviral-infected yet concealed cells to occur in multiple tissues before the hosts' immune system becomes aware of the presence of the foreign antigens.

Monocyte and macrophage infections may be critical for viral persistence and dissemination in various tissues and for transmission of infection. In this regard, the predominant cell types in many bodily fluids (e.g., alveolar fluid, colostrum, semen, vaginal secretions) are monocyte/macrophage. In semen, for example, the numbers of macrophage exceed those of lymphocytes by more than 20-fold (Meltzer *et al.*, 1990; Meltzer and Gendelman, 1992). Macrophage secretory products are directly implicated in the HIV-1 pathogenesis of central nervous system (CNS) and in the control of viral latency in HIV-infected T-cells. Hence, in the 'Trojan horse' model of HIV-1 infection and encephalitis, the virus is proposed to initially enter the CNS carried by monocyte/macrophage. Subsequently, virus replication is activated and additional monocyte/macrophage are recruited into the CNS via cytokine/chemokine networks and endothelial-cell–leucocyte interactions at the blood–brain barrier (Kim *et al.*, 2003; Verani *et al.*, 2005; Alexaki and Wigdahl, 2008). It seems likely that the ongoing seeding of other tissues may depend on replenishment by latently infected monocytes that are not recognized by established virus-specific immune responses.

## Concluding remarks

Lentiviruses are a diverse yet uniquely cohesive group of retroviruses. As a group, their combined impact on the health and well being of animals and humans alike for over a century has been, and continues to be, enormous. One of the most unifying features of the lentivirus group is the infection of monocyte/macrophage cells. Non-productive monocyte infections provide a valuable mechanism for lentiviruses to continually seed target tissues with infected cells, even amidst robust, virus-specific, host immune responses. In this regard, it would appear that latently infected monocytes might represent essentially the principal mechanism of lentiviral persistence for this intriguing family of pathogens.

## References

Ackley, C.D., Yamamoto, J.K., Levy, N., Pedersen, N.C., and Cooper, M.D. (1990). Immunologic abnormalities in pathogen-free cats experimentally infected with feline immunodeficiency virus. J. Virol. *64*, 5652–5655.

Adams, D.S., Oliver, R.E., Ameghino, E., DeMartini, J.C., Verwoerd, D.W., Houwers, D.J., Waghela, S., Gorham, J.R., Hyllseth, B., Dawson, M., *et al.* (1984). Global survey of serological evidence of caprine arthritis-encephalitis virus infection. Vet. Rec. *115*, 493–495.

Aldrovandi, G.M., Gao, L., Bristol, G., and Zack, J.A. (1998). Regions of human immunodeficiency virus type 1 nef required for function *in vivo*. J. Virol. *72*, 7032–7039.

Alexaki, A., and Wigdahl, B. (2008). HIV-1 Infection of bone marrow hematopoietic progenitor cells and their role in Trafficking and Viral Dissemination. PLoS Pathog. *4*, e1000215.

Anderson, S., Shugars, D.C., Swanstrom, R., and Garcia, J.V. (1993). Nef from primary isolates of human immunodeficiency virus type 1 suppresses surface CD4 expression in human and mouse T-cells. J. Virol. *67*, 4923–4931.

Assogba, B.D., Leavell, S., Porter, K., and Burkhard, M.J. (2007). Mucosal administration of low dose cell associated feline immunodeficiency virus promotes viral latency. J. Infect. Dis. *195*, 1184–1188.

Bakhanashvili, M., and Hizi, A. (1993). Fidelity of DNA synthesis exhibited *in vitro* by the reverse transcriptase of the lentivirus equine infectious anemia virus. Biochemistry *32*, 7559–7567.

Barber, S.A., Bruett, L., and Clements, J.E. (2000). Involvement of a membrane-associated serine/threonine kinase complex in cellular binding of visna virus. Virology *274*, 321–330.

Barboric, M., Taube, R., Nekrep, N., Fujinaga, K., and Peterlin, B.M. (2000). Binding of Tat to TAR and recruitment of positive transcription elongation factor b occur independently in bovine immunodeficiency virus. J. Virol. *74*, 6039–6044.

Barker, E., Barnett, S.W., Stamatatos, L., and Levy, J.A. (1995). The Human Immunodeficiency Viruses, J.A. Levy, ed. (New York: Plenum Press), pp. 1–96.

Barre-Sinoussi, F., Chermann, J.C., Rey, F., Nugeyre, M.T., Chamaret, S., Gruest, J., Dauguet, C., Axler-Blin, C., Vezinet-Brun, F., Rouzioux, C., *et al.* (1983). Isolation of a T-lymphotropic retrovirus from a patient at risk for acquired immune deficiency syndrome (AIDS). Science *220*, 868–871.

Bendinelli, M., Pistello, M., Lombardi, S., Poli, A., Garzelli, C., Matteucci, D., Ceccherini-Nelli, L., Malvaldi, G., and Tozzini, F. (1995). Feline immunodeficiency virus: an interesting model for AIDS studies and an important cat pathogen. Clin. Microbiol. Rev. *8*, 87–112.

Bertoni, G., Zahno, M.L., Zanoni, R., Vogt, H.R., Peterhans, E., Ruff, G., Cheevers, W.P., Sonigo, P., and Pancino, G. (1994). Antibody reactivity to the immunodominant epitopes of the caprine arthritis-encephalitis virus gp38 transmembrane protein associates with the development of arthritis. J. Virol. *68*, 7139–7147.

Bertoni, G., Hertig, C., Zahno, M.L., Vogt, H.R., Dufour, S., Cordano, P., Peterhans, E., Cheevers, W.P., Sonigo, P., and Pancino, G. (2000). B-cell epitopes of the envelope glycoprotein of caprine arthritis-encephalitis virus and antibody response in infected goats. J. Gen. Virol. *81*, 2929–2940.

Bisgrove, D., Lewinski, M., Bushman, F., and Verdin, E. (2005). Molecular mechanisms of HIV-1 proviral latency. Expert Rev. Anti Infec.t Ther. *3*, 805–814.

Blankson, J.N., Persaud, D., and Siliciano, R.F. (2002). The challenge of viral reservoirs in HIV-1 infection. Annu. Rev. Med. *53*, 557–593.

Bouhamdan, M., Benichou, S., Rey, F., Navarro, J.M., Agostini, I., Spire, B., Camonis, J., Slupphaug, G., Vigne, R., Benarous, R., and Sire, J. (1996). Human immunodeficiency virus type 1 Vpr protein binds to the uracil DNA glycosylase DNA repair enzyme. J. Virol. *70*, 697–704.

Brodie, S.J., Pearson, L.D., Zink, M.C., Bickle, H.M., Anderson, B.C., Marcom, K.A., and DeMartini, J.C. (1995). Ovine lentivirus expression and disease. Virus replication, but not entry, is restricted to macrophages of specific tissues. Am. J. Pathol. *146*, 250–263.

Brown, E.W., Yuhki, N., Packer, C., and O'Brien, S.J. (1994). A lion lentivirus related to feline immunodeficiency virus: epidemiologic and phylogenetic aspects. J. Virol. *68*, 5953–5968.

Brown, W.C., Bissey, L., Logan, K.S., Pedersen, N.C., Elder, J.H., and Collisson, E.W. (1991). Feline immunodeficiency virus infects both CD4$^+$ and CD8+ T-lymphocytes. J. Virol. *65*, 3359–3364.

Bruett, L., Barber, S.A., and Clements, J.E. (2000). Characterization of a membrane-associated protein implicated in visna virus binding and infection. Virology *271*, 132–141.

Budiarso, I.T., and Hardjosworo, S. (1976). Jembrana disease in Bali cattle. Austral. Vet. J. *52*, 97–97.

Callanan, J.J., Hosie, M.J., and Jarrett, O. (1991). Transmission of feline immunodeficiency virus from mother to kitten. Vet. Rec. *128*, 332–333.

Carpenter, M.A., Brown, E.W., Culver, M., Johnson, W.E., Pecon-Slattery, J., Brousset, D., and O'Brien, S.J. (1996). Genetic and phylogenetic divergence of feline immunodeficiency virus in the puma (*Puma concolor*). J. Virol. *70*, 6682–6693.

Carpenter, S., Miller, L.D., Alexandersen, S., Whetstone, C.A., VanDerMaaten, M.J., Viuff, B., Wannemuehler, Y., Miller, J.M., and Roth, J.A. (1992). Characterization of early pathogenic effects after experimental infection of calves with bovine immunodeficiency-like virus. J. Virol. *66*, 1074–1083.

Carter, C.A., and Ehrlich, L.S. (2008). Cell biology of HIV-1 infection of macrophages. Annu. Rev. Microbiol. *62*, 425–443.

Casey, J.M., Kim, Y., Andersen, P.R., Watson, K.F., Fox, J.L., and Devare, S.G. (1985). Human T-cell lymphotropic virus type III: immunologic characterization and primary structure analysis of the major internal protein, p24. J. Virol. *55*, 417–423.

Chadwick, B.J., Coelen, R.J., Sammels, L.M., Kertayadnya, G., and Wilcox, G.E. (1995). Genomic sequence analysis identifies Jembrana disease virus as a new bovine lentivirus. J. Gen. Virol. *76*, 189–192.

Chadwick, B.J., Desport, M., Brownlie, J., Wilcox, G.E., and Dharma, D.M. (1998). Detection of Jembrana disease virus in spleen, lymph nodes, bone marrow and other tissues by *in situ* hybridization of paraffin-embedded sections. J. Gen. Virol. *79 (Pt 1)*, 101–106.

Chakrabarti, L.A. (2004). The paradox of simian immunodeficiency virus infection in sooty mangabeys: active viral replication without disease progression. Front. Biosci. *9*, 521–539.

Charman, H.P., Bladen, S., Gilden, R.V., and Coggins, L. (1976). Equine infectious anemia virus: evidence favoring classification as a retravirus. J. Virol. *19*, 1073–1079.

Chebloune, Y., Karr, B., Sheffer, D., Leung, K., and Narayan, O. (1996). Variations in lentiviral gene expression in monocyte-derived macrophages from naturally infected sheep. J. Gen. Virol. *77*, 2037–2051.

Chen, B.K., Gandhi, R.T., and Baltimore, D. (1996). CD4 down-modulation during infection of human T-cells with human immunodeficiency virus type 1 involves independent activities of vpu, env, and nef. J. Virol. *70*, 6044–6053.

Chen, R., Wang, H., and Mansky, L.M. (2002). Roles of uracil-DNA glycosylase and dUTPase in virus replication. J. Gen. Virol. *83*, 2339–2345.

Chun, T.W., Carruth, L., Finzi, D., Shen, X., DiGiuseppe, J.A., Taylor, H., Hermankova, M., Chadwick, K., Margolick, J., Quinn, T.C., *et al.* (1997). Quantification of latent tissue reservoirs and total body viral load in HIV-1 infection. Nature *387*, 183–188.

Cichutek, K., and Norley, S. (1993). Lack of immune suppression in SIV-infected natural hosts. Aids 7 Suppl. *1*, S25–35.

Clavel, F., Guetard, D., Brun-Vezinet, F., Chamaret, S., Rey, M.A., Santos-Ferreira, M.O., Laurent, A.G., Dauguet, C., Katlama, C., and Rouzioux, C. (1986). Isolation of a new human retrovirus from West African patients with AIDS. Science *233*, 343–346.

Clements, J.E., Pedersen, F.S., Narayan, O., and Haseltine, W.A. (1980a). Genomic changes associated with antigenic variation of visna virus durig persistent infection. Proc. Natl. Acad. Sci. U.S.A. *77*, 4454–4458.

Clements, J.E., Narayan, O., and Cork, L.C. (1980b). Biochemical characterization of the virus causing leukoencephalitis and arthritis in goats. J. Gen. Virol. *50*, 423–427.

Clements, J.E., Gdovin, S.L., Montelaro, R.C., and Narayan, O. (1988). Antigenic variation in lentiviral diseases. Annu. Rev. Immunol. *6*, 139–159.

Clements, J.E., and Zink, M.C. (1996). Molecular biology and pathogenesis of animal lentivirus infections. Clin. Microbiol. Rev. *9*, 100–117.

Cork, L.C., Hadlow, W.J., Crawford, T.B., Gorham, J.R., and Piper, R.C. (1974). Infectious leukoencephalomyelitis of young goats. J. Infect. Dis. *129*, 134–141.

Craigo, J.K., Leroux, C., Howe, L., Steckbeck, J.D., Cook, S.J., Issel, C.J., and Montelaro, R.C. (2002). Transient immune suppression of inapparent carriers infected with a principal neutralizing domain-deficient equine infectious anaemia virus induces neutralizing antibodies and lowers steady-state virus replication. J. Gen. Virol. *83*, 1353–1359.

Craigo, J.K., and Montelaro, R.C. (2008). Equine Infectious Anemia Virus (Retroviridae), In Encyclopedia of Virology (Academic Press), pp. 167–174.

Crawford, T.B., Adams, D.S., Cheevers, W.P., and Cork, L.C. (1980). Chronic arthritis in goats caused by a retrovirus. Science *207*, 997–999.

Dalziel, R.G., Hopkins, J., Watt, N.J., Dutia, B.M., Clarke, H.A., and McConnell, I. (1991). Identification of a putative cellular receptor for the lentivirus visna virus. J. Gen. Virol. *72*, 1905–1911.

Daniel, M.D., Letvin, N.L., King, N.W., Kannagi, M., Sehgal, P.K., Hunt, R.D., Kanki, P.J., Essex, M., and Desrosiers, R.C. (1985). Isolation of T-cell tropic HTLV-III-like retrovirus from macaques. Science *228*, 1201–1204.

de Parseval, A., and Elder, J.H. (1999). Demonstration that orf2 encodes the feline immunodeficiency virus transactivating (Tat) protein and characterization of a unique gene product with partial rev activity. J. Virol. *73*, 608–617.

de Parseval, A., Chatterji, U., Sun, P., and Elder, J.H. (2004). Feline immunodeficiency virus targets activated CD4+ T-cells by using CD134 as a binding receptor. Proc. Natl. Acad. Sci. U.S.A. *101*, 13044–13049.

Dean, G.A., Reubel, G.H., Moore, P.F., and Pedersen, N.C. (1996). Proviral burden and infection kinetics of feline immunodeficiency virus in lymphocyte subsets of blood and lymph node. J. Virol. *70*, 5165–5169.

Desport, M., Stewart, M.E., Mikosza, A.S., Sheridan, C.A., Peterson, S.E., Chavand, O., Hartaningsih, N., and Wilcox, G.E. (2007). Sequence analysis of Jembrana disease virus strains reveals a genetically stable lentivirus. Virus Res. *126*, 233–244.

Edinger, A.L., Amedee, A., Miller, K., Doranz, B.J., Endres, M., Sharron, M., Samson, M., Lu, Z.H., Clements, J.E., Murphey-Corb, M., *et al.* (1997). Differential utilization of CCR5 by macrophage and T-cell tropic simian immunodeficiency virus strains. Proc. Natl. Acad. Sci. U.S.A. *94*, 4005–4010.

Elder, J.H., Lerner, D.L., Hasselkus-Light, C.S., Fontenot, D.J., Hunter, E., Luciw, P.A., Montelaro, R.C., and Phillips, T.R. (1992). Distinct subsets of retroviruses encode dUTPase. J. Virol. *66*, 1791–1794.

Elder, J.H., Sundstrom, M., de Rozieres, S., de Parseval, A., Grant, C.K., and Lin, Y.-C. (2008). Molecular mechanisms of FIV infection. Vet. Immunol. Immunopathol. *123*, 3–13.

English, R.V., Johnson, C.M., Gebhard, D.H., and Tompkins, M.B. (1993). *In vivo* lymphocyte tropism of feline immunodeficiency virus. J. Virol. *67*, 5175–5186.

Fagerness, A.J., Flaherty, M.T., Perry, S.T., Jia, B., Payne, S.L., and Fuller, F.J. (2006). The S2 accessory gene of equine infectious anemia virus is essential for expression of disease in ponies. Virology *349*, 22–30.

Finzi, D., Hermankova, M., Pierson, T., Carruth, L.M., Buck, C., Chaisson, R.E., Quinn, T.C., Chadwick, K., Margolick, J., Brookmeyer, R., *et al.* (1997). Identification of a Reservoir for HIV-1 in Patients on Highly Active Antiretroviral Therapy. Science *278*, 1295.

Foster, J.L., Anderson, S.J., Frazier, A.L., and Garcia, J.V. (1994). Specific suppression of human CD4 surface expression by Nef from the pathogenic simian immunodeficiency virus SIVmac239open. Virology *201*, 373–379.

Freed, E.O. (2002). Viral Late Domains. J. Virol. *76*, 4679–4687.

Fultz, P.N., Gordon, T.P., Anderson, D.C., and McClure, H.M. (1990). Prevalence of natural infection with simian immunodeficiency virus and simian T-cell leukemia virus type I in a breeding colony of sooty mangabey monkeys. Aids *4*, 619–625.

Gallo, R.C., Sarin, P.S., Gelmann, E.P., Robert-Guroff, M., Richardson, E., Kalyanaraman, V.S., Mann, D., Sidhu, G.D., Stahl, R.E., Zolla-Pazner, S., *et al.* (1983). Isolation of human T-cell leukemia virus in acquired immune deficiency syndrome (AIDS). Science *220*, 865–867.

Gao, F., Bailes, E., Robertson, D.L., Chen, Y., Rodenburg, C.M., Michael, S.F., Cummins, L.B., Arthur, L.O., Peeters, M., Shaw, G.M., *et al.* (1999). Origin of HIV-1 in the chimpanzee Pan troglodytes troglodytes. Nature *397*, 436–441.

Garcia, J.V., and Miller, A.D. (1991). Serine phosphorylation-independent down-regulation of cell-surface CD4 by nef. Nature *350*, 508–511.

Gardner, M.B. (1996). The history of simian AIDS. J. Med. Primatol. *25*, 148–157.

Garvey, K.J., Oberste, M.S., Elser, J.E., Braun, M.J., and Gonda, M.A. (1990). Nucleotide sequence and genome organization of biologically active proviruses of the bovine immunodeficiency-like virus. Virology *175*, 391–409.

Geleziunas, R., Bour, S., and Wainberg, M.A. (1994). Cell surface down-modulation of CD4 after infection by HIV-1. Faseb J 8, 593–600.

Gendelman, H.E., Narayan, O., Molineaux, S., Clements, J.E., and Ghotbi, Z. (1985). Slow, persistent replication of lentiviruses: role of tissue macrophages and macrophage precursors in bone marrow. Proc. Natl. Acad. Sci. U.S.A. *82*, 7086–7090.

Gendelman, H.E., Narayan, O., Kennedy-Stoskopf, S., Kennedy, P.G., Ghotbi, Z., Clements, J.E., Stanley, J., and Pezeshkpour, G. (1986). Tropism of sheep lentiviruses for monocytes: susceptibility to infection and virus gene expression increase during maturation of monocytes to macrophages. J. Virol. *58*, 67–74.

Gojobori, T., and Yokoyama, S. (1985). Rates of evolution of the retroviral oncogene of Moloney murine sarcoma virus and of its cellular homologues. Proc. Natl. Acad. Sci. U.S.A. *82*, 4198–4201.

Gonda, M.A., Charman, H.P., Walker, J.L., and Coggins, L. (1978). Scanning and transmission electron microscopic study of equine infectious anemia virus. Am. J. Vet. Res. *39*, 731–740.

Gonda, M.A., Wong-Staal, F., Gallo, R.C., Clements, J.E., Narayan, O., and Gilden, R.V. (1985). Sequence homology and morphologic similarity of HTLV-III and visna virus, a pathogenic lentivirus. Science *227*, 173–177.

Gonda, M.A., Braun, M.J., Clements, J.E., Pyper, J.M., Wong-Staal, F., Gallo, R.C., and Gilden, R.V. (1986). Human T-cell lymphotropic virus type III shares sequence homology with a family of pathogenic lentiviruses. Proc. Natl. Acad. Sci. U.S.A. *83*, 4007–4011.

Gonda, M.A., Braun, M.J., Carter, S.G., Kost, T.A., Bess, J.W., Jr, Arthur, L.O., and Van der Maaten, M.J. (1987). Characterization and molecular cloning of a bovine lentivirus related to human immunodeficiency virus. Nature *330*, 388–391.

Gonda, M.A. (1992). Bovine immunodeficiency virus. Aids *6*, 759–776.

Gorrell, M.D., Brandon, M.R., Sheffer, D., Adams, R.J., and Narayan, O. (1992). Ovine lentivirus is macrophagetropic and does not replicate productively in T-lymphocytes. J. Virol. *66*, 2679–2688.

Grego, E., Bertolotti, L., Carrozza, M.L., Profiti, M., Mazzei, M., Tolari, F., and Rosati, S. (2005). Genetic and antigenic characterization of the matrix protein of two genetically distinct ovine lentiviruses. Vet. Microbiol. *106*, 179–185.

Grego, E., Bertolotti, L., Quasso, A., Profiti, M., Lacerenza, D., Muz, D., and Rosati, S. (2007). Genetic characterization of small ruminant lentivirus in Italian mixed flocks: evidence for a novel genotype circulating in a local goat population. J. Gen. Virol. *88*, 3423–3427.

Gunnery, S., Green, S.R., and Mathews, M.B. (1992). Tat-responsive region RNA of human immunodeficiency virus type 1 stimulates protein synthesis *in vivo* and *in vitro*: relationship between structure and function. Proc. Natl. Acad. Sci. U.S.A. *89*, 11557–11561.

Guy, B., Kieny, M.P., Riviere, Y., Le Peuch, C., Dott, K., Girard, M., Montagnier, L., and Lecocq, J.P. (1987). HIV F/3′ orf encodes a phosphorylated GTP-binding protein resembling an oncogene product. Nature *330*, 266–269.

Hahn, B.H., Shaw, G.M., De Cock, K.M., and Sharp, P.M. (2000). AIDS as a zoonosis: scientific and public health implications. Science *287*, 607–614.

Hammond, S.A., Li, F., McKeon, B.M., Sr., Cook, S.J., Issel, C.J., and Montelaro, R.C. (2000). Immune responses and viral replication in long-term inapparent carrier ponies inoculated with equine infectious anemia virus. J. Virol. *74*, 5968–5981.

Han, Y., Lassen, K., Monie, D., Sedaghat, A.R., Shimoji, S., Liu, X., Pierson, T.C., Margolick, J.B., Siliciano, R.F., and Siliciano, J.D. (2004). Resting CD4+ T-cells from human immunodeficiency virus type 1 (HIV-1)-infected individuals carry integrated HIV-1 genomes within actively transcribed host genes. J. Virol. *78*, 6122–6133.

Harmache, A., Vitu, C., Russo, P., Bouyac, M., Hieblot, C., Peveri, P., Vigne, R., and Suzan, M. (1995). The caprine arthritis encephalitis virus tat gene is dispensable for efficient viral replication *in vitro* and *in vivo*. J. Virol. *69*, 5445–5454.

Harris, R.S., Sheehy, A.M., Craig, H.M., Malim, M.H., and Neuberger, M.S. (2003). DNA deamination: not just a trigger for antibody diversification but also a mechanism for defense against retroviruses. Nat. Immunol. *4*, 641–643.

Harrold, S.M., Cook, S.J., Cook, R.F., Rushlow, K.E., Issel, C.J., and Montelaro, R.C. (2000). Tissue sites of persistent infection and active replication of equine infectious anemia virus during acute disease and asymptomatic infection in experimentally infected equids. J. Virol. *74*, 3112–3121.

Hartaningsih, N., Wilcox, G.E., Dharma, D.M., and Soetrisno, M. (1993). Distribution of Jembrana disease in cattle in Indonesia. Vet. Microbiol. *38*, 23–29.

Hartmann, K. (1998). Feline immunodeficiency virus infection: an overview. Vet. J. *155*, 123–137.

Henson, J.B., and McGuire, T.C. (1974). Equine infectious anemia. Prog. Med. Virol. *18*, 143–159.

Hirsch, V.M. (2004). What can natural infection of African monkeys with simian immunodeficiency virus tell us about the pathogenesis of AIDS? AIDS Rev. *6*, 40–53.

Hosie, M.J., Robertson, C., and Jarrett, O. (1989). Prevalence of feline leukaemia virus and antibodies to feline immunodeficiency virus in cats in the United Kingdom. Vet. Rec. *125*, 293–297.

Hötzel, I., and Cheevers, W. (2002). Differential receptor usage of small ruminant lentiviruses in ovine and caprine cells: host range but not cytopathic phenotype is determined by receptor usage. Virology *301*, 21–31.

Ishida, T., Washizu, T., Toriyabe, K., Motoyoshi, S., Tomoda, I., and Pedersen, N.C. (1989). Feline immunodeficiency virus infection in cats of Japan. J. Am. Vet. Med. Assoc. *194*, 221–225.

Jolly, P.E., and Narayan, O. (1989). Evidence for interference, coinfections, and intertypic virus enhancement of infection by ovine-caprine lentiviruses. J. Virol. *63*, 4682–4688.

Kanki, P.J., Kurth, R., Becker, W., Dreesman, G., McLane, M.F., and Essex, M. (1985). Antibodies to simian T-lymphotropic retrovirus type III in African green monkeys and recognition of STLV-III viral proteins by AIDS and related sera. Lancet *1*, 1330–1332.

Kent, J.R., Kang, W., Miller, C.G., and Fraser, N.W. (2003). Herpes simplex virus latency-associated transcript gene function. J. Neurovirol. *9*, 285–290.

Kertayadnya, G., Wilcox, G.E., Soeharsono, S., Hartaningsih, N., Coelen, R.J., Cook, R.D., Collins, M.E., and Brownlie, J. (1993). Characteristics of a retrovirus associated with Jembrana disease in Bali cattle. J. Gen. Virol. *74*, 1765–1773.

Kim, W.K., Corey, S., Alvarez, X., and Williams, K. (2003). Monocyte/macrophage traffic in HIV and SIV encephalitis. J. Leukoc. Biol. *74*, 650–656.

Knowles, D.P., Jr, Cheevers, W.P., McGuire, T.C., Brassfield, A.L., Harwood, W.G., and Stem, T.A. (1991). Structure and genetic variability of envelope glycoproteins of two antigenic variants of caprine arthritis-encephalitis lentivirus. J. Virol. *65*, 5744–5750.

Kono, Y., hirasawa, K., Fukunaga, Y., and Taniguchi, T. (1976). Recrudesence of equine infectious anemia by treatment with immunosuppressive drugs. Natl. Inst. Anim. Hlth Q. *16*, 8–15.

Kono, Y. (1988). Antigenic variation of equine infectious anemia virus as detected by virus neutralization. Arch. Virol. *98*, 91–97.

Kuiken, C., Foley, B., Hahn, B.H., McCutchan, F., Mellors, J.W., Mullins, J.I., Wolinsky, S.M., and Korber, B. (1999). Human Retroviruses and AIDS (Los Alamos, NM: Los Alamos National Laboratory).

Kuiken, C.L., Foley, B., Hahn, B., Korber, B., McCutchan, F., Marx, P.A., Mellors, J.W., Mullins, J.I., Sodroski, J., and Wolinksy, S. (2000). Human Retroviruses and AIDS 2000: A Compilation and Analysis of Nucleic Acid and Amino Acid Sequences., In Theoretical Biology and Biophysics Group, Los Alamos National Laboratory (Los Alamos, NM.).

Lackner, A.A., and Veazey, R.S. (2007). Current Concepts in AIDS Pathogenesis: Insights from the SIV/Macaque Model. Annu. Rev. Med. *58*, 461–476.

Lassen, K.G., Bailey, J.R., and Siliciano, R.F. (2004). Analysis of human immunodeficiency virus type 1 transcriptional elongation in resting CD4$^+$ T Cells *In Vivo*. J. Virol. *78*, 9105–9114.

Lavallee, C., Yao, X.J., Ladha, A., Gottlinger, H., Haseltine, W.A., and Cohen, E.A. (1994). Requirement of the Pr55gag precursor for incorporation of the Vpr product into human immunodeficiency virus type 1 viral particles. J. Virol. *68*, 1926–1934.

Leroux, C., Vuillermoz, S., Mornex, J.F., and Greenland, T. (1995). Genomic heterogeneity in the pol region of ovine lentiviruses obtained from bronchoalveolar cells of infected sheep from France. J. Gen. Virol. *76*, 1533–1537.

Leroux, C., Chastang, J., Greenland, T., and Mornex, J.F. (1997a). Genomic heterogeneity of small ruminant lentiviruses: existence of heterogeneous populations in sheep and of the same lentiviral genotypes in sheep and goats. Arch. Virol. *142*, 1125–1137.

Leroux, C., Issel, C., and Montelaro, R.C. (1997b). Novel and dynamic evolution of equine infectious anemia virus genomic quasispecies associated with sequential disease cycles in an experimentally infected pony. J. Virol. *71*, 9627–9639.

Leroux, C., Cadore, J.L., and Montelaro, R.C. (2004). Equine infectious anemia virus (EIAV): what has HIV's country cousin got to tell us? Vet. Res. *35*, 485–512.

Levy, J.A. (2006). HIV pathogenesis: knowledge gained after two decades of research. Adv. Dent. Res. *19*, 10–16.

Lewinski, M.K., Bisgrove, D., Shinn, P., Chen, H., Hoffmann, C., Hannenhalli, S., Verdin, E., Berry, C.C., Ecker, J.R., and Bushman, F.D. (2005). Genome-wide analysis of chromosomal features repressing human immunodeficiency virus transcription. J. Virol. *79*, 6610–6619.

Li, F., Puffer, B.A., and Montelaro, R.C. (1998). The S2 gene of equine infectious anemia virus is dispensable for viral replication *in vitro*. J. Virol. *72*, 8344–8348.

Li, F., Leroux, C., Craigo, J.K., Cook, S.J., Issel, C.J., and Montelaro, R.C. (2000). The S2 gene of equine infectious anemia virus is a highly conserved determinant of viral replication and virulence properties in experimentally infected ponies. J. Virol. *74*, 573–579.

Lichtenstein, D.L., Rushlow, K.E., Cook, R.F., Raabe, M.L., Swardson, C.J., Kociba, G.J., Issel, C.J., and Montelaro, R.C. (1995). Replication *in vitro* and *in vivo* of an equine infectious anemia virus mutant deficient in dUTPase activity. J. Virol. *69*, 2881–2888.

Lichtenstein, D.L., Issel, C.J., and Montelaro, R.C. (1996). Genomic quasispecies associated with the initiation of infection and disease in ponies experimentally infected with equine infectious anemia virus. J. Virol. *70*, 3346–3354.

Lignee, M. (1843). Memoire et observations sur une maladie de sang, connue sous le nom d'anhémie hydrohémie, cachexie aquise du cheval. Rec. Med. Vet. Ec. Alfort. *20*, 30–45.

Lu, Y.L., Spearman, P., and Ratner, L. (1993). Human immunodeficiency virus type 1 viral protein R localization in infected cells and virions. J. Virol. *67*, 6542–6550.

Luttge, B.G., Shehu-Xhilaga, M., Demirov, D.G., Adamson, C.S., Soheilian, F., Nagashima, K., Stephen, A.G., Fisher, R.J., and Freed, E.O. (2008). Molecular characterization of feline immunodeficiency virus budding. J. Virol. *82*, 2106–2119.

McConnel, M.B., Katada, M., McConnell, S., and Moore, R. (1977). Demonstration of equine infectious anemia virus in primary leukocyte cultures by electron microscopy. Am. J. Vet. Res. *38*, 2067–2069.

McGuire, T.C., and Crawford, T.B. (1979). Immunology of a persistent retrovirus infection--equine infectious anemia. Adv. Vet. Sci. Comp. Med. *23*, 137–159.

McNab, W.B., Jacobs, R.M., and Smith, H.E. (1994). A serological survey for bovine immunodeficiency-like virus in Ontario dairy cattle and associations between test results, production records and management practices. Can. J. Vet. Res. *58*, 36–41.

Malmquist, W.A., Van der Maaten, M.J., and Boothe, A.D. (1969). Isolation, immunodiffusion, immunofluorescence, and electron microscopy of a syncytial virus of lymphosarcomatous and apparently normal cattle. Cancer Res. *29*, 188–200.

Mangeat, B., Turelli, P., Caron, G., Friedli, M., Perrin, L., and Trono, D. (2003). Broad antiretroviral defence by human APOBEC3G through lethal editing of nascent reverse transcripts. Nature *424*, 99–103.

Mariani, R., Chen, D., Schrofelbauer, B., Navarro, F., Konig, R., Bollman, B., Munk, C., Nymark-McMahon, H., and Landau, N.R. (2003). Species-specific exclusion of APOBEC3G from HIV-1 virions by Vif. Cell *114*, 21–31.

Marlink, R., Kanki, P., Thior, I., Travers, K., Eisen, G., Siby, T., Traore, I., Hsieh, C.C., Dia, M.C., Gueye, E.H., and et, a. (1994). Reduced rate of disease development after HIV-2 infection as compared to HIV-1. Science *265*, 1587–1590.

Marlink, R.G., Ricard, D., M'Boup, S., Kanki, P.J., Romet-Lemonne, J.L., N'Doye, I., Diop, K., Simpson, M.A., Greco, F., Chou, M.J., and *et al.* (1988). Clinical, hematologic, and immunologic cross-sectional evaluation of individuals exposed to human immunodeficiency virus type-2 (HIV-2). AIDS Res. Hum. Retroviruses *4*, 137–148.

Marsh, H. (1923). Progressive pneumonia in sheep. J. Am. Vet. Med. Assoc. *62*, 458–473.

Marsh, M., Theusner, K., and Pelchen-Matthews, A. (2009). HIV assembly and budding in macrophages. Biochem. Soc. Trans. *37*, 185–189.

Maury, W. (1994). Monocyte maturation controls expression of equine infectious anemia virus. J. Virol. *68*, 6270–6279.

Meas, S., Usui, T., Ohashi, K., Sugimoto, C., and Onuma, M. (2002). Vertical transmission of bovine leukemia virus and bovine immunodeficiency virus in dairy cattle herds. Vet. Microbiol. *84*, 275–282.

Meltzer, M.S., Skillman, D.R., Hoover, D.L., Hanson, B.D., Turpin, J.A., Kalter, D.C., and Gendelman, H.E. (1990). Macrophages and the human immunodeficiency virus. Immunol. Today *11*, 217–223.

Meltzer, M.S., and Gendelman, H.E. (1992). Mononuclear phagocytes as targets, tissue reservoirs, and immunoregulatory cells in human immunodeficiency virus disease. Curr. Top. Microbiol. Immunol. *181*, 239–263.

Montelaro, R.C., Parekh, B., Orrego, A., Issel, C.J., Payne, S., Parekh, B., Montelaro, R.C., Issel, C.J., Montelaro, R.C., West, M., and Issel, C.J. (1984). Antigenic variation during persistent infection by equine infectious anemia virus, a retrovirus. J. Biol. Chem. *259*, 10539–10544.

Montelaro, R.C., Robey, W.G., West, M.D., Issel, C.J., and Fischinger, P.J. (1988). Characterization of the serological cross-reactivity between glycoproteins of the human immunodeficiency virus and equine infectious anaemia virus. J. Gen. Virol. *69*, 1711–1717.

Montelaro, R.C., Ball, J.M., and Rushlow, K. (1993). Equine retroviruses, In The Retroviridae, J.A. Levy, ed. (New York: Plenum Press), pp. 257–360.

Mornex, J.F., Lena, P., Loire, R., Cozon, G., Greenland, T., Guigen, F., Jacquier, M.F., and Cordier, G. (1994). Lentivirus-induced interstitial lung disease: pulmonary pathology in sheep naturally infected by the visna–maedi virus. Vet. Res. *25*, 478–488.

Murphey-Corb, M., Martin, L.N., Rangan, S.R., Baskin, G.B., Gormus, B.J., Wolf, R.H., Andes, W.A., West, M., and Montelaro, R.C. (1986). Isolation of an HTLV-III-related retrovirus from macaques with simian AIDS and its possible origin in asymptomatic mangabeys. Nature *321*, 435–437.

Narayan, O., Clements, J.E., Strandberg, J.D., Cork, L.C., and Griffin, D.E. (1980). Biological characterization of the virus causing leukoencephalitis and arthritis in goats. J. Gen. Virol. *50*, 69–79.

Narayan, O., Wolinsky, J.S., Clements, J.E., Strandberg, J.D., Griffin, D.E., and Cork, L.C. (1982). Slow virus replication: the role of macrophages in the persistence and expression of visna viruses of sheep and goats. J. Gen. Virol. *59*, 345–356.

Narayan, O., Kennedy-Stoskopf, S., Sheffer, D., Griffin, D.E., and Clements, J.E. (1983). Activation of caprine arthritis-encephalitis virus expression during maturation of monocytes to macrophages. Infect. Immun. *41*, 67–73.

Narayan, O., Zink, M., Gorrel, M., Crane, S., Huso, D., and Jolly, P. (1993). The lentiviruses of sheep and goats, In The Retroviridae (New York: Plenum Press), pp. 229–256.

O'Neil, L.L., Burkhard, M.J., Diehl, L.J., and Hoover, E.A. (1995). Vertical transmission of feline immunodeficiency virus. AIDS Res. Hum. Retroviruses *11*, 171–182.

O'Neil, L.L., Burkhard, M.J., and Hoover, E.A. (1996). Frequent perinatal transmission of feline immunodeficiency virus by chronically infected cats. J. Virol. *70*, 2894–2901.

Palsson, P.A. (1976). Meadi and visna in sheep (Amsterdam: North Holland Publishing Company).

Pancio, H.A., Vander Heyden, N., and Ratner, L. (2000). The C-terminal proline-rich tail of human immunodeficiency virus type 2 Vpx is necessary for nuclear localization of the viral preintegration complex in non-dividing cells. J. Virol. *74*, 6162–6167.

Pandrea, I., Onanga, R., Kornfeld, C., Rouquet, P., Bourry, O., Clifford, S., Telfer, P.T., Abernethy, K., White, L.T., Ngari, P., *et al.* (2003). High levels of SIVmnd-1 replication in chronically infected *Mandrillus sphinx*. Virology *317*, 119–127.

Payne, S.L., Salinovich, O., Nauman, S.M., Issel, C.J., and Montelaro, R.C. (1987a). Course and extent of variation of equine infectious anemia virus during parallel persistent infections. J. Virol. *61*, 1266–1270.

Payne, S.L., Fang, F.D., Liu, C.P., Dhruva, B.R., Rwambo, P., Issel, C.J., and Montelaro, R.C. (1987b). Antigenic variation and lentivirus persistence: variations in envelope gene sequences during EIAV infection resemble changes reported for sequential isolates of HIV. Virology *161*, 321–331.

Payne, S.L., Rushlow, K., Dhruva, B.R., Issel, C.J., and Montelaro, R.C. (1989). Localization of conserved and variable antigenic domains of equine infectious anemia virus envelope glycoproteins using recombinant env-encoded protein fragments produced in Escherichia coli. Virology *172*, 609–615.

Payne, S.L., and Elder, J.H. (2001). The role of retroviral dUTPases in replication and virulence. Curr. Protein Pept. Sci. *2*, 381–388.

Pedersen, N. (1993). The feline immunodeficiency virus., In The Retroviridae, J.A. Levy, ed. (New York: Plenum Press), pp. 181–228.

Pedersen, N.C., Ho, E.W., Brown, M.L., and Yamamoto, J.K. (1987). Isolation of a T-lymphotropic virus from domestic cats with an immunodeficiency-like syndrome. Science *235*, 790–793.

Pedersen, N.C., Yamamoto, J.K., Ishida, T., and Hansen, H. (1989). Feline immunodeficiency virus infection. Vet. Immunol. Immunopathol. *21*, 111–129.

Peluso, R., Haase, A., Stowring, L., Edwards, M., and Ventura, P. (1985). A Trojan Horse mechanism for the spread of visna virus in monocytes. Virology *147*, 231–236.

Perk, K. (1999). Concealed locations of lentiviruses in caprine arthritis encephalitis system. Virology *253*, 8–9.

Peterhans, E., Greenland, T., Badiola, J., Harkiss, G., Bertoni, G., Amorena, B., Eliaszewicz, M., Juste, R.A., Krassnig, R., Lafont, J.P., *et al.* (2004). Routes of transmission and consequences of small ruminant lentiviruses (SRLVs) infection and eradication schemes. Vet. Res. *35*, 257–274.

Phillips, T.R., Prospero-Garcia, O., Puaoi, D.L., Lerner, D.L., Fox, H.S., Olmsted, R.A., Bloom, F.E., Henriksen, S.J., and Elder, J.H. (1994). Neurological abnormalities associated with feline immunodeficiency virus infection. J. Gen. Virol. *75*, 979–987.

Phillips, T.R., Prospero-Garcia, O., Wheeler, D.W., Wagaman, P.C., Lerner, D.L., Fox, H.S., Whalen, L.R., Bloom, F.E., Elder, J.H., and Henriksen, S.J. (1996). Neurologic dysfunctions caused by a molecular clone of feline immunodeficiency virus, FIV-PPR. J. Neurovirol. *2*, 388–396.

Pierson, T., McArthur, J., and Siliciano, R.F. (2000). Reservoirs for HIV-1: mechanisms for viral persistence in the presence of antiviral immune responses and antiretroviral therapy. Annu. Rev. Immunol. *18*, 665–708.

Pierson, T.C., Zhou, Y., Kieffer, T.L., Ruff, C.T., Buck, C., and Siliciano, R.F. (2002). Molecular characterization of preintegration latency in human immunodeficiency virus type 1 infection. J. Virol. *76*, 8518.

Piguet, V., and Trono, D. (1999). The Nef protein of primate lentiviruses. Rev. Med. Virol. *9*, 111–120.

Pisoni, G., Quasso, A., and Moroni, P. (2005). Phylogenetic analysis of small-ruminant lentivirus subtype B1 in mixed flocks: evidence for natural transmission from goats to sheep. Virology *339*, 147–152.

Pisoni, G., Bertoni, G., Boettcher, P., Ponti, W., and Moroni, P. (2006). Phylogenetic analysis of the gag region encoding the matrix protein of small ruminant lentiviruses: comparative analysis and molecular epidemiological applications. Virus Res. *116*, 159–167.

Preston, B.D., Poiesz, B.J., and Loeb, L.A. (1988). Fidelity of HIV-1 reverse transcriptase. Science *242*, 1168–1171.

Reina, R., Mora, M.I., Glaria, I., Garcia, I., Solano, C., Lujan, L., Badiola, J.J., Contreras, A., Berriatua, E., Juste, R., *et al.* (2006). Molecular characterization and phylogenetic study of maedi Visna and Caprine Arthritis Encephalitis viral sequences in sheep and goats from Spain. Virus Res. *121*, 189–198.

Richards, R.G., Sowers, L.C., Laszlo, J., and Sedwick, W.D. (1984). The occurrence and consequences of deoxyuridine in DNA. Adv. Enzyme Regul. *22*, 157–185.

Roberts, J.D., Bebenek, K., and Kunkel, T.A. (1988). The accuracy of reverse transcriptase from HIV-1. Science *242*, 1171–1173.

Roizman, B., and Sears, A.E. (1987). An inquiry into the mechanisms of herpes simplex virus latency. Annu. Rev. Microbiol. *41*, 543–571.

Roizman, B., and Sears, A.E. (1995). Herpes simplex viruses and their replication., In Fields virology, B. Fields, D. Knipe, and P. Howley, eds. (Philadelphia: Lippincott-Raven), pp. 2231–2296.

Ross, M.J., and Klotman, P.E. (2004). HIV-associated nephropathy. Aids *18*, 1089–1099.

Rwambo, P.M., Issel, C.J., Adams, W.V., Jr, Hussain, K.A., Miller, M., and Montelaro, R.C. (1990a). Equine infectious anemia virus (EIAV) humoral responses of recipient ponies and antigenic variation during persistent infection. Arch. Virol. *111*, 199–212.

Rwambo, P.M., Issel, C.J., Hussain, K.A., and Montelaro, R.C. (1990b). *In vitro* isolation of a neutralization escape mutant of equine infectious anemia virus (EIAV). Arch. Virol. *111*, 275–280.

Ryan, S., Tiley, L., McConnell, I., and Blacklaws, B. (2000). Infection of dendritic cells by the maedi-visna lentivirus. J. Virol. *74*, 10096–10103.

Salinovich, O., Payne, S.L., Montelaro, R.C., Hussain, K.A., Issel, C.J., and Schnorr, K.L. (1986). Rapid emergence of novel antigenic and genetic variants of equine infectious anemia virus during persistent infection. J. Virol. *57*, 71–80.

Scott, J.V., Stowring, L., Haase, A.T., Narayan, O., and Vigne, R. (1979). Antigenic variation in visna virus. Cell *18*, 321–327.

Sellon, D., Walker, K.M., Russel, K.E., Perry, S.T., and Fuller, F. (1996). Phorbol ester stimulation of equine macrophage cultures alters expression of equine infectious anemia virus. Vet. Microbiol. *52*, 209–221.

Shah, C., Boni, J., Huder, J.B., Vogt, H.R., Muhlherr, J., Zanoni, R., Miserez, R., Lutz, H., and Schupbach, J. (2004a). Phylogenetic analysis and reclassification of caprine and ovine lentiviruses based on 104 new isolates: Evidence for regular sheep-to-goat transmission and worldwide propagation through livestock trade. Virology *319*, 12–26.

Shah, C., Huder, J.B., Boni, J., Schonmann, M., Muhlherr, J., Lutz, H. and Schupbach, J. (2004b). Direct evidence for natural transmission of small-ruminant lentiviruses of subtype A4 from goats to sheep and vice versa. J. Virol. *78*, 7518–7522.

Shannon, R.P. (2001). SIV cardiomyopathy in non-human primates. Trends Cardiovasc. Med. *11*, 242–246.

Sharp, P.M., Shaw, G.M., and Hahn, B.H. (2005). Simian immunodeficiency virus infection of chimpanzees. J. Virol. *79*, 3891–3902.

Sheehy, A.M., Gaddis, N.C., Choi, J.D., and Malim, M.H. (2002). Isolation of a human gene that inhibits HIV-1 infection and is suppressed by the viral Vif protein. Nature *418*, 646–650.

Shen, A., Zink, M.C., Mankowski, J.L., Chadwick, K., Margolick, J.B., Carruth, L.M., Li, M., Clements, J.E., and Siliciano, R.F. (2003). Resting CD4+ T-lymphocytes but not thymocytes provide a latent viral reservoir in a simian immunodeficiency virus–*Macaca nemestrina* model of human immunodeficiency virus type 1-infected patients on highly active antiretroviral therapy. J. Virol. *77*, 4938.

Shimojima, M., Miyazawa, T., Ikeda, Y., McMonagle, E.L., Haining, H., Akashi, H., Takeuchi, Y., Hosie, M.J., and Willett, B.J. (2004). Use of CD134 as a primary receptor by the feline immunodeficiency virus. Science *303*, 1192–1195.

Sigurdsson, B. (1954). maedi, a slow progressive pheumonia of sheep: an epizoological and a pathological study. Br. Vet. J. *110*, 255–260.

Sigurdsson, B., Palsson, P., and Grimsson, H. (1957). Visna, a demylienating transmissible disease of sheep. J. Neuropathol. Exp. Neurol. *16*, 389–403.

Silvestri, G., Sodora, D.L., Koup, R.A., Paiardini, M., O'Neil, S.P., McClure, H.M., Staprans, S.I., and Feinberg, M.B. (2003). Nonpathogenic SIV infection of sooty mangabeys is characterized by limited bystander immunopathology despite chronic high-level viremia. Immunity *18*, 441–452.

Silvestri, G. (2005). Naturally SIV-infected sooty mangabeys: are we closer to understanding why they do not develop AIDS? J. Med. Primatol. *34*, 243–252.

Silvestri, G., Paiardini, M., Pandrea, I., Lederman, M.M., and Sodora, D.L. (2007). Understanding the benign nature of SIV infection in natural hosts. J. Clin. Invest. *117*, 3148–3154.

Simon, V., Ho, D.D., and Abdool Karim, Q. (2006). HIV/AIDS epidemiology, pathogenesis, prevention, and treatment. Lancet *368*, 489–504.

Snider, T.G., Hoyt, P.G., Jenny, B.F., Coats, K.S., Luther, D.G., Storts, R.W., Battles, J.K., and Gonda, M.A. (1997). Natural and experimental bovine immunodeficiency virus infection in cattle. Vet. Clin. North Am. Food Anim. Pract. *13*, 151–176.

Soeharsono, S., Hartaningsih, N., Soetrisno, M., Kertayadnya, G., and Wilcox, G.E. (1990). Studies of experimental Jembrana disease in Bali cattle. I. Transmission and persistence of the infectious agent in ruminants and pigs, and resistance of recovered cattle to re-infection. J. Comp. Pathol. *103*, 49–59.

Soeharsono, S., Wilcox, G.E., Putra, A.A., Hartaningsih, N., Sulistyana, K., and Tenaya, M. (1995a). The transmission of Jembrana disease, a lentivirus disease of Bos javanicus cattle. Epidemiol. Infect. *115*, 367–374.

Soeharsono, S., Wilcox, G.E., Dharma, D.M., Hartaningsih, N., Kertayadnya, G., and Budiantono, A. (1995b). Species differences in the reaction of cattle to Jembrana disease virus infection. J. Comp. Pathol. *112*, 391–402.

Southgate, C.D., and Green, M.R. (1995). Delineating minimal protein domains and promoter elements for transcriptional activation by lentivirus Tat proteins. J. Virol. *69*, 2605–2610.

Sparger, E.E., Shacklett, B.L., Renshaw-Gegg, L., Barry, P.A., Pedersen, N.C., Elder, J.H., and Luciw, P.A. (1992). Regulation of gene expression directed by the long terminal repeat of the feline immunodeficiency virus. Virology *187*, 165–177.

St-Louis, M.-C., Cojocariu, M., and Archambault, D. (2004). The molecular biology of bovine immunodeficiency virus: a comparison with other lentiviruses. Anim. Hlth Rese. Rev. *5*, 125–143.

Stewart, M., Desport, M., Hartaningsih, N., and Wilcox, G. (2005). TaqMan real-time reverse transcription-PCR and JDVp26 antigen capture enzyme-linked immunosorbent assay to quantify Jembrana disease virus load during the acute phase of *in vivo* infection. J. Clin. Microbiol. *43*, 5574–5580.

Stewart, M.E., Desport, M., Setiyaningsih, S., Hartaningsih, N., and Wilcox, G.E. (2008). Analysis of Jembrana disease virus mRNA transcripts produced during acute infection demonstrates a complex transcription pattern. Virus Res. *135*, 336–339.

Subbramanian, R.A., and Cohen, E.A. (1994). Molecular biology of the human immunodeficiency virus accessory proteins. J. Virol. *68*, 6831–6835.

Swiggard, W.J., Baytop, C., Yu, J.J., Dai, J., Li, C., Schretzenmair, R., Theodosopoulos, T., and O'Doherty, U. (2005). Human immunodeficiency virus type 1 can establish latent infection in resting CD4+ T-cells in the absence of activating stimuli. J. Virol. *79*, 14179–14188.

Thorne, C., and Newell, M.L. (2000). Epidemiology of HIV infection in the newborn. Early Hum. Dev. *58*, 1–16.

Threadgill, D.S., Steagall, W.K., Flaherty, M.T., Fuller, F.J., Perry, S.T., Rushlow, K.E., Le Grice, S.F., and Payne, S.L. (1993). Characterization of equine infectious anemia virus dUTPase: growth properties of a dUTPase-deficient mutant. J. Virol. *67*, 2592–2600.

Troyer, J.L., Pecon-Slattery, J., Roelke, M.E., Johnson, W., VandeWoude, S., Vazquez-Salat, N., Brown, M., Frank, L., Woodroffe, R., Winterbach, C., *et al.* (2005). Seroprevalence and genomic divergence of circulating strains of feline immunodeficiency virus among Felidae and Hyaenidae species. J. Virol. *79*, 8282–8294.

Turelli, P., Petursson, G., Guiguen, F., Mornex, J.F., Vigne, R., and Querat, G. (1996). Replication properties of dUTPase-deficient mutants of caprine and ovine lentiviruses. J. Virol. *70*, 1213–1217.

Vallee, H.C.H. (1904). Nature infertieuse de l'anemie du cheval. C. R. Acad. Sci. *139*, 331–333.

Van der Maaten, M.J., Boothe, A.D., and Seger, C.L. (1972). Isolation of a virus from cattle with persistent lymphocytosis. J. Natl. Cancer Inst. *49*, 1649–1657.

Van der Maaten, M.J., Whetstone, C.A., Khramtsov, V.V., and Miller, J.M. (1990). Experimentally-induced infections with bovine immunodeficiency-like virus, a bovine lentivirus. Dev. Biol. Stand. *72*, 91–95.

VandeWoude, S., and Apetrei, C. (2006). Going wild: lessons from naturally occurring T-lymphotropic lentiviruses. Clin. Microbiol. Rev. *19*, 728–762.

Vassylyev, D.G., and Morikawa, K. (1996). Precluding uracil from DNA. Structure *4*, 1381–1385.

Verani, A., Gras, G., and Pancino, G. (2005). Macrophages and HIV-1: dangerous liaisons. Mol. Immunol. *42*, 195–212.

Villet, S., Faure, C., Bouzar, B.A., Morin, T., Verdier, G., Chebloune, Y., and Legras, C. (2003a). Lack of trans-activation function for maedi Visna virus and Caprine arthritis encephalitis virus Tat proteins. Virology *307*, 317–327.

Villet, S., Bouzar, B.A., Morin, T., Verdier, G., Legras, C., and Chebloune, Y. (2003b). maedi-visna virus and caprine arthritis encephalitis virus genomes encode a Vpr-like but no Tat protein. J. Virol. *77*, 9632–9638.

Walder, R., Kalvatchev, Z., Tobin, G.J., Barrios, M.N., Garzaro, D.J., and Gonda, M.A. (1995). Possible role of bovine immunodeficiency virus in bovine paraplegic syndrome: evidence from immunochemical, virological and seroprevalence studies. Res. Virol. *146*, 313–323.

Waters, A.K., De Parseval, A.P., Lerner, D.L., Neil, J.C., Thompson, F.J., and Elder, J.H. (1996). Influence of ORF2 on host cell tropism of feline immunodeficiency virus. Virology *215*, 10–16.

Weiland, F., Matheka, H.D., Coggins, L., and Hatner, D. (1977). Electron microscopic studies on equine infectious anemia virus (EIAV). Brief report. Arch. Virol. *55*, 335–340.

Welsch, S., Keppler, O.T., Habermann, A., Allespach, I., Krijnse-Locker, J., and Krausslich, H.G. (2007). HIV-1 buds predominantly at the plasma membrane of primary human macrophages. PLoS Pathog. *3*, e36.

Whetstone, C.A., VanDerMaaten, M.J., and Black, J.W. (1990). Humoral immune response to the bovine immunodeficiency-like virus in experimentally and naturally infected cattle. J. Virol. *64*, 3557–3561.

Wilcox, G. (1997). Jembrana disease. Austral. Vet. J. *75*, 492–493.

Wilcox, G.E., Kertayadnya, G., Hartaningsih, N., Dharma, D.M., Soeharsono, S., and Robertson, T. (1992). Evidence for a viral aetiology of Jembrana disease in Bali cattle. Vet. Microbiol. *33*, 367–374.

Wilcox, G.E., Chadwick, B.J., and Kertayadnya, G. (1995). Recent advances in the understanding of Jembrana disease. Vet. Microbiol. *46*, 249–255.

Willbold, D., Metzger, A.U., Sticht, H., Gallert, K.C., Voit, R., Dank, N., Bayer, P., Krauss, G., Goody, R.S., and Rosch, P. (1998). Equine infectious anemia virus transactivator is a homeodomain-type protein. J. Mol. Biol. *277*, 749–755.

Willett, B.J., Picard, L., Hosie, M.J., Turner, J.D., Adema, K., and Clapham, P.R. (1997). Shared usage of the chemokine receptor CXCR4 by the feline and human immunodeficiency viruses. J. Virol. *71*, 6407–6415.

Willey, R.L., Maldarelli, F., Martin, M.A., and Strebel, K. (1992). Human immunodeficiency virus type 1 Vpu protein induces rapid degradation of CD4. J. Virol. *66*, 7193–7200.

Williams, S.A., and Greene, W.C. (2005). Host factors regulating post-integration latency of HIV. Trends Microbiol. *13*, 137–139.

Wong, J.K., Hezareh, M., Gunthard, H.F., Havlir, D.V., Ignacio, C.C., Spina, C.A., and Richman, D.D. (1997). Recovery of replication-competent HIV despite prolonged suppression of plasma viremia. Science *278*, 1291.

Wright, S.M., Mleczko, A., and Coats, K.S. (2002). Bovine immunodeficiency virus expression *in vitro* is reduced in the presence of beta-chemokines, MIP-1alpha, MIP-1beta and RANTES. Vet. Res. Commun. *26*, 239–250.

Zanoni, R.G. (1998). Phylogenetic analysis of small ruminant lentiviruses. J. Gen. Virol. *79*, 1951–1961.

Zhang, B., Jin, S., Jin, J., Li, F., and Montelaro, R.C. (2005). A tumor necrosis factor receptor family protein serves as a cellular receptor for the macrophage-tropic equine lentivirus. Proc. Natl. Acad. Sci. U.S.A. *102*, 9918–9923.

Zhang, B., Sun, C., Jin, S., Cascio, M., and Montelaro, R.C. (2008). Mapping of equine lentivirus receptor 1 residues critical for equine infectious anemia virus envelope binding. J. Virol. *82*, 1204–1213.

Zhou, Y., Zhang, H., Siliciano, J.D., and Siliciano, R.F. (2005). Kinetics of human immunodeficiency virus type 1 decay following entry into resting $CD4^+$ T cells. J. Virol. *79*, 2199–2210.

Zink, M.C., Montelaro, R.C., Leroux, C., Cunningham, M.W., and Fujinami, R.S. (2000). Lentivirus infections and the immune system, In Effects of Microbes on the Immune System (Philadlphia: Lippincott Williams & Wilkins).

Zink, M.C., Laast, V.A., Helke, K.L., Brice, A.K., Barber, S.A., Clements, J.E., and Mankowski, J.L. (2006). From mice to macaques – animal models of HIV nervous system disease. Curr. HIV Res. *4*, 293–305.

# A Bird's-eye View of Macrophage Biology

2

Ian Ross

Abstract

Macrophages possess elaborate sensors for pathogens and have evolved complex chemically mediated interactions with other components of the immune system. Macrophage precursors differentiate into multiple end-stage specialized cell types in mammals, each suited for specific roles in homeostasis and defence. This chapter attempts to paint a broad picture of macrophage biology, ranging from embryological origins, through their role in the coordination of immune defences, to specific mechanisms used to address different kinds of threats to the parent organism. Research in macrophage biology is at a particularly exciting stage with new defence pathways still in the process of discovery. At the same time, enough is known to glimpse the complete picture that will one day be available for this staggeringly complex molecular machine which is a central organiser of immune defences. Macrophages are critically involved in disease in humans and animals because of their roles both in chronic inflammatory diseases, and during infections by important pathogens, including lentiviruses. In each case, an understanding of macrophage defensive mechanisms and how they have been maladapted or thwarted in disease states provides a range of opportunities for specific therapeutic intervention.

## Introduction

The focus of this chapter is a broad overview of macrophage biology, to allow specialists in other areas to quickly build a picture of the macrophage in its natural setting, throughout normal development and health as well as diseases of various types. At the same time an effort has been made to put into context current developments in rapidly evolving research topics which have led to recent divergence from our understanding of the textbook macrophage. This entails, at times, some editorializing and opinion, which it is hoped will be constructive. Even a broad coverage must unfortunately omit important and exciting areas including immunosuppression by glucocorticoid receptors and the roles of PPARγ, retinoic acid, prostaglandin and leukotriene biology. Similarly macrophage–tumour interactions and antibody dependent cellular cytotoxicity (ADCC) are only briefly mentioned. Nonetheless, the central role of the macrophage in its natural setting – in the organism and not just in the Petri dish – can be glimpsed here. In particular, recently developing research areas such as antiviral molecular pattern receptors are discussed in sufficient detail to both contextualise them and to provide bearings which will enable the interested reader to launch into the literature with a sense of where each topic fits in the overall picture. Recent work has clarified (or confused) many longstanding assumptions or misunderstandings about macrophage biology (haematopoietic development being an example) and where relevant some of these are addressed so that outdated concepts are not perpetuated, even if the older picture was 'neater' than the current reality. The textbook macrophage is nearly always an oversimplification, as are theoretical synthetic schemes like the Th1/Th2 and M1/M2 polarities, which are usually intended to be guidelines rather than rulebooks. Sticking fast to such models can give rise to (often needless) controversies which usually signify only that 'more work is needed'.

## Historical context

Most general accounts of macrophage biology rightly begin with the observations of Ilya Mechnikov in the 1880s, who introduced the idea that phagocytosis is a hallmark of innate defence (Tauber, 1991; Gordon, 2007). The term 'macrophage' was initially used to describe wandering phagocytes observed across the range of animal systems and could not at that time have had the more restricted definition used today, especially since the concept of immune defence was itself in the early stages. In an age which lacked molecular tools, phagocytosis – the engulfment and destruction of the invading pathogen – provided a readily understood model for immune defence, in comparison to the invisible humoral factors whose source was then unknown. Nevertheless phagocytosis, while undeniably important, is neither the sole province of macrophages (as we now understand them) nor the primary mechanism by which they exert their functions in mammals.

At about the time Mechnikov was working, the presence of humoral immunity was established, in particular due to Paul Erlich's work. As often happens, an unnecessary rivalry emerged between the adherents of phagocytic and humoral immunity, and their relative importance in disease. The impressive ability of vaccination, by then a venerable empirical technique, to produce new humoral defences ('antibodies') which protected against important human pathogens led to the concept of *acquired immunity*, and perhaps the idea that the other defensive mechanisms were rather primitive and inadequate. The subsequent division of immunology into independent areas of 'innate' and 'acquired' immunity has persisted to the present day. Such a division is of course somewhat artificial since both the differences and the similarities between B- and T-cell-mediated immunity are (in terms of molecular mechanisms) as dramatic as those between lymphoid and myeloid cells, while their interdependence has become more evident with time. Nonetheless, T- and B-lymphocytes have always been studied together, with myeloid lineages (until the last decades) usually studied separately. Humans are easily impressed by visual differences, and the striking morphological differences between macrophages and lymphocytes (and the corresponding lack of difference between lymphocytes) may have reinforced the apparent naturalness of the lymphoid–myeloid distinction, while the lack of molecular tools to discriminate cells at the level of surface markers prevented a realization of the true commonality of all haematopoietic cells until the late 1970s and 1980s. In reality, since both these systems co-evolved in vertebrates, there is scarcely a process in acquired immunity in which macrophages are not intimately involved with lymphocytes in the resolution (or sometimes, as we now appreciate, causality) of disease processes.

The primordial idea of the macrophage as the 'big eater' is therefore both a blessing and a curse. On the one hand it underscores the undeniable roles of macrophages in immune defence and their phylogenetic ubiquity, but by caricaturing them as bacteria-eating sentinels it fails to convey the many trophic and homeostatic roles of macrophages during normal tissue development and maintenance which are also important.

## The molecular basis of macrophage functions

Most cellular mechanisms and processes were fixed well before the advent of multicellularity. Multicellular organisms evolved from free living single celled organisms which already possessed phagocytic abilities, sophisticated receptors, signalling networks, gene control, defence mechanisms and the necessary apparatus for the production and secretion of proteins including cytokine-like molecules. Consequently, from an evolutionary perspective the functions of macrophages represent a re-working or elaboration of the capabilities inherited from our single celled precursors. The fact that 'wandering phagocytes' are ubiquitous in multicellular organisms, particularly with respect to defensive functions, reflects this ancient heritage and explains the parallels between the phagocytes of invertebrates and vertebrates, as well as the impressive array of roles carried out by macrophages in vertebrates, which can be seen as the result of innumerable functional adaptations which have utilized an inherited and relatively fixed palette of molecular scaffolds and processes.

Such a general view helps to frame the somewhat byzantine mechanisms by which immune function is realized. It also guides an appreciation of the often surprising parallels between the tools of immune cells and those used by other tissues for quite different purposes; examples include the widespread occurrence of 'immunoglobin' domains in 'non-immune' proteins, the use of Toll receptors for body patterning in *Drosophila* and the relatively few core signalling pathways which are re-used combinatorially by many different receptors. One lesson which has been well learned is that while protein structure is well-conserved, protein function (in a biological context) is relatively malleable. Nonetheless, when it comes to understanding the present immune system in humans, in which we have a very particular interest, the search for useful generalities needs to give way to a detailed enumeration of the specific ways that human or animal macrophages operate. Clearly, some macrophage functions in humans are of relatively recent origin and reflect the relentless selective effect of pathogens on host populations. This is also seen in mice, and will be true for other vertebrate species, so that species-specific macrophage behaviour always needs to be included in the study of animal disease. The ability to distinguish between mechanisms which must necessarily be general (such as conserved signalling pathways) and those which are likely to be almost arbitrary (such as the complex and frequently redundant families of protein messengers) has been part of the legacy of studies of molecular phylogeny, in particular the human genome project and its many progeny.

## Three paradigms

### The reticuloendothelial system (RES)

The earliest classification of macrophages relied upon their wandering phagocytic functions, and without molecular tools no rigorous scheme was possible. The collective functions of the macrophage system, as it was then understood, were characterized by the concept of the reticuloendothelial system (RES) which recognized the intimate relationship between myeloid leucocytes and the endothelial cells of the circulatory system which form a close functional relationship (Aschoff, 1924). In this view, macrophages ('histiocytes') arose from the local mesoderm, and multiple cell types were grouped in terms of their function in different tissues. The immune roles of visible immune tissues (the liver, spleen, mucosal-associated lymphoid tissue and the microglia of the brain) were unified by their functions in detoxification and immunity. The precise lineage relationships of endothelial and histiocytic cells were not a major focus of this classification system.

### The mononuclear phagocyte system (MPS)

The discovery in the 1960s and 1970s that there exist multipotent cells (stem cells) which can give rise to several lineages in colony forming assays marked the rise of the concept of haematopoietic lineages, in conjunction with the rise of the unifying notion (across developmental biology) of the process of differentiation and in particular the occurrence of stem cell niches which continue to replenish cell lineages throughout life. The task of establishing this paradigm was greatly accelerated by the advent of monoclonal antibodies which could characterize and isolate cells which, by all morphological and functional criteria, had previously seemed identical. The process of cellular differentiation is now aptly understood as a tree-like branching process, with few central limbs but many terminal buds. Structurally, this process reflects the reality that multistep branching processes are a natural and evolutionarily conservative way to generate diversity. Simultaneously, numerical expansion of a cell lineage requires multiple rounds of cell division and when this occurs during differentiation, the observed hierarchical expansion series results as a natural consequence.

Like all conceptual metaphors this can be taken to extremes, so that division into further and further subsets with few functional differences eventually serves little useful purpose – ultimately this process of multiplying lineages stops only at the single cell. A pragmatic line must be drawn and this is the basis of describing certain lineages as 'terminally differentiated' even though they can subsequently undergo 'activation' which in reality is no more than a further differentiation. In the case of macrophages, the stimuli for the early steps of differentiation tend to be innate growth factors and hormonal signals, whereas the later ones tend to be environmental signals (foreign molecules) received at unpredictable times (or not at all). Clearly though, the development of the immune system under pathogen-free conditions is abnormal; interaction with the environment is a usual part of the development of immunity.

### Systems biology

More recently, the rise of systems biology has led to a new perspective of cellular differentiation which focuses on the functions encoded by particular genetic programmes rather than solely on the history of the cell type that enacts these functions. In this view, sets of genes are the mediators of a function and these can derive from one, two or several lineages, depending on the situation. Here the focus is on gene networks and not on the linear heritage of the cell type. Cells are assemblages of gene-expressing units and cell populations are not pure lineages uniformly expressing fixed genetic programmes. Instead, it is recognized that it is populations of cells which express genetic programmes while individual cells express genes stochastically, albeit in a biased fashion (Ross *et al.*, 1994b; Kelso *et al.*, 1995; Hume, 2000). In this view, it is meaningful to talk of a Th2 population, but not (strictly) a single Th2 cell. The terms of classification in systems biology are therefore functional and not ontological. While such a perspective has considerable validity and will be useful in systems biology itself, it seems likely to complement rather than displace the well-entrenched view of differentiation as a hierarchical branching, ontological process.

## Macrophage ontogeny and differentiation

### Macrophage functions

The early recognition that macrophages form an important reservoir *in vivo* for the 'T-cell virus' HIV, and that macrophage-tropic viruses evolved *in vivo* helped stimulate a renewed interest in macrophages and the innate immune system generally. The first cytokines (notably the interferons) were discovered in the 1950s, and others were identified in the subsequent decades, but the pace of discovery was slow and the ability to identify gene families required the molecular tools of the 1980s, such as lambda cDNA library screens. However, the general view widened with the discovery of the Toll-like and Nod-like receptors (TLRs) in the late 1990s in conjunction with the bioinformatic capability conferred by genomics. The subsequent explosion of research into the awkwardly named 'pathogen-associated molecular patterns' (PAMPs) initiated a paradigmatic shift in our view of macrophages and their role in immune surveillance and general immunity, autoimmune disease and inflammation in all contexts, both in their helpful physiological responses to infection, and as mediators of chronic disease and tumorigenesis. Today, the involvement of macrophages in disease processes extends to cardiovascular disease, obesity, neurological diseases, cancer biology, chronic inflammatory diseases such as arthritis, bowel disease, chronic obstructive pulmonary disease, osteoporosis and many other pathologies. These complex pathologies reflect the multitude of normal physiological processes in which macrophages are involved, and the ways in which such potent weapons can also harm their owner. Nonetheless, the primary immunological role of macrophages remains the protection of the multicellular organism from external pathogens.

The particular relevance of macrophages for viral disease comes from their central involvement as sentinels of viral infection, initiators of inflammation and educators of the acquired immune system through antigen presentation and cytokine production. From a purely functional point of view, pathogens can be divided broadly into multicellular parasites, fungi, bacteria and viruses, each of which requires a different immune strategy. Macrophages play a central role in deciding which type of response is most appropriate to such threats. As for lymphocytes, macrophages both respond to and produce a bewildering array of cytokines as well as small molecule effectors, to coordinate and direct immune responses.

Immune functions are of course present widely in non-haematopoietic tissues, and it is probably most accurate to say that the immune system is a bodily function which simply reaches particular specialization in the cells of the haematopoietic system. However, in addition to the specialized functions in immunity, the haematopoietic system gives rise to oxygen carrying cells, cells to build and destroy bone, effect blood clotting and wound repair, remodel developing tissues and to carry out a multitude of other functions. Finally, the type of robust immune response which is acceptable in dispensable, rapidly regenerated tissues with a high load of foreign organisms (such as the gut, lungs, hepatic system and mucosal surfaces) would be inappropriate for highly sheltered delicate tissues such as the brain and reproductive system, so that site-dependent variation in response also broadens the repertoire of responses.

Consequently, cells derived from the monocytic lineage encompass a wide array of highly specialized functions and cell types including osteoclasts, dendritic cells and a diverse array of resident macrophages in virtually every tissue, notably including Kupffer cells in the liver, microglia in nervous tissues, osteoclasts and bone-lining macrophages in the bone. Many of these functions relate to homeostatic mechanisms, tissue repair and remodelling or specialized physiology and not just to immune defence. Finally, despite a common morphological and molecular heritage, many differences exist in these populations in terms of cell surface receptors, genetic programming and functional responsiveness, so that the selection of any one cellular marker inevitably cuts such populations into sometimes artificially distinct groupings. Any overview of macrophage biology necessarily suffers from the impossibility of generalizing about specific receptors, effectors and phenotypes which may be shared by only a subset of macrophages and may be contraindicated by some other subset. Thus the devil is in the details, and comprehensive delineation of such details will not be attempted in this overview, except by reference to the literature which contains many fine and detailed reviews in relation to specific aspects of macrophage biology.

## What is a macrophage?

As noted above, scientific observations of macrophages 'wandering immune cells' were initiated by Ilya Mechnikov, who both named these cells and described the phagocytosis of foreign bodies in starfish larvae (Tauber, 1991). In fact, phagocytes occur in virtually all multicellular life forms. Specialized phagocytes occur in the embryos of insects to higher vertebrates, in addition to facultatively phagocytic cells, largely derived from mesenchymal tissues. Whether the differentiation programmes for macrophage ontogeny have stayed constant at the genetic level over this time is not known, but in mature vertebrates at least, macrophages are presently defined mainly by their origin in the haematopoietic system of the bone marrow which also gives rise to blood cells. Keeping in mind that vertebrates (phylogenetically) are a minor and recent branch of the eukaryotes, haematopoietic cells can thus be viewed as a specialized form of the ancient haemolymph phagocytic lineage.

There is of course no absolute definition of the term 'macrophage' even today and even in vertebrates. This reflects the pragmatics of classification as well as the historical development of the field. Early descriptions of macrophages were necessarily morphological and functional, characteristics which do not always reflect genuine lineage development pathways, and it was not until the 1980s that the specifics of haematopoietic lineage development could be put on a firm foundation. In some cases, the specific differentiation events that give rise to some classes of macrophage-like cells are still debated, microglia being a case in point. Functionally, terminal differentiation gives rise to a multitude of macrophage-like cells, not all of which are relevant to a particular function such as immune defence. Finally, many macrophage subsets are defined in terms of surface or enzymatic markers (described below), which often – but not always – reflect functional specialization. Therefore the term 'macrophage' is used broadly here to include embryonic phagocytes, tissue monocyte-derived lineages as well as circulating monocytes and their inflammatory progeny. Once the details of cellular differentiation and lineage commitment are completely understood, a nomenclature based on genuine developmental lineages may replace that defined by functional or marker-based classifications, if it creates a more intelligible and useful system.

## Functional vs. ontological definitions

The relatively recent understanding of macrophage lineage development means that the mononuclear phagocyte system (Hume *et al.*, 2002; Hume, 2006) is a functional, collective term which has only been presumed to be based on common cellular origins. Generally speaking, it is most satisfying to base classifications on ontological development, because no matter how convergent two similar cell types may be, their developmental history inevitably leaves differences which can only be explained by reference to origins. Ontology is thus more fundamental than function and much effort in the last 3 decades (particularly since the advent of monoclonal antibodies) has gone into unravelling the complex pathways involved in haematopoietic differentiation and expansion and functional specialization (van Furth and Cohn, 1968; Hume *et al.*, 2002; Hume, 2006). Ontology is closely bound up with genetic programmes of differentiation – expression of genetic regulators such as transcription factors and growth factor receptors. On the other hand, functional classifications can sometimes be more useful, especially when the function is a specialized one (such as immunity) that cuts across ontological lines. Functional classifications are delineated not only by morphology and location, but by the expression of functional markers – cell surface receptors, enzymes, and behavioural properties such as phagocytosis, which result from them. Thus it makes sense to regard microglia as macrophages regardless of whether they in fact replicate locally or are replaced from the pool of circulating monocytes.

Although it is now clearly understood that the haematopoietic system is the origin of most macrophage-like cells, there is still controversy about the origin of some subsets such as embryonic phagocytes in vertebrates (Hume *et al.*, 1995; Faust *et al.*, 1997; Herbomel *et al.*, 1999, 2001; Lichanska and Hume, 2000).

## Early embryonic phagocytes

In the mouse embryo, a small population of mature maternal macrophages can be demonstrated at 7.5 days post coitus (dpc) that represent the earliest macrophage population, located in and near the yolk sac mesoderm (Bertrand *et al.*, 2005). These may mediate scavenger functions prior to the maturation of the embryo's own phagocytes, and are clearly mature adult macrophages derived from the usual haematopoietic lineage, complete with classic markers such as the F4/80 antibody (Hume and Gordon, 1983). These are likely to be maternally derived placental macrophages. As such, they are not embryonic macrophages proper in the sense of being derived from the embryo itself during organogenesis (Hume and Gordon, 1983; Shepard and Zon, 2000).

Shortly afterwards (8 dpc), true embryonically derived macrophages appear to develop from precursors in the yolk sac mesoderm, emerging directly from precursor cells without the conventional differentiation pathway via monocyte precursors (Palis and Yoder, 2001; Bertrand *et al.*, 2005). The early developmental sequence of these macrophages has been demonstrated to consist of a mesodermal precursor in the primitive streak (the haemangioblast). Until recently it has been uncertain as to whether haemangioblasts give rise independently to endothelial cells and haematopoietic precursors, or whether they differentiate first into endothelial cells and then haematopoietic cells. The truth appears to lie between, in that following the expression of the zinc finger transcription factor Scl (Tal1) they give rise to an immature, incompletely committed endothelial lineage which is Tie2$^{hi}$c-Kit$^{+}$CD41$^{-}$ and which derive from the yolk sac blood islands. Tie2 is a receptor tyrosine kinase for angiopoietin, which is typically expressed on vascular endothelium, whereas c-Kit (CD117) is the receptor for stem cell factor. Although most of these cells continue to form committed Tie2$^{hi}$ endothelial lineages, a small proportion convert to Tie2$^{-}$c-Kit$^{+}$CD45$^{+}$ and with the activation of the key haematopoietic transcription factor Runx1 (AML1) they go on to form haematopoietic lineages which include macrophages (Lancrin *et al.*, 2009). Lineage tracing experiments also support the necessity for an endothelial stage and not a mesodermal origin of haematopoietic precursors (Zovein *et al.*, 2008).

These cells lack classical macrophage markers such as F4/80 (EMR1) and in particular, lack PU.1, the key transcriptional regulator of macrophage lineage commitment in the haematopoietic process (Bonifer and Hume, 2008) which is discussed below. Between 8dpc and 9dpc these 'yolk sac macrophages' migrate to the anterior embryo (Ovchinnikov, 2008) through the developing circulation. Assuming the normally developing embryo to be pathogen-free, the function of these cells is probably to participate in tissue remodelling, in particular the clearance of apoptotic cells which is an important macrophage role in the adult as well as in embryogenesis (Henson and Hume, 2006). These cells persist throughout embryogenesis.

The placenta is also a site of haematopoiesis *in utero*, though placental haematopoiesis does not appear to predate the rise of precursors in the AGM region and placental haematopoiesis is most active in the mid-gestation period. Haematopoietic stem cells appear to rise independently in the placenta and have been shown to give rise to some lymphoid and myeloid lineages, although they have not yet been documented as a source of *bona fide* macrophages (Alvarez-Silva *et al.*, 2003; Mikkola *et al.*, 2005; Rhodes *et al.*, 2008; Barcena *et al.*, 2009a,b). Interestingly it has been shown that interleukin 3 (IL-3), a Runx1 target, is necessary to amplify haematopoietic stem cells from the aorta-gonads-mesonephros (AGM) region, yolk sac and placenta (Robin *et al.*, 2006). This suggests the need for further work to delineate embryonic roles for the growth factors and cytokines which regulate adult haematopoiesis.

### Onset of haematopoiesis proper

In the adult vertebrate, macrophages derive from a progressive sequence of differentiation steps through a monocyte stage in the bone marrow. As the fetus continues to develop, macrophages appear predominantly in the fetal liver. It was long assumed on the basis of timing and location that their appearance in the liver was the result of migration from two sites – an earlier presumed migration from the yolk sac at 8.5–9 dpc and later, more prolific wave at 9.5 dpc from the AGM (Bonifer *et al.*, 1998; Dzierzak *et al.*, 1998) so that by 10 dpc the latter represents the major source of liver embryonic macrophages (Dzierzak *et al.*, 1998; Lichanska and Hume, 2000). Recently, considerable research has established that in fact all fetal haematopoietic precursors originate completely in the embryo proper, in the para-aortic splanchnopleura (PAS) and the AGM at around 9 dpc (Medvinsky and Dzierzak, 1996; Cumano and Godin, 2001, 2007) and probably migrate out into the yolk sac early in embryogenesis prior to the liver becoming the major site of haematopoiesis after 11 dpc (Dzierzak *et al.*, 1998; Cumano and Godin, 2007). A shift from the liver to the bone marrow as the primary site of haematopoiesis happens during neonatal life. Unlike the early embryonic phagocytes, fetal liver macrophages appear to be very similar to adult macrophages, with the expression of similar markers such as PU.1 and F4/80 (Lichanska *et al.*, 1999, 2000).

Given the absence of key macrophage markers, it is questionable to what extent the earliest embryonic macrophages actually resemble macrophages 'as we know them' and to what extent their genetic programmes overlap the familiar biology associated with mature macrophages. Interestingly, studies from zebrafish and *Drosophila* also show evidence of two independent populations of mononuclear phagocytes. The first macrophage precursors in zebra fish similarly appear without classical haematopoietic islands and migrate both into the yolk sac and into the head mesoderm in a CSF-1-dependent process (Herbomel *et al.*, 2001).

In so far as it has been studied, human embryonic development appears to parallel that in the mouse (Tavian *et al.*, 1999, 2001) so that it is probable that similar classes of phagocytic cells occur in the human embryo with similar dynamics.

## Macrophage lineage development in the adult

### Bone marrow haematopoiesis

Traditionally, monocyte precursors in the bone marrow have been characterized as 'monoblasts' and 'promonocytes' according to various empirical criteria such as enzyme staining, and this nomenclature may still be found in clinical texts. A great deal of detailed work on haematopoiesis over the last two decades has led to a much more nuanced (though complicated) view. The current model of haematopoietic lineage development holds that rare haematopoietic stem cells (HSC) in the bone marrow, which normally undergo self-renewal, can also divide asymmetrically to give rise to a short-term HSC which is biased towards differentiation rather than self-renewal (Congdon and Reya, 2008). This uneven division process is viewed as either an occasional random event or else an event produced by the action of growth factors on cell surface receptors in a deterministic way. In fact these approaches are not incompatible because the stochastic process of gene expression means that stem cells will randomly switch on growth factor receptor genes, and the stimulation of these receptors may then (in a deterministic fashion) lead to a fixed developmental sequence (Hume, 2000; Paldi, 2003) based largely on the positive feedback (switching) loops involving key transcription factors such as GATA1 for T-cells, EBF and Pax5 for B-cells (Lukin 2008), and Runx1, C/EBPα and PU.1 for myeloid lineages (Cobaleda and Busslinger, 2008). Nonetheless, until the intertwined action of specific sets of transcription factors, microRNAs and chromatin changes 'lock in' a particular fate, it is likely that some developmental flexibility will remain within a given lineage. The regulation of the transcription of the CSF1R, discussed later, provides a model for such processes (Bonifer and Hume, 2008).

Assuming for the present that short-term HSCs exist as a defined cell type, these are supposed to give rise to multipotent progenitor cells (MPCs) which are the last common ancestor of all haematopoietic lineages. Because no clear branching has occurred to this stage it is difficult to ascertain to what degree these subtypes actually exist as separable or unique entities, rather than as steps on a continuum. In the conventional view, the differentiation process appears to divide at this stage into one of two great lineages, the committed lymphoid precursor (CLP; giving rise to B-cell, T-cells, NK cells and lymphoid dendritic cells) and the committed myeloid precursor (CMP; which results in erythrocytes, platelets, monocytes, granulocytes and myeloid dendritic cells). Recent work suggests that this scheme is too simplistic and that separate precursors may also exist, even at this early stage, for myeloid-lymphoid stem cells and erythroid-lymphoid lineages which give different end-stage lineages (Giebel and Punzel, 2008; Luc *et al.*, 2008a,b).

Cells can certainly be isolated as marker subsets (in the case of myeloid CMPs, $Lin^-$ SCA-$1^-$ $Kit^+$ IL7R$\alpha^-$ Fc$\gamma R^{Lo}$ $CD34^+$ and in the case of CLPs $Lin^-$ SCA-$1^{Lo}$ $Kit^{Lo}$ IL7R$\alpha^+$) which have identified differentiation potentials, but as indicated above, it is not certain that such subsets represent the only ways that these lineages can evolve, with the 'low erythroid' GM-lymphoid precursors identified by Luc *et al.* (2008a) having a $Lin^-$ Sca-$1^+$ $Kit^+$ Flt$3^{hi}$ $Mpl^-$ phenotype. Furthermore, even later in differentiation independent crossover pathways may exist between lymphoid and myeloid cells (Cobaleda and Busslinger, 2008). In particular, close relationships have been observed between B-cells and macrophages which suggest that lineage switching – or alternative commitment models – may occur (Fancke *et al.*, 2008; Feng *et al.*, 2008). Similarly, a number of myeloid tumours have been found to have rearranged immunoglobulin genes, suggesting that they are either derived originally from B-cells or that the B-cell programme for Ig activation has been activated in these cells at some stage. Such controversies underscore the fact that differentiation is essentially a property of the *genome* expressed in a local context, and are a reflection of the activation of particular genetic programmes, and it is likely that such programmes can occur in several contexts. Therefore, while the conventional scheme is certainly one route of development, possibly the major one, it is also probable that reality is more complex. Spectacular recent work has demonstrated the use of as few as 4 transcription factors to reverse the differentiated state of mature tissue cells into totipotent cells capable of giving rise to a new organism (Takahashi and Yamanaka, 2006; Yu *et al.*, 2007). Whether such flexibility is employed as a normal part of haematopoiesis or is purely aberrant in this context, there is at least no doubt that transdifferentiation and de-differentiation is possible. In summary, to fully describe haematopoietic development it is not sufficient to show that a differentiation pathway actually exists; it is also necessary to describe alternative routes, their relative contributions within the range of normal development and the degree of flexibility they hold when abnormal development occurs. In particular, the *in vivo* stem cell niche is likely to be an important influence which purely *in vitro* studies cannot accurately capture (Raaijmakers and Scadden, 2008). It is wise, therefore, not to hold too fast at this stage to simple textbook pictures of haematopoietic stem cell development and the early origin of macrophage lineages.

The picture is not necessarily clearer further down the developmental pathway. In the case of the myeloid lineage, a further branching into a megakaryocyte/erythroid progenitor cell (MkEP) and a granulocyte/monocyte progenitor (GMP) occurs (if it has not occurred already). Supposedly, granulocytes, monocytes and myeloid DCs branch off after this point. Down the monocyte lineage is the monocyte-dendritic cell precursor (MDP) in the bone marrow, which is currently supposed to yield plasmacytoid and conventional DCs as well as monocytes, which then go on to differentiate into tissue and inflammatory macrophages (Fig. 2.1). However, these categories are not as simple as they appear. Firstly, the ontology of myeloid lineage DCs is still a contentious matter. It is by no means clear that 'dendritic cells', despite being morphologically dramatic, are anything more than a terminal differentiation state similar to that occurring in other tissue macrophages. The privileged identification of DCs rests on their branching earlier (i.e. as a 'lineage') than other tissue macrophages which are, in contrast, supposed to be the product of the migration of circulating monocytes to specific tissue niches, followed by differentiation *in situ*. However, DCs neither give rise to other more differentiated cell types nor are they necessarily more differentiated in their function than less visually dramatic tissue macrophages. In this alternative view, the DC morphology and antigen-presenting function is the result of a typical terminal differentiation event which goes through the monocyte lineage, just as for the differentiation of (say) Kupffer cells in the liver or osteoclasts in bone. As such it could arise by activation from several myeloid states. Such a view is supported by studies showing that of the monocytes which enter a tissue in response to a chemotactic signal, around half later re-enter the circulation with a DC phenotype and then migrate to lymphoid tissue contributing to the proliferating categories of DCs from various tissue niches (Hume, 2008). Nonetheless, for at least one DC type (classical spleen DCs) the DC precursors branch from the MDP prior to giving rise to monocytes (Liu *et al.*, 2009), suggesting that an experimental clarification of this debate may be reached in the near future.

In contrast, granulocytes have long been accepted as a completely different cell type to monocyte–macrophages, and one which branches early from circulating precursors. Nonetheless, recent data suggests that granulocytes can themselves mature to form conventional macrophages and

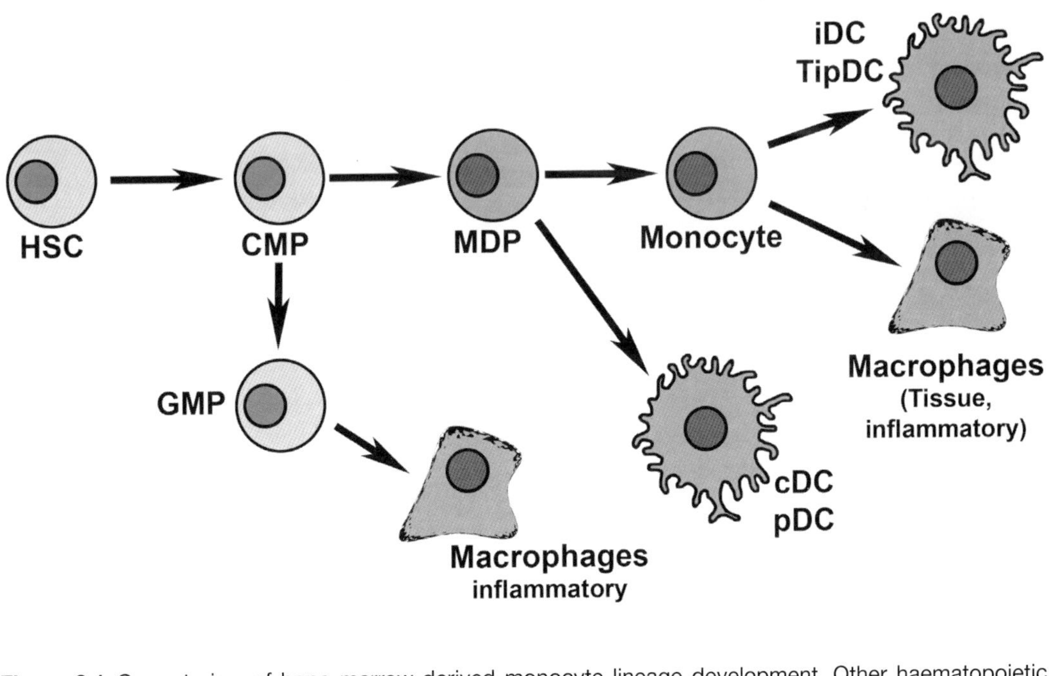


**Figure 2.1** Current view of bone marrow-derived monocyte lineage development. Other haematopoietic cells types are not shown. HSC, haematopoietic stem cell; CMP, common myeloid precursor; MDP, myeloid dendritic precursor; GMP, granulocyte–macrophage precursor; DC, dendritic cell; cDC: classical spleen DC; pDC, plasmacytoid DC; iDC, inflammatory DC; TipDC, TNFα- and iNOS-producing DC.

therefore represent a detour on the macrophage differentiation pathway (Sasmono *et al.*, 2007). In some situations, then, the replacement of granulocytes by macrophages may be partly or wholly due to transdifferentiation rather than turnover.

Finally, terminal differentiation is marked by maximal diversity and less irreversible commitment than earlier branching steps. At which point 'differentiation' transforms into 'activation' is unclear, and the degree of developmental flexibility of 'terminally differentiated' macrophages is still uncertain. For example, transdifferentiation of mature macrophages into endothelial tissue seems to occur. Clearly much work is still needed to clarify these relationships and to define the plasticity or rigidity of terminal differentiation events. The controversial question of to what degree 'terminal' macrophages can transdifferentiate into other tissue cell types (especially endothelial cells) remains to be settled, the experimental techniques requiring great rigour and ingenuity in order to avoid artefacts and alternative explanations.

Regardless of the details of lineage branching at this level, it is clearly possible to isolate populations which we can regard (for experimental purposes) as being specific differentiated subsets of the myeloid lineage. Of these subsets, we can perhaps regard granulocytes (neutrophils) and DCs as terminally differentiated cells which travel via the circulation, whereas other tissue macrophages do not leave the compartment in which they have settled.

### Extramedullary haematopoiesis

To further confound the conventional hierarchical scheme of bone marrow derived haematopoietic lineage development, it is now clear that local regeneration of macrophages occurs in at least three sites: alveolar macrophages in the lung, microglia in the brain and Langerhans cells in the skin. It seems likely that this list will expand with ongoing careful study of smaller tissue sites; the retina, for example, also seems to have a subpopulation of locally replicating (or at least, long-lived) tissue macrophages (Davies *et al.*, 2006; Xu *et al.*, 2007). Whether such local regeneration is a proliferative response on the part of the entire local population, or is due to a small set of local stem cells, is yet to be established. In any case, it is clear that circulating 'terminally differentiated $CD14^+$ monocytes' in fact represent an intermediate step in differentiation.

Finally, the massive number of monocytes produced each day from the bone marrow means that most monocyte–macrophages must die relatively quickly. Generally monocytes have two main fates. Apart from a population which patrols the endothelium (Auffray *et al.*, 2007) they undergo diapedesis and migrate into the tissues in response to a variety of signals. There they either contribute to the pool of local tissue macrophages, or else they undergo activation to form inflammatory macrophages, usually in response to an inflammatory signal that indicates the presence of a pathogen, but which sometimes is an inappropriate activation which forms the foundation of chronic inflammatory diseases. Turnover of tissue macrophages can be quite slow, and residence times of months are commonly found in tracer studies. Inflammatory macrophages, on the other hand, infiltrate in great numbers and must be dispersed equally quickly once an infection is resolved. In this case the only two options are extravasation (into the lymph or circulation) with apoptosis occurring elsewhere, or else apoptosis at the inflammatory site, both of which are known to occur. The relative balance of these processes, however, has not been well defined.

An understanding of haematopoietic stem cell development is of great clinical and research importance, both as a model for understanding other less tractable stem cell developmental pathways, and because of the many clinical implications of bone marrow replacement therapy. In the case of HIV infection, there is also evidence that stem cells can be infected with HIV and thence give rise to an ongoing lineage of constitutively provirally infected blood lineages, with profound implications for understanding viral spread and latency (Alexaki and Wigdahl, 2008).

## Macrophage growth factors and receptors

The isolation in the 1980s of soluble factors (the colony-stimulating factors or CSFs) which enabled the *in vitro* proliferation of bone marrow precursors into colonies of distinct haematopoietic lineages was not only an important step in understanding haematopoiesis but also provided a valuable experimental tool to the investigation of haematopoietic biology. The identification of these key growth factors and their receptors coincided with the discovery of the intracellular signalling pathways which mediate their effects, and the transcription factors which lock in those developmental programmes. This in turn has revealed a multitude of functions in the downstream regulation of mature myeloid cells and the coordination of myeloid functions.

Three key factors in this process are interleukin 3 (IL-3; 'multi-CSF') which acts early in the developmental pathway to stimulate proliferation of myeloid precursors, CSF1 ('macrophage CSF' or MCSF) which is the primordial factor regulating monocyte and tissue macrophage development, and CSF2 ('granulocyte–macrophage CSF' or GM-CSF) which is particularly concerned with the mobilization of inflammatory granulocytes and macrophages during infection. There is also a set of subsequently acting factors including RANKL (receptor activated NFκB ligand) which is necessary (along with CSF1) for osteoclast formation and IL-4 which acts, again in concert with CSF1, to stimulate DC formation. It seems likely that many other factors, most already described, will be shown to act in the maturation of subsets of more terminally differentiated monocytes and tissue macrophages. For the purposes of the present overview, only IL-3, CSF1 and CSF2 will be discussed in detail. One other cytokine acts in this capacity – IL-34, which was recently discovered and shown to signal through the CSR1R (Lin *et al.*, 2008). At present, however, little is known of this growth factor and its biological roles.

## Interleukin 3

Interleukin 3, as the alternative name 'multi-CSF' suggests, is a general stimulator of proliferation of all myeloid cells lineages including the mast cell and erythropoietic lineages. Made primarily by T-cells, IL-3 acts mainly on bone marrow progenitor cells to stimulate myeloid differentiation, partly by up-regulation of the transcription factor Id1 in haematopoietic stem cells (Leeanansaksiri *et al.*, 2005). Mature IL-3 is a 13-kDa monomeric glycosylated four-helix bundle cytokine which is closely related to GM-CSF (Feng *et al.*, 1996).

The human IL-3 receptor is a type I heterodimeric cytokine receptor which consists of an IL-3 specific alpha chain (IL3RA; CD123), primarily responsible for the IL-3 specificity of binding, and an IL-receptor beta chain (IL-3RB; CD131) which confers high affinity binding and which is also utilized in the GM-CSF and IL-5 receptors. In murine IL-3, the β-chain is also IL-3 specific, but there is clear structural homology between the IL-3, GM-CSF and IL-5 cytokines and the α/β-receptor structure (Miyajima *et al.*, 1993). There is also similarity between the β-chain and the IL-6 gp130 subunit. Monomeric β-chain does not bind IL-3. The α chain binds IL-3 with low affinity, whereas the heterodimer binds with high affinity. In contrast, the intracellular signalling triggered by IL-3 binding depends primarily on the β-chain, which means that there are common signalling pathways triggered by IL-3, GM-CSF and IL-5 activation, though the cellular contexts differ.

The heterodimerization of the receptor (and possibly the formation of a higher order complex; (Hansen *et al.*, 2008) produced by IL-3 binding results in signalling primarily (though not exclusively) through the Ras (Satoh *et al.*, 1992; Vojtek *et al.*, 1993) and the JAK (Silvennoinen *et al.*, 1993) pathways, the latter resulting in STAT5 activation. PI3K/Akt pathways lead to effects on the cell cycle, FOXO3a (Park *et al.*, 2009) and NFκB signalling (Lee *et al.*, 2008). Recently it has been shown that the FcRγ chain is also constitutively associated with the IL3R in cells in which it is expressed, and that this affects downstream signalling, leading to IL-4 production in basophils (Hida *et al.*, 2009). Since monocyte–macrophages also express FcRγ signalling effects in these cells may also occur.

The clinical importance of IL-3 stems mainly from its potential to mobilize stem cell precursors from the bone marrow into the circulation, an effect which is exploited during bone marrow transplantation to speed reconstitution of a functional haematopoietic system.

Intracellular signalling downstream of the IL3R provides anti-apoptotic signals which mediate the survival of haematopoietic precursors and their replication so that they contribute to tissue expansion (Yen and Yang-Yen, 2006; Park *et al.*, 2009). As already noted, this activity also appears to be important in embryonic development downstream of Runx1 (Robin *et al.*, 2006).

In addition, IL-3 acts on mature cells, and like GM-CSF can activate cells carrying the IL-3R. IL-3 is needed for the activation of microglia (Natarajan *et al.*, 2004) with implications for neuronal inflammatory diseases including HIV-induced dementia. In Alzheimer's disease IL-3 may be neuroprotective against amyloid β neurotoxicity (Zambrano *et al.*, 2007), while linkage of SNPs at the IL-3, IL-3Rα, IL-3Rβ (and the GM-CSFRα) loci to schizophrenia has been observed (Chen *et al.*, 2007; Sun *et al.*, 2008). There is considerable interest with a view to vaccine development in the role of IL-3 in DC differentiation, as DCs generated via IL-3 and IFNβ appear to prime functional antigen-specific CD8 T-cells (Breckpot *et al.*, 2005; Renneson *et al.*, 2005). Remarkably, IL-3 is capable of antagonizing the pro-inflammatory effects of TNFα by prevented TNFα-induced AP-1 activity, resulting in potent anti-inflammatory effects on arthritis and bone resorption in experimental mouse models (Yogesha *et al.*, 2009).

IL-3 is capable of directing granulocyte–macrophage precursors down the basophil differentiation pathway (Ohmori *et al.*, 2009) and roles for IL-3 in basophil biology, though peripheral to a discussion of macrophage biology, are of great current interest (Burdach *et al.*, 1998; Kowal *et al.*, 2005; Lim *et al.*, 2006) for asthma pathology, but disruption of IL-3 is non-lethal and other cytokines such as GM-CSF appear to be able to substitute for it in most cases. Viruses such as HTLVIII can also induce GM-CSF by $CD34^{+}$TF-1 cells thereby obviating the need for IL-3 (Bovolenta *et al.*, 2002).

A complex literature details multiple, sometimes contradictory effects of IL-3, many of them the result of *in vitro* studies. The importance of some of these effects *in vivo* has yet to be demonstrated. This confusion may be mitigated somewhat by the recent discovery that a second (SP2) isoform of the α subunit of the receptor exists which mediates differential and perhaps antagonistic effects on monocyte differentiation, including effects on the gene expression levels of non-coding microRNAs and of key transcription factors C/EBPα and PU.1 (Chen *et al.*, 2009). Another source of confusion may stem from the fact that the effects of IL-3 may depend on the tissue niche and cooperation with other factors. For example, Zeoli *et al.* (2008) have shown that IL-3 may contribute to the expansion and transdifferentiation of $CD45^{+}$ haematopoietic precursors including incorporation into newly formed blood vessels in organs and tumours. Similarly, effects of IL-3 on erythroid differentiation result from synergism between IL-3 and erythropoietin (Ghinassi *et al.*, 2007). It may be that more subtle effects of IL-3 on macrophage biology require specialized *in vivo* systems and are yet to emerge.

## Colony-stimulating factor 2 (granulocyte–macrophage CSF)

As noted above, CSF2 (GM-CSF) is closely related to IL-3 and IL-5. Although CSF2 is the proper gene name, most of the literature refers to CSF2 as 'GM-CSF', which will be used here. Like IL-3, GM-CSF is a glycosylated four-helix bundle cytokine, though it is present *in vivo* as a dimer of 13-kDa subunits. The GM-CSF receptor is also structurally similar to the heterodimeric IL-3 and IL-5 receptors, comprising a ligand-specific alpha chain structurally homologous to the IL-3 and IL-5 alpha chains, and sharing a common beta subunit. Like the IL-3 and IL-5 receptors, signalling through the GM-CSF receptor is mediated primarily by Ras and JAK-STAT pathways, though a multitude of downstream kinases are also activated (Barreda *et al.*, 2004). Unlike IL-3, downstream signalling is also affected by C/EBPβ, with deficient GM-CSF signalling apparent in C/EBPβ knockout mice (Akagi *et al.*, 2008). GM-CSFR, as well as a splice variant (Pelley *et al.*, 2007; Rosas *et al.*, 2007) can be cleaved at the cell surface to yield a soluble receptor. Recent work suggests that actual receptor activation requires formation of a dodecameric higher order complex which (by structural analogy) may also apply to the IL-3 and IL-5 receptors (Hansen *et al.*, 2008).

GM-CSF is produced in response to immune stimuli by a variety of cell types including T-cells, endothelial cells and fibroblasts and a variety of other tissues (Barreda *et al.*, 2004). It is also produced in both autocrine and paracrine fashion by monocyte–macrophages to modulate the local environment and to recruit circulating neutrophils, monocytes and lymphocytes (Shi *et al.* 2006). Although the highest expression of murine GM-CSFR *in vivo* was found on monocytes and granulocytes, with variable expression on tissue macrophages (Rosas *et al.*, 2007), GM-CSF also acts on other cell types; tolerance-inducing actions on $T_{reg}$ cells (Kared *et al.*, 2008), anti-tumour effects, survival of Ag-stimulated CTLs (Selleri *et al.*, 2008) and effects on eosinophils to modulate allergic responses (Muessel *et al.*, 2008). It also promotes differentiation of neural stem cells in the bone marrow into neurons (Choi *et al.*, 2007; Kruger *et al.*, 2007) and gives neuroprotection against apoptosis (Huang *et al.*, 2007). Immunologically, GM-CSF contributes to DC differentiation (Conti and Gessani, 2008) and augments the ability of macrophages to stimulate T-cells during infection (Makino *et al.*, 2007). There is also evidence that GM-CSF is a key part of Dectin-1 responses to β-glucans by macrophages (Harada and Ohno, 2008).

Like IL-3, GM-CSF has a multitude of biological effects beyond the scope of the present discussion. Classically, however, it acts both to promote granulocyte and/or macrophage differentiation from haematopoietic precursors, with mixed granulocyte and macrophage colonies being produced by cultivation of bone marrow with GM-CSF. On mature cells it generally acts to promote inflammatory activation. Its effects on differentiation, particularly the granulocyte–macrophage ratio, are dependent on its concentration and the presence of other cytokines such as CSF1.

GM-CSF is particularly important in alveolar macrophage development and function. GM-CSF knockout mice display a phenotype resembling the human disease *pulmonary alveolar proteinosis* in which anti-GM-CSF antibodies result in very low levels of functional GM-CSF, displaying defects in neutrophils infiltrating the lung (Uchida *et al.*, 2007) and in alveolar macrophages including the generation of foamy lipid-filled macrophages and impaired surfactant clearance (Thomassen *et al.*, 2007). The maturation of alveolar macrophages which is promoted by GM-CSF appears to be mediated mainly by up-regulation of PU.1 expression in alveolar macrophage precursors as a result of GM-CSF signalling (Bonfield *et al.*, 2003).

Responses of alveolar macrophages to LPS are mediated by GM-CSF as is maintenance of alveolar macrophage function during hyperoxic stress (Baleeiro *et al.*, 2006). The respiratory (oxidative) burst is maintained in alveolar macrophages by GM-CSF, and this is opposed by the inhibitory cytokine IL-10 (Dang *et al.*, 2006).

GM-CSF has been implicated in chronic inflammatory conditions especially chronic obstructive pulmonary disease (COPD) (Hamilton and Anderson, 2004; Vlahos *et al.*, 2006) and rheumatoid arthritis (RA) and there is evidence from experimental models (Plater-Zyberk *et al.*, 2008) to support blockade of GM-CSF as a therapeutic strategy for rheumatoid arthritis. Lari *et al.* (2007) describe the differentiation of GM-CSF-derived macrophages into phenotypically distinct osteoclasts in the presence of RANKL + CSF1, which is particularly relevant to bone resorption in rheumatoid arthritis. This is supported by Nomura *et al.* (2008) and is in accord with the observations of Park *et al.* (2007) that GM-CSF induces osteoclastogenesis in breast cancer.

In contrast to the effects of IL-3 in Alzheimer's disease, it has been suggested that GM-CSF operates to modulate Aβ production by influencing amyloid precursor protein (APP) trafficking (Volmar *et al.*, 2008). In experimental allergic encephalomyelitis (EAE), a model for multiple sclerosis (MS), GM-CSF contributed to activation of microglia and transition to frank disease (Ponomarev *et al.*, 2007).

There is a large literature on the use of GM-CSF for anti-tumour immunotherapy (Hege *et al.*, 2006; Chang *et al.*, 2007; Kaspar *et al.*, 2007; Simmons *et al.*, 2007; Toubaji *et al.*, 2007). GM-CSF-mediated macrophage activation may be important in the development of atherosclerosis (Stojakovic *et al.*, 2007) and may also have a role in the inflammation associated with adiposity (Kim *et al.*, 2008) and perhaps metabolic syndrome.

Although GM-CSF has not been widely used as an antiviral agent, viral mechanisms to inhibit GM-CSF action are not unknown (Deane *et al.*, 2009). More commonly, GM-CSF has been used clinically as an adjuvant (Hartoonian *et al.*, 2009), for example against HIV (Spearman *et al.*, 2009); although it did not in this case augment the antiviral activity of an anti-HIV vaccine. Despite early optimism (Kedzierska *et al.*, 1998), GM-CSF does not appear to be particularly useful for controlling primary HIV infection, though it has been used clinically to support neutrophil function in HIV-related secondary infections (Heit *et al.*, 2006; de Silva *et al.*, 2007).

### Colony-stimulating factor 1 (CSF1)

Colony-stimulating factor 1 is the central factor responsible for the development and regulation of myeloid lineages in general, and the possession of the CSF1 receptor (CSF1R, CD115) is a key marker for myeloid cells, as well as many (though not all) of the terminal cell types to which this lineage leads. *In vitro* culture of bone marrow with CSF1 as a sole additional growth factor leads to the outgrowth over a week of a relatively pure population of bone marrow derived macrophages which can be alternatively differentiated into osteoclasts, dendritic cells, or activated macrophages.

The major circulating isoform of CSF1 is a 34-kDa glycosylated four-helix bundle cytokine which is cleaved from a membrane-bound precursor. Circulating CSF1 is thought to be primarily due to endothelial cells (Roth and Stanley, 1992) but other sources include leucocytes, fibroblasts, osteoblasts and bone marrow stromal cells as well as autocrine production by monocytes (Sweet and Hume, 2003). It binds to the CSF1 receptor (CSF1R; CD115) which is the product of the *c-fms* protooncogene. The CSF1R, a dimeric tyrosine kinase, is confined to cells of the monocyte lineage and binding of CSF1 to the receptor leads to activation of multiple independent intracellular signalling cascades including effects on cytokinesis, cell proliferation and cell survival.

CSF1 was first characterized as the factor responsible for monocyte differentiation from bone marrow, and this function has been further demonstrated by administration of CSF1 to mice, resulting in a 10-fold increase in monocyte levels in the circulation, liver, spleen and peritoneal cavity (Hume *et al.*, 1988). Rising circulatory CSF1 levels due to infection (Francois *et al.*, 1997) or LPS exposure (Engelhardt *et al.*, 1991) act to increase monocyte production in humans and similar observations have been made for mice (Cheers *et al.*, 1988a; 1988b; Cenci *et al.*, 1991; Gregory *et al.*, 1992; Roth *et al.*, 1997).

It exists in several forms although the best studied is the soluble end-to-end disulfide-linked homodimer which is cleaved from a membrane-bound precursor. It also exists as a secreted proteoglycan and as the uncleaved membrane-bound glycoprotein (Price *et al.*, 1992; Stanley *et al.*, 1997). A much studied model of CSF1 function is the *op/op* (osteopetrotic) mouse which has a natural truncation of the CSF1 gene, and which is marked by greatly reduced monocyte numbers, low body weight, loss of peritoneal macrophages, osteopetrosis due to lack of osteoclasts, failure of tooth eruption, and reduced fertility. They have greatly elevated levels of GM-CSF which is presumed to be due to compensatory responses to low monocyte numbers. Failure to fully restore a normal phenotype with supplementation by soluble CSF1 suggests that the other CSF1 isoforms are also biologically important. This is supported by transgenic studies showing that the chondroitin sulfate glycosaminoglycan form of CSF1 is specifically required to correct several defects in *op/op* mice (Nandi *et al.*, 2006).

The presence of CSF1 is required for monocyte and macrophage survival except following activation by inflammatory stimuli such as LPS and TNFα which independently activate NFκB-dependent anti-apoptotic programmes (Sester *et al.*, 2005). Survival is also maintained by JNK signalling (Himes *et al.*, 2005; 2006) which targets AML1 (Runx1) which is already known to be a key factor in macrophage development. Macrophage survival requires constant signalling to inhibit apoptosis, the onset of which is prevented by short lived anti-apoptotic proteins, particularly $Bcl_{XL}$ and Mpl1. These must be constantly produced as a result of the receipt of survival signals by the cell, and processes that trigger macrophage apoptosis also frequently target these proteins.

The action of CSF1, via the CSF1R, is immensely complex, activating multiple cellular pathways and having a plethora of cellular effects which impact on the physiological roles of monocyte–macrophages (Pixley and Stanley, 2004; Yu *et al.*, 2008). Numerous intracellular signalling pathways diverge from the activated CSF1R, and consequently CSF1 effects include a central role in the differentiation of monocyte–macrophage lineages, immediate effects on cellular motility and chemotaxis, the mediation of tissue homing and macrophage activation, and priming for cytokine release. CSF1 is always present at active levels (tens of ng per cubic centimetre) in the tissue milieu, so it is normal for macrophages to be exposed to CSF1 (though the levels vary).

Although CSF1 is sometimes thought of primarily as a differentiation factor, disease models support an important role for CSF1/CSF1R in primary immune responses, with up- or down-regulation of CSF1 levels affecting responses to model pathogens such as *Listeria monocytogenes* (Kayashima *et al.*, 1991; Gregory *et al.*, 1992; Bernier *et al.*, 2001) and *Candida albicans* (Cenci *et al.*, 1991). CSF1 both activates macrophages and enhances cytokine production.

Signalling via CSF1R up-regulates mRNA for key cytokines including IL-6, GM-CSF, IL-1α and IL-1β (Evans *et al.*, 1995; Kamdar *et al.*, 1996), and primes for the production of IL-12 and TNFα, the release of which can be triggered by LPS in both murine and human monocytes (Warren and Ralph, 1986; Evans *et al.*, 1995; Kamdar *et al.*, 1996; 1997). In human monocytes, IL-8 secretion was also enhanced by CSF1 (Hashimoto *et al.*, 1996).

Priming is at least partly TLR-mediated and repressive effects are also possible. For example, while CSF1 primes for LPS effects (mediated by TLR4), it represses CpG signalling (which operates via TLR9) by down-regulation of TLR9 mRNA (Sweet *et al.*, 2002). Some priming may also be effected by autocrine GM-CSF production, as GM-CSF knockout mice showed reduced priming for IL-6 release (Evans *et al.*, 1998).

The CSF1R is a Class III receptor tyrosine kinase (RTK) and consists of an extracellular portion composed of five immunoglobulin domains, the distal three of which bind CSF1, while the two domains closest to the membrane mediate dimerization of the receptor and the initiation of signalling. The intracellular domain is a complex tyrosine kinase signalling domain comprising an ATPase-binding domain, a kinase insert which binds signalling factors, the kinase domain and finally a cytoplasmic tail which is also involved in signalling. There are at least seven tyrosines which are phosphorylated during the initiation of various signalling pathways which include at least 28 potential signalling partners (Pixley and Stanley, 2004). Major signalling pathways include those regulating cell survival and apoptosis, cell cycle control, cell motility and cytoskeletal reorganization including phagocytosis, transcriptional control and genome programming leading to differentiation, and receptor internalization and processing. The delineation of these complex pathways, and their crosstalk with other pathways, has been a longstanding research programme which is an ongoing process. It was only very recently that the actual binding of CSF1 to the CSF1R has been characterized (Chen *et al.*, 2009).

Key pathways include c-Src binding to activate PI3K/Akt signalling (differentiation, cytokinesis and survival), Lyn binding to activate cell cycle and proliferation, Grb2 binding to activate the Ras and PI3K pathways (proliferation, differentiation), PLCγ2 binding to activate phospholipase C (PLC) pathways, Cbl binding to activate receptor internalization and processing and a range of negative regulators on these and other pathways. The end-results of these pathways depend on many factors including the levels of ligand and receptor, the competency of the cell for the various signalling factors involved, the presence of competing or cooperative signals and crosstalk from other receptor pathways or internal processes. The major outcome is macrophage survival, proliferation and differentiation down a multitude of phenotypic pathways to generate end-stage tissue macrophages of numerous types. CSF1 also produces immediate effects including ruffling of lamellipodia and increased cell motility through immediate cytokinetic signalling events (Pixley and Stanley, 2004; Yu *et al.*, 2008), which affects monocyte recruitment to sites of infection (Rovida *et al.*, 2005; Wheeler *et al.*, 2006).

Recruitment and the release of inflammatory cytokines is thought to be of central importance in a variety of chronic inflammatory conditions including atherosclerosis, RA, inflammatory bowel disease (IBD), renal disease, and COPD which are all conditions in which macrophage responses are a central component of the chronic pathology and tissue destruction. A somewhat similar situation exists for some cancers, including prostate (Richardsen *et al.*, 2008) and breast cancer (Lawicki *et al.*, 2006), where monocyte recruitment to the tumour and dysregulated signalling may play pathogenic roles (Hernandez *et al.*, 2009). The central role for macrophages in immunity and physiological processes means that disruption of CSF1 signalling is likely to impact on many diseases.

In the case of viral infections, CSF1 is an important regulator of monocyte–macrophage responses. In the case of HIV-1 infection, CSF1 levels rise leading to increased monocyte levels and increased HIV-1 replication within monocytes. Importantly, CSF1 leads to induction of CCR5 (Wang *et al.*, 1998) which may increase the ability of HIV to replicate within macrophages. At the same time, the *Nef* gene product from HIV interferes with CSF1R signalling (Hiyoshi *et al.*, 2008) and prevents differentiation and activation of macrophages. In the case of other viruses, CSF1 may increase resistance, as observed for Sendai virus (Matsuzawa *et al.*, 1997).

## Signalling pathways and transcription factors in macrophage development

Macrophages utilize a vast array of signalling pathways and transcription factors to initiate and regulate maturational programmes, receptor and effector functions and the coordination of immune responses. Most of these are used in common with other tissue types and generally speaking it is the ways that signalling pathways are used combinatorially rather than the possession of unique pathways which distinguish one tissue from another. It is not possible, therefore, to enumerate the many intracellular pathways used by macrophages, though particular attention is due to the set of signalling pathways which act as coordinators of inflammatory responses. These are the MyD88, IRF, MAP kinase, NFκB and caspase-dependent pathways which together act as central coordinators for pattern recognition receptor mediated responses which are described later.

In contrast, the set of transcription factors which are centrally important in establishing the myeloid haematopoietic lineage development is relatively circumscribed. The role of AML1 (Runx1) in the early formation of embryonic myeloid precursors has already been described, and this factor is also a key determinant of haematopoiesis in the adult. Although it has many targets in the cell, the key target in myelopoiesis is the PU.1 promoter, which it regulates. PU.1, an Ets family transcription factor, is in turn a key positive feedback switch in differentiation, regulating not only its own expression but a vast array of myeloid genes. PU.1 is present in both B-cells and myeloid cells, but with different combinations of other factors. In the case of B-cells, Pax5 acts to close off myeloid gene expression, acting as a repressor, for example, for the CSF1R gene by binding to the PU.1 sites in competition with PU.1 (Tagoh *et al.*, 2006). In myeloid cells, Runx1, C/EBPα and PU.1 act as master regulators for myeloid differentiation pathways. In the case of myelopoiesis, the simultaneous activation of PU.1 and subsequently CSF1R in a sense defines myeloid cells and circumscribes the subsequent patterns of gene expression of which the myeloid lineages are capable. Runx1 and C/EBPα activate PU.1, which switches on very early in myeloid lineage development and rapidly up-regulates the CSF1R. The presence of CSF1 thereafter maintains the viability of myeloid precursors and also guides their subsequent development and behaviour. In this sense, the pathway from stem cell to committed myeloid cell is conceptually short, despite the complexity of the subsequent gene expression and regulation and PU.1 therefore deserves the somewhat overused epithet of being a 'master regulator'.

Given the far reaching influence of PU.1 and CSF1 on myeloid function, the fundamental importance of myeloid cells in disease processes (especially chronic diseases) and the potential importance of the CSF1R as a therapeutic target, it is not surprising that much effort has been devoted to understanding the transcriptional regulation of the PU.1 and CSF1R promoters. As is usually the case, what appears to be conceptually simple is, in practice, convoluted and complex, and after two decades of study, many details still require clarification. Nonetheless, the general pattern is clear.

Gene expression in haematopoietic stem cells is somewhat fluid, and many of the 'downstream' genes or receptors are expressed at low and variable levels, with the relevant chromatin being actively held in an open conformation. The process of differentiation begins with the gradual down regulation of genes whose function is to maintain this fluid, open state and the up-regulation of factors which will go on to become important in cell fate determination. It seems likely that at this stage, cell fates are still not 'locked in' and this may be responsible for the difficulty in defining rigid 'canonical' pathways of hierarchical differentiation. Bonifer *et al.* (2008) provide an excellent review of this transitional process and the subsequent gene expression events leading to the genetic regulation of macrophage specific genes. In the case of the myeloid lineage, Runx1 is probably the earliest myeloid-specific key gene to bias haematopoietic stem cells towards the myeloid lineages, and in this sense it recapitulates the situation in embryonic development discussed above. Runx1 and CBFβ together start to lock in the chromatin state of the cell, as well as transcriptionally activating key genes such as PU.1 (Huang *et al.*, 2008). Expression of PU.1 feeds back upon its own promoter which is also bound by C/EBPα, Elf-1 and Atf2. In B-cells, Elf-1 and Atf2 are bound instead by Oct1 and Bob1. In the absence of Pax5, Runx1, PU.1 and C/EBPα together activate the CSF1R promoter which then provides the canonical CSF1R protein on the surface of the cell and opens the cell to external regulation by CSF1. At the same time, these key transcription factors up-regulate other gene products needed to lock in high level CSF1R expression, leading to a stepwise increase in the level and robustness of CSF1R gene expression. Downstream signalling from the CSF1R then directs the expression of suites of other macrophage-specific genes.

Interestingly, observations that 'macrophage' cell lines possess rearranged immunoglobulin genes (Ross *et al.*, 1994a) and appear to be derived from B-cell tumours argued for a close relationship between B-cell and macrophage gene expression. These observations have been borne out by subsequent studies (Xie *et al.*, 2004) that showed the relatively few differences between B-cells and macrophages in terms of key transcription factors and suggest that inactivation of a very few genes, including Pax5, is sufficient to switch on the CSF1R gene and confer a macrophage phenotype on a B-cell. In this view, B-lymphocytes are (phylogenetically speaking) modified macrophages.

This brief description may suggest that few genes are required for macrophage development, but nothing could be further from the truth. Many other transcription factors are also involved; but these are either not tissue specific in the way PU.1 and Pax5 are, and consequently do not force the direction of differentiation, or else they are regulated by other factors in a cascade.

Like most critical genes, the CSF1R gene is highly regulated at multiple levels. Transcription is not only under the control of chromatin organization and tissue-specific transcription factors, but there are two separate promoters, for expression in trophoblasts and macrophage lineages. A tissue-specific enhancer known as the FIRE element switches on subsequent to promoter activation, and regulation occurs at the level of transcriptional termination, translational regulation, post translational modification, trafficking control to regulate surface expression, protease-dependent shedding of the extracellular domain, internalization and recycling and a host of other controls on the subsequent signalling pathways.

The presence of locked in CSF1R expression is a critical step on the myeloid differentiation pathway. The further ramification of macrophage lineage differentiation states is provided for by the activation of downstream signalling events and their subsequent genetic switching consequences. Although there may be dozens of these differentiation pathways, the best known are the conversion via CSF1 + IL-4 or GM-CSF signalling of monocytes to dendritic cells, and the differentiation of monocytes under the influence of RANKL + CSF1 into osteoclasts, via MITF, NFATc and NFκB pathways. For these transitions, the relevant signalling events and downstream transcription factors have been examined in some detail. Other, less studied transformations include the conversion of monocytes into the various tissue macrophage subsets. Furthermore, as previously noted there is a case to be made that macrophage activation is also a differentiation event, albeit a complex one, since (apart from in textbooks) there is no neat dividing line between differentiation and the morphological and genetic changes that we call 'activation'. The main conceptual difference is that differentiation is continuous and inevitable, resulting from internal cues (e.g. growth factors), whereas activation is irregular and contingent on external cues (e.g. TLR ligands). However, feedback from pathogenic signals affects the differentiation process, and activation can also be produced by internal cues in the absence of pathogens. While the discrimination between differentiation and activation is useful in a general sense, its utility is limited when considering the detailed specifics of a given cellular response.

## Macrophage diversity in the body

While it is not possible to comprehensively cover the literature with respect to all tissue macrophages, an introduction is provided here to the major types of macrophage subsets encountered in vertebrates. Numerically, circulating monocytes, Kupffer cells and alveolar macrophages represent the biggest subsets, but the pathophysiological importance of some other subsets warrants attention.

### Circulating monocyte subsets

The complex origin of circulating monocytes from bone marrow precursors has already been discussed, although the simpler and older terminology of 'monoblasts' and 'promonocytes' is still used, particularly in clinical contexts. Morphologically, blood monocytes are somewhat heterogeneous, which has traditionally been taken as evidence of differing stages of maturation. Monocytes are relatively small (12–15 μm) spherical granular cells with irregular nuclei. They contain cytoplasmic granules some of which are peroxidase positive, and which are the precursors of the lysosomes in the mature macrophage. Other typical monocyte histochemical markers include non-specific esterase, myeloperoxidase, lysozyme and N-acetylglucosaminidase. They adhere to glass and are phagocytic. Monocytes display the CSF1R (CD115) which is necessary for signalling by the key macrophage growth factor CSF1. Because of their medical importance and accessibility, blood monocytes and *in vitro* monocyte-derived macrophages are well described in the literature in comparison to tissue macrophages which are generally harder to study. Following their genesis in the bone marrow, human monocytes enter the circulation between 13–26 h. Once circulating, their half life has been estimated at ~70 h (Whitelaw, 1972) similar to the half-life of 39 h in the rabbit (Issekutz *et al.*, 1981). Monocyte production is around $9.4 \times 10^8$ cells per day which is well in excess of what is needed for maintenance of tissue levels. Following the onset of infection, the inflamed site is rapidly infiltrated by monocytes and neutrophils, though the large numbers of neutrophils means that these short-lived cells predominate in the first 12 h, with monocyte–macrophages being the major cell type only after 12–24 h (Issekutz *et al.*, 1981).

A longstanding question concerns to what extent circulating tissue monocytes are a homogenous population or whether (at the other extreme) there exist multiple subsets of circulating monocytes with specific homing properties. Recently it has become clear that functional monocyte subsets do exist (Geissmann *et al.*, 2003). Circulating murine monocytes can be broadly divided into two sets, 'inflammatory' monocytes ($Gr1^+$, $CCR2^+$ and L-selectin$^+$) which enter inflamed areas after neutrophils and which eventually become dendritic cells and 'resident' monocytes ($Gr1^-$ $CX_3CR1^+$ $LFA1^+$). Gr1, a murine-specific marker, is also known as Ly6C. In humans, similar subsets are characterized by $CD14^+CD16^-$ 'inflammatory' monocytes vs. $CD14^{lo}CD16^+$ which resemble the murine 'resident' phenotype. Auffray *et al.* (2007) showed using *in vivo* videomicroscopy that the resident monocytes actively patrol small and defined areas of the endothelium, both with and against the blood flow, so that the entire endothelial surface is constantly surveyed by these monocytes. Upon tissue injury or *Listeria monocytogenes* infection, these monocytes immediately (within 1–2 h) extravasate and begin differentiation into macrophages, releasing TNFα and other cytokines well before the arrival of either neutrophils or classical infiltrating monocytes. However, neutrophils secrete factors including LL-37 and heparin binding protein (HBP) which accelerate the arrival and activation of these inflammatory monocytes (Soehnlein *et al.*, 2008). It is proposed that these cells eventually give rise to 'M2' alternatively activated macrophages (discussed later) whereas $Gr1^+$ cells give rise to some tissue macrophage subsets, TipDCs and macrophage suppressor cells (Auffray *et al.*, 2009a). Our understanding of these subsets and their functions – particularly with respect to the development of tissue macrophages – is rapidly evolving. Ongoing work suggests that fractalkine receptor ($CX_3CR1$) provides a survival signal for patrolling monocytes and that this may be important for the development of vascular inflammation including atherosclerosis (Combadiere *et al.*, 2008; Landsman *et al.*, 2009). Since it has been shown that CD16$^+$ monocytes are greatly expanded during HIV1 infection (Ancuta *et al.*, 2004), it seems likely that these subsets will be important in many immune functions including lentiviral infections. Notably, there are significant differences between murine and human macrophage biology at this level of detail so differences with other vertebrates are to be expected.

## Tissue macrophages

Although mostly derived from a common pool (or at least similar pools) of circulating monocytes, tissue macrophages in different locations must adapt to the local tissue niche, interact specifically with the cells which make up that tissue and perform the varied tasks required of that tissue. The intercalated phenotype often seen in tissue macrophages means that macrophage–tissue interactions are both structural and functional, and consequently it is difficult to meaningfully study isolated tissue macrophages *in vitro*. Furthermore, in normal (non-infectious) circumstances, tissue macrophages are often characterized by severely repressed inflammatory capabilities which means that most of the hallmarks of classic macrophages (especially surface markers; CD4 for example) are down-regulated. This does not mean, however, that there is no commonality between tissue macrophages and inflammatory macrophages. Once an inflammatory reaction begins, tissue macrophages may (though do not always) transform into inflammatory cells. The following survey of tissue macrophage types constitutes a brief survey of the major subsets. Each constitutes a major field of study in itself, and each site has implications for the pathology of viral diseases. In the case of HIV, the roles of Langerhans cells, microglia and alveolar macrophages are particularly pertinent to the pathology of the disease. Dendritic cells, while not (as a class) specific to one tissue, make up critical parts of the monocyte repertoire and therefore require consideration as well.

### *Dendritic cells*

Establishing the origin and function of dendritic cells (DC) has been one of the more contentious chapters in the history of the innate immune system. Dendritic cells develop generally from circulating monocytes following diapedesis into the tissues, and are marked primarily by their role in antigen presentation. The striking morphology and marked antigen-presenting activity of DCs led to their early classification in the 1970s separate from macrophages, a move which was controversial at the time but which is more generally accepted today. DC-like antigen presenting cells were soon classified into two broad classes – a population thought to be derived from lymphoid progenitors (plasmacytoid DCs) and a population derived from monocytes (myeloid DCs). A third class (follicular DCs) is now not thought to be haematopoietic in origin at all and was classified with DCs largely on the basis of morphology. Myeloid DCs differentiate from monocytes under the influence of CSF1 and IL-4 but also (in an inflammatory situation) other activators such as GM-CSF. Flt3 ligand is also an important cytokine in DC development. Typical signs include the acquisition of a dendritic morphology, up-regulation (in humans) of CD11c, MHC class II, and lectins such as CD1a and DC-SIGN (CD209). Typically, DCs are phagocytic, can present antigen (including viral antigens) to T-cells, are responsive to cytokines such as interferon-$\gamma$ and can induce tolerance under some circumstances. As noted, inflammatory DCs which produce TNF$\alpha$ and inducible nitric oxide synthase (iNOS) are referred to as 'Tip DCs'. Other designations include the 'resident DCs' (both CD8$^+$ and CD8$^-$) which occupy specific tissue niches, and 'interstitial DCs' which generally infiltrate into and leave tissues (Naik, 2008). However, the plethora of subsets, tissue variants, surface markers and behaviours is too complex to cover here. Surprisingly, plasmacytoid DCs derived from the lymphoid lineage respond to CSF1 (Fancke *et al.*, 2008) and presumably express CSF1R, supporting the idea that reprogramming of lymphoid cells to a macrophage-like genetic programme is possible. Since the expression of PU.1 and C/EBP$\alpha$ is sufficient to confer a macrophage-like phenotype on fibroblasts (Feng *et al.*, 2008), this may not require extensive changes.

A current controversy regarding DCs centres on whether the multiple subsets identified by surface marker analysis represent meaningful categories and differentiation pathways, or whether they simply represent the ramifications of macrophage flexibility. A recent review of macrophage subsets and their relationship to DCs is provided by Auffrey *et al.* (2009b). Proponents of DCs as a radically separate population, distinct from macrophages, point to the apparently specific functions or behaviours of the many subsets and seek to define the reasons for such a complex set of differentiation pathways (Naik, 2008). Others (Ferenbach and Hughes, 2008; Hume, 2008) argue that DCs are simply members of a flexible spectrum of macrophage behaviour and that DCs as a distinct class do not exist, pointing to the fact that many DCs (or DC-like cells) are derived from monocytes just as for the other tissue macrophages such as osteoclasts or Kupffer cells, that there is no marker which is confined to DCs, and no known transcription factors or pattern of gene regulation which locks differentiation recognizably into a 'DC programme' of differentiation. Nonetheless, recent results suggest that there are at least some specific DC subsets which do not pass through a monocyte stage but that are derived from a pre-monocyte bone marrow precursor (Liu *et al.*, 2009). Further experimentation should elaborate whether this observation is robust and whether it represents the general case. Although somewhat of a semantic argument, there is a rationale behind the dispute. If the plethora of DC subsets is the result of an adaptable and malleable response to local environmental factors then the number of subsets will multiply almost indefinitely as more and more surface markers are drawn in to delineate the sets. This will bring confusion rather than clarity to the field. On the other hand, if DC subsets are genuinely hard-wired into myeloid biology, then it is worthwhile to understand the reasons for such a proliferation of end-states, and DC types may be discrete and therefore finite. Although there are still puzzles to be solved, it seems best at this stage to regard DCs as a macrophage lineage subset akin to osteoclasts and Langerhans cells, which are after all, dramatically different in their own ways to their parent monocytes or monocyte lineage precursors (MLPs). In this view, incursion of MLPs into each tissue will lead to the generation of a spectrum of phenotypic responses in which DC-like antigen-presenting cells will be at one end of the spectrum. It may be that the genetic programme which supports antigen presentation necessarily confers DC-like properties on cells which switch on this programme, or it may be that the tissue environment selectively recruits MLPs which have DC potential. In the long term, this issue may only be resolved by painstaking, lineage-based characterization of multiple tissue subsets which are currently lumped together as 'DCs'.

### Kupffer cells

Derived from circulating monocytes, Kupffer cells are stellate macrophages lining the hepatic sinusoids, adhering to the fenestrated sinusoidal epithelium and extending numerous filopodial and lamellipodial projections. As the portal circulation reaches the liver, the Kupffer cells are exposed to gut-derived toxins and bacteria, and play a vital role in removing these components prior to the passage of the circulation to the rest of the body. The largest tissue macrophage subtype, they clearly have a sentinel role in protecting the wider circulatory system from gut-derived pathogens and debris. They also have local tissue roles including the processing of haem breakdown products and red cells and are important in liver injury and disease (Bilzer *et al.*, 2006). Clearance of senescent red cells and red cell fragments is mediated by scavenger receptor and leads to bilirubin excretion via the bile duct (Naito *et al.*, 2004; Willekens *et al.*, 2005). Kupffer cells also act as antigen presenting cells within the liver (Maemura *et al.*, 2005). In mice they strongly stain with the antibody F4/80 (anti-EMR1) and also with the antibody KiM2R which targets an unknown 45-kDa marker present on 'resident macrophages' particularly in the spleen and liver.

Kupffer cells are the most numerous tissue macrophages, since they comprise approximately 15% of liver cells and at least 50% of the total body complement of tissue macrophages (Wisse, 1974a,b). Although studies of Kupffer cell residence times have conflicted, the consensus is that they are long-lived cells with residence times of the order of months (Naito *et al.*, 2004). This is supported by a recent study which revealed both short- and long-lived populations, the latter primarily having a non-inflammatory phenotype

(Klein *et al.*, 2007). Some Kupffer cells leave the liver by migrating to the hepatic lymph nodes, and *in situ* apoptotic Kupffer cells are probably phagocytosed by others, but it is not clear exactly how the population numbers in the liver are regulated. Although embryonic haematopoiesis occurs in the liver, it migrates to the bone marrow shortly after birth and in the adult, all Kupffer cells appear to be derived from the circulating monocyte pool (Naito *et al.*, 2004). The main growth factor for Kupffer cells appears to be CSF1, and severe deficiencies in Kupffer cell numbers are seen in the *op/op* mouse (which has a truncation of the CSF1 gene) following Kupffer cell depletion (Naito *et al.*, 2004; Yamamoto *et al.*, 2008).

Their clearance role means that scavenger receptors, FcR and complement receptors are important surface molecules for Kupffer cells. In particular, a recently identified complement receptor CRIg confers on Kupffer cells the ability to effectively clear C3-opsonized pathogens from the circulation (Helmy *et al.*, 2006). Adenovirus is effectively cleared from the circulation by Kupffer cells for which SR and CR are the predominate receptors (Xu *et al.*, 2008). Lentiviral vectors are also sequestered by Kupffer cells, but not by sinusoidal epithelial cells suggesting that they represent the primary defence for the liver against viral infection (van Til *et al.*, 2005).

The fact that Kupffer cells screen the portal circulation means that they are constantly exposed to bacterial products including lipopolysaccharide (LPS) from commensal bacteria and food antigens. Clearly an inflammatory response to such molecules is undesirable. At the same time, the liver clears viruses from the systemic circulation and mounts effective inflammatory defences against them using a collaborative Kupffer cell-NK cell mechanism. Tu *et al.* (2008) demonstrate that this apparent paradox can be explained by the fact that bacterial products activate the MyD88-dependent signalling pathway (described below) and hence produce the inhibitory cytokine IL-10, whereas viral ligands primarily activate the MyD88-independent interferon-β pathway (via TLR3) which fails to produce significant IL-10. At the same time, stimulated Kupffer cells produce NK activators (IL-12 and IL-18). In the first case, IL-10 suppresses NK responses but in its absence (during viral clearance) an inflammatory NK response ensues. They suggest that successful hepatic viruses including Hepatitis B and C are able to subvert this response. Distant viral infections can also activate Kupffer cells, being responsible for the mild non-infectious hepatitis that can accompany influenza infection (Polakos *et al.*, 2006).

Given that the liver filters the entire blood supply, pathogens entering via other sites must also escape Kupffer cell surveillance and this can be relevant to distant organs and sites of infection. Remarkably, malaria sporozoites have been shown to enter hepatic cells by first travelling through Kupffer cells (Baer *et al.*, 2007). In order to protect themselves from intracellular killing, sporozoites produce a protein which is able to inhibit the respiratory burst (Usynin *et al.*, 2007).

The ability to endocytose and kill bacteria has been demonstrated to be important for survival of *Listeria*-infected mice, at least partly by the recruitment of neutrophils. Defence against *Salmonella* is also dependent on activation of Kupffer cells via TLR4 (Vazquez-Torres *et al.*, 2004). Although less frequently studied than other macrophage subtypes, it is to be expected that Kupffer cells will coordinate and direct immune responses within the liver.

The influence of Kupffer cells is also apparent in larger physiological processes. They are centrally involved in endotoxic shock and shock following trauma and haemorrhage. Activated Kupffer cells produce mediators, particularly MCP1, which produce distant organ dysfunction (Hildebrand *et al.*, 2006). The presence of scavenger receptors also enables Kupffer cells to regulate the uptake of lipoprotein and they have an important role in the biology of cholesterol uptake. As with macrophages in other organs, Kupffer cells have been implicated in chronic inflammatory conditions; including alcoholic steatosis, atherosclerosis and cancer.

A multitude of other factors produced by activated Kupffer cells, including prostaglandins, cytokines and chemokines, affect local hepatic cells, circulating lymphocytes and distant organ sites, and have broad physiological effects. The role of Kupffer cells in the liver is consequently not a simple 'tissue macrophage' effect, but rather an integral part of the physiological role of the liver itself.

## Alveolar and lung interstitial macrophages

The lung, like the gut, comprises a vast mucosal surface which necessarily deals with large numbers of potential pathogens, both viral and bacterial, as well as physical particles, both inert and chemically reactive. Three monocyte derived populations are found in the lung; alveolar and interstitial macrophages which line the alveoli and are found in the interstitial tissue respectively, and interstitial dendritic cells which are primarily involved in antigen presentation (Zaslona *et al.*, 2009). It is likely that interstitial macrophages are an intermediate on the pathway to alveolar macrophages (Landsman and Jung, 2007).

Resident alveolar and interstitial macrophages synthesize and secrete a broad array of cytokines and effector molecules including the key cytokines IL-1, IL-6 and TNFα as well as chemokines (including IL-8) and prostaglandins (Maus *et al.*, 2002). Their functions include the phagocytosis and removal of particles and microorganisms, the recruitment of neutrophils when required, and the removal of necrotic and apoptotic cells, both infected and uninfected. Alveolar macrophages, in direct contact with incoming particles or pathogens, release more inflammatory mediators including reactive oxygen species (ROS) and are more phagocytic than interstitial macrophages. As appropriate for 'frontline' cells they can express scavenger receptor, chemokine and cytokine receptors, GPCRs (G protein coupled receptors), FcR (Fc receptors) and TLRs (Gwinn and Vallyathan, 2006). Interstitial macrophages express C3R, ICAM1 and Ia, and secrete more IL-1 and IL-6 suggesting a role for coordination of immune responses rather than direct anti-pathogen attack (Barnes, 2004).

Apart from the cytokine-mediated coordination of inflammation, alveolar macrophages initiate a range of toxic effector mechanisms, particularly the respiratory burst which is a source of ROS (Gwinn and Vallyathan, 2006). The respiratory burst, initiated by an inducible NADPH oxidase, is important for killing fungal pathogens as illustrated by the genetic condition chronic granulomatous disease (CGD) in which a subunit of the respiratory burst oxidase is missing. Although such patients are more prone to bacterial pathogens, they are particularly susceptible to fungal pathogens such as *Aspergillus* (Segal, 2007).

The lungs are naturally exposed to a range of viral, bacterial and fungal organisms with few adverse effects. Although healthy individuals can be colonized by successful pathogens such as *Mycobacterium tuberculosis*, influenza virus and respiratory syncytial virus, it is the presence of a compromise in lung function or immunity that most often results in potentially life-threatening lung conditions. Notable examples include *Pseudomonas aeruginosa* in cystic fibrosis patients and *Pneumocystis carinii* during HIV infection. Successful pathogens have usually evolved mechanisms to inactivate or reduce the activity of alveolar macrophages, such as the PB1-F2 virulence protein of Influenza A which destroys alveolar macrophages and enables rapid colonization by bacterial infections (Coleman, 2007). As with most tissues, some pathogens (*Bacillus anthracis*, *Mycobacterium tuberculosis*) have learned to survive within macrophages and to even use them as a central part of the pathogenic strategy (Guidi-Rontani, 2002). In such cases, macrophage survival is actually detrimental and mechanisms to induce macrophage apoptosis can be helpful. The critical determinant of macrophage apoptosis is the short-lived Bcl family member Mcl1 (Marriott and Dockrell, 2007) which must be constantly renewed and acts as a 'fuse' to trigger apoptosis in the absence of a 'stay alive' signal such as that delivered by the CSF1R.

Like Kupffer cells, alveolar macrophages have a long residence time in the lung, with half-lives estimated to be several months (Thomas *et al.*, 1976; Barnes, 2004). In addition, although the bulk of alveolar and interstitial macrophages are probably replaced from the circulating monocyte pool, there is evidence that to some extent they replicate *in situ* (probably from the interstitial macrophage population) and thus represent a pool of naturally self-replenishing tissue macrophages. In contrast, the lung DC population in the interstitium is turned over every few days, is poorly phagocytic and is primarily concerned with antigen presentation (Vermaelen and Pauwels, 2005).

The central role for alveolar macrophages in the clearance of particles and the resistance to pathogens means that they need to mount a rapid, effective and if necessary prolonged inflammatory response to potential threats. At the same time, the delicate nature of lung tissue means that alveolar macrophages need to be as quiescent as possible during normal lung function while still able to remove inert particulates and other non-infectious substances including apoptotic cells. This knife-edge balance between activation and repression of inflammation is particularly acute in the lung, and the outcome of long term unresolved inflammation is an often fatal chronic lung disease, significant examples including chronic obstructive pulmonary disease, silicosis or asbestosis.

The need both for repression of inflammation when infection is not present, and for an active anti-inflammatory response once the response to an infection is achieved, requires a robust network of checks and balances. Normally, repression of inflammation is achieved at least in part by anti-inflammatory fatty acid derivatives acting through peroxisome proliferator-activated receptor gamma (PPARγ). PPARγ represses the NFκB-mediated cytokine gene activation which is downstream of the CSF1R and the secretion of anti-inflammatory fatty acid derivatives is known to be important for normal lung homeostasis (Reddy, 2008). Phagocytosis of non-infected apoptotic cells also down-regulates macrophage inflammatory characteristics, promoting the release of anti-inflammatory factors such as TGFβ and IL-10. Local (not systemic) production of TGFβ also seems important for maintaining alveolar macrophages in a repressed state until IFNγ priming and TLR activation cooperatively signal the need for an activated phenotype (Lambrecht, 2006).

Following infection, an active process of resolution must include the removal of apoptotic cells (including neutrophils) and debris and the re-establishment of the normal anti-inflammatory environment.

Without an appropriate programme of resolution, lung infections take longer to resolve and produce more damage including scarring and alveolar collapse. Such processes are typical of chronic inflammatory lung diseases such as COPD, silicosis and asbestosis, the pathology of which can best be understood as that of unresolved inflammation. Sometimes even during infections the inflammatory response rather than the pathogen itself seems to be the main cause of pathological damage, as seems to be the case for *Pneumocystis carinii* (Gigliotti and Wright, 2005). Lung cancer is another long term sequelae of unresolved inflammation. Although it arises in non-macrophage tissues, there is a good case that many of the actual mutagenic events are produced by the inflammatory environment produced by alveolar macrophages. It has been shown for example that mesothelial cells die in the presence of asbestos and can only form tumours in the presence of high local levels of anti-apoptotic TNFα which is secreted by alveolar macrophages (Yang *et al.*, 2006).

Although the pathology of chronic inflammation may seem to be opposite to that of lung infection, they are closely related. For example, the key environmental factor in COPD is of course smoking. One might predict that the presence of chronic inflammation – while damaging – would reduce the chances of exogenous infection, but this is not the case. Although cigarette smoke has some immunosuppressive effects, it up-regulates host defence cytokines including TNFα, MIP2, IL-6 and MCP1. However, Gualano *et al.* (2008) showed that despite this apparently robust defence, smoked animals were more susceptible to influenza A. This is important for COPD management because exacerbations of COPD – which are the major cause of morbidity and disease progression in COPD – are precipitated by respiratory viruses.

Another observation in chronic inflammation is that unlike most other macrophages, alveolar macrophages in COPD are resistant to glucocorticoid immunosuppression, via histone deacetylase-mediated inactivation of glucocorticoid repression of NFκB-mediated inflammatory pathways rendering these agents of little use for COPD treatment (Barnes, 2004). In summary, alveolar macrophages are both necessary for survival of acute infections and also fundamentally involved in many chronic inflammatory pathologies through complex mechanisms which are not completely understood.

### Microglia

Microglia are tissue macrophages in the brain and nervous systems and make up about 10% of the cells within the brain, though the proportion varies from 0.5% to 15% in different locations (Lawson *et al.*, 1990). They carry out several key functions apart from immunological roles. They form a sentinel network, have a major role in phagocytosing apoptotic cells and debris (Neumann *et al.*, 2009), are involved in neuronal pruning and reconfiguration, the generation of new synapses and the modulation of neuronal function such as neuropathic pain (Marin-Teva *et al.*, 2004; Coull *et al.*, 2005; Stevens *et al.*, 2007). However, when activated by pathogenic signals they are capable of becoming activated in a classical macrophage manner as well as transforming into antigen-presenting cells.

Several microglial morphologies are found within the brain. So called quiescent, resting or ramified microglia typically have long ramified processes forming an active sentinel network throughout the brain which sample the local environment (Nimmerjahn *et al.*, 2005). These are essentially non-inflammatory cells which lack MHC class II, do not secrete cytokines, and have low levels of CD45 and FcR. Upon receipt of an activating signal, the ramified processes contract and the cell can transform to a typical activated phagocytic macrophage phenotype, in which case it is referred to as an amoeboid microglial cell. In fact, depending on the stimuli the actual phenotypes of such cells may vary considerably, as seen with other tissue macrophages. As with other macrophages, the specific activation stimulus (e.g. TLR4 vs. TLR9) determines the resultant phenotype. For example, TLR4 stimulation by LPS up-regulates typical proinflammatory cytokines such as IFN$\alpha$, TNF$\alpha$ and IL-6, phagocytosis and MHC (Ito *et al.*, 2002; Kielian *et al.*, 2002), whereas TLR9 stimulation by CpG DNA leads to secretion of IL-12 (Dalpke *et al.*, 2002). Transformation to a DC-like phenotype with up-regulation of MHCII and antigen presentation abilities may also occur.

In addition to genuine microglia, other macrophage populations occur, notably the perivascular macrophages in and around the blood vessels of the brain (which appear microglia-like but which are clearly derived from circulating monocytes), meningeal macrophages, and choroid plexus macrophages. These sites do not involve intimate contact with neurons and glial cells, and effectively only genuine microglia are normally found within the brain tissue itself. Perivascular macrophages have been shown to be derived from $CD68^+Gr1^-$ 'patrolling' macrophages presumably similar to the ($Gr1^-$ $CX_3CR1^+$ $LFA1^+$) 'resident' cells discussed above (Audoy-Remus *et al.*, 2008).

The issue of whether microglia are a local population or whether they enter from the circulating monocyte pool has been highly contentious. During early embryogenesis, macrophages from the liver and yolk sac colonize the developing nervous system as part of a general movement throughout the embryo, where undoubtedly they play a major role in shaping the developing tissue and absorbing apoptotic cells (Ovchinnikov, 2008). Around birth and before the establishment of the blood–brain barrier there is a second wave of colonization of the brain. During one or both of these processes, a local population is established which includes $CD34^+$ cells that can proliferate locally and which obviate the need for a continuous replenishment of myeloid precursors from the circulation. Upon local inflammatory activation these cells can produce focal proliferation (microgliosis). Isolated microglial cells proliferate in response to CSF1, so microglial differentiation appears to be a typical CSF1R-mediated myeloid differentiation process.

Many studies carried out in the last 15 years have indicated that circulating monocytes can infiltrate the brain and contribute to the microglial population, so that the conventional wisdom up to around 2005 was that microglia are indeed continuously turned over from bone marrow precursors. However, these studies utilized whole body irradiation and bone marrow replacement, and it has recently become clear that artefacts of this process contributed to the conclusions. More refined analyses have now shown that in the normal, undiseased situation microglia are entirely regenerated within the brain and circulating monocytes do not enter the brain (Chan *et al.*, 2007). However, certain pathologies change this situation and circulating monocytes can then enter the brain and differentiate (probably under the influence of astrocytes) into microglia.

Careful analysis has revealed the preconditions required for this process. The first is that some kind of inflammatory insult must occur which leads to CCL2 (MCP-1) release as a chemoattractant signal. Two types of circulating monocytes can then enter the brain. The first type is what is known as 'inflammatory monocytes' (see above) which are Ly6C$^{hi}$CCR2$^{+}$CD34$^{-}$ CX3CR1$^{lo}$ cells presumably similar to those which enter other inflammatory sites. These cells can enter the brain and differentiate into microglia but they cannot proliferate further and therefore have relatively short residence times. The other type is composed of CD34$^{+}$ cells in the circulation which are released from the bone marrow. These also differentiate into microglia but can proliferate within the brain, thereby genuinely contributing to the resident microglial population. Whole animal irradiation and bone marrow replacement leads to release of these CD34$^{+}$ cells into the circulation and subsequent incorporation into the brain, but in the absence of brain irradiation or of a CCL2 signal (e.g. in CCL2 knockout animals) the observed entry of circulating monocytes does not occur in the normal undiseased situation.

In diseases where there is a major inflammatory insult or where the blood–brain barrier is weakened, CCL2 signals may attract circulating monocytes, and this can be important for either resolution or exacerbation of pathology. For example, in West Nile Virus infections in the brain, inflammatory monocytes from the circulation are recruited to the inflammatory site and appear to worsen disease pathology (Getts *et al.*, 2008). On the other hand, incursive inflammatory monocytes appear to limit formation of amyloid β plaques and disease progression (Simard *et al.*, 2006; El Khoury *et al.*, 2007).

Interestingly, D'Mello *et al.* (2009) showed that the inflammatory site does not have to actually be in the brain. Using a murine model of hepatic inflammation, the release of TNFα from the hepatic site induced CCL2 from microglia as well as increasing circulating CCR2$^{+}$ monocytes. This led to infiltration of circulating monocytes into the brain.

In short, while circulating monocytes do not infiltrate the brain during normal life, this cannot be assumed to be the case during disease (even when the brain itself is not infected). Clearly this has implications for many diseases but especially for neurodegenerative conditions including multiple sclerosis, amyotrophic lateral sclerosis, Parkinson's disease, Alzheimer's disease, neural tumours, meningitis, prion diseases, ageing, viral infections (especially HIV), and many others.

In some HIV-infected individuals, HIV-associated dementia develops along with other symptoms of AIDS. Microglia are a major reservoir for HIV and appear to be the only cells in the brain which are actively infected. HIV enters microglia via the usual CD4$^{+}$ coreceptor route (typically CCR5). Cytokines can affect HIV uptake: IL-4 can up-regulate microglial CD4, IL-10 up-regulates CCR5, while cytokine agonists of CCR5 can block viral entry (Rock *et al.*, 2004). The extensive literature on HIV and microglia will not be covered here beyond some concluding remarks, but the reviews by Rock *et al.* (2004) and Kramer-Hämmerle *et al.* (2005) provide a detailed overview of this complex area.

### Dermal macrophages

The skin contains several monocyte-derived populations including Langerhans cells, dermal dendritic cells (DC) and dermal macrophages. In the epidermis, Langerhans cells (LC) are monocyte-derived antigen-presenting skin cells which form a regular network of sentinels. They act to initiate inflammatory responses and present antigen, typically migrating to local lymph nodes to stimulate a lymphocytic response. The origin and lifespan of LC has been a matter of much debate. It is now established that they can both replicate *in situ* from local stem cells (the normal case in uninflamed skin) and can also be replenished from circulating precursors (monocytes and/or DC precursors). Consequently they are normally long lived with residence times of months or even longer, but upon depletion (e.g. by UV irradiation) they can be replaced by CD14$^{+}$Ly6G$^{+}$ monocytes in a relatively short time (although this seems to take many days, unlike typical DC populations).

Langerhans cells are characterized by their location in the epidermis, their local proliferation, their expression of the protein langerin (CD207) and by the presence of tennis-racquet shaped Birbeck granules which are formed following ligation of langerin and which seem to represent a non-classical antigen-presenting pathway. Classically, LC migrate to draining lymph nodes and present antigen to CD8$^{+}$ T$_{h}$ cells and this is thought to be important for viral immunity, although cytolytic viruses such as HSV and CMV seem to subvert this process. Cross-presentation (i.e. antigen presentation to $CD8^+$ cytotoxic T-cells via the MHC I rather than MHC II pathway) following phagocytosis of lysed LCs is thought to assist in antigen presentation in this situation. LC have long been known to be important for HIV infection, not only as a route of primary infection via breaks in the skin barrier (especially in foreskin and vaginal mucosa), but also as reservoirs of virus (de Witte *et al.*, 2007).

HIV entry into subepidermal DCs and transmission to local lymph nodes is known to be mediated by CD209 (DC-SIGN) which like other lectin receptors such as CD1a can bind mannose on the HIV envelope. Importantly, entry of HIV-1 into LC is not DC-SIGN, CD1a or CD4-dependent. However, binding of gp120 by langerin is also known to occur and this was assumed to be a route of infection not only of LC but also of T-cells following migration and antigen presentation. Recent work (de Witte *et al.*, 2007; de Jong *et al.*, 2008a,b) has suggested that langerin actually *blocks* HIV infection and cross-infection of T-cells, and that it is only following activation and down-regulation of langerin by LPS or TNFα (or specific Ab) that LC infection and T-cell infection, or alternatively, the presence of excess virus, occurs. On the other hand, Kawamura *et al.* (2008) argue that in a human skin model system, LC and not skin DCs are mainly responsible for HIV transmission to T-cells. These differences are possibly due to differences in experimental conditions.

In addition to Langerhans cells there is a $langerin^+$ population of DCs which is short-lived and occupies the dermal layer. These DCs do not proliferate locally and are replenished via a conventional route from circulating common DC precursors. They readily migrate to draining lymph nodes to present antigen. Other non-langerin containing DC are also present in the dermis along with conventional monocyte-derived macrophages. Ochoa *et al.* (2008) show that the dermal DC population consists of $CD1^+$ DCs and $DC\text{-}SIGN^+$ (CD209) macrophages.

Finally, although langerin-expressing Langerhans cells are classically present in the skin, it is not very surprising that langerin-containing LC-like cells are also present in other organs such as cornea (Mayer *et al.*, 2007) and the intestinal mucosal surfaces of the gastrointestinal tract (Merad *et al.*, 2008). Unfortunately it is not possible to do justice here to the intricacies of differentiation and phenotypes of dermal macrophage subsets, especially since the literature is growing rapidly.

### Osteal macrophages and osteoclasts

Apart from bone marrow cells which constitute the stem cell niche for haematopoiesis, bone contains at least three populations of monocyte derived cells. Firstly trabecular bone has a population of macrophages which in the mouse are F4/80 positive and negative for tartrate resistant acid phosphatase (TRAP), a classic osteoclast marker. Another population is TRAP positive and F4/80 low, including those destined to become osteoclasts. Osteoclasts are giant multinucleated cells formed by the fusion of large numbers of monocyte-derived precursors, which function to resorb bone. Osteoclasts overlie the bone surface and form a tightly sealed acidified resorption pit into which they secrete proteins (including TRAP) which resorb bone, contributing to the continuous dynamic remodelling of the bone surface. Of great importance in joint replacement, osteoporosis and other bone diseases, osteoclast differentiation from monocytes is well studied and is mediated by CSF1 in combination with RANKL. A key transcription factor in this process is NFATc1.

Finally, there is a recently described population of bone lining macrophages or osteal macrophages which are at least structurally separate and which form a canopy over the bone-forming osteoblast layer which lines the bone surface (Chang *et al.*, 2008; Pettit *et al.*, 2008). Osteal macrophages (or *osteomacs*) are hypothesised to collaborate with osteoblasts in a close relationship, to regulate bone maintenance and reformation following osteoclastic resorption. The balance between resorption and replacement is a key process in bone turnover, and chronic inflammatory processes in the bone (for example surrounding artificial joint replacements and during osteoporosis) lead to excessive bone resorption and fragility especially in trabecular bone.

### Other tissue macrophage populations

It is not possible here to provide more than a thumbnail sketch of the tissue macrophage populations already mentioned, or to do more than mention the many other populations which are known to exist. Splenic sinusoidal macrophages, like Kupffer cells, play a sentinel role and are frequent targets for viral infection including dengue virus, hepatitis and lentiviruses such as human, simian and feline immunodeficiency viruses. Tissue macrophage populations are also present in the lamina propria of the intestine, in close collaboration with other immune cells in Peyer's patches and the lymphoid organs, and in the peritoneum where they form a residential population which is important in endometriosis and reproductive biology. During disease processes, the formation of granulomas involves the conversion of macrophages to giant cells at the centre of the granuloma. Virtually every tissue contains a macrophage subpopulation, often referred to as histiocytes, and it is likely that new macrophage derived or related populations remain to be discovered. For example, it was recently shown that Paneth cells in the crypts of the small intestine express CSF1R and are CSF1 dependent (Huynh *et al.*, 2009). These cells express TLR4, are LPS-sensitive and secrete antimicrobial peptides and are hypothesised to be involved in the pathology of Crohn's disease (Wehkamp and Stange, 2006). A common source of experimental confusion results from contamination of supposedly pure tissue preparations with a monocyte derived subpopulation which gives rise to unexpected expression of immune-related genes in these samples.

## Experimental macrophage systems

A range of experimental systems have evolved to study primary macrophage cell biology, the most widely used of which are (in the mouse) the bone marrow derived macrophage (BMDM or BMM), the elicited peritoneal macrophages and the ES cell derived macrophage, whereas in humans, the conversion of blood monocytes to a macrophage-like cell is effected by culturing freshly isolated blood monocytes on an adherent substrate (or preferably, by negative sorting of other compartments). BMDM are produced by *in vitro* culturing of bone marrow in the presence of CSF1, which produces a relatively non-inflammatory macrophage somewhat representative of a generic tissue macrophage, but which can be activated using INF$\gamma$ and/or LPS. Thioglycollate elicited peritoneal macrophages are used as a model of highly activated phagocytic inflammatory macrophages and are produced following the injection of thioglycollate into the peritoneal cavity of a mouse. In addition to these primary cell systems, a range of immortalized cell lines are used which have various strengths and weaknesses. The most widely used are the murine RAW264.7 cell line, derived from an Abelson virus-induced tumour, and the human 'monocyte-like' THP1 and MonoMac6 lines. The human cell line HL60 is often used as a monocyte model but is a relatively poor model lacking many monocytic attributes.

### Macrophage cell surface proteins

Cell surface proteins can be looked at from several perspectives. They are expressed on the surface for a reason, and the presence or absence of a cell surface protein can confer a functional property or behaviour on a cell, so the functional perspective is the first and most important. A particular behaviour may require the expression of several unrelated proteins, including intracellular proteins, which form a circuit or mechanism, as do the multiple proteins involved in phagocytosis for example. Other key macrophage functions include antigen presentation proteins, growth factor receptors, cytokine receptors and cell adhesion molecules. Since proteins fall into classes, noting the class of a protein can convey a great deal of information even if that protein is the only member of its class expressed on the cell surface – the Toll-like receptors, interleukin receptors or GPCRs provide instructive examples. Surface proteins are often used for classification and cell sorting, in which case the function may be less important than the level of expression. Thus monocytes can be differentiated into subsets using CD markers. Finally, since approximately 2000 different proteins are typically expressed on the surface of a given cell, coverage of the complete range of cell surface proteins expressed by macrophages is more suitably achieved with a database than a review. The emphasis below, therefore, is on selecting a few proteins or protein classes which exemplify the distinctive character of macrophages as opposed to other cell types.

Key macrophage growth factor receptors are described elsewhere in this chapter. Receptors for pathogen-associated molecular patterns (PAMPS) are particularly important for macrophage immune functions and govern much macrophage biology and hence warrant a section of their own. Differentiation and activation markers give a perspective on the state and purpose of a given macrophage, so some notable CD markers are described briefly, with some comments on their functional roles. Cytokine receptors provide an understanding of the information flowing between macrophages and other tissues and will also be covered in outline. Each protein has an extensive biochemistry and the literature contains reviews of the structure and function of most of these important proteins. While it is not possible to cover the structure and function of these in detail, it is hoped that a general orientation will be achieved as to how these proteins fit into macrophage functionality. Finally, cell surface proteins are used by viruses, including lentiviruses as docking and entry mechanisms, and some important viral receptors are briefly described.

## The classification of macrophages by cell surface markers

Cell surface markers are principally used for two purposes – firstly to sort leucocyte subsets for purification by flow cytometry or magnetic bead based fractionation, and secondly to characterize functional subsets of cells such as monocyte subclasses. Although internal antigens can be used for this purpose, the necessary fixing step prevents the isolation of live cells. Therefore cell surface markers which can be bound by antibody without affecting cell viability have become very important. Since several antibodies may be isolated which bind to a single cell surface antigen, the cluster of differentiation (CD) marker scheme aims to identify unique antigens and assign them a CD number, and is administered by the Human Cell Differentiation Molecules Council. The characterization of monocyte–macrophages by CD markers is thus a pragmatic trade-off between the need to sort cells in a manageable way on the one hand, by using minimal markers, and on the other, the desire to completely define functional subsets in maximum detail by using as many markers as are relevant. Consequently the marker subsets used to define monocytes and macrophages vary depending on the purpose at hand. Not surprisingly, CD markers are usually closely aligned with the functional characteristics of the cell type. Obviously for histological studies the use of a much wider variety of markers, including internal proteins such as transcription factors (such as PU.1) is possible.

Monocytes are classically $CD45^{+}CD14^{+}$ cells. CD45, present on most leucocytes, is a protein tyrosine phosphatase (PTP1c) and also known as B220 or Leucocyte Common Antigen (LCA). CD14 is the monocyte co-receptor for LPS and is generally restricted to monocyte–macrophages. Because cell surface receptors are up- and down-regulated according to the needs of the cell, there is no 'ideal' marker which will unambiguously identify all cells of a certain type nor exclude all cells of other types. Other cells may express CD14, and expression of CD14 is also labile, being readily down-regulated by LPS and also by pathogenic stimuli, for example HIV gp120.

When a single tissue is being considered, a monocyte–macrophage marker specific to that tissue may be the best characteristic, as for the KiM2R antibody for Kupffer cells. Examples of key cell monocyte- and macrophage-specific markers are noted in the course of discussing the different tissue subsets.

The CSF1R (CD115) itself is an excellent and biologically relevant general marker for resting monocytes and macrophages, although in activated macrophages, surface expression of the CSF1R is lost due to internalization of the receptor. Generally speaking, the best and most discriminative macrophage marker in mice, though not in humans, is the F4/80 monoclonal antibody target EMR1 (EGF module-containing mucin-like hormone receptor 1). It is expressed in virtually no other tissue, and is maintained by most macrophage subsets throughout development. Unfortunately, the human homologue of murine EMR1 (CD97) is accompanied by at least 2 other isoforms (EMR2 and EMR3) which have related and non-overlapping functions. Perhaps as a result, the human EMR proteins are not as narrowly constrained and are also expressed on activated lymphocytes and granulocytes. The use of a panel of markers (with careful thought as to the purpose at hand) is therefore the most effective way to define macrophages as a subset.

Sometimes the function of a cell type (e.g. antigen presentation) is of more interest than the cell identity *per se*. Many functional markers are up and down-regulated as macrophages mature and differentiate, and this reflects key macrophage functions: for example, MHCI and MHCII are highly expressed in dendritic cells, moderately expressed in many other macrophage subsets, yet completely absent from some macrophages especially tissue macrophages which need to remain immunologically quiescent in their normal tissue niches. Phagocytic macrophages express receptors for opsonized (FcγR, complement receptor) and non-opsonized (scavenger receptor, mannose receptor) targets. They also express cell surface effector molecules such as proteases (e.g. urokinase plasminogen activator or uPA), protease inhibitors (PAI2) and an array of signalling molecules (including the toll-like receptors). CD33, which is a SIGLEC family sialoglycoprotein receptor, is expressed early in myelopoiesis and is often used in studies of bone marrow. Particularly relevant to lentiviral expression is the coexpression of CD4 and CCR5 which mediate the uptake of HIV1, and DC-SIGN which can also be used by HIV as a means of infection.

## Macrophage receptors for pathogen-associated molecular patterns (PAMPS)

The ability to identify and respond appropriately to other potentially dangerous organisms probably arose at a time before multicellular organisms existed. At this time, foreign and potentially dangerous organisms consisted of bacteria, viruses and other eukaryotes. Plants and animals share classes of PAMP receptors such as the nucleotide oligomerization domain (NOD)-like receptors (NLR) which are homologous to the R (resistance) proteins in plants (Wilmanski *et al.*, 2008). The advent of multicellularity required the maintenance of a generally sterile internal environment and a greater investment in longevity, and this in turn requires much more robust defences against foreign organisms, whether opportunistic or pathogenic. The multiplication of defensive mechanisms and the gathering of accurate and detailed intelligence regarding potential threats must have started very early after the adoption of multicellularity as a way of life. Macrophage PAMP receptors therefore result from some 2 billion years of co-evolution with pathogens, and it is not surprising that they are complicated, redundant and robust, nor that the mechanisms used by pathogens to evade them are sometimes astonishing in their subtlety. The elucidation of these mechanisms over a single decade has been a *tour de force* of modern biology which mirrors the rapid description of the key intracellular signalling pathways in the 1990s.

Although the acronym is new, the concept of PAMPs is not. The presence of molecules on the surface of immune cells which bind to pathogens and mediate immune responses has long been known, examples apart from LPS including mannose receptors, scavenger receptors and GPCRs such as the 'f-Met-Leu-Pro' formyl peptide receptor (FPR). Nonetheless, the discovery of the Toll-like receptors (TLRs) advanced the field in several ways. Firstly it identified a whole class of receptors whose function was primarily a signalling function against highly specific pathogen motifs, and it uncovered a key signalling pathway in this process. Most importantly, it focused the attention of the immunology community on the specificity of the innate immune response to different pathogens, a specificity which had been less clearly seen with the more general purpose receptors. This led to a paradigmatic shift for which the discovery of the TLRs was a lever.

### Toll-like receptors

Macrophages had long been known to detect bacterial products such as LPS, and CD14 was known to be involved in this process, but it was clearly not a free-standing LPS receptor, and the nature of LPS detection and signalling remained elusive. At this stage, in 1996, nothing was known of PAMP receptors although TLR1 had already been cloned. The discovery was made that Toll, a *Drosophila* developmental pathway receptor which utilized an interleukin-type signalling domain, was important for pathogen defence via the well-known NFκB pathway (Lemaitre *et al.*, 1996). Similarly, Medzhitov *et al.* (1997) cloned a human Toll homologue and showed that a constitutively active mutant induced both NFκB and NFκB-induced genes including IL-1, IL-6, IL-8 and B7.1. Although the ligand was still unknown, the implications for immunity were clearly understood, in particular the role of the innate immune system in triggering lymphocyte activation and responses. It was soon shown that mice with a mutation on TLR4 were defective in LPS signalling and that the Toll homologue TLR4 therefore played a vital role in LPS detection (Poltorak *et al.*, 1998). Within a few years, 10 Toll-like proteins had been identified by homology searching within the human genome and their functions identified by painstaking biochemical studies (O'Neill, 2008a). All are expressed by monocyte–macrophage lineage cells.

The TLRs are integral membrane proteins present either on the surface (TLRs 1–2, 4–6) or on intracellular lysosomal membranes (TLRs 3, 7, 8 and 9). They are composed of an extracellular domain which contains multiple repeats of an alpha helical motif rich in leucine. This leucine-rich repeat (LRR) region forms a curved or horseshoe-shaped domain consisting of a hydrophobic inner side and a hydrophilic outer surface. LRR domains are not unique to TLRs, and form a variety of binding surfaces for ligands or for other proteins. The inner membrane (cytoplasmic) portion of TLRs is composed of a domain which is homologous to the interleukin receptors – the Toll-Interleukin Receptor (TIR) domain. This domain has access to the factors in the MyD88 adapter pathway which leads to signalling through NFκB and the TBK1 adapter 'MyD88 independent' pathway which leads to signalling through IRF3. The former initiates cytokine gene expression, typically with a peak expression at around 8–12 hours after signalling. The latter activates the interferon response elements and leads to the secretion of interferons α and β. Since macrophages typically have interferon receptors as well as TLRs, a common observation from TLR signalling is a second wave of interferon response genes some 12 hours later than the initial TLR response. There are also signalling options to apoptosis pathways. The elucidation of the complex pathways and consequences of TIR signalling is one of the most-reviewed areas in immunology in the last decade and O'Neill's recent (2008) overview provides an introduction to the field.

The cell surface TLRs are mainly concerned with the detection of products unique to bacteria, including specific subtypes of bacteria, or to host cell components which are produced in response to cell damage (heparan sulfate, HSP70). Thus, bacterial cell wall components include lipopolysaccharide (TLR4), lipotechoic acids and other cell wall glycolipids or lipoproteins (TLR2), triacyl lipopeptides (TLR1), Mycoplasma-specific diacyl lipopeptides (TLR6). In addition, the bacterial flagellar protein flagellin is detected by TLR5. The presence of bacteria, whether Gram (+) or Gram (–) is then easily detectable by extracellular receptors which signal via the intracellular adapter MyD88 (although little is known of TLR10 function to date).

The intracellular TLRs are responsive to viral PAMPs especially double stranded (TLR3) and single stranded (TLR7/8) RNA as well as unmethylated (i.e. non-mammalian) CpG containing DNA (TLR9). These, then, are the TLRs specifically concerned with viral defence. In particular, TLR7 has been identified as important for HIV pathology. It has been argued that HIV RNA itself can trigger a full-blown immunodeficiency (Baenziger *et al.*, 2009) which is supported by results showing that primate hosts such as sooty mangabeys which fail to secrete high levels of interferon α in response to TLR7 and 9 stimulation do not progress to SIV pathology. The argument is that it is the host response to HIV/SIV that drives the pathology.

TLRs are not primarily focussed on immediate action by the macrophage such as phagocytosis, though this may form part of the subsequent responses. The primary function of TLRs is to trigger the release of specific sets of cytokines which coordinate an appropriate and holistic response by all elements of the immune system. The events subsequent to viral TLR triggers are different to those triggered by bacteria and this reflects the core differences between these pathogen types. Bacterial responses are generally focussed outwards towards destruction of an external pathogen. Anti-viral responses are focussed on warning other cells and the destruction (or apoptosis) of the infected cell, to minimize viral replication.

### Nod-like receptors (NLRs)

Although not the first member of the family identified, the prototypical Nod receptor is Nod2, which was initially identified as a protein implicated in Crohn's disease susceptibility. The combination of cell biology tools and a disease phenotype concentrated research on this protein. The archetypal domain structure of Nod2 consists of a C-terminal ligand binding LRR domain, the eponymous nucleotide oligomerization domain (NOD) and an N-terminal caspase activating domain (CARD). The recently discovered role of LRRs as ligand binding domains for TLRs invited comparison with the LRRs of these intracellular sensors. Unlike the TLRs which are all membrane embedded with the TIR domain facing the cytoplasm, Nod proteins are true intracellular cytoplasmic sensors.

It appears that these proteins are present in an inactive form where the LRR domain folds back on the NOD domain and prevents it from being activated. Binding of the ligand induces a conformational change so that the NOD domain is exposed or otherwise activated. This domain binds ATP and interacts with other proteins to form a cyclic structure. This complex enables the binding of further partners to the N-terminal interaction domains which are primarily of two kinds – NOD domains as in NOD2, or PYRIN domains as in NALP3 (cryopyrin). The nature of these domains determines the subsequent signalling route. Either signalling goes via the binding of RIP2 and thence through NFκB and p38 MAP kinase (as for most of the TLRs), thereby initiating cytokine production, or else it forms a so-called ASC complex which activates caspases, leading to the production of interleukin 1β (in the case of caspase 1) or initiates apoptosis. In the case of NOD2, the ligand is muramyl dipeptide, a peptidoglycan (bacterial cell wall) fragment which has long been known as a potent inflammatory stimulant and adjuvant.

NLR proteins containing pyrin domains are even more extensive than NOD proteins and the activation of the so-called 'inflammasome' can be initiated by a wide array of stimuli including sterile compounds (uric acid crystals in the case of gout, aluminium, asbestos, silica) UV irradiation, abnormal ion fluxes (potassium, calcium), extracellular ATP binding to cellular GPCRs and a variety of bacterial products, particularly relating to pathogens which invade the cell such as *Listeria, Salmonella,* and *Mycoplasma.*

An understanding of the NLR proteins is providing clarity to the biology of a host of inflammatory conditions including many which are due to mutations or allelic variations of NLR family members. This includes gout, asbestosis, silicosis, inflammatory bowel disease and numerous congenital autoinflammatory conditions. It is to be expected that successful pathogens have devised mechanisms to interfere with or stimulate the activation of these proteins, providing fertile ground for the investigation of pathogenic mechanisms as well as a potential therapeutic route for manipulating the immune system in the context of specific infections.

The NLR family is complex and extensive, and this brief outline simply introduces the general principles. To date, over 20 members of this family are known and extensive studies have only been conducted on a few of these. Even with this limited analysis to date, the field is complex and the reader is referred to recent reviews for further detail (Church *et al.,* 2008; Wilmanski *et al.,* 2008).

#### *Rig-I-like receptors (RLRs)*

To date, three members of this family are known; Rig-I, Mda5 and LGP2. Although structurally different to the NLRs and TLRs they are conceptually related in that they detect specific PAMPs (typically viral RNA) and trigger pathogen-specific responses (interferon signalling pathways). Structurally these proteins are RNA helicases which exist in the cytoplasm and detect single- (ss) and double-stranded (ds) viral RNA, though bacterial (*Helicobacter*) RNA is also implicated (Rad 2009). Rig-I and Mda5 possess CARD domains and signal through the adapter IPS-1 which is localized to mitochondria. This results in the activation of antiviral responses via IRF3 and p38 MAPK/NFκB in a manner similar to TLR3 (Mikkelsen 2009). Though LGP2 can bind dsRNA (Li *et al.,* 2009) it has no CARD domains and its purpose and manner of action is less clear, possibly involving modulation of responses to Rig-I and/or Mda5. Rig-I and Mda5 detect different features of viral RNA and therefore show specificity to different sets of viruses. Rig-I exists in a complex with HSP90 (Matsumiya *et al.*, 2009) and binds to RNA molecules which contains appropriate motifs and activates interferon pathways. These motifs include a 5′-triphosphate (present in influenza RNA) but also other signals (Myong *et al.*, 2009; Uzri and Gehrke, 2009). The RLRs are active helicases but this does not seem necessary for their action (Bamming and Horvath, 2009). Although the mechanisms by which Rig-I and Mda5 activate antiviral responses are still being worked out, it is clear that this system is important for antiviral responses and very likely to lentiviral infections. It is to be expected that viruses have evolved mechanisms for inactivating RLR; Gerlier and Valentin (2009) review these mechanisms for measles viruses. Similar mechanisms have been noted for RSV (Ling *et al.*, 2009), Ebola (Prins *et al.*, 2009), vaccinia (Zhang *et al.*, 2009), rinderpest virus (Boxer *et al.*, 2009), bovine viral diarrhoea virus (Baril *et al.*, 2009), paramyxovirus (Childs *et al.*, 2009), hepatitis C virus (Liang *et al.*, 2008), hantavirus (Alff *et al.*, 2008), metapneumavirus (Bao *et al.*, 2008), and doubtless many others. The potential to interfere with such viral defences is clearly of great therapeutic interest (Bowie and Unterholzner, 2008).

Not unexpectedly, RLRs are tied in with other inflammatory mechanisms – for example, TNFα up-regulates Rig-I mRNA in synoviocytes which may be relevant to joint inflammation in viral diseases (Imaizumi *et al.*, 2009). Similarly, both TLR3 and the RLRs may work cooperatively for effective antiviral responses (Negishi *et al.*, 2008).

## HIN-200 family proteins

It has long been known that cytoplasmic dsDNA initiates an apoptotic response in primary macrophages (Stacey *et al.*, 1996), suggesting that a cytoplasmic system exists which detects cytoplasmic dsDNA which will also probably comprise part of the antiviral defence system, in this case for viruses which either retrotranscribe their genomes into dsDNA or which are dsDNA viruses to begin with, such as adenovirus. Recently this sensor has been identified as the Hin-200 protein AIM2 (Fernandes-Alnemri *et al.*, 2009; Roberts *et al.*, 2009) which in response to the detection of cytoplasmic dsDNA activates the caspase pathway and leads to cell death by apoptosis. AIM2 is also antagonized by the related protein p202, revealing a typical intrinsic regulatory feedback mechanism (Roberts *et al.*, 2009). Since there are at least 12 Hin-200 protein in mice and at least four in humans, it is to be expected that a complex and important biology of the Hin-200 proteins will be revealed in the next few years, including cross-talk with the other innate immune responses. Hin-200 proteins are implicated in systemic lupus erythematosus (SLE) and as for many other chronic inflammatory pathologies, the elucidation of this pathway should lead to a much better understanding of the origin and potential treatment of this condition. Similarly, successful viruses undoubtedly inactivate or compromise this pathway and there may be therapeutic opportunities in opposing these viral strategies.

## Pathogen-binding PAMP receptors

Several other classes of molecules exist which detect the presence of pathogens and mediate responses to them. Most of these have long been known although much remains to be learned about them. Primarily these do not signal, unlike the TLRs and NLRs, but act as immediate effectors. Often they simply enable pathogen binding and therefore the potentiation of other macrophage functions. A brief introduction to these molecules is provided below.

### *Scavenger receptors*

The primary function of scavenger receptors is to recognize and bind pathogenic molecules which – assuming they are attached to the pathogen – enable the internalization and destruction of potential pathogens. Although conventionally grouped by their 'scavenger function' they are in reality composed of several different protein classes and therefore do not form a unified group in the way the TLRs do. The name is partly due to the broad ligand specificity which often includes host as well as pathogen ligands, specifically host low density lipoproteins (LDLs) such as oxidized and acetylated LDLs, which makes them important in the development of atherosclerosis. The best known are the scavenger receptor A class (SR-A) which bind bacterial LPS, LTA, CpG DNA and bacterial cells as well as modified LDLs.

SR-AI is homotrimeric and comprises membrane-anchored coiled-coil and collagenous domain proteins with a cysteine-rich C-terminal binding domain. Others in the class include SRCL-I/II in which the C-terminal domain is a C-type lectin domain, and MARCO which is important for the clearance of pneumococcal infections (Arredouani *et al.*, 2004; Arredouani and Kobzik, 2004). CD36 is class B scavenger receptor which unusually has transmembrane domains at either terminus with an unstructured extracellular loop in between. CD36 recognizes diacylated lipopeptide but also extracellular matrix. Although CD68 (macrosialin) is a structurally related molecule it is not known, at this stage, to bind pathogen components.

As binding molecules, scavenger receptors may act as coreceptors for other complexes. For example, it has been reported that CD36 may act as a coreceptor for TLR2, to broaden the detection specificity to include agonists such as diacylated lipopeptide which would otherwise not be recognized by TLR2. This suggests that some of the specificities of the TLR ligands may be mediated by coreceptors such as scavenger receptors.

In this context it is worth recalling that LPS binding by TLR4 requires both CD14 and MD2, even though TLR4 forms the actual signalling domain. CD14 is not a scavenger receptor, but a GPI-anchored extracellular molecule, although it also exists as a soluble LPS-binding protein.

*C-type lectin domain (CTLD) receptors*

Lectins are proteins which bind carbohydrate moieties, often utilizing calcium in the process. The best known macrophage C-type lectins are the Mannose Receptor (MR; CD206), DC-SIGN (CD209; 'dendritic cell specific ICAM-3 grabbing non-integrin') and Dectin-1 (CLEC7A). Type I CTLD proteins have multiple CTLD domains and include DC-SIGN, MR, Dectin-2, DEC-205, PLA2 receptor and Endo180. Type II CTLs are small molecules with a single CTLD and include CD69 (which is not generally expressed on macrophages), and Dectin-1, the receptor for fungal β-glucans, which is a macrophage specific protein. Other C-type lectins which cannot be covered in detail here include Dectin-2, responsible for leukotriene generation in response to house dust mite (Barrett *et al.*, 2009). Because of their importance, a brief overview is provided for the biology of MR, DC-SIGN and Dectin-1.

*Mannose receptor*

Because terminal mannose is not typically present on mammalian proteins, but is frequently present on microorganisms, the detection of terminal mannose, fucose, glucose or N-acetylglucosamine by the MR acts as a PAMP signal. The MR (MRC1) comprises an N-terminal membrane anchor, eight CTL domains, a fibronectin linker and a C-terminal cysteine-rich carbohydrate binding domain. MRC1 is up-regulated by IL-4 and is typical of IL-4-activated ('M2') macrophages (described below). Typically, MRC1 binding initiates phagocytic uptake, a process which is subverted by some pathogens including *Mycobacterium tuberculosis* and *Clostridium perfringens*, to provide entry to the host macrophage intracellular compartment which is where these pathogens reside. The MR has long been known to be important in the phagocytosis of microorganisms including fungal pathogens such as yeast and is a classical macrophage marker.

*DC-SIGN*

As the name suggests, DC-SIGN is highly expressed by dendritic cells as well as by some macrophage subsets including those in the lung. The receptor itself is a small membrane-anchored molecule with a single CTLD. It forms a tetrameric complex which binds a range of high-mannose and fucose-containing carbohydrates including α1–3 and α1–6-mannotriose residues (Appelmelk *et al.*, 2003), lipoarabinomannan (ManLAM) (Geijtenbeek *et al.*, 2003) and phosphatidylinositol mannosides (PIM) (Torrelles *et al.*, 2006). Pathogens containing these groups include *Streptococcus pneumoniae, Helicobacter pylori* and *Mycobacterium tuberculosis,* the last appearing to subvert DC-SIGN binding to arrest DC maturation (Geijtenbeek *et al.*, 2003). Importantly, DC-SIGN also mediates HIV-binding and enables HIV to avoid antigen presentation by dendritic cells, as well as mediating transmission of the virus to T-cells (Geijtenbeek and van Kooyk, 2003). DC-SIGN is thus an important HIV receptor and the pattern of DC-SIGN expression in macrophages and DCs is of great interest.

*Dectin-1*

Dectin-1 is the receptor for fungal β-glucans found in yeast cell walls. Dectin-1 is unusual in having an intracellular signal transduction domain which initiates a tyrosine kinase signalling pathway via Syk and CARD9 to NFκB and which also activates the oxidative burst, a critical antifungal defence, for example for *Candida* (Skrzypek *et al.*, 2009) or *Aspergillus* (Werner *et al.*, 2009) species as well as a range of other fungal diseases. In addition, Dectin-1 stimulation increases TNFα production. In combination with TLR2, Dectin-1 signalling results in up-regulation of IL-6 and IL-23 while simultaneously leading to the down-regulation of IL-12, thus generating a Th17 response which is appropriate for fungal pathogens (Dennehy *et al.*, 2009; Huang *et al.*, 2009; Manicassamy *et al.*, 2009). The interaction between Dectin-1 and TLR2 pathways emphasizes the previous point about cooperation between different PAMP receptors and the need to consider signalling crosstalk and the complex phenotypes that can be generated by context-dependent signalling (O'Neill, 2008b).

*Formyl peptide receptor (FPR)*

The f-Met-Leu-Pro (formyl peptide; FPR) receptors are GPCRs which detect a common N-terminal motif from bacterial (or mitochondrial) proteins, triggering MAP kinase and NFκB signalling. Although often considered neutrophil receptors, FPRs are also present on monocytes and FPR stimulation results in chemotaxis, respiratory burst activation and prostaglandin release (Migeotte *et al.*, 2006). Interestingly, it was shown that FPR stimulation resulted in down-regulation of CCR5, with implications for HIV uptake (Shen *et al.*, 2000).

*Complement receptors*

The activation, breakdown and binding to pathogens of complement molecules, particularly C3 and its product C3b, creates an alarm signal for macrophages which is detected via cell surface complement receptors (CR) especially CR1 and CR3. CR1 is a large extracellular membrane protein which contains the antigenic sites making up the Knops blood group. It is not confined to monocyte/macrophages but is also expressed on erythrocytes and other leucocytes. CR1 mediates binding to C3b and C4b complement proteins, activating phagocytosis of complement-opsonized particles. It also acts as a negative feedback control on complement activation. CR3 ($\alpha_M$ $\beta_2$) is a heterodimer of ITGAM (integrin alpha M; $\alpha_M$; CD11b) and integrin beta 2 ($\beta_2$; CD18) and is the antigen for the Mac-1 antibody, a prominent cell surface marker for leucocytes especially monocytes, NK cells and neutrophils. C3R binds (among other ligands) inactivated C3b (iC3b) and mediates monocyte adhesion and spreading and also activation (e.g. respiratory burst) and chemotaxis. Complement receptors can phagocytose and destroy opsonized viral particles but because complement can also bind to the activated Fc region of immunoglobulins, it has been implicated also in antibody-dependent enhancement (ADE) of viral infections, which is discussed later. CR4 (CD11c/CD18; Itgax) also binds iC3b and mediates similar functions.

*Immunoglobulin receptors*

FcR bind the Fc region of immunoglobulins when the binding region is exposed by Ig binding to a ligand. This enables the recognition of active immunoglobulin complexes such as opsonized virus or bacteria. There are numerous FcR which bind different Ig classes but for monocyte/macrophages the most important are the Fcγ Receptors FcγRI (CD64), FcγRII (CD32) and FcγRIII (CD16) which bind to and enhance phagocytosis of activated IgG. FcR signal through an intracellular ITAM (immunoreceptor tyrosine activation motif) located on the receptor itself or a co-receptor, which is phosphorylated upon receptor activation and initiates a kinase cascade and intracellular signalling. FcR also mediate an important macrophage- and NK-dependent response known as antibody dependent cell-mediated cytotoxicity (ADCC) whereby a cytotoxic response (secretion of cytotoxic molecules such as perforin) is generated to cells containing bound IgG. This response is particularly important as an anti-tumour response, when tumours display unfamiliar antigens such as fetal or mutated proteins. Macrophages can also express

FcαRI (CD89) which binds IgA, for example on mucosal surfaces where IgA is important.

## Cytokine production by macrophages

Cytokines and chemokines represent major pathways of information flow between macrophages and the rest of the body, especially other immune cells. Not surprisingly, these patterns of information flow are immensely complex. Here a few particularly important cytokines are described, with passing reference to others. Although the nomenclature has been confused by historical processes, two significant structural families of cytokines exist. Many (but not all) interleukins are four-helix bundle ligands which bind to IL-like signalling receptors, whereas the chemokine ligands are cysteine-stabilized proteins with receptors which are GPCRs.

For simplicity these have been divided into cytokines made by, and acting upon macrophages, though in reality some fall into both categories (autocrine) and others are produced only in specific situations. In addition, chemokines are considered separately both because their role is more circumscribed and because they form a family which shares close structural homology.

### Key cytokines produced by macrophages

*TNFα (tumour necrosis factor α)*

As its name suggests, TNFα (cachexin) is capable of killing tumour cells. Discovered in 1968 and cloned in 1985, TNFα is a member of a large family of structurally and functionally related proteins. Produced mainly but not exclusively by macrophages, TNFα is a critical cytokine in the mediation of immune responses especially acute inflammatory responses, and is capable of both activating cells and causing their death by triggering apoptosis.

Soluble TNFα exists as a 51-kDa homotrimer of processed polypeptides which are cleaved from the macrophage cell surface by ADAM17, also known as TACE (TNFα-converting enzyme). TNFα also signals as a membrane-bound cell surface complex. Structurally, TNFα is composed of β-pleated jellyroll domains. It binds to the receptors TNFR1, which is widely distributed and TNFR2, which is confined primarily to immune cells. The receptors contain a domain (the death domain or DD) which upon ligand binding is de-repressed and bound by TRADD (TNF receptor-associated death domain) which then acts to nucleate several signalling complexes leading variously to NFκB, JNK/AP1 and caspase 3 apoptotic pathways (Grivennikov *et al.*, 2006; Li and Lin, 2008). Macrophages have TNFα receptors and are stimulated by TNFα secretion in an autocrine fashion. Since NFκB frequently has antiapoptotic effects, including independence from growth factor requirements, TNFα signalling can lead to either apoptosis or resistance to it, and this balance is exploited by successful macrophage pathogens such as *Salmonella, Yersinia* and *Mycobacteria* to either destroy macrophages or to keep them alive as hosts. Although JNK signalling is generally viewed as stress-related proapoptotic pathway, it is crucial for survival in macrophages (Himes *et al.*, 2006). Extensive crosstalk with other intracellular signalling pathways guarantees that the effects of TNFα signalling are highly dependent upon the type of cell and its state. TNFα may also be able to exploit the production of ROS to kill cells including virally infected cells (Morgan *et al.*, 2008).

TNFα release is triggered in response to endogenous (IL-1) and exogenous (LPS) mediators, the mechanisms being highly regulated at every level from chromatin structure through to membrane trafficking and cell surface cleavage. In macrophages, IFNγ priming leads to the accumulation of TNFα mRNA, but not to translation and release of the protein, and similar effects are seen in response to some PAMP receptors such as NOD2 following MDP stimulation (Wolfert *et al.*, 2002). Consequently, subsequent TLR4 stimulation leads to a large and rapid release of TNFα which has the appearance of being synergistic. Macrophages are capable of producing significant amounts of TNFα within a half-hour of TLR signalling so that TNFα acts as an acute response signal to infection. Like IL-1 and IL-6 it mediates fever and acute phase responses such as C-reactive protein (CRP) release from hepatocytes, and also leads to the classic local inflammatory signs of oedema, heat and vasodilation. Downstream effects include prostaglandin release and insulin resistance. Chronic effects of TNFα include cachexia which is seen in cancer.

TNFα is a central mediator of the inflammation in chronic inflammatory diseases including inflammatory bowel disease and rheumatoid arthritis, and TNFα blockade using therapeutic antibodies has been used successfully to treat these diseases (Alonso-Ruiz *et al.*, 2008; Behm and Bickston, 2008) although not without side effects due to the loss of this important cytokine, including the reactivation of latent tuberculosis and hepatitis B/C (Nathan *et al.*, 2006; Jacobs *et al.*, 2007). TNFα has significant metabolic effects including insulin resistance and it is involved in the pro-inflammatory pathology of diabetes and obesity.

TNFα generally has antiviral effects and can act synergistically with interferons to produce effective antiviral responses, which may be possible to exploit therapeutically (Bartee *et al.*, 2008, 2009). The biology of TNFα and TNF family members is vast and complex and readers are referred to the literature for more details (Kruglov *et al.*, 2008).

*IL-1β (interleukin 1β)*

The first discovered interleukin, IL-1 acts upon the brain stem to produce fever, which led to an early name of 'endogenous pyrogen' and lymphocyte activating factor (LAF). IL-1 activates the acquired immune system, including natural killer cells, and initiates the maturation of B-lymphocytes. IL-1 activity is actually produced by two closely related polypeptides, IL-1α and IL-1β, the latter being an archetypal macrophage-secreted proinflammatory cytokine, having a wide array of effects on numerous tissues. Both are produced as proproteins which are processed prior to release, IL-1α by calpain and IL-1β by the caspase-1-containing inflammasome; caspase-1 was originally known as ICE (interleukin 1-converting enzyme). Both bind to the IL-1R, a bipartite interleukin receptor family receptor which is present particularly on lymphocytes but also in many other tissues. Closely related cytokines include IL-18 and IL-33 which are also processed by caspase 1 and are reviewed by Arend *et al.* (2008).

The secretion of IL-1β lies downstream of many PAMP receptor signalling pathways, especially the intracellular signals such as the NLRs and intracellular TLRs. For example, the NALP3 (cryopyrin) inflammasome pathways activate caspase-1 and hence lead to the release of IL-1β. Inappropriate activation and secretion of IL-1β has also been proposed to be a pathological consequence of mutations in Nod2 which contribute to the pathogenesis of Crohn's disease. As a consequence, although it is a critical component of general inflammation, IL-1 and its close relatives are also centrally involved in many disease processes especially those related to chronic inflammation (Barksby *et al.*, 2007) This includes any disease in which macrophages play a significant role, including (but certainly not limited to) rheumatoid arthritis (Kay and Calabrese, 2004), neurological disease (Shaftel *et al.*, 2008), cancer (Apte *et al.*, 2006) stress responses (Goshen and Yirmiya, 2009) and HIV dementia (Brabers and Nottet, 2006).

*IL-6 (interleukin 6)*

IL-6 is a typical four-helix bundle cytokine secreted primarily, although not exclusively, by macrophages. Originally described as a factor secreted by T-cells in the maturation of B-cells into plasma cells, IL-6 is now recognized as a pleiotropic cytokine with complex and widespread effects, including both pro- and anti-inflammatory actions (Kishimoto, 2005). Other tissues secreting IL-6 include adipocytes and muscle, where IL-6 secretion is postulated to mediate the anti-inflammatory effects of moderate exercise, possibly by down-regulating TNFα expression, but is also implicated in the aetiology of exercise-related stress fatigue (Robson-Ansley *et al.*, 2007) and chronic fatigue syndrome – though it is not the only one, other candidates including IL-1, IFNα and CRP. The IL-6 receptor is widely expressed, so that IL-6 secretion affects most tissues.

Expression of IL-6 is regulated by NFκB and IL-6 production is a consequence of pro-inflammatory signalling by PAMP receptors such as TLR4. The presence of LPS in particular leads to IL-6 secretion. IL-6 acts as a positive feedback regulator of myelopoiesis, accelerates the maturation of acquired immune responses and like IL-1, mediates fever and the acute phase response by stimulating hepatocytes to release CRP (Ohsugi, 2007). It also has effects on glucose metabolism and has been implicated in diabetes pathology.

Like IL-1 and TNFα, it is involved in a range of chronic inflammatory diseases (Naugler and Karin, 2008). Although a recent therapeutic approach, the blockade of IL-6 has been shown to be successful as an alternative to anti-TNFα therapy in treating chronic inflammatory diseases such as Crohn's disease, reflecting this widespread importance of IL-6 for inflammation (Sebba, 2008).

*IL-8 (interleukin 8)*

IL-8 (CXCL8) is actually a chemokine which binds to classical GPCR-type chemokine receptors (CXCR1/2). Mainly secreted by macrophages and endothelial cells, but also by T-cells, NK cells, neutrophils, fibroblasts and epithelial cells, the primary function of IL-8 is in neutrophil chemotaxis, to enable phagocytosis and killing of potential pathogens. It also stimulates leukotriene release. Release of IL-8 is accentuated by TLR stimulation via NFκB and thus IL-8 release forms part of a Th1/M1 pattern of classical macrophage activation, with secretion usually downstream of IL-1 or TNFα signalling as part of the proinflammatory response to these cytokines, both in bacterial and viral infections. IL-8 secretion by macrophages is antagonized by the macrophage-suppressive cytokine IL-10. The importance of neutrophil infiltration in lung disease has led to a particular interest in IL-8 biology in this setting (Mukaida, 2003) although it is also implicated in tumour biology and other chronic inflammatory conditions. Because IL-8 can inhibit IFNα/β activity it may exacerbate some viral infections including CMV and adenovirus.

*IL-12 (interleukin 12)*

IL-12 is a key proinflammatory cytokine associated with the orientation of an immune response towards the Th1 type, acting primarily on T-cells in a manner antagonistic to IL-4. It enhances $CD8^+$ (cytotoxic) T-cell activation and NK cell activity.

IL-12 is a heterodimeric four-helix bundle cytokine composed of IL-12α ('p35') and IL-12β ('p40'). Since IL-12β can also form an inactive homodimer, the transcriptional regulation and biological effects of IL-12 are complex (Zhang and Wang, 2008). In addition, IL-12 shares a common chain (p40) with IL-23 which has a different spectrum of activity, notably the 'Th17' response to fungal pathogens (Lyakh *et al.*, 2008), due to Dectin-1/TLR2 signalling (by fungal PAMPs) or Nod2 signalling (by *Mycobacteria* spp.). IL-27 is another cytokine related to IL-12 and is coordinated in a related manner (Goriely *et al.*, 2008). An important activity of IL-12 is its action on T-cells and NK cells to stimulate IFNγ and TNFα release, which are the key priming factors for macrophage activation, so this represents a T-cell-macrophage positive feedback loop. Stimulation of MyD88 signalling TLRs leads to IL-12 release, and this response is used as an indicator of the type of macrophage activation present in a given situation. The balance of responses between different IL-12 family members in response to pathogen signals is delicately controlled (Goriely *et al.*, 2008). IL-12 responses are often repressed by successful tumours and the use of IL-12 family cytokines in cancer therapy is an active area of interest (Weiss *et al.*, 2007).

Because IL-12 production by macrophages coordinates NK cells to produce IFNγ and TNFα, and stimulates the production of cytotoxic T-cells, it is important in antiviral responses. IL-12 is also a key player in the pathology of chronic inflammation and anti-IL-12 therapies are of great interest for therapeutic applications (Paunovic *et al.*, 2008).

## Key cytokines which act upon macrophages

*IFNγ (gamma interferon)*

The homodimeric type II interferon IFNγ acts as an important priming factor for macrophage activation and is a central player in antiviral defences (Schroder *et al.*, 2004; Young, 2006; Schoenborn and Wilson, 2007a,b). Macrophages express the interferon gamma receptor (IFGR1) which signals through the JAK-STAT pathway. Secreted primarily by Th1 T-cells, cytotoxic T-cells and NK cells, particularly in response to macrophage secretion of IL-12, IFNγ acts to prime macrophages for TLR activation, up-regulates MHC class II to enable antigen presentation and enhances MHC class I presentation (Strehl *et al.*, 2005). In the presence of TLR stimulation, for example with LPS stimulation via TLR4 or dsRNA stimulation through TLR3, IFNγ produces a synergistic activation of classical macrophage activities including phagocytosis, and the respiratory burst. The release of IFNγ antagonizes IL-4, to inhibit Th2 and M2 development.

*IL-10 (interleukin 10)*
IL-10 is one of two anti-inflammatory regulators responsible for limiting Th1 type inflammation in a large number of tissues mediated by macrophages and T-cells (the other is TGFβ). On monocyte/macrophages it inhibits macrophage activation and knockout studies indicate an important role in the control of gut inflammation (Gasche *et al.*, 2003). A four-helix bundle cytokine, it acts as a homodimer and is secreted mainly by monocytes, but also by Th2 cells, mast cells and epithelial cells. The receptor is a JAK-STAT (STAT3) interleukin receptor consisting of two subunits, IL-10R1 and IL-10R2, which is a common chain for several other cytokines including IL-22, IL-26, IL-28α/β and IL-29 (Donnelly *et al.*, 2004). The powerful advantages of an anti-inflammatory cytokine are exploited by some viruses such as EBV and CMV which secrete their own version (cmvIL-10; Gruber *et al.*, 2008). IL-10 secretion can also become a cause of failure to clear pathogens; IL-10 blockade has been used experimentally to clear chronic viral infections in mice (Ejrnaes *et al.*, 2006).

*IL-4/13 (interleukins 4 and 13)*
IL-4 is a four-helix bundle cytokine which binds to a typical interleukin JAK-STAT receptor, the IL4R, which acts in macrophages primarily through STAT6. It stimulates the development of T0 cells down a Th2 pathway, which subsequently also produce IL-4. On B-cells, IL-4 stimulates class switching to IgE, and hence is involved in allergic responses to parasites. The action of IL-4 on macrophages is primarily to induce an 'alternatively activated macrophage' phenotype (sometimes referred to as an M2 phenotype) which is discussed later. Conceptually, IL-4 can be viewed as the primary antagonist of the Th1 promoting cytokines IL-12 and IFNγ. IL-13 shares a receptor chain with IL-4 and has an overlapping spectrum of activity. The action of IL-4 on monocytes is important in the further differentiation of monocytes into DCs and IL-4 up-regulates a set of genes which include fusion proteins (enhancing the formation of multinucleated giant cells), and the mannose receptor. It also promotes endocytosis which increases competence for antigen presentation. As a key cytokine in the generation of 'alternatively activated macrophages' IL-4 is also involved in allergic responses and allergic diseases including asthma. Conversely, the importance of Th1 responses in viral immunity means that IL-4 is generally not associated with enhanced viral resistance. The biology of IL-4 and IL-13 is discussed further during consideration of alternative macrophage activation.

*TGFβ (transforming growth factor beta)*
The biology of TGFβ and its related family members is extensive and not confined to monocyte–macrophages, being important in fibrosis, cancer, vasculogenesis and a host of other conditions. It is mentioned here primarily because TGFβ has a multitude of modulatory effects on macrophages. At the same time, it is also produced by macrophages and can act in an autocrine fashion. It is generally viewed as an anti-inflammatory cytokine which opposes the action of the pro-inflammatory TNFα, IL-1, IL-6, IL-12 and IFNγ. Its actions are complex and not always anti-inflammatory; however, together IL-10 produced by T-cells and TGFβ (from numerous sources) can be viewed as regulatory checks on the proinflammatory cytokines (Wan and Flavell, 2007; Li and Flavell, 2008).

Mature soluble TGFβ comprises a 25-kDa dimer which is produced by proteolytic cleavage from a large pro-protein, though signalling via uncleaved membrane-bound TGFβ is also important especially in the context of regulatory T-cell ($T_{Reg}$) generation. Structurally, the mature molecule is a cysteine knot protein related to a wide family of proteins including the bone morphogenetic proteins (BMPs) and myostatin (which regulates muscle growth). The TGFβ receptor signals via the Smad or DAXX signal transduction pathways (Schmierer and Hill, 2007). In the case of Smad signalling, heterodimeric Smad3&4 acts as a promoter of signalling while Smad7 acts as a signalling inhibitor. In many cells Smad3/4 signalling activates an apoptotic programme but in macrophages it leads to a phenotype which can be described as a 'wound healing' or fibrotic response, including the secretion of platelet derived growth factor-B (PDGF-B), extracellular matrix components. TGFβ also promotes the trophic roles of osteoblasts, while inhibiting osteoclast formation, the respiratory burst and other pathways of macrophage activation. Thus while its effects appear to be immunosuppressive, TGFβ stimulation does in fact represent a kind of macrophage differentiation, and by promoting the Th17 response is involved in at least some forms of immune response (Wahl, 2007). Another complication with TGFβ signalling is that it can be local and context dependent such that the effects of local TGFβ expression (especially at the cell surface) can be very different to the administration of systemic TGFβ.

### Chemokines

As the name suggests, chemokines are primarily concerned with the trafficking and activation of leucocytes to sites of infection, by acting as chemoattractants. Chemokines and their receptors form a large superfamily with common structural motifs. They are generally small proteins of 8–10 kDa with two or four conserved cysteines that form stabilizing disulfide bonds in a Greek key structural motif. A unified 'C' nomenclature has been developed to reduce the confusion due to the multiple historical names. Their receptors are GPCRs which trigger protein kinase C (PKC)-dependent signalling pathways, activating MAP kinase cascades and subsequently cellular migration, expression of integrins and other mechanisms that lead to appropriate chemotactic responses.

Chemokines fall into four related classes which are distinguished structurally according to the configuration of the cysteine residues. The C chemokines have only a single pair of conserved cysteines which bridge their N- and C-terminal regions and comprise only 2 members, XCL1 (lymphotactin-α) and XCL2 (lymphotactin-β). The CC chemokines have the central cysteines adjacent to each other in the middle of the protein – these bridge outwards to the N- and C-termini. The CXC chemokines have a single residue between the central cysteines and the $CX_3C$ chemokine fractalkine ($CX_3CL1$) has three internal residues. These motifs reflect the evolutionary structure of the families and provide a useful basis for dividing structures and functions. Most chemokines are in the CC (CCL1–28; 24 human members) and CXC (21 members) classes.

The patterns of chemokine and chemokine receptor expression in macrophages determine which subsets of macrophages will be attracted to specific sites of inflammation or competent for activities such as phagocytosis. $CX_3CL1$ (fractalkine) plays an important role in the behaviour of circulating monocytes which is discussed elsewhere. In the context of HIV infection, the actions of the T-cell-expressed chemokine CCL5 (RANTES) and its receptor CCR5 are particularly important since CCR5 is an important co-receptor for HIV entry to macrophages. CCL3 (MIP-1α) and CCL4 (MIP-1β) are also CCR5 ligands and are also relevant in this context. When chemokines bind to their receptors, the receptor–ligand complex is internalized and degraded and so the presence of CCL5 will tend to down-regulate the surface levels of CCR5, reducing HIV susceptibility. CXCR4 is also capable of allowing HIV entry but is not generally expressed on macrophages. Individuals with specific CCR5 mutations are generally protected from HIV progression to AIDS. Chemokine receptors expressed on monocytes include CCR1 and CCR2, which are able to bind many if not most chemokines. Neutrophil migration is particularly controlled by chemokine secretion. RANTES, MCP1, MCP3 and eotaxin are particularly important as monocyte attractants, however monocytes do not necessarily respond to chemokines to the same extent as neutrophils.

Given the complexity of the chemokine family, the delineation of which cytokines regulate monocyte–macrophage behaviour *in vivo* is still in progress. Furthermore, some viruses modulate chemokine secretion, or even produce their own chemokines, to modify host immune behaviour to their own advantage, examples including measles and CMV (Abt *et al.*, 2009; Frascaroli *et al.*, 2009).

## Macrophage activation

### Classical activation

Activation is the classic response to microbial infection. Usually, an infiltrating monocyte/macrophage has already encountered cellular signals indicating the need to prepare for activation. Of these, one of the most central and potent is gamma interferon (IFNγ), typically made by T-cells ($CD4^+$ Th1 and $CD8^+$ Tc1 cells). Priming of macrophages with IFNγ leads to subsequent up-regulation of proinflammatory cytokines especially IL-12, IL-6 and TNFα as well as initiation of phagocytic capacity suitable for dealing with bacterial pathogens.

IFNγ does not immediately produce a dramatic change in morphology but the intracellular changes are profound. IFNγ puts a macrophage into a state of readiness, up-regulating multiple genes, though not necessarily permitting the translation of their gene products. Such a condition is referred to as priming. When a primed macrophage encounters lipopolysaccharide or some other pathogen signal, it rapidly transforms into the activated state. Phagocytosis, cytokine production, respiratory burst, cell spreading and motility appear in short order. In rough outline the following events occur: Signalling pathways, particularly through the MyD88 adaptor system, activate the production of interferon beta (IFNβ) which is secreted. NFκB activation combined with p38 MAP kinase pathways stimulates the expression of a vast array of inflammatory genes including interleukins, cytokines and others. TNFα switches into translation and large amounts are trafficked to the cell surface and released. These changes typically occur from 30 min (TNFα secretion) through to a peak at around 8–10 h, after which they decline. Feedback from TNFα and from autocrine IFNβ acting on macrophage interferon alpha/beta receptors (IFNαR) leads to a late induction of another vast array of interferon-inducible genes. Typically these peak at 12–24 h and while most decline, there are some irreversible changes in gene regulation which means that the state of the post-activation macrophage is not a return to the pre-activated phenotype. In particular there are changes in surface receptor genes and in the likelihood of apoptosis.

A critically important aspect of macrophage activation is the balance between survival and apoptosis which is regulated by NFκB/p38/JNK (pro-survival) and the levels of apoptotic proteins especially $Bcl_{XL}$/Mpl1 (the loss of which initiates apoptosis). This is because in the case of many activating stimuli it is desirable for the macrophage to continue phagocytosis or cell killing as long as possible. However, some pathogens – particularly viruses, but also intracellular bacteria such as *Listeria, Salmonella* and *Mycobacteria* – actually utilize the activated macrophage as a host and in these cases it can be preferable that the macrophage enter apoptosis relatively quickly. Successful macrophage pathogens have mechanisms to prevent apoptosis, typically by inducing antiapoptotic genes or producing proteins which complex with and inactivate inducers of apoptosis.

It should be stressed that the phenomenon of macrophage activation is a general principle or description, not a single process. It was first described at a time when morphological and behavioural changes (cell spreading, phagocytosis) were the primary methods of characterizing cells. In fact, activation is a terminal differentiation process which occurs with some commonalities in a range of already differentiated macrophage populations. As such, the actual programme of activation is different for each cell type and is suited for the niche in which the cell finds itself. Consequently the details of which markers or properties are up- or down-regulated during activation will vary substantially between, say, Kupffer cells, microglia and alveolar macrophages. Furthermore, different pattern recognition receptors (PRR) are designed to react differently according to the stimulus, so that dsRNA activation through TLR3 (a signal for RNA viruses) will produce a different pattern of activation from TLR4 activation despite the overlap produced by common intracellular signalling pathways. So a further discrimination must be made in terms of the pattern of activation in response to different signals. In this case, it is useful to note similarities in PRR-specific responses across a range of macrophage cell types which can provide guiding principles for understanding the types of behaviour that each tissue niche may impose upon its resident macrophages, as well as the responses of infiltrating blood monocytes to each tissue environment. It is clearly neither possible nor useful to use cellular markers to definitively subdivide macrophage populations into innumerable transient subsets. Although a detailed description of tissue responses is necessary, the search for guiding principles is more helpful for understanding the evolutionary rationale behind activation responses.

The complexity of macrophage activation illustrates that the so-called 'non-specific' responses of the innate immune system are actually highly specific to the cellular niche in which a PAMP is encountered, rather than to a specific antigen.

### Alternative activation

Macrophage activation is traditionally thought of as an inflammatory response to potentially pathogenic organisms, with a heavy emphasis on viral and bacterial responses. The recognition that some classes of stimuli result in 'activation' (morphological, behavioural and transcriptional changes) with a quite different character, led to the concept of 'alternatively activated' macrophages, also referred to as M2 macrophages (with classical activation described as M1 by analogy to the Th1 vs. Th2 models for T-lymphocyte activation). As for the Th1/Th2 concept, this is probably an over-simplifying characterization, but it does convey the idea that different pathogen classes require very different responses. In particular, the engulfment of dying cells by macrophages is a homeostatic function which tends to dampen inflammation, and perhaps represents a resolution-stage function, while responses to multicellular parasites clearly require a different approach to viruses or bacteria which can be entirely engulfed by a single macrophage.

The discovery of IL-4 in 1982 helped provide the basis for the paradigm of the activation by APC of naïve T-cells into T-helper cells type 1 (Th1) vs. type 2 (Th2) which are mutually inhibitory subsets. According to this model, Th1 secrete IFNγ and IL-2 and TNFβ which promote classical macrophage activation, whereas Th2 secrete IL-4, IL-10, and IL-13 which bias the immune response towards eosinophil, basophil and mast cell responses needed for fungal and parasitic infections. The concept of 'alternative activation' of macrophages arose as a counterpart to the Th1/Th2 model of T-cell activation, with the designations 'M1' (classically activated) and 'M2' (alternatively activated). However, reality is much more complex (even for T-cells) and it is now recognized that activated macrophages do not fall into two neat categories. Attempted modifications to the scheme proposed M1 vs. three categories of M2 cells (M2a, b and c). Other modifying agents include glucocorticoids which act via the intracellular glucocorticoid receptor (GR), and TGFβ which acts on the TGFβR and IL-10 (IL-10R).

As one might expect, IL-4 and IL-13 are secreted by eosinophils, basophils, mast cells as well as Th2 cells, thereby drawing macrophages into a suitable activation profile following the stimulation (by fungal, protozoan or helminthic pathogens) of these cell types. Other cell types such as adipocytes can also contribute. The IL-4R shares a common β-chain with the IL-13R and the two cytokines have overlapping, though not identical, activities. As with the IL-3R and GM-CSFR, signalling is mediated by the PI3K and JAK/STAT pathways, the latter, in this case, through STAT6. There is also crosstalk with other signalling pathways especially the PPARγ, PPARδ and IL-10 pathways which can modify the phenotype substantially. The generally anti-inflammatory effects of TGFβ, which are mediated by TGFβR on macrophages, can be nullified by down-regulation of the TGFβR by IL-4, though simultaneous activation with glucocorticoids up-regulates TGFβR thereby maintaining an additional control on these cells. Differences between IL-4 and IL-13 signalling result in overlapping but separate gene expression signatures, and this is when studied in 'pure' models. It is likely that in the context of 'real world' situations, a very complex interlocking set of activation profiles may occur.

Phenotypes which differ between M1 and M2 cells include an increased focus on antigen presentation in M2 (leading ultimately to a DC phenotype), increased pinocytosis, and increased fusion to form giant cells, presumably to allow the destruction of pathogens too large to be ingested by single macrophages (Varin and Gordon, 2009).

The molecular consequences of IL-4 stimulation include the up-regulation of markers such as arginase (ARG1) which shifts macrophage metabolism away from iNOS activation and nitric oxide (NO) production (a significant M1 response in mice) and towards the production of polyamines such as those used in wound healing.

Receptors important for fungal pathogens are up-regulated including Dectin-1, DC-SIGN, mannose receptor, scavenger receptor A, GPCRs for sensing nucleotides, and a behavioural repertoire which includes antigen presentation, cell fusion and an inhibition of autophagy.

In contrast to the dramatic suite of proinflammatory cytokines secreted by classically activated macrophages, M2 alternatively activated macrophages generally do not produce the canonical proinflammatory cytokines IL-12, IL-8, TNFα, and IL-1β, which requires a down-regulation of the TLR-NFκB-inflammasome pathways.

Part of the signature of alternatively activated macrophages is a tendency to enhance wound-repair signals including fibrosis, by the secretion of fibronectin, PDGFc, insulin-like growth factor 1 (IGF1) and coagulation factors (Martinez *et al.*, 2009).

How valid is the concept of alternatively activated macrophages? There is no doubt that the Th2–M2 phenotype produced by IL-4/IL-13 stimulation is different from (and antagonistic to) the proinflammatory phenotype of classically activated macrophages. However, the proposition that these two phenotypes are polar opposites is perhaps a simplification. Relatively recent increases in global population (over the last 2000–1000 years) have led to advantageous conditions for epidemic bacterial and viral diseases and the focus of immunology in the last century has been largely on these pathogens. In contrast, immune challenges during the period of human evolution would have included many diverse stimuli, and contagious diseases spread by contact may have been less significant than local endemic parasitic and protozoan diseases. It seems likely, then, that many 'alternative' macrophage activation states exist which are tailored to various pathogenic events. Thus 'classical activation' may represent only one (albeit important) activation state of a significant number of possibilities. In particular, significant and common pathogens for early humans will have included nematodes such as roundworms, platyhelminths such as cestodes (tapeworms) and trematodes (flukes, including schistosomes), protozoans such as malarial and amoeboid parasites, and a range of fungal diseases.

Furthermore, since foreign organisms may have been encountered by early humans particularly in the context of physical injury, the fibrotic wound healing responses that seem to commonly accompany macrophage activation may form part of a general strategy of healing juxtaposed with immune defences.

Mosser and Edwards (2008) argue that the 'M2' category has evolved into an 'everything else' category which negates its explanatory usefulness. They bring into the argument the well-known homeostatic roles of macrophages and suggest that instead of an M1–M2 polarity, macrophage activation exists on a 'colour wheel' spectrum, a continuum between three primary poles analogous to the three primary colours – inflammatory, regulatory and wound healing. They suggest that IL-4-stimulated 'M2' macrophages, while important for dealing with some helminth infections, are not primarily immune defence macrophages, but are aimed at repairing tissue damage i.e. are wound healing macrophages. They also discuss the concept of regulatory macrophages or macrophage-derived suppressor cells (MDSCs; (Gabrilovich and Nagaraj, 2009).

MDSCs are proposed to have an immunosuppressive phenotype and evolve into tumour-associated macrophages (TAMs). Their surface phenotype is $Lin^{-}HLA^{-}DR^{-}CD33^{+}$ or $CD11b^{+}CD14^{-}CD33^{+}$ and they are characterized by constant STAT3 activation and production of NO (in mice) ROS and peroxynitrite. Clinically their most significant feature is suppression of T-cell responses, including the promotion of $T_{Reg}$ cells which enables tumour proliferation, but they are conceived of as being part of a negative regulatory part of the innate immune system. There is controversy about whether these cells truly exist as a coherent functional group in normal immunity, which will only be resolved by further study. However, as a feature of tumour-associated macrophages they may well be important in tumour biology.

The study of tumour-associated macrophages (TAMs), the foam cells of atherosclerosis and the uric acid crystal-laden macrophages of gout provide examples of 'alternatively activated' macrophages which are undoubtedly of clinical importance. Nonetheless, it is arguable that regardless of their clinical importance today, these phenotypes are unlikely to have been selected for by evolution, and that their characteristics are the result of the serendipitous adaptation of an existing immune defence to an emerging tumour, excessive oxidized LDL or the predisposing pathology for gout and other modern diseases of excess.

Martinez and Gordon (2009), in contrast, support an M1/M2 classification scheme, though recognizing the complexity of the real situation. Given that the Th1/Th2 polarity is well established in the literature, the M1/M2 scheme seems a reasonable first approximation as long as it is remembered that this is a simplified picture of the true situation, more an *aide-memoire* or conceptual framework than a dogma.

The above discussion is merely a brief introduction to the study of macrophage activation. An understanding of macrophage activation represents an important attempt to synthesize a vast array of phenotypic observations into a coherent model of macrophage responses to potential pathogens and tissue damage, as well as encompassing the homeostatic functions which represent the 'bread and butter' of macrophage biology. Since macrophage activation involves changes in the regulation of a substantial fraction of the genes in the human genome (Wells *et al.*, 2003) it is not surprising that this understanding has some way to go before it is really clear.

## Macrophage tropic viruses and antibody-dependent enhancement (ADE)

Macrophages are commonly considered as an important defence against viruses, and in general this is the case. Sometimes macrophages can be infected by viruses with a broad host range, largely because they are closely associated with virally infected cells, but in these cases they are not the primary target cell type. Examples include poliovirus, measles, hepatitis A (which mainly infect hepatocytes but with some Kupffer cell involvement) and hepatitis C for which evidence of monocyte infection *in vivo* is controversial. CMV and EBV are broad host range viruses known to infect monocytes at times.

However, in at least some cases macrophages are important or primary reservoirs or mediators of viral pathology. Historically the first examples are the flaviviruses, especially dengue virus, West Nile virus, yellow fever and Japanese encephalitis virus. Dengue virus and the filoviruses Ebola and Marburg virus cause haemorrhagic fever which is a consequence of the activation of the innate immune system during the course of viral pathology. Measles, influenza, rabies and the arenaviruses Lassa and Junin virus also target macrophages.

Dengue virus was the virus in which the mechanism of antibody-dependent enhancement (ADE) of viral infection was first described. Although neutralizing antibodies confer protection upon the host to the primary serotype, dengue fever is characterized by a marked increase in severity upon re-infection by a different serotype, a response which is still incompletely understood. ADE has been proposed as a contributor to this phenomenon. Halstead and O'Rourke (1977) described a paradoxical enhancement of viral infectivity for dengue virus in the presence of antiserum. Instead of the inactivation of virus by antibody-dependent blocking of infectivity, or by phagocytosis and lysosomal destruction, such viruses are capable of being drawn into the macrophage via antibody-dependent uptake mechanisms, remaining viable, and then initiating a productive infection. The overall effect is an enhanced infectivity from a low titre viral stock. This process can be mediated by cellular receptors which bind the IgG Fc region, including Fc$\gamma$R on macrophages.

A number of potential mechanisms may explain ADE. The simplest explanations consider the ability of antibody dependent binding to increase the effective concentration of viruses at the host cell surface. Once IgG is bound to the virus surface, either FcR or activated C1q bound to the IgG Fc region may bring virus within close range of the normal cellular uptake receptor. Alternatively these mechanisms may simply bring the virus into the cell via the endocytic pathway. If the antibody binding is relatively weak (which may result between serotypes) it may be released early in this process enabling live virus to commence the infection. It may be that some or all of these mechanisms occur in different infections, as well as others. For example, in feline infectious peritonitis virus, ADE still occurs in the absence of the normal viral uptake receptor, feline aminopeptidase N (fAPN)(Takano *et al.*, 2008a,b). This argues that viral uptake is more likely to be endocytic. At the same time C1q can inhibit dengue virus ADE instead of enhancing it (Mehlhop *et al.*, 2007) whereas ADE in Zaire Ebola virus is C1q-dependent (Takada *et al.*, 2007). In the case of HIV-1, both complement receptor and CD4 are required for ADE suggesting that in this case, the 'proximity' model for ADE is relevant (Robinson, 2006).

It seems likely that more complex mechanisms may contribute as well. FcγR signalling leads to JAK-STAT activation with downstream consequences on intracellular signalling events. This could result in the enhancement of viral transcription, as well as repression of cellular mechanisms which mediate responses to viral infection such as TLR3 and Rig-I like receptors. Rodrigo *et al.* (2006) showed that signalling-incompetent FcγR led to compromised ADE in dengue virus infection. Mahalingam and Lidbury (2002) also showed that in macrophages NFκB- and STAT-1-dependent antiviral responses (including TNFα, IFNβ and IP-10) induced by LPS were suppressed by ADE of Ross River Virus (RRV), although macrophages are not the primary host of this virus. Interestingly, RRV infections in an animal model are abolished in the absence of signalling-competent complement receptor CR3 (Morrison *et al.*, 2008) although this may not be directly related to ADE. Another suggestion is that ADE may result in secretion of IL-10 and a switch to a Th2/M2 phenotype which would interfere with the normal Th1/M1 antiviral responses (Suhrbier and La Linn, 2003). It seems that ADE is a phenomenon which needs to be assessed in the context of individual viruses.

## Conclusion

Macrophages are both a central cell in immune responses to infection and at times a specifically targeted viral host. Successful pathogens have clearly evolved mechanisms to subvert the numerous defence mechanisms presented by macrophages (in conjunction with the rest of the immune system). The variation inherent in human and animal physiology and genomes means that no pathogen can be successful in all members of a species, and the natural variation of the course of infection between individuals provides insight into the specific mechanisms of viral pathology, well exemplified by the discovery of CCR5 mutations in humans. A full understanding of these viral pathogenic mechanisms presents many therapeutic opportunities to intervene in viral (or other) pathologies. It is difficult, perhaps overwhelming, to keep in mind all aspects of macrophage biology when investigating one aspect of disease. At the same time, it is valuable to have a broad knowledge of many apparently unrelated facets of immunity (including macrophage biology) in order to be prepared for the opportunities that observation brings occasionally in the course of research.

## References


Abt, M., Gassert, E., and Schneider-Schaulies, S. (2009). Measles virus modulates chemokine release and chemotactic responses of dendritic cells. J. Gen. Virol. *90*, 909–914.

Akagi, T., Saitoh, T., O'Kelly, J., Akira, S., Gombart, A.F., and Koeffler, H.P. (2008). Impaired response to GM-CSF and G-CSF, and enhanced apoptosis in C/EBPbeta-deficient hematopoietic cells. Blood *111*, 2999–3004.

Alexaki, A., and Wigdahl, B. (2008). HIV-1 infection of bone marrow hematopoietic progenitor cells and their role in trafficking and viral dissemination. PLoS Pathog. *4*, e1000215.

Alff, P.J., Sen, N., Gorbunova, E., Gavrilovskaya, I.N., and Mackow, E.R. (2008). The NY-1 hantavirus Gn cytoplasmic tail coprecipitates TRAF3 and inhibits cellular interferon responses by disrupting TBK1–TRAF3 complex formation. J. Virol. *82*, 9115–9122.

Alonso-Ruiz, A., Pijoan, J.I., Ansuategui, E., Urkaregi, A., Calabozo, M., and Quintana, A. (2008). Tumor necrosis factor alpha drugs in rheumatoid arthritis: systematic review and metaanalysis of efficacy and safety. BMC Musculoskelet. Disord. *9*, 52.

Alvarez-Silva, M., Belo-Diabangouaya, P., Salaun, J., and Dieterlen-Lievre, F. (2003). Mouse placenta is a major hematopoietic organ. Development *130*, 5437–5444.

Ancuta, P., Moses, A., and Gabuzda, D. (2004). Transendothelial migration of CD16+ monocytes in response to fractalkine under constitutive and inflammatory conditions. Immunobiology *209*, 11–20.

Appelmelk, B.J., van Die, I., van Vliet, S.J., Vandenbroucke-Grauls, C.M., Geijtenbeek, T.B., and van Kooyk, Y. (2003). Cutting edge: carbohydrate profiling identifies new pathogens that interact with dendritic cell-specific ICAM-3-grabbing nonintegrin on dendritic cells. J. Immunol. *170*, 1635–1639.

Apte, R.N., Dotan, S., Elkabets, M., White, M.R., Reich, E., Carmi, Y., Song, X., Dvozkin, T., Krelin, Y., and Voronov, E. (2006). The involvement of IL-1 in tumorigenesis, tumor invasiveness, metastasis and tumor–host interactions. Cancer Metastasis Rev. *25*, 387–408.

Arend, W.P., Palmer, G., and Gabay, C. (2008). IL-1, IL-*18*, and IL-33 families of cytokines. Immunol. Rev. *223*, 20–38.

Arredouani, M., Yang, Z., Ning, Y., Qin, G., Soininen, R., Tryggvason, K., and Kobzik, L. (2004). The scavenger receptor MARCO is required for lung defense against pneumococcal pneumonia and inhaled particles. J. Exp. Med. *200*, 267–272.

Arredouani, M.S., and Kobzik, L. (2004). The structure and function of marco, a macrophage class a scavenger receptor. Cell Mol Biol (Noisy-le-grand) *50 Online Pub*, OL657–665.

Aschoff, L. (1924). Das reticulo-endotheliale system. Ergebnisse der inneren Medizin und Kinderheilkunde *26*, 1–118.

Audoy-Remus, J., Richard, J.F., Soulet, D., Zhou, H., Kubes, P., and Vallieres, L. (2008). Rod-Shaped monocytes patrol the brain vasculature and give rise to perivascular macrophages under the influence of proinflammatory cytokines and angiopoietin-2. J. Neurosci. *28*, 10187–10199.

Auffray, C., Fogg, D., Garfa, M., Elain, G., Join-Lambert, O., Kayal, S., Sarnacki, S., Cumano, A., Lauvau, G., and Geissmann, F. (2007). Monitoring of blood vessels and tissues by a population of monocytes with patrolling behavior. Science *317*, 666–670.

Auffray, C., Fogg, D.K., Narni-Mancinelli, E., Senechal, B., Trouillet, C., Saederup, N., Leemput, J., Bigot, K., Campisi, L., Abitbol, M., *et al.* (2009a). CX3CR1+ CD115+ CD135+ common macrophage/DC precursors and the role of CX3CR1 in their response to inflammation. J. Exp. Med. *206*, 595–606.

Auffray, C., Sieweke, M.H., and Geissmann, F. (2009b). Blood monocytes: development, heterogeneity, and relationship with dendritic cells. Annu. Rev. Immunol. *27*, 669–692.

Baenziger, S., Heikenwalder, M., Johansen, P., Schlaepfer, E., Hofer, U., Miller, R.C., Diemand, S., Honda, K., Kundig, T.M., Aguzzi, A., and Speck, R.F. (2009). Triggering TLR7 in mice induces immune activation and lymphoid system disruption, resembling HIV-mediated pathology. Blood *113*, 377–388.

Baer, K., Roosevelt, M., Clarkson, A.B., Jr, van Rooijen, N., Schnieder, T., and Frevert, U. (2007). Kupffer cells are obligatory for *Plasmodium yoelii* sporozoite infection of the liver. Cell. Microbiol. *9*, 397–412.

Baleeiro, C.E., Christensen, P.J., Morris, S.B., Mendez, M.P., Wilcoxen, S.E., and Paine, R., 3rd (2006). GM-CSF and the impaired pulmonary innate immune response following hyperoxic stress. Am. J. Physiol. Lung Cell Mol. Physiol. *291*, L1246–1255.

Bamming, D., and Horvath, C.M. (2009). Regulation of signal transduction by enzymatically inactive antiviral RNA helicase proteins MDA5, RIG-I, and LGP2. J. Biol. Chem. *284*, 9700–9712.

Bao, X., Liu, T., Shan, Y., Li, K., Garofalo, R.P., and Casola, A. (2008). Human metapneumovirus glycoprotein G inhibits innate immune responses. PLoS Pathog. *4*, e1000077.

Barcena, A., Kapidzic, M., Muench, M.O., Gormley, M., Scott, M.A., Weier, J.F., Ferlatte, C., and Fisher, S.J. (2009a). The human placenta is a hematopoietic organ during the embryonic and fetal periods of development. Dev. Biol. 327, 24–33.

Barcena, A., Muench, M.O., Kapidzic, M., and Fisher, S.J. (2009b). A new role for the human placenta as a hematopoietic site throughout gestation. Reprod. Sci. *16*, 178–187.

Baril, M., Racine, M.E., Penin, F., and Lamarre, D. (2009). MAVS dimer is a crucial signaling component of innate immunity and the target of hepatitis C virus NS3/4A protease. J. Virol. *83*, 1299–1311.

Barksby, H.E., Lea, S.R., Preshaw, P.M., and Taylor, J.J. (2007). The expanding family of interleukin-1 cytokines and their role in destructive inflammatory disorders. Clin. Exp. Immunol. *149*, 217–225.

Barnes, P.J. (2004). Alveolar macrophages as orchestrators of COPD. Copd *1*, 59–70.

Barreda, D.R., Hanington, P.C., and Belosevic, M. (2004). Regulation of myeloid development and function by colony stimulating factors. Dev. Comp. Immunol. *28*, 509–554.

Barrett, N.A., Maekawa, A., Rahman, O.M., Austen, K.F., and Kanaoka, Y. (2009). Dectin-2 recognition of house dust mite triggers cysteinyl leukotriene generation by dendritic cells. J. Immunol. *182*, 1119–1128.

Bartee, E., Mohamed, M.R., and McFadden, G. (2008). Tumor necrosis factor and interferon: cytokines in harmony. Curr. Opin. Microbiol. *11*, 378–383.

Bartee, E., Mohamed, M.R., Lopez, M.C., Baker, H.V., and McFadden, G. (2009). The addition of tumor necrosis factor plus beta interferon induces a novel synergistic antiviral state against poxviruses in primary human fibroblasts. J. Virol. *83*, 498–511.

Behm, B.W., and Bickston, S.J. (2008). Tumor necrosis factor-alpha antibody for maintenance of remission in Crohn's disease. Cochrane Database Syst Rev, CD006893.

Bernier, T., Halter, R., Pau, D., Rittinghausen, S., and Emmendorffer, A. (2001). M-CSF transgenic mice: role of M-CSF in infection and autoimmunity. Exp. Toxicol. Pathol. *53*, 165–173.

Bertrand, J.Y., Jalil, A., Klaine, M., Jung, S., Cumano, A., and Godin, I. (2005). Three pathways to mature macrophages in the early mouse yolk sac. Blood *106*, 3004–3011.

Bilzer, M., Roggel, F., and Gerbes, A.L. (2006). Role of Kupffer cells in host defense and liver disease. Liver Int. *26*, 1175–1186.

Bonfield, T.L., Raychaudhuri, B., Malur, A., Abraham, S., Trapnell, B.C., Kavuru, M.S., and Thomassen, M.J. (2003). PU.1 regulation of human alveolar macrophage differentiation requires granulocyte–macrophage colony-stimulating factor. Am. J. Physiol. Lung Cell Mol. Physiol. *285*, L1132–1136.

Bonifer, C., Faust, N., Geiger, H., and Muller, A.M. (1998). Developmental changes in the differentiation capacity of haematopoietic stem cells. Immunol. Today *19*, 236–241.

Bonifer, C., and Hume, D.A. (2008). The transcriptional regulation of the Colony-Stimulating Factor 1 Receptor (csf1r) gene during hematopoiesis. Front. Biosci. *13*, 549–560.

Bonifer, C., Hoogenkamp, M., Krysinska, H., and Tagoh, H. (2008). How transcription factors program chromatin--lessons from studies of the regulation of myeloid-specific genes. Semin. Immunol. *20*, 257–263.

Bovolenta, C., Pilotti, E., Mauri, M., Turci, M., Ciancianaini, P., Fisicaro, P., Bertazzoni, U., Poli, G., and Casoli, C. (2002). Human T-cell leukemia virus type 2 induces survival and proliferation of CD34(+) TF-1 cells through activation of STAT1 and STAT5 by secretion of interferon-gamma and granulocyte macrophage-colony-stimulating factor. Blood *99*, 224–231.

Bowie, A.G., and Unterholzner, L. (2008). Viral evasion and subversion of pattern-recognition receptor signalling. Nat. Rev. Immunol. *8*, 911–922.

Boxer, E.L., Nanda, S.K., and Baron, M.D. (2009). The rinderpest virus non-structural C protein blocks the induction of type 1 interferon. Virology *385*, 134–142.

Brabers, N.A., and Nottet, H.S. (2006). Role of the pro-inflammatory cytokines TNF-alpha and IL-1beta in HIV-associated dementia. Eur J. Clin. Invest. *36*, 447–458.

Breckpot, K., Corthals, J., Bonehill, A., Michiels, A., Tuyaerts, S., Aerts, C., Heirman, C., and Thielemans, K. (2005). Dendritic cells differentiated in the presence of IFN-β and IL-3 are potent inducers of an antigen-specific CD8+ T-cell response. J. Leukoc. Biol. *78*, 898–908.

Burdach, S., Nishinakamura, R., Dirksen, U., and Murray, R. (1998). The physiologic role of interleukin-3, interleukin-5, granulocyte–macrophage colony-stimulating factor, and the beta c receptor system. Curr. Opin. Hematol. *5*, 177–180.

Cenci, E., Bartocci, A., Puccetti, P., Mocci, S., Stanley, E.R., and Bistoni, F. (1991). Macrophage colony-stimulating factor in murine candidiasis: serum and tissue levels during infection and protective effect of exogenous administration. Infect. Immun. *59*, 868–872.

Chan, W.Y., Kohsaka, S., and Rezaie, P. (2007). The origin and cell lineage of microglia: new concepts. Brain Res. Rev. *53*, 344–354.

Chang, C.J., Chen, Y.H., Huang, K.W., Cheng, H.W., Chan, S.F., Tai, K.F., and Hwang, L.H. (2007). Combined GM-CSF and IL-12 gene therapy synergistically suppresses the growth of orthotopic liver tumors. Hepatology *45*, 746–754.

Chang, M.K., Raggatt, L.J., Alexander, K.A., Kuliwaba, J.S., Fazzalari, N.L., Schroder, K., Maylin, E.R., Ripoll, V.M., Hume, D.A., and Pettit, A.R. (2008). Osteal tissue macrophages are intercalated throughout human and mouse bone lining tissues and regulate osteoblast function *in vitro* and *in vivo*. J. Immunol. *181*, 1232–1244.

Cheers, C., Haigh, A.M., Kelso, A., Metcalf, D., Stanley, E.R., and Young, A.M. (1988). Production of colony-stimulating factors (CSFs) during infection: separate determinations of macrophage-, granulocyte-, granulocyte-macrophage-, and multi-CSFs. Infect. Immun. *56*, 247–251.

Cheers, C., and Stanley, E.R. (1988). Macrophage production during murine listeriosis: colony-stimulating factor 1 (CSF-1) and CSF-1-binding cells in genetically resistant and susceptible mice. Infect. Immun. *56*, 2972–2978.

Chen, J., Olsen, J., Ford, S., Mirza, S., Walker, A., Murphy, J.M., and Young, I.G. (2009). A new isoform of interleukin-3 receptor α with novel differentiation activity and high affinity binding mode. J. Biol. Chem. *284*, 5763–5773.

Chen, X., Wang, X., Hossain, S., O'Neill, F.A., Walsh, D., van den Oord, E., Fanous, A., and Kendler, K.S. (2007). Interleukin 3 and schizophrenia: the impact of sex and family history. Mol. Psychiatry *12*, 273–282.

Childs, K.S., Andrejeva, J., Randall, R.E., and Goodbourn, S. (2009). Mechanism of mda-5 Inhibition by paramyxovirus V proteins. J. Virol. *83*, 1465–1473.

Choi, J.K., Choi, B.H., Ha, Y., Park, H., Yoon, S.H., Park, H.C., and Park, S.R. (2007). Signal transduction pathways of GM-CSF in neural cell lines. Neurosci. Lett. *420*, 217–222.

Church, L.D., Cook, G.P., and McDermott, M.F. (2008). Primer: inflammasomes and interleukin 1beta in inflammatory disorders. Nat. Clin. Pract. Rheumatol. *4*, 34–42.

Cobaleda, C., and Busslinger, M. (2008). Developmental plasticity of lymphocytes. Curr. Opin. Immunol. *20*, 139–148.

Coleman, J.R. (2007). The PB1-F2 protein of Influenza A virus: increasing pathogenicity by disrupting alveolar macrophages. Virol. J. *4*, 9.

Combadiere, C., Potteaux, S., Rodero, M., Simon, T., Pezard, A., Esposito, B., Merval, R., Proudfoot, A., Tedgui, A., and Mallat, Z. (2008). Combined inhibition of CCL2, CX3CR1, and CCR5 abrogates Ly6C(hi) and Ly6C(lo) monocytosis and almost abolishes atherosclerosis in hypercholesterolemic mice. Circulation *117*, 1649–1657.

Congdon, K.L., and Reya, T. (2008). Divide and conquer: how asymmetric division shapes cell fate in the hematopoietic system. Curr. Opin. Immunol. *20*, 302–307.

Conti, L., and Gessani, S. (2008). GM-CSF in the generation of dendritic cells from human blood monocyte precursors: recent advances. Immunobiology *213*, 859–870.

Coull, J.A., Beggs, S., Boudreau, D., Boivin, D., Tsuda, M., Inoue, K., Gravel, C., Salter, M.W., and De Koninck, Y. (2005). BDNF from microglia causes the shift in neuronal anion gradient underlying neuropathic pain. Nature *438*, 1017–1021.

Cumano, A., and Godin, I. (2001). Pluripotent hematopoietic stem cell development during embryogenesis. Curr. Opin. Immunol. *13*, 166–171.

Cumano, A., and Godin, I. (2007). Ontogeny of the hematopoietic system. Annu. Rev. Immunol. *25*, 745–785.

D'Mello, C., Le, T., and Swain, M.G. (2009). Cerebral microglia recruit monocytes into the brain in response to tumor necrosis factoralpha signaling during peripheral organ inflammation. J. Neurosci. *29*, 2089–2102.

Dalpke, A.H., Schafer, M.K., Frey, M., Zimmermann, S., Tebbe, J., Weihe, E., and Heeg, K. (2002). Immunostimulatory CpG-DNA activates murine microglia. J. Immunol. *168*, 4854–4863.

Dang, P.M., Elbim, C., Marie, J.C., Chiandotto, M., Gougerot-Pocidalo, M.A., and El-Benna, J. (2006). Anti-inflammatory effect of interleukin-10 on human neutrophil respiratory burst involves inhibition of GM-CSF-induced p47PHOX phosphorylation through a decrease in ERK1/2 activity. Faseb J. *20*, 1504–1506.

Davies, M.H., Eubanks, J.P., and Powers, M.R. (2006). Microglia and macrophages are increased in response to ischemia-induced retinopathy in the mouse retina. Mol. Vis. *12*, 467–477.

de Jong, M.A., de Witte, L., Oudhoff, M.J., Gringhuis, S.I., Gallay, P., and Geijtenbeek, T.B. (2008). TNF-alpha and TLR agonists increase susceptibility to HIV-1 transmission by human Langerhans cells ex vivo. J. Clin. Invest. *118*, 3440–3452.

de Silva, T.I., Cope, A., Goepel, J., and Greig, J.M. (2007). The use of adjuvant granulocyte–macrophage colony-stimulating factor in HIV-related disseminated atypical mycobacterial infection. J. Infect. *54*, e207–210.

de Witte, L., Nabatov, A., Pion, M., Fluitsma, D., de Jong, M.A., de Gruijl, T., Piguet, V., van Kooyk, Y., and Geijtenbeek, T.B. (2007). Langerin is a natural barrier to HIV-1 transmission by Langerhans cells. Nat. Med. *13*, 367–371.

de Witte, L., Nabatov, A., and Geijtenbeek, T.B. (2008). Distinct roles for DC-SIGN+-dendritic cells and Langerhans cells in HIV-1 transmission. Trends Mol. Med. *14*, 12–19.

Deane, D., Ueda, N., Wise, L.M., Wood, A.R., Percival, A., Jepson, C., Inglis, N.F., Fleming, S.B., Mercer, A.A., and McInnes, C.J. (2009). Conservation and variation of the parapoxvirus GM-CSF-inhibitory factor (GIF) proteins. J. Gen. Virol. *90*, 970–977.

Dennehy, K.M., Willment, J.A., Williams, D.L., and Brown, G.D. (2009). Reciprocal regulation of IL-23 and IL-12 following co-activation of Dectin-1 and TLR signaling pathways. Eur J. Immunol. *39*, 1379–1386.

Donnelly, R.P., Sheikh, F., Kotenko, S.V., and Dickensheets, H. (2004). The expanded family of class II cytokines that share the IL-10 receptor-2 (IL-10R2) chain. J Leukoc. Biol. *76*, 314–321.

Dzierzak, E., Medvinsky, A., and de Bruijn, M. (1998). Qualitative and quantitative aspects of haematopoietic cell development in the mammalian embryo. Immunol. Today *19*, 228–236.

Ejrnaes, M., Filippi, C.M., Martinic, M.M., Ling, E.M., Togher, L.M., Crotty, S., and von Herrath, M.G. (2006). Resolution of a chronic viral infection after interleukin-10 receptor blockade. J. Exp. Med. *203*, 2461–2472.

El Khoury, J., Toft, M., Hickman, S.E., Means, T.K., Terada, K., Geula, C., and Luster, A.D. (2007). Ccr2 deficiency impairs microglial accumulation and accelerates progression of Alzheimer-like disease. Nat. Med. *13*, 432–438.

Engelhardt, R., Mackensen, A., and Galanos, C. (1991). Phase I trial of intravenously administered endotoxin (*Salmonella abortus equi*) in cancer patients. Cancer Res. *51*, 2524–2530.

Evans, R., Kamdar, S.J., Fuller, J.A., and Krupke, D.M. (1995). The potential role of the macrophage colony-stimulating factor, CSF-1, in inflammatory responses: characterization of macrophage cytokine gene expression. J. Leukoc. Biol. *58*, 99–107.

Evans, R., Shultz, L.D., Dranoff, G., Fuller, J.A., and Kamdar, S.J. (1998). CSF-1 regulation of Il6 gene expression by murine macrophages: a pivotal role for GM-CSF. J. Leukoc. Biol. *64*, 810–816.

Fancke, B., Suter, M., Hochrein, H., and O'Keeffe, M. (2008). M-CSF: a novel plasmacytoid and conventional dendritic cell poietin. Blood *111*, 150–159.

Faust, N., Huber, M.C., Sippel, A.E., and Bonifer, C. (1997). Different macrophage populations develop from embryonic/fetal and adult hematopoietic tissues. Exp. Hematol. *25*, 432–444.

Feng, R., Desbordes, S.C., Xie, H., Tillo, E.S., Pixley, F., Stanley, E.R., and Graf, T. (2008). PU.1 and C/EBPalpha/beta convert fibroblasts into macrophage-like cells. Proc. Natl. Acad. Sci. U.S.A. *105*, 6057–6062.

Feng, Y., Klein, B.K., and McWherter, C.A. (1996). Three-dimensional solution structure and backbone dynamics of a variant of human interleukin-3. J. Mol. Biol. *259*, 524–541.

Ferenbach, D., and Hughes, J. (2008). Macrophages and dendritic cells: what is the difference? Kidney Int. *74*, 5–7.

Fernandes-Alnemri, T., Yu, J.W., Datta, P., Wu, J., and Alnemri, E.S. (2009). AIM2 activates the inflammasome and cell death in response to cytoplasmic DNA. Nature *458*, 509–513.

Francois, B., Trimoreau, F., Vignon, P., Fixe, P., Praloran, V., and Gastinne, H. (1997). Thrombocytopenia in the sepsis syndrome: role of hemophagocytosis and macrophage colony-stimulating factor. Am. J. Med. *103*, 114–120.

Frascaroli, G., Varani, S., Blankenhorn, N., Pretsch, R., Bacher, M., Leng, L., Bucala, R., Landini, M.P., and Mertens, T. (2009). Human cytomegalovirus paralyzes macrophage motility through down-regulation of chemokine receptors, reorganization of the cytoskeleton, and release of macrophage migration inhibitory factor. J. Immunol. *182*, 477–488.

Gabrilovich, D.I., and Nagaraj, S. (2009). Myeloid-derived suppressor cells as regulators of the immune system. Nat. Rev. Immunol. *9*, 162–174.

Gasche, C., Grundtner, P., Zwirn, P., Reinisch, W., Shaw, S.H., Zdanov, A., Sarma, U., Williams, L.M., Foxwell, B.M., and Gangl, A. (2003). Novel variants of the IL-10 receptor 1 affect inhibition of monocyte TNF-alpha production. J. Immunol. *170*, 5578–5582.

Geijtenbeek, T.B., Van Vliet, S.J., Koppel, E.A., Sanchez-Hernandez, M., Vandenbroucke-Grauls, C.M., Appelmelk, B., and Van Kooyk, Y. (2003). Mycobacteria target DC-SIGN to suppress dendritic cell function. J. Exp. Med. *197*, 7–17.

Geijtenbeek, T.B., and van Kooyk, Y. (2003). DC-SIGN: a novel HIV receptor on DCs that mediates HIV-1 transmission. Curr. Top. Microbiol. Immunol. *276*, 31–54.

Geissmann, F., Jung, S., and Littman, D.R. (2003). Blood monocytes consist of two principal subsets with distinct migratory properties. Immunity *19*, 71–82.

Gerlier, D., and Valentin, H. (2009). Measles virus interaction with host cells and impact on innate immunity. Curr. Top. Microbiol. Immunol. *329*, 163–191.

Getts, D.R., Terry, R.L., Getts, M.T., Muller, M., Rana, S., Shrestha, B., Radford, J., Van Rooijen, N., Campbell, I.L., and King, N.J. (2008). Ly6c+ 'inflammatory monocytes' are microglial precursors recruited in a pathogenic manner in West Nile virus encephalitis. J. Exp. Med. *205*, 2319–2337.

Ghinassi, B., Verrucci, M., Jelicic, K., Di Noia, A., Migliaccio, G., and Migliaccio, A.R. (2007). Interleukin-3 and erythropoietin cooperate in the regulation of the expression of erythroid-specific transcription factors during erythroid differentiation. Exp. Hematol. *35*, 735–747.

Giebel, B., and Punzel, M. (2008). Lineage development of hematopoietic stem and progenitor cells. Biol. Chem. *389*, 813–824.

Gigliotti, F., and Wright, T.W. (2005). Immunopathogenesis of Pneumocystis carinii pneumonia. Expert Rev. Mol. Med. *7*, 1–16.

Gordon, S. (2007). The macrophage: past, present and future. Eur J. Immunol. *37 Suppl. 1*, S9–17.

Goriely, S., Neurath, M.F., and Goldman, M. (2008). How microorganisms tip the balance between interleukin-12 family members. Nat. Rev. Immunol. *8*, 81–86.

Goshen, I., and Yirmiya, R. (2009). Interleukin-1 (IL-1): a central regulator of stress responses. Front. Neuroendocrinol. *30*, 30–45.

Gregory, S.H., Wing, E.J., Tweardy, D.J., Shadduck, R.K., and Lin, H.S. (1992). Primary listerial infections are exacerbated in mice administered neutralizing antibody to macrophage colony-stimulating factor. J. Immunol. *149*, 188–193.

Grivennikov, S.I., Kuprash, D.V., Liu, Z.G., and Nedospasov, S.A. (2006). Intracellular signals and events activated by cytokines of the tumor necrosis factor superfamily: From simple paradigms to complex mechanisms. Int. Rev. Cytol. *252*, 129–161.

Gruber, S.G., Gloria Luciani, M., Grundtner, P., Zdanov, A., and Gasche, C. (2008). Differential signaling of cmvIL-10 through common variants of the IL-10 receptor 1. Eur. J. Immunol. *38*, 3365–3375.

Gualano, R.C., Hansen, M.J., Vlahos, R., Jones, J.E., Park-Jones, R.A., Deliyannis, G., Turner, S.J., Duca, K.A., and Anderson, G.P. (2008). Cigarette smoke worsens lung inflammation and impairs resolution of influenza infection in mice. Resp. Res. *9*, 53.

Guidi-Rontani, C. (2002). The alveolar macrophage: the Trojan horse of *Bacillus anthracis*. Trends Microbiol. *10*, 405–409.

Gwinn, M.R., and Vallyathan, V. (2006). Respiratory burst: role in signal transduction in alveolar macrophages. J. Toxicol. Environ. Health B Crit. Rev. *9*, 27–39.

Halstead, S.B., and O'Rourke, E.J. (1977). Antibody-enhanced dengue virus infection in primate leukocytes. Nature *265*, 739–741.

Hamilton, J.A., and Anderson, G.P. (2004). GM-CSF Biology. Growth Factors *22*, 225–231.

Hansen, G., Hercus, T.R., McClure, B.J., Stomski, F.C., Dottore, M., Powell, J., Ramshaw, H., Woodcock, J.M., Xu, Y., Guthridge, M., *et al.* (2008). The structure of the GM-CSF receptor complex reveals a distinct mode of cytokine receptor activation. Cell *134*, 496–507.

Harada, T., and Ohno, N. (2008). Contribution of dectin-1 and granulocyte macrophage-colony stimulating factor (GM-CSF) to immunomodulating actions of beta-glucan. Int. Immunopharmacol. *8*, 556–566.

Hartoonian, C., Ebtekar, M., Soleimanjahi, H., Karami, A., Mahdavi, M., Rastgoo, N., and Azadmanesh, K. (2009). Effect of immunological adjuvants: GM-CSF (granulocyte-monocyte colony stimulating factor) and IL-23 (interleukin-23) on immune responses generated against hepatitis C virus core DNA vaccine. Cytokine *46*, 43–50.

Hashimoto, S., Yoda, M., Yamada, M., Yanai, N., Kawashima, T., and Motoyoshi, K. (1996). Macrophage colony-stimulating factor induces interleukin-8 production in human monocytes. Exp. Hematol. *24*, 123–128.

Hege, K.M., Jooss, K., and Pardoll, D. (2006). GM-CSF gene-modifed cancer cell immunotherapies: of mice and men. Int. Rev. Immunol. *25*, 321–352.

Heit, B., Jones, G., Knight, D., Antony, J.M., Gill, M.J., Brown, C., Power, C., and Kubes, P. (2006). HIV and other lentiviral infections cause defects in neutrophil chemotaxis, recruitment, and cell structure: immunorestorative effects of granulocyte–macrophage colony-stimulating factor. J. Immunol. *177*, 6405–6414.

Helmy, K.Y., Katschke, K.J., Jr, Gorgani, N.N., Kljavin, N.M., Elliott, J.M., Diehl, L., Scales, S.J., Ghilardi, N., and van Lookeren Campagne, M. (2006). CRIg: a macrophage complement receptor required for phagocytosis of circulating pathogens. Cell *124*, 915–927.

Henson, P.M., and Hume, D.A. (2006). Apoptotic cell removal in development and tissue homeostasis. Trends Immunol. *27*, 244–250.

Herbomel, P., Thisse, B., and Thisse, C. (1999). Ontogeny and behaviour of early macrophages in the zebrafish embryo. Development *126*, 3735–3745.

Herbomel, P., Thisse, B., and Thisse, C. (2001). Zebrafish early macrophages colonize cephalic mesenchyme and developing brain, retina, and epidermis through a M-CSF receptor-dependent invasive process. Dev. Biol. *238*, 274–288.

Hernandez, L., Smirnova, T., Kedrin, D., Wyckoff, J., Zhu, L., Stanley, E.R., Cox, D., Muller, W.J., Pollard, J.W., Van Rooijen, N., and Segall, J.E. (2009). The EGF/CSF-1 Paracrine invasion loop can be triggered by heregulin β1 and CXCL12. Cancer Res. *69*, 3221–3227.

Hida, S., Yamasaki, S., Sakamoto, Y., Takamoto, M., Obata, K., Takai, T., Karasuyama, H., Sugane, K., Saito, T., and Taki, S. (2009). Fc receptor gamma-chain, a constitutive component of the IL-3 receptor, is required for IL-3-induced IL-4 production in basophils. Nat. Immunol. *10*, 214–222.

Hildebrand, F., Hubbard, W.J., Choudhry, M.A., Frink, M., Pape, H.C., Kunkel, S.L., and Chaudry, I.H. (2006). Kupffer cells and their mediators: the culprits in producing distant organ damage after trauma-hemorrhage. Am. J. Pathol. *169*, 784–794.

Himes, S.R., Cronau, S., Mulford, C., and Hume, D.A. (2005). The Runx1 transcription factor controls CSF-1-dependent and -independent growth and survival of macrophages. Oncogene *24*, 5278–5286.

Himes, S.R., Sester, D.P., Ravasi, T., Cronau, S.L., Sasmono, T., and Hume, D.A. (2006). The JNK are important for development and survival of macrophages. J. Immunol. *176*, 2219–2228.

Hiyoshi, M., Suzu, S., Yoshidomi, Y., Hassan, R., Harada, H., Sakashita, N., Akari, H., Motoyoshi, K., and Okada, S. (2008). Interaction between Hck and HIV-1 Nef negatively regulates cell surface expression of M-CSF receptor. Blood *111*, 243–250.

Huang, G., Zhang, P., Hirai, H., Elf, S., Yan, X., Chen, Z., Koschmieder, S., Okuno, Y., Dayaram, T., Growney, J.D., *et al.* (2008). PU.1 is a major downstream target of AML1 (RUNX1) in adult mouse hematopoiesis. Nat. Genet. *40*, 51–60.

Huang, H., Ostroff, G.R., Lee, C.K., Wang, J.P., Specht, C.A., and Levitz, S.M. (2009). Distinct patterns of dendritic cell cytokine release stimulated by fungal β-glucans and Toll-like receptor agonists. Infect. Immun. *77*, 1774–1781.

Huang, X., Choi, J.K., Park, S.R., Ha, Y., Park, H., Yoon, S.H., Park, H.C., Park, J.O., and Choi, B.H. (2007). GM-CSF inhibits apoptosis of neural cells via regulating the expression of apoptosis-related proteins. Neurosci. Res. *58*, 50–57.

Hume, D.A., and Gordon, S. (1983). Mononuclear phagocyte system of the mouse defined by immunohistochemical localization of antigen F4/80. Identification of resident macrophages in renal medullary and cortical interstitium and the juxtaglomerular complex. J. Exp. Med. *157*, 1704–1709.

Hume, D.A., Pavli, P., Donahue, R.E., and Fidler, I.J. (1988). The effect of human recombinant macrophage colony-stimulating factor (CSF-1) on the murine mononuclear phagocyte system *in vivo*. J. Immunol. *141*, 3405–3409.

Hume, D.A., Monkley, S.J., and Wainwright, B.J. (1995). Detection of c-fms protooncogene in early mouse embryos by whole mount *in situ* hybridization indicates roles for macrophages in tissue remodelling. Br. J. Haematol *90*, 939–942.

Hume, D.A. (2000). Probability in transcriptional regulation and its implications for leukocyte differentiation and inducible gene expression. Blood *96*, 2323–2328.

Hume, D.A., Ross, I.L., Himes, S.R., Sasmono, R.T., Wells, C.A., and Ravasi, T. (2002). The mononuclear phagocyte system revisited. J. Leukoc. Biol. *72*, 621–627.

Hume, D.A. (2006). The mononuclear phagocyte system. Curr. Opin. Immunol. *18*, 49–53.

Hume, D.A. (2008). Macrophages as APC and the dendritic cell myth. J. Immunol. *181*, 5829–5835.

Huynh, D., Dai, X.M., Nandi, S., Lightowler, S., Trivett, M., Chan, C.K., Bertoncello, I., Ramsay, R.G., and Stanley, E.R. (2009). CSF-1 dependence of Paneth cell development in the mouse small intestine. Gastroenterology *137*, 136–144.

Imaizumi, T., Matsumiya, T., Yoshida, H., Naraoka, T., Uesato, R., Ishibashi, Y., Ota, K., Toh, S., Fukuda, S., and Satoh, K. (2009). Tumor-necrosis factor-alpha induces retinoic acid-inducible gene-I in rheumatoid fibroblast-like synoviocytes. Immunol. Lett. *122*, 89–93.

Issekutz, T.B., Issekutz, A.C., and Movat, H.Z. (1981). The *in vivo* quantitation and kinetics of monocyte migration into acute inflammatory tissue. Am. J. Pathol. *103*, 47–55.

Ito, T., Amakawa, R., and Fukuhara, S. (2002). Roles of toll-like receptors in natural interferon-producing cells as sensors in immune surveillance. Hum. Immunol. *63*, 1120–1125.

Jacobs, M., Samarina, A., Grivennikov, S., Botha, T., Allie, N., Fremond, C., Togbe, D., Vasseur, V., Rose, S., Erard, F., *et al.* (2007). Reactivation of tuberculosis by tumor necrosis factor neutralization. Eur. Cytokine Netw. *18*, 5–13.

Kamdar, S.J., Chapoval, A.I., Phelps, J., Fuller, J.A., and Evans, R. (1996). Differential sensitivity of mouse mononuclear phagocytes to CSF-1 and LPS: the potential *in vivo* relevance of enhanced IL-6 gene expression. Cell. Immunol. *174*, 165–172.

Kamdar, S.J., Fuller, J.A., Nishikawa, S.I., and Evans, R. (1997). Priming of mouse macrophages with the macrophage colony-stimulating factor (CSF-1) induces a variety of pathways that regulate expression of the interleukin 6 (Il6) and granulocyte–macrophage colony-stimulating factor (Csfgm) genes. Exp. Cell. Res. *235*, 108–116.

Kared, H., Leforban, B., Montandon, R., Renand, A., Layseca Espinosa, E., Chatenoud, L., Rosenstein, Y., Schneider, E., Dy, M., and Zavala, F. (2008). Role of GM-CSF in tolerance induction by mobilized hematopoietic progenitors. Blood *112*, 2575–2578.

Kaspar, M., Trachsel, E., and Neri, D. (2007). The antibody-mediated targeted delivery of interleukin-15 and GM-CSF to the tumor neovasculature inhibits tumor growth and metastasis. Cancer Res. *67*, 4940–4948.

Kawamura, T., Koyanagi, Y., Nakamura, Y., Ogawa, Y., Yamashita, A., Iwamoto, T., Ito, M., Blauvelt, A., and Shimada, S. (2008). Significant virus replication in Langerhans cells following application of HIV to abraded skin: relevance to occupational transmission of HIV. J. Immunol. *180*, 3297–3304.

Kay, J., and Calabrese, L. (2004). The role of interleukin-1 in the pathogenesis of rheumatoid arthritis. Rheumatology (Oxford) *43* Suppl. 3, iii2–iii9.

Kayashima, S., Tsuru, S., Shinomiya, N., Katsura, Y., Motoyoshi, K., Rokutanda, M., and Nagata, N. (1991). Effects of macrophage colony-stimulating factor on reduction of viable bacteria and survival of mice during Listeria monocytogenes infection: characteristics of monocyte subpopulations. Infect. Immun. *59*, 4677–4680.

Kedzierska, K., Rainbird, M.A., Lopez, A.F., and Crowe, S.M. (1998). Effect of GM-CSF on HIV-1 replication in monocytes/macrophages *in vivo* and *in vitro*: a review. Vet. Immunol. Immunopathol. *63*, 111–121.

Kelso, A., Groves, P., Troutt, A.B., and Francis, K. (1995). Evidence for the stochastic acquisition of cytokine profile by $CD4^+$ T-cells activated in a T helper type 2-like response *in vivo*. Eur J. Immunol. *25*, 1168–1175.

Kielian, T., Mayes, P., and Kielian, M. (2002). Characterization of microglial responses to *Staphylococcus aureus:* effects on cytokine, costimulatory molecule, and Toll-like receptor expression. J. Neuroimmunol. *130*, 86–99.

Kim, D.H., Sandoval, D., Reed, J.A., Matter, E.K., Tolod, E.G., Woods, S.C., and Seeley, R.J. (2008). The role of GM-CSF in adipose tissue inflammation. Am J. Physiol. Endocrinol. Metab. *295*, E1038–1046.

Kishimoto, T. (2005). Interleukin-6: from basic science to medicine--40 years in immunology. Annu. Rev. Immunol. *23*, 1–21.

Klein, I., Cornejo, J.C., Polakos, N.K., John, B., Wuensch, S.A., Topham, D.J., Pierce, R.H., and Crispe, I.N. (2007). Kupffer cell heterogeneity: functional properties of bone marrow derived and sessile hepatic macrophages. Blood *110*, 4077–4085.

Kowal, K., Nolte, H., Skov, P.S., and DuBuske, L.M. (2005). Effect of allergen-specific immunotherapy on recombinant human interleukin 3-mediated amplification of allergen-induced basophil histamine release. Allergy Asthma Proc. *26*, 456–462.

Kramer-Hammerle, S., Rothenaigner, I., Wolff, H., Bell, J.E., and Brack-Werner, R. (2005). Cells of the central nervous system as targets and reservoirs of the human immunodeficiency virus. Virus Res. *111*, 194–213.

Kruger, C., Laage, R., Pitzer, C., Schabitz, W.R., and Schneider, A. (2007). The hematopoietic factor GM-CSF (granulocyte–macrophage colony-stimulating factor) promotes neuronal differentiation of adult neural stem cells *in vitro*. BMC Neurosci. *8*, 88.

Kruglov, A.A., Kuchmiy, A., Grivennikov, S.I., Tumanov, A.V., Kuprash, D.V., and Nedospasov, S.A. (2008). Physiological functions of tumor necrosis factor and the consequences of its pathologic overexpression or blockade: mouse models. Cytokine Growth Factor Rev. *19*, 231–244.

Lambrecht, B.N. (2006). Alveolar macrophage in the driver's seat. Immunity *24*, 366–368.

Lancrin, C., Sroczynska, P., Stephenson, C., Allen, T., Kouskoff, V., and Lacaud, G. (2009). The haemangioblast generates haematopoietic cells through a haemogenic endothelium stage. Nature *457*, 892–895.

Landsman, L., and Jung, S. (2007). Lung macrophages serve as obligatory intermediate between blood monocytes and alveolar macrophages. J. Immunol. *179*, 3488–3494.

Landsman, L., Bar-On, L., Zernecke, A., Kim, K.W., Krauthgamer, R., Shagdarsuren, E., Lira, S.A., Weissman, I.L., Weber, C., and Jung, S. (2009). CX3CR1 is required for monocyte homeostasis and atherogenesis by promoting cell survival. Blood *113*, 963–972.

Lari, R., Fleetwood, A.J., Kitchener, P.D., Cook, A.D., Pavasovic, D., Hertzog, P.J., and Hamilton, J.A. (2007). Macrophage lineage phenotypes and osteoclastogenesis--complexity in the control by GM-CSF and TGF-beta. Bone *40*, 323–336.

Lawicki, S., Szmitkowski, M., and Wojtukiewicz, M. (2006). The pretreatment plasma level and diagnostic utility of M-CSF in benign breast tumor and breast cancer patients. Clin. Chim. Acta *371*, 112–116.

Lawson, L.J., Perry, V.H., Dri, P., and Gordon, S. (1990). Heterogeneity in the distribution and morphology of microglia in the normal adult mouse brain. Neuroscience *39*, 151–170.

Lee, H.Y., Youn, S.W., Kim, J.Y., Park, K.W., Hwang, C.I., Park, W.Y., Oh, B.H., Park, Y.B., Walsh, K., Seo, J.S., and Kim, H.S. (2008). FOXO3a turns the tumor necrosis factor receptor signaling towards apoptosis through reciprocal regulation of c-Jun N-terminal kinase and NF-kappaB. Arterioscler. Thromb. Vasc. Biol. *28*, 112–120.

Leeanansaksiri, W., Wang, H., Gooya, J.M., Renn, K., Abshari, M., Tsai, S., and Keller, J.R. (2005). IL-3 induces inhibitor of DNA-binding protein-1 in hemopoietic progenitor cells and promotes myeloid cell development. J. Immunol. *174*, 7014–7021.

Lemaitre, B., Nicolas, E., Michaut, L., Reichhart, J.M., and Hoffmann, J.A. (1996). The dorsoventral regulatory gene cassette spatzle/Toll/cactus controls the potent antifungal response in *Drosophila* adults. Cell *86*, 973–983.

Li, H., and Lin, X. (2008). Positive and negative signaling components involved in TNFαlpha-induced NF-kappaB activation. Cytokine *41*, 1–8.

Li, M.O., and Flavell, R.A. (2008). Contextual regulation of inflammation: a duet by transforming growth factor-beta and interleukin-10. Immunity *28*, 468–476.

Li, X., Ranjith-Kumar, C.T., Brooks, M.T., Dharmaiah, S., Herr, A.B., Kao, C., and Li, P. (2009). The RIG-I like receptor LGP2 recognizes the termini of double-stranded RNA. J. Biol. Chem. *284*, 13881–13891.

Liang, Y., Ishida, H., Lenz, O., Lin, T.I., Nyanguile, O., Simmen, K., Pyles, R.B., Bourne, N., Yi, M., Li, K., and Lemon, S.M. (2008). Antiviral suppression vs restoration of RIG-I signaling by hepatitis C protease and polymerase inhibitors. Gastroenterology *135*, 1710–1718, e1712.

Lichanska, A.M., Browne, C.M., Henkel, G.W., Murphy, K.M., Ostrowski, M.C., McKercher, S.R., Maki, R.A., and Hume, D.A. (1999). Differentiation of the mononuclear phagocyte system during mouse embryogenesis: the role of transcription factor PU.1. Blood *94*, 127–138.

Lichanska, A.M., and Hume, D.A. (2000). Origins and functions of phagocytes in the embryo. Exp. Hematol. *28*, 601–611.

Lim, L.H., Burdick, M.M., Hudson, S.A., Mustafa, F.B., Konstantopoulos, K., and Bochner, B.S. (2006). Stimulation of human endothelium with IL-3 induces selective basophil accumulation *in vitro*. J. Immunol. *176*, 5346–5353.

Lin, H., Lee, E., Hestir, K., Leo, C., Huang, M., Bosch, E., Halenbeck, R., Wu, G., Zhou, A., Behrens, D., *et al.* (2008). Discovery of a cytokine and its receptor by functional screening of the extracellular proteome. Science *320*, 807–811.

Ling, Z., Tran, K.C., and Teng, M.N. (2009). Human respiratory syncytial virus nonstructural protein NS2 antagonizes the activation of beta interferon transcription by interacting with RIG-I. J. Virol. *83*, 3734–3742.

Liu, K., Victora, G.D., Schwickert, T.A., Guermonprez, P., Meredith, M.M., Yao, K., Chu, F.F., Randolph, G.J., Rudensky, A.Y., and Nussenzweig, M. (2009). *In vivo* analysis of dendritic cell development and homeostasis. Science *324*, 392–397.

Luc, S., Anderson, K., Kharazi, S., Buza-Vidas, N., Boiers, C., Jensen, C.T., Ma, Z., Wittmann, L., and Jacobsen, S.E. (2008a). Down-regulation of Mpl marks the transition to lymphoid-primed multipotent progenitors with gradual loss of granulocyte-monocyte potential. Blood *111*, 3424–3434.

Luc, S., Buza-Vidas, N., and Jacobsen, S.E. (2008b). Delineating the cellular pathways of hematopoietic lineage commitment. Semin. Immunol. *20*, 213–220.

Lyakh, L., Trinchieri, G., Provezza, L., Carra, G., and Gerosa, F. (2008). Regulation of interleukin-12/interleukin-23 production and the T-helper 17 response in humans. Immunol. Rev. *226*, 112–131.

Maemura, K., Zheng, Q., Wada, T., Ozaki, M., Takao, S., Aikou, T., Bulkley, G.B., Klein, A.S., and Sun, Z. (2005). Reactive oxygen species are essential mediators in antigen presentation by Kupffer cells. Immunol. Cell Biol. *83*, 336–343.

Mahalingam, S., and Lidbury, B.A. (2002). Suppression of lipopolysaccharide-induced antiviral transcription factor (STAT-1 and NF-kappa B) complexes by antibody-dependent enhancement of macrophage infection by Ross River virus. Proc. Natl. Acad. Sci. U.S.A. *99*, 13819–13824.

Makino, M., Maeda, Y., Fukutomi, Y., and Mukai, T. (2007). Contribution of GM-CSF on the enhancement of the T-cell-stimulating activity of macrophages. Microbes Infect. *9*, 70–77.

Manicassamy, S., Ravindran, R., Deng, J., Oluoch, H., Denning, T.L., Kasturi, S.P., Rosenthal, K.M., Evavold, B.D., and Pulendran, B. (2009). Toll-like receptor 2-dependent induction of vitamin A-metabolizing enzymes in dendritic cells promotes T regulatory responses and inhibits autoimmunity. Nat. Med. *15*, 401–409.

Marin-Teva, J.L., Dusart, I., Colin, C., Gervais, A., van Rooijen, N., and Mallat, M. (2004). Microglia promote the death of developing Purkinje cells. Neuron *41*, 535–547.

Marriott, H.M., and Dockrell, D.H. (2007). The role of the macrophage in lung disease mediated by bacteria. Exp Lung Res. *33*, 493–505.

Martinez, F.O., Helming, L., and Gordon, S. (2009). Alternative activation of macrophages: an immunologic functional perspective. Annu. Rev. Immunol. *27*, 451–483.

Matsumiya, T., Imaizumi, T., Yoshida, H., Satoh, K., Topham, M.K., and Stafforini, D.M. (2009). The levels of retinoic acid-inducible gene I are regulated by heat shock protein 90-alpha. J. Immunol. *182*, 2717–2725.

Matsuzawa, K., Yoo, Y.C., Fukushima, A., Yoshimatsu, K., Arikawa, J., and Azuma, I. (1997). Protective effect of mucosal administration of recombinant human macrophage colony-stimulating factor (rhM-CSF) on mucosal infection of Sendai virus in mice. Vaccine *15*, 85–89.

Maus, U.A., Koay, M.A., Delbeck, T., Mack, M., Ermert, M., Ermert, L., Blackwell, T.S., Christman, J.W., Schlondorff, D., Seeger, W., and Lohmeyer, J. (2002). Role of resident alveolar macrophages in leukocyte traffic into the alveolar air space of intact mice. Am J. Physiol. Lung Cell Mol. Physiol. *282*, L1245–1252.

Mayer, W.J., Irschick, U.M., Moser, P., Wurm, M., Huemer, H.P., Romani, N., and Irschick, E.U. (2007). Characterization of antigen-presenting cells in fresh and cultured human corneas using novel dendritic cell markers. Invest Ophthalmol. Vis. Sci. *48*, 4459–4467.

Medvinsky, A., and Dzierzak, E. (1996). Definitive hematopoiesis is autonomously initiated by the AGM region. Cell *86*, 897–906.

Medzhitov, R., Preston-Hurlburt, P., and Janeway, C.A., Jr (1997). A human homologue of the Drosophila Toll protein signals activation of adaptive immunity. Nature *388*, 394–397.

Mehlhop, E., Ansarah-Sobrinho, C., Johnson, S., Engle, M., Fremont, D.H., Pierson, T.C., and Diamond, M.S. (2007). Complement protein C1q inhibits antibody-dependent enhancement of flavivirus infection in an IgG subclass-specific manner. Cell Host Microbe 2, 417–426.

Merad, M., Ginhoux, F., and Collin, M. (2008). Origin, homeostasis and function of Langerhans cells and other langerin-expressing dendritic cells. Nat. Rev. Immunol. *8*, 935–947.

Migeotte, I., Communi, D., and Parmentier, M. (2006). Formyl peptide receptors: a promiscuous subfamily of G protein-coupled receptors controlling immune responses. Cytokine Growth Factor Rev. *17*, 501–519.

Mikkola, H.K., Gekas, C., Orkin, S.H., and Dieterlen-Lievre, F. (2005). Placenta as a site for hematopoietic stem cell development. Exp. Hematol. 33, 1048–1054.

Miyajima, A., Mui, A.L., Ogorochi, T., and Sakamaki, K. (1993). Receptors for granulocyte–macrophage colony-stimulating factor, interleukin-3, and interleukin-5. Blood 82, 1960–1974.

Morgan, M.J., Kim, Y.S., and Liu, Z.G. (2008). TNFαlpha and reactive oxygen species in necrotic cell death. Cell Res. *18*, 343–349.

Morrison, T.E., Simmons, J.D., and Heise, M.T. (2008). Complement receptor 3 promotes severe ross river virus-induced disease. J. Virol. *82*, 11263–11272.

Mosser, D.M., and Edwards, J.P. (2008). Exploring the full spectrum of macrophage activation. Nat. Rev. Immunol. *8*, 958–969.

Muessel, M.J., Scott, K.S., Friedl, P., Bradding, P., and Wardlaw, A.J. (2008). CCL11 and GM-CSF differentially use the Rho GTPase pathway to regulate motility of human eosinophils in a three-dimensional microenvironment. J. Immunol. *180*, 8354–8360.

Mukaida, N. (2003). Pathophysiological roles of interleukin-8/CXCL8 in pulmonary diseases. Am J. Physiol. Lung Cell Mol. Physiol. *284*, L566–577.

Myong, S., Cui, S., Cornish, P.V., Kirchhofer, A., Gack, M.U., Jung, J.U., Hopfner, K.P., and Ha, T. (2009). Cytosolic viral sensor RIG-I is a 5′-triphosphate-dependent translocase on double-stranded RNA. Science *323*, 1070–1074.

Naik, S.H. (2008). Demystifying the development of dendritic cell subtypes, a little. Immunol. Cell Biol. *86*, 439–452.

Naito, M., Hasegawa, G., Ebe, Y., and Yamamoto, T. (2004). Differentiation and function of Kupffer cells. Med. Electron Microsc. *37*, 16–28.

Nandi, S., Akhter, M.P., Seifert, M.F., Dai, X.M., and Stanley, E.R. (2006). Developmental and functional significance of the CSF-1 proteoglycan chondroitin sulfate chain. Blood *107*, 786–795.

Natarajan, C., Sriram, S., Muthian, G., and Bright, J.J. (2004). Signaling through JAK2-STAT5 pathway is essential for IL-3-induced activation of microglia. Glia *45*, 188–196.

Nathan, D.M., Angus, P.W., and Gibson, P.R. (2006). Hepatitis B and C virus infections and anti-tumor necrosis factor-alpha therapy: guidelines for clinical approach. J. Gastroenterol. Hepatol. *21*, 1366–1371.

Naugler, W.E., and Karin, M. (2008). The wolf in sheep's clothing: the role of interleukin-6 in immunity, inflammation and cancer. Trends Mol. Med. *14*, 109–119.

Negishi, H., Osawa, T., Ogami, K., Ouyang, X., Sakaguchi, S., Koshiba, R., Yanai, H., Seko, Y., Shitara, H., Bishop, K., *et al.* (2008). A critical link between Toll-like receptor 3 and type II interferon signaling pathways in antiviral innate immunity. Proc. Natl. Acad. Sci. U.S.A. *105*, 20446–20451.

Neumann, H., Kotter, M.R., and Franklin, R.J. (2009). Debris clearance by microglia: an essential link between degeneration and regeneration. Brain *132*, 288–295.

Nielsen, A.R., and Pedersen, B.K. (2007). The biological roles of exercise-induced cytokines: IL-6, IL-8, and IL-15. Appl. Physiol. Nutr. Metab. *32*, 833–839.

Nimmerjahn, A., Kirchhoff, F., and Helmchen, F. (2005). Resting microglial cells are highly dynamic surveillants of brain parenchyma *in vivo*. Science *308*, 1314–1318.

Nomura, K., Kuroda, S., Yoshikawa, H., and Tomita, T. (2008). Inflammatory osteoclastogenesis can be induced by GM-CSF and activated under TNF immunity. Biochem. Biophys. Res. Commun. *367*, 881–887.

O'Neill, L.A. (2008a). The interleukin-1 receptor/Toll-like receptor superfamily: 10 years of progress. Immunol. Rev. *226*, 10–18.

O'Neill, L.A. (2008b). When signaling pathways collide: positive and negative regulation of toll-like receptor signal transduction. Immunity *29*, 12–20.

Ochoa, M.T., Loncaric, A., Krutzik, S.R., Becker, T.C., and Modlin, R.L. (2008). 'Dermal dendritic cells' comprise two distinct populations: CD1+ dendritic cells and CD209+ macrophages. J. Invest. Dermatol. *128*, 2225–2231.

Ohmori, K., Luo, Y., Jia, Y., Nishida, J., Wang, Z., Bunting, K.D., Wang, D., and Huang, H. (2009). IL-3 induces basophil expansion *in vivo* by directing granulocyte-monocyte progenitors to differentiate into basophil lineage-restricted progenitors in the bone marrow and by increasing the number of basophil/mast cell progenitors in the spleen. J. Immunol. *182*, 2835–2841.

Ohsugi, Y. (2007). Recent advances in immunopathophysiology of interleukin-6: an innovative therapeutic drug, tocilizumab (recombinant humanized anti-human interleukin-6 receptor antibody), unveils the mysterious etiology of immune-mediated inflammatory diseases. Biol. Pharm. Bull. *30*, 2001–2006.

Ovchinnikov, D.A. (2008). Macrophages in the embryo and beyond: much more than just giant phagocytes. Genesis *46*, 447–462.

Paldi, A. (2003). Stochastic gene expression during cell differentiation: order from disorder? Cell. Mol. Life Sci. *60*, 1775–1778.

Palis, J., and Yoder, M.C. (2001). Yolk-sac hematopoiesis: the first blood cells of mouse and man. Exp. Hematol. *29*, 927–936.

Park, B.K., Zhang, H., Zeng, Q., Dai, J., Keller, E.T., Giordano, T., Gu, K., Shah, V., Pei, L., Zarbo, R.J., *et al.* (2007). NF-kappaB in breast cancer cells promotes osteolytic bone metastasis by inducing osteoclastogenesis via GM-CSF. Nat. Med. *13*, 62–69.

Park, S., Guo, J., Kim, D., and Cheng, J.Q. (2009). Identification of 24p3 as a direct target of Foxo3a regulated by interleukin-3 through the phosphoinositide 3-kinase/Akt pathway. J. Biol. Chem. *284*, 2187–2193.

Paunovic, V., Carroll, H.P., Vandenbroeck, K., and Gadina, M. (2008). Signalling, inflammation and arthritis: crossed signals: the role of interleukin (IL)-*12*, -*17*, -23 and -27 in autoimmunity. Rheumatology (Oxford) *47*, 771–776.

Pelley, J.L., Nicholls, C.D., Beattie, T.L., and Brown, C.B. (2007). Discovery and characterization of a novel splice variant of the GM-CSF receptor alpha subunit. Exp. Hematol. 35, 1483–1494.

Pettit, A.R., Chang, M.K., Hume, D.A., and Raggatt, L.J. (2008). Osteal macrophages: a new twist on coupling during bone dynamics. Bone *43*, 976–982.

Pixley, F.J., and Stanley, E.R. (2004). CSF-1 regulation of the wandering macrophage: complexity in action. Trends Cell Biol. *14*, 628–638.

Plater-Zyberk, C., Joosten, L.A., Helsen, M.M., Koenders, M.I., Baeuerle, P.A., and van den Berg, W.B. (2008). Combined blockade of GM-CSF and IL-17 pathways potently suppresses chronic destructive arthritis in a TNFαindependent mouse model. Ann. Rheum. Dis. *68*, 721–728.

Polakos, N.K., Cornejo, J.C., Murray, D.A., Wright, K.O., Treanor, J.J., Crispe, I.N., Topham, D.J., and Pierce, R.H. (2006). Kupffer cell-dependent hepatitis occurs during influenza infection. Am. J. Pathol. 168, 1169–1178; quiz 1404–1165.

Poltorak, A., He, X., Smirnova, I., Liu, M.Y., Van Huffel, C., Du, X., Birdwell, D., Alejos, E., Silva, M., Galanos, C., *et al.* (1998). Defective LPS signaling in C3H/HeJ and C57BL/10ScCr mice: mutations in Tlr4 gene. Science *282*, 2085–2088.

Ponomarev, E.D., Shriver, L.P., Maresz, K., Pedras-Vasconcelos, J., Verthelyi, D., and Dittel, B.N. (2007). GM-CSF production by autoreactive T-cells is required for the activation of microglial cells and the onset of experimental autoimmune encephalomyelitis. J. Immunol. *178*, 39–48.

Price, L.K., Choi, H.U., Rosenberg, L., and Stanley, E.R. (1992). The predominant form of secreted colony stimulating factor-1 is a proteoglycan. J. Biol. Chem. *267*, 2190–2199.

Prins, K.C., Cardenas, W.B., and Basler, C.F. (2009). Ebola virus protein VP35 impairs the function of interferon regulatory factor-activating kinases IKKepsilon and TBK-1. J. Virol. *83*, 3069–3077.

Raaijmakers, M.H., and Scadden, D.T. (2008). Evolving concepts on the microenvironmental niche for hematopoietic stem cells. Curr Opin Hematol *15*, 301–306.

Reddy, R.C. (2008). Immunomodulatory role of PPAR-gamma in alveolar macrophages. J. Invest. Med. *56*, 522–527.

Renneson, J., Salio, M., Mazouz, N., Goldman, M., Marchant, A., and Cerundolo, V. (2005). Mature dendritic cells differentiated in the presence of interferon-beta and interleukin-3 prime functional antigen-specific CD8 T-cells. Clin. Exp. Immunol. *139*, 468–475.

Rhodes, K.E., Gekas, C., Wang, Y., Lux, C.T., Francis, C.S., Chan, D.N., Conway, S., Orkin, S.H., Yoder, M.C., and Mikkola, H.K. (2008). The emergence of hematopoietic stem cells is initiated in the placental vasculature in the absence of circulation. Cell Stem Cell *2*, 252–263.

Richardsen, E., Uglehus, R.D., Due, J., Busch, C., and Busund, L.T. (2008). The prognostic impact of M-CSF, CSF-1 receptor, CD68 and CD3 in prostatic carcinoma. Histopathology *53*, 30–38.

Roberts, T.L., Idris, A., Dunn, J.A., Kelly, G.M., Burnton, C.M., Hodgson, S., Hardy, L.L., Garceau, V., Sweet, M.J., Ross, I.L., *et al.* (2009). HIN-200 proteins regulate caspase activation in response to foreign cytoplasmic DNA. Science *323*, 1057–1060.

Robin, C., Ottersbach, K., Durand, C., Peeters, M., Vanes, L., Tybulewicz, V., and Dzierzak, E. (2006). An unexpected role for IL-3 in the embryonic development of hematopoietic stem cells. Dev. Cell *11*, 171–180.

Robinson, W.E. (2006). Mechanism for complement-mediated, antibody-dependent enhancement of human immunodeficiency virus type 1 infection in MT2 cells is enhanced entry through CD4, CD*21*, and CXCR4 chemokine receptors. Viral Immunol. *19*, 434–447.

Robson-Ansley, P.J., Blannin, A., and Gleeson, M. (2007). Elevated plasma interleukin-6 levels in trained male triathletes following an acute period of intense interval training. Eur. J. Appl. Physiol. *99*, 353–360.

Rock, R.B., Gekker, G., Hu, S., Sheng, W.S., Cheeran, M., Lokensgard, J.R., and Peterson, P.K. (2004). Role of microglia in central nervous system infections. Clin. Microbiol. Rev. *17*, 942–964.

Rodrigo, W.W., Jin, X., Blackley, S.D., Rose, R.C., and Schlesinger, J.J. (2006). Differential enhancement of dengue virus immune complex infectivity mediated by signaling-competent and signaling-incompetent human Fcgamma RIA (CD64) or FcgammaRIIA (CD32). J. Virol. *80*, 10128–10138.

Rosas, M., Gordon, S., and Taylor, P.R. (2007). Characterisation of the expression and function of the GM-CSF receptor alpha-chain in mice. Eur J. Immunol. *37*, 2518–2528.

Ross, I.L., Dunn, T.L., Yue, X., Roy, S., Barnett, C.J., and Hume, D.A. (1994a). Comparison of the expression and function of the transcription factor PU.1 (Spi-1 proto-oncogene) between murine macrophages and B-lymphocytes. Oncogene *9*, 121–132.

Ross, I.L., Browne, C.M., and Hume, D.A. (1994b). Transcription of individual genes in eukaryotic cells occurs randomly and infrequently. Immunol. Cell Biol. *72*, 177–185.

Roth, P., and Stanley, E.R. (1992). The biology of CSF-1 and its receptor. Curr. Top. Microbiol. Immunol. *181*, 141–167.

Roth, P., Bartocci, A., and Stanley, E.R. (1997). Lipopolysaccharide induces synthesis of mouse colony-stimulating factor-1 *in vivo*. J. Immunol. *158*, 3874–3880.

Rovida, E., Lugli, B., Barbetti, V., Giuntoli, S., Olivotto, M., and Dello Sbarba, P. (2005). Focal adhesion kinase is redistributed to focal complexes and mediates cell spreading in macrophages in response to M-CSF. Biol. Chem. *386*, 919–929.

Sasmono, R.T., Ehrnsperger, A., Cronau, S.L., Ravasi, T., Kandane, R., Hickey, M.J., Cook, A.D., Himes, S.R., Hamilton, J.A., and Hume, D.A. (2007). Mouse neutrophilic granulocytes express mRNA encoding the macrophage colony-stimulating factor receptor (CSF-1R) as well as many other macrophage-specific transcripts and can transdifferentiate into macrophages *in vitro* in response to CSF-1. J. Leukoc. Biol. *82*, 111–123.

Satoh, T., Uehara, Y., and Kaziro, Y. (1992). Inhibition of interleukin 3 and granulocyte–macrophage colony-stimulating factor stimulated increase of active ras. GTP by herbimycin A, a specific inhibitor of tyrosine kinases. J. Biol. Chem. *267*, 2537–2541.

Schmierer, B., and Hill, C.S. (2007). TGFβeta-SMAD signal transduction: molecular specificity and functional flexibility. Nat. Rev. Mol. Cell Biol. *8*, 970–982.

Schoenborn, J.R., and Wilson, C.B. (2007). Regulation of interferon-gamma during innate and adaptive immune responses. Adv. Immunol. *96*, 41–101.

Schroder, K., Hertzog, P.J., Ravasi, T., and Hume, D.A. (2004). Interferon-gamma: an overview of signals, mechanisms and functions. J. Leukoc. Biol. *75*, 163–189.

Sebba, A. (2008). Tocilizumab: the first interleukin-6-receptor inhibitor. Am. J. Health Syst. Pharm. *65*, 1413–1418.

Segal, B.H. (2007). Role of macrophages in host defense against aspergillosis and strategies for immune augmentation. Oncologist *12* Suppl. *2*, 7–13.

Selleri, S., Deola, S., Pos, Z., Jin, P., Worschech, A., Slezak, S.L., Rumio, C., Panelli, M.C., Maric, D., Stroncek, D.F., *et al.* (2008). GM-CSF/IL-3/IL-5 receptor common beta chain (CD131) expression as a biomarker of antigen-stimulated CD8+ T-cells. J. Transl. Med. *6*, 17.

Sester, D.P., Trieu, A., Brion, K., Schroder, K., Ravasi, T., Robinson, J.A., McDonald, R.C., Ripoll, V., Wells, C.A., Suzuki, H., *et al.* (2005). LPS regulates a set of genes in primary murine macrophages by antagonising CSF-1 action. Immunobiology *210*, 97–107.

Shaftel, S.S., Griffin, W.S., and O'Banion, M.K. (2008). The role of interleukin-1 in neuroinflammation and Alzheimer disease: an evolving perspective. J. Neuroinflammation *5*, 7.

Shen, W., Li, B., Wetzel, M.A., Rogers, T.J., Henderson, E.E., Su, S.B., Gong, W., Le, Y., Sargeant, R., Dimitrov, D.S., *et al.* (2000). Down-regulation of the chemokine receptor CCR5 by activation of chemotactic formyl peptide receptor in human monocytes. Blood *96*, 2887–2894.

Shepard, J.L., and Zon, L.I. (2000). Developmental derivation of embryonic and adult macrophages. Curr. Opin. Hematol. *7*, 3–8.

Silvennoinen, O., Witthuhn, B.A., Quelle, F.W., Cleveland, J.L., Yi, T., and Ihle, J.N. (1993). Structure of the murine Jak2 protein-tyrosine kinase and its role in interleukin 3 signal transduction. Proc. Natl. Acad. Sci. U.S.A. *90*, 8429–8433.

Simard, A.R., Soulet, D., Gowing, G., Julien, J.P., and Rivest, S. (2006). Bone marrow-derived microglia play a critical role in restricting senile plaque formation in Alzheimer's disease. Neuron *49*, 489–502.

Simmons, A.D., Li, B., Gonzalez-Edick, M., Lin, C., Moskalenko, M., Du, T., Creson, J., VanRoey, M.J., and Jooss, K. (2007). GM-CSF-secreting cancer immunotherapies: preclinical analysis of the mechanism of action. Cancer Immunol. Immunother. *56*, 1653–1665.

Skrzypek, F., Cenci, E., Pietrella, D., Rachini, A., Bistoni, F., and Vecchiarelli, A. (2009). Dectin-1 is required for human dendritic cells to initiate immune response to *Candida albicans* through Syk activation. Microbes Infect. *11*, 661–670.

Soehnlein, O., Zernecke, A., Eriksson, E.E., Rothfuchs, A.G., Pham, C.T., Herwald, H., Bidzhekov, K., Rottenberg, M.E., Weber, C., and Lindbom, L. (2008). Neutrophil secretion products pave the way for inflammatory monocytes. Blood *112*, 1461–1471.

Spearman, P., Kalams, S., Elizaga, M., Metch, B., Chiu, Y.L., Allen, M., Weinhold, K.J., Ferrari, G., Parker, S.D., McElrath, M.J., *et al.* (2009). Safety and immunogenicity of a CTL multiepitope peptide vaccine for HIV with or without GM-CSF in a phase I trial. Vaccine *27*, 243–249.

Stacey, K.J., Sweet, M.J., and Hume, D.A. (1996). Macrophages ingest and are activated by bacterial DNA. J. Immunol. *157*, 2116–2122.

Stanley, E.R., Berg, K.L., Einstein, D.B., Lee, P.S., Pixley, F.J., Wang, Y., and Yeung, Y.G. (1997). Biology and action of colony-stimulating factor-1. Mol. Reprod. Dev. *46*, 4–10.

Stevens, B., Allen, N.J., Vazquez, L.E., Howell, G.R., Christopherson, K.S., Nouri, N., Micheva, K.D., Mehalow, A.K., Huberman, A.D., Stafford, B., *et al.* (2007). The classical complement cascade mediates CNS synapse elimination. Cell *131*, 1164–1178.

Stojakovic, M., Krzesz, R., Wagner, A.H., and Hecker, M. (2007). CD154-stimulated GM-CSF release by vascular smooth muscle cells elicits monocyte activation--role in atherogenesis. J. Mol. Med. *85*, 1229–1238.

Strehl, B., Seifert, U., Kruger, E., Heink, S., Kuckelkorn, U., and Kloetzel, P.M. (2005). Interferon-gamma, the functional plasticity of the ubiquitin–proteasome system, and MHC class I antigen processing. Immunol. Rev. *207*, 19–30.

Suhrbier, A., and La Linn, M. (2003). Suppression of antiviral responses by antibody-dependent enhancement of macrophage infection. Trends Immunol. *24*, 165–168.

Sun, S., Wang, F., Wei, J., Cao, L.Y., Wu, G.Y., Lu, L., Kosten, T.A., Kosten, T.R., and Zhang, X.Y. (2008). Association between interleukin-3 receptor alpha polymorphism and schizophrenia in the Chinese population. Neurosci. Lett. *440*, 35–37.

Sweet, M.J., Campbell, C.C., Sester, D.P., Xu, D., McDonald, R.C., Stacey, K.J., Hume, D.A., and Liew, F.Y. (2002). Colony-stimulating factor-1 suppresses responses to CpG DNA and expression of toll-like receptor 9 but enhances responses to lipopolysaccharide in murine macrophages. J. Immunol. *168*, 392–399.

Sweet, M.J., and Hume, D.A. (2003). CSF-1 as a regulator of macrophage activation and immune responses. Arch. Immunol. Ther. Exp. (Warsz) *51*, 169–177.

Tagoh, H., Ingram, R., Wilson, N., Salvagiotto, G., Warren, A.J., Clarke, D., Busslinger, M., and Bonifer, C. (2006). The mechanism of repression of the myeloid-specific c-fms gene by Pax5 during B lineage restriction. Embo J. *25*, 1070–1080.

Takada, A., Ebihara, H., Feldmann, H., Geisbert, T.W., and Kawaoka, Y. (2007). Epitopes required for antibody-dependent enhancement of Ebola virus infection. J. Infect. Dis. *196* Suppl. *2*, S347–356.

Takahashi, K., and Yamanaka, S. (2006). Induction of pluripotent stem cells from mouse embryonic and adult fibroblast cultures by defined factors. Cell *126*, 663–676.

Takano, T., Katada, Y., Moritoh, S., Ogasawara, M., Satoh, K., Satoh, R., Tanabe, M., and Hohdatsu, T. (2008a). Analysis of the mechanism of antibody-dependent enhancement of feline infectious peritonitis virus infection: aminopeptidase N is not important and a process of acidification of the endosome is necessary. J. Gen. Virol. *89*, 1025–1029.

Takano, T., Kawakami, C., Yamada, S., Satoh, R., and Hohdatsu, T. (2008b). Antibody-dependent enhancement occurs upon re-infection with the identical serotype virus in feline infectious peritonitis virus infection. J. Vet. Med. Sci. *70*, 1315–1321.

Tauber, A.I. (1991). The immunological self: a centenary perspective. Perspect. Biol. Med. *35*, 74–86.

Tavian, M., Cortes, F., Charbord, P., Labastie, M.C., and Peault, B. (1999). Emergence of the haematopoietic system in the human embryo and foetus. Haematologica *84* Suppl. EHA-4, 1–3.

Tavian, M., Robin, C., Coulombel, L., and Peault, B. (2001). The human embryo, but not its yolk sac, generates lympho-myeloid stem cells: mapping multipotent hematopoietic cell fate in intraembryonic mesoderm. Immunity *15*, 487–495.

Thomas, E.D., Ramberg, R.E., Sale, G.E., Sparkes, R.S., and Golde, D.W. (1976). Direct evidence for a bone marrow origin of the alveolar macrophage in man. Science *192*, 1016–1018.

Thomassen, M.J., Barna, B.P., Malur, A.G., Bonfield, T.L., Farver, C.F., Malur, A., Dalrymple, H., Kavuru, M.S., and Febbraio, M. (2007). ABCG1 is deficient in alveolar macrophages of GM-CSF knockout mice and patients with pulmonary alveolar proteinosis. J. Lipid Res. *48*, 2762–2768.

Torrelles, J.B., Azad, A.K., and Schlesinger, L.S. (2006). Fine discrimination in the recognition of individual species of phosphatidyl-myo-inositol mannosides from *Mycobacterium tuberculosis* by C-type lectin pattern recognition receptors J. Immunol. *177*, 1805–1816.

Toubaji, A., Hill, S., Terabe, M., Qian, J., Floyd, T., Simpson, R.M., Berzofsky, J.A., and Khleif, S.N. (2007). The combination of GM-CSF and IL-2 as local adjuvant shows synergy in enhancing peptide vaccines and provides long term tumor protection. Vaccine 25, 5882–5891.

Tu, Z., Bozorgzadeh, A., Pierce, R.H., Kurtis, J., Crispe, I.N., and Orloff, M.S. (2008). TLR-dependent cross talk between human Kupffer cells and NK cells. J. Exp. Med. *205*, 233–244.

Uchida, K., Beck, D.C., Yamamoto, T., Berclaz, P.Y., Abe, S., Staudt, M.K., Carey, B.C., Filippi, M.D., Wert, S.E., Denson, L.A., *et al.* (2007). GM-CSF autoantibodies and neutrophil dysfunction in pulmonary alveolar proteinosis. N. Engl. J. Med. *356*, 567–579.

Usynin, I., Klotz, C., and Frevert, U. (2007). Malaria circumsporozoite protein inhibits the respiratory burst in Kupffer cells. Cell. Microbiol. *9*, 2610–2628.

Uzri, D., and Gehrke, L. (2009). Nucleotide sequences and modifications that determine RIG-I/RNA binding and signaling activities. J. Virol. *83*, 4174–4184.

van Furth, R., and Cohn, Z.A. (1968). The origin and kinetics of mononuclear phagocytes. J. Exp. Med. *128*, 415–435.

van Til, N.P., Markusic, D.M., van der Rijt, R., Kunne, C., Hiralall, J.K., Vreeling, H., Frederiks, W.M., Oude-Elferink, R.P., and Seppen, J. (2005). Kupffer cells and not liver sinusoidal endothelial cells prevent lentiviral transduction of hepatocytes. Mol. Ther. *11*, 26–34.

Varin, A., and Gordon, S. (2009). Alternative activation of macrophages: Immune function and cellular biology. Immunobiology *7*, 630–641.

Vazquez-Torres, A., Vallance, B.A., Bergman, M.A., Finlay, B.B., Cookson, B.T., Jones-Carson, J., and Fang, F.C. (2004). Toll-like receptor 4 dependence of innate and adaptive immunity to Salmonella: importance of the Kupffer cell network. J. Immunol. *172*, 6202–6208.

Vermaelen, K., and Pauwels, R. (2005). Pulmonary dendritic cells. Am. J. Respir. Crit. Care Med. *172*, 530–551.

Vlahos, R., Bozinovski, S., Hamilton, J.A., and Anderson, G.P. (2006). Therapeutic potential of treating chronic obstructive pulmonary disease (COPD) by neutralising granulocyte macrophage-colony stimulating factor (GM-CSF). Pharmacol. Ther. *112*, 106–115.

Vojtek, A.B., Hollenberg, S.M., and Cooper, J.A. (1993). Mammalian Ras interacts directly with the serine/threonine kinase Raf. Cell *74*, 205–214.

Volmar, C.H., Ait-Ghezala, G., Frieling, J., Paris, D., and Mullan, M.J. (2008). The granulocyte macrophage colony stimulating factor (GM-CSF) regulates amyloid beta (Abeta) production. Cytokine *42*, 336–344.

Wahl, S.M. (2007). Transforming growth factor-beta: innately bipolar. Curr. Opin. Immunol. *19*, 55–62.

Wan, Y.Y., and Flavell, R.A. (2007). 'Yin-Yang' functions of transforming growth factor-beta and T regulatory cells in immune regulation. Immunol. Rev. *220*, 199–213.

Wang, J., Roderiquez, G., Oravecz, T., and Norcross, M.A. (1998). Cytokine regulation of human immunodeficiency virus type 1 entry and replication in human monocytes/macrophages through modulation of CCR5 expression. J. Virol. *72*, 7642–7647.

Warren, M.K., and Ralph, P. (1986). Macrophage growth factor CSF-1 stimulates human monocyte production of interferon, tumor necrosis factor, and colony stimulating activity. J. Immunol. *137*, 2281–2285.

Wehkamp, J., and Stange, E.F. (2006). Paneth cells and the innate immune response. Curr. Opin. Gastroenterol. *22*, 644–650.

Weiss, J.M., Subleski, J.J., Wigginton, J.M., and Wiltrout, R.H. (2007). Immunotherapy of cancer by IL-12-based cytokine combinations. Expert Opin. Biol. Ther. *7*, 1705–1721.

Wells, C.A., Ravasi, T., Sultana, R., Yagi, K., Carninci, P., Bono, H., Faulkner, G., Okazaki, Y., Quackenbush, J., Hume, D.A., and Lyons, P.A. (2003). Continued discovery of transcriptional units expressed in cells of the mouse mononuclear phagocyte lineage. Genome Res. *13*, 1360–1365.

Werner, J.L., Metz, A.E., Horn, D., Schoeb, T.R., Hewitt, M.M., Schwiebert, L.M., Faro-Trindade, I., Brown, G.D., and Steele, C. (2009). Requisite role for the dectin-1 beta-glucan receptor in pulmonary defense against Aspergillus fumigatus. J. Immunol. *182*, 4938–4946.

Wheeler, A.P., Smith, S.D., and Ridley, A.J. (2006). CSF-1 and PI 3-kinase regulate podosome distribution and assembly in macrophages. Cell Motil. Cytoskeleton *63*, 132–140.

Whitelaw, D.M. (1972). Observations on human monocyte kinetics after pulse labeling. Cell Tissue Kinet. *5*, 311–317.

Willekens, F.L., Werre, J.M., Kruijt, J.K., Roerdinkholder-Stoelwinder, B., Groenen-Dopp, Y.A., van den Bos, A.G., Bosman, G.J., and van Berkel, T.J. (2005). Liver Kupffer cells rapidly remove red blood cell-derived vesicles from the circulation by scavenger receptors. Blood *105*, 2141–2145.

Wilmanski, J.M., Petnicki-Ocwieja, T., and Kobayashi, K.S. (2008). NLR proteins: integral members of innate immunity and mediators of inflammatory diseases. J. Leukoc. Biol. *83*, 13–30.

Wisse, E. (1974a). Observations on the fine structure and peroxidase cytochemistry of normal rat liver Kupffer cells. J. Ultrastruct. Res. *46*, 393–426.

Wisse, E. (1974b). Kupffer cell reactions in rat liver under various conditions as observed in the electron microscope. J. Ultrastruct. Res. *46*, 499–520.

Wolfert, M.A., Murray, T.F., Boons, G.J., and Moore, J.N. (2002). The origin of the synergistic effect of muramyl dipeptide with endotoxin and peptidoglycan. J. Biol. Chem. *277*, 39179–39186.

Xie, H., Ye, M., Feng, R., and Graf, T. (2004). Stepwise reprogramming of B-cells into macrophages. Cell *117*, 663–676.

Xu, H., Chen, M., Mayer, E.J., Forrester, J.V., and Dick, A.D. (2007). Turnover of resident retinal microglia in the normal adult mouse. Glia 55, 1189–1198.

Xu, Z., Tian, J., Smith, J.S., and Byrnes, A.P. (2008). Clearance of adenovirus by Kupffer cells is mediated by scavenger receptors, natural antibodies, and complement. J. Virol. *82*, 11705–11713.

Yamamoto, T., Kaizu, C., Kawasaki, T., Hasegawa, G., Umezu, H., Ohashi, R., Sakurada, J., Jiang, S., Shultz, L., and Naito, M. (2008). Macrophage colony-stimulating factor is indispensable for repopulation and differentiation of Kupffer cells but not for splenic red pulp macrophages in osteopetrotic (op/op) mice after macrophage depletion. Cell Tissue Res. *332*, 245–256.

Yang, H., Bocchetta, M., Kroczynska, B., Elmishad, A.G., Chen, Y., Liu, Z., Bubici, C., Mossman, B.T., Pass, H.I., Testa, J.R., *et al.* (2006). TNF-alpha inhibits asbestos-induced cytotoxicity via a NF-kappaB-dependent pathway, a possible mechanism for asbestos-induced oncogenesis. Proc. Natl. Acad. Sci. U.S.A. *103*, 10397–10402.

Yen, J.J., and Yang-Yen, H.F. (2006). Transcription factors mediating interleukin-3 survival signals. Vitam. Horm. *74*, 147–163.

Yogesha, S.D., Khapli, S.M., Srivastava, R.K., Mangashetti, L.S., Pote, S.T., Mishra, G.C., and Wani, M.R. (2009). IL-3 inhibits TNF-alpha-induced bone resorption and prevents inflammatory arthritis. J. Immunol. *182*, 361–370.

Young, H.A. (2006). Unraveling the pros and cons of interferon-gamma gene regulation. Immunity *24*, 506–507.

Young, H.A., and Bream, J.H. (2007). IFN-γamma: recent advances in understanding regulation of expression, biological functions, and clinical applications. Curr. Top. Microbiol. Immunol. *316*, 97–117.

Yu, J., Vodyanik, M.A., Smuga-Otto, K., Antosiewicz-Bourget, J., Frane, J.L., Tian, S., Nie, J., Jonsdottir, G.A., Ruotti, V., Stewart, R., *et al.* (2007). Induced pluripotent stem cell lines derived from human somatic cells. Science *318*, 1917–1920.

Yu, W., Chen, J., Xiong, Y., Pixley, F.J., Dai, X.M., Yeung, Y.G., and Stanley, E.R. (2008). CSF-1 receptor structure/function in MacCsf1r-/- macrophages: regulation of proliferation, differentiation, and morphology. J. Leukoc. Biol. *84*, 852–863.

Zambrano, A., Otth, C., Mujica, L., Concha, I.I., and Maccioni, R.B. (2007). Interleukin-3 prevents neuronal death induced by amyloid peptide. BMC Neurosci *8*, 82.

Zaslona, Z., Wilhelm, J., Cakarova, L., Marsh, L.M., Seeger, W., Lohmeyer, J., and von Wulffen, W. (2009). Transcriptome profiling of primary murine monocytes, lung macrophages and lung dendritic cells reveals a distinct expression of genes involved in cell trafficking. Respir Res. *10*, 2.

Zeoli, A., Dentelli, P., Rosso, A., Togliatto, G., Trombetta, A., Damiano, L., di Celle, P.F., Pegoraro, L., Altruda, F., and Brizzi, M.F. (2008). Interleukin-3 promotes expansion of hemopoietic-derived CD45+ angiogenic cells and their arterial commitment via STAT5 activation. Blood *112*, 350–361.

Zhang, P., Langland, J.O., Jacobs, B.L., and Samuel, C.E. (2009). Protein kinase PKR-dependent activation of mitogen activated protein kinases occurs through mitochondrial adapter IPS-1 and is antagonized by vaccinia virus E3L. J. Virol. *83*, 5718–5725.

Zhang, S., and Wang, Q. (2008). Factors determining the formation and release of bioactive IL-12: regulatory mechanisms for IL-12p70 synthesis and inhibition. Biochem. Biophys. Res. Commun. *372*, 509–512.

Zovein, A.C., Hofmann, J.J., Lynch, M., French, W.J., Turlo, K.A., Yang, Y., Becker, M.S., Zanetta, L., Dejana, E., Gasson, J.C., *et al.* (2008). Fate tracing reveals the endothelial origin of hematopoietic stem cells. Cell Stem Cell 3, 625–636.

# The Macrophage in HIV-1 Infection and Disease Progression

# 3

Paul R. Gorry, Jasminka Sterjovski and Melissa J. Churchill

Abstract

Most individuals who progress to late stages of HIV-1 infection do so whilst harbouring CCR5-restricted (R5) viral strains, which has been associated with enhanced tropism for cells of the monocyte/macrophage lineage (enhanced M-tropism). However, the molecular mechanisms contributing to enhanced M-tropism by R5 HIV-1 strains, and precisely how these HIV-1 variants cause immunodeficiency are unknown. This chapter examines the relationship between viral coreceptor usage, M-tropism, and pathogenicity of HIV-1. We highlight studies suggesting that enhanced M-tropism of R5 HIV-1 may arise from HIV-1 variants that have acquired an increased ability to scavenge low levels of CD4 and CCR5 expressed on the macrophage cell surface. These viral variants may have decreased sensitivity to HIV-1 entry inhibitors and increased ability to cause cytopathic effects in CD4+ T-cells, and influence disease progression in subjects who exclusively harbour R5 HIV-1 strains at late stages of infection.

## Introduction

Human immunodeficiency virus type 1 (HIV-1) infection is characterized by the progressive loss of CD4+ T lymphocytes, leading to the development of opportunistic infections and cancers defining AIDS (Barre-Sinoussi *et al.*, 1983; Gallo *et al.*, 1984). Studies in animal models suggest that high level virus replication *per se* is not sufficient for CD4+ T-cell depletion (Chakrabarti *et al.*, 2000), indicating that specific viral determinants and/or host responses are major contributors to HIV-1 pathogenicity. While viral cytopathic effects and immunological clearance contribute to the death of infected cells (Ho *et al.*, 1995; Wei *et al.*, 1995), more recent studies have identified chronic immune activation as a principal determinant of immune exhaustion and subsequent depletion of uninfected CD4+ T-cells (Brenchley *et al.*, 2006; Brenchley *et al.*, 2008; Brenchley and Douek, 2008). The viral envelope glycoproteins (Env) are significant mediators of acute cytotoxicity accompanying virus production by the infected cell (Sodroski *et al.*, 1986). In addition to CD4+ T-lymphocytes, HIV-1 also infects cells of the monocyte/macrophage lineage including monocytes in peripheral blood, macrophages in lymphoid tissues, and microglia in the central nervous system (CNS) (Gartner *et al.*, 1986b; Price, 1996). These cells may be involved in HIV-1 transmission and persistence, and their long lifespan compromises effective antiretroviral therapy (Crowe and Sonza, 2000; Pierson *et al.*, 2000b; Schrager and D'Souza, 1998; Sonza *et al.*, 2001). HIV-1 disease progression is associated with enhanced tropism for macrophages (M-tropism) (Li *et al.*, 1999; Tuttle *et al.*, 2002). Thus, M-tropism of HIV-1 is important for HIV-1 pathogenesis. However, the mechanisms involved in the enhancement of M-tropism and how M-tropic viral variants cause CD4+ T-cell depletion in HIV-1 infected subjects are poorly understood.

## HIV-1 entry

The gp120 envelope glycoproteins of HIV-1 Env, which exist as spikes on the exterior of the virus particle, are responsible for attachment between the virus and the target cells (reviewed in Gorry *et al.*, 2004; Gorry *et al.*, 2005). Viral attachment to cells involves binding of gp120 to cellular CD4, and then to a chemokine receptor as a cellular coreceptor (reviewed in Wyatt and Sodroski, 1998) (Fig. 3.1). Initial binding of gp120 to CD4 occurs with high affinity, and triggers a conformational change in gp120 that exposes the binding site for the secondary coreceptor. The subsequent binding of gp120 to the coreceptor then leads to further conformational changes in its structure which expose the N-terminus of the transmembrane glycoprotein of Env, gp41, which contains a fusion peptide that embeds into the host cell membrane. Structural studies have revealed that the gp41 ectodomain forms an extended coiled coil conformation and a six helix bundle structure (Chan *et al.*, 1997). Formation of the six helical bundle brings together the viral and target cell membranes, allowing fusion between and entry of the virion into the cell.

## HIV-1 coreceptors and cellular tropism

The tropism of HIV-1 is influenced, but not determined, by the choice of coreceptor used for viral entry (reviewed in Gorry *et al.*, 2004; Gorry *et al.*, 2005) (Fig. 3.2). CCR5 is the principal coreceptor used by M-tropic viruses, which can replicate in CD4+ T-cells and macrophages (Alkhatib *et al.*, 1996; Choe *et al.*, 1996; Deng *et al.*, 1996; Doranz *et al.*, 1996; Dragic *et al.*, 1996), and these are referred to as R5 viruses. CXCR4 mediates entry of T-cell tropic viruses, which are capable of replicating in CD4+ T-cells and T-cell lines (Feng *et al.*, 1996), and these are referred to as X4 viruses. Dual-tropic HIV-1 viruses can use both coreceptors for HIV-1 entry and can replicate in CD4+ T-cells, macrophages and T-cell lines, and are referred to as R5X4 viruses (Collman *et al.*, 1992). Some viral variants can use alternative coreceptors for HIV-1 entry in transfected cells, including CCR2b, CCR3, CCR8, CX3CR1, Gpr1, Gpr15, Strl33, Apj and Rdc1 (Choe *et al.*, 1996; Choe *et al.*, 1998; Combadiere *et al.*, 1998; Deng *et al.*, 1997; Doranz *et al.*, 1996; Edinger *et al.*, 1998a; Edinger *et al.*, 1998b; Farzan *et al.*, 1997; Horuk *et al.*, 1998; Jinno *et al.*, 1998; Liao *et al.*, 1997; Rucker *et al.*, 1997; Samson *et al.*, 1998; Shimizu *et al.*, 1999). However, the role of alternative coreceptors *in vivo* remains uncertain, and usage of coreceptors other than CCR5 and CXCR4

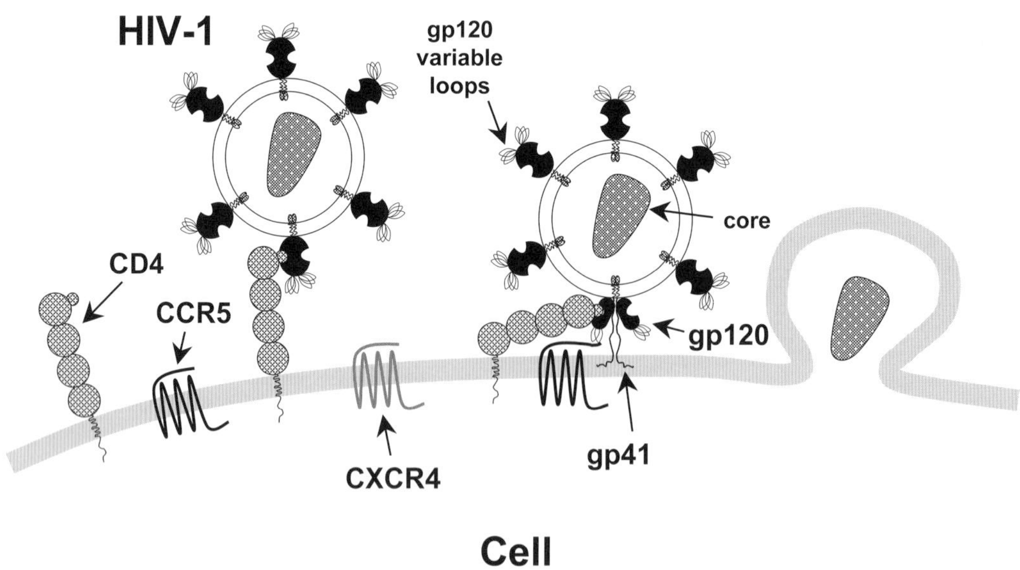


**Figure 3.1** Mechanism of HIV-1 entry. Upon binding to CD4, HIV-1 gp120 undergoes a conformational change in the variable loop regions to expose the coreceptor binding site. Binding of gp120 to a coreceptor, usually CCR5 or CXCR4, then triggers the Env to undergo further conformational changes resulting in the insertion of the fusion peptide of gp41 into the target cell membrane. This interaction brings the viral and cellular membranes together allowing fusion and entry of the virion core into the cell. Adapted from Gorry *et al.* (2004).

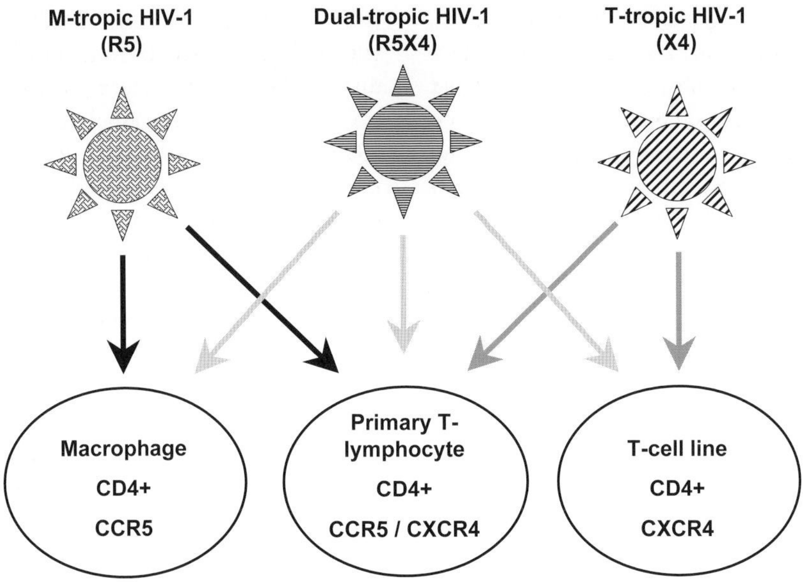


**Figure 3.2** HIV-1 coreceptors and cellular tropism. M-tropic HIV-1 strains can infect macrophages and primary T-lymphocytes, and typically use CCR5 as a coreceptor for entry. T-tropic HIV-1 strains can infect continuous T-cell lines and primary T-lymphocytes, and use CXCR4 as a coreceptor for entry. Dual tropic HIV-1 strains can infect all three cell types, and can use both CCR5 and CXCR4 as coreceptors for entry. Adapted from Gorry *et al.* (2004).

for entry into primary cells is considered rare. Although coreceptor preference by HIV-1 influences cell tropism, the degree of M-tropism is not always predicted by CCR5 usage (Cheng-Mayer *et al.*, 1997; Cunningham *et al.*, 2000; Dittmar *et al.*, 1997; Hung *et al.*, 1999). Thus, determinants that underlie cellular tropism of HIV-1 are likely to be more complex than coreceptor usage alone.

To this end, studies that specifically examined mechanisms involved in cellular tropism of HIV-1 showed that CCR5 usage of primary HIV-1 isolates is, in fact, neither necessary nor sufficient for M-tropism (Gorry *et al.*, 2001; Li *et al.*, 1999; Peters *et al.*, 2004; Peters *et al.*, 2006). Non-M-tropic R5 HIV-1 strains that are replication competent in primary CD4+ T-cells are frequently unable to infect macrophages and microglia due to a block in virus entry or an early post-entry step. The bottleneck to M-tropism by these viral strains is presently unclear, but may be related to differences in dependence on CCR5 conformation (Doms, 2000; Hill *et al.*, 1998), and/or posttranslational modifications such as sulfation (Farzan *et al.*, 1999) or O-linked glycosylation (Carlsson *et al.*, 1986; Farzan *et al.*, 1999; Fukuda *et al.*, 1986) that exist between macrophages, resting T cells, and activated T cells. Adding further evidence for the dissociation between cellular tropism and coreceptor usage *per se*, a subset of highly M-tropic primary HIV-1 strains use only CXCR4 for entry macrophage lineage cells (Gorry *et al.*, 2001). Thus, the cellular tropism of HIV-1 can be restricted at the entry level or during an early post-entry step by a mechanism independent of coreceptor specificity.

## Relationship between macrophage tropism and disease progression

There are many host and viral factors that may affect the rate of disease progression during HIV-1 infection (Buchacz *et al.*, 1998; Fauci, 1996; Mikhail *et al.*, 2003; Rowland-Jones *et al.*, 2001). However, the enhanced ability of primary HIV-1 strains to replicate in macrophage-lineage cells is associated with viral variants that emerge at late stages of disease (Li *et al.*, 1999; Tuttle *et al.*, 2002), suggesting that M-tropism contributes to HIV-1 pathogenesis (reviewed in Gorry *et al.*, 2004; Gorry *et al.*, 2005). Previous studies have shown that accelerated CD4+ T-cell depletion is frequently caused by viral variants that have aquired the ability to use CXCR4 or both CCR5 and CXCR4 for HIV-1 entry, which occurs in approximately 40–50% of infected individuals who progress to AIDS (Bjorndal *et al.*, 1997; Connor *et al.*, 1997). Since CXCR4 is widely expressed in cells within peripheral blood and lymphoid tissues, rapid depletion of CD4+ T-cells by X4 or R5X4 viral variants most likely results from increased availability of target cells. However, R5 M-tropic HIV-1 variants cause CD4+ T-cell depletion allowing progression to AIDS in most HIV-1 infected individuals (Bjorndal *et al.*, 1997; Connor *et al.*, 1997). Moreover, because it is frequently believed that CXCR4 expressed on macrophages does not support productive HIV-1 infection (Connor *et al.*, 1996; Karita *et al.*, 1997; Schuitemaker and Miedema, 1996), it is not known whether late-emerging primary X4 and R5X4 viruses are M-tropic and mediate pathogenicity via productive macrophage infection. The cell surface expression levels of CD4, CCR5 and CXCR4 are generally lower on macrophages compared to CD4+ T-cells (Lewin *et al.*, 1996; Ometto *et al.*, 1999; Wang *et al.*, 2002). The underlying molecular mechanisms responsible for the acquisition or enhancement of M-tropism by R5 viruses are unknown, but previous studies suggests that low CD4 levels restricts infection of macrophages by certain non-M-tropic R5 HIV-1 and X4 SIV strains (Bannert *et al.*, 2000; Mori *et al.*, 2000). In addition, CCR5 expression levels as well as the density of CCR5 at the cell surface can also restrict infection by R5 HIV-1 strains (Fear *et al.*, 1998; Kuhmann *et al.*, 2000; Rana *et al.*, 1997; Reynes *et al.*, 2001; Tuttle *et al.*, 1998). Together, these studies suggest that differences between the affinity of HIV-1 gp120 for CD4 and/or coreceptor may influence M-tropism of primary HIV-1 strains.

## Monocyte/macrophages are important for HIV-1 pathogenesis

Macrophages are a significant cellular reservoir that support the long term persistence of HIV-1, and are involved in the dissemination of virus to other cells (Gartner *et al.*, 1986b; Gartner and Popovic, 1990; Igarashi *et al.*, 2001) (reviewed in Gorry *et al.*, 2004; Gorry *et al.*, 2005). Tissues where persistently infected macrophages reside include the brain, lung, lymph node, spleen, bone marrow and gastrointestinal tract. Evidence suggests that CD16+ monocytes may be precursors to infected tissue macrophages, and that they are preferentially infected with HIV-1 and promote highly efficient replication of virus upon differentiation to macrophages (Ancuta *et al.*, 2003; Ancuta *et al.*, 2004; Ancuta *et al.*, 2006a; Ancuta *et al.*, 2006b; Ellery *et al.*, 2007). Macrophages are relatively long-lived cells, and do not undergo virus-induced cytopathic effects as readily as CD4+ T-cells (Crowe *et al.*, 1987; Gartner *et al.*, 1986a; Gartner *et al.*, 1986b; Ho *et al.*, 1986; Meltzer *et al.*, 1990), so they may continue to shed virus for sustained periods without undergoing cell death. Macrophages are important for regulating HIV-1 disease progression, and secrete neurotoxic toxic factors that are involved in the development of AIDS dementia (Brew *et al.*, 1995; Kerr *et al.*, 1998; Levy, 1993; Meltzer *et al.*, 1990; Meltzer and Gendelman, 1992; Pemberton *et al.*, 1997; Smith *et al.*, 2001). Furthermore, since most drugs used in highly active antiretroviral therapy (HAART) have relatively poor CNS penetrance (Pierson *et al.*, 2000b; Schrager and D'Souza, 1998), HIV-1 may be protected within macrophages and microglia in brains of patients receiving HAART. This viral compartmentalization contributes to the inability of HAART to effectively eradicate HIV-1 from the body, and may contribute to the viral rebound that occurs upon cessation of HAART (Belmonte *et al.*, 2003; Collman *et al.*, 2003; Crowe *et al.*, 2003; Lambotte *et al.*, 2000; Lambotte *et al.*, 2003; Smith *et al.*, 2003; Sonza and Crowe, 2001). In addition, previous studies using a chimeric simian-human immunodeficiency virus (SHIV) model demonstrated that tissue macrophages are the principal source of high levels of plasma virus in monkeys when CD4+ T-lymphocytes are no longer measurable in peripheral blood (Igarashi *et al.*, 2001; Igarashi *et al.*, 2003). HIV-1 infection impairs macrophage function, particularly their ability to phagocytose pathogens such as *Candida albicans, Toxoplasma gondii* and *Mycobacterium avium*, which commonly cause HIV-1-related opportunistic infections (Crowe *et al.*, 1994; Kedzierska *et al.*, 2000; Kedzierska *et al.*, 2002). Moreover, R5 HIV-1 strains, which are potentially M-tropic, persist almost exclusively at early stages of HIV-1 infection, suggesting that they may be preferentially transmitted from person to person. Together, these studies provide evidence that macrophages are important for HIV-1 transmission, persistence and pathogenicity.

## Evolution of macrophage tropic HIV-1

Early studies on primary HIV-1 infection showed that newly acquired infections are typically characterized by a relatively homogenous viral population (Pang *et al.*, 1992; Wolinsky *et al.*, 1992; Zhang *et al.*, 1993; Zhu *et al.*, 1993), suggesting that particular viral variants are transmitted (reviewed in Gorry *et al.*, 2004; Gorry *et al.*, 2005). However, later studies showed that initial infection was associated with the presence of multiple HIV-1 variants, but that most variants were sequestered at seroconversion upon the development of HIV-1 specific antibodies (Learn *et al.*, 2002; Lawson *et al.*, 2002). These studies suggest a role for the humoral immune response in the selection of viral strains during early HIV-1 infection. However, more recent studies using single genome amplification techniques showed that productive, clinical infection occurs in most cases by infection with a single viral variant (Keele *et al.*, 2008).

Studies examining viral evolution during HIV-1 disease progression after seroconversion have also yielded conflicting results (reviewed in Gorry *et al.*, 2004; Gorry *et al.*, 2005). Some reports showed that progression of HIV-1 infection is associated with a narrowing of viral diversification (Delwart *et al.*, 1994; Delwart *et al.*, 1997; Ganeshan *et al.*, 1997; Liu *et al.*, 1997; Lukashov *et al.*, 1995; Shioda *et al.*, 1997; Wolfs *et al.*, 1990; Wolinsky *et al.*, 1996), whereas other studies showed that disease progression is associated with increasing rates of viral diversity (Goodenow *et al.*, 1989; Markham *et al.*, 1998; McNearney *et al.*, 1992). It is unclear why these studies yielded discrepant results, but possibilities may include differences in the frequency of sampling, and differences in the region of the viral genome used for phylogenetic analyses. However, a later study helped reconcile the conflicting studies by analysing the evolution of HIV-1 *env* sequences during infection, over the course of 6 to 12 years in a cohort of 9 subjects, after seroconversion until the development of advanced disease (Shankarappa *et al.*, 1999). This study identified 3 distinct phases during chronic HIV-1 infection leading to development of AIDS; an early phase with increases (approximately 1% per year) in both viral divergence and diversity; an intermediate phase characterized by a further increase in viral divergence but with a stabilization or decline in viral diversity; and a late phase characterized by a stabilization of viral divergence and a continued stability or decline in viral diversity. X4 viruses frequently appeared during the transition between the early and intermediate phases. However, studies that examined virus evolution in subjects exclusively harbouring R5 viral variants to late stages of HIV-1 infection showed increases in viral diversity at late stages of infection (Li *et al.*, 1999). Thus, selection pressures involved in HIV-1 evolution during disease progression may differ between subjects harbouring R5 viral variants and those who develop X4 or R5X4 viral variants.

## Cytopathicity of macrophage tropic HIV-1

R5 HIV-1 viral variants are frequently thought to be less cytopathic than X4 viruses (reviewed in Gorry *et al.*, 2004; Gorry *et al.*, 2005), which is a belief that arose from *ex vivo* studies comparing the cytopathicity of R5 and X4 strains in human lymphoid tissue culture systems, and SCID/hu mouse studies that showed increased cytopathicity by X4 viruses (Glushakova *et al.*, 1995; Glushakova *et al.*, 1998; Picchio *et al.*, 1998). However, these studies could not explain why R5 HIV-1 variants cause CD4+ T-cell depletion in most subjects who progress to late stages of HIV-1 infection. Helping reconcile this issue, a later study showed that R5 and X4 viruses are equally cytopathic for their respective T-cell targets, but that CD4+/CCR5+ T-cells were less abundant in the *ex vivo* systems than CD4+/CXCR4+ T-cells (Grivel and Margolis, 1999). Although these studies shed light on mechanisms involved in the accelerated pathogenicity of X4 viruses, they could not explain why R5 viruses also cause significant CD4+ T-cell loss at late stages of infection if their target cells constitute a small proportion of the T-cell pool. Further studies provided some clarification. Experiments using SCID-hu Thy/Liv mice demonstrated two distinct phases of R5 virus-mediated T-cell depletion in thymus; an initial non-pathogenic phase of virus replication in medullary stromal cells, and after 3 to 4 weeks a cytopathic phase of virus replication in thymocytes (Berkowitz *et al.*, 1998; Scoggins *et al.*, 2000). This was distinct to that observed for X4 viruses, which was characterized by rapid replication in medullary and cortical thymocytes and rapid depletion. Thus, R5 and X4 strains have distinct cytopathic consequences. In support of these studies, Fais et. al., (1999) conducted experiments with SCID mice reconstituted with human peripheral blood lymphocytes and showed a significant delay of CD4+ T-cell loss when infected with R5 virus, whereas X4 viruses caused very rapid depletion of CD4+ T-cells. In this study, the permissiveness of thymocytes to R5 HIV-1 infection coincided with the emergence of an activated CD45RO+ memory phenotype, suggesting that the ability of R5 HIV-1 to deplete T-cells may depend on the activation state of CCR5+ T-cells. In support of this contention, CCR5 is mainly expressed on the surface of memory T-cells whereas CXCR4 is expressed mainly on naïve CD4+ T-cells (Bleul *et al.*, 1997). Similar results were obtained in studies using rhesus macaques infected with R5 or X4 SHIV variants (Harouse *et al.*, 1999). In these studies, R5 SHIV caused profound depletion of CD4+ intestinal T-cells, which are mainly CD45RO+ memory T-cells, and then gradual depletion of peripheral CD4+ T-cells, similar to human AIDS. In contrast, X4 SHIV cause profound depletion of peripheral CD4+ T-cells that was not evident in the intestine and not typical of human AIDS.

The molecular mechanisms contributing to R5 virus-mediated T-cell depletion are poorly understood (reviewed in Gorry *et al.*, 2004; Gorry *et al.*, 2005). However, the ability of R5 viruses to deplete CD4+ T-cells is more pronounced in viruses isolated from patients with AIDS compared to viruses isolated from the same individuals at earlier stages of infection (Scoggins *et al.*, 2000), suggesting that M-tropic R5 viruses emerging at late stages of infection have augmented cytopathicity. Furthermore, R5 HIV-1 strains isolated from subjects with AIDS are more cytotoxic to CD4+ T-cells (Kwa *et al.*, 2003) and are less inhibited by natural ligands of CCR5 (Jansson *et al.*, 1996; Jansson *et al.*, 1999; Koning *et al.*, 2003) compared to R5 HIV-1 strains isolated at earlier stages of infection. The persistence of R5 HIV-1 strains at late stages of HIV-1 infection may be promoted by the predominance of memory CD4+ T-cells in mucosal tissues (De Maria *et al.*, 1993), thus contributing to CD4+ T-cell depletion at later stages of HIV-1 infection. Together, these studies demonstrate that R5 M-tropic HIV-1 variants are intrinsically cytopathic, and have a mode of pathogenicity that is distinct from X4 HIV-1 strains.

## Envelope–receptor interactions important for macrophage tropism

Previous studies that showed low levels of cell surface CD4 or CCR5 expression on macrophages can restrict infection of certain R5 strains (Bannert *et al.*, 2000; Fear *et al.*, 1998; Kuhmann *et al.*, 2000; Mori *et al.*, 2000; Rana *et al.*, 1997; Reynes *et al.*, 2001; Tuttle *et al.*, 1998) led to further studies that investigated Env–receptor interactions between different R5 viruses (reviewed in Gorry *et al.*, 2004; Gorry *et al.*, 2005). A comparison between cytopathic and non-cytopathic R5 HIV-1 strains demonstrated increased CCR5 affinity and increased ability to scavenge low levels of cell surface expressed CD4 and CCR5 by Env clones derived from a cytopathic R5 HIV-1 strain (UK1br) that induced high levels of macrophage cytopathicity and induced neuronal apoptosis in primary brain cultures, compared to Env clones derived from a non-cytopathic R5 HIV-1 strain (MACS2br) that did not exhibit these cytopathic effects (Gorry *et al.*, 2002). These results suggest that the cytopathicity of R5 HIV-1 is associated with an enhanced ability of Env to bind to and utilize its entry receptors. Consistent with this contention, an X4 SHIV variant that was adapted for growth in macaque macrophages *in vivo* developed increased Env-mediated membrane fusing capacity, as a result of increased binding of Env to CXCR4 (Liu *et al.*, 1999). Together, these studies suggest that enhancement of M-tropism, as a result of increased Env-receptor binding, may be a mechanism by which HIV-1 viruses augment pathogenicity.

Primary HIV-1 strains with increased Env-CCR5 affinity may have an increased ability to scavenge low levels of CD4 and CCR5 for HIV-1 entry (reviewed in Gorry *et al.*, 2004; Gorry *et al.*, 2005). To this end, a number of studies have shown that primary HIV-1 isolates and Env clones with augmented CCR5 affinity can enter and/or mediate cell-to-cell fusion with cells expressing very low levels of CCR5, whereas isolates and Env clones without increased Env-CCR5 affinity have an increased requirement for higher CD4 and CCR5 levels (Gorry *et al.*, 2001; Gorry *et al.*, 2002; Martin *et al.*, 2001; Martin-Garcia *et al.*, 2005a; Martin-Garcia *et al.*, 2005b; Peters *et al.*, 2004; Thomas *et al.*, 2007). These results suggest that development or enhancement of M-tropism by R5 HIV-1 strains during HIV-1 infection may be associated with variants that have accquired increased Env-CCR5 affinity that can scavenge low levels of CD4 and CCR5 expressed on the macrophage cell surface.

In addition, other investigators have shown that SIV strains with enhanced M-tropism have an increased ability to scavenge low levels of CD4 expressed on macrophages for virus entry (Bannert *et al.*, 2000; Means *et al.*, 2001; Mori *et al.*, 2000), which was most likely due to increased Env-CD4 affinity. Thus, an increased affinity of SIV for CD4 may permit macrophages infection. Together, the data suggest that late emerging viral variants with increased ability to scavenge low levels of CD4 and/or CCR5 for virus entry and increased CD4 and/or CCR5 affinity may be important for HIV-1 disease progression, especially in subjects exclusively harbouring R5 viral variants.

The mechanisms underlying increased CCR5 affinity of R5 viruses are presently unclear (reviewed in Gorry *et al.*, 2004; Gorry *et al.*, 2005). However, these viral strains have increased *in vitro* replication competence (Blaak *et al.*, 1998), decreased sensitivity to inhibition by natural ligands of CCR5 (Koning *et al.*, 2003), increased Env-mediated fusogenicity (Gorry *et al.*, 2002) and kinetics of virus-cell fusion (Reeves *et al.*, 2002; Sterjovski *et al.*, 2007), increased ability to cause CD4+ T-cell cytopathicity (LaBonte *et al.*, 2000), and increased ability to cause bystander cell apoptosis (Holm *et al.*, 2004) compared to R5 HIV-1 variants that do not have enhanced affinity for CCR5. Selection pressure imposed by β-chemokines may contribute to the development of these phenotypes. For example, although selection by β-chemokines has been associated with development of X4 or R5X4 viral variants (Jansson *et al.*, 1996; Scarlatti *et al.*, 1997), it may also select for R5 variants that are less sensitive to their inhibition, by acquiring an increased ability to compete for CCR5 binding, possibly due to increased or more appropriate exposure of the gp120 CCR5 binding domain. Such conformational alterations may contribute to the enhanced ability of R5 HIV-1 strains to replicate in macrophages and enhanced cytopathicity.

The regions in gp120 that are associated with enhanced CCR5 affinity are probably strain specific and/or context dependent. However, single amino acid changes in the gp120 bridging sheet, which forms an important structural component of the CCR5 binding domain after gp120 has interacted with CD4, have been shown to significantly alter CCR5 affinity (Rizzuto *et al.*, 1998). In addition, alterations in N-linked glycosylation patterns that better expose the coreceptor binding site may also allow enhanced CCR5 binding (Dehghani *et al.*, 2003; Gorry *et al.*, 2002; Hoffman *et al.*, 1999; Kolchinsky *et al.*, 1999; Kolchinsky *et al.*, 2001).

## Env determinants in HIV-1 contributing to macrophage tropism

Whilst a number of studies have investigated mechanisms contributing to pathogenicity of X4 and R5X4 viruses that arise during late stages of HIV-1 infection (Glushakova *et al.*, 1995; Glushakova *et al.*, 1998; Picchio *et al.*, 1998), the molecular mechanisms responsible for pathogenicity of M-tropic R5 HIV-1 strains are incompletely understood (Gorry *et al.*, 2005; Gorry *et al.*, 2005). In one study (Repits *et al.*, 2008), a shift towards a more positive net charge of gp120 was associated with late-emerging R5 virus and was localized to the V2 and V4 variable loop regions of gp120. A more positive charge resulted in enhanced cell attachment in a receptor-independent mechanism.

In another recent study, changes in the R5 Env occurring over the course of HIV-1 infection included changes in the length of variable loops and position of glycans, as well as single amino acids across conserved and variable regions of gp120 and gp41 which were driven by changes in the neutralizing humoral response of the individual (Bunnik *et al.*, 2008). However, many of these changes were virus-specific and further studies are required to identify common changes associated with disease progression across a range of patient cohorts.

While the exact molecular mechanisms that lead to the increased pathogenicity of late stage M-tropic R5 HIV-1 are not well defined, valuable insights can be gained from studies that examine the emergence of high affinity R5 variants in immune privileged sites. Previous studies have shown that neurovirulent R5 HIV-1 strains isolated from the brain have greater CCR5 affinity and reduced dependence on the levels of CCR5 or CD4 for virus entry (Gorry *et al.*, 2002; Martin-Garcia *et al.*, 2006; Peters *et al.*, 2004). Together, these findings may explain why R5 variants with greater CCR5 affinity, reduced CCR5/CD4 dependence, faster fusion kinetics and greater susceptibility to neutralizing antibodies arise during later stages of infection, when reduced selection pressure due to a weakening immune system allows HIV-1 to adopt a more open Env conformation.

The immune response against HIV-1 is continually evolving over the course of infection, which in turn forces HIV-1 to adapt in order to escape neutralization (Richman *et al.*, 2003). A study that examined the emergence of escape virus in response to neutralizing antibodies reported that neutralization escape mutations occurred predominantly in N-linked glycosylation sites, giving rise to the 'evolving glycan shield' model of HIV-1 escape from immune selection pressures (Wei *et al.*, 2003). Changes in the glycosylation pattern led to changes in glycan packing that allowed neutralizing antibody escape without inhibiting interactions with entry receptors.

Late emerging M-tropic R5 viruses have demonstrated increased sensitivity to neutralizing antibodies, particularly those that target the CD4 binding site (Gray *et al.*, 2005; Peters *et al.*, 2008; Repits *et al.*, 2005). During earlier stages of HIV-1 infection, a strong immune response leads to generation of antibodies against many of the HIV-1 proteins, including Env (Burton *et al.*, 2005; Kimata, 2006; Pantophlet and Burton, 2006; Richman *et al.*, 2003). In order to escape the immune response, the Env proteins need to exist in a conformation or have a glycan shield that protects key receptor binding sites and other neutralizing epitopes from neutralizing antibodies (Parren *et al.*, 1999). *In vitro* studies have demonstrated that the virus can afford to adopt a less defensive Env conformation in the absence of selection pressure from neutralizing antibodies, such that receptor binding sites are more readily accessible and fusion can occur at a more rapid pace (Moore *et al.*, 1995; Moore and Ho, 1995; Moore and Doms, 2003). Late-emerging R5 viral variants that have increased sensitivity to antibody neutralization may be able to replicate and persist *in vivo* during late stages of HIV-1 infection when the humoral immune response is waning (Gorry *et al.*, 2005). Reduced selection pressure from antibodies may allow the evolution of R5 variants with increased CD4/CCR5 interactions, similar to that which occurs in immune privileged anatomical sites such as the brain (Gorry *et al.*, 2001; Gorry *et al.*, 2002; Martin *et al.*, 2001; Peters *et al.*, 2004; Shieh *et al.*, 2000; Strizki *et al.*, 1996). This is consistent with the long held view that protection of HIV-1 from neutralizing antibodies carries the price of a less efficient interaction with entry receptors (Moore and Ho, 1995). Recently, this conjecture was further supported by a study that showed that changes in the quality and magnitude of the neutralizing antibody response over the course of HIV-1 infection exerted significant selection pressure on R5 HIV-1 Env glycoproteins and coincided with changes in gp160 length and number of glycosylation sites (Bunnik *et al.*, 2008). Together, these studies highlight a significant role for the changing host immune response in the evolution of the late emerging, M-tropic R5 HIV-1 phenotype.

The apparent increase in resistance to inhibition by natural CCR5 ligands that is associated with late emerging M-tropic R5 HIV-1 suggests that ligand competition may also influence R5 HIV-1 evolution over the course of infection. The HIV-1 gp120 protein and natural CC chemokine ligands are mutually competitive for CCR5 binding (Hill *et al.*, 1997; Trkola *et al.*, 1996; Wu *et al.*, 1996). Compared with other natural CCR5 ligands, RANTES demonstrates the most potent and broad inhibition of HIV-1 infection (Trkola *et al.*, 1998), despite almost 10-fold less affinity for CCR5 compared to another CCR5 ligand, MIP-1α (Aramori *et al.*, 1997). This suggests that a RANTES specific mechanism may be responsible for ligand-mediated selection pressures against HIV-1 gp120. RANTES is also most effective at down-modulating CCR5 expression compared with other CCR5 ligands (Trkola *et al.*, 1998). Given that CCR5 down-modulation and direct blocking via CCR5 ligands is effective in preventing HIV-1 infection (Alkhatib *et al.*, 1997; Trkola *et al.*, 1998), there may be a selective advantage for R5 HIV-1 strains to evolve alternative mechanisms of engaging CCR5 that out compete RANTES binding. Therefore, ligand competition at the CCR5 receptor and/or changes in the cytokine balance over the course of infection may also contribute to the emergence of M-tropic R5 HIV-1 strains.

## Host factors influencing M-tropism of HIV-1 and disease progression

Host genetic factors have been associated with delayed onset of AIDS and prolonged survival, as well as increased susceptibility to HIV-1 infection. Polymorphisms in chemokines and chemokine receptors associated with changes in the susceptibility of HIV-1 infection and disease progression include chemokine receptors CCR5 and CCR2, the CXCR4 ligand SDF1, the leucocyte chemotactic and adhesion receptor $CX_3CR1$, and the CCR5 ligand RANTES (reviewed in Reiche *et al.*, 2007). One of the earliest and most widely studied polymorphism to be associated with susceptibility and severity of HIV-1 infection is the CCR5 Δ32 polymorphism. The CCR5 Δ32 allele contains a 32-nucleotide deletion in the open reading frame of the gene that results in a truncated, non-functional form of the protein (Liu *et al.*, 1996). Homozygosity at the CCR5 Δ32 allele is most common in Caucasians and is observed in ~1% of the HIV-1 uninfected population, while heterozygosity is found in ~15 to 20% (Huang *et al.*, 1996; Samson *et al.*, 1996; Zimmerman *et al.*, 1997). Individuals homozygous for the CCR5 Δ32 allele are highly resistant to HIV-1 transmission apart from rare cases (Gray *et al.*, 2006; Kuipers *et al.*, 1999; Liu *et al.*, 1996; Michael *et al.*, 1998; Naif *et al.*, 2002; Samson *et al.*, 1996), whereas heterozygotes are susceptible but typically have delayed onset of AIDS compared to individuals homozygous for the wild-type allele (de Roda Husman *et al.*, 1997; Dean *et al.*, 1996; Michael *et al.*, 1997; Zimmerman *et al.*, 1997). The level of expression of CCR5 on CD4+ T-cells and macrophages from CCR5 Δ32 homozygous and heterozygous donors restricts replication of HIV-1 *in vitro* and *in vivo* (Benkirane *et al.*, 1997; Blaak *et al.*, 2000; Picchio *et al.*, 1998; Venkatesan *et al.*, 2002; Wu *et al.*, 1997). Reduced levels of CCR5 expression in heterozygous individuals may promote the persistence of R5 viruses with enhanced M-tropism, which have reduced CCR5-dependence (Gorry *et al.*, 2007). Polymorphisms in the CCR5 gene promoter have also been associated with reduced susceptibility to infection with HIV-1 and delayed progression to AIDS (Ometto *et al.*, 2001; Yang *et al.*, 2003). The protection from HIV-1 infection conferred by CCR5 polymorphisms and the absence of deleterious effects in the individual has made the CCR5 protein an important target in HIV-1 chemotherapy.

Polymorphisms of the human major histocompatibility complex (MHC) have also been widely studied for their association with susceptibility and severity of HIV infection. The MHC maps to chromosome six and contains the highly variable HLA class I and class II genes, the products of which are fundamental to the specificity and magnitude of the acquired immune response (Carrington and O'Brien, 2003). The MHC class I HLA loci, HLA-A, HLA-B and HLA-C, encode molecules that bind and present primarily intracellular antigens to CD8+ T lymphocytes and regulate the cytotoxic T-lymphocyte (CTL) response. Gene products from class II loci, HLA-DR, HLA-DQ and HLA-DP, bind and present primarily extracellular antigens to CD4+ T-cells and help trigger CD4+ helper T-cell activation of the antibody immune response. Both MHC class I and class II polymorphisms have been associated with susceptibility to HIV-1 infection and the rate of disease progression (Anastassopoulou and Kostrikis, 2003; Carrington and O'Brien,

2003; Itescu *et al.*, 1994; Lockett *et al.*, 2001; MacDonald *et al.*, 2000; MacDonald *et al.*, 2001; Malhotra *et al.*, 2001; Tiwari and Terasaki, 1981).

Class I homozygosity is associated with rapid disease progression to AIDS (Carrington *et al.*, 1999; Tang *et al.*, 1999), which may be due to a more restricted range of peptide recognition that limits the immune response to HIV-1. Specific class I alleles, such as HLA-B*27 and HLA-B*57, have been associated with delayed progression to AIDS (Gao *et al.*, 2001; Hendel *et al.*, 1999; Kaslow *et al.*, 1996; Magierowska *et al.*, 1999; Migueles *et al.*, 2000), and have also been associated with an immunodominant response against HIV-1 Gag and RT epitopes (Kelleher *et al.*, 2001; Klein *et al.*, 1998; Nietfield *et al.*, 1995). Other HLA types correlate with increased risk of disease progression to AIDS, such as HLA-B*35 and HLA-B*08 (Carrington *et al.*, 1999; Gao *et al.*, 2001; Itescu *et al.*, 1992; Klein *et al.*, 1994; Sahmoud *et al.*, 1993; Scorza Smeraldi *et al.*, 1988). Combinations of HLA alleles, or haplotypes, can also contribute to disease progression in a co-operative manner. In general, combinations of HLA-B molecules with Bw4 epitopes confer protective effects against disease progression to AIDS by regulating natural killer (NK) cell response to HIV-1 infection through the KIR3DL1 NK cell receptor, whereas HLA-B molecules with Bw6 specificity are associated with more rapid progression to AIDS (Carrington and O'Brien, 2003; Gaudieri *et al.*, 2005; Gumperz *et al.*, 1995).

Together, these studies demonstrate that while the Env glycoproteins contribute significantly to the pathogenic effects of HIV-1 infection via enhanced M-tropism, syncytia formation and activation of cell death signals, a multitude of other gene products and host cells factors contribute to the overall pathogenesis of HIV-1 and disease progression.

## Implications for HIV-1 entry inhibitors

Development of HIV-1 entry inhibitors represents a new era in the treatment of HIV-1 infection. While many compounds are still in the developmental stage, others such as the CCR5 antagonist maraviroc and the T-20 fusion inhibitor have been licensed for clinical use (Dolin, 2008; Fatkenheuer *et al.*, 2008; Lalezari *et al.*, 2003; Lazzarin *et al.*, 2003; Qian *et al.*, 2008). However, there are several challenges currently facing the development of entry inhibitors; these include oral availability, toxicity and the highly variable and evolving nature of the M-tropic HIV-1 Env proteins. Perhaps one of the greatest challenges confronting the successful use of entry inhibitors in the treatment of HIV-1 is the emergence of M-tropic R5 variants during later stages of disease that have demonstrated reduced sensitivity to current co-receptor and fusion inhibitors in development or clinical use (reviewed in Sterjovski *et al.*, 2006). These findings may impact significantly on the therapeutic use of entry inhibitors.

While late-emerging M-tropic HIV-1 variants are associated with reduced sensitivity to some entry inhibitors, they have also been associated with greater sensitivity to inhibition by neutralizing antibodies. Selection of HIV-1 variants due to immunological pressure is thought to underlie alterations in the mode or efficiency of coreceptor usage during later stages of HIV-1 infection (Bunnik *et al.*, 2008; Gray *et al.*, 2005). This, in turn, may affect sensitivity of HIV-1 to inhibition by natural ligands of CCR5, and may also have considerable impact on the efficacy of newly developed HIV-1 entry inhibitors (Jansson *et al.*, 1999; Karlsson *et al.*, 2004; Scarlatti *et al.*, 1997) (reviewed in Sterjovski *et al.*, 2006).

Evolution of pathogenic M-tropic R5 HIV-1 variants involves an increase in the efficiency of HIV-1 entry. Previous studies have shown that R5 viruses isolated from AIDS patients have an increased ability to scavenge low CCR5 levels for HIV-1 entry compared to R5 viruses isolated from early stages of HIV-1 infection, suggesting increased Env–CCR5 affinity (Gorry *et al.*, 2002; Gray *et al.*, 2005), which has also been linked to increased kinetics of virus-cell fusion (Reeves *et al.*, 2002). Importantly, changes in Env-coreceptor affinity may potentially affect sensitivity to inhibition by a number of different HIV-1 entry inhibitors; for example, recent studies have reported reductions in the sensitivity of late emerging R5 HIV-1 variants to inhibition by CCR5 inhibitors as well as the fusion inhibitor T-20 (Gray *et al.*, 2005; Reeves *et al.*, 2002; Repits *et al.*, 2005). This may be a result of increased competition between gp120 and the CCR5 inhibitor, or by increased kinetics of virus-cell fusion which could potentially reduce the opportunity for fusion inhibitors to act. Together, these studies underscore the importance of inhibiting late-emerging M-tropic R5 HIV-1 viral variants in the design and clinical use of HIV-1 entry inhibitors.

Cellular and anatomical HIV-1 reservoirs subvert virus eradication by re-seeding the body with unique HIV-1 variants (Saksena and Smit, 2005; Schrager and D'Souza, 1998; Sonza and Crowe, 2001) (reviewed in Sterjovski *et al.*, 2006). For example, viral variants within the CNS evolve in a unique manner due to compartmentalization in an environment that is not subjected to normal immune selection pressures (Churchill *et al.*, 2004; Ohagen *et al.*, 2003; Thompson *et al.*, 2004), and frequently results in enhanced M-tropism (reviewed in Gorry *et al.*, 2005). As discussed above, late-emerging R5 HIV-1 strains show enhanced tropism for cells of macrophage lineage and may contribute to the viral load in cellular reservoirs of HIV-1. Thus, compartmentalization and emergence of viral variants from tissue reservoirs contributes to genetic diversification of the viral quasispecies, which probably imposes a significant challenge to the success of antiretroviral therapy. R5 viruses may also be selected in individuals undergoing treatment with a CXCR4 inhibitor (Este *et al.*, 1999). While stringent screening of the viral phenotype of patients potentially undergoing CXCR4 inhibitor treatment (should CXCR4 inhibitors eventually become clinically useful) will identify those who will benefit most from treatment, it does not eliminate the possibility of selecting R5 virus harboured in tissue reservoirs (Crowe and Sonza, 2000; Pierson *et al.*, 2000a; Pierson *et al.*, 2000b; Schrager and D'Souza, 1998; Sonza and Crowe, 2001).

## Implications for HIV-1 vaccine design

While the exact structure of the late emerging, high affinity and neutralization sensitive M-tropic Env is unknown, the hypothesized open conformation has therapeutic potential as a possible target against late emerging M-tropic R5 variants. One of the major challenges of HIV-1 vaccine design is associated with high diversity of the HIV-1 *env* gene. Variation of the *env* genes both within and between patients, as well as between different subtypes of HIV-1 may prevent the development of a vaccine that results in broadly neutralizing antibodies across all HIV-1 strains (Zwick and Burton, 2007). This obstacle can be circumvented by targeting highly conserved epitopes of the Env protein that are involved in key functional mechanisms, such as receptor binding and creation of the fusion complex. Peripheral blood and CNS-derived M-tropic R5 HIV-1 strains display increased sensitivity to neutralizing antibodies that target the CD4 binding site of gp120 (Gray *et al.*, 2005; Peters *et al.*, 2008), suggesting that they have evolved Env conformations with constitutively greater exposure of the CD4 binding site. This has significance in the development of both therapeutic and preventative vaccines.

The emergence of HIV-1 variants at later stages of disease that are sensitive to neutralization by antibodies directed against the CD4 binding site suggests that anti-CD4 binding site antibodies are not present at the time of their emergence. Thus, using M-tropic R5 HIV-1 Envs as vaccine candidates may lead the way to development of therapeutic vaccines that have the potential to avert the emergence of pathogenic R5 HIV-1 strains later in infection and slow, and possibly prevent, the progression to AIDS in the majority of HIV-1 infected individuals. However, anti-HIV-1 neutralizing antibodies rarely, if ever, impose sufficient selection pressure to prevent viral escape from occurring altogether (Poignard *et al.*, 1999; Quakkelaar *et al.*, 2007; Zwick and Burton, 2007). Thus, a vaccine candidate must induce sufficiently high titres of broadly reactive neutralizing antibodies to make a significant impact on viral load and disease progression.

Investigation of late emerging, M-tropic R5 Envs as potential vaccine candidates may also have significant implications in the development of vaccines designed to prevent potential transmission of HIV-1 and dissemination of infection in the new host. A diverse population of HIV-1 strains, referred to as a 'viral swarm', is transmitted to the new host and begins to replicate shortly after transmission (Learn *et al.*, 2002). Selection of viruses best suited to replication in the context of the early immune response to HIV-1 infection leads to a rapidly purifying selection and homogenization of the viral population (Learn *et al.*, 2002; Long *et al.*, 2002; Mosier, 2005; Ritola et al., 2004; Shankarappa et al., 1999). Tissue-specific selection is also thought to contribute to early selection pressures on viral strains (van't Wout et al., 1998). R5 HIV-1 strains are more efficient at crossing epithelial barriers since activated CD4+ T-cells that express CCR5 are the most permissive target cell at mucosal sites of transmission and lymph nodes (Carreno et al., 2002; Hladik et al., 1999; Meng et al., 2002; Schacker et al., 2001; Zhang et al., 1999). Early R5 selection variants have shorter V2 regions and fewer potential N-linked glycosylation sites (PNGS) compared with R5 viruses found during chronic infection (Derdeyn et al., 2004; Fouchier et al., 1995; Nabatov et al., 2004; Sagar et al., 2006). Early R5 transmission variants with shorter V1/V2 loops and less PNGS are also more easily neutralized by antibodies and have a greater propensity to utilize low CD4 concentrations on target cells to mediate entry compared to R5 HIV-1 isolated from chronic stages of HIV-1 infection (Sagar et al., 2006). Thus, early transmission R5 variants have similar genetic and functional features as late stage AIDS R5 strains, such as shorter V2 loops, less PNGS and greater sensitivity to neutralization by antibodies. The reversion to increased neutralization sensitivity of R5 HIV-1 at later stages of infection is likely to be a result of the evolving humoral immune response to HIV-1 infection (Bunnik et al., 2008; Wei et al., 2003; Wyatt et al., 1998). In a study by Derdeyn et al. (2004), recipient viruses were found to be highly sensitive to neutralization by sera from the transmitting partner, suggesting that exposure of neutralizing epitopes that are lost during adaptive evolution are favoured during transmission of virus to a new host. M-tropic R5 HIV-1 with increased sensitivity to antibody neutralization may evolve during later stages of infection when the immune response is less effective or robust and the viral entry proteins can afford to adopt similar conformations to those that were advantageous during transmission (Bunnik et al., 2008; Gray et al., 2005; Peters et al., 2008). Thus, investigation of M-tropic R5 Env as potential vaccine candidates could potentially provide an opportunity to circumvent or slow the initial transmission of HIV-1 and the emergence of pathogenic R5 strains during later stages of infection.

## Summary

M-tropism of HIV-1 results from a multitude of factors that permit efficient virus entry into macrophages, not solely the coreceptor specificity of the virus, and is an important viral determinant contributing to pathogenesis of HIV-1. M-tropism of HIV-1 strains occurs more frequently at late stages of infection compared to earlier stages, which, in part is due to an increased ability of the virus to interact with relatively low levels of CD4 and CCR5 expressed on the macrophage cell surface. Thus, alterations in the affinity of HIV-1 gp120 for its entry receptors during HIV-1 infection may underpin the acquisition or enhancement of M-tropism by R5 viruses. Importantly, M-tropic R5 viral strains appear to have reduced sensitivity to HIV-1 entry inhibitors, and an increased ability to cause CD4+ T-cell loss. A better understanding of late emerging, M-tropic R5 HIV-1 strains will advance our knowledge of HIV-1 pathogenesis as well as the development of new therapeutics to treat HIV-1 infection.

## Acknowledgements

PRG is the recipient of an Australian National Health and Medical Research Council (NHMRC) R. Douglas Wright Biomedical Career Development Award.

## References

Alkhatib, G., Combadiere, C., Broder, C.C., Feng, Y., Kennedy, P.E., Murphy, P.M., and Berger, E.A. (1996). CC CKR5: a RANTES, MIP-1alpha, MIP-1beta receptor as a fusion cofactor for macrophage-tropic HIV-1. Science *272*, 1955–1958.

Alkhatib, G., Locati, M., Kennedy, P.E., Murphy, P.M., and Berger, E.A. (1997). HIV-1 coreceptor activity of CCR5 and its inhibition by chemokines: independence from G protein signaling and importance of coreceptor downmodulation. Virology *234*, 340–348.

Anastassopoulou, C.G., and Kostrikis, L.G. (2003). The impact of human allelic variation on HIV-1 disease. Curr. HIV Res. *1*, 185–203.

Ancuta, P., Rao, R., Moses, A., Mehle, A., Shaw, S.K., Luscinskas, F.W., and Gabuzda, D. (2003). Fractalkine preferentially mediates arrest and migration of CD16+ monocytes. J. Exp. Med. *197*, 1701–1707.

Ancuta, P., Moses, A., and Gabuzda, D. (2004). Transendothelial migration of CD16+ monocytes in response to fractalkine under constitutive and inflammatory conditions. Immunobiology *209*, 11–20.

Ancuta, P., Autissier, P., Wurcel, A., Zaman, T., Stone, D., and Gabuzda, D. (2006a). CD16+ monocyte-derived macrophages activate resting T cells for HIV infection by producing CCR3 and CCR4 ligands. J. Immunol. *176*, 5760–5771.

Ancuta, P., Kunstman, K.J., Autissier, P., Zaman, T., Stone, D., Wolinsky, S.M., and Gabuzda, D. (2006b). CD16+ monocytes exposed to HIV promote highly efficient viral replication upon differentiation into macrophages and interaction with T cells. Virology *344*, 267–276.

Aramori, I., Ferguson, S.S., Bieniasz, P.D., Zhang, J., Cullen, B., and Cullen, M.G. (1997). Molecular mechanism of desensitization of the chemokine receptor CCR-5: receptor signaling and internalization are dissociable from its role as an HIV-1 co-receptor. EMBO J. *16*, 4606–4616.

Bannert, N., Schenten, D., Craig, S., and Sodroski, J. (2000). The level of CD4 expression limits infection of primary rhesus monkey macrophages by a T-tropic simian immunodeficiency virus and macrophage-tropic human immunodeficiency viruses. J. Virol. *74*, 10984–10993.

Barre-Sinoussi, F., Chermann, J.C., Rey, F., Nugeyre, M.T., Chamaret, S., Gruest, J., Dauguet, C., Axler-Blin, C., Vezinet-Brun, F., Rouzioux, C., *et al.* (1983). Isolation of a T-lymphotropic retrovirus from a patient at risk for acquired immune deficiency syndrome (AIDS). Science *220*, 868–871.

Belmonte, L., Bare, P., de Bracco, M.M., and Ruibal-Ares, B.H. (2003). Reservoirs of HIV replication after successful combined antiretroviral treatment. Curr. Med. Chem. *10*, 303–312.

Benkirane, M., Jin, D.Y., Chun, R.F., Koup, R.A., and Jeang, K.T. (1997). Mechanism of transdominant inhibition of CCR5-mediated HIV-1 infection by ccr5delta32. J. Biol. Chem. *272*, 30603–30606.

Berkowitz, R.D., Alexander, S., Bare, C., Linquist-Stepps, V., Bogan, M., Moreno, M.E., Gibson, L., Wieder, E.D., Kosek, J., Stoddart, C.A., *et al.* (1998). CCR5- and CXCR4-utilizing strains of human immunodeficiency virus type 1 exhibit differential tropism and pathogenesis *in vivo*. J. Virol. *72*, 10108–10117.

Bjorndal, A., Deng, H., Jansson, M., Fiore, J.R., Colognesi, C., Karlsson, A., Albert, J., Scarlatti, G., Littman, D.R., and Fenyo, E.M. (1997). Coreceptor usage of primary human immunodeficiency virus type 1 isolates varies according to biological phenotype. J. Virol. *71*, 7478–7487.

Blaak, H., Brouwer, M., Ran, L.J., de Wolf, F., and Schuitemaker, H. (1998). *In vitro* replication kinetics of human immunodeficiency virus type 1 (HIV-1) variants in relation to virus load in long-term survivors of HIV-1 infection. J. Infect. Dis. *177*, 600–610.

Blaak, H., Ran, L.J., Rientsma, R., and Schuitemaker, H. (2000). Susceptibility of *in vitro* stimulated PBMC to infection with NSI HIV-1 is associated with levels of CCR5 expression and beta-chemokine production. Virology *267*, 237–246.

Bleul, C.C., Wu, L., Hoxie, J.A., Springer, T.A., and Mackay, C.R. (1997). The HIV coreceptors CXCR4 and CCR5 are differentially expressed and regulated on human T lymphocytes. Procedings of the National Academy of Sciences USA *94*, 1925–1930.

Brenchley, J.M., Price, D.A., Schacker, T.W., Asher, T.E., Silvestri, G., Rao, S., Kazzaz, Z., Bornstein, E., Lambotte, O., Altmann, D., *et al.* (2006). Microbial translocation is a cause of systemic immune activation in chronic HIV infection. Nat. Med. *12*, 1365–1371.

Brenchley, J.M., and Douek, D.C. (2008). HIV infection and the gastrointestinal immune system. Mucosal Immunol. *1*, 23–30.

Brenchley, J.M., Paiardini, M., Knox, K.S., Asher, A.I., Cervasi, B., Asher, T.E., Scheinberg, P., Price, D.A., Hage, C.A., Kholi, L.M., *et al.* (2008). Differential Th17 CD4 T-cell depletion in pathogenic and nonpathogenic lentiviral infections. Blood *112*, 2826–2835.

Brew, B.J., Corbeil, J., Pemberton, L., Evans, L., Saito, K., Penny, R., Cooper, D.A., and Heyes, M.P. (1995). Quinolinic acid production is related to macrophage tropic isolates of HIV-1. Journal of Neurovirology *1*, 369–374.

Buchacz, K.A., Wilkinson, D.A., Krowka, J.F., Koup, R.A., and Padian, N.S. (1998). Genetic and immunological host factors associated with susceptibility to HIV-1 infection. Aids *12 Suppl A*, S87–94.

Bunnik, E.M., Pisas, L., van Nuenen, A.C., and Schuitemaker, H. (2008). Autologous neutralizing humoral immunity and evolution of the viral envelope in the course of subtype B human immunodeficiency virus type 1 infection. J. Virol. *82*, 7932–7941.

Burton, D.R., Stanfield, R.L., and Wilson, I.A. (2005). Antibody vs. HIV in a clash of evolutionary titans. Proc. Natl. Acad. Sci. U.S.A. *102*, 14943–14948.

Carlsson, S.R., Sasaki, H., and Fukuda, M. (1986). Structural variations of O-linked oligosaccharides present in leukosialin isolated from erythroid, myeloid, and T-lymphoid cell lines. J. Biol. Chem. *261*, 12787–12795.

Carreno, M.P., Krieff, C., Irinopoulou, T., Kazatchkine, M.D., and Belec, L. (2002). Enhanced transcytosis of R5-tropic human immunodeficiency virus across tight monolayer of polarized human endometrial cells under pro-inflammatory conditions. Cytokine *20*, 289–294.

Carrington, M., Nelson, G.W., Martin, M.P., Kissner, T., Vlahov, D., Goedert, J.J., Kaslow, R., Buchbinder, S., Hoots, K., and O'Brien, S.J. (1999). HLA and HIV-1: heterozygote advantage and B*35-Cw*04 disadvantage. Science *283*, 1748–1752.

Carrington, M., and O'Brien, S.J. (2003). The influence of HLA genotype on AIDS. Annu Rev Med *54*, 535–551.

Chakrabarti, L.A., Lewin, S.R., Zhang, L., Gettie, A., Luckay, A., Martin, L.N., Skulsky, E., Ho, D.D., Cheng-Mayer, C., and Marx, P.A. (2000). Normal T-cell turnover in sooty mangabeys harboring active simian immunodeficiency virus infection. J. Virol. *74*, 1209–1223.

Chan, D.C., Fass, D., Berger, J.M., and Kim, P.S. (1997). Core structure of gp41 from the HIV envelope glycoprotein. Cell *89*, 263–273.

Cheng-Mayer, C., Liu, R., Landau, N.R., and Stamatatos, L. (1997). Macrophage tropism of human immunodeficiency virus type 1 and utilization of the CC-CKR5 coreceptor. J. Virol. *71*, 1657–1661.

Choe, H., Farzan, M., Sun, Y., Sullivan, N., Rollins, B., Ponath, P.D., Wu, L., Mackay, C.R., LaRosa, G.,

Newman, W., *et al.* (1996). The beta-chemokine receptors CCR3 and CCR5 facilitate infection by primary HIV-1 isolates. Cell *85*, 1135–1148.

Choe, H., Farzan, M., Konkel, M., Martin, K., Sun, Y., Marcon, L., Cayabyab, M., Berman, M., Dorf, M.E., Gerard, N., *et al.* (1998). The orphan seven-transmembrane receptor apj supports the entry of primary T-cell-line-tropic and dualtropic human immunodeficiency virus type 1. J. Virol. *72*, 6113–6118.

Churchill, M., Sterjovski, J., Gray, L., Cowley, D., Chatfield, C., Learmont, J., Sullivan, J.S., Crowe, S.M., Mills, J., Brew, B.J., *et al.* (2004). Longitudinal analysis of nef/long terminal repeat-deleted HIV-1 in blood and cerebrospinal fluid of a long-term survivor who developed HIV-associated dementia. J. Infect. Dis. *190*, 2181–2186.

Collman, R., Balliet, J.W., Gregory, S.A., Friedman, H., Kolson, D.L., Nathanson, N., and Srinivasan, A. (1992). An infectious molecular clone of an unusual macrophage-tropic and highly cytopathic strain of human immunodeficiency virus type 1. J. Virol. *66*, 7517–7521.

Collman, R.G., Perno, C.F., Crowe, S.M., Stevenson, M., and Montaner, L.J. (2003). HIV and cells of macrophage/dendritic lineage and other non-T cell reservoirs: new answers yield new questions. J. Leukoc. Biol. *74*, 631–634.

Combadiere, C., Salzwedel, K., Smith, E.D., Tiffany, H.L., Berger, E.A., and Murphy, P.M. (1998). Identification of CX3CR1. A chemotactic receptor for the human CX3C chemokine fractalkine and a fusion coreceptor for HIV-1. Journal of Biological Chemistry *273*, 23799–23804.

Connor, R.I., Paxton, W.A., Sheridan, K.E., and Koup, R.A. (1996). Macrophages and CD4+ T lymphocytes from two multiply exposed, uninfected individuals resist infection with primary non-syncytium-inducing isolates of human immunodeficiency virus type 1. J. Virol. *70*, 8758–8764.

Connor, R.I., Sheridan, K.E., Ceradini, D., Choe, S., and Landau, N.R. (1997). Change in coreceptor use coreceptor use correlates with disease progression in HIV-1--infected individuals. J. Exp. Med. *185*, 621–628.

Crowe, S.M., Mills, J., and McGrath, M.S. (1987). Quantitative immunocytofluorographic analysis of CD4 surface antigen expression and HIV infection of human peripheral blood monocyte/macrophages. AIDS Res. Hum. Retroviruses 3, 135–145.

Crowe, S.M., Vardaxis, N.J., Kent, S.J., Maerz, A.L., Hewish, M.J., McGrath, M.S., and Mills, J. (1994). HIV infection of monocyte-derived macrophages *in vitro* reduces phagocytosis of Candida albicans. J. Leukoc. Biol. *56*, 318–327.

Crowe, S.M., and Sonza, S. (2000). HIV-1 can be recovered from a variety of cells including peripheral blood monocytes of patients receiving highly active antiretroviral therapy: a further obstacle to eradication. Journal of Leukocyte Biology *68*, 345–350.

Crowe, S., Zhu, T., and Muller, W.A. (2003). The contribution of monocyte infection and trafficking to viral persistence, and maintenance of the viral reservoir in HIV infection. J. Leukoc. Biol. *74*, 635–641.

Cunningham, A.L., Li, S., Juarez, J., Lynch, G., Alali, M., and Naif, H. (2000). The level of HIV infection of macrophages is determined by interaction of viral and host cell genotypes. Journal of Leukocyte Biology *68*, 311–317.

De Maria, R., Fais, S., Silvestri, M., Frati, L., Pallone, F., Santoni, A., and Testi, R. (1993). Continuous *in vivo* activation and transient hyporesponsiveness to TcR/CD3 triggering of human gut lamina propria lymphocytes. European Journal of Immunology *23*, 3104–3108.

de Roda Husman, A.M., Koot, M., Cornelissen, M., Keet, I.P., Brouwer, M., Broersen, S.M., Bakker, M., Roos, M.T., Prins, M., de Wolf, F., *et al.* (1997). Association between CCR5 genotype and the clinical course of HIV-1 infection. Ann. Intern. Med. *127*, 882–890.

Dean, M., Carrington, M., Winkler, C., Huttley, G.A., Smith, M.W., Allikmets, R., Goedert, J.J., Buchbinder, S.P., Vittinghoff, E., Gomperts, E., *et al.* (1996). Genetic restriction of HIV-1 infection and progression to AIDS by a deletion allele of the CKR5 structural gene. Hemophilia Growth and Development Study, Multicenter AIDS Cohort Study, Multicenter Hemophilia Cohort Study, San Francisco City Cohort, ALIVE Study. Science *273*, 1856–1862.

Dehghani, H., Puffer, B.A., Doms, R.W., and Hirsch, V.M. (2003). Unique pattern of convergent envelope evolution in simian immunodeficiency virus-infected rapid progressor macaques: association with CD4-independent usage of CCR5. J. Virol. *77*, 6405–6418.

Delwart, E.L., Sheppard, H.W., Walker, B.D., Goudsmit, J., and Mullins, J.I. (1994). Human immunodeficiency virus type 1 evolution *in vivo* tracked by DNA heteroduplex mobility assays. J. Virol. *68*, 6672–6683.

Delwart, E.L., Pan, H., Sheppard, H.W., Wolpert, D., Neumann, A.U., Korber, B., and Mullins, J.I. (1997). Slower evolution of human immunodeficiency virus type 1 quasispecies during progression to AIDS. J. Virol. *71*, 7498–7508.

Deng, H., Liu, R., Ellmeier, W., Choe, S., Unutmaz, D., Burkhart, M., Di Marzio, P., Marmon, S., Sutton, R.E., Hill, C.M., *et al.* (1996). Identification of a major co-receptor for primary isolates of HIV-1. Nature *381*, 661–666.

Deng, H.K., Unutmaz, D., KewalRamani, V.N., and Littman, D.R. (1997). Expression cloning of new receptors used by simian and human immunodeficiency viruses. Nature *388*, 296–300.

Derdeyn, C.A., Decker, J.M., Bibollet-Ruche, F., Mokili, J.L., Muldoon, M., Denham, S.A., Heil, M.L., Kasolo, F., Musonda, R., Hahn, B.H., *et al.* (2004). Envelope-constrained neutralization-sensitive HIV-1 after heterosexual transmission. Science *303*, 2019–2022.

Dittmar, M.T., McKnight, A., Simmons, G., Clapham, P.R., Weiss, R.A., and Simmonds, P. (1997). HIV-1 tropism and co-receptor use. Nature *385*, 495–496.

Dolin, R. (2008). A new class of anti-HIV therapy and new challenges. N. Engl. J. Med. *359*, 1509–1511.

Doms, R.W. (2000). Beyond receptor expression: the influence of receptor conformation, density, and affinity in HIV-1 infection. Virology *276*, 229–237.

Doranz, B.J., Rucker, J., Yi, Y., Smyth, R.J., Samson, M., Peiper, S.C., Parmentier, M., Collman, R.G., and Doms, R.W. (1996). A dual-tropic primary HIV-1 isolate that uses fusin and the beta-chemokine receptors CKR-5, CKR-3, and CKR-2b as fusion cofactors. Cell *85*, 1149–1158.

Dragic, T., Litwin, V., Allaway, G.P., Martin, S.R., Huang, Y., Nagashima, K.A., Cayanan, C., Maddon, P.J., Koup, R.A., Moore, J.P., *et al.* (1996). HIV-1 entry into CD4+ cells is mediated by the chemokine receptor CC-CKR-5. Nature *381*, 667–673.

Edinger, A.L., Hoffman, T.L., Sharron, M., Lee, B., O'Dowd, B., and Doms, R.W. (1998a). Use of GPR1, GPR15, and STRL33 as coreceptors by diverse human immunodeficiency virus type 1 and simian immunodeficiency virus envelope proteins. Virology *249*, 367–378.

Edinger, A.L., Hoffman, T.L., Sharron, M., Lee, B., Yi, Y., Choe, W., Kolson, D.L., Mitrovic, B., Zhou, Y., Faulds, D., *et al.* (1998b). An orphan seven-transmembrane domain receptor expressed widely in the brain functions as a coreceptor for human immunodeficiency virus type 1 and simian immunodeficiency virus. J. Virol. *72*, 7934–7940.

Ellery, P.J., Tippett, E., Chiu, Y.L., Paukovics, G., Cameron, P.U., Solomon, A., Lewin, S.R., Gorry, P.R., Jaworowski, A., Greene, W.C., *et al.* (2007). The CD16+ Monocyte Subset Is More Permissive to Infection and Preferentially Harbors HIV-1 *In Vivo*. J. Immunol. *178*, 6581–6589.

Este, J.A., Cabrera, C., Blanco, J., Gutierrez, A., Bridger, G., Henson, G., Clotet, B., Schols, D., and De Clercq, E. (1999). Shift of clinical human immunodeficiency virus type 1 isolates from X4 to R5 and prevention of emergence of the syncytium-inducing phenotype by blockade of CXCR4. J. Virol. *73*, 5577–5585.

Fais, S., Lapenta, C., Santini, S.M., Spada, M., Parlato, S., Logozzi, M., Rizza, P., and Belardelli, F. (1999). Human immunodeficiency virus type 1 strains R5 and X4 induce different pathogenic effects in hu-PBL-SCID mice, depending on the state of activation/differentiation of human target cells at the time of primary infection. J. Virol. *73*, 6453–6459.

Farzan, M., Choe, H., Martin, K., Marcon, L., Hofmann, W., Karlsson, G., Sun, Y., Barrett, P., Marchand, N., Sullivan, N., *et al.* (1997). Two orphan seven-transmembrane segment receptors which are expressed in CD4-positive cells support simian immunodeficiency virus infection. Journal of Experimental Medicine *186*, 405–411.

Farzan, M., Mirzabekov, T., Kolchinsky, P., Wyatt, R., Cayabyab, M., Gerard, N.P., Gerard, C., Sodroski, J., and Choe, H. (1999). Tyrosine sulfation of the amino terminus of CCR5 facilitates HIV-1 entry. Cell *96*, 667–676.

Fatkenheuer, G., Nelson, M., Lazzarin, A., Konourina, I., Hoepelman, A.I., Lampiris, H., Hirschel, B., Tebas, P., Raffi, F., Trottier, B., *et al.* (2008). Subgroup analyses of maraviroc in previously treated R5 HIV-1 infection. N. Engl. J. Med. *359*, 1442–1455.

Fauci, A.S. (1996). Host factors and the pathogenesis of HIV-induced disease. Nature *384*, 529–534.

Fear, W.R., Kesson, A.M., Naif, H., Lynch, G.W., and Cunningham, A.L. (1998). Differential tropism and chemokine receptor expression of human immunodeficiency virus type 1 in neonatal monocytes, monocyte-derived macrophages, and placental macrophages. J. Virol. *72*, 1334–1344.

Feng, Y., Broder, C.C., Kennedy, P.E., and Berger, E.A. (1996). HIV-1 entry cofactor: functional cDNA cloning of a seven-transmembrane, G protein-coupled receptor. Science *272*, 872–877.

Fouchier, R.A., Broersen, S.M., Brouwer, M., Tersmette, M., Van't Wout, A.B., Groenink, M., and Schuitemaker, H. (1995). Temporal relationship between elongation of the HIV type 1 glycoprotein 120 V2 domain and the conversion toward a syncytium-inducing phenotype. AIDS Res. Hum. Retroviruses *11*, 1473–1478.

Fukuda, M., Carlsson, S.R., Klock, J.C., and Dell, A. (1986). Structures of O-linked oligosaccharides isolated from normal granulocytes, chronic myelogenous leukemia cells, and acute myelogenous leukemia cells. J. Biol. Chem. *261*, 12796–12806.

Gallo, R.C., Salahuddin, S.Z., Popovic, M., Shearer, G.M., Kaplan, M., Haynes, B.F., Palker, T.J., Redfield, R., Oleske, J., Safai, B., *et al.* (1984). Frequent detection and isolation of cytopathic retroviruses (HTLV-III) from patients with AIDS and at risk for AIDS. Science *224*, 500–503.

Ganeshan, S., Dickover, R.E., Korber, B.T., Bryson, Y.J., and Wolinsky, S.M. (1997). Human immunodeficiency virus type 1 genetic evolution in children with different rates of development of disease. J. Virol. *71*, 663–677.

Gao, X., Nelson, G.W., Karacki, P., Martin, M.P., Phair, J., Kaslow, R., Goedert, J.J., Buchbinder, S., Hoots, K., Vlahov, D., *et al.* (2001). Effect of a single amino acid change in MHC class I molecules on the rate of progression to AIDS. N. Engl. J. Med. *344*, 1668–1675.

Gartner, S., Markovits, P., Markovitz, D.M., Betts, R.F., and Popovic, M. (1986a). Virus isolation from and identification of HTLV-III/LAV-producing cells in brain tissue from a patient with AIDS. Journal of the American Medical Association *256*, 2365–2371.

Gartner, S., Markovits, P., Markovitz, D.M., Kaplan, M.H., Gallo, R.C., and Popovic, M. (1986b). The role of mononuclear phagocytes in HTLV-III/LAV infection. Science *233*, 215–219.

Gartner, S., and Popovic, M. (1990). Macrophage tropism of HIV-1. AIDS Research and Human Retroviruses *6*, 1017–1021.

Gaudieri, S., Nolan, D., McKinnon, E., Witt, C.S., Mallal, S., and Christiansen, F.T. (2005). Associations between KIR epitope combinations expressed by HLA-B/-C haplotypes found in an HIV-1 infected study population may influence NK mediated immune responses. Mol Immunol *42*, 557–560.

Glushakova, S., Baibakov, B., Margolis, L.B., and Zimmerberg, J. (1995). Infection of human tonsil histocultures: a model for HIV pathogenesis. Nature Med *1*, 1320–1322.

Glushakova, S., Grivel, J.C., Fitzgerald, W., Sylwester, A., Zimmerberg, J., and Margolis, L.B. (1998). Evidence for the HIV-1 phenotype switch as a causal factor in acquired immunodeficiency. Nature Med *4*, 346–349.

Goodenow, M., Huet, T., Saurin, W., Kwok, S., Sninsky, J., and Wain-Hobson, S. (1989). HIV-1 isolates are rapidly evolving quasispecies: evidence for viral mixtures and preferred nucleotide substitutions. J. Acquir. Immune Defic. Syndr. *2*, 344–352.

Gorry, P.R., Bristol, G., Zack, J.A., Ritola, K., Swanstrom, R., Birch, C.J., Bell, J.E., Bannert, N., Crawford, K., Wang, H., *et al.* (2001). Macrophage Tropism of Human Immunodeficiency Virus Type 1 Isolates from Brain and Lymphoid Tissues Predicts Neurotropism Independent of Coreceptor Specificity. J. Virol. *75*, 10073–10089.

Gorry, P.R., Taylor, J., Holm, G.H., Mehle, A., Morgan, T., Cayabyab, M., Farzan, M., Wang, H., Bell, J.E., Kunstman, K., *et al.* (2002). Increased CCR5 affinity and reduced CCR5/CD4 dependence of a neurovirulent primary human immunodeficiency virus type 1 isolate. J. Virol. *76*, 6277–6292.

Gorry, P.R., Sterjovski, J., Churchill, M., Witlox, K., Gray, L., Cunningham, A., and Wesselingh, S. (2004). The role of viral coreceptors and macrophage tropism in human immunodeficiency virus type 1 disease progression. Sex Health *1*, 23–34.

Gorry, P.R., Churchill, M., Crowe, S.M., Cunningham, A.L., and Gabuzda, D. (2005). Pathogenesis of macrophage tropic HIV. Curr. HIV Res. *3*, 53–60.

Gorry, P.R., Dunfee, R.L., Mefford, M.E., Kunstman, K., Morgan, T., Moore, J.P., Mascola, J.R., Agopian, K., Holm, G.H., Mehle, A., *et al.* (2007). Changes in the V3 region of gp120 contribute to unusually broad coreceptor usage of an HIV-1 isolate from a CCR5 Delta32 heterozygote. Virology *362*, 163–178.

Gray, L., Sterjovski, J., Churchill, M., Ellery, P., Nasr, N., Lewin, S.R., Crowe, S.M., Wesselingh, S.L., Cunningham, A.L., and Gorry, P.R. (2005). Uncoupling coreceptor usage of human immunodeficiency virus type 1 (HIV-1) from macrophage tropism reveals biological properties of CCR5-restricted HIV-1 isolates from patients with acquired immunodeficiency syndrome. Virology *337*, 384–398.

Gray, L., Churchill, M.J., Keane, N., Sterjovski, J., Ellett, A.M., Purcell, D.F., Poumbourios, P., Kol, C., Wang, B., Saksena, N.K., *et al.* (2006). Genetic and functional analysis of R5X4 human immunodeficiency virus type 1 envelope glycoproteins derived from two individuals homozygous for the CCR5delta32 allele. J. Virol. *80*, 3684–3691.

Grivel, J.C., and Margolis, L.B. (1999). CCR5- and CXCR4-tropic HIV-1 are equally cytopathic for their T-cell targets in human lymphoid tissue. Nature Med *5*, 344–346.

Gumperz, J.E., Litwin, V., Phillips, J.H., Lanier, L.L., and Parham, P. (1995). The Bw4 public epitope of HLA-B molecules confers reactivity with natural killer cell clones that express NKB1, a putative HLA receptor. J. Exp. Med. *181*, 1133–1144.

Harouse, J.M., Gettie, A., Tan, R.C., Blanchard, J., and Cheng-Mayer, C. (1999). Distinct pathogenic sequela in rhesus macaques infected with CCR5 or CXCR4 utilizing SHIVs. Science *284*, 816–819.

Hendel, H., Caillat-Zucman, S., Lebuanec, H., Carrington, M., O'Brien, S., Andrieu, J.M., Schachter, F., Zagury, D., Rappaport, J., Winkler, C., *et al.* (1999). New class I and II HLA alleles strongly associated with opposite patterns of progression to AIDS. J. Immunol. *162*, 6942–6946.

Hill, C.M., Deng, H., Unutmaz, D., Kewalramani, V.N., Bastiani, L., Gorny, M.K., Zolla-Pazner, S., and Littman, D.R. (1997). Envelope glycoproteins from human immunodeficiency virus types 1 and 2 and simian immunodeficiency virus can use human CCR5 as a coreceptor for viral entry and make direct CD4-dependent interactions with this chemokine receptor. J. Virol. *71*, 6296–6304.

Hill, C.M., Kwon, D., Jones, M., Davis, C.B., Marmon, S., Daugherty, B.L., DeMartino, J.A., Springer, M.S., Unutmaz, D., and Littman, D.R. (1998). The amino terminus of human CCR5 is required for its function as a receptor for diverse human and simian immunodeficiency virus envelope glycoproteins. Virology *248*, 357–371.

Hladik, F., Lentz, G., Akridge, R.E., Peterson, G., Kelley, H., McElroy, A., and McElrath, M.J. (1999). Dendritic cell–T-cell interactions support coreceptor-independent human immunodeficiency virus type 1 transmission in the human genital tract. J. Virol. *73*, 5833–5842.

Ho, D.D., Rota, T.R., and Hirsch, M.S. (1986). Infection of monocyte/macrophages by human T lymphotropic virus type III. Journal of Clinical Investigation *77*, 1712–1715.

Hoffman, T.L., LaBranche, C.C., Zhang, W., Canziani, G., Robinson, J., Chaiken, I., Hoxie, J.A., and Doms, R.W. (1999). Stable exposure of the coreceptor-binding site in a CD4-independent HIV-1 envelope protein. Procedings of the National Academy of Sciences USA *96*, 6359–6364.

Ho, D.D., Neumann, A.U., Perelson, A.S., Chen, W., Leonard, J.M., and Markowitz, M. (1995). Rapid turnover of plasma virions and CD4 lymphocytes in HIV-1 infection. Nature *373*, 123–126.

Holm, G.H., Zhang, C., Gorry, P.R., Peden, K., Schols, D., De Clercq, E., and Gabuzda, D. (2004). Apoptosis of bystander T cells induced by human immunodeficiency virus type 1 with increased envelope/receptor affinity and coreceptor binding site exposure. J. Virol. *78*, 4541–4551.

Horuk, R., Hesselgesser, J., Zhou, Y., Faulds, D., Halks-Miller, M., Harvey, S., Taub, D., Samson, M., Parmentier, M., Rucker, J., *et al.* (1998). The CC chemokine I-309 inhibits CCR8-dependent infection by diverse HIV-1 strains. Journal of Biological Chemistry *273*, 386–391.

Huang, Y., Paxton, W.A., Wolinsky, S.M., Neumann, A.U., Zhang, L., He, T., Kang, S., Ceradini, D., Jin, Z., Yazdanbakhsh, K., *et al.* (1996). The role of a mutant CCR5 allele in HIV-1 transmission and disease progression. Nat. Med. *2*, 1240–1243.

Hung, C.S., Pontow, S., and Ratner, L. (1999). Relationship between productive HIV-1 infection of macrophages and CCR5 utilization. Virology *264*, 278–288.

Igarashi, T., Brown, C.R., Endo, Y., Buckler-White, A., Plishka, R., Bischofberger, N., Hirsch, V., and Martin, M.A. (2001). Macrophage are the principal reservoir and sustain high virus loads in rhesus macaques after the depletion of CD4+ T cells by a highly pathogenic simian immunodeficiency virus/HIV type 1 chimera (SHIV): Implications for HIV-1 infections of humans. Procedings of the National Academy of Sciences USA *98*, 658–663.

Igarashi, T., Imamichi, H., Brown, C.R., Hirsch, V.M., and Martin, M.A. (2003). The emergence and characterization of macrophage-tropic SIV/HIV chimeric viruses (SHIVs) present in CD4+ T cell-depleted rhesus monkeys. J. Leukoc. Biol. *74*, 772–780.

Itescu, S., Mathur-Wagh, U., Skovron, M.L., Brancato, L.J., Marmor, M., Zeleniuch-Jacquotte, A., and Winchester, R. (1992). HLA-B35 is associated with accelerated progression to AIDS. J. Acquir. Immune Defic. Syndr. *5*, 37–45.

Itescu, S., Rose, S., Dwyer, E., and Winchester, R. (1994). Certain HLA-DR5 and -DR6 major histocompatibility complex class II alleles are associated with a CD8 lymphocytic host response to human immunodeficiency virus type 1 characterized by low lymphocyte viral strain heterogeneity and slow disease progression. Proc. Natl. Acad. Sci. U.S.A. *91*, 11472–11476.

Jansson, M., Popovic, M., Karlsson, A., Cocchi, F., Rossi, P., Albert, J., and Wigzell, H. (1996). Sensitivity to inhibition by beta-chemokines correlates with biological phenotypes of primary HIV-1 isolates. Proc. Natl. Acad. Sci. U.S.A. 93, 15382–15387.

Jansson, M., Backstrom, E., Bjorndal, A., Holmberg, V., Rossi, P., Fenyo, E.M., Popovic, M., Albert, J., and Wigzell, H. (1999). Coreceptor usage and RANTES sensitivity of non-syncytium-inducing HIV-1 isolates obtained from patients with AIDS. J Hum Virol *2*, 325–338.

Jinno, A., Shimizu, N., Soda, Y., Haraguchi, Y., Kitamura, T., and Hoshino, H. (1998). Identification of the chemokine receptor TER1/CCR8 expressed in brain-derived cells and T cells as a new coreceptor for HIV-1 infection. Biochemical and Biophysical Research Communications *243*, 497–502.

Karita, E., Nkengasong, J.N., Willems, B., Vanham, G., Fransen, K., Heyndrickx, L., Janssens, W., Piot, P., and van der Groen, G. (1997). Macrophage-tropism of HIV-1 isolates of different genetic subtypes. Aids *11*, 1303–1304.

Karlsson, I., Antonsson, L., Shi, Y., Oberg, M., Karlsson, A., Albert, J., Olde, B., Owman, C., Jansson, M., and Fenyo, E.M. (2004). Coevolution of RANTES sensitivity and mode of CCR5 receptor use by human immunodeficiency virus type 1 of the R5 phenotype. J. Virol. *78*, 11807–11815.

Kaslow, R.A., Carrington, M., Apple, R., Park, L., Munoz, A., Saah, A.J., Goedert, J.J., Winkler, C., O'Brien, S.J., Rinaldo, C., *et al.* (1996). Influence of combinations of human major histocompatibility complex genes on the course of HIV-1 infection. Nat. Med. *2*, 405–411.

Kedzierska, K., Mak, J., Mijch, A., Cooke, I., Rainbird, M., Roberts, S., Paukovics, G., Jolley, D., Lopez, A., and Crowe, S.M. (2000). Granulocyte-macrophage colony-stimulating factor augments phagocytosis of Mycobacterium avium complex by human immunodeficiency virus type 1-infected monocytes/macrophages *in vitro* and *in vivo*. Journal of Infectious Diseases *181*, 390–394.

Kedzierska, K., Ellery, P., Mak, J., Lewin, S.R., Crowe, S.M., and Jaworowski, A. (2002). HIV-1 down-modulates gamma signaling chain of Fc gamma R in human macrophages: a possible mechanism for inhibition of phagocytosis. Journal of Immunology *168*, 2895–2903.

Keele, B.F., Giorgi, E.E., Salazar-Gonzalez, J.F., Decker, J.M., Pham, K.T., Salazar, M.G., Sun, C., Grayson, T., Wang, S., Li, H., *et al.* (2008). Identification and characterization of transmitted and early founder virus envelopes in primary HIV-1 infection. Proc. Natl. Acad. Sci. U.S.A. *105*, 7552–7557.

Kelleher, A.D., Long, C., Holmes, E.C., Allen, R.L., Wilson, J., Conlon, C., Workman, C., Shaunak, S., Olson, K., Goulder, P., *et al.* (2001). Clustered mutations in HIV-1 gag are consistently required for escape from HLA-B27-restricted cytotoxic T lymphocyte responses. J. Exp. Med. *193*, 375–386.

Kerr, S.J., Armati, P.J., Guillemin, G.J., and Brew, B.J. (1998). Chronic exposure of human neurons to quinolinic acid results in neuronal changes consistent with AIDS dementia complex. Aids *12*, 355–363.

Kimata, J.T. (2006). HIV-1 fitness and disease progression: insights from the SIV-macaque model. Curr. HIV Res. *4*, 65–77.

Klein, M.R., Keet, I.P., D'Amaro, J., Bende, R.J., Hekman, A., Mesman, B., Koot, M., de Waal, L.P., Coutinho, R.A., and Miedema, F. (1994). Associations between HLA frequencies and pathogenic features of human immunodeficiency virus type 1 infection in seroconverters from the Amsterdam cohort of homosexual men. J. Infect. Dis. *169*, 1244–1249.

Klein, M.R., van der Burg, S.H., Hovenkamp, E., Holwerda, A.M., Drijfhout, J.W., Melief, C.J., and Miedema, F. (1998). Characterization of HLA-B57-restricted human immunodeficiency virus type 1 Gag- and RT-specific cytotoxic T lymphocyte responses. J. Gen. Virol. *79 (Pt 9)*, 2191–2201.

Kolchinsky, P., Mirzabekov, T., Farzan, M., Kiprilov, E., Cayabyab, M., Mooney, L.J., Choe, H., and Sodroski, J. (1999). Adaptation of a CCR5-using, primary human immunodeficiency virus type 1 isolate for CD4-independent replication. J. Virol. *73*, 8120–8126.

Kolchinsky, P., Kiprilov, E., Bartley, P., Rubinstein, R., and Sodroski, J. (2001). Loss of a single N-linked glycan allows CD4-independent human immunodeficiency virus type 1 infection by altering the position of the gp120 V1/V2 variable loops. J. Virol. *75*, 3435–3443.

Koning, F.A., Kwa, D., Boeser-Nunnink, B., Dekker, J., Vingerhoed, J., Hiemstra, H., and Schuitemaker, H. (2003). Decreasing sensitivity to RANTES (regulated on activation, normally T cell-expressed and -secreted) neutralization of CC chemokine receptor 5-using, non-syncytium-inducing virus variants in the course of human immunodeficiency virus type 1 infection. J. Infect. Dis. *188*, 864–872.

Kuhmann, S.E., Platt, E.J., Kozak, S.L., and Kabat, D. (2000). Cooperation of multiple CCR5 coreceptors is required for infections by human immunodeficiency virus type 1. J. Virol. *74*, 7005–7015.

Kuipers, H., Workman, C., Dyer, W., Geczy, A., Sullivan, J., and Oelrichs, R. (1999). An HIV-1-infected individual homozygous for the CCR-5 delta32 allele and the SDF-1 3'A allele. Aids *13*, 433–434.

Kwa, D., Vingerhoed, J., Boeser, B., and Schuitemaker, H. (2003). Increased *In Vitro* Cytopathicity of CC Chemokine Receptor 5-Restricted Human Immunodeficiency Virus Type 1 Primary Isolates Correlates with a Progressive Clinical Course of Infection. J. Infect. Dis. *187*, 1397–1403.

LaBonte, J.A., Patel, T., Hofmann, W., and Sodroski, J. (2000). Importance of membrane fusion mediated by human immunodeficiency virus envelope glycoproteins for lysis of primary CD4-positive T cells. J. Virol. *74*, 10690–10698.

Lalezari, J.P., Henry, K., O'Hearn, M., Montaner, J.S., Piliero, P.J., Trottier, B., Walmsley, S., Cohen, C., Kuritzkes, D.R., Eron, J.J., Jr, *et al.* (2003). Enfuvirtide, an HIV-1 fusion inhibitor, for drug-resistant HIV infection in North and South America. N. Engl. J. Med. *348*, 2175–2185.

Lambotte, O., Taoufik, Y., de Goer, M.G., Wallon, C., Goujard, C., and Delfraissy, J.F. (2000). Detection of infectious HIV in circulating monocytes from patients on prolonged highly active antiretroviral therapy. J. Acquir. Immune Defic. Syndr. *23*, 114–119.

Lambotte, O., Deiva, K., and Tardieu, M. (2003). HIV-1 persistence, viral reservoir, and the central nervous system in the HAART era. Brain Pathology *13*, 95–103.

Lawson, V.A., Oelrichs, R., Guillon, C., Imrie, A.A., Cooper, D.A., Deacon, N.J., and McPhee, D.A. (2002). Adaptive changes after human immunodeficiency virus type 1 transmission. AIDS Res. Hum. Retroviruses *18*, 545–556.

Lazzarin, A., Clotet, B., Cooper, D., Reynes, J., Arasteh, K., Nelson, M., Katlama, C., Stellbrink, H.J., Delfraissy, J.F., Lange, J., *et al.* (2003). Efficacy of enfuvirtide in patients infected with drug-resistant HIV-1 in Europe and Australia. N. Engl. J. Med. *348*, 2186–2195.

Learn, G.H., Muthui, D., Brodie, S.J., Zhu, T., Diem, K., Mullins, J.I., and Corey, L. (2002). Virus population homogenization following acute human immunodeficiency virus type 1 infection. J. Virol. *76*, 11953–11959.

Levy, J.A. (1993). Pathogenesis of human immunodeficiency virus infection. Microbiological Reviews *57*, 183–289.

Lewin, S.R., Sonza, S., Irving, L.B., McDonald, C.F., Mills, J., and Crowe, S.M. (1996). Surface CD4 is critical to *in vitro* HIV infection of human alveolar macrophages. AIDS Research and Human Retroviruses *12*, 877–883.

Li, S., Juarez, J., Alali, M., Dwyer, D., Collman, R., Cunningham, A., and Naif, H.M. (1999). Persistent CCR5 utilization and enhanced macrophage tropism by primary blood human immunodeficiency virus type 1 isolates from advanced stages of disease and comparison to tissue-derived isolates. J. Virol. *73*, 9741–9755.

Liao, F., Alkhatib, G., Peden, K.W., Sharma, G., Berger, E.A., and Farber, J.M. (1997). STRL33, A novel chemokine receptor-like protein, functions as a fusion cofactor for both macrophage-tropic and T cell line-tropic HIV-1. Journal of Experimental Medicine *185*, 2015–2023.

Liu, R., Paxton, W.A., Choe, S., Ceradini, D., Martin, S.R., Horuk, R., MacDonald, M.E., Stuhlmann, H., Koup, R.A., and Landau, N.R. (1996). Homozygous defect in HIV-1 coreceptor accounts for resistance of some multiply-exposed individuals to HIV-1 infection. Cell *86*, 367–377.

Liu, S.L., Schacker, T., Musey, L., Shriner, D., McElrath, M.J., Corey, L., and Mullins, J.I. (1997). Divergent patterns of progression to AIDS after infection from the same source: human immunodeficiency virus type 1 evolution and antiviral responses. J. Virol. *71*, 4284–4295.

Liu, Z.Q., Muhkerjee, S., Sahni, M., McCormick-Davis, C., Leung, K., Li, Z., Gattone, V.H., 2nd, Tian, C., Doms, R.W., Hoffman, T.L., *et al.* (1999). Derivation and biological characterization of a molecular clone of SHIV(KU-2) that causes AIDS, neurological disease, and renal disease in rhesus macaques. Virology *260*, 295–307.

Lockett, S.F., Robertson, J.R., Brettle, R.P., Yap, P.L., Middleton, D., and Leigh Brown, A.J. (2001). Mismatched human leukocyte antigen alleles protect against heterosexual HIV transmission. J. Acquir. Immune Defic. Syndr. *27*, 277–280.

Long, E.M., Rainwater, S.M., Lavreys, L., Mandaliya, K., and Overbaugh, J. (2002). HIV type 1 variants transmitted to women in Kenya require the CCR5 coreceptor for entry, regardless of the genetic complexity of the infecting virus. AIDS Res. Hum. Retroviruses *18*, 567–576.

Lukashov, V.V., Kuiken, C.L., and Goudsmit, J. (1995). Intrahost human immunodeficiency virus type 1 evolution is related to length of the immunocompetent period. J. Virol. *69*, 6911–6916.

MacDonald, K.S., Fowke, K.R., Kimani, J., Dunand, V.A., Nagelkerke, N.J., Ball, T.B., Oyugi, J., Njagi, E., Gaur, L.K., Brunham, R.C., *et al.* (2000). Influence of HLA supertypes on susceptibility and resistance to human immunodeficiency virus type 1 infection. J. Infect. Dis. *181*, 1581–1589.

MacDonald, K.S., Embree, J.E., Nagelkerke, N.J., Castillo, J., Ramhadin, S., Njenga, S., Oyug, J., Ndinya-Achola, J., Barber, B.H., Bwayo, J.J., *et al.* (2001). The HLA A2/6802 supertype is associated with reduced risk of perinatal human immunodeficiency virus type 1 transmission. J. Infect. Dis. *183*, 503–506.

Magierowska, M., Theodorou, I., Debre, P., Sanson, F., Autran, B., Riviere, Y., Charron, D., and Costagliola, D. (1999). Combined genotypes of CCR5, CCR2, SDF1, and HLA genes can predict the long-term nonprogressor status in human immunodeficiency virus-1-infected individuals. Blood 93, 936–941.

Malhotra, U., Holte, S., Dutta, S., Berrey, M.M., Delpit, E., Koelle, D.M., Sette, A., Corey, L., and McElrath, M.J.

(2001). Role for HLA class II molecules in HIV-1 suppression and cellular immunity following antiretroviral treatment. J. Clin. Invest. *107*, 505–517.

Markham, R.B., Wang, W.C., Weisstein, A.E., Wang, Z., Munoz, A., Templeton, A., Margolick, J., Vlahov, D., Quinn, T., Farzadegan, H., *et al.* (1998). Patterns of HIV-1 evolution in individuals with differing rates of CD4 T cell decline. Proc. Natl. Acad. Sci. U.S.A. 95, 12568–12573.

Martin, J., LaBranche, C.C., and Gonzalez-Scarano, F. (2001). Differential CD4/CCR5 utilization, gp120 conformation, and neutralization sensitivity between envelopes from a microglia-adapted human immunodeficiency virus type 1 and its parental isolate. J. Virol. 75, 3568–3580.

Martin-Garcia, J., Cao, W., Varela-Rohena, A., Plassmeyer, M.L., and Gonzalez-Scarano, F. (2005a). HIV-1 tropism for the central nervous system: Brain-derived envelope glycoproteins with lower CD4 dependence and reduced sensitivity to a fusion inhibitor. Virology *346*, 169–179.

Martin-Garcia, J., Cocklin, S., Chaiken, I.M., and Gonzalez-Scarano, F. (2005b). Interaction with CD4 and antibodies to CD4-induced epitopes of the envelope gp120 from a microglial cell-adapted human immunodeficiency virus type 1 isolate. J. Virol. *79*, 6703–6713.

Martin-Garcia, J., Cao, W., Varela-Rohena, A., Plassmeyer, M.L., and Gonzalez-Scarano, F. (2006). HIV-1 tropism for the central nervous system: Brain-derived envelope glycoproteins with lower CD4 dependence and reduced sensitivity to a fusion inhibitor. Virology *346*, 169–179.

McNearney, T., Hornickova, Z., Markham, R., Birdwell, A., Arens, M., Saah, A., and Ratner, L. (1992). Relationship of human immunodeficiency virus type 1 sequence heterogeneity to stage of disease. Proc. Natl. Acad. Sci. U.S.A. *89*, 10247–10251.

Means, R.E., Matthews, T., Hoxie, J.A., Malim, M.H., Kodama, T., and Desrosiers, R.C. (2001). Ability of the V3 loop of simian immunodeficiency virus to serve as a target for antibody-mediated neutralization: correlation of neutralization sensitivity, growth in macrophages, and decreased dependence on CD4. J. Virol. 75, 3903–3915.

Meltzer, M.S., Nakamura, M., Hansen, B.D., Turpin, J.A., Kalter, D.C., and Gendelman, H.E. (1990). Macrophages as susceptible targets for HIV infection, persistent viral reservoirs in tissue, and key immunoregulatory cells that control levels of virus replication and extent of disease. AIDS Research and Human Retroviruses *6*, 967–971.

Meltzer, M.S., and Gendelman, H.E. (1992). Mononuclear phagocytes as targets, tissue reservoirs, and immunoregulatory cells in human immunodeficiency virus disease. Current Topics in Microbiology and Immunology *181*, 239–263.

Meng, G., Wei, X., Wu, X., Sellers, M.T., Decker, J.M., Moldoveanu, Z., Orenstein, J.M., Graham, M.F., Kappes, J.C., Mestecky, J., *et al.* (2002). Primary intestinal epithelial cells selectively transfer R5 HIV-1 to CCR5+ cells. Nat. Med. *8*, 150–156.

Michael, N.L., Chang, G., Louie, L.G., Mascola, J.R., Dondero, D., Birx, D.L., and Sheppard, H.W. (1997). The role of viral phenotype and CCR-5 gene defects in HIV-1 transmission and disease progression. Nat. Med. 3, 338–340.

Michael, N.L., Nelson, J.A., KewalRamani, V.N., Chang, G., O'Brien, S.J., Mascola, J.R., Volsky, B., Louder, M., White, G.C., 2nd, Littman, D.R., *et al.* (1998). Exclusive and persistent use of the entry coreceptor CXCR4 by human immunodeficiency virus type 1 from a subject homozygous for CCR5 delta32. J. Virol. *72*, 6040–6047.

Migueles, S.A., Sabbaghian, M.S., Shupert, W.L., Bettinotti, M.P., Marincola, F.M., Martino, L., Hallahan, C.W., Selig, S.M., Schwartz, D., Sullivan, J., *et al.* (2000). HLA B*5701 is highly associated with restriction of virus replication in a subgroup of HIV-infected long term nonprogressors. Proc. Natl. Acad. Sci. U.S.A. *97*, 2709–2714.

Mikhail, M., Wang, B., and Saksena, N.K. (2003). Mechanisms involved in non-progressive HIV disease. AIDS Reviews, 5, 230–244.

Moore, J.P., Cao, Y., Qing, L., Sattentau, Q.J., Pyati, J., Koduri, R., Robinson, J., Barbas, C.F., 3rd, Burton, D.R., and Ho, D.D. (1995). Primary isolates of human immunodeficiency virus type 1 are relatively resistant to neutralization by monoclonal antibodies to gp120, and their neutralization is not predicted by studies with monomeric gp120. J. Virol. *69*, 101–109.

Moore, J.P., and Ho, D.D. (1995). HIV-1 neutralization: the consequences of viral adaptation to growth on transformed T cells. Aids *9 Suppl A*, S117–136.

Moore, J.P., and Doms, R.W. (2003). The entry of entry inhibitors: a fusion of science and medicine. Proc. Natl. Acad. Sci. U.S.A. *100*, 10598–10602.

Mori, K., Rosenzweig, M., and Desrosiers, R.C. (2000). Mechanisms for adaptation of simian immunodeficiency virus to replication in alveolar macrophages. J. Virol. *74*, 10852–10859.

Mosier, D.E. (2005). HIV-1 envelope evolution and vaccine efficacy. Curr Drug Targets Infect Disord 5, 171–177.

Nabatov, A.A., Pollakis, G., Linnemann, T., Kliphius, A., Chalaby, M.I., and Paxton, W.A. (2004). Intrapatient alterations in the human immunodeficiency virus type 1 gp120 V1V2 and V3 regions differentially modulate coreceptor usage, virus inhibition by CC/CXC chemokines, soluble CD4, and the b12 and 2G12 monoclonal antibodies. J. Virol. *78*, 524–530.

Naif, H.M., Cunningham, A.L., Alali, M., Li, S., Nasr, N., Buhler, M.M., Schols, D., de Clercq, E., and Stewart, G. (2002). A human immunodeficiency virus type 1 isolate from an infected person homozygous for CCR5Delta32 exhibits dual tropism by infecting macrophages and MT2 cells via CXCR4. J. Virol. *76*, 3114–3124.

Nietfield, W., Bauer, M., Fevrier, M., Maier, R., Holzwarth, B., Frank, R., Maier, B., Riviere, Y., and Meyerhans, A. (1995). Sequence constraints and recognition by CTL of an HLA-B27-restricted HIV-1 gag epitope. J. Immunol. *154*, 2189–2197.

Ohagen, A., Devitt, A., Kunstman, K.J., Gorry, P.R., Rose, P.P., Korber, B., Taylor, J., Levy, R., Murphy, R.L., Wolinsky, S.M., *et al.* (2003). Genetic and functional analysis of full-length human immunodeficiency virus type 1 env genes derived from brain and blood of patients with AIDS. J. Virol. *77*, 12336–12345.

Ometto, L., Zanchetta, M., Cabrelle, A., Esposito, G., Mainardi, M., Chieco-Bianchi, L., and De Rossi, A. (1999). Restriction of HIV type 1 infection in macrophages heterozygous for a deletion in the CC-chemokine receptor 5 gene. AIDS Research and Human Retroviruses *15*, 1441–1452.

Ometto, L., Bertorelle, R., Mainardi, M., Zanchetta, M., Tognazzo, S., Rampon, O., Ruga, E., Chieco-Bianchi, L., and De Rossi, A. (2001). Polymorphisms in the CCR5 promoter region influence disease progression in perinatally human immunodeficiency virus type 1-infected children. J. Infect. Dis. *183*, 814–818.

Pang, S., Shlesinger, Y., Daar, E.S., Moudgil, T., Ho, D.D., and Chen, I.S. (1992). Rapid generation of sequence variation during primary HIV-1 infection. Aids *6*, 453–460.

Pantophlet, R., and Burton, D.R. (2006). GP120: target for neutralizing HIV-1 antibodies. Annu. Rev. Immunol. *24*, 739–769.

Parren, P.W., Moore, J.P., Burton, D.R., and Sattentau, Q.J. (1999). The neutralizing antibody response to HIV-1: viral evasion and escape from humoral immunity. Aids *13 Suppl A*, S137–162.

Pemberton, L.A., Kerr, S.J., Smythe, G., and Brew, B.J. (1997). Quinolinic acid production by macrophages stimulated with IFN-gamma, TNF-alpha, and IFN-alpha. Journal of Interferon and Cytokine Research *17*, 589–595.

Peters, P.J., Bhattacharya, J., Hibbitts, S., Dittmar, M.T., Simmons, G., Bell, J., Simmonds, P., and Clapham, P.R. (2004). Biological analysis of human immunodeficiency virus type 1 R5 envelopes amplified from brain and lymph node tissues of AIDS patients with neuropathology reveals two distinct tropism phenotypes and identifies envelopes in the brain that confer an enhanced tropism and fusigenicity for macrophages. J. Virol. *78*, 6915–6926.

Peters, P.J., Sullivan, W.M., Duenas-Decamp, M.J., Bhattacharya, J., Ankghuambom, C., Brown, R., Luzuriaga, K., Bell, J., Simmonds, P., Ball, J., *et al.* (2006). Non-macrophage-tropic human immunodeficiency virus type 1 R5 envelopes predominate in blood, lymph nodes, and semen: implications for transmission and pathogenesis. J. Virol. *80*, 6324–6332.

Peters, P.J., Duenas-Decamp, M.J., Sullivan, W.M., Brown, R., Ankghuambom, C., Luzuriaga, K., Robinson, J., Burton, D.R., Bell, J., Simmonds, P., *et al.* (2008). Variation in HIV-1 R5 macrophage-tropism correlates with sensitivity to reagents that block envelope: CD4 interactions but not with sensitivity to other entry inhibitors. Retrovirology 5, 5.

Picchio, G.R., Gulizia, R.J., Wehrly, K., Chesebro, B., and Mosier, D.E. (1998). The cell tropism of human immunodeficiency virus type 1 determines the kinetics of plasma viremia in SCID mice reconstituted with human peripheral blood leukocytes. J. Virol. *72*, 2002–2009.

Pierson, T., Hoffman, T.L., Blankson, J., Finzi, D., Chadwick, K., Margolick, J.B., Buck, C., Siliciano, J.D., Doms, R.W., and Siliciano, R.F. (2000a). Characterization of chemokine receptor utilization of viruses in the latent reservoir for human immunodeficiency virus type 1. J. Virol. *74*, 7824–7833.

Pierson, T., McArthur, J., and Siliciano, R.F. (2000b). Reservoirs for HIV-1: mechanisms for viral persistence in the presence of antiviral immune responses and antiretroviral therapy. Annual Reviews in Immunology *18*, 665–708.

Poignard, P., Sabbe, R., Picchio, G.R., Wang, M., Gulizia, R.J., Katinger, H., Parren, P.W., Mosier, D.E., and Burton, D.R. (1999). Neutralizing antibodies have limited effects on the control of established HIV-1 infection *in vivo*. Immunity *10*, 431–438.

Price, R.W. (1996). Neurological complications of HIV infection. Lancet *348*, 445–452.

Qian, K., Morris-Natschke, S.L., and Lee, K.H. (2008). HIV entry inhibitors and their potential in HIV therapy. Med Res. Rev. *29*, 369–393.

Quakkelaar, E.D., van Alphen, F.P., Boeser-Nunnink, B.D., van Nuenen, A.C., Pantophlet, R., and Schuitemaker, H. (2007). Susceptibility of recently transmitted subtype B human immunodeficiency virus type 1 variants to broadly neutralizing antibodies. J. Virol. *81*, 8533–8542.

Rana, S., Besson, G., Cook, D.G., Rucker, J., Smyth, R.J., Yi, Y., Turner, J.D., Guo, H.H., Du, J.G., Peiper, S.C., *et al.* (1997). Role of CCR5 in infection of primary macrophages and lymphocytes by macrophage-tropic strains of human immunodeficiency virus: resistance to patient-derived and prototype isolates resulting from the delta ccr5 mutation. J. Virol. *71*, 3219–3227.

Reeves, J.D., Gallo, S.A., Ahmad, N., Miamidian, J.L., Harvey, P.E., Sharron, M., Pohlmann, S., Sfakianos, J.N., Derdeyn, C.A., Blumenthal, R., *et al.* (2002). Sensitivity of HIV-1 to entry inhibitors correlates with envelope/coreceptor affinity, receptor density, and fusion kinetics. Procedings of the National Academy of Sciences USA 99, 16249–16254.

Reiche, E.M., Bonametti, A.M., Voltarelli, J.C., Morimoto, H.K., and Watanabe, M.A. (2007). Genetic polymorphisms in the chemokine and chemokine receptors: impact on clinical course and therapy of the human immunodeficiency virus type 1 infection (HIV-1). Curr. Med. Chem. *14*, 1325–1334.

Repits, J., Oberg, M., Esbjornsson, J., Medstrand, P., Karlsson, A., Albert, J., Fenyo, E.M., and Jansson, M. (2005). Selection of human immunodeficiency virus type 1 R5 variants with augmented replicative capacity and reduced sensitivity to entry inhibitors during severe immunodeficiency. J. Gen. Virol. *86*, 2859–2869.

Repits, J., Sterjovski, J., Badia-Martinez, D., Mild, M., Gray, L., Churchill, M.J., Purcell, D.F., Karlsson, A., Albert, J., Fenyo, E.M., *et al.* (2008). Primary HIV-1 R5 isolates from end-stage disease display enhanced viral fitness in parallel with increased gp120 net charge. Virology *379*, 125–134.

Reynes, J., Portales, P., Segondy, M., Baillat, V., Andre, P., Avinens, O., Picot, M.C., Clot, J., Eliaou, J.F., and Corbeau, P. (2001). CD4 T cell surface CCR5 density as a host factor in HIV-1 disease progression. Aids *15*, 1627–1634.

Richman, D.D., Wrin, T., Little, S.J., and Petropoulos, C.J. (2003). Rapid evolution of the neutralizing antibody response to HIV type 1 infection. Proc. Natl. Acad. Sci. U.S.A. *100*, 4144–4149.

Ritola, K., Pilcher, C.D., Fiscus, S.A., Hoffman, N.G., Nelson, J.A., Kitrinos, K.M., Hicks, C.B., Eron, J.J., Jr, and Swanstrom, R. (2004). Multiple V1/V2 env variants are frequently present during primary infection with human immunodeficiency virus type 1. J. Virol. *78*, 11208–11218.

Rizzuto, C.D., Wyatt, R., Hernandez-Ramos, N., Sun, Y., Kwong, P.D., Hendrickson, W.A., and Sodroski, J. (1998). A conserved HIV gp120 glycoprotein structure involved in chemokine receptor binding. Science *280*, 1949–1953.

Rowland-Jones, S., Pinheiro, S., and Kaul, R. (2001). New insights into host factors in HIV-1 pathogenesis. Cell *104*, 473–476.

Rucker, J., Edinger, A.L., Sharron, M., Samson, M., Lee, B., Berson, J.F., Yi, Y., Margulies, B., Collman, R.G., Doranz, B.J., *et al.* (1997). Utilization of chemokine receptors, orphan receptors, and herpesvirus-encoded receptors by diverse human and simian immunodeficiency viruses. J. Virol. *71*, 8999–9007.

Sagar, M., Wu, X., Lee, S., and Overbaugh, J. (2006). Human immunodeficiency virus type 1 V1-V2 envelope loop sequences expand and add glycosylation sites over the course of infection, and these modifications affect antibody neutralization sensitivity. J. Virol. *80*, 9586–9598.

Sahmoud, T., Laurian, Y., Gazengel, C., Sultan, Y., Gautreau, C., and Costagliola, D. (1993). Progression to AIDS in French haemophiliacs: association with HLA-B35. Aids *7*, 497–500.

Saksena, N.K., and Smit, T.K. (2005). HAART & the molecular biology of AIDS dementia complex. Indian J Med Res. *121*, 256–269.

Samson, M., Libert, F., Doranz, B.J., Rucker, J., Liesnard, C., Farber, C.M., Saragosti, S., Lapoumeroulie, C., Cognaux, J., Forceille, C., *et al.* (1996). Resistance to HIV-1 infection in caucasian individuals bearing mutant alleles of the CCR-5 chemokine receptor gene. Nature *382*, 722–725.

Samson, M., Edinger, A.L., Stordeur, P., Rucker, J., Verhasselt, V., Sharron, M., Govaerts, C., Mollereau, C., Vassart, G., Doms, R.W., *et al.* (1998). ChemR23, a putative chemoattractant receptor, is expressed in monocyte-derived dendritic cells and macrophages and is a coreceptor for SIV and some primary HIV-1 strains. European Journal of Immunology *28*, 1689–1700.

Scarlatti, G., Tresoldi, E., Bjorndal, A., Fredriksson, R., Colognesi, C., Deng, H.K., Malnati, M.S., Plebani, A., Siccardi, A.G., Littman, D.R., *et al.* (1997). *In vivo* evolution of HIV-1 co-receptor usage and sensitivity to chemokine-mediated suppression. Nat. Med. *3*, 1259–1265.

Schacker, T., Little, S., Connick, E., Gebhard, K., Zhang, Z.Q., Krieger, J., Pryor, J., Havlir, D., Wong, J.K., Schooley, R.T., *et al.* (2001). Productive infection of T cells in lymphoid tissues during primary and early human immunodeficiency virus infection. J. Infect. Dis. *183*, 555–562.

Schrager, L.K., and D'Souza, M.P. (1998). Cellular and anatomical reservoirs of HIV-1 in patients receiving potent antiretroviral combination therapy. JAMA *280*, 67–71.

Schuitemaker, H., and Miedema, F. (1996). Viral and cellular determinants of HIV-1 replication in macrophages. Aids *10 Suppl A*, S25–32.

Scoggins, R.M., Taylor, J.R., Jr, Patrie, J., van't Wout, A.B., Schuitemaker, H., and Camerini, D. (2000). Pathogenesis of primary R5 human immunodeficiency virus type 1 clones in SCID-hu mice. J. Virol. *74*, 3205–3216.

Scorza Smeraldi, R., Fabio, G., Lazzarin, A., Eisera, N., Uberti Foppa, C., Moroni, M., and Zanussi, C. (1988). HLA-associated susceptibility to AIDS: HLA B35 is a major risk factor for Italian HIV-infected intravenous drug addicts. Hum Immunol *22*, 73–79.

Shankarappa, R., Margolick, J.B., Gange, S.J., Rodrigo, A.G., Upchurch, D., Farzadegan, H., Gupta, P., Rinaldo, C.R., Learn, G.H., He, X., *et al.* (1999). Consistent viral evolutionary changes associated with the progression of human immunodeficiency virus type 1 infection. J. Virol. *73*, 10489–10502.

Shieh, J.T., Martin, J., Baltuch, G., Malim, M.H., and Gonzalez-Scarano, F. (2000). Determinants of syncytium formation in microglia by human immunodeficiency virus type 1: role of the V1/V2 domains. J. Virol. *74*, 693–701.

Shimizu, N., Soda, Y., Kanbe, K., Liu, H.Y., Jinno, A., Kitamura, T., and Hoshino, H. (1999). An orphan G protein-coupled receptor, GPR1, acts as a coreceptor to allow replication of human immunodeficiency virus types 1 and 2 in brain-derived cells. J. Virol. *73*, 5231–5239.

Shioda, T., Oka, S., Xin, X., Liu, H., Harukuni, R., Kurotani, A., Fukushima, M., Hasan, M.K., Shiino, T., Takebe, Y., *et al.* (1997). *In vivo* sequence variability of human immunodeficiency virus type 1 envelope gp120: association of V2 extension with slow disease progression. J. Virol. *71*, 4871–4881.

Smith, D.G., Guillemin, G.J., Pemberton, L., Kerr, S., Nath, A., Smythe, G.A., and Brew, B.J. (2001). Quinolinic acid is produced by macrophages stimulated by platelet activating factor, Nef and Tat. Journal of Neurovirology *7*, 56–60.

Smith, P.D., Meng, G., Salazar-Gonzalez, J.F., and Shaw, G.M. (2003). Macrophage HIV-1 infection and the gastrointestinal tract reservoir. J. Leukoc. Biol. *74*, 642–649.

Sodroski, J., Goh, W.C., Rosen, C., Campbell, K., and Haseltine, W.A. (1986). Role of the HTLV-III/LAV envelope in syncytium formation and cytopathicity. Nature *322*, 470–474.

Sonza, S., Maerz, A., Deacon, N., Meanger, J., Mills, J., and Crowe, S. (1996). Human immunodeficiency virus type 1 replication is blocked prior to reverse transcription and integration in freshly isolated peripheral blood monocytes. J. Virol. *70*, 3863–3869.

Sonza, S., and Crowe, S.M. (2001). Reservoirs for HIV infection and their persistence in the face of undetectable viral load. AIDS Patient Care STDS *15*, 511–518.

Sonza, S., Mutimer, H.P., Oelrichs, R., Jardine, D., Harvey, K., Dunne, A., Purcell, D.F., Birch, C., and Crowe, S.M. (2001). Monocytes harbour replication-competent, non-latent HIV-1 in patients on highly active antiretroviral therapy. Aids *15*, 17–22.

Sterjovski, J., Churchill, M., Wesselingh, S.L., and Gorry, P.R. (2006). HIV-1 entry inhibitors: classes, applications and factors affecting potency. Curr. HIV Res. *4*, 387–400.

Sterjovski, J., Churchill, M.J., Ellett, A., Gray, L.R., Roche, M.J., Dunfee, R.L., Purcell, D.F., Saksena, N., Wang, B., Sonza, S., *et al.* (2007). Asn 362 in gp120 contributes to enhanced fusogenicity by CCR5-restricted HIV-1 envelope glycoprotein variants from patients with AIDS. Retrovirology *4*, 89.

Strizki, J.M., Albright, A.V., Sheng, H., O'Connor, M., Perrin, L., and Gonzalez-Scarano, F. (1996). Infection of primary human microglia and monocyte-derived macrophages with human immunodeficiency virus type 1 isolates: evidence of differential tropism. J. Virol. *70*, 7654–7662.

Tang, J., Costello, C., Keet, I.P., Rivers, C., Leblanc, S., Karita, E., Allen, S., and Kaslow, R.A. (1999). HLA class I homozygosity accelerates disease progression in human immunodeficiency virus type 1 infection. AIDS Res. Hum. Retroviruses *15*, 317–324.

Thomas, E.R., Dunfee, R.L., Stanton, J., Bogdan, D., Taylor, J., Kunstman, K., Bell, J.E., Wolinsky, S.M., and Gabuzda, D. (2007). Macrophage entry mediated by HIV Envs from brain and lymphoid tissues is determined by the capacity to use low CD4 levels and overall efficiency of fusion. Virology *360*, 105–119.

Thompson, K.A., Churchill, M.J., Gorry, P.R., Sterjovski, J., Oelrichs, R.B., Wesselingh, S.L., and McLean, C.A. (2004). Astrocyte specific viral strains in HIV dementia. Ann. Neurol. *56*, 873–877.

Tiwari, J.L., and Terasaki, P.I. (1981). HLA-DR and disease associations. Prog. Clin. Biol. Res. *58*, 151–163.

Triques, K., and Stevenson, M. (2004). Characterization of restrictions to human immunodeficiency virus type 1 infection of monocytes. J. Virol. *78*, 5523–5527.

Trkola, A., Dragic, T., Arthos, J., Binley, J.M., Olson, W.C., Allaway, G.P., Cheng-Mayer, C., Robinson, J., Maddon, P.J., and Moore, J.P. (1996). CD4-dependent, antibody-sensitive interactions between HIV-1 and its co-receptor CCR-5. Nature *384*, 184–187.

Trkola, A., Paxton, W.A., Monard, S.P., Hoxie, J.A., Siani, M.A., Thompson, D.A., Wu, L., Mackay, C.R., Horuk, R., and Moore, J.P. (1998). Genetic subtype-independent inhibition of human immunodeficiency virus type 1 replication by CC and CXC chemokines. J. Virol. 72, 396–404.

Tuttle, D.L., Harrison, J.K., Anders, C., Sleasman, J.W., and Goodenow, M.M. (1998). Expression of CCR5 increases during monocyte differentiation and directly mediates macrophage susceptibility to infection by human immunodeficiency virus type 1. J. Virol. *72*, 4962–4969.

Tuttle, D.L., Anders, C.B., Aquino-De Jesus, M.J., Poole, P.P., Lamers, S.L., Briggs, D.R., Pomeroy, S.M., Alexander, L., Peden, K.W., Andiman, W.A., *et al.* (2002). Increased replication of non-syncytium-inducing HIV type 1 isolates in monocyte-derived macrophages is linked to advanced disease in infected children. AIDS Res. Hum. Retroviruses *18*, 353–362.

van't Wout, A.B., Ran, L.J., Kuiken, C.L., Kootstra, N.A., Pals, S.T., and Schuitemaker, H. (1998). Analysis of the temporal relationship between human immunodeficiency virus type 1 quasispecies in sequential blood samples and various organs obtained at autopsy. J. Virol. *72*, 488–496.

Venkatesan, S., Petrovic, A., Van Ryk, D.I., Locati, M., Weissman, D., and Murphy, P.M. (2002). Reduced cell surface expression of CCR5 in CCR5Delta 32 heterozygotes is mediated by gene dosage, rather than by receptor sequestration. J. Biol. Chem. *277*, 2287–2301.

Wang, J., Crawford, K., Yuan, M., Wang, H., Gorry, P.R., and Gabuzda, D. (2002). Regulation of CC chemokine receptor 5 and CD4 expression and human immunodeficiency virus type 1 replication in human macrophages and microglia by T helper type 2 cytokines. J. Infect. Dis. *185*, 885–897.

Wei, X., Ghosh, S.K., Taylor, M.E., Johnson, V.A., Emini, E.A., Deutsch, P., Lifson, J.D., Bonhoeffer, S., Nowak, M.A., Hahn, B.H., *et al.* (1995). Viral dynamics in human immunodeficiency virus type 1 infection. Nature *373*, 117–122.

Wei, X., Decker, J.M., Wang, S., Hui, H., Kappes, J.C., Wu, X., Salazar-Gonzalez, J.F., Salazar, M.G., Kilby, J.M., Saag, M.S., *et al.* (2003). Antibody neutralization and escape by HIV-1. Nature *422*, 307–312.

Wolfs, T.F., de Jong, J.J., Van den Berg, H., Tijnagel, J.M., Krone, W.J., and Goudsmit, J. (1990). Evolution of sequences encoding the principal neutralization epitope of human immunodeficiency virus 1 is host dependent, rapid, and continuous. Proc. Natl. Acad. Sci. U.S.A. *87*, 9938–9942.

Wolinsky, S.M., Wike, C.M., Korber, B.T., Hutto, C., Parks, W.P., Rosenblum, L.L., Kunstman, K.J., Furtado, M.R., and Munoz, J.L. (1992). Selective transmission of human immunodeficiency virus type-1 variants from mothers to infants. Science *255*, 1134–1137.

Wolinsky, S.M., Korber, B.T., Neumann, A.U., Daniels, M., Kunstman, K.J., Whetsell, A.J., Furtado, M.R., Cao, Y., Ho, D.D., and Safrit, J.T. (1996). Adaptive evolution of human immunodeficiency virus-type 1 during the natural course of infection. Science *272*, 537–542.

Wu, L., Gerard, N.P., Wyatt, R., Choe, H., Parolin, C., Ruffing, N., Borsetti, A., Cardoso, A.A., Desjardin, E., Newman, W., *et al.* (1996). CD4-induced interaction of primary HIV-1 gp120 glycoproteins with the chemokine receptor CCR-5. Nature *384*, 179–183.

Wu, L., Paxton, W.A., Kassam, N., Ruffing, N., Rottman, J.B., Sullivan, N., Choe, H., Sodroski, J., Newman, W., Koup, R.A., *et al.* (1997). CCR5 levels and expression pattern correlate with infectability by macrophage-tropic HIV-1, *in vitro*. J. Exp. Med. *185*, 1681–1691.

Wyatt, R., Kwong, P.D., Desjardins, E., Sweet, R.W., Robinson, J., Hendrickson, W.A., and Sodroski, J.G. (1998). The antigenic structure of the HIV gp120 envelope glycoprotein. Nature *393*, 705–711.

Wyatt, R., and Sodroski, J. (1998). The HIV-1 envelope glycoproteins: fusogens, antigens, and immunogens. Science *280*, 1884–1888.

Yang, C., Li, M., Limpakarnjanarat, K., Young, N.L., Hodge, T., Butera, S.T., McNicholl, J.M., Mastro, T.D., and Lal, R.B. (2003). Polymorphisms in the CCR5 coding and noncoding regions among HIV type 1-exposed, persistently seronegative female sex-workers from Thailand. AIDS Res. Hum. Retroviruses *19*, 661–665.

Zhang, L.Q., MacKenzie, P., Cleland, A., Holmes, E.C., Brown, A.J., and Simmonds, P. (1993). Selection for specific sequences in the external envelope protein of human immunodeficiency virus type 1 upon primary infection. J. Virol. *67*, 3345–3356.

Zhang, Z., Schuler, T., Zupancic, M., Wietgrefe, S., Staskus, K.A., Reimann, K.A., Reinhart, T.A., Rogan, M., Cavert, W., Miller, C.J., *et al.* (1999). Sexual transmission and propagation of SIV and HIV in resting and activated CD4+ T cells. Science *286*, 1353–1357.

Zhu, T., Mo, H., Wang, N., Nam, D.S., Cao, Y., Koup, R.A., and Ho, D.D. (1993). Genotypic and phenotypic characterization of HIV-1 patients with primary infection. Science *261*, 1179–1181.

Zimmerman, P.A., Buckler-White, A., Alkhatib, G., Spalding, T., Kubofcik, J., Combadiere, C., Weissman, D., Cohen, O., Rubbert, A., Lam, G., *et al.* (1997). Inherited resistance to HIV-1 conferred by an inactivating mutation in CC chemokine receptor 5: studies in populations with contrasting clinical phenotypes, defined racial background, and quantified risk. Mol. Med. *3*, 23–36.

Zwick, M.B., and Burton, D.R. (2007). HIV-1 neutralization: mechanisms and relevance to vaccine design. Curr. HIV Res. *5*, 608–624.

# Post-entry Restrictions to Lentiviral Replication

4

Jenny L Anderson and Gilda Tachedjian


## Abstract

Host organisms contain numerous defences to protect themselves against invading pathogens like lentiviruses. These defence systems include intracellular inhibitory proteins termed 'restriction factors', which have recently risen to prominence with the discovery of APOBEC3 and TRIM5 cellular proteins as key factors restricting invading retroviruses in host cells. Understanding how these and other intracellular restriction factors block retroviruses is providing new insights into barriers of retroviral replication and the cross-species transmission of lentiviruses. Moreover, these restriction factors have important ramifications for identifying new targets for antiviral therapy, developing better models of HIV-1/AIDS and selecting retroviral vectors for gene therapy. Thus, restriction factors have emerged as a dynamic, important area of lentivirus research. This chapter reviews rapid advances in understanding intracellular restriction factors targeting retroviruses after entry into host cells. These restriction factors include TRIM5α, TRIMCyp, cyclophilin A, Fv1, APOBEC3, TRIM28, ZAP, tetherin and calcium-modulating cyclophilin ligand. Lentiviral countermeasures to circumvent these intracellular restrictions are also canvassed. Delineating the biology of these intracellular restrictions appears promising for developing novel antiviral therapies to curb lentiviruses as effectively as the naturally occurring intracellular restriction factors.


## Introduction

The complex relationship between host organisms and their pathogens is heavily influenced by host defence systems and pathogen countermeasures. For lentiviruses and other retroviral pathogens, they must contend with innate and adaptive immune mechanisms in the host. Recently, the role of intracellular defences have emerged as critical components in the antiviral armour of the host following the discovery of various intracellular proteins that potently restrict the replication of invading retroviruses (Gao *et al.*, 2002; Sheehy *et al.*, 2002; Stremlau *et al.*, 2004; Wolf and Goff, 2007; Barr *et al.*, 2008; Neil *et al.*, 2008; Varthakavi *et al.*, 2008). As these intracellular host proteins block or 'restrict' retroviral replication, they are thus referred to as 'restriction factors'.

The identification of intracellular restriction factors targeting retroviruses occurred around 40 years ago with the report of inbred mice having different susceptibility to Friend strains of murine leukaemia virus (MLV) (Lilly, 1967). Among the series of mouse genes found to influence this viral susceptibility, the cellular Friend virus susceptibility 1 (Fv1) gene (Lilly, 1970; Pincus *et al.*, 1971) was shown to target the capsid core of invading MLV strains and restrict viral infection (Hopkins *et al.*, 1977; DesGroseillers and Jolicoeur, 1983; Ou *et al.*, 1983). Similar restrictions to MLV strains in human cells, and then to lentiviruses in primate cells were described in the late 1990s to early 2000s (Hofmann *et al.*, 1999;

Towers *et al.*, 2000; Besnier *et al.*, 2002; Cowan *et al.*, 2002; Munk *et al.*, 2002; Towers *et al.*, 2002; Hatziioannou *et al.*, 2003). The subsequent discovery of cellular APOBEC3G and TRIM5α in 2002 and 2004 respectively as key determinants of cell susceptibility to retroviruses sparked tremendous interest in the role of intracellular proteins in regulating retroviral infection and cross-species transmission (Sheehy *et al.*, 2002; Stremlau *et al.*, 2004). The recent discovery that tetherin and calcium-modulating cyclophilin ligand (CAML) can regulate late stages of retroviral replication broadened the scope of restriction factors (Neil *et al.*, 2008; Varthakavi *et al.*, 2008). Moreover, new restrictions are emerging in the literature that await further characterization (Schmitz *et al.*, 2004; Pineda *et al.*, 2007; Han *et al.*, 2008; Kahl *et al.*, 2008; Yan and Kozak, 2008), and may also emerge from large screens of cellular proteins impacting HIV-1 replication (Brass *et al.*, 2008; Konig *et al.*, 2008; Zhou *et al.*, 2008). Thus, restriction factors appear to have great influence on lentiviral replication *in vivo* and are important factors to consider during lentiviral replication in macrophages.

Restriction factors can target diverse members of the *Retroviridae*. These viruses reverse-transcribe their viral RNA genome into complementary, double stranded DNA (cDNA) via their reverse transcriptase (RT) enzyme while replicating in host cells. Some restriction factors also target *Hepadnaviridae* (i.e. hepatitis B viruses), endogenous retroviruses and retrotransposons in the larger collection of retro-transcribing viruses, as well as other virus families. Thus, restriction factors can target diverse pathogens.

Restriction factors can target both early stages (binding to integration) and late stages (gene expression to particle release) of retrovirus replication in host cells (lentivirus replication is reviewed in Chapter 1). For instance, TRIM5α, TRIMCyp, cyclophilin A (CypA) and Fv1 target the capsid shell of invading retroviral cores to restrict early viral replication in cells (DesGroseillers and Jolicoeur, 1983; Franke *et al.*, 1994; Thali *et al.*, 1994; Nisole *et al.*, 2004; Sayah *et al.*, 2004; Stremlau *et al.*, 2004). Conversely, tetherin and CAML target lentiviral release from cells late in retroviral replication (Neil *et al.*, 2008; Varthakavi *et al.*, 2008). Some restriction factors like the APOBEC3 proteins have an unusual mode of action where they must be present in virus producer cells and packaged into virions late in replication, but restrict the early replication of this virus in target cells (Sheehy *et al.*, 2002). Of note, many of these restriction factors target early viral replication indicating this stage may be particularly susceptible to therapeutic intervention. Thus, restriction factors can have diverse targets and consequences in retroviral replication, perhaps facilitating their collective ability to block retrovirus replication *in vivo*.

Understanding how these intracellular restriction factors impede retroviruses may enhance our understanding of retroviral replication and transmission across animal species. Moreover, they may unveil new retroviral targets and/or strategies for therapeutic intervention, lead to better animal models of HIV/AIDS and guide better choices of retroviral vectors for gene therapy in different animals. Thus restriction factors are an important area of retrovirus research and advances in this rapidly moving field are reviewed here. Restriction factors targeting retroviruses after entry into host cells will be discussed according to their retroviral target and stages of retroviral replication they impact (summarized in Table 4.1). This includes restriction of: viral capsid/uncoating by Fv1, TRIM5α, TRIMCyp and CypA; viral genomes/reverse transcription/integration by APOBEC3; viral gene expression by TRIM28, ZAP and RNA silencing; and virion release by tetherin, CAML, UBP, β-catenin and E-cadherin.

## Restrictions of lentiviral capsid cores

### Fv1

A role for host cell factors targeting the incoming capsid core of invading retroviruses to restrict infection first emerged from experiments with gammaretroviruses (Lilly, 1967). Studies of mice susceptibility to Friend strains of murine leukaemia virus (MLV) revealed a series of loci including Friend virus susceptibility 1 (Fv1) that regulates infection in inbred mice (Lilly, 1970; Pincus *et al.*, 1971). While Fv1 targets gammaretroviruses as opposed to lentiviruses, many aspects of Fv1 restriction relate to lentivirus restriction by TRIM5 proteins warranting the discussion of Fv1 restriction here.

Two predominant Fv1 alleles were identified in inbred mice as mediating Fv1 restriction: *Fv1ⁿ* and *Fv1ᵇ*. Moreover, the ability of MLV strains to replicate in the presence of these alleles allowed their classification as N-tropic, B-tropic or NB-tropic. For instance, *Fv1ⁿ* in NIH Swiss mice allows the replication of N-tropic MLV (N-MLV) but restricts B-tropic MLV (B-MLV). Conversely, *Fv1ᵇ* in Balb/c mice permits B-MLV replication but blocks N-MLV. NB-tropic MLV is not restricted by endogenous *Fv1ⁿ* or *Fv1ᵇ* (Hartley *et al.*, 1970). An *Fv1⁰* 'null' allele is also present in wild mice and allows N-, B- and NB-tropic strains of MLV to replicate (Hartley and Rowe, 1975; Kozak, 1985). An alternative $Fv1^{nr}$ allele allows some N-MLV strains to replicate but restricts other N-MLV and B-MLV strains (Kozak, 1985). Finally, restriction mediated by Fv1 is codominant, with heterozygous mice expressing *Fv1ⁿ* and *Fv1ᵇ* able to restrict both N-MLV and B-MLV (Pincus *et al.*, 1971; Rowe and Hartley, 1972).

Fv1 restriction of MLV is manifest in the target cell early during virus replication, allowing reverse transcription of the MLV RNA genome into cDNA in the cytoplasm but formation of nuclear viral cDNA circles and provirus is impaired. This implies that Fv1 blocks the nuclear entry of MLV cDNA and/or subsequent integration (Jolicoeur and Baltimore, 1976; Pryciak and Varmus, 1992). Fv1 targets the MLV capsid (CA) protein to restrict virus infection (Hopkins *et al.*, 1977; DesGroseillers and Jolicoeur, 1983; Ou *et al.*, 1983), with amino acid 110 in the MLV CA the primary determinant for *Fv1ⁿ* or *Fv1ᵇ* restriction (Kozak and Chakraborti, 1996). Restriction can be overcome by high titres of sensitive virus (Pincus *et al.*, 1975; Duran-Troise *et al.*, 1977), perhaps owing to the low endogenous expression of Fv1 (Yap and Stoye, 2003). However, restriction is not abrogated by CA in a Gag precursor context or when expressed alone in target cells before infection (Dodding *et al.*, 2005). This infers that Fv1 targets the mature CA shell of invading MLV cores for restriction permitting reverse transcription but preventing viral cDNA accumulation in the nucleus (Jolicoeur and Baltimore, 1976; Pryciak and Varmus, 1992).

A significant advance in understanding Fv1 restriction arose when Stoye and colleagues used positional cloning to isolate the Fv1 gene (Best *et al.*, 1996). Analysing genomic DNA from different animal species indicates the Fv1 gene is limited to murine cells (Best *et al.*, 1996). Intriguingly, the Fv1 gene exhibits sequence homology to the human and murine ERV-L family of endogenous retroviruses that corresponds to a *gag* gene based on the position of homology within ERV-L (Best *et al.*, 1996; Benit *et al.*, 1997). Moreover, Fv1 contains a major homology region (MHR) (Benit *et al.*, 1997) characteristic of orthoretroviral CA proteins and its integrity is essential for Fv1 restriction (Bishop *et al.*, 2001). Hence, Fv1 appears to encode a CA-like protein.

The Fv1ⁿ and Fv1ᵇ proteins differ at three positions towards the C-terminus that combine to determine the specificity of restriction (Bishop *et al.*, 2001). Studies of purified, recombinant Fv1ⁿ (Bishop *et al.*, 2006a) reveal that it has protease-resistant N-terminal and C-terminal domains separated by a protease-sensitive, central linker region. The N-terminal domain contains a strongly predicted, coiled-coil motif [known to mediate the oligomerization of TRIM proteins, (Reymond *et al.*, 2001)] and this purified domain can form extended helical dimers (Bishop *et al.*, 2006b). This indicates the N-terminal domain allows Fv1 oligomerization consistent with earlier data showing overexpressed Fv1 aggregates in cells (Yap and Stoye, 2003). The C-terminal domain contains the MHR region (Bishop *et al.*, 2006a) with the three restriction specificity determinants located towards the far end of the C-terminus (Bishop *et al.*, 2001). Thus, the N-terminal domain appears to allow Fv1 dimerization while the C-terminal domain encodes the specificity determinant for targeting retroviruses.

This structural organization of Fv1 is analogous to the TRIM5α and TRIMCyp restriction factors that target lentiviruses. These TRIM5 proteins contain an N-terminal RING, B-box and coiled-coil motifs (the latter two domains enabling oligomerization), and a C-terminal B30.2 or cyclophilin A (CypA) domain mediating specific viruses targeted for restriction (*see below*). In fact, the Fv1 N-terminal domain can functionally replace the TRIM5 RING, B-box and coiled-coil motifs when fused to the CypA

**Table 4.1** Host cell factors restricting retroviral pathogens after entry

| Restriction factor | Host[a] | Viruses restricted | Retroviral target | Retroviral replication block | Retroviral aversion |
|---|---|---|---|---|---|
| Fv1 | Mice | Retroviruses<br>Gamma | CA shell of invading virion | Nuclear viral cDNA | Mutate viral CA sequence |
| TRIM5α | Primates<br>Cow<br>Rabbit | Retroviruses<br>Lenti<br>Gamma<br>Beta<br>Spuma | CA shell of invading virion | *Most instances*<br>Viral CA uncoating and reverse transcription<br>*Select instances*<br>Nuclear viral cDNA | Mutate viral CA sequence |
| TRIMCyp | *Aotus* monkeys<br>Macaques<br>Rhesus<br>Pig-tailed<br>Cynomologous | Retroviruses<br>Lenti<br>Gamma | CA | Viral CA uncoating and reverse transcription | Mutate viral CA sequence |
| CypA | Primates | Retroviruses<br>Lenti | CA | *Target cell OWM CypA restricting HIV-1*<br>Viral cDNA | |
| | | | | *Intravirion human and agm CypA restricting SIV*<br>Viral reverse transcription | SIV Vif |
| APOBEC3[b] | Primates<br>Mice<br>Rat<br>Hamster<br>Horse<br>Rabbit | Retroviruses<br>Lenti<br>Alpha<br>Beta<br>Gamma<br>Delta<br>Spuma<br>Endogenous retroviruses<br>LTR and non-LTR retrotransposons<br>Hepatitis B virus<br>Adeno-associated virus<br>Human papillomavirus | ssDNA | *Virion-associated A3 in target cell*<br>Hypermutate reverse transcribing viral cDNA<br>Impair viral gene expression<br>Decrease viral cDNA<br>impair reverse transcription<br>impair integration<br>*Target cell LMM A3*<br>Decrease viral cDNA<br>*Target cell HMM A3*<br>Usurp retrotransposon RNA and impair retrotransposition | Lentiviral Vif<br>Prevent A3 virion incorporation<br>Reduce A3G translation |

| | | | | | |
|---|---|---|---|---|---|
| TRIM28 | Mice | Retroviruses<br>Lenti<br>Beta<br>Gamma<br>Spuma | Provirus<br>Repressor binding site<br>Via ZFP809 cell protein | Transcription | |
| ZAP | Rat<br>Human | Retroviruses<br>Gamma<br>Alphaviruses<br>Filoviruses | Viral mRNA<br>Zap-responsive element | Degrades viral mRNA | |
| Tetherin | Primates<br>Mice | Retroviruses<br>Lenti<br>Alpha<br>Beta<br>Delta<br>Spuma<br>Filoviruses<br>Lassa virus | Virion lipid envelope | Virion release | HIV-1 Vpu |
| CAML | Human | Retroviruses<br>Lenti<br>Gamma | HIV-1 Vpu | Virion release | HIV-1 Vpu<br>HIV-2 Env |

a Host animals whose restriction factor is defined to restrict retroviral pathogens.

b Encompasses APOBEC3 (A3) proteins varying in number and properties between mammals.

agm, African green monkey; CA, capsid; cDNA, complementary DNA; Env, envelope; HIV-1 and HIV-2, human immunodeficiency virus type 1 and type 2; HMM, high molecular mass; LMM, low molecular mass; NWM, New world monkeys, OWM, Old world monkeys; SIV, simian immunodeficiency virus; ssDNA, single-stranded DNA; Vif, virion infectivity factor; Vpu, viral protein U.

specificity determinant and restrict HIV-1 and feline immunodeficiency virus (FIV) lentiviruses (Schaller *et al.*, 2007b; Yap *et al.*, 2007). This indicates the Fv1 N-terminal domain and TRIM5 N-terminal RING, B-box and coiled-coil motifs may share overlapping attributes for restriction, such as oligomerization.

Despite these many insights into Fv1 restriction, the precise mechanism used by Fv1 to restrict MLV remains elusive. One theory is that the Fv1 C-terminal specificity domain interacts directly or indirectly with the outer CA shell of invading MLV cores. This is consistent with the ability of excess sensitive virus to abrogate restriction (Pincus *et al.*, 1975; Duran-Troise *et al.*, 1977). Moreover, structural models place the MLV CA residues mediating Fv1 restriction on the outer surface of CA lattices that form the shell of invading retroviral cores, and thus in an accessible position for Fv1 interaction (Bishop *et al.*, 2006a). However, direct interaction between Fv1 and MLV CA has yet to be demonstrated probably because Fv1 interacts with the CA shell of invading MLV cores, which can be challenging to handle. Fv1 might multimerize via its N-terminal domain (Bishop *et al.*, 2006b) either before or after capsid interaction for restriction. Studies with a TRIM5α-21RING fusion protein indicate that TRIM5α preferentially interacts with retroviral CA shells as a dimer under stringent conditions of reduced restriction factor levels (Langelier *et al.*, 2008). This may favour the weakly expressed Fv1 similarly interacting with MLV CA shells as a dimer. Fv1 interaction with MLV allows reverse transcription but probably perturbs this viral cDNA accessing the nucleus and/or cellular DNA for integration (Jolicoeur and Baltimore, 1976; Pryciak and Varmus, 1992). Therefore, Fv1 interaction with invading MLV cores might sterically perturb and/or mask key motifs in these viral complexes impeding trafficking into the nucleus and/or integration to ultimately block viral infection. The Fv1 MHR located in the C-terminal domain is essential for restriction (Bishop *et al.*, 2001) but its role in restriction also remains unclear. Clues may lie in studies of MHR function in orthoretroviral CA proteins, which is implicated in immature and mature CA assembly, including potential conformational reorganization of CA for mature CA assembly (Purdy *et al.*, 2008). Perhaps the MHR of Fv1 also induces Fv1 conformational changes, possibly assisting Fv1 assembly with invading viral cores, or perturbing the CA shell of MLV cores to ultimately impair nuclear entry and integration despite complete reverse transcription. Defining the role of the Fv1 MHR should facilitate a detailed understanding of the Fv1 restriction mechanism, as will continuing comparison to TRIM5α and TRIMCyp restriction.

## TRIM5α, TRIMCyp and cyclophilin A

### *Discovery of TRIM5α*

The discovery of Fv1 (Lilly, 1970; Pincus *et al.*, 1971) set a precedent for the expression of intrinsic, antiviral proteins in host cells that target the CA of invading retroviruses to block infection. While the Fv1 gene appears limited to murine cells, a similar Fv1-like restriction blocking N-MLV replication was subsequently identified in human and other mammalian cells (Towers *et al.*, 2000). Like Fv1, this restriction was dominant, saturable by excess amounts of affected virus, and mapped to the same N-MLV CA residue at amino acid 110 (Towers *et al.*, 2000; Towers *et al.*, 2002; Hatziioannou *et al.*, 2003). However unlike Fv1, this restriction blocked the accumulation of N-MLV reverse transcription products (Towers *et al.*, 2000) whereas Fv1 blocks N-MLV after reverse transcription at nuclear entry and/or integration (Jolicoeur and Baltimore, 1976; Pryciak and Varmus, 1992). In human cells, this Fv1-like restriction blocking N-MLV was termed Resistance Factor 1 [Ref1, (Towers *et al.*, 2000)].

In addition to MLV CA, evidence of a similar restriction targeting lentiviral CA also emerged from studies seeking animal models of HIV/AIDS. These studies found HIV-1 only infected a limited host range of chimpanzees (Alter *et al.*, 1984), gibbon apes (Lusso *et al.*, 1988) and pig-tailed macaques (Agy *et al.*, 1992). Conversely, most Old World monkey (OWM) species like rhesus macaques or their cells were refractory to HIV-1 infection (Shibata *et al.*, 1991; Hofmann *et al.*, 1999). HIV-1 restriction in the rhesus cells was not due to a block in HIV-1 entry or its reverse transcriptase (RT) as chimeric SIV containing the HIV-1 envelope protein (Shibata *et al.*, 1991) or RT (Balzarini *et al.*, 1997) still infected rhesus cells. Rather, HIV-1 reverse transcription products were impaired in rhesus cells (Shibata *et al.*, 1995) indicating a post-entry block early in HIV-1 replication. SIV and HIV-1 chimeric viruses mapped this HIV-1 restriction to the CA-p2 of the HIV-1 *gag* gene (Dorfman and Gottlinger, 1996). Therefore, rhesus cells contain a post-entry restriction of HIV-1 replication mapping to the HIV-1 CA protein, similar to Fv1 and Ref1 restriction of N-MLV.

Examining the post-entry ability of HIV-1 and SIV from rhesus macaques ($SIV_{mac}$) to infect various mammalian cell types further revealed that cells from OWM in Africa and Asia generally restrict HIV-1 but not $SIV_{mac}$ (Hofmann *et al.*, 1999). Conversely, cells from New World monkeys (NWM) in Central and South America generally restrict $SIV_{mac}$ but not HIV-1. These restrictions correlated with the species of the cells rather than tissue type, implying widely expressed, species-specific cellular factors were mediating the post-entry restriction of these primate lentiviruses (Hofmann *et al.*, 1999). These and other lentivirus restrictions in primate cells also shared many features of Fv1 and Ref1 restriction of N-MLV. That is, these lentivirus restrictions were (a) saturable with excess amounts of affected virus, (b) dominant in heterokaryons of restricted and permissive cells and (c) in the case of HIV-1, restriction mapped to the CA-p2 domain (Besnier *et al.*, 2002; Cowan *et al.*, 2002; Munk *et al.*, 2002; Hatziioannou *et al.*, 2003). Given the similarity to Fv1 restriction, this lentivirus restriction in primate cells was termed lentivirus susceptibility factor 1 (Lv1; Cowan *et al.*, 2002). However, unlike Fv1 that restricts after reverse transcription at nuclear entry and/or integration, Lv1 restricted lentiviral replication at reverse transcription (Besnier *et al.*, 2002; Cowan *et al.*, 2002; Munk *et al.*, 2002) similar to Ref1 (Towers *et al.*, 2000). Cross-saturation experiments indicated restriction-sensitive but divergent retroviruses could saturate the restriction of each other. This inferred that primate cells use the same cell factor/s to restrict divergent retroviruses (Hatziioannou *et al.*, 2003). However, the identity of the cellular factors mediating Lv1 and Ref1 restriction was unknown.

This seminal advance arose in 2004 when Stremlau and colleagues used a genetic screen to identify the rhesus macaque TRIM5α protein (rhTRIM5α) as restricting HIV-1 replication in rhesus macaque cells (Stremlau *et al.*, 2004). Depleting endogenous rhTRIM5α in rhesus cells using siRNA prevented HIV-1 restriction, and expressing rhTRIM5α in permissive human cells was sufficient to restrict HIV-1. Therefore, rhTRIM5α was necessary and sufficient for restricting HIV-1. RhTRIM5α also exhibited other hallmarks expected of the cellular factor restricting HIV-1 in rhesus cells from earlier work. This included blocking HIV-1 replication after entry at reverse transcription; restriction mapping to the HIV CA-p2 sequences; and rhTRIM5α having minor impact on $SIV_{mac}$ from rhesus macaques but potently restricting HIV-1. Therefore, rhTRIM5α was mediating the restriction of HIV-1 in rhesus macaque cells.

### *TRIM protein family*

TRIM5α belongs to a large family of TRIM proteins characterized by a tripartite motif of a RING (really interesting new gene), one or two B-boxes and a coiled-coil domain [reviewed in Reymond *et al.* (2001) and Nisole *et al.* (2005)]. TRIM proteins are also known as RBCC proteins given their tripartite motif arrangement. The RING domain is a zinc-finger that coordinates two zinc atoms and has E3 ubiquitin ligase activity in some TRIM proteins like TRIM5δ (Xu *et al.*, 2003). B-box domains are unique to TRIM proteins and are of two types, B-box 1 and B-box 2, which differ in the number and spacing of their conserved cysteine and histidine residues. B-box 1 always precedes B-Box 2 if present in the same TRIM protein. B-box domains also coordinate zinc but their function in TRIM proteins lacks clarity. The coiled-coil domain oligomerizes with other coiled-coil domains. This allows homo- and hetero-oligomerization of TRIM proteins, and arrangement into higher-order structures or 'bodies' that can define discrete subcellular compartments in the cytoplasm and/or nucleus of cells (Reymond *et al.*, 2001). TRIM proteins can also contain variable C-terminal domains, like the B30.2 (PRYSPRY) domain of TRIM5α. TRIM5α is generated from the TRIM5 gene. The mRNA transcribed from the TRIM5 gene alternatively splices into different isoforms. The TRIM5α isoform translates the B30.2 domain, giving TRIM5α the characteristic structure of a RING, B-box 2, coiled-coil and B30.2 domain (Nisole *et al.*, 2005).

*TRIM5 as a new class of intracellular restriction factors*

Following the discovery of rhTRIM5α (Stremlau *et al.*, 2004), analogous TRIM5α proteins in various primate species were quickly found to underlie the post-entry, species-specific, Lv1 and Ref1 restrictions of lentiviruses and N-MLV respectively (Hatziioannou *et al.*, 2004b; Keckesova *et al.*, 2004; Perron *et al.*, 2004; Yap *et al.*, 2004). This rapidly expanding field has since identified TRIM5α proteins in many primate species including humans, apes, OWM and NWM (Song *et al.*, 2005b; Ylinen *et al.*, 2005; Newman *et al.*, 2006; Kaiser *et al.*, 2007; Kratovac *et al.*, 2008). These TRIM5α proteins target specific retroviruses that commonly originate from another animal species. They can also restrict diverse types of retroviruses including lentiviruses and gammaretroviruses (Hatziioannou *et al.*, 2004b; Keckesova *et al.*, 2004; Perron *et al.*, 2004; Stremlau *et al.*, 2004; Yap *et al.*, 2004; Saenz *et al.*, 2005; Ylinen *et al.*, 2005), and more recently spumaretroviruses (Yap *et al.*, 2008) and betaretroviruses (Diehl *et al.*, 2008). Restriction by these TRIM5α proteins maps to the retroviral CA protein (Hatziioannou *et al.*, 2004a; Perron *et al.*, 2004; Stremlau *et al.*, 2004; Sebastian and Luban, 2005; Ylinen *et al.*, 2005; Stremlau *et al.*, 2006a; Perron *et al.*, 2007; Kratovac *et al.*, 2008; Langelier *et al.*, 2008). In most instances, restriction results in an early viral replication block that prevents the accumulation of retroviral reverse transcription products (Keckesova *et al.*, 2004; Perron *et al.*, 2004; Stremlau *et al.*, 2004; Ylinen *et al.*, 2005; Anderson *et al.*, 2006; Wu *et al.*, 2006). However, like Fv1 restriction, exceptions that block after reverse transcription do occur such as squirrel monkey TRIM5α restriction of $SIV_{mac}$ (Ylinen *et al.*, 2005).

In addition to TRIM5α, a novel TRIM5 variant fusing the TRIM5 RBCC domains to a CypA domain, termed TRIMCyp, was also identified in the *Aotus* species of NWM and 3 different macaque species of OWM. This TRIMCyp variant was first identified in NWM owl monkey cells (Nisole *et al.*, 2004; Sayah *et al.*, 2004), which specifically restrict HIV-1 in contrast to most other NWM cells (Hofmann *et al.*, 1999). This specific HIV-1 restriction arises from the retrotransposition of a CypA pseudogene between exons 7 and 8 in the TRIM5 locus of owl monkeys. mRNA transcribed from this gene splices to the CypA insert instead of B30.2 sequence, thus encoding a novel TRIMCyp protein where the TRIM5 RBCC domains fuse to CypA instead of B30.2. As CypA binds HIV-1 CA protein (Luban *et al.*, 1993), owl monkey TRIMCyp thus uses its CypA domain to bind the CA shell of invading HIV-1 cores for restriction (Nisole *et al.*, 2004; Diaz-Griffero *et al.*, 2006b). Owl monkey TRIMCyp appears to use the same method to bind and restrict FIV, SIV from African green monkeys ($SIV_{agm}$) and other primate lentiviral CA (Diaz-Griffero *et al.*, 2006b; Lin and Emerman, 2006; Zhang *et al.*, 2006; Kratovac *et al.*, 2008). This is because like HIV-1, CypA binds FIV CA (Lin and Emerman, 2006) and TRIMCyp restriction of these viruses is blocked by cyclosporine A (CsA), which competitively binds CypA and blocks CA binding (Diaz-Griffero *et al.*, 2006b; Lin and Emerman, 2006; Zhang *et al.*, 2006). To date, TRIMCyp is only found in the *Aotus* species of NWMs (Ribeiro *et al.*, 2005).

TRIMCyp was also recently identified in three OWM macaque species: *Macaca nemestrina* (pig-tailed macaques), *Macaca mulatta* (rhesus macaques) and *Macaca fascicularis* (cynomologous macaques) (Liao *et al.*, 2007; Brennan *et al.*, 2008; Newman *et al.*, 2008; Virgen *et al.*, 2008; Wilson *et al.*, 2008a). These OWM TRIMCyps also fuse the TRIM5 RBCC domains to CypA. However unlike NWM TRIMCyp, these OWM TRIMCyps arose from the insertion of the CypA pseudogene into the 3′ untranslated region of the TRIM5 locus accompanied by a splice acceptor mutation (Brennan *et al.*, 2008; Newman *et al.*, 2008; Wilson *et al.*, 2008a) that seemingly favours alternative splicing to the CypA. These OWM TRIMCyps thus represent a second independent generation of TRIMCyp to the NWM *Aotus* species. The convergent evolution and maintenance of these TRIMCyps from two independent retrotransposition events in NWM and OWM indicates these TRIM5 RBCC fusion proteins binding retroviral CA must confer key selective advantages to the host.

The OWM TRIMCyps also differ from NWM TRIMCyp in restriction specificity. For instance, NWM owl monkey TRIMCyp restricts HIV-1 but all three OWM macaque TRIMCyps do not (Nisole *et al.*, 2004; Sayah *et al.*, 2004; Liao *et al.*, 2007; Brennan *et al.*, 2008; Kratovac *et al.*, 2008; Newman *et al.*, 2008; Virgen *et al.*, 2008; Wilson *et al.*, 2008a). This different HIV-1 restriction in NWM owl monkey versus OWM pig-tailed macaque TRIMCyps maps to a single amino acid change in the CypA domain that alters pig-tailed macaque TRIMCyp binding to HIV-1 CA (Kratovac *et al.*, 2008; Virgen *et al.*, 2008). This same amino acid change also exists in the CypA domain of cynomologous macaque (Brennan *et al.*, 2008) and rhesus macaque (rh) TRIMCyps (Wilson *et al.*, 2008a). This amino acid change therefore probably explains why all three OWM TRIMCyps fail to restrict HIV-1. While rhTRIMCyp does not restrict HIV-1 or Equine infectious anaemia virus (EIAV), it restricts HIV-2 and FIV more potently than rhTRIM5α (Wilson *et al.*, 2008a). Moreover, the rhTRIMCyp allele is common in Indian but not Chinese rhesus macaques (Wilson *et al.*, 2008a). Perhaps different retroviral evolutionary pressure in distinct geographic locations is selecting for distinct TRIM5 alleles in animals at these different locations. The rhTRIMCyp allele may be selected for in Indian rhesus macaques for better restriction of lentiviruses like HIV-2 and FIV, at the expense of restricting lentiviruses like HIV-1 and EIAV.

In addition to rhTRIMCyp, recent work reveals multiple TRIM5α alleles (TRIM5 RBCC fused to B30.2) exist in rhesus macaques and sooty mangabeys (Newman *et al.*, 2006). These TRIM5α alleles within the same species can have different retroviral specificities. For instance, the rhTRIM5α alleles differ in their ability to restrict HIV-2 (Wilson *et al.*, 2008b). Therefore, TRIM5 restriction specificity differs *within* an animal species as well as *between* animal species (Newman *et al.*, 2006). Most non-synonymous polymorphisms between alleles cluster in the coiled-coil and B30.2 domains indicating evolutionary selection (Newman *et al.*, 2006). Moreover, the different ability of rhTRIM5α alleles to restrict HIV-2 maps to a TFP339Q mutation in the B30.2 domain (Wilson *et al.*, 2008b). This indicates the coiled-coil and B30.2 domains are both under evolutionary selection and determine the retroviral specificity of TRIM5 proteins.

Pig-tailed macaques also contain two novel TRIM5 isoforms, TRIM5θ and TRIM5η, in addition to TRIMCyp, which probably arise by alternative splicing of mRNA from the TRIM5 locus (Brennan *et al.*, 2007). TRIM5θ encodes the TRIM5 RBCC only and thus lacks a recruitment domain to bind retroviruses for restriction. Conversely, TRIM5η encodes the TRIM5 RBCC and most of the B30.2 but lacks 9 amino acids coded by exon 7 and does not restrict any of 10 diverse retroviruses tested (Brennan *et al.*, 2007; Kratovac *et al.*, 2008). Thus, the biological purpose of these isoforms is unclear. Different TRIM proteins heteromultimerize via their coiled-coil domains and coexpression of TRIM5α with short, mutated, splice variants or different TRIM5 alleles or other TRIMs (e.g. TRIM34) can be dominant-negative and perturb TRIM5α restriction (Stremlau *et al.*, 2004; Berthoux *et al.*, 2005a; Passerini *et al.*, 2006; Zhang *et al.*, 2006; Schaller *et al.*, 2007a; Wilson *et al.*, 2008a,b). Therefore, TRIM5θ and/or TRIM5η might be expressed via alternative splicing under certain conditions in pig-tailed macaque cells to act as a dominant negative and regulate TRIMCyp restriction.

Apart from primates, TRIM genes capable of restricting retroviruses are also found in cows (Si *et al.*, 2006; Ylinen *et al.*, 2006) and rabbits (Schaller *et al.*, 2007a). TRIM5-like genes have also been recently reported in mice and rats (Tareen *et al.*, 2009). While the antiviral activity of these rodent TRIM5-like genes remains unknown, they show evidence of positive selection indicative of evolutionary conflict and potential antiviral function like other mammalian TRIM5 genes (Tareen *et al.*, 2009).

TRIM5α and TRIMCyp proteins (hereafter also termed TRIM5 proteins given their common TRIM5 RBCC domains) typically restrict retroviruses from other animal species, with minor impact on retroviruses originating from the same species. Therefore, TRIM5 proteins constitute a new class of intracellular antiviral proteins that help protect hosts against the cross-species transmission of retroviruses. Overexpressing TRIM5 and Fv1 proteins can also expand their antiviral specificity (Bock *et al.*, 2000; Ylinen *et al.*, 2005; Passerini *et al.*, 2006; Ylinen *et al.*, 2006;

Schaller *et al.*, 2007a). This may arise *in vivo* as human interferons elevate TRIM5 expression in cell culture (Asaoka *et al.*, 2005; Sakuma *et al.*, 2007b; Carthagena *et al.*, 2008), and expand owl monkey TRIMCyp restriction to include N-MLV and B-MLV (Carthagena *et al.*, 2008). Notably, human and African green monkey TRIM1 also restricts N-MLV, albeit less potently than TRIM5α (Yap *et al.*, 2004). Moreover, other TRIM proteins have antiviral function including: TRIM19 (promyelocytic leukaemia/PML), TRIM22, TRIM25, TRIM28 and TRIM34 (Nisole *et al.*, 2005; Zhang *et al.*, 2006; Towers, 2007; Gack *et al.*, 2007; Wolf and Goff, 2007; Barr *et al.*, 2008). Many TRIM proteins are also inducible by type I and type II interferons that are important for pathogen resistance (Orimo *et al.*, 2000; Rhodes *et al.*, 2002; Asaoka *et al.*, 2005; Rajsbaum *et al.*, 2008). Therefore, TRIM proteins may play a wider role in immunity to pathogen invasion (reviewed in Ozato *et al.*, 2008).

*Function of TRIM5 domains*

TRIM5α and TRIMCyp contain the common N-terminal RBCC domains fused to either a B30.2 or CypA domain at the C-terminus respectively (Fig. 4.1). While all of these domains can contribute to restriction, their relative importance and absolute function has in some instances been controversial.

The *B30.2 (PRYSPRY)* or *CypA domain* of TRIM5 proteins bind the CA shell of invading retroviral cores and thus determine the specific retroviruses restricted by the various TRIM5 proteins (Sawyer *et al.*, 2005; Sebastian and Luban, 2005; Song *et al.*, 2005a; Stremlau *et al.*, 2005, 2006a; Yap *et al.*, 2005; Perron *et al.*, 2007; Kar *et al.*, 2008; Langelier *et al.*, 2008). B30.2 domains contain a PRY and SPRY domain (Rhodes *et al.*, 2005) and structural studies indicate SPRY/B30.2 domains form protein binding interfaces and thus are protein interaction domains (Woo *et al.*, 2006; James *et al.*, 2007). Initial rhTRIM5α studies demonstrated the B30.2 domain is crucial for restriction and subsequent mutagenesis studies revealed the B30.2 domain mediates the specific retroviruses restricted by the various TRIM5α proteins (Stremlau *et al.*, 2004, 2005; Sawyer *et al.*, 2005; Song *et al.*, 2005a; Yap *et al.*, 2005). Phylogenetic analyses of different primate TRIM5α indicates the B30.2 domain has encountered strong positive selection during evolution, defined by a substantial increase in non-synonymous versus synonymous mutations in this domain (Sawyer *et al.*, 2005). Moreover, the B30.2 domain contains a 13-amino-acid 'patch' harbouring a cluster of positively selected residues confirmed through mutagenesis as a major determinant of TRIM5α retroviral specificity. The B30.2 domain contains multiple variable regions with species-specific amino acid polymorphism and polymorphism in length (Song *et al.*, 2005a). Variable region 1 encompasses the 'patch' and polymorphism at amino acid 332 in this region regulates primate TRIM5α restriction of retroviruses like HIV-1, $SIV_{mac}$ and chimeric virus containing extinct *Pan troglodytes* endogenous retrovirus (PtERV1) CA (Stremlau *et al.*, 2005; Yap *et al.*, 2005; Li *et al.*, 2006; Ohkura *et al.*, 2006; Kaiser *et al.*, 2007). However, variable regions 1, 2

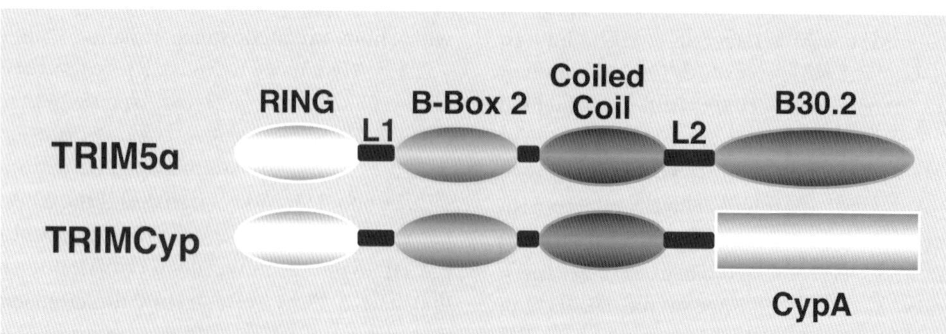


**Figure 4.1** Schematic diagram of the domain composition of TRIM5α and TRIMCyp. These proteins contain amino terminal RING, B-Box 2 and Coiled-coil domains joined by linker 2 (L2) protein sequence to either a B30.2 or CypA domain at the carboxy terminus. L1, linker 1 protein sequence.

and 3 all determine TRIM5α retroviral specificity (Ohkura *et al.*, 2006).

TRIM5α and TRIMCyp restriction targets retroviral CA protein assembled into the polymeric lattice of mature retroviral cores (Dodding *et al.*, 2005; Forshey *et al.*, 2005; Sebastian and Luban, 2005; Stremlau *et al.*, 2006a; Diaz-Griffero *et al.*, 2007; Perron *et al.*, 2007; Kar *et al.*, 2008; Langelier *et al.*, 2008). The CypA domain of owl monkey TRIMCyp binds HIV-1 and FIV CA, and enables TRIMCyp to bind the CA shell of invading HIV-1 cores for restriction (Luban *et al.*, 1993; Nisole *et al.*, 2004; Diaz-Griffero *et al.*, 2006b; Lin and Emerman, 2006). As the B30.2 domain of TRIM5α mediates retroviral specificity (Sawyer *et al.*, 2005; Song *et al.*, 2005a; Stremlau *et al.*, 2005; Yap *et al.*, 2005) and replaces the CypA domain of owl monkey TRIMCyp, the TRIM5α B30.2 domain was also thought to bind the CA shell of invading retroviral cores for restriction. However, it has been challenging to demonstrate TRIM5α B30.2-mediated binding to retroviral CA. This is because TRIM5α targets retroviral CA assembled into the polymeric lattice shell of invading retroviral cores (Dodding *et al.*, 2005; Forshey *et al.*, 2005) that can be difficult to handle due to their instability. Multiple lines of evidence now demonstrate that TRIM5α binds invading retroviral cores via their B30.2 domain. Firstly, using detergent to strip the membrane envelope from MLV virions reveals human (hu) TRIM5α in cell lysates binds to N-MLV cores but not B-MLV cores via the B30.2 domain (Sebastian and Luban, 2005). Secondly, primate TRIM5α variants or owl monkey TRIMCyp in cell lysates bind synthetic, *in vitro* assembled HIV-1 capsid-nucleocapsid (CA-NC) lattices that recapitulate the hexagonal CA lattice structure of HIV-1 cores, in a B30.2 species-specific or CypA-dependent manner (Diaz-Griffero *et al.*, 2006b; Stremlau *et al.*, 2006a). Finally, recent studies with purified, chimeric TRIM5-21R (TRIM5α substituted with a TRIM21 RING domain to stop rapid protein turnover for purification) reveal TRIM5-21R directly binds HIV-1 CA lattices or native EIAV cores via the B30.2 domain in the absence of other cellular proteins (Kar *et al.*, 2008; Langelier *et al.*, 2008). This collectively confirms the B30.2 domain of TRIM5α proteins directly bind the CA shell of invading retroviral cores for restriction.

Therefore, the B30.2 domain of TRIM5α and the CypA domain of TRIMCyp allow TRIM5 proteins to bind the CA shell of invading retroviral cores for restriction, and variation in the sequence of these domains determines the retroviral specificity of these TRIM5 proteins. The TRIM5α B30.2 domain structure has been modelled based on recently solved structures for SPRY/B30.2 domains of GUSTAVAS (Woo *et al.*, 2006), PRYSPRY-19q13.4.1 (Grutter *et al.*, 2006), SSB-2 (Masters *et al.*, 2006) and TRIM21 (James *et al.*, 2007). This modelling predicts the TRIM5α B30.2 variable regions are located in exposed loops protruding from the B30.2 β-sandwich core and form a protein binding interface (Ohkura *et al.*, 2006; Woo *et al.*, 2006; James *et al.*, 2007). As the B30.2 domain mediates the retroviral specificity of TRIM5α and the retroviral CA determines restriction, these structural models suggest the TRIM5α B30.2 variable region loops form a binding site for retroviral CA. Moreover, amino acid or length changes in these loops probably allow TRIM5α proteins to selectively recognize and restrict different retroviruses. A recent study also identifies a patch predicted to lie on the B30.2 domain surface that is involved in restriction but is dispensable for binding retroviral CA (Sebastian *et al.*, 2009). Thus, the B30.2 domain may have a secondary function in addition to binding retroviral CA, such as binding cofactors of restriction.

The TRIM5 *coiled-coil domain* is essential for restriction and enables TRIM5 proteins to oligomerize with themselves or other TRIM proteins (Reymond *et al.*, 2001; Xu *et al.*, 2003; Mische *et al.*, 2005; Perez-Caballero *et al.*, 2005a; Song *et al.*, 2005c; Javanbakht *et al.*, 2006; Yap *et al.*, 2007). In fact, TRIM5α oligomerization with short, mutated, splice variants or different TRIM5 alleles, or other TRIM proteins like TRIM34, can be dominant-negative and block TRIM5α restriction (Stremlau *et al.*, 2004; Berthoux *et al.*, 2005a; Passerini *et al.*, 2006; Zhang *et al.*, 2006; Schaller *et al.*, 2007a; Wilson *et al.*, 2008a,b). Phylogenetic comparison of primate TRIM5α sequences (Sawyer *et al.*, 2005) and TRIM5α alleles in rhesus macaques and sooty mangabeys (Newman *et al.*, 2006) reveal non-synonymous polymorphisms cluster in the coiled-coil as well as the B30.2 domain. This suggests TRIM5 oligomerization may assist retroviral CA recognition. TRIM5α was initially reported to preferentially oligomerize into trimers based on the gel mobility of chemically cross-linked TRIM5α from cell lysates (Mische *et al.*, 2005). However, recent studies demonstrate that purified chimeric TRIM5α substituted with a TRIM21 RING preferentially exists as a dimer by analytical ultracentrifugation and mass spectrometry, with unusually slow mobility by gel electrophoresis giving the misleading impression of a trimer (Kar *et al.*, 2008; Langelier *et al.*, 2008). These chimeric dimers bind synthetic CA-NC lattices more efficiently than the monomeric protein under stringent conditions, supporting TRIM5α oligomerization into dimers for efficient binding to retroviral cores (Langelier *et al.*, 2008). This is consistent with reports that TRIM5α oligomerization increases binding avidity for retroviral CA and restriction potency (Javanbakht *et al.*, 2005; Mische *et al.*, 2005; Javanbakht *et al.*, 2006). Fusing CypA to specific oligomerization domains can also generate artificial restriction factors and enhance restriction (Javanbakht *et al.*, 2007; Yap *et al.*, 2007). These TRIM5 dimers may enable multiple contacts of their affiliated B30.2 domains with the hexameric CA lattice shell of retroviral cores for efficient binding and restriction.

TRIM5α and TRIMCyp can also oligomerize into higher order complexes termed 'bodies' found primarily in the cytoplasm of host cells (Reymond *et al.*, 2001; Xu *et al.*, 2003; Perez-Caballero *et al.*, 2005b; Song *et al.*, 2005c; Diaz-Griffero *et al.*, 2006a; Diaz-Griffero *et al.*, 2006b; Wu *et al.*, 2006; Campbell *et al.*, 2007). These bodies are dynamic and continuously exchange TRIM5 protein between the bodies and the cytosolic pool indicating they are not dead end protein aggregations but instead have functional significance. These bodies are also highly mobile and traffic along the microtubule network (Campbell *et al.*, 2007). Moreover, they are sensitive to proteasome inhibition and heat shock, where they can accumulate into large aggregations (Diaz-Griffero *et al.*, 2006a; Wu *et al.*, 2006; Anderson *et al.*, 2008; Campbell *et al.*, 2008). A potential role for these TRIM5α cytoplasmic bodies in restriction has been the subject of much controversy, as discussed below.

The *linker 2 region* connecting the coiled-coil domain to the B30.2 or CypA domain in TRIM5 proteins also impacts TRIM5α oligomerization (Javanbakht *et al.*, 2006). Mutating this linker region impairs TRIM5α oligomerization, retroviral capsid binding and restriction (Mische *et al.*, 2005; Javanbakht *et al.*, 2006). It is possible the spacing provided by this linker 2 region between the TRIM5 coiled-coil and CA-binding domains enables optimal and/or flexible TRIM5 multimerization and CA binding, enhancing TRIM5 binding and restriction of different retroviral CA cores.

B-Box domains are zinc-coordinating motifs unique to TRIM proteins (Nisole *et al.*, 2005). The role of the *B-box 2 domain* in TRIM5 proteins has been perplexing given B-box 2 mutations can have variable impact on restriction. For instance, deleting or mutating conserved cysteine and histidine residues in B-box 2 impairs rhTRIM5α and huTRIM5α restriction, but has little impact on owl monkey TRIMCyp or chimeric rhTRIMCyp restriction (Javanbakht *et al.*, 2005; Perez-Caballero *et al.*, 2005a; Diaz-Griffero *et al.*, 2007; Yap *et al.*, 2007). Moreover, conditionally defective B-box mutations exist that perturb restriction of one retrovirus but not another (Li *et al.*, 2007a). For example, mutating rhTRIM5α B-box 2 residues 120 and 121 prevents N-MLV restriction but not HIV-1 restriction (Li and Sodroski, 2008). This conditionally defective phenotype and overall requirement for B-box 2 during restriction correlates with the B30.2 domain and restriction potency. That is, the B-box 2 is required in examples of weak restriction by TRIM5α but is less important for potent restriction by TRIM5α or TRIMCyp (Li *et al.*, 2007a; Li and Sodroski, 2008). CA binding assays indicate the B-box 2 enhances the CA-binding efficiency of TRIM5α proteins (Li *et al.*, 2007a). Moreover, recent studies reveal the B-box 2 promotes the higher-order oligomerization of TRIM5α beyond initial dimerization, which enables more efficient TRIM5α binding to HIV-1 CA lattices (Li and Sodroski, 2008). Therefore, the B-box 2 promotes the higher-order oligomerization of TRIM5α to enhance binding to the CA shell of retroviral cores for restriction, a particular advantage when retrovirus restriction (and thus B30.2 interaction with CA) is weak. When restriction is potent, high affinity interaction between the TRIM5α B30.2 domain and retroviral CA probably obviates the need for B-box 2 higher-order oligomerization for efficient TRIM5α binding to viral CA cores. Thus, B-box 2 higher-order oligomerization and B30.2 CA binding appear to be complementary mechanisms to achieve efficient TRIM5α interaction with retroviral cores for restriction.

The *RING domain* at the N-terminus of TRIM5 restriction factors is a zinc-coordinating motif (Nisole *et al.*, 2005). RING domains define a major family of E3 ubiquitin ligases acting in the ubiquitin–proteasome system. E3 ubiquitin ligases complex with E2 ubiquitin-conjugating enzymes and transfer ubiquitin to bound substrates resulting in substrate monoubiquitylation or polyubiquitylation. Polyubiquitylated substrates are degraded by proteasomes (Liu *et al.*, 2005b). As RING domains function as E3 ubiquitin ligases in some TRIM proteins like TRIM5δ (Xu *et al.*, 2003) it was speculated that it may similarly function as an E3 ubiquitin ligase in TRIM5α (Stremlau *et al.*, 2004). Consistent with this notion, TRIM5α and owl monkey TRIMCyp are rapidly turned over in cells mediated by the RING and B-box 2 or RING respectively, with the RING domain facilitating TRIM5 polyubiquitylation (Diaz-Griffero *et al.*, 2006a). Recent work confirms that TRIM5α is a RING-finger-type E3 ubiquitin ligase that self-ubiquitylates or can be ubiquitylated by Ro52 E3 ubiquitin ligase (Yamauchi *et al.*, 2008). Purified chimeric TRIM5α substituted with the TRIM21 RING also self-ubiquitylates (Kar *et al.*, 2008; Langelier *et al.*, 2008). Thus, the TRIM5 RING has E3 ubiquitin ligase activity allowing TRIM5 self-ubiquitylation. Recent experiments show TRIM5 proteins undergo proteasome-dependent degradation upon exposure to restriction-sensitive retroviral CA cores during restriction (Rold and Aiken, 2008). Moreover, adding proteasome inhibitors rescues viral reverse transcription products from restriction and the clearance of virions from restricted cells is prevented (Anderson *et al.*, 2006; Wu *et al.*, 2006; Campbell *et al.*, 2008). This collectively implicates the TRIM5 RING domain, ubiquitylation and the ubiquitin–proteasome system in TRIM5 restriction. However, mutagenesis studies show the RING domain enhances restriction potency but is not essential for restriction (Stremlau *et al.*, 2004; Javanbakht *et al.*, 2005; Perez-Caballero *et al.*, 2005b). Therefore, the absolute role of the TRIM5 RING, ubiquitylation and ubiquitin–proteasome function during restriction remains controversial as discussed below. Notably, a H43Y non-synonymous mutation in the huTRIM5α RING domain is reported in 43% of Central and South Americans, which impairs huTRIM5α restriction (Sawyer *et al.*, 2006). Thus, this naturally occurring mutation infers the huTRIM5α RING domain and its E3 ubiquitin ligase activity may be involved in restricting retroviral infection *in vivo*.

*Restriction mechanism of TRIM5 proteins*

Despite rapid advances in understanding many aspects of restriction by TRIM5α and TRIMCyp proteins, their precise restriction mechanism remains controversial. What is clear is that TRIM5 restriction initiates with direct binding of the TRIM5 restriction factor to the CA shell of invading retroviral cores in target cells. This is supported by experiments showing TRIM5 restriction requires stable polymeric CA (Dodding *et al.*, 2005; Forshey *et al.*, 2005). Furthermore, purified TRIM5-21R chimeric protein or cell lysate-derived TRIM5α and TRIMCyp directly bind synthetic HIV-1 CA lattices, detergent stripped retroviral cores and/or pelletable CA (Nisole *et al.*, 2004; Sebastian and Luban, 2005; Stremlau *et al.*, 2006a; Diaz-Griffero *et al.*, 2007; Perron *et al.*, 2007; Kar *et al.*, 2008; Langelier *et al.*, 2008). TRIM5 proteins use their C-terminal B30.2 or CypA domain to target retroviral CA residues particularly present in exposed loops on the CA shell of invading retroviral cores (Hatziioannou *et al.*, 2004a; Nisole *et al.*, 2004; Perron *et al.*, 2004; Stremlau *et al.*, 2004; Ylinen *et al.*, 2005; Diaz-Griffero *et al.*, 2006b; Kratovac *et al.*, 2008; Langelier *et al.*, 2008). TRIM5 interaction with the CA shell of retroviral cores also requires the TRIM5 coiled-coil oligomerization domain (Mische *et al.*, 2005; Perez-Caballero *et al.*, 2005a; Song *et al.*, 2005c; Javanbakht *et al.*, 2006) and B-box 2 higher order oligomerization when TRIM5α B30.2 interaction with retroviral CA is weak (Li and Sodroski, 2008). This is consistent with the notion that sufficient amounts of TRIM5 must oligomerize with retroviral cores for restriction. *In situ* fluorescence microscopy also supports this notion by showing cytosolic HIV-1 complexes localize with rhTRIM5α cytoplasmic bodies during restriction (Campbell *et al.*, 2008).

However, a role for TRIM5 cytoplasmic bodies during restriction remains controversial. Some consider cytoplasmic bodies irrelevant for restriction as drug treatments or cell lines lacking cytoplasmic bodies *before* infection do not prevent TRIM5α restriction (Perez-Caballero *et al.*, 2005b; Song *et al.*, 2005c). However, these studies do not investigate the presence of cytoplasmic bodies in the cells *after* virus addition, where the virus may form a central point around which cytoplasmic bodies can still form. While these studies illustrate that pre-formed cytoplasmic bodies are not required for restriction, recent experiments imaging cells *after* virus addition indicate that invading HIV-1 cores localize with rhTRIM5α cytoplasmic bodies during restriction (Campbell *et al.*, 2008). Moreover, other data shows that restriction requires the TRIM5 coiled-coil oligomerization domain and under some circumstances, B-box 2 higher-order oligomerization (Mische *et al.*, 2005; Perez-Caballero *et al.*, 2005a; Song *et al.*, 2005c; Javanbakht *et al.*, 2006; Li and Sodroski, 2008). Thus, these data collectively indicate that a critical amount of TRIM5 is likely to accumulate with retroviral cores for restriction.

After sufficient quantities of TRIM5 protein have bound to the CA shell of invading retroviral cores, the consequence of this TRIM5 interaction for restriction remains a further point of contention. Some propose from CA sedimentation studies that TRIM5 binding leads to accelerated uncoating of CA from invading retroviral cores (Stremlau *et al.*, 2006a; Diaz-Griffero *et al.*, 2007; Perron *et al.*, 2007). Other studies illustrate the rate of CA uncoating from retroviral cores is critical for HIV-1 reverse transcription, with unstable CA mutations (that mimic accelerated uncoating) blocking HIV-1 reverse transcription products and infection (Forshey *et al.*, 2002). Thus, TRIM5 binding and accelerated uncoating of CA from invading viral cores is proposed to restrict viral reverse transcription products and replication (Stremlau *et al.*, 2006a). Although, precisely how TRIM5 binding leads to accelerated uncoating of CA from retroviral cores remains unclear.

Separate studies with proteasome inhibitors suggest a multi-step model for TRIM5 restriction (Anderson *et al.*, 2006; Wu *et al.*, 2006; Campbell *et al.*, 2008). These studies find that proteasome inhibitors perturb TRIM5 restriction, rescuing viral reverse transcription products from restriction but not viral nuclear cDNA or infection. Moreover, fluorescence microscopy reveals proteasome inhibitors reduce HIV-1 turnover and allow HIV-1 to accumulate in large rhTRIM5α cytoplasmic bodies, which probably obscure their movement into the nucleus explaining why viral nuclear cDNA and infection remain blocked. As proteasome inhibitors rescue one step of viral replication from restriction (reverse transcription), but not all steps of viral replication (viral nuclear cDNA and infection remain blocked), these studies indicate TRIM5 restriction involves multiple steps. Moreover, as the restriction of viral reverse transcription products is sensitive to proteasome inhibitors, this indicates proteasomes are involved in restricting viral reverse transcription products (Anderson *et al.*, 2006; Wu *et al.*, 2006; Campbell *et al.*, 2008).

This impact of proteasome inhibitors on TRIM5 restriction is similar to Fv1 restriction of MLV, squirrel monkey TRIM5α restriction of $SIV_{mac}$ and bovine TRIM restriction of HIV-2 (Jolicoeur and Baltimore, 1976; Pryciak and Varmus, 1992; Ylinen *et al.*, 2005; Ylinen *et al.*, 2006). In all cases, reverse transcription occurs but nuclear viral cDNA and infection are impaired. Moreover, artificial fusion of different TRIM RBCC domains with CypA generates novel restriction factors that can block retroviruses before or after reverse transcription (Yap *et al.*, 2006). These examples set a precedent for TRIM5 proteins being able to restrict retroviruses at multiple stages of viral replication. Consequently, a multi-step model is proposed for TRIM5 restriction in which TRIM5 binds and accumulates with retroviral cores (step 1) and rapidly blocks viral reverse transcription products through proteasome-mediated activity (step 2). When step 2 is blocked by proteasome inhibitors, step 1 is stabilized and reverse transcription products accumulate, but as the viral cores are trapped in large TRIM5 bodies blocking their movement into the nucleus, viral nuclear cDNA and infection remain blocked (Anderson *et al.*, 2006; Wu *et al.*, 2006; Campbell *et al.*, 2008). In this model, TRIM5 accumulation with retroviral cores is sufficient to induce restriction (step 1) but TRIM5 proteins have the added ability to rapidly block reverse transcription products via proteasome-mediated activity (step 2). Perhaps these two steps in TRIM5 restriction are more effective at restricting virus *in vivo*, consistent with reports the TRIM5 RING domain is not essential for restriction, but enhances restriction potency (Stremlau *et al.*, 2004; Javanbakht *et al.*, 2005; Perez-Caballero *et al.*, 2005b).

How this multi-step model relates to the accelerated uncoating model for TRIM5 restriction is unclear. A potential connection between these two models may lie in a recent study showing TRIM5 proteins are rapidly degraded in a proteasome-sensitive manner during restriction (Rold and Aiken, 2008). Perhaps TRIM5 accumulates with invading retroviral capsids (step 1) and this TRIM5 is then rapidly degraded by proteasomes, causing the CA shell of viral cores to fall apart (accelerated uncoating), restricting reverse transcription products (step 2) and viral infection. This explanation would encompass all available data published thus far (Fig. 4.2).

Finally, it is formally possible that TRIM5 accumulation with retroviral cores may have multiple outcomes during restriction via mechanisms not necessarily connected. For instance, TRIM5 interaction with retroviral cores may result in: (i) proteasome-dependent loss of TRIM5 protein and/or (ii) accelerated CA uncoating to ultimately impede reverse transcription products, or (iii) permit reverse transcription but sterically obscure viral cDNA trafficking to the nucleus and/or integration to ultimately restrict viral replication. Perhaps various mechanisms apply depending on the particular TRIM5 protein binding to the particular retrovirus, the protein residues involved, the binding affinity, the extent of oligomerization and the stability of the bound TRIM5 protein. Such a flexible model might account for the select instances where TRIM5 restriction results in a block *after* viral reverse transcription (Ylinen *et al.*, 2005, 2006; Yap *et al.*, 2006) rather than *at* reverse transcription as per most TRIM5 restriction examples (Keckesova *et al.*, 2004; Perron *et al.*, 2004; Stremlau *et al.*, 2004; Ylinen *et al.*, 2005; Anderson *et al.*, 2006; Wu *et al.*, 2006). A recent study suggests that the stability of the restriction factor interacting with retroviral cores influences the restriction outcome (Schaller *et al.*, 2007b). Replacing the TRIM5 RBCC of TRIMCyp with the Fv1 N-terminal domain generates an Fv1-CypA restriction factor with the retroviral specificity of TRIMCyp, but with the Fv1 antiviral outcome of restricting *after* reverse transcription, rather than *at* reverse transcription like native TRIMCyp. As Fv1 has a longer half-life than TRIMCyp, this may afford Fv1-restricted virus the time to reverse transcribe leading to a block later in viral replication (Schaller *et al.*, 2007b) compared to TRIMCyp and TRIM5α that have rapid half-lives (Diaz-Griffero *et al.*, 2006a) and usually restrict at reverse transcription. Thus, the stability of the restriction factor bound to the retroviral CA may determine the restriction outcome. Ultimately, while the precise restriction mechanism of TRIM proteins requires greater clarity, it is clear these proteins must initially accumulate with invading retroviral cores for restriction, and these proteins can restrict retroviruses via multiple steps, including a proteasome-sensitive step (Fig. 4.2).

In addition to this primary activity of TRIM5 proteins that bind and restrict invading retroviruses in target cells, rhTRIM5α is also reported to block HIV-1 production from virus producer cells through inducing rapid degradation of the HIV-1 Gag precursor protein (Sakuma *et al.*, 2007a). This antiviral activity maps to the rhTRIM5α RBCC domain, as opposed to the B30.2 domain that determines retroviral specificity during TRIM5 restriction in target cells. Thus, rhTRIM5α appears to recognize HIV-1 Gag in virus producer cells via a different mechanism than post-entry restriction in target cells. Moreover, while rhTRIM5α can induce this restriction in producer cells, huTRIM5α does not (Sakuma *et al.*, 2007a). However, this producer cell restriction of rhTRIM5α remains controversial as others find no effect of endogenous rhTRIM5α on HIV-1 virion production (Zhang *et al.*, 2008a). Furthermore, others find excess expression of both rhTRIM5α and huTRIM5α is required to efficiently impair HIV-1 production and thus question whether endogenous rhTRIM5α restricts HIV-1 virion production *in vivo* (Zhang *et al.*, 2008a). The different results in these studies may arise from technical differences between experiments or the finding that this rhTRIM5α producer cell restriction is

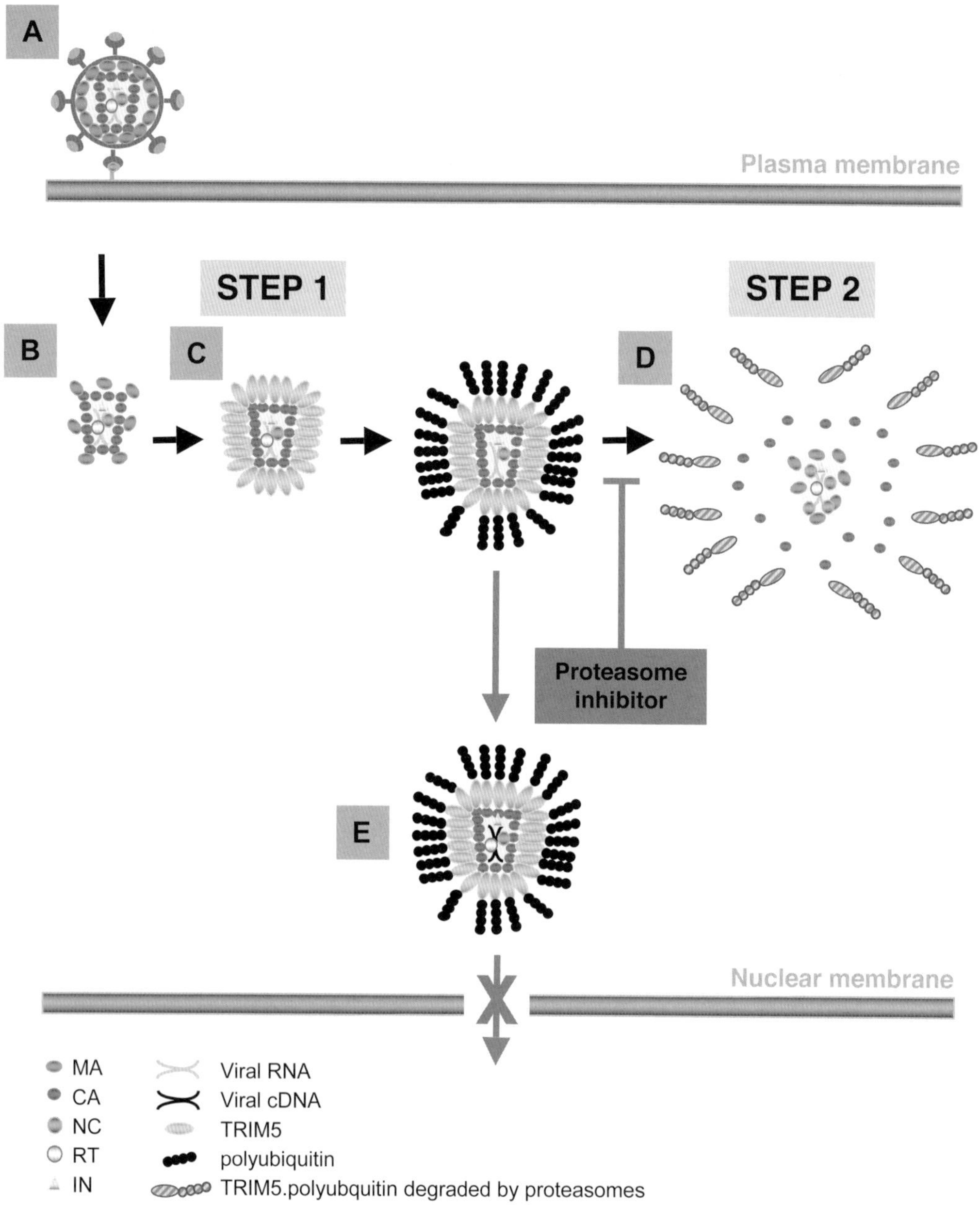


**Figure 4.2** Potential model of TRIM5 restriction of retroviruses. TRIM5 proteins restrict retroviruses during early viral replication in target cells. When retroviruses invade target cells, virion envelope glycoproteins bind receptors on the cell plasma membrane (A) and mediate fusion of virion and cell membranes, releasing the internal virion core into the cell cytoplasm (B). Cellular TRIM5 proteins target the capsid (CA) shell of these invading retroviral cores for restriction. Current data indicates that TRIM5 proteins restrict retroviruses via multiple steps. The following model encompasses available data published thus far. Firstly, TRIM5 proteins bind and accumulate with the CA shell of invading retroviral cores (step 1, C). Secondly, TRIM5 proteins are polyubiquitylated and rapidly degraded by proteasomes during restriction. This potentially causes the associated CA shell of viral cores to rapidly fall apart (accelerated uncoating) and stop reverse transcription product formation (step 2, D). Thus viral infection is restricted. If proteasome inhibitors are added to halt step 2, step 1 is stabilized and reverse transcription products now accumulate but these viral cores remain trapped in TRIM5 bodies that obscure movement into the nucleus, thus blocking viral nuclear cDNA and infection (E). In this model, TRIM5 accumulation with retroviral cores is sufficient to induce restriction (step 1) but TRIM5 proteins have the added ability to rapidly block reverse transcription products via proteasome activity (step 2). These multiple steps may enable TRIM5 to restrict retroviruses more effectively *in vivo*. CA, capsid; cDNA, complementary DNA; IN, integrase; MA, matrix; NC, nucleocapsid; RT, reverse transcriptase. A colour version of this figure is located in the plate section at the back of the book.

cell line dependent (Sakuma *et al.*, 2008). As interferons enhance rhTRIM5α, huTRIM5α, agmTRIM5α and owl monkey TRIMCyp expression (Asaoka *et al.*, 2005; Sakuma *et al.*, 2007b; Carthagena *et al.*, 2008; Rajsbaum *et al.*, 2008), it is possible interferon-induced elevation of TRIM5α levels might restrict HIV-1 production if sufficient rhTRIM5α was expressed in cells. However, whether this occurs *in vivo* is unknown. Thus, the role of rhTRIM5α in restricting HIV-1 virion production from producer cells remains controversial.

*Arsenic trioxide and heat shock impact TRIM5 restriction*

TRIM5α restriction of retroviruses in target cells is sensitive to arsenic trioxide and other drugs that disrupt mitochondria membrane potential (Berthoux *et al.*, 2003; Berthoux *et al.*, 2004; Keckesova *et al.*, 2004; Sayah and Luban, 2004; Sebastian *et al.*, 2006). Treating restrictive cells with these drugs stimulates viral cDNA synthesis and counteracts huTRIM5α/Ref-1 restriction of N-MLV and EIAV (Berthoux *et al.*, 2003; Keckesova *et al.*, 2004; Sayah and Luban, 2004). Moreover, treating restrictive cells with these drugs also stimulates HIV-1 cDNA synthesis and perturbs rhTRIM5α, agmTRIM5α and TRIMCyp restriction of HIV-1 in specific cell types (Berthoux *et al.*, 2004; Saenz *et al.*, 2005; Sebastian *et al.*, 2006). In contrast to proteasome inhibitors that do not rescue viral infection (Anderson *et al.*, 2006; Wu *et al.*, 2006), these drugs can rescue retroviral infection from restriction (Berthoux *et al.*, 2003; Berthoux *et al.*, 2004; Keckesova *et al.*, 2004; Sayah and Luban, 2004; Saenz *et al.*, 2005; Sebastian *et al.*, 2006). RNA interference experiments map this positive effect of arsenic trioxide to perturbing huTRIM5α restriction (Sebastian *et al.*, 2006). The ability of arsenic trioxide to rescue HIV-1 infection from rhTRIM5α, agmTRIM5α and TRIMCyp restriction is cell type dependent (Sebastian *et al.*, 2006), suggesting cell-specific factors regulate the effect of this drug on TRIM5 restriction. This cell type dependency may also explain why separate work finds arsenic trioxide effects on lentiviral restriction complicated and not clearly TRIM5α dependent (Saenz *et al.*, 2005). Ultimately, how arsenic trioxide and other drugs perturbing mitochondrial membrane potential rescue retroviral cDNA synthesis and infection from TRIM5α restriction remains unclear.

Exposing cells to a different type of stress imparted by prolonged exposure at high temperatures, a condition known to induce the heat shock response, also perturbs rhTRIM5α and TRIMCyp restriction of HIV-1 (Anderson *et al.*, 2008). Unlike proteasome inhibitors that efficiently rescue reverse transcription products from restriction but not nuclear viral cDNA, this heat treatment rescues both viral reverse transcription products and nuclear viral cDNA from restriction. Viral infection also remains blocked in the case of rhTRIM5α but is partially rescued in the case of TRIMCyp. The heat-induced change in rhTRIM5α restriction correlates with changes in the morphology of rhTRIM5α cytoplasmic bodies, where very large bodies can accumulate or the bodies disappear in some cells (Anderson *et al.*, 2008). Similarly, proteasome inhibitors that perturb TRIM5α restriction also change the morphology of TRIM5α cytoplasmic bodies (Wu *et al.*, 2006). This is consistent with the possibility that TRIM5 cytoplasmic bodies influence restriction, but ultimately how heat shock perturbs TRIM5 restriction remains unclear.

*Effects on the cross-species transmission of retroviruses*

Primate TRIM5α proteins have encountered strong positive selection during evolution (Sawyer *et al.*, 2005; Song *et al.*, 2005a), particularly in the B30.2 domain that interacts with the CA shell of retroviral cores and mediates retroviral specificity. This suggests TRIM5α proteins have evolved under the influence of antagonistic interactions with retroviruses. That is, retroviruses have probably evolved to evade TRIM5α restriction, thus imposing strong selective pressure on TRIM5 proteins to co-evolve to block these undesirable viruses and protect the host, thereby placing selective pressure back on the retroviruses and so on. This antagonistic pathogen-host scenario is commonly referred to as the Red Queen hypothesis [reviewed in (Towers, 2007)]. This strong positive selection of primate TRIM5α proteins during evolution therefore supports the likelihood that TRIM5α proteins are important barriers to the cross-species transmission of retroviruses.

The independent, convergent evolution of TRIMCyp in NWM and OWM further supports the binding of TRIM5 proteins to the CA shell of retroviral cores as important barriers to the cross-species transmission of retroviruses (Nisole *et al.*, 2004; Sayah *et al.*, 2004; Liao *et al.*, 2007; Brennan *et al.*, 2008; Newman *et al.*, 2008; Virgen *et al.*, 2008; Wilson *et al.*, 2008a). Lower-order primate TRIM5α frequently restrict a diverse array of retroviruses, as seen for rhTRIM5α, agmTRIM5α and owl monkey TRIMCyp (Kratovac *et al.*, 2008). This suggests that lower order primate TRIM5 proteins may be significant barriers to cross-species transmission of primate lentiviruses. Moreover, many primate TRIM5 proteins share a common ability to restrict N-MLV (Hatziioannou *et al.*, 2004b; Keckesova *et al.*, 2004; Perron *et al.*, 2004; Yap *et al.*, 2004; Song *et al.*, 2005b; Newman *et al.*, 2006; Kratovac *et al.*, 2008). This conserved ability to restrict retroviruses like N-MLV suggests it is a selective advantage in many primate species and thus of importance *in vivo*. Therefore, these findings collectively suggest TRIM5 proteins have evolved to play an important role in restricting specific retroviruses.

Conversely, recent work indicates that higher-order primate TRIM5α from humans, chimpanzees and gorillas are generally *not* able to inhibit diverse retroviral CA inserted into HIV-1 based, chimeric constructs (Kratovac *et al.*, 2008). This suggests these higher-order primate TRIM5α proteins are not a major barrier to infection by other primate lentiviruses and instead, may have permitted coinfection by multiple lentiviruses allowing recombinations and new pathogenic lentiviruses to evolve, leading to the current HIV pandemic today (Kratovac *et al.*, 2008). This coincides with separate work suggesting the selection pressure to restrict an ancient endogenous retrovirus PtERV1 and/or similar retroviruses during human evolution may have left humans more susceptible to other retroviruses like HIV-1 (Kaiser *et al.*, 2007). Additionally, primate TRIM5 proteins do not restrict chimeric MLV-based viruses containing various ancient primate or mouse endogenous gammaretroviral CA sequences (Perez-Caballero *et al.*, 2008). This again suggests primate TRIM5α proteins are unlikely to be major barriers to the cross-species transmission of endogenous gammaretroviruses. Instead, APOBEC3 cytidine deaminases played a role in restricting the ancient primate gammaretroviruses (Perez-Caballero *et al.*, 2008). Consequently, these recent studies suggest the overall ability of primate TRIM5 proteins to *solely* prevent the cross-species transmission of ancient gammaretroviruses (Perez-Caballero *et al.*, 2008), or higher-order primate TRIM5α to restrict modern primate lentiviruses (Kratovac *et al.*, 2008) is unlikely.

Therefore, while TRIM5 proteins probably contribute to preventing the cross-species transmission of retroviruses to endure strong positive selection, they are unlikely to act alone. Instead, TRIM5 proteins probably act in concert with other antiviral mechanisms such as APOBEC3 proteins to collectively prevent the cross-species transmission of specific retroviruses.

*A role for cyclophilin A in restriction*

In the search for host cell proteins that bind retroviral CA, a yeast two-hybrid screen unveiled CypA as binding HIV-1 CA both in isolation and in the context of Pr55Gag precursor protein (Luban *et al.*, 1993). CypA is a ubiquitous, cytosolic enzyme first identified as binding the cyclosporine A (CsA) immunosuppressant drug in lymphoid cells (Handschumacher *et al.*, 1984). CypA is a peptidyl-prolyl isomerase that binds proteins and assists with protein folding (reviewed in Barik, 2006). In the context of HIV-1, CypA binds an exposed proline-rich loop spanning Pro85-Pro93 in HIV-1 CA, binding primarily at CA G89 and P90 residues (Franke *et al.*, 1994; Thali *et al.*, 1994; Gamble *et al.*, 1996; Yoo *et al.*, 1997). Moreover, CypA catalyses the *cis–trans* isomerization of a peptidyl prolyl bond at Gly89-Pro90 in HIV-1 CA (Bosco *et al.*, 2002; Bosco and Kern, 2004). CypA is also packaged into HIV-1 virions through binding the proline rich loop in the CA domain of viral Pr55Gag precursor protein (Franke *et al.*, 1994; Thali *et al.*, 1994; Gamble *et al.*, 1996; Yoo *et al.*, 1997).

In human cells, CypA stimulates HIV-1 infection by stimulating the accumulation of viral cDNA products early during replication (Braaten *et al.*, 1996). CypA depletion experiments show that target cell CypA stimulates HIV-1 infection rather than CypA from virus producer cells (Sokolskaja *et al.*, 2004; Hatziioannou *et al.*, 2005). Moreover, CypA reportedly protects HIV-1 from a human cell restriction to stimulate HIV-1 infection in human target cells although how this occurs is unclear (Towers *et al.*, 2003; Sayah and Luban, 2004). Initial N-MLV saturation studies suggested CypA binding to the CA shell of invading HIV-1 cores might mask Ref-1 (huTRIM5α) interaction with the viral cores to protect HIV-1 from restriction (Sayah and Luban, 2004). However, subsequent studies found CypA protected HIV-1 from restriction in human cells independent of huTRIM5α restriction (Hatziioannou *et al.*, 2005; Keckesova *et al.*, 2006; Sokolskaja *et al.*, 2006; Stremlau *et al.*, 2006b). Therefore, human cells are proposed to contain antiviral factor(s) distinct from huTRIM5α that restrict HIV-1, and CypA protects HIV-1 from restriction by these unknown antiviral factor(s) in human cells permitting HIV-1 infection [reviewed in (Luban, 2007; Towers, 2007)]. However, the identity of these proposed antiviral factor(s) remains unknown and thus, the precise manner in which CypA protects HIV-1 from restriction in human target cells awaits clarification.

While CypA protects HIV-1 from restriction in human cells, studies show CypA can have the opposite effect and promote HIV-1 restriction in owl monkey, rhesus macaque and African green monkey (agm) cells (Towers *et al.*, 2003; Sayah and Luban, 2004; Berthoux *et al.*, 2005b; Keckesova *et al.*, 2006; Sokolskaja *et al.*, 2006; Stremlau *et al.*, 2006b). The discovery of TRIMCyp explained how CypA promotes HIV-1 restriction in owl monkey cells (Nisole *et al.*, 2004; Sayah *et al.*, 2004). Owl monkey TRIMCyp directly binds HIV-1 CA via the CypA domain (Nisole *et al.*, 2004; Diaz-Griffero *et al.*, 2006b) similar to human CypA (Franke *et al.*, 1994; Thali *et al.*, 1994; Gamble *et al.*, 1996; Yoo *et al.*, 1997). Therefore, the CypA domain of TRIMCyp recruits the entire owl monkey TRIMCyp restriction factor to invading HIV-1 CA shells for restriction. This same CypA recruitment mechanism also probably applies to owl monkey TRIMCyp restriction of FIV and SIV from the tantalus species of African green monkey ($SIV_{agm}$TAN). This is because TRIMCyp restriction of both viruses is sensitive to CsA inhibition of CypA, and FIV binds human CypA (Diaz-Griffero *et al.*, 2006b; Lin and Emerman, 2006; Zhang *et al.*, 2006).

Conversely, how CypA promotes HIV-1 restriction in rhesus macaque and agm cells is unclear. Some studies report CypA requires TRIM5α to promote HIV-1 restriction in these OWM cells, implying CypA sensitizes HIV-1 to TRIM5α-mediated restriction in these cells (Berthoux *et al.*, 2005b; Keckesova *et al.*, 2006). This led to a model proposing CypA binds the CA shell of invading HIV-1 cores and alters CA conformation via isomerization, promoting TRIM5α binding and restriction (Berthoux *et al.*, 2005b; Keckesova *et al.*, 2006). However, separate work finds CypA is not required but rather enhances TRIM5α-mediated restriction in these cells (Stremlau *et al.*, 2006b; Lin and Emerman, 2008). This infers TRIM5α has restriction activity not requiring CypA. Therefore, OWM cells have CypA-dependent and CypA-independent TRIM5α activity restricting HIV-1 (Lin and Emerman, 2008). The CypA-dependent TRIM5α restriction activity maps to two loops in HIV-1 CA, has minor impact on reverse transcription products but acts mostly after reverse transcription (Berthoux *et al.*, 2004; Lin and Emerman, 2008). This coincides with HIV-1 CA studies suggesting CypA influences CA uncoating from invading HIV-1 cores and perturbs events after reverse transcription (Qi *et al.*, 2008; Ylinen *et al.*, 2009). As OWM TRIM5α also targets the CA shell of invading HIV-1 cores, CypA may possibly alter the CA uncoating of HIV-1 cores to enhance TRIM5α restriction in OWM cells. This possibility is supported by the study of HIV-1 CA mutants whose stability influences owl monkey TRIMCyp restriction (Forshey *et al.*, 2005). Recent work inactivating TRIM5α with dominant-negative TRIM5 RBCC indicates CypA can also restrict HIV-1 independent of TRIM5α in OWM cells (Nakayama *et al.*, 2008). While the mechanism is not clear, this CypA antiviral effect appears specific to OWM cells in the limited cell types tested. Thus, OWM cells may have a specific restriction factor that uses CypA to restrict HIV-1, in addition to CypA enhancement of TRIM5α restriction (Nakayama *et al.*,

2008). Collectively, these studies reveal HIV-1 restriction in OWM cells is complex. Restriction may involve antiviral activity from CypA and TRIM5α, acting in combination and in isolation. However, their individual antiviral mechanisms and relative importance to overall HIV-1 restriction in OWM cells requires clarification.

All the abovementioned studies concern the effects of CypA on restrictions targeting HIV-1 CA in target cells. However, CypA can also have antiviral activity mediated by the virus producer cells, restricting $SIV_{agm}$ and $SIV_{mac}$ by a mechanism counteracted by the SIV accessory protein, virion infectivity factor [Vif, (Takeuchi *et al.*, 2007)]. This CypA restriction activity correlates with CypA in human or agm producer cells and efficient CypA incorporation into Vif-deficient $SIV_{agm}$ or $SIV_{mac}$ virions. However, restriction of these viruses is manifest in the target cell, impairing Vif-deficient $SIV_{agm}$ or $SIV_{mac}$ infection. These features of CypA restriction are similar to APOBEC3-mediated restriction (*see below*). However, the CypA restriction of Vif-deficient SIV infection shows no evidence of cytidine deaminase hypermutation, indicating this infection block is not the product of APOBEC3 antiviral activity. Instead, CypA restriction of Vif-deficient $SIV_{agm}$ impairs reverse transcription product levels, similar to TRIM5 restriction. However, RNA interference studies indicate this CypA-mediated restriction is independent of huTRIM5α in human target cells. Therefore, precisely how intra-virion CypA impedes reverse transcription products to impair Vif-deficient SIV infection of target cells during restriction remains unclear and requires further experimentation. $SIV_{agm}$ and $SIV_{mac}$ Vif in virus producer cells both counteract this restriction activity of CypA, preventing its incorporation into virions. These SIV Vif proteins do not alter the intracellular stability of CypA in virus producer cells and the mechanism that prevents CypA incorporation into SIV virions is unknown (Takeuchi *et al.*, 2007). Whether this antiviral activity of virion-associated CypA counteracted by Vif impacts more retroviruses than these two SIV isolates requires testing. Ultimately these diverse antiviral roles of CypA indicate that, like TRIM5 proteins, this ubiquitous cytosolic protein is an influential component of the antiviral armour in host cells to defend against invading retroviruses.

## Restrictions of lentiviral genomes and early replication events

### APOBEC3 proteins

#### *Discovery of APOBEC3G*

The ability of APOBEC3 (apolipoprotein B mRNA-editing enzyme catalytic polypeptide-like 3) proteins to restrict reverse-transcribing pathogens like lentiviruses emerged from studies of the HIV-1 accessory protein, Virion infectivity factor (Vif). HIV-1 Vif is packaged into virions (Liu *et al.*, 1995) and is essential for producing highly infectious virions from HIV-infected primary cells (Fisher *et al.*, 1987; Strebel *et al.*, 1987). However, cell types differ in requiring Vif for HIV-1 replication. Some cell types are *permissive* and do not need Vif for HIV-1 replication, like HeLa, 293T, Jurkat, SupT1 and COS cell lines. Conversely, other cell types, including human $CD4^+$ T-cells and macrophages infected by HIV-1 *in vivo*, are *restrictive* and Vif is required for HIV-1 replication (reviewed in Bour and Strebel, 2000). In restrictive cells, the replication of Vif-deficient HIV-1 (HIV-1ΔVif) is restored when Vif is added to the virus-producing cells but not target cells, indicating Vif acts in virus producing cells (von Schwedler *et al.*, 1993). However, Vif enhances viral infectivity in target cells, promoting the accumulation of reverse transcribed viral cDNA early in viral replication (von Schwedler *et al.*, 1993). Heterokaryon experiments fusing permissive and restrictive cells demonstrate that restriction is dominant (Madani and Kabat, 1998; Simon *et al.*, 1998). This indicates restrictive cells contain a dominant inhibitory factor that blocks HIV-1 replication in the absence of Vif, and that Vif can overcome this inhibitory factor.

In 2002, Sheehy and colleagues identified this dominant inhibitory factor using subtractive hybridization of cDNA from a closely related permissive and restrictive cell line (Sheehy *et al.*, 2002). This technique unveiled a human cellular protein initially called CEM15, and soon identified as Apolipoprotein B mRNA-editing enzyme catalytic polypeptide-like 3G (APOBEC3G, A3G), which fitted all the criteria expected of the dominant inhibitory factor counteracted by Vif. Firstly, A3G was present in all restrictive cells and absent from all permissive cells tested. Secondly, expressing A3G in permissive cells converted them to a restrictive phenotype indicating this protein was a dominant inhibitor. Moreover, titrating increasing amounts of A3G into permissive virus producer cells induced a dose dependent loss in the infectivity of HIV-1ΔVif produced from these cells. Thirdly, co-expressing Vif in these producer cells overcame the infectivity loss, demonstrating that Vif expression in virus producer cells overcomes the dominant A3G restriction. Finally, in the absence of Vif, A3G was packaged into HIV-1 virions providing an explanation for its subsequent antiviral activity in target cells. Therefore, A3G fulfilled all of the criteria expected of the dominant inhibitory factor, and thus was defined as a cellular restriction factor blocking HIV-1ΔVif infection that is overcome by Vif (Sheehy *et al.*, 2002). Moreover, this seminal study revealed that HIV-1 Vif functions to overcome A3G restriction and permit HIV-1 replication in human cells.

*Cytidine deaminase family*

A3G belongs to a family of cellular cytidine deaminase enzymes founded by APOBEC1 in 1993 (Teng *et al.*, 1993). APOBEC1 is involved in lipid metabolism, where it deaminates a cytosine at nucleotide 6666 in apolipoprotein B mRNA (Teng *et al.*, 1993). In addition to APOBEC1, the family also includes activation-induced deaminase (AID), APOBEC2, APOBEC3A-H and APOBEC4 (Liao *et al.*, 1999; Muramatsu *et al.*, 1999; Jarmuz *et al.*, 2002; Conticello *et al.*, 2005; Rogozin *et al.*, 2005; OhAinle *et al.*, 2006). AID impacts adaptive and innate immunity by mediating somatic hypermutation and class switch recombination to diversify the antibody repertoire, and inhibiting L1 retrotransposition (Muramatsu *et al.*, 1999, 2000; Macduff *et al.*, 2009). Cytidine deaminase enzymes are characterized by their ability to hydrolyse or 'deaminate' an amino group in cytosine bases (C) to generate uracil bases (U), in a DNA and/or RNA context (Teng *et al.*, 1993; Liao *et al.*, 1999; Muramatsu *et al.*, 1999; Harris *et al.*, 2002; Jarmuz *et al.*, 2002; Petersen-Mahrt *et al.*, 2002; Harris *et al.*, 2003; Lecossier *et al.*, 2003; Zhang *et al.*, 2003; Bishop *et al.*, 2004a; Rogozin *et al.*, 2005). In DNA, this deamination promotes guanine (G) to adenine (A) base mutations in the complementary strand (Harris *et al.*, 2003; Lecossier *et al.*, 2003; Mangeat *et al.*, 2003; Zhang *et al.*, 2003).

The APOBEC3 (A3) genes are found in mammals and are located between the conserved *CBX6* and *CBX7* genes (Conticello *et al.*, 2005; LaRue *et al.*, 2009). Mammals vary in their number of A3 genes. For instance, the human A3 locus on chromosome 22 encodes 7 distinct A3 genes: A3A, A3B, A3C, A3DE, A3F, A3G and A3H (Jarmuz *et al.*, 2002; Conticello *et al.*, 2005; LaRue *et al.*, 2009). Conversely, rodents have one A3 gene (Mariani *et al.*, 2003; LaRue *et al.*, 2009). Phylogenetic comparison of APOBEC sequences in different primates indicates that the majority of APOBEC genes, including most A3 genes, have encountered strong positive selection given their elevated non-synonymous to synonymous mutation ratio (Sawyer *et al.*, 2004). Therefore, it is possible evolutionary pressure imparted by reverse transcribing pathogens targeted by A3 proteins may have driven the expansion of A3 genes in mammals with multiple A3 genes. Interestingly, multiple A3 proteins restrict a reconstructed, ancient human endogenous retrovirus K (HERV-K), and two inactivated HERV-K proviruses in the human germline contain signature G-A hypermutation mediated by human A3G (Lee *et al.*, 2008). This supports the possibility that HERV-K and other ancient reverse transcribing pathogens may have provided pressure on the A3 locus to expand during primate evolution to counteract these invading pathogens.

The APOBEC proteins catalyse deamination via a catalytic domain (CTD or Z domain) that coordinates zinc. This domain has a hallmark active site containing the His-X-Glu-$X_{aa23-28}$-Pro-Cys-$X_{aa2-4}$-Cys motif, where the histidine and two cysteines bind zinc and the glutamate enables hydrolysis during deamination [reviewed in (Harris and Liddament, 2004)]. These Z domains can be present in one or two copies in APOBEC proteins. For instance, human A3A, A3C and A3H are single domain deaminases whilst A3B, A3DE, A3F and A3G are double domain deaminases (Jarmuz *et al.*, 2002; LaRue *et al.*, 2009). Moreover, these Z domains can be classified as 3 types (recently proposed to be termed Z1, Z2 and

Z3) based on their unique sequence features and thus distinct phylogenetic clustering (Conticello 2005). While the arrangement of A3 genes among primates align relatively closely, these A3 genes differ considerably in number, arrangement and Z-domain composition in other mammals like horses, cats and artiodactyls (LaRue *et al.*, 2009). As a result, naming A3 genes in other mammals based on the terminology of the human A3 genes is inaccurate and a new nomenclature is proposed for non-primate A3 genes (LaRue *et al.*, 2009).

### *APOBEC3 proteins as a new class of restriction factors*

Since the discovery of A3G as a restriction factor counteracted by Vif in 2002 (Sheehy *et al.*, 2002), the A3 proteins are now reported to restrict a broad range of reverse transcribing pathogens. These pathogens include lentiviruses, alpharetroviruses, betaretroviruses, gammaretroviruses, deltaretroviruses, spumaviruses, hepadnaviruses, endogenous retroviruses, long terminal repeat (LTR)-retrotransposons and non-LTR retrotransposons (Harris *et al.*, 2003; Mangeat *et al.*, 2003; Bishop *et al.*, 2004b; Turelli *et al.*, 2004; Yu *et al.*, 2004a; Dutko *et al.*, 2005; Lochelt *et al.*, 2005; Russell *et al.*, 2005; Sasada *et al.*, 2005; Bogerd *et al.*, 2006b; Chen *et al.*, 2006; Doehle *et al.*, 2006; Jonsson *et al.*, 2007; Okeoma *et al.*, 2007; Wiegand and Cullen, 2007; Bogerd *et al.*, 2008; Lee *et al.*, 2008; Munk *et al.*, 2008). Moreover, recent studies demonstrate these proteins also target DNA viruses, including adeno-associated virus (Chen *et al.*, 2006) and human papillomavirus (Vartanian *et al.*, 2008). Therefore, these A3 proteins have broad activity in restricting primarily reverse transcribing pathogens.

Focusing on lentiviruses that feature in this book, HIV-1 is potently restricted by human (hu) A3G and huA3F, and modestly restricted by huA3B (Sheehy *et al.*, 2002; Bishop *et al.*, 2004b; Wiegand *et al.*, 2004; Yu *et al.*, 2004a). While huA3G and huA3F are counteracted by HIV-1 Vif, huA3B restricts HIV-1 independently of Vif (Bishop *et al.*, 2004b; Yu *et al.*, 2004a). The expression of both huA3G and huA3F that counteract Vif in $CD4^+$ T-cells suggests that they primarily restrict HIV-1 *in vivo* (Bishop *et al.*, 2004b; Wiegand *et al.*, 2004). Moreover, HIV-1 sequences also contain hypermutations at characteristic nucleotide motifs of huA3G and huA3F deamination (Bishop *et al.*, 2004b) consistent with impacting HIV-1 replication *in vivo*. HIV-1 is also modestly restricted by huA3DE in a Vif-sensitive manner. As huA3DE is expressed in relevant target cells of HIV-1 infection and some HIV-1 clinical isolates have hypermutations at sequences targeted by huA3DE, it is likely that huA3DE may also contribute to HIV-1 restriction *in vivo* (Dang *et al.*, 2006). Mutations occur in human and chimpanzee (cpz) A3H that impair their A3H expression, but reversing these mutations back to sequences found in other primate A3H restores efficient huA3H and cpzA3H expression, and HIV-1 restriction (Dang *et al.*, 2008). Polymorphisms occur in the huA3H gene producing multiple huA3H variants that differ in their expression and restriction of reverse transcribing pathogens (OhAinle *et al.*, 2008; Harari *et al.*, 2009). Why A3H variants persist in the human population is unclear but they may influence susceptibility to retroviral infection and retrotransposition. A3 proteins in other animal species also restrict HIV-1. For instance, cpzA3G restricts HIV-1 in a Vif sensitive manner, while rhA3G, rhA3H, agmA3G, mouse (m) A3, mA1, rat A1, hamster A1 and rabbit A1 all restrict HIV-1 in a Vif-independent manner (Mariani *et al.*, 2003; Bishop *et al.*, 2004b; OhAinle *et al.*, 2006; Ikeda *et al.*, 2008). Equine A3F1 also potently restricts HIV-1ΔVif (Bogerd *et al.*, 2008).

Multiple A3 proteins also restrict the OWM retroviruses, $SIV_{mac}$ and $SIV_{agm}$. $SIV_{mac}$ is restricted by huA3C, huA3DE, huA3F, huA3G, cpzA3G, rhA3G and agmA3G in a Vif-sensitive manner, and by huA3B, mA3 and rabbit A1 in a Vif-insensitive manner (Mariani *et al.*, 2003; Yu *et al.*, 2004a; Dang *et al.*, 2006; Ikeda *et al.*, 2008). Conversely, $SIV_{agm}$ is restricted by huA3C, agmA3G, rhA3G and rhA3H in a Vif-sensitive manner, and by huA3B, huA3DE, huA3F, huA3G, cpzA3G, mA3 and rabbit A1 in a Vif-insensitive manner (Mariani *et al.*, 2003; Yu *et al.*, 2004a; Dang *et al.*, 2006; OhAinle *et al.*, 2006; Ikeda *et al.*, 2008). Specific feline A3 (Munk *et al.*, 2008) and rabbit A1 proteins (Ikeda *et al.*, 2008) restrict FIV. EIAV, which is unique among lentiviruses in lacking a *vif* gene (Oberste and Gonda, 1992), is potently restricted by huA3G and weakly restricted by specific equine A3 proteins (Mangeat *et al.*, 2003; Bogerd *et al.*, 2008). Collectively, these examples illustrate that A3 proteins from specific, often broad animal species can restrict lentiviruses. This chapter focuses on lentiviral restrictions but recent reviews elsewhere canvass the broad restriction of other retroviruses, retroelements and DNA viruses by various A3 proteins (Chiu and Greene, 2008; Goila-Gaur and Strebel, 2008).

The natural cellular functions of the A3 proteins are unknown. However, as A3 proteins collectively have broad antiviral activity and have expanded in vertebrate lineages, this indicates A3 proteins may naturally exist to defend host organisms against primarily reverse transcribing pathogens. A3 knockout mice have no obvious phenotypic differences other than an impaired ability to restrict mouse mammary tumour virus (Okeoma *et al.*, 2007; Low *et al.*, 2009). This supports the theory that the primary function of A3 proteins is to protect cells from invading reverse transcribing pathogens.

#### *A prototype for restriction by A3 proteins: huA3G restriction of HIV-1*

Following the discovery of HIV-1ΔVif restriction by huA3G (Sheehy *et al.*, 2002), much research has focused on defining how huA3G restricts HIV-1ΔVif replication, and how HIV-1 Vif counteracts huA3G. The depth of research in this area has unveiled multiple antiviral activities of huA3G common to many other examples of A3 restriction and viral evasion mechanisms (Fig. 4.3). Thus, huA3G restriction of HIV-1 serves as a prototype for understanding many aspects of A3 restriction and will be discussed in detail here to illustrate many of these concepts. The restriction mechanism of A3 proteins targeting other reverse transcribing pathogens and their diverse countermeasures are reviewed elsewhere (Chiu and Greene, 2008).

#### *Virion-associated huA3G restriction of HIV-1*

In the absence of Vif, huA3G in virus producing cells is efficiently packaged into assembling HIV-1 virions. HuA3G is packaged via interaction with the nucleocapsid (NC) domain of HIV-1 Gag precursor proteins and single-stranded nucleic acid, like HIV-1 RNA that binds NC and huA3G (Mariani *et al.*, 2003; Alce and Popik, 2004; Cen *et al.*, 2004; Luo *et al.*, 2004; Schafer *et al.*, 2004; Svarovskaia *et al.*, 2004; Zennou *et al.*, 2004; Khan *et al.*, 2007; Soros *et al.*, 2007; Bogerd and Cullen, 2008). Recent studies indicate newly synthesized huA3G is packaged into virions and associates with virion cores (Soros *et al.*, 2007). Moreover, huA3G is inactivated by the viral RNA until digestion of the viral RNA template during reverse transcription, which dually activates huA3G to deaminate and provides the single-stranded viral cDNA template for deamination (Soros *et al.*, 2007).

In the search for defining how intravirion huA3G restricts HIV-1ΔVif replication in target cells, as huA3G is a cytidine deaminase packaged into HIV-1ΔVif virions, intravirion huA3G was initially speculated to deaminate the HIV-1 genome during replication in target cells. This was confirmed in studies showing huA3G causes hypermutation of reverse transcribed HIV-1ΔVif or MLV cDNA in target cells (Harris *et al.*, 2003; Lecossier *et al.*, 2003; Mangeat *et al.*, 2003; Zhang *et al.*, 2003). Moreover, huA3G deaminates single-stranded viral cDNA during reverse transcription, which is mostly the minus strand of the viral cDNA. This results in G to A hypermutation of the complementary DNA strand, which is mainly the positive strand of the viral cDNA (Harris *et al.*, 2003; Lecossier *et al.*, 2003; Mangeat *et al.*, 2003; Zhang *et al.*, 2003; Yu *et al.*, 2004b; Iwatani *et al.*, 2006). Two gradients of hypermutation are induced by huA3G along HIV-1 cDNA (Yu *et al.*, 2004b; Suspene *et al.*, 2006) correlating with the length of time the viral cDNA is probably single-stranded before reverse transcription of the complementary DNA strand (Suspene *et al.*, 2006). However, these hypermutation gradients may also be explained by the 3′ to 5′ directionality of processive huA3G deamination on single-stranded DNA (Chelico *et al.*, 2006; Furukawa *et al.*, 2009). The huA3G C-terminal deaminase domain mediates hypermutation (Hache *et al.*, 2005; Navarro *et al.*, 2005; Newman *et al.*, 2005; Iwatani *et al.*, 2006). Recent structural studies reveal this domain folds into a β-sheet core and multiple α-helices forming an active centre structurally similar to APOBEC2 (Chen *et al.*, 2008b; Holden *et al.*, 2008; Furukawa *et al.*, 2009). APOBEC proteins target and deaminate preferred nucleotide sequences (Beale *et al.*, 2004; Bishop

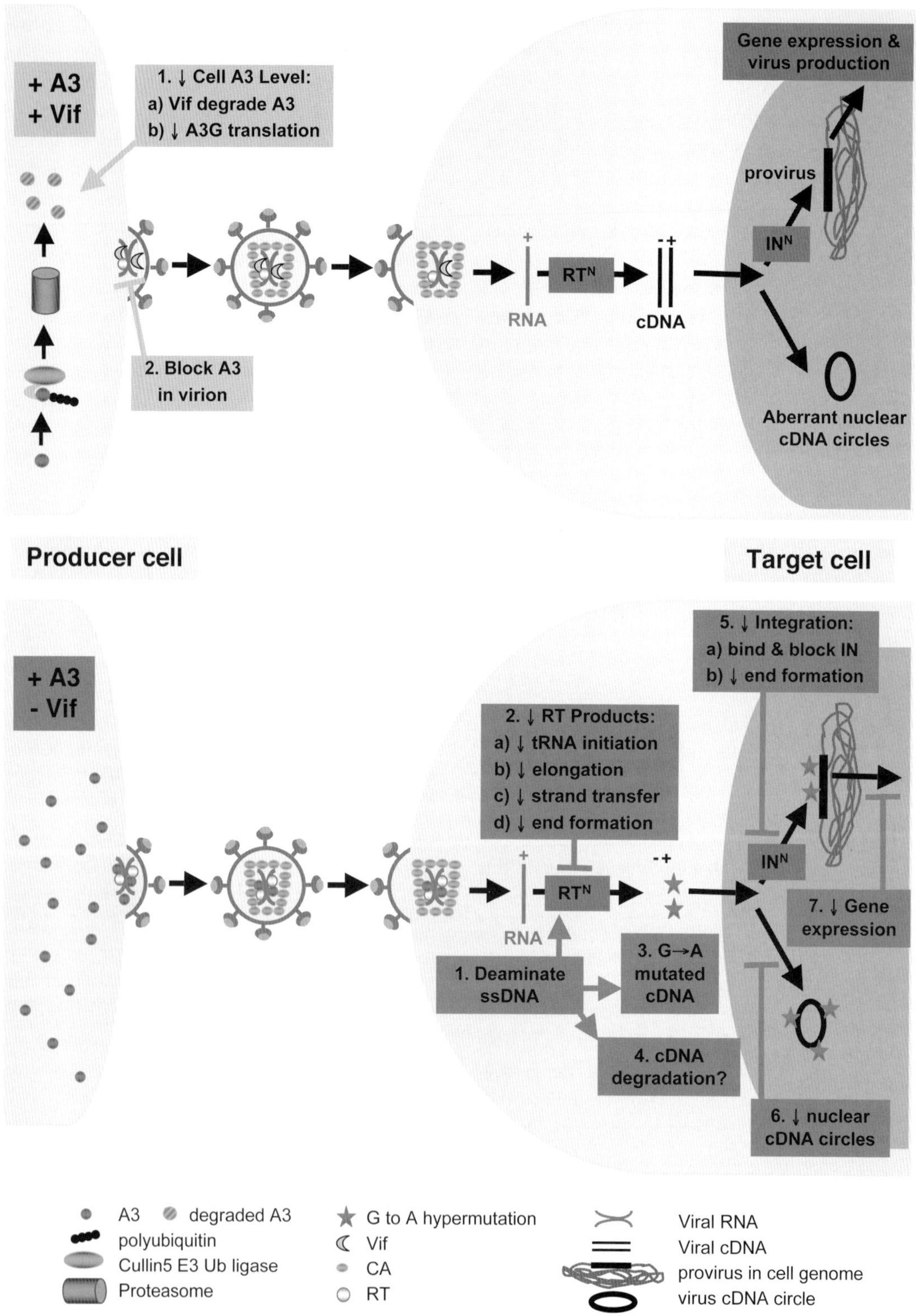


**Figure 4.3** Virion-associated huA3G restricts HIV-1 and HIV-1 avoids huA3G via multiple mechanisms. HIV-1 restriction by virion-associated huA3G serves as a prototype for understanding mechanisms used by virion-associated A3 proteins to restrict retroviruses, and retroviral countermeasures. **Upper panel:** To avoid virion-associated huA3G, HIV-1 uses multiple strategies to prevent huA3G incorporation into virions (*upper*

*et al.*, 2004b; Liddament *et al.*, 2004; Wiegand *et al.*, 2004; Yu *et al.*, 2004b; Dang *et al.*, 2006). For instance, A3G favours 5′-CC-3′ dinucleotide motifs on single-stranded DNA (Harris *et al.*, 2003; Beale *et al.*, 2004; Bishop *et al.*, 2004b; Wiegand *et al.*, 2004; Yu *et al.*, 2004b), with the 3′ underlined cytidine favoured for deamination due to the 3′ to 5′ directionality of this deaminase (Chelico *et al.*, 2006; Furukawa *et al.*, 2009). In contrast, A3F and A3B favour 5′-TC-3′ dinucleotide motifs (Bishop *et al.*, 2004b; Liddament *et al.*, 2004; Wiegand *et al.*, 2004). Various APOBEC proteins can hypermutate the single-stranded DNA of specific pathogens (Harris *et al.*, 2003; Bishop *et al.*, 2004b; Yu *et al.*, 2004a; Lochelt *et al.*, 2005; Bogerd *et al.*, 2006a; Dang *et al.*, 2006; OhAinle *et al.*, 2006). The extensive hypermutation of viral cDNA by APOBEC proteins probably perturbs viral sequence and translation codons, as per huA3G restriction of HIV-1ΔVif (Yu *et al.*, 2004b), thereby impairing viral protein expression and restricting replication.

Apart from deamination-mediated restriction of viral protein expression, intravirion huA3G clearly restricts the replication of HIV-1ΔVif in target cells via additional mechanisms which occur before protein expression. This is because huA3G impairs the formation of HIV-1ΔVif cDNA products in the cytoplasm and nucleus of target cells earlier in replication (Mangeat *et al.*, 2003; Mariani *et al.*, 2003; Aires da Silva *et al.*, 2004; Bishop *et al.*, 2006b; Guo *et al.*, 2006; Luo *et al.*, 2007; Mbisa *et al.*, 2007; Anderson and Hope, 2008). Specifically, huA3G induces modest declines in early reverse transcription products, a larger decline in late reverse transcription products and the greatest decline is in nuclear viral cDNA, including 2-LTR circles and integrated provirus (Mangeat *et al.*, 2003; Mariani *et al.*, 2003; Aires da Silva *et al.*, 2004; Luo *et al.*,

---

*left cell*) and thus avoid intravirion huA3G restricting early HIV replication in target cells. Firstly, HIV-1 limits intracellular levels of huA3G in virus producer cells by: a) expressing its Vif accessory protein that binds huA3G and an E3 ubiquitin ligase (Cullin 5, Elongin C and Elongin B), which polyubiquitylates and targets huA3G for proteasome-mediated degradation; and b) impeding huA3G translation. Secondly, HIV-1 may directly prevent huA3G incorporation into virions, as observed for other retroviruses that do not encode Vif accessory proteins but directly prevent A3 virion incorporation to avoid A3 antiviral activity. This collective ability of HIV-1 to limit the huA3G pool in virus producer cells and directly prevent huA3G packaging impedes huA3G incorporation into assembling virions, thereby escaping subsequent huA3G restriction in target cells. As a result, these HIV-1 virions can enter target cells, reverse transcribe their viral RNA genomes into cDNA, which then move into the nucleus where they either integrate into the cell genome leading to productive infection (viral gene expression and virion production) or form defective, nuclear viral cDNA circles (upper right cell). **Lower panel:** If HIV-1 lacks Vif and/or is unable to prevent huA3G incorporation into virions in the virus producer cell (*lower left cell*), this virion-associated huA3G can have multiple antiviral activities on early HIV replication in target cells to restrict viral infection (*lower right cell*). For instance, during reverse transcription ($RT^N$) of the viral plus strand RNA genome into cDNA, intravirion huA3G deaminates single-stranded viral cDNA (1) resulting in G to A hypermutation of the complementary viral DNA strand (3). This deamination may lead to degradation of the viral cDNA (4), although this possibility remains controversial given opposing results from studies examining whether cell base excision repair enzymes cause viral cDNA loss. Alternatively, huA3G also perturbs the reverse transcription process (2) possibly in multiple ways (decreasing tRNA initiation, elongation, strand transfer and complete cDNA end formation), which leads to collectively greater losses in viral reverse transcription product levels. HuA3G also further perturbs integration ($IN^N$) of viral cDNA into the cell genome in the nucleus (5) potentially through binding and obscuring the function of viral integrase protein, and/or decreasing complete reverse transcription of the viral cDNA ends, which are both essential for integration. HuA3G also depletes levels of aberrant viral cDNA circles formed in the nucleus (6) to a greater magnitude than viral cDNA products formed in the cytoplasm. Finally, any hypermutated viral cDNA that avoids earlier huA3G restrictions and integrates into the cell genome is probably defective for gene expression (7) given the extent of A3G-mediated hypermutation that perturbs nucleotide sequence and translation of open reading frames. These multiple antiviral activities of virion-associated huA3G probably enhance the potency of A3 restriction to more efficiently restrict retroviral pathogens *in vivo*. Note that target cell huA3G can also form high molecular mass and low molecular mass complexes that usurp retrotransposon RNA and prevent retrotransposition, or may block HIV-1 reverse transcription in certain human target cells respectively (*not shown*). A3, APOBEC3; CA, capsid; cDNA, complementary DNA; RT, reverse transcriptase/transcription; ssDNA, single-stranded DNA. A colour version of this figure is located in the plate section at the back of the book.

2007; Mbisa *et al.*, 2007; Anderson and Hope, 2008). While the magnitude of these effects differs between studies, probably due to the titre of huA3G used to make virus (Anderson and Hope, 2008), the decline in viral cDNA products follows this general trend.

The huA3G-mediated losses in viral cDNA were initially proposed to arise from deamination and the recruitment of enzymes from the cellular base excision repair pathway (Harris *et al.*, 2003). It was speculated that huA3G-mediated deamination and the generation of uracil bases in single-stranded viral cDNA could recruit cellular uracil DNA glycosylases such as UNG-2 or SMUG, which cleave off the aberrant uracil bases. This produces abasic sites that could be recognized by apurinic/apyrimidinic endonuclease (APE), which subsequently cleaves the viral DNA backbone during attempted base excision repair. However, if the complementary DNA strand is not yet reverse transcribed to complete repair, the viral DNA could fall apart. Thus, huA3G deamination might cause the viral cDNA to degrade leading to the loss of viral cDNA products during restriction. In support of this model, UNG-2, SMUG and APE are packaged into HIV-1 virions (Priet *et al.*, 2005; Schrofelbauer *et al.*, 2005; Yang *et al.*, 2007a). Moreover, inhibiting UNG-2 or APE in virus producing cells partially rescues HIV-1ΔVif cDNA production and infection from huA3G-mediated restriction (Yang *et al.*, 2007a). However, separate studies find UNG-2 and SMUG dispensable for huA3G restriction of various reverse transcribing pathogens and thus dispute this model (Langlois *et al.*, 2005; Kaiser and Emerman, 2006; Mbisa *et al.*, 2007; Schumacher *et al.*, 2008). Therefore, a role for deamination-mediated degradation of viral cDNA during restriction remains controversial.

Instead, the ability of huA3G to decrease various HIV-1ΔVif cDNA products during restriction has been attributed to its effects on multiple steps of viral replication that alter the synthesis of these viral cDNA products, rather than induce their degradation. For instance, huA3G and huA3F perturb annealing of tRNA primer that initiates reverse transcription potentially accounting for the modest loss in early reverse transcription products (Guo *et al.*, 2006, 2007a; Yang *et al.*, 2007b). Secondly, they perturb multiple steps during reverse transcription including elongation, minus and plus strand transfer potentially contributing to losses in late reverse transcription products (Iwatani *et al.*, 2007; Li *et al.*, 2007b; Mbisa *et al.*, 2007; Bishop *et al.*, 2008). Thirdly, they perturb proviral integration, potentially via interaction with viral integrase and/or effects on tRNA primer processing/removal during reverse transcription that alters the viral cDNA ends required for integration, potentially causing the loss in nuclear integrated provirus (Luo *et al.*, 2007; Mbisa *et al.*, 2007). The explanation for the loss in nuclear, viral 2-LTR circles remains unclear and may also arise from the alteration of cDNA ends by huA3G (Mbisa *et al.*, 2007). Alternatively, huA3G perturbs the sedimentation and *in vitro* integration activity of cytoplasmic HIV-1 complexes (Carr *et al.*, 2006; Mbisa *et al.*, 2007) and acts primarily in the cytoplasm for its antiviral effects on viral cDNA products (Anderson and Hope, 2008). Thus, huA3G may also perturb the structural integrity and/or trafficking of viral complexes into the nucleus to help deplete nuclear viral 2-LTR circles and integrated provirus. Ultimately, huA3G and huA3F probably have deleterious effects on multiple steps early in HIV-1ΔVif replication in addition to perturbing viral protein coding sequence through deamination to collectively restrict viral infection (Fig. 4.3). These various antiviral mechanisms may ensure that invading reverse transcribing pathogens are efficiently restricted *in vivo*.

The precise role of deamination in these antiviral activities of virion-associated huA3G remains controversial as various studies suggest that A3 proteins can restrict their target pathogens in the absence of deamination. Firstly, multiple examples exist where different A3 proteins restrict viral pathogens with minimal, if any, detectable hypermutation. This includes: huA3G and huA3F restriction of hepatitis B virus (Turelli *et al.*, 2004; Rosler *et al.*, 2005); huA3G restriction of human T-cell leukaemia virus type 1 (Sasada *et al.*, 2005); huA3A, huA3B and huA3F restriction of L1 retrotransposons (Bogerd *et al.*, 2006b; Stenglein and Harris, 2006); huA3A restriction of retrotransposons and adeno-associated virus (Bogerd *et al.*, 2006a; Chen *et al.*, 2006); mouse A3 restriction of mouse mammary tumour virus (Okeoma *et al.*, 2007); and efficiently expressed huA3H restriction of HIV-1 (Dang *et al.*, 2008). Moreover, huA3G in resting CD4$^+$ T-cells can restrict invading HIV-1, depleting viral cDNA with minimal hypermutation (Chiu *et al.*, 2005). These examples therefore set a precedent for A3 proteins having deaminase-independent restriction activities. However, deaminated viral cDNA may not be detected in these studies if it is rapidly degraded. Secondly, A3 proteins containing catalytically inactive deaminase domains can still restrict reverse transcribing pathogens (Turelli *et al.*, 2004; Newman *et al.*, 2005; Sasada *et al.*, 2005; Bishop *et al.*, 2006b; Bogerd *et al.*, 2006b; Stenglein and Harris, 2006; Holmes *et al.*, 2007; Iwatani *et al.*, 2007; Li *et al.*, 2007b; Okeoma *et al.*, 2007; Yang *et al.*, 2007b; Bishop *et al.*, 2008). Furthermore, huA3G and huA3F restriction correlates with losses in reverse transcribed HIV-1ΔVif cDNA rather than deamination (Bishop *et al.*, 2006b; Holmes *et al.*, 2007). However, recent work suggests that the ability of these deamination-defective A3 proteins to restrict reverse transcribing pathogens arises from overexpression. If these A3 mutants are instead expressed at wild-type protein levels (Miyagi *et al.*, 2007) and lower physiological levels (Schumacher *et al.*, 2008; Browne *et al.*, 2009), their restriction activity is impaired. Therefore the latter findings indicate deamination is important for efficient A3 restriction (Miyagi *et al.*, 2007; Schumacher *et al.*, 2008; Browne *et al.*, 2009). Consequently, the importance and contribution of deamination-dependent versus deamination-independent antiviral activities to overall restriction by A3 proteins ultimately remains unclear. However, being able to restrict pathogens via multiple means, deamination-dependent and deamination-independent, probably enhances the potency of restriction by these A3 proteins and may also expand the range of pathogens they inhibit.

### *Averting virion-associated A3 restriction*

To counteract the antiviral activity of virion-associated huA3G, HIV-1 encodes the Vif accessory protein that must be present in the virus producer cell (Sheehy *et al.*, 2002). In virus-producing cells, Vif binds to both huA3G and an E3 ubiquitin ligase containing Cullin 5, Elongin B, Elongin C, Rbx1 and Nedd8 (Yu *et al.*, 2003; Mehle *et al.*, 2004b). This promotes polyubiquitylation of huA3G and targets huA3G for rapid degradation by cellular proteasomes (Conticello *et al.*, 2003; Marin *et al.*, 2003; Sheehy *et al.*, 2003; Stopak *et al.*, 2003; Yu *et al.*, 2003; Mehle *et al.*, 2004b; Mehle *et al.*, 2004a). The loss in intracellular huA3G coincides with a loss of huA3G incorporated into virions allowing the virions to infect target cells due to the loss of virion-associated huA3G (Conticello *et al.*, 2003; Marin *et al.*, 2003; Sheehy *et al.*, 2003; Stopak *et al.*, 2003; Yu *et al.*, 2003; Mehle *et al.*, 2004a,b). Therefore Vif effectively acts as a linker, targeting huA3G for proteasome-mediated degradation and impairing huA3G levels in producer cells, diminishing huA3G incorporation into virions to avoid restriction by virion-associated huA3G.

HIV-1 Vif also counteracts other human A3 proteins by recruiting cellular Cullin 5, Elongin C and Elongin B, and targeting these A3 proteins for proteasome-mediated turnover in virus producer cells, including huA3F and huA3DE (Liu *et al.*, 2005a; Dang *et al.*, 2006; Zhang *et al.*, 2008d,e). Most lentiviruses encode Vif proteins except EIAV probably reflecting the importance of Vif for counteracting A3 restriction *in vivo*. Like HIV-1 Vif, other primate lentiviral Vif proteins also recruit Cullin 5-derived E3 ubiquitin ligases (Yu *et al.*, 2003; Luo *et al.*, 2005; Zhang *et al.*, 2008e). In fact, primate lentiviral Vif proteins act as substrate receptors for Cullin5-derived E3 ubiquitin ligases, using their viral BC-box (SLQYLA motif) and HCCH motif to engage Elongin C/Elongin B and Cullin 5 respectively (Mehle *et al.*, 2004b; Yu *et al.*, 2004c; Luo *et al.*, 2005; Mehle *et al.*, 2006). Non-primate lentiviral Vif from visna–maedi virus (VMV) and bovine immunodeficiency virus also recruit Elongin C and Elongin B, but bind Cullin 5 poorly consistent with lacking the HCCH motif. Instead, VMV Vif binds Cullin 2 efficiently (Luo *et al.*, 2005). Therefore, lentiviral Vif proteins appear to link A3 proteins to Cullin-derived E3 ubiquitin ligases and induce proteasome-mediated degradation of the bound A3 protein as a common mechanism to evade A3 restriction.

HIV-1 Vif uses common and distinct residues to bind various A3 proteins. For instance, the YXXL motif highly conserved among lentiviral Vif proteins is required for Vif to bind A3G, A3F and A3C (Pery *et al.*, 2009). However the HIV-1

Vif YRHHY at residues 40–44 is specific for binding huA3G, while the QVDRMR at residues 12–17 and residues 74–79 are specific for binding huA3F (Mehle *et al.*, 2007; Russell and Pathak, 2007; He *et al.*, 2008; Pery *et al.*, 2009). The huA3G and huA3F proteins also use distinct domains to bind HIV-1 Vif. While both contain two deaminase domains, huA3G uses residues in the N-terminal domain while huA3F uses residues in the C-terminal domain to bind and induce Vif degradation (Huthoff and Malim, 2007; Russell and Pathak, 2007; Zhang *et al.*, 2008b).

The ability of lentiviral Vif proteins to restrict primate A3 proteins is also species-specific. For instance, HIV-1 Vif counteracts huA3G and cpzA3G but not agmA3G or rhA3G. Conversely, agm Vif counteracts agmA3G and rhA3G, but not huA3G or cpzA3G (Mariani *et al.*, 2003). This species specificity maps to A3G residue 128 that is aspartic acid in huA3G and lysine in agmA3G (Bogerd *et al.*, 2004; Mangeat *et al.*, 2004; Schrofelbauer *et al.*, 2004; Xu *et al.*, 2004). Switching this residue between huA3G and agmA3G switches their lentiviral Vif sensitivity and this residue is key for lentiviral Vif interaction and A3G exclusion from virions (Bogerd *et al.*, 2004; Mangeat *et al.*, 2004; Schrofelbauer *et al.*, 2004; Xu *et al.*, 2004). A3G residue 128 is located in the N-terminal domain of A3G and on this domain's surface in structural models derived from the A3G C-terminal domain structure (Furukawa *et al.*, 2009). Thus, A3G residue 128 is in an accessible position for specifically binding lentiviral Vif proteins. Moreover, the surface potential surrounding residue 128 varies between huA3G and agmA3G models (Furukawa *et al.*, 2009). This is consistent with the charge difference between the aspartic acid in huA3G and the basic lysine at residue 128 in agmA3G being important for recognition of different lentiviral Vif proteins (Bogerd *et al.*, 2004; Mangeat *et al.*, 2004; Schrofelbauer *et al.*, 2004; Xu *et al.*, 2004).

In addition to binding, degrading and thus depleting the intracellular pool of various A3 proteins, lentiviral Vif proteins may also evade A3 restriction by other means (Kao *et al.*, 2003; Stopak *et al.*, 2003; Kao *et al.*, 2004; Santa-Marta *et al.*, 2005). For instance, HIV-1 Vif may also reduce huA3G translation to contribute to the depletion of huA3G levels in producer cells (Kao *et al.*, 2003; Stopak *et al.*, 2003). Moreover, HIV-1 Vif prevents virion packaging and restriction by a huA3G C97A mutant resistant to Vif-mediated degradation (Opi *et al.*, 2007). This suggests that the effects of HIV-1 Vif on A3G packaging into virions and degradation can be separated. Thus, HIV-1 Vif may counteract A3 proteins in multiple ways, impeding A3 packaging into virions as well as depleting intracellular A3 levels in producer cells via proteasome degradation and reducing A3 translation (Fig. 4.3) (Conticello *et al.*, 2003; Kao *et al.*, 2003; Marin *et al.*, 2003; Sheehy *et al.*, 2003; Stopak *et al.*, 2003; Yu *et al.*, 2003; Mehle *et al.*, 2004b; Mehle *et al.*, 2004a; Luo *et al.*, 2005).

Retroviruses that lack Vif must use alternative mechanisms to evade A3-mediated restriction. These viruses lacking Vif commonly obscure A3 packaging into virions to evade restriction by virion-associated A3. For instance, HTLV-I (Derse *et al.*, 2007), MLV (Doehle *et al.*, 2005; Abudu *et al.*, 2006; Zhang *et al.*, 2008c), foamy viruses (Lochelt *et al.*, 2005; Russell *et al.*, 2005) and Mason-Pfizer Monkey virus (Doehle *et al.*, 2006) all block A3 incorporation into virions via various strategies without inducing A3 degradation. Additionally, EIAV and MLV appear to perturb restriction by specific A3 proteins after A3 virion packaging via novel mechanism/s (Abudu *et al.*, 2006; Bogerd *et al.*, 2008). Therefore, in addition to Vif, retroviruses can employ multiple tactics to counteract restriction by A3 proteins. This points to the important role these A3 proteins play in limiting the spread of reverse transcribing pathogens *in vivo* given viruses have evolved multiple tactics to evade A3 restriction.

### *huA3G complexes and post-entry restriction of HIV-1*

In addition to the antiviral activities of virion-associated A3 proteins, huA3G is also reported to have antiviral activity in specific target cells. This emerged from the report that huA3G in unstimulated peripheral blood CD4$^+$ T-cells and monocytes assembles into low molecular mass (LMM) complexes that restrict invading HIV-1 (Chiu *et al.*, 2005). Target cell LMM huA3G is reported to block HIV-1 after entry, impairing the level of reverse transcribed viral cDNA. Moreover, despite the LMM complexes being active for deamination *in vitro*, only minor G-to-A hypermutation of HIV-1 cDNA is observed in cells suggesting LMM huA3G primarily restricts HIV-1 via non-deamination mechanisms. Activation of peripheral blood CD4$^+$ T-cells and differentiation of monocytes into macrophages both promote huA3G assembly into high molecular mass (HMM) complexes that are inactive for *in vitro* deamination and correlate with a loss of HIV-1 restriction (Chiu *et al.*, 2005). This is proposed to explain why HIV-1 replicates in activated peripheral blood CD4$^+$ T-cells and macrophages because huA3G is assembled into the inactive HMM form. Conversely, HIV-1 presumably fails to replicate in unstimulated peripheral blood CD4$^+$ T-cells and monocytes due to restriction by active LMM huA3G. Notably, HIV-1 Vif does not influence this LMM huA3G restriction. This is presumably because not enough Vif is present in virions to overcome target cell LMM huA3G, and Vif is not expressed until late in viral replication after the post-entry restriction by LMM huA3G would occur (Chiu *et al.*, 2005).

This assembly of huA3G into inactive HMM or active LMM complexes also correlates with the susceptibility of other primary cells to HIV-1 infection. For instance, in contrast to resting CD4$^+$ cells in peripheral blood that contain LMM huA3G and resist HIV-1 infection, resting CD4$^+$ cells in lymphoid tissue contain inactive HMM huA3G and are permissive for HIV-1 infection (Chiu *et al.*, 2005; Kreisberg *et al.*, 2006). Moreover, treating resting peripheral blood CD4$^+$ T-cells with IL-2, IL-7 or IL-15 shifts huA3G from active LMM into inactive HMM form, possibly explaining why cytokines make resting CD4$^+$ cells permissive to HIV-1 infection (Stopak *et al.*, 2007). The poor ability of HIV-1 to infect immature dendritic cells (iDCs) also correlates with huA3G and huA3F expression, and the small subset of iDCs infected noticeably lack huA3G expression (Pion *et al.*, 2006). Maturation of these DCs resists HIV-1 infection further and also correlates with elevated huA3G expression (Pion *et al.*, 2006) that assembles into active LMM form (Stopak *et al.*, 2007). Finally, while the majority of peripheral blood monocytes (CD14$^{high}$CD16-) contain active LMM huA3G and resist HIV-1 infection, a minor CD16$^+$ subset of monocytes are permissive to HIV-1 infection correlating with huA3G in the inactive HMM form (Ellery *et al.*, 2007). Thus, the presence of huA3G in active LMM or inactive HMM forms correlates with the susceptibility of different human cell types to HIV-1 infection, suggesting target cell huA3G may play a role in restricting HIV-1 *in vivo*.

However, in contrast to initial findings (Chiu *et al.*, 2005), recent work has failed to demonstrate that target cell huA3G restricts HIV-1 in unstimulated peripheral CD4$^+$ T-cells (Kamata *et al.*, 2009). Despite using similar methods and reagents to deplete huA3G from unstimulated CD4$^+$ T-cells via siRNA, recent work found huA3G depletion did not increase HIV-1 infection of unstimulated CD4$^+$ T-cells, in opposition to earlier work. Thus, this recent study argues against target cell huA3G as primarily responsible for HIV-1 restriction in unstimulated CD4$^+$ T-cells. Therefore, the ability of target cell huA3G to restrict HIV-1 at least in the context of peripheral blood unstimulated CD4$^+$ T-cells is unclear (Kamata *et al.*, 2009).

Macrophages and activated CD4$^+$ T-cells are the major cell types infected by HIV-1 *in vivo* and contain inactive HMM huA3G complexes (Chiu *et al.*, 2005). However, HIV-1 still requires Vif to replicate in these cells (Fisher *et al.*, 1987; Strebel *et al.*, 1987; Bour and Strebel, 2000) because huA3G can incorporate into virions produced from these cells and subsequently restrict HIV-1 infection of target cells. Therefore, HIV-1 is still susceptible to virion-associated huA3G when replicating in these cells that contain huA3G in inactive HMM complexes. This raises a key question of how huA3G that assembles into HMM complexes in these cells is also packaged into assembling virions in an active form. Recent work indicates that assembling HIV-1 virions in producer cells package newly synthesized huA3G, rather than huA3G already in inactive HMM complexes (Soros *et al.*, 2007). Moreover, this virion-associated huA3G complexed with the viral RNA remains inactive until the RNaseH activity of the HIV-1 RT enzyme removes the viral RNA genome during reverse transcription. This RT activity has the dual effect of activating the packaged huA3G for deamination in concert with synthesizing the single stranded viral cDNA target of deamination. Therefore, by packaging newly synthesized huA3G and activating its deamination during viral reverse transcription in target cells, virion-associated huA3G can restrict HIV-1 generated from these cells that also contain inactive HMM huA3G complexes (Soros *et al.*, 2007).

If active LMM huA3G protects cells from infection by invading retroviruses like HIV-1, this raises the issue of why cells assemble huA3G into inactive HMM complexes at all. Biochemical and subcellular localization analyses of huA3G complexes indicate it associates with various cytoplasmic ribonucleoprotein (RNP) complexes and RNA-rich sites of RNA processing, function and storage, including processing bodies, stress granules, Staufen granules, polysomes and RoRNPs (Chiu *et al.*, 2006; Kozak *et al.*, 2006; Wichroski *et al.*, 2006; Gallois-Montbrun *et al.*, 2007). HuA3F also accumulates with various cytoplasmic RNPs and the RNA-rich sites of processing bodies and stress granules (Wichroski *et al.*, 2006; Gallois-Montbrun *et al.*, 2008). Therefore, it is conceivable that huA3G and huA3F assemble into inactive HMM and active LMM forms to regulate their restriction activity in cells. Sequestration into HMM complexes may prevent these constitutively expressed, nucleic acid editing proteins from being permanently activated in the cell where they might induce deleterious editing of cellular nucleic acids. Perhaps these proteins are activated by specific conditions to engage undesirable targets. Alternatively, huA3G or huA3F may influence the function of the associated cytoplasmic RNPs and RNA-rich sites. Conversely, huA3G HMM complexes sequester Alu and human small Y (hY) retroelement RNAs and prevent retrotransposition (Chiu *et al.*, 2006; Gallois-Montbrun *et al.*, 2008). Moreover, huA3F also complexes with Alu retroelement RNA and suppresses retrotransposition, although it lacks the ability to sequester hY retroelement RNA in contrast to huA3G (Gallois-Montbrun *et al.*, 2008). Therefore an alternative, more likely explanation for the HMM complexes is these complexes function to sequester retroelement RNAs into inactive complexes to prevent deleterious retrotransposition in cells.

These studies collectively indicate that huA3G, huA3F and potentially other A3 proteins have multiple cellular activities to restrict diverse reverse transcribing pathogens. In the case of huA3G and huA3F, this includes (a) sequestering specific retroelement RNAs into inactive HMM complexes to block deleterious retrotransposition, (b) forming active LMM complexes that potentially block invading retroviruses directly entering target cells, and (c) packaging newly synthesized huA3 into assembling virions to block the subsequent replication of these virions in target cells by multiple means. To circumvent these antiviral effects and successfully replicate in host species, retroviruses like HIV-1 potentially infect target cells with inactive HMM huA3s to avoid restriction by active LMM huA3s, and encode Vif accessory proteins to degrade newly synthesized huA3 in producer cells or adopt other virion exclusion methods to prevent subsequent restriction by virion-associated A3s. Therefore, A3 proteins are clearly a powerful arm of the host defences given the diversity of reverse transcribing pathogens they restrict, and the elaborate countermeasures pathogens have evolved to avoid A3 restrictions, such as the Vif accessory protein.

## Restrictions of lentivirus gene expression

### TRIM28 restriction of retroviral transcription

Among the TRIM family members implicated in protecting hosts from viruses (Nisole *et al.*, 2005; Barr *et al.*, 2008), TRIM28 (also known as Kap-1 and Tif1-beta) is defined as a transcriptional co-repressor of Moloney murine leukaemia virus (Wolf and Goff, 2007). While this restriction is largely described for gammaretroviruses, it also impacts lentiviruses (Ellis, 2005; Wolf *et al.*, 2008a).

This transcriptional restriction emerged over 30 years ago in mouse embryonic carcinoma (EC) and embryonic stem (ES) cells that restrict MLV replication (Teich *et al.*, 1977; Barklis *et al.*, 1986). This restriction targets the integrated provirus and promotes transcriptional silencing of proviral gene expression. Silencing is mediated to some extent by reduced binding of transcription factors to viral enhancers present in the retroviral LTR (Linney *et al.*, 1984; Hilberg *et al.*, 1987; Akgun *et al.*, 1991). However, trans-acting repressors appear to play a more prominent role (Flanagan *et al.*, 1989; Tsukiyama *et al.*, 1989). A major target for trans-acting repressors is the repressor binding site (RBS) DNA sequence in MLV provirus that derives from the $tRNA^{Pro}$ primer binding site (PBS) in the viral RNA genome (Barklis *et al.*, 1986; Loh *et al.*, 1987; Feuer *et al.*, 1989; Kempler *et al.*, 1993). During virus assembly, the cellular $tRNA^{Pro}$ anneals to the 5′-end of the viral genome and primes minus strand strong stop DNA synthesis during reverse transcription in the target cell (Goff, 1990). Since the RBS is present in the DNA sequence, its function is expected to be distinct to that of the PBS present in the viral RNA genome. Transcriptional repression by RBS is independent of its position or orientation in the cell genome (Harada *et al.*, 1979; Petersen *et al.*, 1991). Studies demonstrate that a single point mutation (termed the B2 mutation) in the RBS abrogates MLV transcriptional repression in embryonic cells indicating the factor(s) responsible for repression no longer bind their DNA target (Harada *et al.*, 1979; Barklis *et al.*, 1986; Petersen *et al.*, 1991; Wolf and Goff, 2007).

Recently, TRIM28 was identified as a transcriptional co-repressor of MLV (Wolf and Goff, 2007; Wolf *et al.*, 2008b; Wolf *et al.*, 2008a). TRIM28 was identified as part of a high molecular weight complex present in nuclear extracts of EC and ES cells that binds a DNA RBS probe corresponding to the MLV PBS sequence (Wolf and Goff, 2007). This TRIM28 containing complex was not detected using a probe expressing the B2 mutation nor in nuclear lysates from cells known to support MLV replication. Furthermore, siRNA knockdown of TRIM28 abrogates transcriptional repression (Wolf and Goff, 2007). Although TRIM28 is important in retroviral gene silencing, it does not bind DNA directly but instead binds to ZFP809, a Kruppel-associated box (KRAB) zinc finger DNA-binding protein that binds the RBS (Friedman *et al.*, 1996; Wolf and Goff, 2009). TRIM28 recruits other factors known to be involved in transcriptional silencing and heterochromatin formation. This includes heterochromatin protein 1, which is required for RBS mediated restriction of MLV in EC cells but is dispensable for the TRIM28 containing protein complex binding the RBS (Wolf *et al.*, 2008b). Interestingly, NIH3T3 cells form TRIM28 containing complexes that lack silencing activity, suggesting these cells fail to recruit a downstream factor for silencing (Wolf *et al.*, 2008b). TRIM28-dependent, PBS targeted, transcriptional silencing in embryonic cells has also been described for retroviruses that use the $tRNA^{Lys1,2}$ primer. This includes Mason-Pfizer monkey virus, spumavirus and VMV (Wolf *et al.*, 2008a). While restriction is less potent than that mediated by the $tRNA^{Pro}$ PBS of MLV, these studies indicate TRIM28 restricts a broad range of retroviruses.

Whether TRIM28-dependent transcriptional silencing restricts other lentiviruses apart from VMV remains unclear. HIV-1 based vectors can be transcriptionally silenced, particularly under circumstances where integrated provirus numbers are limited (Ellis, 2005). Transcriptional silencing has also been observed in other mammalian cells including human haematopoietic stem cells (Haas *et al.*, 2003). Moreover, KRAB box transcription factors engineered to target the HIV-1 PBS mediate transcriptional silencing of HIV-1 provirus (Eberhardy *et al.*, 2006). Thus, in addition to VMV, potentially other lentiviruses may encounter TRIM28-dependent transcriptional repression.

## ZAPping viral mRNA stability

Using a cDNA screen to detect cellular factors that restrict retroviral replication, cellular Zinc-finger antiviral protein (ZAP) was identified as restricting retroviruses by decreasing viral mRNA stability post-transcription (Gao *et al.*, 2002). The original antiviral cDNA from Rat2 fibroblasts encoded a fusion protein containing the first 254 codons of ZAP with a C-terminal zeocin resistance gene (NZAP-Zeo). Overexpressing NZAP-Zeo or full-length ZAP restricts MLV infection (Gao *et al.*, 2002). Moreover, rat ZAP (rZAP) also restricts Sinbis virus and other alphaviruses, plus Marburg and Ebola filoviruses (Bick *et al.*, 2003; Muller *et al.*, 2007). Despite ZAP's activity against retroviruses, alphaviruses and filoviruses, it lacks true broad-spectrum antiviral activity as it fails to inhibit HIV-1, herpes simplex virus type I, yellow fever virus, and vesicular stomatitis virus in cell culture (Bick *et al.*, 2003; Kerns *et al.*, 2008). While ZAP is not currently known to inhibit lentiviruses, its capacity to restrict gammaretroviruses and its novel mechanism of action warrants inclusion here.

ZAP expresses four unusual CCCH type zinc fingers at its N-terminus (Gao *et al.*, 2002). Similar motifs are found in trisethraprolin (TTP), which binds to AU-rich elements (ARE) and recruits exosomes to degrade ARE-containing RNAs (Lai *et al.*, 1999; Chen *et al.*, 2001). Consistent with the function of these novel CCCH type zinc fingers, rZAP restricts MLV replication in target cells by specifically degrading cytoplasmic MLV mRNA rather than nuclear MLV mRNA (Gao *et al.*, 2002) through directly binding specific viral recognition sequences (ZAP-responsive element, ZRE) in the MLV 3′ LTR (Guo *et al.*, 2004). ZAP also targets Sinbis virus RNA but the Sinbis ZRE maps to multiple regions and its sequence is different to the MLV ZRE (Guo *et al.*, 2004). Like TTP-mediated RNA degradation, ZAP binding to MLV ZRE also recruits exosomes (Guo *et al.*, 2007b). Optimal ZAP function also requires the p72 DEAD box RNA helicase, presumably to remove RNA secondary structure for more efficient RNA degradation (Chen *et al.*, 2008a). Thus, ZAP is proposed to degrade viral mRNAs by binding the viral mRNA ZRE via its zinc finger motif, and then recruiting poly(A)-specific ribonuclease (PARN) to remove the mRNA poly(A) tail, the exosome, and the p72 RNA helicase to restructure the RNA for efficient degradation (Zhu and Gao, 2008).

Human ZAP exists as short and long isoforms and restricts MLV and Sinbis virus more potently than the original rat NZAP-Zeo (Kerns *et al.*, 2008). Moreover, the long isoform contains a C-terminal poly(ADP-ribose) polymerase (PARP)-like domain absent in rZAP. The ZAP gene has endured recurrent positive selection indicating it confers a selective advantage to the host. Surprisingly, positive selection occurs in the PARP but not the CCCH RNA-binding domain (Kerns *et al.*, 2008). However, the role of the PARP domain in ZAP-mediated viral restriction awaits definition. Importantly, ZAP expression is up-regulated by type I interferons, Sinbis virus and CMV infection (Ryman *et al.*, 2005; DeFilippis *et al.*, 2006). This supports a role for ZAP in host innate defences against viruses.

## Post-transcription restriction by RNA silencing

RNA silencing is an ancient form of intracellular immunity that confers protection against virus, transgenes and retrotransposons in plants and invertebrates by post-transcriptional gene regulation (Ding and Voinnet, 2007). The role of RNA silencing in defending mammalian cells from viral infections under physiological conditions has been controversial (Cullen, 2006; Lin and Cullen, 2007; Yeung *et al.*, 2007). RNA silencing, also termed RNA interference (RNAi), is mediated by small non-coding microRNAs (miRNAs) that are key to the post-transcriptional regulation of eukaryotic gene expression by either degrading target mRNA or arresting translation. MicroRNAs derive from longer non-coding primary miRNA (pri-miRNA) by a stepwise maturation process (Lee *et al.*, 2002; Bushati and Cohen, 2007). The pri-miRNA is processed in the cell nucleus by the RNase III endonuclease, Drosha, to form a pre-miRNA containing a stem-loop structure around 60 base pairs. This pre-miRNA is exported from the nucleus to the cytoplasm by exportin 5 (Yi *et al.*, 2003) where it is then cleaved by another RNase III endonuclease, Dicer, yielding a miRNA duplex of 19–24 base pairs. The miRNA duplex is incorporated into the RNA-induced silencing complex where one strand of the miRNA duplex is degraded, enabling the other miRNA strand to anneal to its mRNA target. Perfect sequence homology of the miRNA with its target results in mRNA degradation. Imperfect sequence complementarity results in mRNA translation arrest, which is the main mechanism of miRNA-mediated post-transcriptional silencing in mammalian cells (Nilsen, 2007). This form of RNA silencing usually involves miRNA targeting the 3′ untranslated regions (UTR) of mRNA. Imperfect base pairing of a single miRNA can potentially recognize and inhibit the translation of nearly 100 mRNAs in the cell (Brennecke *et al.*, 2005). Currently, over 800 human miRNAs are described (http://microrna.sanger.ac.uk) (Kumar and Jeang, 2008). Apart from cellular miRNA, viral RNA can also adopt stem loop structures recognized by Drosha (in the nucleus) or Dicer (in the cytoplasm) and be processed to form viral miRNA (Corbeau, 2008). These viral miRNAs may restrict viral replication by targeting viral or cellular mRNAs with sequence homology to the viral miRNA. Conversely, cellular miRNAs may target viral mRNAs or cellular mRNAs expressing cellular proteins required for viral replication to restrict invading viruses (Corbeau, 2008).

If RNA interference legitimately constitutes part of the mammalian innate intracellular immune response against viral infections, it is proposed that several criteria should be satisfied (Cullen, 2006). Firstly, viral infections should result in the production of viral miRNAs. Secondly, inhibiting the RNA interference machinery should enhance viral replication. Finally, the virus should have evolved a mechanism to evade the antiviral response mediated by RNA interference.

Concerning a role for RNA interference in restricting HIV-1, the existence of HIV-1 miRNAs remains controversial despite several studies supporting the first criteria (Cullen, 2009). Bioinformatic analysis of the HIV-1 genome predicts candidate pre-miRNAs that, upon maturation to miRNAs, could potentially interact with the 3′ UTR of human mRNAs (Bennasser *et al.*, 2004). This includes cellular mRNAs expressing CD4 or IL-2 involved in HIV-1 infection (Couturier and Root-Bernstein, 2005). A viral nef-derived miRNA, miR-N367, has been detected in HIV-1 infected and *nef* transduced cells, which acts post-transcriptionally by binding the viral 3′ UTR or down-regulating transcription from the HIV-1 LTR (Omoto *et al.*, 2004; Omoto and Fujii, 2005). The HIV-1 trans-activation response element is also reportedly processed by Dicer to yield viral miRNAs detected in HIV-1 infected cells that block gene transcription (Klase *et al.*, 2007; Ouellet *et al.*, 2008). *In silico* analysis also identifies two HIV-1 siRNAs mapping to the viral Rev-responsive element, which are detected in HIV-1 infected cells and target the cognate *env* mRNA (Bennasser *et al.*, 2005). However, other studies question the existence of these viral miRNAs and have failed to detect HIV-1 miRNAs in infected cells (Pfeffer *et al.*, 2005; Cullen, 2006; Lin and Cullen, 2007). Therefore, while viral miRNAs clearly exist for herpesvirus family members like Epstein Barr virus, human cytomegalovirus, herpes simplex virus type 1, and Kaposi's sarcoma-associated virus, more evidence is required to support the notion that HIV-1 miRNAs are produced in infected cells (Cullen, 2009).

Despite the controversy over the existence of HIV-1 miRNAs, other studies support the second criteria that blocking RNA interference enhances HIV-1 replication. Inhibiting the RNA interference machinery by siRNA knockdown of Dicer or Drosha increases HIV-1 replication thus supporting the role of RNA interference in HIV-1 restriction (Triboulet *et al.*, 2007; Qian *et al.*, 2009). Inhibiting the RNA interference machinery decreases the biogenesis of cellular miRNAs like cellular miR-17–5p and miR-20a that restrict the translation of cellular p300/CREB Binding protein-associated factor (PCAF) involved in HIV-1 gene expression (Triboulet *et al.*, 2007). Moreover, down-regulating Dicer rescues abundant HIV-1 *gag* mRNA translation (Qian *et al.*, 2009). This collectively supports a role for RNA interference in restricting HIV-1 replication.

If RNA interference restricts viral replication, the virus would be expected to evolve a mechanism to evade this antiviral response. In support of this third criterion, the HIV-1 Tat protein reportedly has RNA silencing suppressor (RSS) activity and counteracts the repression of HIV-1 replication (Bennasser *et al.*, 2005; Qian *et al.*, 2009). This RSS activity of Tat was initially reported to block Dicer's ability to process mature miRNAs (Bennasser *et al.*, 2005; Bennasser and Jeang, 2006). However, recent work indicates Tat's RSS activity is independent of RNA processing and restricts mRNA translation presumably by sequestering mature miRNAs (Qian *et al.*, 2009). However, others question a role for Tat as an RSS given experiments are typically not performed using physiological levels of this viral regulatory protein (Cullen, 2006; Lin and Cullen, 2007). Tat over-expression can lead to non-specific activation of heterologous promoters used in the various studies, confounding data interpretation (Cullen, 2006; Lin and Cullen, 2007). Irrespective of whether Tat is a bona fide RSS, the impact of the Tat RSS activity is modest and ultimately a role for HIV-1 factors with RSS activity to support a role for RNA interference in HIV-1 restriction remains controversial.

Cellular miRNAs also appear to impact HIV-1 latency. Cellular miR-28, miR-125b, miR-150, miR-223 and miR-382 are enriched in resting compared to activated CD4+ T-cells (Huang *et al.*, 2007). Moreover, specific inhibitors of these miRNAs increase HIV-1 protein translation in resting cells thus supporting the role of cellular miRNAs and RNA interference in HIV-1 latency (Huang *et al.*, 2007). Recently, cellular miR-28, miR-150, miR-223 and miR-382 expression correlated with HIV-1 restriction in monocytes purified from peripheral blood mononuclear cells (PBMCs) (Wang *et al.*, 2009). In comparison, these cellular miRNAs are poorly expressed in macrophages that are permissive for HIV-1 replication. Hence, these data collectively indicate that HIV-1 restriction by the RNA silencing machinery probably facilitates HIV-1 latency in PBMCs and helps restrict HIV-1 replication in monocytes. Therefore, while the role of RNA interference in restricting HIV-1 is controversial, evidence is mounting to support a role for miRNA in modulating the HIV-1 lentivirus in host cells.

## Restrictions of lentivirus particle release

### Vpu abrogates restrictions of lentiviral particle release

HIV-1 Vpu abrogates several cellular restriction factors including tetherin (CD317, BST2 or HM1.24), calcium-modulating cyclophilin ligand (CAML) and Vpu binding protein (UBP) that are all implicated in perturbing viral particle release from cells (Callahan *et al.*, 1998; Neil *et al.*, 2008; Van Damme *et al.*, 2008; Varthakavi *et al.*, 2008). HIV-1 Vpu is an accessory protein expressed late during HIV-1 replication in cells from biscistronic mRNA also encoding viral Env (Cohen *et al.*, 1988; Strebel *et al.*, 1988; Schwartz *et al.*, 1990). Vpu is limited to HIV-1, $SIV_{cpz}$ and SIV from greater spot-nosed monkeys, and absent in HIV-2 and other lentiviruses (Huet *et al.*, 1990; Courgnaud *et al.*, 2002). However, HIV-2 Env has similar functions to Vpu in overcoming the restriction of viral particle release (Bour *et al.*, 1996; Bour and Strebel, 1996; Ritter *et al.*, 1996; Schubert *et al.*, 1999; Abada *et al.*, 2005). Vpu is not essential for HIV-1 replication *in vitro* but is important for *in vivo* pathogenesis, as inactivating Vpu in SHIV chimeric viruses decreases viral load and CD4+ T-cell depletion in macaques (Singh *et al.*, 2003; Hout *et al.*, 2004a,b; Hill *et al.*, 2008).

Vpu is a type I integral membrane protein with N-terminal transmembrane and cytoplasmic domains (Maldarelli *et al.*, 1993). Vpu is ascribed several functions including down-regulating cellular CD4, MHC class I and II, plus enhancing viral particle release, ion channel activity and apoptosis (Bour and Strebel, 2003; Hout *et al.*, 2004b; Nomaguchi *et al.*, 2008). Consistent with these functions, Vpu localizes to various cellular compartments including the endoplasmic reticulum (ER), Golgi, pericentriolar recycling endosome and plasma membrane (Klimkait *et al.*, 1990; Pacyniak *et al.*, 2005; Varthakavi *et al.*, 2006; Neil *et al.*, 2008; Van Damme and Guatelli, 2008). The ability of Vpu to induce CD4 degradation in the ER is well described. Vpu targets CD4 in the ER for degradation by binding the human beta transducin repeat-containing protein (βTrCP), thereby recruiting a cullin-RING ubiquitin ligase complex to the cytoplasmic tail of CD4 (Margottin *et al.*, 1998; Malim and Emerman, 2008). This promotes polyubiquitylation and proteosomal degradation of CD4 but not Vpu. CD4 degradation also allows the viral Env protein to escape premature binding to its cellular CD4 receptor in the ER and move to the cell surface for incorporation into assembling virions to produce infectious virions (Margottin *et al.*, 1998; Malim and Emerman, 2008). The impact of Vpu on CD4 degradation and viral particle release is functionally distinct and maps primarily to the cytoplasmic and transmembrane domains respectively (Yao *et al.*, 1992; Schubert and Strebel, 1994; Friborg *et al.*, 1995; Schubert *et al.*, 1996a; Margottin *et al.*, 1998; Paul *et al.*, 1998). The ability of Vpu to oligomerize into ion channels and enhance viral particle release may be related but awaits clarification (Maldarelli *et al.*, 1993; Ewart *et al.*, 1996; Schubert *et al.*, 1996b). As CD4 disrupts Vpu oligomerization at the cell surface, CD4 degradation may indirectly effect Vpu-mediated virus particle release (Bour *et al.*, 1999).

Vpu clearly enhances HIV-1 viral particle release from host cells (Strebel *et al.*, 1989; Terwilliger *et al.*, 1989; Gendelman *et al.*, 1990; Klimkait *et al.*, 1990; Gottlinger *et al.*, 1993; Balliet *et al.*, 1994; Geraghty *et al.*, 1994; Schubert *et al.*, 1995). Vpu also prevents endocytosis of newly formed viral particles in intracellular vacuoles, which were previously considered to be sites of virus budding (Harila *et al.*, 2006; Neil *et al.*, 2006; Van Damme and Guatelli, 2008). Infection of cells with HIV-1 lacking functional Vpu causes viral particles to accumulate at the cell surface as observed by electron microscopy (Neil *et al.*, 2008). While this phenotype appears similar to defects conferred by mutations in the HIV-1 Gag p6 late domain, which is also critical for viral particle egress, significant differences exist between the two. In contrast to the p6 late domain phenotype, Vpu-defective HIV-1 have no defects in Gag expression or processing, the virions that accumulate at the cell surface are mature, and the viral particles can be released from the cell surface by subtilisin protease treatment (Strebel *et al.*, 1989; Terwilliger *et al.*, 1989; Neil *et al.*, 2006). Consequently, these studies indicate that the Vpu defect arises at a stage after fission of viral and cell membranes (Klimkait *et al.*, 1990; Gottlinger *et al.*, 1993; Neil *et al.*, 2006).

The ability of Vpu to enhance viral particle release is also host cell species and cell type dependent (Strebel *et al.*, 1989; Terwilliger *et al.*, 1989; Geraghty *et al.*, 1994; Sakai *et al.*, 1995; Varthakavi *et al.*, 2003; Neil *et al.*, 2006). Vpu enhances HIV-1 particle release in most human cells including primary CD4$^{+}$ T-cells and monocyte derived macrophages, but not cells from non-human primates (Gendelman *et al.*, 1990; Balliet *et al.*, 1994; Schubert *et al.*, 1995; Deora and Ratner, 2001). Remarkably, in cells that require Vpu for enhancing virion release (referred to as Vpu dependent), Vpu also enhances the egress of other retroviruses normally lacking a Vpu gene suggesting Vpu's effect is independent of specific viral sequences (Gottlinger *et al.*, 1993; Jouvenet *et al.*, 2008).

Heterokaryon experiments fusing cells that are Vpu-dependent (HeLa) and Vpu-independent (COS7) reveal that Vpu antagonizes a host cell factor that restricts HIV-1 particle release (Varthakavi *et al.*, 2003). Moreover, expression of the restriction factor can be induced by interferon α (IFN-α) treatment suggesting the factor is part of the innate immune response (Neil *et al.*, 2007). Tetherin, also known as CD317, BST2 or HM1.24 (Neil *et al.*, 2008; Van Damme *et al.*, 2008), is one of several cellular restriction factors reported to decrease HIV-1 particle release that is counteracted by Vpu.

## Tetherin (CD317, BST2 or HM1.24)

Tetherin was identified as a host cell restriction factor counteracted by Vpu using microarray analysis of cellular mRNA and a set of stringent criteria: (1) its expected presence and absence in cells that display or do not display Vpu-dependent defects in viral particle release; (2) a requirement for Vpu enhancement of virion release in cells after treatment of Vpu independent cells with IFN-α; and (3) the factor is either associated with the cell membrane or secreted due to its capacity to prevent fission of virion and cell membranes (Neil *et al.*, 2008). While there were several candidates that conformed to these stringent criteria, tetherin was considered the most likely candidate based on its mRNA expression levels in Vpu-dependent and -independent cell lines. The role of tetherin was confirmed by its overexpression in 293T-cells (that are normally Vpu independent), which recapitulated the virion release defect for Vpu-defective HIV but not wild-type virus. Additionally, depleting tetherin expression in Vpu-dependent cells negated the requirement of Vpu for virion release (Neil *et al.*, 2008). These experiments were performed using single cycle HIV-1 replication assays in cells lacking CD4 to avoid confounding effects of Vpu on CD4 expression, which can also impact on virion infectivity and egress (Lama *et al.*, 1999; Levesque *et al.*, 2003; Tanaka *et al.*, 2003). Tetherin expression also caused virions to accumulate at the cell surface and in endosomal compartments while concomitant Vpu expression abrogated these effects (Neil *et al.*, 2008). The role of tetherin as an intracellular restriction factor counteracted by Vpu was independently confirmed by others (Van Damme *et al.*, 2008), who focused on it as the putative restriction factor after a proteomic analysis indicated Vpu modulated tetherin expression at the plasma membrane (Bartee *et al.*, 2006).

The exact mechanism by which tetherin inhibits viral particle release remains unclear although its structure provides some clues. Tetherin is a glycosylated integral membrane protein containing an N-terminal cytoplasmic tail, transmembrane domain, a putative extracellular coiled-coil domain and putative glycosyl phosphatidylinositol GPI anchor at the C-terminus (Goto *et al.*, 1994; Kupzig *et al.*, 2003; Blasius *et al.*, 2006; Rollason *et al.*, 2007). Tetherin is predicted to localize to lipid rafts at the plasma membrane (Kupzig *et al.*, 2003) leading to speculation that it may form links between the plasma membrane and viral membranes to physically prevent virion release (Neil *et al.*, 2008; Van Damme *et al.*, 2008). Tetherin interaction with membranes is consistent with its broad-spectrum inhibition of retrovirus and filovirus particle release (Jouvenet *et al.*, 2008; Sakuma *et al.*, 2009). As tetherin can form dimers, several topologies are suggested to explain how viral particles tether to each other and to the plasma membrane (Neil *et al.*, 2008). This proposed mechanism probably requires tetherin incorporation into viral particles (Gottlinger, 2008).

The mechanism by which Vpu abrogates the effects of tetherin on viral particle release also remains unclear. One possibility is that Vpu may degrade tetherin by proteasomes. In support of this possibility, the Kaposi's sarcoma-associated herpes virus K5 protein is an ubiquitin ligase that down-regulates tetherin (Bartee *et al.*, 2006). As Vpu also down-regulates tetherin, Vpu may degrade tetherin via proteosomes in a manner similar to K5 (Bartee *et al.*, 2006; Gottlinger, 2008). Alternatively, Vpu may alter tetherin function and/or its localization in host cells. This is supported by data showing Vpu produced from HIV-1 infection or exogenous expression down-regulates tetherin from the cell surface and co-localizes with tetherin. In contrast, in the absence of Vpu, virions associate with tetherin at the cell surface and in cell compartments (Bartee *et al.*, 2006; Neil *et al.*, 2008; Van Damme *et al.*, 2008). Moroever, the pericentriolar recycling endosome is required for Vpu-mediated enhancement of HIV-1 particle release (Varthakavi *et al.*, 2006). However, more research is ultimately required to determine how Vpu abrogates tetherin restriction.

Human tetherin is a host cell restriction factor induced by IFN-α with broad-spectrum activity against enveloped viruses including retroviruses and filoviruses (Jouvenet *et al.*, 2008; Neil *et al.*, 2008; Van Damme *et al.*, 2008; Sakuma *et al.*, 2009). Tetherin orthologues exist in other mammalian species, which raises the possibility that tetherin in other species may also have an antiviral role. Interestingly, Vpu is not required for HIV-1 particle release from agm-derived COS cells and IFN-α treatment does not stimulate HIV-1 release (Geraghty *et al.*, 1994; Neil *et al.*, 2007). Therefore, Vpu specifically abrogates restriction of viral particle release by human tetherin. Understanding the mechanism by which tetherin restricts viral particle release from host cells and how Vpu counteracts this restriction may reveal new knowledge to develop novel therapeutic agents to treat not only HIV/AIDS but other enveloped viruses which infect humans.

## Calcium-modulating cyclophilin ligand (CAML)

Following the discovery of tetherin restriction, calcium-modulating cyclophilin ligand (CAML) was reported to bind Vpu and function as a Vpu-responsive human cell restriction factor (Varthakavi *et al.*, 2008). CAML was identified in a yeast two-hybrid screen of a HeLa cDNA library using Vpu as the bait. Direct interaction between CAML and Vpu was confirmed by co-immunoprecipitation studies with lysates from HIV infected cells using antibodies to endogenous CAML. CAML is absent in agm cells consistent with their Vpu-independent phenotype for viral particle release (Varthakavi *et al.*, 2008). Expressing human CAML in agm cells appears to restrict virus particle release and this appears to be reversed by coexpressing Vpu or HIV-2 Env that both counteract restriction of viral particle release (Bour *et al.*, 1996; Abada *et al.*, 2005). While cellular CAML expression correlates with the Vpu-dependent phenotype, the ability of IFN-α to up-regulate CAML expression was not tested (Varthakavi *et al.*, 2008).

The exact mechanism by which CAML restricts viral particle release and its relationship with tetherin remains unclear. However, as CAML and Vpu are present in recycling endosomes and co-localize within perinuclear compartments, their interaction in these compartments may decrease CAML reaching the plasma membrane (Tran *et al.*, 2003; Varthakavi *et al.*, 2006; Van Damme and Guatelli, 2008). CAML may in turn alter the recycling of another factor (*e.g.* tetherin) required for virion retention at the cell surface. Determining whether CAML affects the cellular localization and activity of tetherin may assist in understanding whether there is interplay between these two Vpu-responsive cellular restriction factors.

### Vpu-binding protein (UBP)

A cellular protein termed Vpu-binding protein (UBP) was also identified in early yeast two-hybrid screens using Vpu as bait (Callahan *et al.*, 1998). UBP contains four copies of a tetratricopeptide repeat (TPR) sequence involved in protein–protein interactions in other proteins. Proteins from the tetratricopeptide family have diverse functions including immunophilins (Callebaut *et al.*, 1992; Ratajczak *et al.*, 1993), organelle targeting proteins (Hase *et al.*, 1983; Dodt *et al.*, 1995) and proteins involved in mitosis (Hirano *et al.*, 1990; Sikorski *et al.*, 1990). UBP RNA is ubiquitously expressed in human tissues and UBP directly binds Vpu and HIV-1 Gag in cell lysates (Callahan *et al.*, 1998). The TPR domain is the minimal region for UBP binding to Vpu (Dutta and Tan, 2008). Overexpressing UBP in HeLa cells (that require Vpu for virus release) reduces viral particle release 5-fold in the presence of Vpu, and a further 2- to 3-fold in the absence of Vpu (Callahan *et al.*, 1998). Hence, UBP reduces viral particle release even in the presence of Vpu, suggesting that Vpu may not completely counteract UBP restriction. Vpu expression in HeLa cells redistributes UBP and Gag to the cell periphery suggesting a role in modulating UBP effects on viral particle release (Handley *et al.*, 2001). While Vpu appears to reduce restriction by UBP, the impact of UBP on viral egress is probably modest compared to CAML (Varthakavi *et al.*, 2008). Ultimately, how UBP reduces viral particle release and whether IFN-α up-regulates UBP as part of the innate intracellular immune response requires further investigation.

### β-Catenin and E-cadherin

β-catenin and E-cadherin complexes may also contribute to the Vpu-dependent restriction of HIV-1 particle release from macrophages and dendritic cells (Salim and Ratner, 2008). Vpu stabilizes several substrates of bTrCP including β-catenin, and Vpu-mediated sequestration of βTrCP up-regulates E-cadherin transcription (Bour *et al.*, 2001; Estrabaud *et al.*, 2007; Salim and Ratner, 2008). Overexpressing E-cadherin or maintaining E-cadherin/β-catenin complexes decreases virion release while down-regulating E-cadherin or disrupting E-cadherin/β-catenin complexes enhances virion release (Salim and Ratner, 2008). As E-cadherin/β-catenin complexes are anchored to the cytokeleton via E-cadherin, disruption of the underlying α-catenin-actin cytoskeleton is speculated to underlie the Vpu-mediated inhibition of HIV-1 release, although this awaits experimental confirmation (Salim and Ratner, 2008). E-cadherin is present at high levels in macrophages and dendritic cells but not T-lymphocytes, and Vpu enhances HIV-1 release from macrophages up to 1000-fold (Balliet *et al.*, 1994; Deora and Ratner, 2001). Therefore, regulation of E-cadherin/β-catenin interactions may represent another restriction mechanism counteracted by Vpu. Of note, β-catenin has also been implicated in restricting HIV-1 replication in astrocytes and human peripheral blood mononuclear cells (Kumar *et al.*, 2008) potentially via a downstream effector of β-catenin signalling, TCF-4, which represses Tat-mediated HIV-1 transcription (Wortman *et al.*, 2002; Rossi *et al.*, 2006; Kumar *et al.*, 2008). However, the details of this potential β-catenin restriction targeting HIV-1 transcription require further elucidation.

## Future perspectives

Intracellular host cell restrictions have revolutionized the retrovirology field in recent years. Identifying TRIM5, APOBEC3, and other cellular proteins as restricting lentiviruses and other pathogens (Gao *et al.*, 2002; Sheehy *et al.*, 2002; Stremlau *et al.*, 2004; Wolf and Goff, 2007; Barr *et al.*, 2008; Neil *et al.*, 2008; Varthakavi *et al.*, 2008) has led to important insights into understanding barriers of retroviral replication and the cross-species transmission of retroviruses. It has also garnered intense interest in how this information can be harnessed to develop new antiviral therapies, animal models of HIV-1/AIDS and facilitate gene therapy.

### Antiviral therapies

With respect to developing new antiviral therapies, the potent TRIM5 and APOBEC3 restriction factors, which are enduring positive selection and evolving to restrict various retroviruses *in vivo* (Sawyer *et al.*, 2004; Sawyer *et al.*, 2005), all target early events in retroviral replication. This infers that this stage of replication is particularly susceptible to therapeutic intervention, and may contain effective targets for developing new therapies that emulate restriction as effectively as the naturally occurring cell restriction proteins.

Considering the TRIM5 and CypA axis of intracellular immunity, these cellular factors target the CA shell of invading retroviral cores for restriction, and retroviruses evade these proteins by changing their viral CA sequence (Accola *et al.*, 1998; Hatziioannou *et al.*, 2004a; Dodding *et al.*, 2005; Forshey *et al.*, 2005; Sebastian and Luban, 2005; Stremlau *et al.*, 2006a). This indicates the viral CA shell is sensitive to inhibition. Therefore, one antiviral approach is to develop new therapeutics that mimic TRIM5 proteins, binding invading retroviral CA shells to perturb their subsequent behaviour. HIV-1 could be particularly sensitive to such inhibitors given optimal uncoating of its CA shell is required for infection (Forshey *et al.*, 2002). Fusing the HIV-1 CA binding sequence, CypA, to certain protein multimerization domains can perturb HIV-1 in cell culture (Javanbakht *et al.*, 2007; Yap *et al.*, 2007), providing a precedent that such factors can be developed and are antiviral. However, protein-based inhibitors require an effective delivery system and if CypA is used to target the lentiviral CA, potential deleterious effects on host cells containing highly conserved CypA must be considered. Moreover, great care must be exercised when introducing protein substrates capable of higher order oligomerization, as overexpression and/or failure to turn over such substrates could lead to large protein aggregates in the cell with toxic consequences. Searching for small molecule inhibitors that perturb invading retroviral CA shells is another possibility that may be less deleterious to host cells. Alternatively, molecules that change the spectrum of retroviruses that the existing host TRIM5 protein can restrict are possibilities. However, a major complication to any approaches seeking to mimic TRIM5 restriction is the fact these proteins interact with the 3 dimensional CA shell of invading retroviruses (Dodding *et al.*, 2005; Forshey *et al.*, 2005; Sebastian and Luban, 2005; Stremlau *et al.*, 2006a). Thus, high throughput assays need to be developed to screen potential inhibitors of the CA shell of invading retrovirus cores. Moreover, defining the mechanism of action of such inhibitors requires working with 3 dimensional retroviral CA structures that is technically demanding (Stremlau *et al.*, 2006a; Langelier *et al.*, 2008).

As TRIM5 proteins are interferon inducible (Asaoka *et al.*, 2005; Sakuma *et al.*, 2007b; Carthagena *et al.*, 2008; Rajsbaum *et al.*, 2008) and up-regulating their expression can expand their antiviral specificity (Bock *et al.*, 2000; Ylinen *et al.*, 2005; Passerini *et al.*, 2006; Ylinen *et al.*, 2006; Schaller *et al.*, 2007a), a different approach could involve up-regulating the expression of existing TRIM5 proteins in the host to expand retroviral specificity to other circulating retroviruses. Delivering a TRIM5 protein with the appropriate restriction specificity, such as rhTRIM5α or huTRIM5α substituted with the rhTRIM5α specificity determinant, potentially using transgenic stem cells is another possibility to impede HIV-1 replication (Anderson and Akkina, 2005, 2008). However, this approach has added issues of potential deleterious effects from activating oncogenes during gene therapy, and/or overexpressing a TRIM5 protein that may form higher order, toxic aggregates and/or bind other TRIM proteins via their coiled-coil domains perturbing TRIM protein function.

Alternatively, the APOBEC3 proteins target diverse reverse transcribing pathogens and lentiviruses primarily subvert A3 proteins via their Vif proteins (Sheehy *et al.*, 2002; Conticello *et al.*, 2003; Marin *et al.*, 2003; Sheehy *et al.*, 2003; Stopak *et al.*, 2003; Yu *et al.*, 2003; Mehle *et al.*, 2004b; Mehle *et al.*, 2004a; Liu *et al.*, 2005a; Luo *et al.*, 2005; Dang *et al.*, 2006; Zhang *et al.*, 2008d,e). Therefore, a different therapeutic approach could be to develop inhibitors that block lentiviral Vif interaction with specific A3 proteins thereby allowing host A3 proteins to regain their ability to restrict specific lentiviruses. Inhibitors that bind specific lentiviral Vif and/or obscure its interaction with host A3 proteins could be valuable. Such an approach could be used to prevent HIV-1 Vif interaction with huA3G and huA3F thereby allowing these human A3 proteins to restrict invading HIV-1 again. Vif peptides and antibodies that antagonize Vif-huA3G binding have been described (Mehle *et al.*, 2007). More recently, a small molecule inhibitor, RN-18, which targets HIV-1 Vif–huA3G interaction and inhibits HIV-1 replication in restrictive cells has been developed (Nathans *et al.*, 2008). This provides a proof-of-principle example of this approach and these inhibitors offer promise for development into new antiviral therapies, which are required due to resistance problems and long term, adverse effects of current HIV-1 drugs.

As Vif and A3 interactions are dose-dependent (Sheehy *et al.*, 2002), an alternative approach may be to stimulate high level expression of specific A3 proteins in patients to overcome the level of lentiviral Vif expressed in virally infected cells and restrict subsequent rounds of infection. Alternatively, delivering exogenous huA3G fused to a Vpr peptide for virion incorporation also curbs HIV-1 replication in primary cells irrespective of HIV-1 Vif expression (Ao *et al.*, 2008). Delivering huA3G stabilized from proteasome-mediated degradation also impedes HIV-1 infectivity (Li *et al.*, 2008). However, these approaches may be problematic as up-regulated expression of other APOBEC proteins, APOBEC1 and AID, in mice can lead to various cancers (Yamanaka *et al.*, 1995; Okazaki *et al.*, 2003), probably due to undesirable deamination of cellular nucleic acid. An alternative approach for huA3G and huA3F may be to stimulate their assembly into antiviral LMM complexes as opposed to inactive HMM complexes, which correlates with restriction to HIV-1 infection in various human target cells (Chiu *et al.*, 2005; Kreisberg *et al.*, 2006; Ellery *et al.*, 2007; Stopak *et al.*, 2007; Trapp *et al.*, 2009). However, the HMM complexes usurp retroelement RNA and impede Alu retrotransposition so stimulating LMM A3 complexes to form is likely to enhance Alu retrotransposition in the host, as observed in cell culture (Chiu *et al.*, 2006; Gallois-Montbrun *et al.*, 2008). Therefore, this approach may have deleterious genetic consequences on the host from enhanced retrotransposition and thus also be problematic. Consequently, of all these possible therapeutic approaches, targeting the Vif molecule and blocking its interaction with specific A3 proteins seems least likely to have deleterious effects on the host and be most promising for therapeutic development in theory. Although, the ability of lentiviruses like HIV-1 to develop resistance to such Vif inhibitors will be an important consideration.

Aside from inhibitors targeting TRIM5-viral CA or APOBEC-lentiviral Vif, new susceptible viral targets may also emerge from defining other host cell restriction factors. For instance, inhibitors targeting Vpu to curb its ability to block tetherin would allow tetherin to restrict HIV-1 particle release again, and thus may be therapeutically effective *in vivo*. A cholesterol binding compound, amphotericin B methyl ester, partially blocks Vpu from counteracting tetherin restriction of HIV-1 particle release (Waheed *et al.*, 2008). This provides a precedent that chemotherapeutic agents that prevent Vpu from counteracting intracellular restrictions of viral particle release may be effective. Therefore, the study of restriction factors is an exciting field inspiring new possibilities of novel antivirals.

## Animal models of HIV-1/AIDS

As well as requiring new inhibitors to curb the current HIV-1 pandemic, another concern is the lack of a suitable, effective *in vivo* model of HIV-1/AIDS to study pathogenesis and test the efficacy of new inhibitors or potential vaccines prior to testing in humans. Initial HIV-1 studies revealed that HIV-1 replicates in few animal species and the limited options of chimpanzees (Alter *et al.*, 1984) and gibbon apes (Lusso *et al.*, 1988) are costly, endangered and often there is no disease induction. As a result, simian AIDS models have emerged as a widely used alternative, which employs macaques infected with SIV or SHIV chimeric viruses (primarily SIV substituted with various HIV-1 genes) [reviewed in Ambrose *et al.* (2007) and Chapter 6]. Despite providing many important insights, these simian AIDS models still have various limitations. For instance, certain HIV-1 drugs cannot be tested in these models before human trials due to differences between HIV-1 and SIV enzymes targeted by the drugs. Moreover, new drugs developed to target specific accessory proteins that differ between HIV-1 and SIV may not be tested properly in this system. The efficacy of various HIV-1 vaccines is also difficult to test in these models due to differences in HIV-1 and SIV sequences, and the elicited immune responses (reviewed in Ambrose *et al.*, 2007). Many of these issues may be avoided if primarily HIV-1 derived virus could be made to replicate in these animal models.

The discovery of the TRIM5 and APOBEC3 proteins as potent restrictors of invading retroviruses that generally originate from different animal species, has led to new strategies for producing HIV-1 that replicates in macaque models. TRIM5 restriction targets the retroviral CA and APOBEC3 restriction is counteracted by lentiviral Vif. By substituting the CA and Vif sequence in HIV-1 for the SIV counterpart, a primarily HIV-1 derived virus is generated that replicates in macaque cells (Hatziioannou *et al.*, 2006; Kamada *et al.*, 2006). One of these viruses also replicates transiently in macaques (Igarashi *et al.*, 2007) although follow up studies showing that these viruses actually cause disease are lacking. A separate approach has employed HIV-1 substituted for the SIV Vif gene (HIV-1/SIV Vif) to avoid A3 restriction, and used this virus to infect pig-tailed macaques whose TRIMCyp does not restrict HIV-1, thereby avoiding TRIM5 restriction in this animal (Hatziioannou *et al.*, 2009). Importantly, this virus robustly and persistently infected pig-tailed macaques for several months analogous to HIV-1, before infection was controlled. This virus and animal model was also sensitive to current HIV-1 drugs that are normally ineffective against other SIV and SHIV isolates. Administering a drug regimen currently used for HIV-1 patients to these pig-tailed macaques as a pre-exposure prophylaxis also protected them from a subsequent high dose challenge of this HIV-1/SIV Vif chimera. This demonstrates the utility of this animal model for testing new anti-HIV drugs and administration parameters (Hatziioannou *et al.*, 2009). Therefore, this choice of pig-tailed macaque and substituting the HIV-1 Vif gene for an SIV counterpart enabled HIV-1 to replicate in another primate species. This not only unveils an animal model for HIV-1 derived virus, but also shows that TRIM5 and A3 restrictions constitute key barriers to retroviral infections in other animal species. Of note, rhesus macaques homozygous for *Mamu-7*, which expresses TRIMCyp that does not restrict HIV-1 (Wilson *et al.*, 2008a), may also be a suitable animal model for the HIV-1/SIV Vif, similar to pig-tailed macaques encoding TRIMCyp (Hatziioannou *et al.*, 2009). Therefore, the discovery of TRIM5 and APOBEC3 restriction has led to significant advances in developing animal models of HIV-1. Although notably, the pig-tailed macaque model using the HIV-1/SIV Vif did not induce disease (Hatziioannou *et al.*, 2009) indicating this model requires further refinement to develop a model that mimics HIV-1 induced AIDS.

## Restriction factors and gene therapy

Intracellular restriction factors are also important to consider in gene therapy applications. For instance, when considering gene therapy vectors for different animal hosts, as both TRIM5 and APOBEC3 in the animal host may restrict the replication of viral vector, the origin of the viral vector should be carefully chosen to circumvent these restrictions where possible and improve transduction. The viral Gag-Pol packaging constructs should be selected such that the viral CA is not restricted by TRIM5 in the intended host species to enhance transduction efficiency. Moreover, cells lacking A3 restriction should be used to produce the viral vectors to avoid the deleterious effects of virion-associated A3 during transduction.

## Emerging restrictions

While this chapter has focused on mostly well-described, cellular restriction factors like TRIM5, APOBEC3 and tetherin, new cellular restrictions are emerging in the literature that await clear definition. For instance, heterokaryon studies indicate human macrophages harbour a dominant restriction targeting SIV from sooty mangabeys ($SIV_{sm}$) that is circumvented by the SIVSM Vpx accessory protein (Sharova *et al.*, 2008). Expressing Vpx in macrophages *in trans* before infection evades the restriction of Vpx-deleted $SIV_{sm}$ and HIV-1 that lacks Vpx (Sharova *et al.*, 2008). This is consistent with studies where expression of $SIV_{mac}$ Vpx *in trans* in DCs promotes subsequent infection by Vpx-deleted $SIV_{mac}$ or HIV-2, and by HIV-1 and FIV that both lack Vpx (Goujon *et al.*, 2007). The Vpx-mediated rescue of viral infection in DCs rescued reverse transcribed viral cDNA and could be recapitulated by treating cells with a proteasome inhibitor, suggesting that Vpx and the proteasome inhibitor act by similar mechanisms and that Vpx counteracts a proteasome-dependent restriction. This is reminiscent of TRIM5 restriction of lentiviruses targeting the viral CA but this Vpx-sensitive restriction appears distinct from TRIM5 restriction (Goujon *et al.*, 2007). While the identity of the putative cellular restriction factor/s remains unknown, Vpx also requires cellular damage DNA binding protein 1 to counteract the macrophage restriction and ubiquitylation. Moreover, proteasomes are needed for Vpx activity (Sharova *et al.*, 2008). An arsenic-sensitive restriction of primate lentiviruses in human DCs that diminishes reverse transcription products and is independent of TRIM5α and A3G has also been reported (Pion *et al.*, 2007). However the underlying cellular restriction factor/s, mechanism and how this relates to the above Vpx-sensitive restriction await definition.

Type I interferons (IFN-α and IFN-β) are cytokines activating the Jak-Stat signalling pathway that results in the transcription of numerous interferon stimulated genes that are part of the innate antiviral response (Takaoka and Yanai, 2006). Two IFN-β inducible cellular restrictions are also implicated in restricting HIV-1 particle release in a Vpu-independent manner, namely the interferon-induced gene 15 (ISG-15) (Okumura *et al.*, 2006) and TRIM22 (Barr *et al.*, 2008). ISG-15 is stimulated by IFN-β, inhibits HIV-1 virion release and siRNA depletion of ISG-15 reverses the ability of IFN-β to modestly decrease HIV-1 particle release (Okumura *et al.*, 2006). While the underlying restriction mechanism of ISG-15 lacks clarity, ISG-15 is an ubiquitin-like protein that inhibits Gag and Tsg101 ubiquitylation, potentially blocking the Gag p6 late domain from hijacking the endosomal sorting complex for HIV-1 egress from host cells (Bieniasz, 2006). IFN-β also up-regulates TRIM22 in human osteosarcoma (HOS) cells and TRIM22 overexpression in HOS and HeLa cells inhibits HIV-1 particle output without dramatic loss in intracellular Gag (Barr *et al.*, 2008). This impact on viral particle release is specific for HIV-1 Gag but not MLV or EIAV Gag. The mechanism of TRIM22 restriction is unclear but requires its E3 ligase catalytic site. TRIM22 expression and IFN-β treatment also increase diffuse cytoplasmic Gag-GFP localization rather than punctate fluorescence at the plasma membrane indicative of virus particles, suggesting TRIM22 perturbs Gag trafficking to the cell surface (Barr *et al.*, 2008). TRIM22 is also reported to restrict HIV-1 replication in COS, 293T and human macrophages by repressing HIV-1 transcription (Tissot and Mechti, 1995; Bouazzaoui *et al.*, 2006). Therefore, TRIM22 appears to restrict HIV-1 by at least two different mechanisms that await further investigation.

Additional lentiviral restrictions, Lv2 and Lv3, distinct from Lv1/Ref1 (TRIM5) are also reported. Lv2 restriction impedes HIV-2 after reverse transcription but before nuclear entry in certain human cells, and is mediated by the viral envelope as well as CA (Schmitz *et al.*, 2004). Lv3 restriction blocks HIV-1 in a rhesus sMAGI cell line in a TRIM5-independent manner that depends on the viral entry pathway (Pineda *et al.*, 2007). Studies of a CypA-independent, HIV-1 CA mutant also suggest that tissue specific factors can impede lentiviruses independent of TRIM5 and CypA (Kahl *et al.*, 2008). A novel restriction distinct from Fv1 also exists in African pygmy mice, which restricts AKV-type MLV by targeting the viral CA and impairing reverse transcription products (Yan and Kozak, 2008). However, the identity of all these cellular restriction factors remains unknown. Finally, three recent studies using RNA interference and high throughput analyses to identify host cell proteins that enhance or restrict HIV-1 infection at early and/or late stages of replication (Brass *et al.*, 2008; Konig *et al.*, 2008; Zhou *et al.*, 2008) may also unveil new host cell restriction factors of HIV-1. It will be important to validate any of these potential cellular restriction factors in biologically relevant cell types and determine their mechanism of action in comprehensive follow up studies, particularly given the small overlap of cell factors emerging from these screens (Zhou *et al.*, 2008). However, this method provides a new approach for identifying cellular restrictions of lentiviruses and may unveil more novel targets for developing new antiviral therapies in the future.

In addition to cellular protein factors that restrict lentiviruses, cellular structures may also pose as barriers to lentiviral infection. For instance, cortical actin underlying the cell plasma membrane may impede the trafficking of retroviruses at the plasma membrane, which in the case of HIV-1 appears to be circumvented by Nef (Campbell *et al.*, 2004). Moreover, the viscous cellular cytoplasm is likely to limit the diffusion of particles (Luby-Phelps, 2000) like lentiviruses through the cell cytoplasm unless microtubules and dynein (McDonald *et al.*, 2002) and possibly other cellular filaments and motors are engaged to aid cytoplasmic movement towards the nucleus for efficient infection. The nuclear membrane itself also constitutes a major barrier to many retroviruses such as MLV, which only infects dividing cells presumably as the virus only gains access to the nucleus during membrane dissolution (Roe *et al.*, 1993; Lewis and Emerman, 1994). This is less of an issue for lentiviruses like HIV-1 that infect non-dividing and dividing cells and thus can traverse the nuclear membrane barrier in non-dividing cells (Weinberg *et al.*, 1991; Bukrinsky *et al.*, 1992; Lewis *et al.*, 1992). However, imaging studies reveal only a small fraction of intracellular HIV-1 complexes localize to the nucleus (Albanese *et al.*, 2008) indicating the nuclear membrane is still a barrier to HIV-1 infection. Condensed chromatin in the nucleus may also act as a barrier given that HIV-1 preferentially localizes to areas of decondensed chromatin (Albanese *et al.*, 2008) and HIV-1 movement is limited once in the nucleus (Arhel *et al.*, 2006). Therefore, cellular structures in addition to cellular restriction factors appear to all serve to collectively restrict invading lentiviruses *in vivo*.

## Concluding remarks

Cellular restriction factors have thus emerged as pivotal components in the antiviral armour of host organisms. They are providing new insights into barriers of lentiviral replication and are aiding our understanding of the cross-species transmission of retroviruses. They are also guiding the choice of better gene therapy vectors for different animal species. Understanding these factors is enabling better models of HIV-1/AIDS to be developed and most importantly, is identifying promising new targets for antiviral therapy. While many unanswered questions and potential new factors remain to be discovered, an exciting future awaits this dynamic research field.

### Acknowledgements

We wish to acknowledge the important studies of numerous researchers contributing to this dynamic field of host cell restriction factors, whose many research advances could not all be cited here. This work was supported by the National Health and Medical Research Council of Australia (NHMRC). J.L.A. was supported by NHMRC Project Grant 488225 and G.T. was supported by NHMRC Senior Research Fellowship 543105 and NHMRC project grants 381705 and 433903.

### Abbreviations

agm, African green monkey; AID, activation-induced deaminase; AIDS, acquired immunodeficiency syndrome; APE, apurinic/apyrimidinic endonuclease; APOBEC3, apolipoprotein B mRNA-editing enzyme catalytic polypepetide-like 3; ARE, AU-rich element; βTrCP, beta transducin repeat-containing protein; CA, capsid; CAML, calcium-modulating cyclophilin ligand; cDNA, complementary DNA; cpz, chimpanzee; CsA, cyclosporine A; CypA, cyclophilin A; DC, dendritic cell; DNA, deoxyribonucleic acid; EC, embryonic carcinoma; EIAV, equine infectious anaemia virus; ER, endoplasmic reticulum; ES, embryonic stem; FIV, feline immunodeficiency virus; Fv1, Friend virus susceptibility 1; HERV-K, human endogenous retrovirus K; HIV-1, human immunodeficiency virus type 1; HMM, high molecular mass; HOS, human osteosarcoma; hu, human; hY, human small Y; iDC, immature dendritic cell; IFN, interferon; ISG-15, interferon-induced gene 15; KRAB, Kruppel-associated box; LMM, low molecular mass; LTR, long terminal repeat; Lv1, lentivirus susceptibility factor 1; m, mouse; MHR, major homology region; miRNA, microRNA; MLV, murine leukaemia virus; NC, nucleocapsid; NWM, New World monkey; NZAP, N terminal zinc-finger antiviral protein; OWM, Old World monkey; PARN, poly(A)-specific ribonuclease; PARP, poly(ADP-ribose)polymerase; PBMC, peripheral blood mononuclear cells; PBS, primer binding site; PCAF, p300/CREB binding protein-associated factor; pri-miRNA, primary microRNA; PtERV1, *Pan troglodytes* endogenous retrovirus; RBCC, RING, B-box and coiled-coil; RBS, repressor binding site; Ref1, resistance factor 1; rh, rhesus macaque; RNA, ribonucleic acid; RNAi, RNA interference; RNP, ribonucleoprotein; RSS, RNA silencing suppressor; RT, reverse transcriptase; SGT, small glutamine-rich tetratricopeptide repeat containing protein; siRNA, short interfering RNA; SIV, simian immunodeficiency virus; TAR, trans-activation response element; Tat, transcriptional activator; TPR, tetratricopeptide repeat; TRIM5α, tripartite motif 5 alpha; TRBP, TAR RNA-binding protein; TRIMCyp, Tripartite motif cyclophilin A fusion protein; tRNA, transfer RNA; TTP, trisethraprolin; UBP, Vpu binding protein; UTR, untranslated region; Vif, Virion infectivity factor; VMV, visna–maedi virus; Vpu, viral protein U; ZAP, zinc-finger antiviral protein; Zeo, zeocin resistance gene; ZRE, ZAP-responsive element.

## References


Abada, P., Noble, B., and Cannon, P.M. (2005). Functional domains within the human immunodeficiency virus type 2 envelope protein required to enhance virus production. J. Virol. *79*, 3627–3638.

Abudu, A., Takaori-Kondo, A., Izumi, T., Shirakawa, K., Kobayashi, M., Sasada, A., Fukunaga, K., and Uchiyama, T. (2006). Murine retrovirus escapes from murine APOBEC3 via two distinct novel mechanisms. Curr. Biol. *16*, 1565–1570.

Accola, M.A., Höglund, S., and Göttlinger, H.G. (1998). A putative α-helical structure which overlaps the capsid-p2 boundary in the human immunodeficiency virus type 1 Gag precursor is crucial for viral particle assembly. J. Virol. *72*, 2072–2078.

Agy, M.B., Frumkin, L.R., Corey, L., Coombs, R.W., Wolinsky, S.M., Koehler, J., Morton, W.R., and Katze, M.G. (1992). Infection of Macaca nemestrina by human immunodeficiency virus type-1. Science *257*, 103–106.

Aires da Silva, F., Santa-Marta, M., Freitas-Vieira, A., Mascarenhas, P., Barahona, I., Moniz-Pereira, J., Gabuzda, D., and Goncalves, J. (2004). Camelized rabbit-derived VH single-domain intrabodies against Vif strongly neutralize HIV-1 infectivity. J. Mol. Biol. *340*, 525–542.

Akgun, E., Ziegler, M., and Grez, M. (1991). Determinants of retrovirus gene expression in embryonal carcinoma cells. J. Virol. *65*, 382–388.

Albanese, A., Arosio, D., Terreni, M., and Cereseto, A. (2008). HIV-1 pre-integration complexes selectively target decondensed chromatin in the nuclear periphery. PLoS ONE 3, e2413.

Alce, T.M., and Popik, W. (2004). APOBEC3G is incorporated into virus-like particles by a direct interaction with HIV-1 Gag nucleocapsid protein. J. Biol. Chem. *279*, 34083–34086.

Alter, H.J., Eichberg, J.W., Masur, H., Saxinger, W.C., Gallo, R., Macher, A.M., Lane, H.C., and Fauci, A.S. (1984). Transmission of HTLV-III infection from human plasma to chimpanzees: an animal model for AIDS. Science *226*, 549–552.

Ambrose, Z., KewalRamani, V.N., Bieniasz, P.D., and Hatziioannou, T. (2007). HIV/AIDS: in search of an animal model. Trends Biotechnol. *25*, 333–337.

Anderson, J., and Akkina, R. (2005). TRIM5alpharh expression restricts HIV-1 infection in lentiviral vector-transduced CD34+-cell-derived macrophages. Mol. Ther. *12*, 687–696.

Anderson, J., and Akkina, R. (2008). Human immunodeficiency virus type 1 restriction by human-rhesus chimeric tripartite motif 5alpha (TRIM 5alpha) in CD34(+) cell-derived macrophages *in vitro* and in T-cells *in vivo* in severe combined immunodeficient (SCID-hu) mice transplanted with human fetal tissue. Hum. Gene Ther. *19*, 217–228.

Anderson, J.L., Campbell, E.M., Wu, X., Vandegraaff, N., Engelman, A., and Hope, T.J. (2006). Proteasome inhibition reveals that a functional preintegration complex intermediate can be generated during restriction by diverse TRIM5 proteins. J. Virol. *80*, 9754–9760.

Anderson, J.L., and Hope, T.J. (2008). APOBEC3G restricts early HIV-1 replication in the cytoplasm of target cells. Virology *375*, 1–12.

Anderson, J.L., Campbell, E.M., Figueiredo, A., and Hope, T.J. (2008). Heat shock perturbs TRIM5alpha restriction of human immunodeficiency virus type 1. J. Virol. *82*, 2575–2579.

Ao, Z., Yu, Z., Wang, L., Zheng, Y., and Yao, X. (2008). Vpr14–88-Apobec3G fusion protein is efficiently incorporated into Vif-positive HIV-1 particles and inhibits viral infection. PLoS ONE 3, e1995.

Arhel, N., Genovesio, A., Kim, K.A., Miko, S., Perret, E., Olivo-Marin, J.C., Shorte, S., and Charneau, P. (2006). Quantitative four-dimensional tracking of cytoplasmic and nuclear HIV-1 complexes. Nat. Methods *3*, 817–824.

Asaoka, K., Ikeda, K., Hishinuma, T., Horie-Inoue, K., Takeda, S., and Inoue, S. (2005). A retrovirus restriction factor TRIM5alpha is transcriptionally regulated by interferons. Biochem. Biophys. Res. Commun. *338*, 1950–1956.

Balliet, J.W., Kolson, D.L., Eiger, G., Kim, F.M., McGann, K.A., Srinivasan, A., and Collman, R. (1994). Distinct effects in primary macrophages and lymphocytes of the human immunodeficiency virus type 1 accessory genes vpr, vpu, and nef: mutational analysis of a primary HIV-1 isolate. Virology *200*, 623–631.

Balzarini, J., De Clercq, E., and Uberla, K. (1997). SIV/HIV-1 hybrid virus expressing the reverse transcriptase gene of HIV-1 remains sensitive to HIV-1-specific reverse transcriptase inhibitors after passage in rhesus macaques. J. Acquir. Immune Defic. Syndr. Hum. Retrovirol. *15*, 1–4.

Barik, S. (2006). Immunophilins: for the love of proteins. Cell. Mol. Life Sci. *63*, 2889–2900.

Barklis, E., Mulligan, R.C., and Jaenisch, R. (1986). Chromosomal position or virus mutation permits retrovirus expression in embryonal carcinoma cells. Cell *47*, 391–399.

Barr, S.D., Smiley, J.R., and Bushman, F.D. (2008). The interferon response inhibits HIV particle production by induction of TRIM22. PLoS Pathog. *4*, e1000007.

Bartee, E., McCormack, A., and Fruh, K. (2006). Quantitative membrane proteomics reveals new cellular targets of viral immune modulators. PLoS Pathog. *2*, e107.

Beale, R.C., Petersen-Mahrt, S.K., Watt, I.N., Harris, R.S., Rada, C., and Neuberger, M.S. (2004). Comparison of the differential context-dependence of DNA deamination by APOBEC enzymes: correlation with mutation spectra *in vivo*. J. Mol. Biol. *337*, 585–596.

Benit, L., De Parseval, N., Casella, J.F., Callebaut, I., Cordonnier, A., and Heidmann, T. (1997). Cloning of a new murine endogenous retrovirus, MuERV-L, with strong similarity to the human HERV-L element and with a gag coding sequence closely related to the Fv1 restriction gene. J. Virol. *71*, 5652–5657.

Bennasser, Y., Le, S.Y., Yeung, M.L., and Jeang, K.T. (2004). HIV-1 encoded candidate micro-RNAs and their cellular targets. Retrovirology *1*, 43.

Bennasser, Y., Le, S.Y., Benkirane, M., and Jeang, K.T. (2005). Evidence that HIV-1 encodes an siRNA and a suppressor of RNA silencing. Immunity *22*, 607–619.

Bennasser, Y., and Jeang, K.T. (2006). HIV-1 Tat interaction with Dicer: requirement for RNA. Retrovirology *3*, 95.

Berthoux, L., Towers, G.J., Gurer, C., Salomoni, P., Pandolfi, P.P., and Luban, J. (2003). As(2)O(3) enhances retroviral reverse transcription and counteracts Ref1 antiviral activity. J. Virol. *77*, 3167–3180.

Berthoux, L., Sebastian, S., Sokolskaja, E., and Luban, J. (2004). Lv1 inhibition of human immunodeficiency virus type 1 is counteracted by factors that stimulate synthesis or nuclear translocation of viral cDNA. J. Virol. *78*, 11739–11750.

Berthoux, L., Sebastian, S., Sayah, D.M., and Luban, J. (2005a). Disruption of human TRIM5alpha antiviral activity by nonhuman primate orthologues. J. Virol. *79*, 7883–7888.

Berthoux, L., Sebastian, S., Sokolskaja, E., and Luban, J. (2005b). Cyclophilin A is required for TRIM5{alpha}-mediated resistance to HIV-1 in Old World monkey cells. Proc. Natl. Acad. Sci. U.S.A. *102*, 14849–14853.

Besnier, C., Takeuchi, Y., and Towers, G. (2002). Restriction of lentivirus in monkeys. Proc. Natl. Acad. Sci. U.S.A. *99*, 11920–11925.

Best, S., Le Tissier, P., Towers, G., and Stoye, J.P. (1996). Positional cloning of the mouse retrovirus restriction gene Fv1. Nature *382*, 826–829.

Bick, M.J., Carroll, J.W., Gao, G., Goff, S.P., Rice, C.M., and MacDonald, M.R. (2003). Expression of the zinc-finger antiviral protein inhibits alphavirus replication. J. Virol. *77*, 11555–11562.

Bieniasz, P.D. (2006). Late budding domains and host proteins in enveloped virus release. Virology *344*, 55–63.

Bishop, K.N., Bock, M., Towers, G., and Stoye, J.P. (2001). Identification of the regions of Fv1 necessary for murine leukemia virus restriction. J. Virol. *75*, 5182–5188.

Bishop, K.N., Holmes, R.K., Sheehy, A.M., and Malim, M.H. (2004a). APOBEC-mediated editing of viral RNA. Science *305*, 645.

Bishop, K.N., Holmes, R.K., Sheehy, A.M., Davidson, N.O., Cho, S.J., and Malim, M.H. (2004b). Cytidine deamination of retroviral DNA by diverse APOBEC proteins. Curr. Biol. *14*, 1392–1396.

Bishop, K.N., Mortuza, G.B., Howell, S., Yap, M.W., Stoye, J.P., and Taylor, I.A. (2006a). Characterization of an amino-terminal dimerization domain from retroviral restriction factor Fv1. J. Virol. *80*, 8225–8235.

Bishop, K.N., Holmes, R.K., and Malim, M.H. (2006b). Antiviral potency of APOBEC proteins does not correlate with cytidine deamination. J. Virol. *80*, 8450–8458.

Bishop, K.N., Verma, M., Kim, E.Y., Wolinsky, S.M., and Malim, M.H. (2008). APOBEC3G inhibits elongation of HIV-1 reverse transcripts. PLoS Pathog. *4*, e1000231.

Blasius, A.L., Giurisato, E., Cella, M., Schreiber, R.D., Shaw, A.S., and Colonna, M. (2006). Bone marrow stromal cell antigen 2 is a specific marker of type I IFN-producing cells in the naive mouse, but a promiscuous cell surface antigen following IFN stimulation. J. Immunol. *177*, 3260–3265.

Bock, M., Bishop, K.N., Towers, G., and Stoye, J.P. (2000). Use of a transient assay for studying the genetic determinants of Fv1 restriction. J. Virol. *74*, 7422–7430.

Bogerd, H.P., Doehle, B.P., Wiegand, H.L., and Cullen, B.R. (2004). A single amino acid difference in the host APOBEC3G protein controls the primate species specificity of HIV type 1 virion infectivity factor. Proc. Natl. Acad. Sci. U.S.A. *101*, 3770–3774.

Bogerd, H.P., Wiegand, H.L., Doehle, B.P., Lueders, K.K., and Cullen, B.R. (2006a). APOBEC3A and APOBEC3B are potent inhibitors of LTR-retrotransposon function in human cells. Nucleic Acids Res. *34*, 89–95.

Bogerd, H.P., Wiegand, H.L., Hulme, A.E., Garcia-Perez, J.L., O'Shea, K.S., Moran, J.V., and Cullen, B.R. (2006b). Cellular inhibitors of long interspersed element 1 and Alu retrotransposition. Proc. Natl. Acad. Sci. U.S.A. *103*, 8780–8785.

Bogerd, H.P., and Cullen, B.R. (2008). Single-stranded RNA facilitates nucleocapsid: APOBEC3G complex formation. RNA *14*, 1228–1236.

Bogerd, H.P., Tallmadge, R.L., Oaks, J.L., Carpenter, S., and Cullen, B.R. (2008). Equine infectious anemia virus resists the antiretroviral activity of equine APOBEC3 proteins through a packaging-independent mechanism. J. Virol. *82*, 11889–11901.

Bosco, D.A., Eisenmesser, E.Z., Pochapsky, S., Sundquist, W.I., and Kern, D. (2002). Catalysis of cis/trans isomerization in native HIV-1 capsid by human cyclophilin A. Proc. Natl. Acad. Sci. U.S.A. *99*, 5247–5252.

Bosco, D.A., and Kern, D. (2004). Catalysis and binding of cyclophilin A with different HIV-1 capsid constructs. Biochemistry *43*, 6110–6119.

Bouazzaoui, A., Kreutz, M., Eisert, V., Dinauer, N., Heinzelmann, A., Hallenberger, S., Strayle, J., Walker, R., Rubsamen-Waigmann, H., Andreesen, R., and von Briesen, H. (2006). Stimulated trans-acting factor of 50 kDa (Staf50) inhibits HIV-1 replication in human monocyte-derived macrophages. Virology *356*, 79–94.

Bour, S., Schubert, U., Peden, K., and Strebel, K. (1996). The envelope glycoprotein of human immunodeficiency virus type 2 enhances viral particle release: a Vpu-like factor? J. Virol. *70*, 820–829.

Bour, S., and Strebel, K. (1996). The human immunodeficiency virus (HIV) type 2 envelope protein is a functional complement to HIV type 1 Vpu that enhances particle release of heterologous retroviruses. J. Virol. *70*, 8285–8300.

Bour, S., Perrin, C., and Strebel, K. (1999). Cell surface CD4 inhibits HIV-1 particle release by interfering with Vpu activity. J. Biol. Chem. *274*, 33800–33806.

Bour, S., and Strebel, K. (2000). HIV accessory proteins: multifunctional components of a complex system. Adv. Pharmacol. *48*, 75–120.

Bour, S., Perrin, C., Akari, H., and Strebel, K. (2001). The human immunodeficiency virus type 1 Vpu protein inhibits NF-kappa B activation by interfering with beta TrCP-mediated degradation of Ikappa B. J. Biol. Chem. *276*, 15920–15928.

Bour, S., and Strebel, K. (2003). The HIV-1 Vpu protein: a multifunctional enhancer of viral particle release. Microbes Infect. *5*, 1029–1039.

Braaten, D., Franke, E.K., and Luban, J. (1996). Cyclophilin A is required for an early step in the life cycle of human immunodeficiency virus type 1 before the initiation of reverse transcription. J. Virol. *70*, 3551–3560.

Brass, A.L., Dykxhoorn, D.M., Benita, Y., Yan, N., Engelman, A., Xavier, R.J., Lieberman, J., and Elledge, S.J. (2008). Identification of host proteins required for HIV infection through a functional genomic screen. Science *319*, 921–926.

Brennan, G., Kozyrev, Y., Kodama, T., and Hu, S.L. (2007). Novel TRIM5 isoforms expressed by Macaca nemestrina. J. Virol. *81*, 12210–12217.

Brennan, G., Kozyrev, Y., and Hu, S.L. (2008). TRIMCyp expression in Old World primates Macaca nemestrina and Macaca fascicularis. Proc. Natl. Acad. Sci. U.S.A. *105*, 3569–3574.

Brennecke, J., Stark, A., Russell, R.B., and Cohen, S.M. (2005). Principles of microRNA-target recognition. PLoS Biol. *3*, e85.

Browne, E.P., Allers, C., and Landau, N.R. (2009). Restriction of HIV-1 by APOBEC3G is cytidine deaminase-dependent. Virology *387*, 313–321.

Bukrinsky, M.I., Sharova, N., Dempsey, M.P., Stanwick, T.L., Bukrinskaya, A.G., Haggerty, S., and Stevenson, M. (1992). Active nuclear import of human immunodeficiency virus type 1 preintegration complexes. Proc. Natl. Acad. Sci. U.S.A. *89*, 6580–6584.

Bushati, N., and Cohen, S.M. (2007). microRNA functions. Annu Rev Cell Dev. Biol. *23*, 175–205.

Callahan, M.A., Handley, M.A., Lee, Y.H., Talbot, K.J., Harper, J.W., and Panganiban, A.T. (1998). Functional interaction of human immunodeficiency virus type 1 Vpu and Gag with a novel member of the tetratricopeptide repeat protein family. J. Virol. *72*, 5189–5197.

Callebaut, I., Renoir, J.M., Lebeau, M.C., Massol, N., Burny, A., Baulieu, E.E., and Mornon, J.P. (1992). An immunophilin that binds M(r) *90*,000 heat shock protein: main structural features of a mammalian p59 protein. Proc. Natl. Acad. Sci. U.S.A. *89*, 6270–6274.

Campbell, E.M., Nunez, R., and Hope, T.J. (2004). Disruption of the actin cytoskeleton can complement the ability of Nef to enhance human immunodeficiency virus type 1 infectivity. J. Virol. *78*, 5745–5755.

Campbell, E.M., Dodding, M.P., Yap, M.W., Wu, X., Gallois-Montbrun, S., Malim, M.H., Stoye, J.P., and Hope, T.J. (2007). TRIM5{alpha} Cytoplasmic Bodies Are Highly Dynamic Structures. Mol. Biol. Cell *18*, 2102–2111.

Campbell, E.M., Perez, O., Anderson, J.L., and Hope, T.J. (2008). Visualization of a proteasome-independent intermediate during restriction of HIV-1 by rhesus TRIM5alpha. J. Cell Biol. *180*, 549–561.

Carr, J.M., Davis, A.J., Coolen, C., Cheney, K., Burrell, C.J., and Li, P. (2006). Vif-deficient HIV reverse transcription complexes (RTCs) are subject to structural changes and mutation of RTC-associated reverse transcription products. Virology *351*, 80–91.

Carthagena, L., Parise, M.C., Ringeard, M., Chelbi-Alix, M.K., Hazan, U., and Nisole, S. (2008). Implication of TRIM alpha and TRIMCyp in interferon-induced anti-retroviral restriction activities. Retrovirology *5*, 59.

Cen, S., Guo, F., Niu, M., Saadatmand, J., Deflassieux, J., and Kleiman, L. (2004). The interaction between HIV-1 Gag and APOBEC3G. J. Biol. Chem. *279*, 33177–33184.

Chelico, L., Pham, P., Calabrese, P., and Goodman, M.F. (2006). APOBEC3G DNA deaminase acts processively 3' --> 5' on single-stranded DNA. Nat. Struct. Mol. Biol. *13*, 392–399.

Chen, C.Y., Gherzi, R., Ong, S.E., Chan, E.L., Raijmakers, R., Pruijn, G.J., Stoecklin, G., Moroni, C., Mann, M., and Karin, M. (2001). AU binding proteins recruit the exosome to degrade ARE-containing mRNAs. Cell *107*, 451–464.

Chen, G., Guo, X., Lv, F., Xu, Y., and Gao, G. (2008a). p72 DEAD box RNA helicase is required for optimal function of the zinc-finger antiviral protein. Proc. Natl. Acad. Sci. U.S.A. *105*, 4352–4357.

Chen, H., Lilley, C.E., Yu, Q., Lee, D.V., Chou, J., Narvaiza, I., Landau, N.R., and Weitzman, M.D. (2006). APOBEC3A is a potent inhibitor of adeno-associated virus and retrotransposons. Curr. Biol. *16*, 480–485.

Chen, K.M., Harjes, E., Gross, P.J., Fahmy, A., Lu, Y., Shindo, K., Harris, R.S., and Matsuo, H. (2008b). Structure of the DNA deaminase domain of the HIV-1 restriction factor APOBEC3G. Nature *452*, 116–119.

Chiu, Y.L., Soros, V.B., Kreisberg, J.F., Stopak, K., Yonemoto, W., and Greene, W.C. (2005). Cellular APOBEC3G restricts HIV-1 infection in resting $CD4^+$ T-cells. Nature *435*, 108–114.

Chiu, Y.L., Witkowska, H.E., Hall, S.C., Santiago, M., Soros, V.B., Esnault, C., Heidmann, T., and Greene, W.C. (2006). High-molecular-mass APOBEC3G complexes restrict Alu retrotransposition. Proc. Natl. Acad. Sci. U.S.A. *103*, 15588–15593.

Chiu, Y.L., and Greene, W.C. (2008). The APOBEC3 cytidine deaminases: an innate defensive network opposing exogenous retroviruses and endogenous retroelements. Annu. Rev. Immunol. *26*, 317–353.

Cohen, E.A., Terwilliger, E.F., Sodroski, J.G., and Haseltine, W.A. (1988). Identification of a protein encoded by the vpu gene of HIV-1. Nature *334*, 532–534.

Conticello, S.G., Harris, R.S., and Neuberger, M.S. (2003). The Vif protein of HIV triggers degradation of the human antiretroviral DNA deaminase APOBEC3G. Curr. Biol. *13*, 2009–2013.

Conticello, S.G., Thomas, C.J., Petersen-Mahrt, S.K., and Neuberger, M.S. (2005). Evolution of the AID/APOBEC family of polynucleotide (deoxy)cytidine deaminases. Mol. Biol. Evol. *22*, 367–377.

Corbeau, P. (2008). Interfering RNA and HIV: reciprocal interferences. PLoS Pathog. *4*, e1000162.

Courgnaud, V., Salemi, M., Pourrut, X., Mpoudi-Ngole, E., Abela, B., Auzel, P., Bibollet-Ruche, F., Hahn, B., Vandamme, A.M., Delaporte, E., and Peeters, M. (2002). Characterization of a novel simian immunodeficiency virus with a vpu gene from greater spot-nosed monkeys (*Cercopithecus nictitans*) provides new insights into simian/human immunodeficiency virus phylogeny. J. Virol. *76*, 8298–8309.

Couturier, J.P., and Root-Bernstein, R.S. (2005). HIV may produce inhibitory microRNAs (miRNAs) that block production of CD28, CD4 and some interleukins. J. Theor. Biol. *235*, 169–184.

Cowan, S., Hatziioannou, T., Cunningham, T., Muesing, M.A., Gottlinger, H.G., and Bieniasz, P.D. (2002). Cellular inhibitors with Fv1-like activity restrict human and simian immunodeficiency virus tropism. Proc. Natl. Acad. Sci. U.S.A. *99*, 11914–11919.

Cullen, B.R. (2006). Is RNA interference involved in intrinsic antiviral immunity in mammals? Nat. Immunol. *7*, 563–567.

Cullen, B.R. (2009). Viral and cellular messenger RNA targets adn viral microRNAs. Nature *457*, 421–425.

Dang, Y., Wang, X., Esselman, W.J., and Zheng, Y.H. (2006). Identification of APOBEC3DE as another antiretroviral factor from the human APOBEC family. J. Virol. *80*, 10522–10533.

Dang, Y., Siew, L.M., Wang, X., Han, Y., Lampen, R., and Zheng, Y.H. (2008). Human cytidine deaminase APOBEC3H restricts HIV-1 replication. J. Biol. Chem. *283*, 11606–11614.

DeFilippis, V.R., Robinson, B., Keck, T.M., Hansen, S.G., Nelson, J.A., and Fruh, K.J. (2006). Interferon regulatory factor 3 is necessary for induction of antiviral genes during human cytomegalovirus infection. J. Virol. *80*, 1032–1037.

Deora, A., and Ratner, L. (2001). Viral protein U (Vpu)-mediated enhancement of human immunodeficiency virus type 1 particle release depends on the rate of cellular proliferation. J. Virol. *75*, 6714–6718.

Derse, D., Hill, S.A., Princler, G., Lloyd, P., and Heidecker, G. (2007). Resistance of human T-cell leukemia virus type 1 to APOBEC3G restriction is mediated by elements in nucleocapsid. Proc. Natl. Acad. Sci. U.S.A. *104*, 2915–2920.

DesGroseillers, L., and Jolicoeur, P. (1983). Physical mapping of the Fv-1 tropism host range determinant of BALB/c murine leukemia viruses. J. Virol. *48*, 685–696.

Diaz-Griffero, F., Li, X., Javanbakht, H., Song, B., Welikala, S., Stremlau, M., and Sodroski, J. (2006a). Rapid turnover and polyubiquitylation of the retroviral restriction factor TRIM5. Virology *349*, 300–315.

Diaz-Griffero, F., Vandegraaff, N., Li, Y., McGee-Estrada, K., Stremlau, M., Welikala, S., Si, Z., Engelman, A., and Sodroski, J. (2006b). Requirements for capsid-binding and an effector function in TRIMCyp-mediated restriction of HIV-1. Virology *351*, 404–419.

Diaz-Griffero, F., Kar, A., Lee, M., Stremlau, M., Poeschla, E., and Sodroski, J. (2007). Comparative requirements for the restriction of retrovirus infection by TRIM5alpha and TRIMCyp. Virology *369*, 400–410.

Diehl, W.E., Stansell, E., Kaiser, S.M., Emerman, M., and Hunter, E. (2008). Identification of postentry restrictions to Mason-Pfizer monkey virus infection in New World monkey cells. J. Virol. *82*, 11140–11151.

Ding, S.W., and Voinnet, O. (2007). Antiviral immunity directed by small RNAs. Cell *130*, 413–426.

Dodding, M.P., Bock, M., Yap, M.W., and Stoye, J.P. (2005). Capsid processing requirements for abrogation of Fv1 and Ref1 restriction. J. Virol. *79*, 10571–10577.

Dodt, G., Braverman, N., Wong, C., Moser, A., Moser, H.W., Watkins, P., Valle, D., and Gould, S.J. (1995). Mutations in the PTS1 receptor gene, PXR1, define complementation group 2 of the peroxisome biogenesis disorders. Nat. Genet. *9*, 115–125.

Doehle, B.P., Schafer, A., Wiegand, H.L., Bogerd, H.P., and Cullen, B.R. (2005). Differential sensitivity of murine leukemia virus to APOBEC3-mediated inhibition is governed by virion exclusion. J. Virol. *79*, 8201–8207.

Doehle, B.P., Bogerd, H.P., Wiegand, H.L., Jouvenet, N., Bieniasz, P.D., Hunter, E., and Cullen, B.R. (2006). The betaretrovirus Mason-Pfizer monkey virus selectively excludes simian APOBEC3G from virion particles. J. Virol. *80*, 12102–12108.

Dorfman, T., and Gottlinger, H.G. (1996). The human immunodeficiency virus type 1 capsid p2 domain confers sensitivity to the cyclophilin-binding drug SDZ NIM 811. J. Virol. *70*, 5751–5757.

Duran-Troise, G., Bassin, R.H., Rein, A., and Gerwin, B.I. (1977). Loss of Fv-1 restriction in Balb/3T3 cells following infection with a single N tropic murine leukemia virus particle. Cell *10*, 479–488.

Dutko, J.A., Schafer, A., Kenny, A.E., Cullen, B.R., and Curcio, M.J. (2005). Inhibition of a yeast LTR retrotransposon by human APOBEC3 cytidine deaminases. Curr. Biol. *15*, 661–666.

Dutta, S., and Tan, Y.J. (2008). Structural and functional characterization of human SGT and its interaction with Vpu of the human immunodeficiency virus type 1. Biochemistry *47*, 10123–10131.

Eberhardy, S.R., Goncalves, J., Coelho, S., Segal, D.J., Berkhout, B., and Barbas, C.F., 3rd (2006). Inhibition of human immunodeficiency virus type 1 replication with artificial transcription factors targeting the highly conserved primer-binding site. J. Virol. *80*, 2873–2883.

Ellery, P.J., Tippett, E., Chiu, Y.L., Paukovics, G., Cameron, P.U., Solomon, A., Lewin, S.R., Gorry, P.R., Jaworowski, A., Greene, W.C., *et al.* (2007). The CD16+ monocyte subset is more permissive to infection and preferentially harbors HIV-1 *in vivo*. J. Immunol. *178*, 6581–6589.

Ellis, J. (2005). Silencing and variegation of gammaretrovirus and lentivirus vectors. Hum. Gene Ther. *16*, 1241–1246.

Estrabaud, E., Le Rouzic, E., Lopez-Verges, S., Morel, M., Belaidouni, N., Benarous, R., Transy, C., Berlioz-Torrent, C., and Margottin-Goguet, F. (2007). Regulated degradation of the HIV-1 Vpu protein through a betaTrCP-independent pathway limits the release of viral particles. PLoS Pathog. *3*, e104.

Ewart, G.D., Sutherland, T., Gage, P.W., and Cox, G.B. (1996). The Vpu protein of human immunodeficiency virus type 1 forms cation-selective ion channels. J. Virol. *70*, 7108–7115.

Feuer, G., Taketo, M., Hanecak, R.C., and Fan, H. (1989). Two blocks in Moloney murine leukemia virus expression in undifferentiated F9 embryonal carcinoma cells as determined by transient expression assays. J. Virol. *63*, 2317–2324.

Fisher, A.G., Ensoli, B., Ivanoff, L., Chamberlain, M., Petteway, S., Ratner, L., Gallo, R.C., and Wong-Staal, F. (1987). The sor gene of HIV-1 is required for efficient virus transmission *in vitro*. Science *237*, 888–893.

Flanagan, J.R., Krieg, A.M., Max, E.E., and Khan, A.S. (1989). Negative control region at the 5' end of murine leukemia virus long terminal repeats. Mol. Cell Biol. *9*, 739–746.

Forshey, B.M., von Schwedler, U., Sundquist, W.I., and Aiken, C. (2002). Formation of a human immunodeficiency virus type 1 core of optimal stability is crucial for viral replication. J. Virol. *76*, 5667–5677.

Forshey, B.M., Shi, J., and Aiken, C. (2005). Structural requirements for recognition of the human immunodeficiency virus type 1 core during host restriction in owl monkey cells. J. Virol. *79*, 869–875.

Franke, E.K., Yuan, H.E., and Luban, J. (1994). Specific incorporation of cyclophilin A into HIV-1 virions. Nature *372*, 359–362.

Friborg, J., Ladha, A., Gottlinger, H., Haseltine, W.A., and Cohen, E.A. (1995). Functional analysis of the phosphorylation sites on the human immunodeficiency virus type 1 Vpu protein. J. Acquir. Immune Defic. Syndr. Hum. Retrovirol. *8*, 10–22.

Friedman, J.R., Fredericks, W.J., Jensen, D.E., Speicher, D.W., Huang, X.P., Neilson, E.G., and Rauscher, F.J., 3rd (1996). KAP-1, a novel corepressor for the highly conserved KRAB repression domain. Genes Dev. *10*, 2067–2078.

Furukawa, A., Nagata, T., Matsugami, A., Habu, Y., Sugiyama, R., Hayashi, F., Kobayashi, N., Yokoyama, S., Takaku, H., and Katahira, M. (2009). Structure, interaction and real-time monitoring of the enzymatic reaction of wild-type APOBEC3G. Embo J. *28*, 440–451.

Gack, M.U., Shin, Y.C., Joo, C.H., Urano, T., Liang, C., Sun, L., Takeuchi, O., Akira, S., Chen, Z., Inoue, S., and Jung, J.U. (2007). TRIM25 RING-finger E3 ubiquitin ligase is essential for RIG-I-mediated antiviral activity. Nature *446*, 916–920.

Gallois-Montbrun, S., Kramer, B., Swanson, C.M., Byers, H., Lynham, S., Ward, M., and Malim, M.H. (2007). Antiviral protein APOBEC3G localizes to ribonucleoprotein complexes found in P bodies and stress granules. J. Virol. *81*, 2165–2178.

Gallois-Montbrun, S., Holmes, R.K., Swanson, C.M., Fernandez-Ocana, M., Byers, H.L., Ward, M.A., and Malim, M.H. (2008). Comparison of cellular ribonucleoprotein complexes associated with the APOBEC3F and APOBEC3G antiviral proteins. J. Virol. *82*, 5636–5642.

Gamble, T.R., Vajdos, F.F., Yoo, S., Worthylake, D.K., Houseweart, M., Sundquist, W.I., and Hill, C.P. (1996). Crystal structure of human cyclophilin A bound to the amino-terminal domain of HIV-1 capsid. Cell *87*, 1285–1294.

Gao, G., Guo, X., and Goff, S.P. (2002). Inhibition of retroviral RNA production by ZAP, a CCCH-type zinc finger protein. Science *297*, 1703–1706.

Gendelman, H.E., Baca, L.M., Turpin, J., Kalter, D.C., Hansen, B., Orenstein, J.M., Dieffenbach, C.W., Friedman, R.M., and Meltzer, M.S. (1990). Regulation of HIV replication in infected monocytes by IFN-alpha. Mechanisms for viral restriction. J. Immunol. *145*, 2669–2676.

Geraghty, R.J., Talbot, K.J., Callahan, M., Harper, W., and Panganiban, A.T. (1994). Cell type-dependence for Vpu function. J. Med. Primatol. *23*, 146–150.

Goff, S.P. (1990). Retroviral reverse transcriptase: synthesis, structure, and function. J. Acquir. Immune Defic. Syndr. *3*, 817–831.

Goila-Gaur, R., and Strebel, K. (2008). HIV-1 Vif, APOBEC, and intrinsic immunity. Retrovirology *5*, 51.

Goto, T., Kennel, S.J., Abe, M., Takishita, M., Kosaka, M., Solomon, A., and Saito, S. (1994). A novel membrane antigen selectively expressed on terminally differentiated human B-cells. Blood *84*, 1922–1930.

Gottlinger, H.G., Dorfman, T., Cohen, E.A., and Haseltine, W.A. (1993). Vpu protein of human immunodeficiency virus type 1 enhances the release of capsids produced by gag gene constructs of widely divergent retroviruses. Proc. Natl. Acad. Sci. U.S.A. *90*, 7381–7385.

Gottlinger, H.G. (2008). HIV/AIDS: virus kept on a leash. Nature *451*, 406–408.

Goujon, C., Riviere, L., Jarrosson-Wuilleme, L., Bernaud, J., Rigal, D., Darlix, J.L., and Cimarelli, A. (2007). SIVSM/HIV-2 Vpx proteins promote retroviral escape from a proteasome-dependent restriction pathway present in human dendritic cells. Retrovirology *4*, 2.

Grutter, C., Briand, C., Capitani, G., Mittl, P.R., Papin, S., Tschopp, J., and Grutter, M.G. (2006). Structure of the PRYSPRY-domain: implications for autoinflammatory diseases. FEBS Lett. *580*, 99–106.

Guo, F., Cen, S., Niu, M., Saadatmand, J., and Kleiman, L. (2006). Inhibition of formula-primed reverse transcription by human APOBEC3G during human immunodeficiency virus type 1 replication. J. Virol. *80*, 11710–11722.

Guo, F., Cen, S., Niu, M., Yang, Y., Gorelick, R.J., and Kleiman, L. (2007a). The interaction of APOBEC3G with human immunodeficiency virus type 1 nucleocapsid inhibits tRNA3Lys annealing to viral RNA. J. Virol. *81*, 11322–11331.

Guo, X., Carroll, J.W., Macdonald, M.R., Goff, S.P., and Gao, G. (2004). The zinc finger antiviral protein directly binds to specific viral mRNAs through the CCCH zinc finger motifs. J. Virol. *78*, 12781–12787.

Guo, X., Ma, J., Sun, J., and Gao, G. (2007b). The zinc-finger antiviral protein recruits the RNA processing exosome to degrade the target mRNA. Proc. Natl. Acad. Sci. U.S.A. *104*, 151–156.

Haas, D.L., Lutzko, C., Logan, A.C., Cho, G.J., Skelton, D., Jin Yu, X., Pepper, K.A., and Kohn, D.B. (2003). The Moloney murine leukemia virus repressor binding site represses expression in murine and human hematopoietic stem cells. J. Virol. *77*, 9439–9450.

Hache, G., Liddament, M.T., and Harris, R.S. (2005). The retroviral hypermutation specificity of APOBEC3F and APOBEC3G is governed by the C-terminal DNA cytosine deaminase domain. J. Biol. Chem. *280*, 10920–10924.

Handley, M.A., Paddock, S., Dall, A., and Panganiban, A.T. (2001). Association of Vpu-binding protein with microtubules and Vpu-dependent redistribution of HIV-1 Gag protein. Virology *291*, 198–207.

Handschumacher, R.E., Harding, M.W., Rice, J., Drugge, R.J., and Speicher, D.W. (1984). Cyclophilin: a specific cytosolic binding protein for cyclosporin A. Science *226*, 544–547.

Harada, F., Peters, G.G., and Dahlberg, J.E. (1979). The primer tRNA for Moloney murine leukemia virus DNA synthesis. Nucleotide sequence and aminoacylation of tRNAPro. J. Biol. Chem. *254*, 10979–10985.

Harari, A., Ooms, M., Mulder, L.C., and Simon, V. (2009). Polymorphisms and splice variants influence the antiretroviral activity of human APOBEC3H. J. Virol. *83*, 295–303.

Harila, K., Prior, I., Sjoberg, M., Salminen, A., Hinkula, J., and Suomalainen, M. (2006). Vpu and Tsg101 regulate intracellular targeting of the human immunodeficiency virus type 1 core protein precursor Pr55gag. J. Virol. *80*, 3765–3772.

Harris, R.S., Petersen-Mahrt, S.K., and Neuberger, M.S. (2002). RNA editing enzyme APOBEC1 and some of its homologs can act as DNA mutators. Mol. Cell *10*, 1247–1253.

Harris, R.S., Bishop, K.N., Sheehy, A.M., Craig, H.M., Petersen-Mahrt, S.K., Watt, I.N., Neuberger, M.S., and Malim, M.H. (2003). DNA deamination mediates innate immunity to retroviral infection. Cell *113*, 803–809.

Harris, R.S., and Liddament, M.T. (2004). Retroviral restriction by APOBEC proteins. Nat. Rev. Immunol. *4*, 868–877.

Hartley, J.W., Rowe, W.P., and Huebner, R.J. (1970). Host-range restrictions of murine leukemia viruses in mouse embryo cell cultures. J. Virol. *5*, 221–225.

Hartley, J.W., and Rowe, W.P. (1975). Clonal cells lines from a feral mouse embryo which lack host-range restrictions for murine leukemia viruses. Virology *65*, 128–134.

Hase, T., Riezman, H., Suda, K., and Schatz, G. (1983). Import of proteins into mitochondria: nucleotide sequence of the gene for a 70-kd protein of the yeast mitochondrial outer membrane. Embo J. *2*, 2169–2172.

Hatziioannou, T., Cowan, S., Goff, S.P., Bieniasz, P.D., and Towers, G.J. (2003). Restriction of multiple divergent retroviruses by Lv1 and Ref1. Embo J. *22*, 385–394.

Hatziioannou, T., Cowan, S., Von Schwedler, U.K., Sundquist, W.I., and Bieniasz, P.D. (2004a). Species-specific tropism determinants in the human immunodeficiency virus type 1 capsid. J. Virol. *78*, 6005–6012.

Hatziioannou, T., Perez-Caballero, D., Yang, A., Cowan, S., and Bieniasz, P.D. (2004b). Retrovirus resistance factors Ref1 and Lv1 are species-specific variants of TRIM5alpha. Proc. Natl. Acad. Sci. U.S.A. *101*, 10774–10779.

Hatziioannou, T., Perez-Caballero, D., Cowan, S., and Bieniasz, P.D. (2005). Cyclophilin interactions with incoming human immunodeficiency virus type 1 capsids with opposing effects on infectivity in human cells. J. Virol. *79*, 176–183.

Hatziioannou, T., Princiotta, M., Piatak, M., Jr, Yuan, F., Zhang, F., Lifson, J.D., and Bieniasz, P.D. (2006). Generation of simian-tropic HIV-1 by restriction factor evasion. Science *314*, 95.

Hatziioannou, T., Ambrose, Z., Chung, N.P., Piatak, M., Jr, Yuan, F., Trubey, C.M., Coalter, V., Kiser, R., Schneider, D., Smedley, J., *et al.* (2009). A macaque model of HIV-1 infection. Proc. Natl. Acad. Sci. U.S.A. *106*, 4425–4429.

He, Z., Zhang, W., Chen, G., Xu, R., and Yu, X.F. (2008). Characterization of conserved motifs in HIV-1 Vif required for APOBEC3G and APOBEC3F interaction. J. Mol. Biol. *381*, 1000–1011.

Hilberg, F., Stocking, C., Ostertag, W., and Grez, M. (1987). Functional analysis of a retroviral host-range mutant: altered long terminal repeat sequences allow expression in embryonal carcinoma cells. Proc. Natl. Acad. Sci. U.S.A. *84*, 5232–5236.

Hill, M.S., Ruiz, A., Pacyniak, E., Pinson, D.M., Culley, N., Yen, B., Wong, S.W., and Stephens, E.B. (2008). Modulation of the severe $CD4^+$ T-cell loss caused by a pathogenic simian-human immunodeficiency virus by replacement of the subtype B vpu with the vpu from a subtype C HIV-1 clinical isolate. Virology *371*, 86–97.

Hirano, T., Kinoshita, N., Morikawa, K., and Yanagida, M. (1990). Snap helix with knob and hole: essential repeats in S. pombe nuclear protein nuc2+. Cell *60*, 319–328.

Hofmann, W., Schubert, D., LaBonte, J., Munson, L., Gibson, S., Scammell, J., Ferrigno, P., and Sodroski, J. (1999). Species-specific, postentry barriers to primate immunodeficiency virus infection. J. Virol. *73*, 10020–10028.

Holden, L.G., Prochnow, C., Chang, Y.P., Bransteitter, R., Chelico, L., Sen, U., Stevens, R.C., Goodman, M.F., and Chen, X.S. (2008). Crystal structure of the antiviral APOBEC3G catalytic domain and functional implications. Nature *456*, 121–124.

Holmes, R.K., Koning, F.A., Bishop, K.N., and Malim, M.H. (2007). APOBEC3F can inhibit the accumulation of HIV-1 reverse transcription products in the absence of hypermutation. Comparisons with APOBEC3G. J. Biol. Chem. *282*, 2587–2595.

Hopkins, N., Schindler, J., and Hynes, R. (1977). Six-NB-tropic murine leukemia viruses derived from a

B-tropic virus of BALB/c have altered p30. J. Virol. *21*, 309–318.

Hout, D.R., Gomez, M.L., Pacyniak, E., Mulcahy, E.R., Gomez, L.M., Jackson, M., Flick, M., Fegley, B., McCormick, C., Wisdom, B.J., *et al.* (2004a). Fusion of the upstream vpu sequences to the env of simian human immunodeficiency virus (SHIV(KU-1bMC33)) results in the synthesis of two envelope precursor proteins, increased numbers of virus particles associated with the cell surface and is pathogenic for pig-tailed macaques. Virology *323*, 91–107.

Hout, D.R., Mulcahy, E.R., Pacyniak, E., Gomez, L.M., Gomez, M.L., and Stephens, E.B. (2004b). Vpu: a multifunctional protein that enhances the pathogenesis of human immunodeficiency virus type 1. Curr HIV Res. *2*, 255–270.

Huang, J., Wang, F., Argyris, E., Chen, K., Liang, Z., Tian, H., Huang, W., Squires, K., Verlinghieri, G., and Zhang, H. (2007). Cellular microRNAs contribute to HIV-1 latency in resting primary CD4+ T-lymphocytes. Nat. Med. *13*, 1241–1247.

Huet, T., Cheynier, R., Meyerhans, A., Roelants, G., and Wain-Hobson, S. (1990). Genetic organization of a chimpanzee lentivirus related to HIV-1. Nature *345*, 356–359.

Huthoff, H., and Malim, M.H. (2007). Identification of amino acid residues in APOBEC3G required for regulation by human immunodeficiency virus type 1 Vif and Virion encapsidation. J. Virol. *81*, 3807–3815.

Igarashi, T., Iyengar, R., Byrum, R.A., Buckler-White, A., Dewar, R.L., Buckler, C.E., Lane, H.C., Kamada, K., Adachi, A., and Martin, M.A. (2007). Human immunodeficiency virus type 1 derivative with 7% simian immunodeficiency virus genetic content is able to establish infections in pig-tailed macaques. J. Virol. *81*, 11549–11552.

Ikeda, T., Ohsugi, T., Kimura, T., Matsushita, S., Maeda, Y., Harada, S., and Koito, A. (2008). The antiretroviral potency of APOBEC1 deaminase from small animal species. Nucleic Acids Res. *36*, 6859–6871.

Iwatani, Y., Takeuchi, H., Strebel, K., and Levin, J.G. (2006). Biochemical activities of highly purified, catalytically active human APOBEC3G: correlation with antiviral effect. J. Virol. *80*, 5992–6002.

Iwatani, Y., Chan, D.S., Wang, F., Maynard, K.S., Sugiura, W., Gronenborn, A.M., Rouzina, I., Williams, M.C., Musier-Forsyth, K., and Levin, J.G. (2007). Deaminase-independent inhibition of HIV-1 reverse transcription by APOBEC3G. Nucleic Acids Res. *35*, 7096–7108.

James, L.C., Keeble, A.H., Khan, Z., Rhodes, D.A., and Trowsdale, J. (2007). Structural basis for PRYSPRY-mediated tripartite motif (TRIM) protein function. Proc. Natl. Acad. Sci. U.S.A. *104*, 6200–6205.

Jarmuz, A., Chester, A., Bayliss, J., Gisbourne, J., Dunham, I., Scott, J., and Navaratnam, N. (2002). An anthropoid-specific locus of orphan C to U RNA-editing enzymes on chromosome 22. Genomics *79*, 285–296.

Javanbakht, H., Diaz-Griffero, F., Stremlau, M., Si, Z., and Sodroski, J. (2005). The contribution of RING and B-box 2 domains to retroviral restriction mediated by monkey TRIM5alpha. J. Biol. Chem. *280*, 26933–26940.

Javanbakht, H., Yuan, W., Yeung, D.F., Song, B., Diaz-Griffero, F., Li, Y., Li, X., Stremlau, M., and Sodroski, J. (2006). Characterization of TRIM5alpha trimerization and its contribution to human immunodeficiency virus capsid binding. Virology *353*, 234–246.

Javanbakht, H., Diaz-Griffero, F., Yuan, W., Yeung, D.F., Li, X., Song, B., and Sodroski, J. (2007). The ability of multimerized cyclophilin A to restrict retrovirus infection. Virology *367*, 19–29.

Jolicoeur, P., and Baltimore, D. (1976). Effect of Fv-1 gene product on proviral DNA formation and integration in cells infected with murine leukemia viruses. Proc. Natl. Acad. Sci. U.S.A. *73*, 2236–2240.

Jonsson, S.R., LaRue, R.S., Stenglein, M.D., Fahrenkrug, S.C., Andresdottir, V., and Harris, R.S. (2007). The restriction of zoonotic PERV transmission by human APOBEC3G. PLoS ONE *2*, e893.

Jouvenet, N., Bieniasz, P.D., and Simon, S.M. (2008). Imaging the biogenesis of individual HIV-1 virions in live cells. Nature *454*, 236–240.

Kahl, C.A., Cannon, P.M., Oldenburg, J., Tarantal, A.F., and Kohn, D.B. (2008). Tissue-specific restriction of cyclophilin A-independent HIV-1- and SIV-derived lentiviral vectors. Gene Ther. *15*, 1079–1089.

Kaiser, S.M., and Emerman, M. (2006). Uracil DNA glycosylase is dispensable for human immunodeficiency virus type 1 replication and does not contribute to the antiviral effects of the cytidine deaminase Apobec3G. J. Virol. *80*, 875–882.

Kaiser, S.M., Malik, H.S., and Emerman, M. (2007). Restriction of an extinct retrovirus by the human TRIM5alpha antiviral protein. Science *316*, 1756–1758.

Kamada, K., Igarashi, T., Martin, M.A., Khamsri, B., Hatcho, K., Yamashita, T., Fujita, M., Uchiyama, T., and Adachi, A. (2006). Generation of HIV-1 derivatives that productively infect macaque monkey lymphoid cells. Proc. Natl. Acad. Sci. U.S.A. *103*, 16959–16964.

Kamata, M., Nagaoka, Y., and Chen, I.S. (2009). Reassessing the role of APOBEC3G in human immunodeficiency virus type 1 infection of quiescent CD4+ T-cells. PLoS Pathog. *5*, e1000342.

Kao, S., Khan, M.A., Miyagi, E., Plishka, R., Buckler-White, A., and Strebel, K. (2003). The human immunodeficiency virus type 1 Vif protein reduces intracellular expression and inhibits packaging of APOBEC3G (CEM15), a cellular inhibitor of virus infectivity. J. Virol. *77*, 11398–11407.

Kao, S., Miyagi, E., Khan, M.A., Takeuchi, H., Opi, S., Goila-Gaur, R., and Strebel, K. (2004). Production of infectious human immunodeficiency virus type 1 does not require depletion of APOBEC3G from virus-producing cells. Retrovirology *1*, 27.

Kar, A.K., Diaz-Griffero, F., Li, Y., Li, X., and Sodroski, J. (2008). Biochemical and biophysical characterization of a chimeric TRIM21-TRIM5alpha protein. J. Virol. *82*, 11669–11681.

Keckesova, Z., Ylinen, L.M., and Towers, G.J. (2004). The human and African green monkey TRIM5alpha genes encode Ref1 and Lv1 retroviral restriction factor activities. Proc. Natl. Acad. Sci. U.S.A. *101*, 10780–10785.

Keckesova, Z., Ylinen, L.M., and Towers, G.J. (2006). Cyclophilin A renders human immunodeficiency virus type 1 sensitive to Old World monkey but not human TRIM5 alpha antiviral activity. J. Virol. *80*, 4683–4690.

Kempler, G., Freitag, B., Berwin, B., Nanassy, O., and Barklis, E. (1993). Characterization of the Moloney murine leukemia virus stem cell-specific repressor binding site. Virology *193*, 690–699.

Kerns, J.A., Emerman, M., and Malik, H.S. (2008). Positive selection and increased antiviral activity associated with the PARP-containing isoform of human zinc-finger antiviral protein. PLoS Genet. *4*, e21.

Khan, M.A., Goila-Gaur, R., Opi, S., Miyagi, E., Takeuchi, H., Kao, S., and Strebel, K. (2007). Analysis of the contribution of cellular and viral RNA to the packaging of APOBEC3G into HIV-1 virions. Retrovirology *4*, 48.

Klase, Z., Kale, P., Winograd, R., Gupta, M.V., Heydarian, M., Berro, R., McCaffrey, T., and Kashanchi, F. (2007). HIV-1 TAR element is processed by Dicer to yield a viral micro-RNA involved in chromatin remodeling of the viral LTR. BMC Mol. Biol. *8*, 63.

Klimkait, T., Strebel, K., Hoggan, M.D., Martin, M.A., and Orenstein, J.M. (1990). The human immunodeficiency virus type 1-specific protein vpu is required for efficient virus maturation and release. J. Virol. *64*, 621–629.

Konig, R., Zhou, Y., Elleder, D., Diamond, T.L., Bonamy, G.M., Irelan, J.T., Chiang, C.Y., Tu, B.P., De Jesus, P.D., Lilley, C.E., *et al.* (2008). Global analysis of host-pathogen interactions that regulate early-stage HIV-1 replication. Cell *135*, 49–60.

Kozak, C.A. (1985). Analysis of wild-derived mice for Fv-1 and Fv-2 murine leukemia virus restriction loci: a novel wild mouse Fv-1 allele responsible for lack of host range restriction. J. Virol. *55*, 281–285.

Kozak, C.A., and Chakraborti, A. (1996). Single amino acid changes in the murine leukemia virus capsid protein gene define the target of Fv1 resistance. Virology *225*, 300–305.

Kozak, S.L., Marin, M., Rose, K.M., Bystrom, C., and Kabat, D. (2006). The anti-HIV-1 editing enzyme APOBEC3G binds HIV-1 RNA and messenger RNAs that shuttle between polysomes and stress granules. J. Biol. Chem. *281*, 29105–29119.

Kratovac, Z., Virgen, C.A., Bibollet-Ruche, F., Hahn, B.H., Bieniasz, P.D., and Hatziioannou, T. (2008). Primate lentivirus capsid sensitivity to TRIM5 proteins. J. Virol. *82*, 6772–6777.

Kreisberg, J.F., Yonemoto, W., and Greene, W.C. (2006). Endogenous factors enhance HIV infection of tissue naive CD4 T-cells by stimulating high molecular mass APOBEC3G complex formation. J. Exp. Med. *203*, 865–870.

Kumar, A., Zloza, A., Moon, R.T., Watts, J., Tenorio, A.R., and Al-Harthi, L. (2008). Active beta-catenin signaling is an inhibitory pathway for human immunodeficiency virus replication in peripheral blood mononuclear cells. J. Virol. *82*, 2813–2820.

Kumar, A., and Jeang, K.T. (2008). Insights into cellular microRNAs and human immunodeficiency virus type 1 (HIV-1). J. Cell. Physiol. *216*, 327–331.

Kupzig, S., Korolchuk, V., Rollason, R., Sugden, A., Wilde, A., and Banting, G. (2003). Bst-2/HM1.24 is a raft-associated apical membrane protein with an unusual topology. Traffic *4*, 694–709.

Lai, W.S., Carballo, E., Strum, J.R., Kennington, E.A., Phillips, R.S., and Blackshear, P.J. (1999). Evidence that tristetraprolin binds to AU-rich elements and promotes the deadenylation and destabilization of tumor necrosis factor alpha mRNA. Mol. Cell Biol. *19*, 4311–4323.

Lama, J., Mangasarian, A., and Trono, D. (1999). Cell-surface expression of CD4 reduces HIV-1 infectivity by blocking Env incorporation in a Nef- and Vpu-inhibitable manner. Curr. Biol. *9*, 622–631.

Langelier, C.R., Sandrin, V., Eckert, D.M., Christensen, D.E., Chandrasekaran, V., Alam, S.L., Aiken, C., Olsen, J.C., Kar, A.K., Sodroski, J.G., and Sundquist, W.I. (2008). Biochemical characterization of a recombinant TRIM5alpha protein that restricts human immunodeficiency virus type 1 replication. J. Virol. *82*, 11682–11694.

Langlois, M.A., Beale, R.C., Conticello, S.G., and Neuberger, M.S. (2005). Mutational comparison of the single-domained APOBEC3C and double-domained APOBEC3F/G anti-retroviral cytidine deaminases provides insight into their DNA target site specificities. Nucleic Acids Res. *33*, 1913–1923.

LaRue, R.S., Andresdottir, V., Blanchard, Y., Conticello, S.G., Derse, D., Emerman, M., Greene, W.C., Jonsson, S.R., Landau, N.R., Lochelt, M., *et al.* (2009). Guidelines for naming nonprimate APOBEC3 genes and proteins. J. Virol. *83*, 494–497.

Lecossier, D., Bouchonnet, F., Clavel, F., and Hance, A.J. (2003). Hypermutation of HIV-1 DNA in the absence of the Vif protein. Science *300*, 1112.

Lee, Y., Jeon, K., Lee, J.T., Kim, S., and Kim, V.N. (2002). MicroRNA maturation: stepwise processing and subcellular localization. Embo J. *21*, 4663–4670.

Lee, Y.N., Malim, M.H., and Bieniasz, P.D. (2008). Hypermutation of an ancient human retrovirus by APOBEC3G. J. Virol. *82*, 8762–8770.

Levesque, K., Zhao, Y.S., and Cohen, E.A. (2003). Vpu exerts a positive effect on HIV-1 infectivity by down-modulating CD4 receptor molecules at the surface of HIV-1-producing cells. J. Biol. Chem. *278*, 28346–28353.

Lewis, P., Hensel, M., and Emerman, M. (1992). Human immunodeficiency virus infection of cells arrested in the cell cycle. Embo J. *11*, 3053–3058.

Lewis, P.F., and Emerman, M. (1994). Passage through mitosis is required for oncoretroviruses but not for the human immunodeficiency virus. J. Virol. *68*, 510–516.

Li, L., Liang, D., Li, J.Y., and Zhao, R.Y. (2008). APOBEC3G-UBA2 fusion as a potential strategy for stable expression of APOBEC3G and inhibition of HIV-1 replication. Retrovirology *5*, 72.

Li, X., Song, B., Xiang, S.H., and Sodroski, J. (2007a). Functional interplay between the B-box 2 and the

B30.2(SPRY) domains of TRIM5alpha. Virology *366*, 234–244.

Li, X., and Sodroski, J. (2008). The TRIM5alpha B-box 2 domain promotes cooperative binding to the retroviral capsid by mediating higher-order self-association. J. Virol. *82*, 11495–11502.

Li, X.Y., Guo, F., Zhang, L., Kleiman, L., and Cen, S. (2007b). APOBEC3G inhibits DNA strand transfer during HIV-1 reverse transcription. J. Biol. Chem. *282*, 32065–32074.

Li, Y., Li, X., Stremlau, M., Lee, M., and Sodroski, J. (2006). Removal of arginine 332 allows human TRIM5alpha to bind human immunodeficiency virus capsids and to restrict infection. J. Virol. *80*, 6738–6744.

Liao, C.H., Kuang, Y.Q., Liu, H.L., Zheng, Y.T., and Su, B. (2007). A novel fusion gene, TRIM5-Cyclophilin A in the pig-tailed macaque determines its susceptibility to HIV-1 infection. Aids *21* Suppl. *8*, S19–26.

Liao, W., Hong, S.H., Chan, B.H., Rudolph, F.B., Clark, S.C., and Chan, L. (1999). APOBEC-2, a cardiac- and skeletal muscle-specific member of the cytidine deaminase supergene family. Biochem. Biophys. Res. Commun. *260*, 398–404.

Liddament, M.T., Brown, W.L., Schumacher, A.J., and Harris, R.S. (2004). APOBEC3F properties and hypermutation preferences indicate activity against HIV-1 *in vivo*. Curr. Biol. *14*, 1385–1391.

Lilly, F. (1967). Susceptibility to two strains of Friend leukemia virus in mice. Science *155*, 461–462.

Lilly, F. (1970). Fv-2: identification and location of a second gene governing the spleen focus response to Friend leukemia virus in mice. J. Natl. Cancer Inst. *45*, 163–169.

Lin, J., and Cullen, B.R. (2007). Analysis of the interaction of primate retroviruses with the human RNA interference machinery. J. Virol. *81*, 12218–12226.

Lin, T.Y., and Emerman, M. (2006). Cyclophilin A interacts with diverse lentiviral capsids. Retrovirology 3, 70.

Lin, T.Y., and Emerman, M. (2008). Determinants of cyclophilin A-dependent TRIM5 alpha restriction against HIV-1. Virology *379*, 335–341.

Linney, E., Davis, B., Overhauser, J., Chao, E., and Fan, H. (1984). Non-function of a Moloney murine leukaemia virus regulatory sequence in F9 embryonal carcinoma cells. Nature *308*, 470–472.

Liu, B., Sarkis, P.T., Luo, K., Yu, Y., and Yu, X.F. (2005a). Regulation of Apobec3F and human immunodeficiency virus type 1 Vif by Vif-Cul5-ElonB/C E3 ubiquitin ligase. J. Virol. *79*, 9579–9587.

Liu, H., Wu, X., Newman, M., Shaw, G.M., Hahn, B.H., and Kappes, J.C. (1995). The Vif protein of human and simian immunodeficiency viruses is packaged into virions and associates with viral core structures. J. Virol. *69*, 7630–7638.

Liu, Y.C., Penninger, J., and Karin, M. (2005b). Immunity by ubiquitylation: a reversible process of modification. Nat. Rev. Immunol. 5, 941–952.

Lochelt, M., Romen, F., Bastone, P., Muckenfuss, H., Kirchner, N., Kim, Y.B., Truyen, U., Rosler, U., Battenberg, M., Saib, A., *et al.* (2005). The antiretroviral activity of APOBEC3 is inhibited by the foamy virus accessory Bet protein. Proc. Natl. Acad. Sci. U.S.A. *102*, 7982–7987.

Loh, T.P., Sievert, L.L., and Scott, R.W. (1987). Proviral sequences that restrict retroviral expression in mouse embryonal carcinoma cells. Mol. Cell Biol. *7*, 3775–3784.

Low, A., Okeoma, C.M., Lovsin, N., de las Heras, M., Taylor, T.H., Peterlin, B.M., Ross, S.R., and Fan, H. (2009). Enhanced replication and pathogenesis of Moloney murine leukemia virus in mice defective in the murine APOBEC3 gene. Virology *385*, 455–463.

Luban, J., Bossolt, K.L., Franke, E.K., Kalpana, G.V., and Goff, S.P. (1993). Human immunodeficiency virus type 1 Gag protein binds to cyclophilins A and B. Cell *73*, 1067–1078.

Luban, J. (2007). Cyclophilin A, TRIM5, and resistance to human immunodeficiency virus type 1 infection. J. Virol. *81*, 1054–1061.

Luby-Phelps, K. (2000). Cytoarchitecture and physical properties of cytoplasm: volume, viscosity, diffusion, intracellular surface area. Int Rev Cytol *192*, 189–221.

Luo, K., Liu, B., Xiao, Z., Yu, Y., Yu, X., Gorelick, R., and Yu, X.F. (2004). Amino-terminal region of the human immunodeficiency virus type 1 nucleocapsid is required for human APOBEC3G packaging. J. Virol. *78*, 11841–11852.

Luo, K., Xiao, Z., Ehrlich, E., Yu, Y., Liu, B., Zheng, S., and Yu, X.F. (2005). Primate lentiviral virion infectivity factors are substrate receptors that assemble with cullin 5-E3 ligase through a HCCH motif to suppress APOBEC3G. Proc. Natl. Acad. Sci. U.S.A. *102*, 11444–11449.

Luo, K., Wang, T., Liu, B., Tian, C., Xiao, Z., Kappes, J., and Yu, X.F. (2007). Cytidine deaminases APOBEC3G and APOBEC3F interact with human immunodeficiency virus type 1 integrase and inhibit proviral DNA formation. J. Virol. *81*, 7238–7248.

Lusso, P., Markham, P.D., Ranki, A., Earl, P., Moss, B., Dorner, F., Gallo, R.C., and Krohn, K.J. (1988). Cell-mediated immune response toward viral envelope and core antigens in gibbon apes (Hylobates lar) chronically infected with human immunodeficiency virus-1. J. Immunol. *141*, 2467–2473.

McDonald, D., Vodicka, M.A., Lucero, G., Svitkina, T.M., Borisy, G.G., Emerman, M., and Hope, T.J. (2002). Visualization of the intracellular behavior of HIV in living cells. J. Cell Biol. *159*, 441–452.

Macduff, D.A., Demorest, Z.L., and Harris, R.S. (2009). AID can restrict L1 retrotransposition suggesting a dual role in innate and adaptive immunity. Nucleic Acids Res. *37*, 1854–1867.

Madani, N., and Kabat, D. (1998). An endogenous inhibitor of human immunodeficiency virus in human lymphocytes is overcome by the viral Vif protein. J. Virol. 72, 10251–10255.

Maldarelli, F., Chen, M.Y., Willey, R.L., and Strebel, K. (1993). Human immunodeficiency virus type 1 Vpu protein is an oligomeric type I integral membrane protein. J. Virol. *67*, 5056–5061.

Malim, M.H., and Emerman, M. (2008). HIV-1 accessory proteins--ensuring viral survival in a hostile environment. Cell Host Microbe *3*, 388–398.

Mangeat, B., Turelli, P., Caron, G., Friedli, M., Perrin, L., and Trono, D. (2003). Broad antiretroviral defence by human APOBEC3G through lethal editing of nascent reverse transcripts. Nature *424*, 99–103.

Mangeat, B., Turelli, P., Liao, S., and Trono, D. (2004). A single amino acid determinant governs the species-specific sensitivity of APOBEC3G to Vif action. J. Biol. Chem. *279*, 14481–14483.

Margottin, F., Bour, S.P., Durand, H., Selig, L., Benichou, S., Richard, V., Thomas, D., Strebel, K., and Benarous, R. (1998). A novel human WD protein, h-beta TrCp, that interacts with HIV-1 Vpu connects CD4 to the ER degradation pathway through an F-box motif. Mol. Cell *1*, 565–574.

Mariani, R., Chen, D., Schrofelbauer, B., Navarro, F., Konig, R., Bollman, B., Munk, C., Nymark-McMahon, H., and Landau, N.R. (2003). Species-specific exclusion of APOBEC3G from HIV-1 virions by Vif. Cell *114*, 21–31.

Marin, M., Rose, K.M., Kozak, S.L., and Kabat, D. (2003). HIV-1 Vif protein binds the editing enzyme APOBEC3G and induces its degradation. Nat. Med. *9*, 1398–1403.

Masters, S.L., Yao, S., Willson, T.A., Zhang, J.G., Palmer, K.R., Smith, B.J., Babon, J.J., Nicola, N.A., Norton, R.S., and Nicholson, S.E. (2006). The SPRY domain of SSB-2 adopts a novel fold that presents conserved Par-4-binding residues. Nat. Struct. Mol. Biol. *13*, 77–84.

Mbisa, J.L., Barr, R., Thomas, J.A., Vandegraaff, N., Dorweiler, I.J., Svarovskaia, E.S., Brown, W.L., Mansky, L.M., Gorelick, R.J., Harris, R.S., *et al.* (2007). Human Immunodeficiency Virus Type 1 cDNAs Produced in the Presence of APOBEC3G Exhibit Defects in Plus-Strand DNA Transfer and Integration. J. Virol. *81*, 7099–7110.

Mehle, A., Strack, B., Ancuta, P., Zhang, C., McPike, M., and Gabuzda, D. (2004a). Vif overcomes the innate antiviral activity of APOBEC3G by promoting its degradation in the ubiquitin–proteasome pathway. J. Biol. Chem. *279*, 7792–7798.

Mehle, A., Goncalves, J., Santa-Marta, M., McPike, M., and Gabuzda, D. (2004b). Phosphorylation of a novel SOCS-box regulates assembly of the HIV-1 Vif-Cul5 complex that promotes APOBEC3G degradation. Genes Dev. *18*, 2861–2866.

Mehle, A., Thomas, E.R., Rajendran, K.S., and Gabuzda, D. (2006). A zinc-binding region in Vif binds Cul5 and determines cullin selection. J. Biol. Chem. *281*, 17259–17265.

Mehle, A., Wilson, H., Zhang, C., Brazier, A.J., McPike, M., Pery, E., and Gabuzda, D. (2007). Identification of an APOBEC3G binding site in human immunodeficiency virus type 1 Vif and inhibitors of Vif-APOBEC3G binding. J. Virol. *81*, 13235–13241.

Mische, C.C., Javanbakht, H., Song, B., Diaz-Griffero, F., Stremlau, M., Strack, B., Si, Z., and Sodroski, J. (2005). Retroviral restriction factor TRIM5alpha is a trimer. J. Virol. *79*, 14446–14450.

Miyagi, E., Opi, S., Takeuchi, H., Khan, M., Goila-Gaur, R., Kao, S., and Strebel, K. (2007). Enzymatically active APOBEC3G is required for efficient inhibition of human immunodeficiency virus type 1. J. Virol. *81*, 13346–13353.

Muller, S., Moller, P., Bick, M.J., Wurr, S., Becker, S., Gunther, S., and Kummerer, B.M. (2007). Inhibition of filovirus replication by the zinc finger antiviral protein. J. Virol. *81*, 2391–2400.

Munk, C., Brandt, S.M., Lucero, G., and Landau, N.R. (2002). A dominant block to HIV-1 replication at reverse transcription in simian cells. Proc. Natl. Acad. Sci. U.S.A. *99*, 13843–13848.

Munk, C., Beck, T., Zielonka, J., Hotz-Wagenblatt, A., Chareza, S., Battenberg, M., Thielebein, J., Cichutek, K., Bravo, I.G., O'Brien, S.J., *et al.* (2008). Functions, structure, and read-through alternative splicing of feline APOBEC3 genes. Genome Biol. *9*, R48.

Muramatsu, M., Sankaranand, V.S., Anant, S., Sugai, M., Kinoshita, K., Davidson, N.O., and Honjo, T. (1999). Specific expression of activation-induced cytidine deaminase (AID), a novel member of the RNA-editing deaminase family in germinal center B-cells. J. Biol. Chem. *274*, 18470–18476.

Muramatsu, M., Kinoshita, K., Fagarasan, S., Yamada, S., Shinkai, Y., and Honjo, T. (2000). Class switch recombination and hypermutation require activation-induced cytidine deaminase (AID), a potential RNA editing enzyme. Cell *102*, 553–563.

Nakayama, E.E., Shingai, Y., Kono, K., and Shioda, T. (2008). TRIM5alpha-independent anti-human immunodeficiency virus type 1 activity mediated by cyclophilin A in Old World monkey cells. Virology *375*, 514–520.

Nathans, R., Cao, H., Sharova, N., Ali, A., Sharkey, M., Stranska, R., Stevenson, M., and Rana, T.M. (2008). Small-molecule inhibition of HIV-1 Vif. Nat. Biotechnol. *26*, 1187–1192.

Navarro, F., Bollman, B., Chen, H., Konig, R., Yu, Q., Chiles, K., and Landau, N.R. (2005). Complementary function of the two catalytic domains of APOBEC3G. Virology *333*, 374–386.

Neil, S.J., Eastman, S.W., Jouvenet, N., and Bieniasz, P.D. (2006). HIV-1 Vpu promotes release and prevents endocytosis of nascent retrovirus particles from the plasma membrane. PLoS Pathog. *2*, e39.

Neil, S.J., Sandrin, V., Sundquist, W.I., and Bieniasz, P.D. (2007). An interferon-alpha-induced tethering mechanism inhibits HIV-1 and Ebola virus particle release but is counteracted by the HIV-1 Vpu protein. Cell Host Microbe *2*, 193–203.

Neil, S.J., Zang, T., and Bieniasz, P.D. (2008). Tetherin inhibits retrovirus release and is antagonized by HIV-1 Vpu. Nature *451*, 425–430.

Newman, E.N., Holmes, R.K., Craig, H.M., Klein, K.C., Lingappa, J.R., Malim, M.H., and Sheehy, A.M. (2005). Antiviral function of APOBEC3G can be dissociated from cytidine deaminase activity. Curr. Biol. *15*, 166–170.

Newman, R.M., Hall, L., Connole, M., Chen, G.L., Sato, S., Yuste, E., Diehl, W., Hunter, E., Kaur, A., Miller, G.M., and Johnson, W.E. (2006). Balancing selection and the evolution of functional polymorphism in Old World monkey TRIM5alpha. Proc. Natl. Acad. Sci. U.S.A. *103*, 19134–19139.

Newman, R.M., Hall, L., Kirmaier, A., Pozzi, L.A., Pery, E., Farzan, M., O'Neil, S.P., and Johnson, W. (2008). Evolution of a TRIM5-CypA splice isoform in old world monkeys. PLoS Pathog. *4*, e1000003.

Nilsen, T.W. (2007). Mechanisms of microRNA-mediated gene regulation in animal cells. Trends Genet. *23*, 243–249.

Nisole, S., Lynch, C., Stoye, J.P., and Yap, M.W. (2004). A Trim5-cyclophilin A fusion protein found in owl monkey kidney cells can restrict HIV-1. Proc. Natl. Acad. Sci. U.S.A. *101*, 13324–13328.

Nisole, S., Stoye, J.P., and Saib, A. (2005). TRIM family proteins: retroviral restriction and antiviral defence. Nat. Rev. Microbiol. *3*, 799–808.

Nomaguchi, M., Fujita, M., and Adachi, A. (2008). Role of HIV-1 Vpu protein for virus spread and pathogenesis. Microbes Infect. *10*, 960–967.

Oberste, M.S., and Gonda, M.A. (1992). Conservation of amino-acid sequence motifs in lentivirus Vif proteins. Virus Genes *6*, 95–102.

OhAinle, M., Kerns, J.A., Malik, H.S., and Emerman, M. (2006). Adaptive evolution and antiviral activity of the conserved mammalian cytidine deaminase APOBEC3H. J. Virol. *80*, 3853–3862.

OhAinle, M., Kerns, J.A., Li, M.M., Malik, H.S., and Emerman, M. (2008). Antiretroelement activity of APOBEC3H was lost twice in recent human evolution. Cell Host Microbe *4*, 249–259.

Ohkura, S., Yap, M.W., Sheldon, T., and Stoye, J.P. (2006). All three variable regions of the TRIM5alpha B30.2 domain can contribute to the specificity of retrovirus restriction. J. Virol. *80*, 8554–8565.

Okazaki, I.M., Hiai, H., Kakazu, N., Yamada, S., Muramatsu, M., Kinoshita, K., and Honjo, T. (2003). Constitutive expression of AID leads to tumorigenesis. J. Exp. Med. *197*, 1173–1181.

Okeoma, C.M., Lovsin, N., Peterlin, B.M., and Ross, S.R. (2007). APOBEC3 inhibits mouse mammary tumour virus replication *in vivo*. Nature *445*, 927–930.

Okumura, A., Lu, G., Pitha-Rowe, I., and Pitha, P.M. (2006). Innate antiviral response targets HIV-1 release by the induction of ubiquitin-like protein ISG15. Proc. Natl. Acad. Sci. U.S.A. *103*, 1440–1445.

Omoto, S., Ito, M., Tsutsumi, Y., Ichikawa, Y., Okuyama, H., Brisibe, E.A., Saksena, N.K., and Fujii, Y.R. (2004). HIV-1 nef suppression by virally encoded microRNA. Retrovirology *1*, 44.

Omoto, S., and Fujii, Y.R. (2005). Regulation of human immunodeficiency virus 1 transcription by nef microRNA. J. Gen. Virol. *86*, 751–755.

Opi, S., Kao, S., Goila-Gaur, R., Khan, M.A., Miyagi, E., Takeuchi, H., and Strebel, K. (2007). Human immunodeficiency virus type 1 Vif inhibits packaging and antiviral activity of a degradation-resistant APOBEC3G variant. J. Virol. *81*, 8236–8246.

Orimo, A., Tominaga, N., Yoshimura, K., Yamauchi, Y., Nomura, M., Sato, M., Nogi, Y., Suzuki, M., Suzuki, H., Ikeda, K., *et al.* (2000). Molecular cloning of ring finger protein 21 (RNF21)/interferon-responsive finger protein (ifp1), which possesses two RING-B box-coiled coil domains in tandem. Genomics *69*, 143–149.

Ou, C.Y., Boone, L.R., Koh, C.K., Tennant, R.W., and Yang, W.K. (1983). Nucleotide sequences of gag-pol regions that determine the Fv-1 host range property of BALB/c N-tropic and B-tropic murine leukemia viruses. J. Virol. *48*, 779–784.

Ouellet, D.L., Plante, I., Landry, P., Barat, C., Janelle, M.E., Flamand, L., Tremblay, M.J., and Provost, P. (2008). Identification of functional microRNAs released through asymmetrical processing of HIV-1 TAR element. Nucleic Acids Res. *36*, 2353–2365.

Ozato, K., Shin, D.M., Chang, T.H., and Morse, H.C., 3rd (2008). TRIM family proteins and their emerging roles in innate immunity. Nat. Rev. Immunol. *8*, 849–860.

Pacyniak, E., Gomez, M.L., Gomez, L.M., Mulcahy, E.R., Jackson, M., Hout, D.R., Wisdom, B.J., and Stephens, E.B. (2005). Identification of a region within the cytoplasmic domain of the subtype B Vpu protein of human immunodeficiency virus type 1 (HIV-1) that is responsible for retention in the golgi complex and its absence in the Vpu protein from a subtype C HIV-1. AIDS Res. Hum. Retroviruses *21*, 379–394.

Passerini, L.D., Keckesova, Z., and Towers, G.J. (2006). Retroviral restriction factors Fv1 and TRIM5alpha act independently and can compete for incoming virus before reverse transcription. J. Virol. *80*, 2100–2105.

Paul, M., Mazumder, S., Raja, N., and Jabbar, M.A. (1998). Mutational analysis of the human immunodeficiency virus type 1 Vpu transmembrane domain that promotes the enhanced release of virus-like particles from the plasma membrane of mammalian cells. J. Virol. 72, 1270–1279.

Perez-Caballero, D., Hatziioannou, T., Yang, A., Cowan, S., and Bieniasz, P.D. (2005a). Human tripartite motif 5alpha domains responsible for retrovirus restriction activity and specificity. J. Virol. *79*, 8969–8978.

Perez-Caballero, D., Hatziioannou, T., Zhang, F., Cowan, S., and Bieniasz, P.D. (2005b). Restriction of human immunodeficiency virus type 1 by TRIM-CypA occurs with rapid kinetics and independently of cytoplasmic bodies, ubiquitin, and proteasome activity. J. Virol. *79*, 15567–15572.

Perez-Caballero, D., Soll, S.J., and Bieniasz, P.D. (2008). Evidence for restriction of ancient primate gammaretroviruses by APOBEC3 but not TRIM5alpha proteins. PLoS Pathog. *4*, e1000181.

Perron, M.J., Stremlau, M., Song, B., Ulm, W., Mulligan, R.C., and Sodroski, J. (2004). TRIM5alpha mediates the postentry block to N-tropic murine leukemia viruses in human cells. Proc. Natl. Acad. Sci. U.S.A. *101*, 11827–11832.

Perron, M.J., Stremlau, M., Lee, M., Javanbakht, H., Song, B., and Sodroski, J. (2007). The human TRIM5alpha restriction factor mediates accelerated uncoating of the N-tropic murine leukemia virus capsid. J. Virol. *81*, 2138–2148.

Pery, E., Rajendran, K.S., Brazier, A.J., and Gabuzda, D. (2009). Regulation of APOBEC3 proteins by a novel YXXL motif in human immunodeficiency virus type 1

Vif and simian immunodeficiency virus SIVagm Vif. J. Virol. *83*, 2374–2381.

Petersen, R., Kempler, G., and Barklis, E. (1991). A stem cell-specific silencer in the primer-binding site of a retrovirus. Mol. Cell Biol. *11*, 1214–1221.

Petersen-Mahrt, S.K., Harris, R.S., and Neuberger, M.S. (2002). AID mutates E. coli suggesting a DNA deamination mechanism for antibody diversification. Nature *418*, 99–103.

Pfeffer, S., Sewer, A., Lagos-Quintana, M., Sheridan, R., Sander, C., Grasser, F.A., van Dyk, L.F., Ho, C.K., Shuman, S., Chien, M., *et al.* (2005). Identification of microRNAs of the herpesvirus family. Nat. Methods *2*, 269–276.

Pincus, T., Hartley, J.W., and Rowe, W.P. (1971). A major genetic locus affecting resistance to infection with murine leukemia viruses. I. Tissue culture studies of naturally occurring viruses. J. Exp. Med. *133*, 1219–1233.

Pincus, T., Hartley, J.W., and Rowe, W.P. (1975). A major genetic locus affecting resistance to infection with murine leukemia viruses. IV. Dose-response relationships in Fv-1-sensitive and resistant cell cultures. Virology *65*, 333–342.

Pineda, M.J., Orton, B.R., and Overbaugh, J. (2007). A TRIM5alpha-independent post-entry restriction to HIV-1 infection of macaque cells that is dependent on the path of entry. Virology *363*, 310–318.

Pion, M., Granelli-Piperno, A., Mangeat, B., Stalder, R., Correa, R., Steinman, R.M., and Piguet, V. (2006). APOBEC3G/3F mediates intrinsic resistance of monocyte-derived dendritic cells to HIV-1 infection. J. Exp. Med. *203*, 2887–2893.

Pion, M., Stalder, R., Correa, R., Mangeat, B., Towers, G.J., and Piguet, V. (2007). Identification of an arsenic-sensitive block to primate lentiviral infection of human dendritic cells. J. Virol. *81*, 12086–12090.

Priet, S., Gros, N., Navarro, J.M., Boretto, J., Canard, B., Querat, G., and Sire, J. (2005). HIV-1-Associated Uracil DNA Glycosylase Activity Controls dUTP Misincorporation in Viral DNA and Is Essential to the HIV-1 Life Cycle. Mol. Cell *17*, 479–490.

Pryciak, P.M., and Varmus, H.E. (1992). Fv-1 restriction and its effects on murine leukemia virus integration *in vivo* and *in vitro*. J. Virol. *66*, 5959–5966.

Purdy, J.G., Flanagan, J.M., Ropson, I.J., Rennoll-Bankert, K.E., and Craven, R.C. (2008). Critical role of conserved hydrophobic residues within the major homology region in mature retroviral capsid assembly. J. Virol. *82*, 5951–5961.

Qi, M., Yang, R., and Aiken, C. (2008). Cyclophilin A-dependent restriction of human immunodeficiency virus type 1 capsid mutants for infection of nondividing cells. J. Virol. *82*, 12001–12008.

Qian, S., Zhong, X., Yu, L., Ding, B., de Haan, P., and Boris-Lawrie, K. (2009). HIV-1 Tat RNA silencing suppressor activity is conserved across kingdoms and counteracts translational repression of HIV-1. Proc. Natl. Acad. Sci. U.S.A. *106*, 605–610.

Rajsbaum, R., Stoye, J.P., and O'Garra, A. (2008). Type I interferon-dependent and -independent expression of tripartite motif proteins in immune cells. Eur J. Immunol. *38*, 619–630.

Ratajczak, T., Carrello, A., Mark, P.J., Warner, B.J., Simpson, R.J., Moritz, R.L., and House, A.K. (1993). The cyclophilin component of the unactivated estrogen receptor contains a tetratricopeptide repeat domain and shares identity with p59 (FKBP59). J. Biol. Chem. *268*, 13187–13192.

Reymond, A., Meroni, G., Fantozzi, A., Merla, G., Cairo, S., Luzi, L., Riganelli, D., Zanaria, E., Messali, S., Cainarca, S., *et al.* (2001). The tripartite motif family identifies cell compartments. Embo J. *20*, 2140–2151.

Rhodes, D.A., Ihrke, G., Reinicke, A.T., Malcherek, G., Towey, M., Isenberg, D.A., and Trowsdale, J. (2002). The 52, 000 MW Ro/SS-A autoantigen in Sjogren's syndrome/systemic lupus erythematosus (Ro52) is an interferon-gamma inducible tripartite motif protein associated with membrane proximal structures. Immunology *106*, 246–256.

Rhodes, D.A., de Bono, B., and Trowsdale, J. (2005). Relationship between SPRY and B30.2 protein domains. Evolution of a component of immune defence? Immunology *116*, 411–417.

Ribeiro, I.P., Menezes, A.N., Moreira, M.A., Bonvicino, C.R., Seuanez, H.N., and Soares, M.A. (2005). Evolution of cyclophilin A and TRIMCyp retrotransposition in New World primates. J. Virol. *79*, 14998–15003.

Ritter, G.D., Jr, Yamshchikov, G., Cohen, S.J., and Mulligan, M.J. (1996). Human immunodeficiency virus type 2 glycoprotein enhancement of particle budding: role of the cytoplasmic domain. J. Virol. *70*, 2669–2673.

Roe, T., Reynolds, T.C., Yu, G., and Brown, P.O. (1993). Integration of murine leukemia virus DNA depends on mitosis. EMBO J. *12*, 2099–2108.

Rogozin, I.B., Basu, M.K., Jordan, I.K., Pavlov, Y.I., and Koonin, E.V. (2005). APOBEC4, a new member of the AID/APOBEC family of polynucleotide (deoxy) cytidine deaminases predicted by computational analysis. Cell Cycle *4*, 1281–1285.

Rold, C.J., and Aiken, C. (2008). Proteasomal degradation of TRIM5alpha during retrovirus restriction. PLoS Pathog. *4*, e1000074.

Rollason, R., Korolchuk, V., Hamilton, C., Schu, P., and Banting, G. (2007). Clathrin-mediated endocytosis of a lipid-raft-associated protein is mediated through a dual tyrosine motif. J. Cell Sci. *120*, 3850–3858.

Rosler, C., Kock, J., Kann, M., Malim, M.H., Blum, H.E., Baumert, T.F., and von Weizsacker, F. (2005). APOBEC-mediated interference with hepadnavirus production. Hepatology *42*, 301–309.

Rossi, A., Mukerjee, R., Ferrante, P., Khalili, K., Amini, S., and Sawaya, B.E. (2006). Human immunodeficiency virus type 1 Tat prevents dephosphorylation of Sp1 by TCF-4 in astrocytes. J. Gen. Virol. *87*, 1613–1623.

Rowe, W.P., and Hartley, J.W. (1972). Studies of genetic transmission of murine leukemia virus by AKR mice. II. Crosses with Fv-1 b strains of mice. J. Exp. Med. *136*, 1286–1301.

Russell, R.A., Wiegand, H.L., Moore, M.D., Schafer, A., McClure, M.O., and Cullen, B.R. (2005). Foamy virus Bet proteins function as novel inhibitors of the APOBEC3 family of innate antiretroviral defense factors. J. Virol. 79, 8724–8731.

Russell, R.A., and Pathak, V.K. (2007). Identification of two distinct human immunodeficiency virus type 1 Vif determinants critical for interactions with human APOBEC3G and APOBEC3F. J. Virol. 81, 8201–8210.

Ryman, K.D., Meier, K.C., Nangle, E.M., Ragsdale, S.L., Korneeva, N.L., Rhoads, R.E., MacDonald, M.R., and Klimstra, W.B. (2005). Sindbis virus translation is inhibited by a PKR/RNase L-independent effector induced by alpha/beta interferon priming of dendritic cells. J. Virol. 79, 1487–1499.

Saenz, D.T., Teo, W., Olsen, J.C., and Poeschla, E.M. (2005). Restriction of feline immunodeficiency virus by Ref1, Lv1, and primate TRIM5alpha proteins. J. Virol. 79, 15175–15188.

Sakai, H., Tokunaga, K., Kawamura, M., and Adachi, A. (1995). Function of human immunodeficiency virus type 1 Vpu protein in various cell types. J. Gen. Virol. 76 ( Pt 11), 2717–2722.

Sakuma, R., Noser, J.A., Ohmine, S., and Ikeda, Y. (2007a). Rhesus monkey TRIM5alpha restricts HIV-1 production through rapid degradation of viral Gag polyproteins. Nat. Med. 13, 631–635.

Sakuma, R., Mael, A.A., and Ikeda, Y. (2007b). Alpha interferon enhances TRIM5alpha-mediated antiviral activities in human and rhesus monkey cells. J. Virol. 81, 10201–10206.

Sakuma, R., Ohmine, S., Mael, A.A., Noser, J.A., and Ikeda, Y. (2008). Reply to: No effect of endogenous TRIM5alpha on HIV-1 production. Nat. Med. 14, 236–238.

Sakuma, T., Noda, T., Urata, S., Kawaoka, Y., and Yasuda, J. (2009). Inhibition of Lassa and Marburg virus production by tetherin. J. Virol. 83, 2382–2385.

Salim, A., and Ratner, L. (2008). Modulation of beta-catenin and E-cadherin interaction by Vpu increases human immunodeficiency virus type 1 particle release. J. Virol. 82, 3932–3938.

Santa-Marta, M., da Silva, F.A., Fonseca, A.M., and Goncalves, J. (2005). HIV-1 Vif can directly inhibit apolipoprotein B mRNA-editing enzyme catalytic polypeptide-like 3G-mediated cytidine deamination by using a single amino acid interaction and without protein degradation. J. Biol. Chem. 280, 8765–8775.

Sasada, A., Takaori-Kondo, A., Shirakawa, K., Kobayashi, M., Abudu, A., Hishizawa, M., Imada, K., Tanaka, Y., and Uchiyama, T. (2005). APOBEC3G targets human T-cell leukemia virus type 1. Retrovirology 2, 32.

Sawyer, S.L., Emerman, M., and Malik, H.S. (2004). Ancient adaptive evolution of the primate antiviral DNA-editing enzyme APOBEC3G. PLoS Biol. 2, E275.

Sawyer, S.L., Wu, L.I., Emerman, M., and Malik, H.S. (2005). Positive selection of primate TRIM5{alpha} identifies a critical species-specific retroviral restriction domain. Proc. Natl. Acad. Sci. U.S.A. 102, 2832–2837.

Sawyer, S.L., Wu, L.I., Akey, J.M., Emerman, M., and Malik, H.S. (2006). High-frequency persistence of an impaired allele of the retroviral defense gene TRIM5alpha in humans. Curr. Biol. 16, 95–100.

Sayah, D.M., Sokolskaja, E., Berthoux, L., and Luban, J. (2004). Cyclophilin A retrotransposition into TRIM5 explains owl monkey resistance to HIV-1. Nature 430, 569–573.

Sayah, D.M., and Luban, J. (2004). Selection for loss of Ref1 activity in human cells releases human immunodeficiency virus type 1 from cyclophilin A dependence during infection. J. Virol. 78, 12066–12070.

Schafer, A., Bogerd, H.P., and Cullen, B.R. (2004). Specific packaging of APOBEC3G into HIV-1 virions is mediated by the nucleocapsid domain of the gag polyprotein precursor. Virology 328, 163–168.

Schaller, T., Hue, S., and Towers, G.J. (2007a). An active TRIM5 protein in rabbits indicates a common antiviral ancestor for mammalian TRIM5 proteins. J. Virol. 81, 11713–11721.

Schaller, T., Ylinen, L.M., Webb, B.L., Singh, S., and Towers, G.J. (2007b). Fusion of cyclophilin A to Fv1 enables cyclosporine-sensitive restriction of human and feline immunodeficiency viruses. J. Virol. 81, 10055–10063.

Schmitz, C., Marchant, D., Neil, S.J., Aubin, K., Reuter, S., Dittmar, M.T., and McKnight, A. (2004). Lv2, a novel postentry restriction, is mediated by both capsid and envelope. J. Virol. 78, 2006–2016.

Schrofelbauer, B., Chen, D., and Landau, N.R. (2004). A single amino acid of APOBEC3G controls its species-specific interaction with virion infectivity factor (Vif). Proc. Natl. Acad. Sci. U.S.A. 101, 3927–3932.

Schrofelbauer, B., Yu, Q., Zeitlin, S.G., and Landau, N.R. (2005). Human immunodeficiency virus type 1 Vpr induces the degradation of the UNG and SMUG uracil-DNA glycosylases. J. Virol. 79, 10978–10987.

Schubert, U., and Strebel, K. (1994). Differential activities of the human immunodeficiency virus type 1-encoded Vpu protein are regulated by phosphorylation and occur in different cellular compartments. J. Virol. 68, 2260–2271.

Schubert, U., Clouse, K.A., and Strebel, K. (1995). Augmentation of virus secretion by the human immunodeficiency virus type 1 Vpu protein is cell type dependent and occurs in cultured human primary macrophages and lymphocytes. J. Virol. 69, 7699–7711.

Schubert, U., Bour, S., Ferrer-Montiel, A.V., Montal, M., Maldarell, F., and Strebel, K. (1996a). The two biological activities of human immunodeficiency virus type 1 Vpu protein involve two separable structural domains. J. Virol. 70, 809–819.

Schubert, U., Ferrer-Montiel, A.V., Oblatt-Montal, M., Henklein, P., Strebel, K., and Montal, M. (1996b). Identification of an ion channel activity of the Vpu transmembrane domain and its involvement in the regulation of virus release from HIV-1-infected cells. FEBS Lett. 398, 12–18.

Schubert, U., Bour, S., Willey, R.L., and Strebel, K. (1999). Regulation of virus release by the macrophage-tropic human immunodeficiency virus type 1 AD8 isolate is redundant and can be controlled by either Vpu or Env. J. Virol. 73, 887–896.

Schumacher, A.J., Hache, G., Macduff, D.A., Brown, W.L., and Harris, R.S. (2008). The DNA deaminase activity of human APOBEC3G is required for Ty1, MusD, and human immunodeficiency virus type 1 restriction. J. Virol. *82*, 2652–2660.

Schwartz, S., Felber, B.K., Fenyo, E.M., and Pavlakis, G.N. (1990). Env and Vpu proteins of human immunodeficiency virus type 1 are produced from multiple bicistronic mRNAs. J. Virol. *64*, 5448–5456.

Sebastian, S., and Luban, J. (2005). TRIM5alpha selectively binds a restriction-sensitive retroviral capsid. Retrovirology *2*, 40.

Sebastian, S., Sokolskaja, E., and Luban, J. (2006). Arsenic counteracts human immunodeficiency virus type 1 restriction by various TRIM5 orthologues in a cell type-dependent manner. J. Virol. *80*, 2051–2054.

Sebastian, S., Grutter, C., de Castillia, C.S., Pertel, T., Olivari, S., Grutter, M.G., and Luban, J. (2009). An invariant surface patch on the TRIM5alpha PRYSPRY domain is required for retroviral restriction but dispensable for capsid binding. J. Virol. *83*, 3365–3373.

Sharova, N., Wu, Y., Zhu, X., Stranska, R., Kaushik, R., Sharkey, M., and Stevenson, M. (2008). Primate lentiviral Vpx commandeers DDB1 to counteract a macrophage restriction. PLoS Pathog. *4*, e1000057.

Sheehy, A.M., Gaddis, N.C., Choi, J.D., and Malim, M.H. (2002). Isolation of a human gene that inhibits HIV-1 infection and is suppressed by the viral Vif protein. Nature *418*, 646–650.

Sheehy, A.M., Gaddis, N.C., and Malim, M.H. (2003). The antiretroviral enzyme APOBEC3G is degraded by the proteasome in response to HIV-1 Vif. Nat. Med. *9*, 1404–1407.

Shibata, R., Kawamura, M., Sakai, H., Hayami, M., Ishimoto, A., and Adachi, A. (1991). Generation of a chimeric human and simian immunodeficiency virus infectious to monkey peripheral blood mononuclear cells. J. Virol. *65*, 3514–3520.

Shibata, R., Sakai, H., Kawamura, M., Tokunaga, K., and Adachi, A. (1995). Early replication block of human immunodeficiency virus type 1 in monkey cells. J. Gen. Virol. 76 ( Pt 11), 2723–2730.

Si, Z., Vandegraaff, N., O'Huigin, C., Song, B., Yuan, W., Xu, C., Perron, M., Li, X., Marasco, W.A., Engelman, A., *et al.* (2006). Evolution of a cytoplasmic tripartite motif (TRIM) protein in cows that restricts retroviral infection. Proc. Natl. Acad. Sci. U.S.A. *103*, 7454–7459.

Sikorski, R.S., Boguski, M.S., Goebl, M., and Hieter, P. (1990). A repeating amino acid motif in CDC23 defines a family of proteins and a new relationship among genes required for mitosis and RNA synthesis. Cell *60*, 307–317.

Simon, J.H., Gaddis, N.C., Fouchier, R.A., and Malim, M.H. (1998). Evidence for a newly discovered cellular anti-HIV-1 phenotype. Nat. Med. *4*, 1397–1400.

Singh, D.K., Griffin, D.M., Pacyniak, E., Jackson, M., Werle, M.J., Wisdom, B., Sun, F., Hout, D.R., Pinson, D.M., Gunderson, R.S., *et al.* (2003). The presence of the casein kinase II phosphorylation sites of Vpu enhances the CD4(+) T-cell loss caused by the simian-human immunodeficiency virus SHIV(KU-lbMC33) in pig-tailed macaques. Virology *313*, 435–451.

Sokolskaja, E., Sayah, D.M., and Luban, J. (2004). Target cell cyclophilin A modulates human immunodeficiency virus type 1 infectivity. J. Virol. *78*, 12800–12808.

Sokolskaja, E., Berthoux, L., and Luban, J. (2006). Cyclophilin A and TRIM5alpha independently regulate human immunodeficiency virus type 1 infectivity in human cells. J. Virol. *80*, 2855–2862.

Song, B., Gold, B., O'Huigin, C., Javanbakht, H., Li, X., Stremlau, M., Winkler, C., Dean, M., and Sodroski, J. (2005a). The B30.2(SPRY) domain of the retroviral restriction factor TRIM5alpha exhibits lineage-specific length and sequence variation in primates. J. Virol. *79*, 6111–6121.

Song, B., Javanbakht, H., Perron, M., Park, D.H., Stremlau, M., and Sodroski, J. (2005b). Retrovirus restriction by TRIM5alpha variants from Old World and New World primates. J. Virol. *79*, 3930–3937.

Song, B., Diaz-Griffero, F., Park, D.H., Rogers, T., Stremlau, M., and Sodroski, J. (2005c). TRIM5alpha association with cytoplasmic bodies is not required for antiretroviral activity. Virology *343*, 201–211.

Soros, V.B., Yonemoto, W., and Greene, W.C. (2007). Newly synthesized APOBEC3G is incorporated into HIV virions, inhibited by HIV RNA, and subsequently activated by RNase H. PLoS Pathog. *3*, e15.

Stenglein, M.D., and Harris, R.S. (2006). APOBEC3B and APOBEC3F inhibit L1 retrotransposition by a DNA deamination-independent mechanism. J. Biol. Chem. *281*, 16837–16841.

Stopak, K., de Noronha, C., Yonemoto, W., and Greene, W.C. (2003). HIV-1 Vif blocks the antiviral activity of APOBEC3G by impairing both its translation and intracellular stability. Mol. Cell *12*, 591–601.

Stopak, K.S., Chiu, Y.L., Kropp, J., Grant, R.M., and Greene, W.C. (2007). Distinct patterns of cytokine regulation of APOBEC3G expression and activity in primary lymphocytes, macrophages, and dendritic cells. J. Biol. Chem. *282*, 3539–3546.

Strebel, K., Daugherty, D., Clouse, K., Cohen, D., Folks, T., and Martin, M.A. (1987). The HIV 'A' (sor) gene product is essential for virus infectivity. Nature *328*, 728–730.

Strebel, K., Klimkait, T., and Martin, M.A. (1988). A novel gene of HIV-1, vpu, and its 16-kilodalton product. Science *241*, 1221–1223.

Strebel, K., Klimkait, T., Maldarelli, F., and Martin, M.A. (1989). Molecular and biochemical analyses of human immunodeficiency virus type 1 vpu protein. J. Virol. *63*, 3784–3791.

Stremlau, M., Owens, C.M., Perron, M.J., Kiessling, M., Autissier, P., and Sodroski, J. (2004). The cytoplasmic body component TRIM5alpha restricts HIV-1 infection in Old World monkeys. Nature *427*, 848–853.

Stremlau, M., Perron, M., Welikala, S., and Sodroski, J. (2005). Species-specific variation in the B30.2(SPRY) domain of TRIM5{alpha} determines the potency of human immunodeficiency virus restriction. J. Virol. *79*, 3139–3145.

Stremlau, M., Perron, M., Lee, M., Li, Y., Song, B., Javanbakht, H., Diaz-Griffero, F., Anderson, D.J.,

Sundquist, W.I., and Sodroski, J. (2006a). Specific recognition and accelerated uncoating of retroviral capsids by the TRIM5alpha restriction factor. Proc. Natl. Acad. Sci. U.S.A. *103*, 5514–5519.

Stremlau, M., Song, B., Javanbakht, H., Perron, M., and Sodroski, J. (2006b). Cyclophilin A: an auxiliary but not necessary cofactor for TRIM5alpha restriction of HIV-1. Virology *351*, 112–120.

Suspene, R., Rusniok, C., Vartanian, J.P., and Wain-Hobson, S. (2006). Twin gradients in APOBEC3 edited HIV-1 DNA reflect the dynamics of lentiviral replication. Nucleic Acids Res. *34*, 4677–4684.

Svarovskaia, E.S., Xu, H., Mbisa, J.L., Barr, R., Gorelick, R.J., Ono, A., Freed, E.O., Hu, W.S., and Pathak, V.K. (2004). Human apolipoprotein B mRNA-editing enzyme-catalytic polypeptide-like 3G (APOBEC3G) is incorporated into HIV-1 virions through interactions with viral and nonviral RNAs. J. Biol. Chem. *279*, 35822–35828.

Takaoka, A., and Yanai, H. (2006). Interferon signalling network in innate defence. Cell. Microbiol. *8*, 907–922.

Takeuchi, H., Buckler-White, A., Goila-Gaur, R., Miyagi, E., Khan, M.A., Opi, S., Kao, S., Sokolskaja, E., Pertel, T., Luban, J., and Strebel, K. (2007). Vif counteracts a cyclophilin A-imposed inhibition of simian immunodeficiency viruses in human cells. J. Virol. *81*, 8080–8090.

Tanaka, M., Ueno, T., Nakahara, T., Sasaki, K., Ishimoto, A., and Sakai, H. (2003). Down-regulation of CD4 is required for maintenance of viral infectivity of HIV-1. Virology *311*, 316–325.

Tareen, S.U., Sawyer, S.L., Malik, H.S., and Emerman, M. (2009). An expanded clade of rodent Trim5 genes. Virology *385*, 473–483.

Teich, N.M., Weiss, R.A., Martin, G.R., and Lowy, D.R. (1977). Virus infection of murine teratocarcinoma stem cell lines. Cell *12*, 973–982.

Teng, B., Burant, C.F., and Davidson, N.O. (1993). Molecular cloning of an apolipoprotein B messenger RNA editing protein. Science *260*, 1816–1819.

Terwilliger, E.F., Cohen, E.A., Lu, Y.C., Sodroski, J.G., and Haseltine, W.A. (1989). Functional role of human immunodeficiency virus type 1 vpu. Proc. Natl. Acad. Sci. U.S.A. *86*, 5163–5167.

Thali, M., Bukovsky, A., Kondo, E., Rosenwirth, B., Walsh, C.T., Sodroski, J., and Gottlinger, H. (1994). Functional association of cyclophilin A with HIV-1 virions. Nature *372*, 363–365.

Tissot, C., and Mechti, N. (1995). Molecular cloning of a new interferon-induced factor that represses human immunodeficiency virus type 1 long terminal repeat expression. J. Biol. Chem. *270*, 14891–14898.

Towers, G., Bock, M., Martin, S., Takeuchi, Y., Stoye, J.P., and Danos, O. (2000). A conserved mechanism of retrovirus restriction in mammals. Proc. Natl. Acad. Sci. U.S.A. *97*, 12295–12299.

Towers, G., Collins, M., and Takeuchi, Y. (2002). Abrogation of Ref1 retrovirus restriction in human cells. J. Virol. *76*, 2548–2550.

Towers, G.J., Hatziioannou, T., Cowan, S., Goff, S.P., Luban, J., and Bieniasz, P.D. (2003). Cyclophilin A modulates the sensitivity of HIV-1 to host restriction factors. Nat. Med. *9*, 1138–1143.

Towers, G.J. (2007). The control of viral infection by tripartite motif proteins and cyclophilin A. Retrovirology *4*, 40.

Tran, D.D., Russell, H.R., Sutor, S.L., van Deursen, J., and Bram, R.J. (2003). CAML is required for efficient EGF receptor recycling. Dev. Cell *5*, 245–256.

Trapp, S., Derby, N.R., Singer, R., Shaw, A., Williams, V.G., Turville, S.G., Bess, J.W., Jr, Lifson, J.D., and Robbiani, M. (2009). Double-stranded RNA analog poly(I:C) inhibits human immunodeficiency virus amplification in dendritic cells via type I interferon-mediated activation of APOBEC3G. J. Virol. *83*, 884–895.

Triboulet, R., Mari, B., Lin, Y.L., Chable-Bessia, C., Bennasser, Y., Lebrigand, K., Cardinaud, B., Maurin, T., Barbry, P., Baillat, V., *et al.* (2007). Suppression of microRNA-silencing pathway by HIV-1 during virus replication. Science *315*, 1579–1582.

Tsukiyama, T., Niwa, O., and Yokoro, K. (1989). Mechanism of suppression of the long terminal repeat of Moloney leukemia virus in mouse embryonal carcinoma cells. Mol. Cell Biol. *9*, 4670–4676.

Turelli, P., Mangeat, B., Jost, S., Vianin, S., and Trono, D. (2004). Inhibition of hepatitis B virus replication by APOBEC3G. Science *303*, 1829.

Van Damme, N., Goff, D., Katsura, C., Jorgenson, R.L., Mitchell, R., Johnson, M.C., Stephens, E.B., and Guatelli, J. (2008). The interferon-induced protein BST-2 restricts HIV-1 release and is down-regulated from the cell surface by the viral Vpu protein. Cell Host Microbe *3*, 245–252.

Van Damme, N., and Guatelli, J. (2008). HIV-1 Vpu inhibits accumulation of the envelope glycoprotein within clathrin-coated, Gag-containing endosomes. Cell. Microbiol. *10*, 1040–1057.

Vartanian, J.P., Guetard, D., Henry, M., and Wain-Hobson, S. (2008). Evidence for editing of human papillomavirus DNA by APOBEC3 in benign and precancerous lesions. Science *320*, 230–233.

Varthakavi, V., Smith, R.M., Bour, S.P., Strebel, K., and Spearman, P. (2003). Viral protein U counteracts a human host cell restriction that inhibits HIV-1 particle production. Proc. Natl. Acad. Sci. U.S.A. *100*, 15154–15159.

Varthakavi, V., Smith, R.M., Martin, K.L., Derdowski, A., Lapierre, L.A., Goldenring, J.R., and Spearman, P. (2006). The pericentriolar recycling endosome plays a key role in Vpu-mediated enhancement of HIV-1 particle release. Traffic *7*, 298–307.

Varthakavi, V., Heimann-Nichols, E., Smith, R.M., Sun, Y., Bram, R.J., Ali, S., Rose, J., Ding, L., and Spearman, P. (2008). Identification of calcium-modulating cyclophilin ligand as a human host restriction to HIV-1 release overcome by Vpu. Nat. Med. *14*, 641–647.

Virgen, C.A., Kratovac, Z., Bieniasz, P.D., and Hatziioannou, T. (2008). Independent genesis of chimeric TRIM5-cyclophilin proteins in two primate species. Proc. Natl. Acad. Sci. U.S.A. *105*, 3563–3568.

von Schwedler, U., Song, J., Aiken, C., and Trono, D. (1993). Vif is crucial for human immunodeficiency virus type 1 proviral DNA synthesis in infected cells. J. Virol. *67*, 4945–4955.

Waheed, A.A., Ablan, S.D., Soheilian, F., Nagashima, K., Ono, A., Schaffner, C.P., and Freed, E.O. (2008). Inhibition of human immunodeficiency virus type 1 assembly and release by the cholesterol-binding compound amphotericin B methyl ester: evidence for Vpu dependence. J. Virol. *82*, 9776–9781.

Wang, X., Ye, L., Hou, W., Zhou, Y., Wang, Y.J., Metzger, D.S., and Ho, W.Z. (2009). Cellular microRNA expression correlates with susceptibility of monocytes/macrophages to HIV-1 infection. Blood *113*, 671–674.

Weinberg, J.B., Matthews, T.J., Cullen, B.R., and Malim, M.H. (1991). Productive human immunodeficiency virus type 1 (HIV-1) infection of nonproliferating human monocytes. J. Exp. Med. *174*, 1477–1482.

Wichroski, M.J., Robb, G.B., and Rana, T.M. (2006). Human retroviral host restriction factors APOBEC3G and APOBEC3F localize to mRNA processing bodies. PLoS Pathog. *2*, e41.

Wiegand, H.L., Doehle, B.P., Bogerd, H.P., and Cullen, B.R. (2004). A second human antiretroviral factor, APOBEC3F, is suppressed by the HIV-1 and HIV-2 Vif proteins. Embo J. *23*, 2451–2458.

Wiegand, H.L., and Cullen, B.R. (2007). Inhibition of alpharetrovirus replication by a range of human APOBEC3 proteins. J. Virol. *81*, 13694–13699.

Wilson, S.J., Webb, B.L., Ylinen, L.M., Verschoor, E., Heeney, J.L., and Towers, G.J. (2008a). Independent evolution of an antiviral TRIMCyp in rhesus macaques. Proc. Natl. Acad. Sci. U.S.A. *105*, 3557–3562.

Wilson, S.J., Webb, B.L., Maplanka, C., Newman, R.M., Verschoor, E.J., Heeney, J.L., and Towers, G.J. (2008b). Rhesus macaque TRIM5 alleles have divergent antiretroviral specificities. J. Virol. *82*, 7243–7247.

Wolf, D., and Goff, S.P. (2007). TRIM28 mediates primer binding site-targeted silencing of murine leukemia virus in embryonic cells. Cell *131*, 46–57.

Wolf, D., Hug, K., and Goff, S.P. (2008a). TRIM28 mediates primer binding site-targeted silencing of Lys1,2 tRNA-utilizing retroviruses in embryonic cells. Proc. Natl. Acad. Sci. U.S.A. *105*, 12521–12526.

Wolf, D., Cammas, F., Losson, R., and Goff, S.P. (2008b). Primer binding site-dependent restriction of murine leukemia virus requires HP1 binding by TRIM28. J. Virol. *82*, 4675–4679.

Wolf, D., and Goff, S.P. (2009). Embryonic stem cells use ZFP809 to silence retroviral DNAs. Nature *458*, 1201–1204.

Woo, J.S., Imm, J.H., Min, C.K., Kim, K.J., Cha, S.S., and Oh, B.H. (2006). Structural and functional insights into the B30.2/SPRY domain. Embo J. *25*, 1353–1363.

Wortman, B., Darbinian, N., Sawaya, B.E., Khalili, K., and Amini, S. (2002). Evidence for regulation of long terminal repeat transcription by Wnt transcription factor TCF-4 in human astrocytic cells. J. Virol. *76*, 11159–11165.

Wu, X., Anderson, J.L., Campbell, E.M., Joseph, A.M., and Hope, T.J. (2006). Proteasome inhibitors uncouple rhesus TRIM5alpha restriction of HIV-1 reverse transcription and infection. Proc. Natl. Acad. Sci. U.S.A. *103*, 7465–7470.

Xu, H., Svarovskaia, E.S., Barr, R., Zhang, Y., Khan, M.A., Strebel, K., and Pathak, V.K. (2004). A single amino acid substitution in human APOBEC3G antiretroviral enzyme confers resistance to HIV-1 virion infectivity factor-induced depletion. Proc. Natl. Acad. Sci. U.S.A. *101*, 5652–5657.

Xu, L., Yang, L., Moitra, P.K., Hashimoto, K., Rallabhandi, P., Kaul, S., Meroni, G., Jensen, J.P., Weissman, A.M., and D'Arpa, P. (2003). BTBD1 and BTBD2 colocalize to cytoplasmic bodies with the RBCC/tripartite motif protein, TRIM5delta. Exp. Cell. Res. *288*, 84–93.

Yamanaka, S., Balestra, M.E., Ferrell, L.D., Fan, J., Arnold, K.S., Taylor, S., Taylor, J.M., and Innerarity, T.L. (1995). Apolipoprotein B mRNA-editing protein induces hepatocellular carcinoma and dysplasia in transgenic animals. Proc. Natl. Acad. Sci. U.S.A. *92*, 8483–8487.

Yamauchi, K., Wada, K., Tanji, K., Tanaka, M., and Kamitani, T. (2008). Ubiquitination of E3 ubiquitin ligase TRIM5 alpha and its potential role. Febs J *275*, 1540–1555.

Yan, Y., and Kozak, C.A. (2008). Novel postentry resistance to AKV ecotropic mouse gammaretroviruses in the African pygmy mouse, *Mus minutoides*. J. Virol. *82*, 6120–6129.

Yang, B., Chen, K., Zhang, C., Huang, S., and Zhang, H. (2007a). Virion-associated uracil DNA glycosylase-2 and apurinic/apyrimidinic endonuclease are involved in the degradation of APOBEC3G-edited nascent HIV-1 DNA. J. Biol. Chem. *282*, 11667–11675.

Yang, Y., Guo, F., Cen, S., and Kleiman, L. (2007b). Inhibition of initiation of reverse transcription in HIV-1 by human APOBEC3F. Virology *365*, 92–100.

Yao, X.J., Gottlinger, H., Haseltine, W.A., and Cohen, E.A. (1992). Envelope glycoprotein and CD4 independence of vpu-facilitated human immunodeficiency virus type 1 capsid export. J. Virol. *66*, 5119–5126.

Yap, M.W., and Stoye, J.P. (2003). Intracellular localisation of Fv1. Virology *307*, 76–89.

Yap, M.W., Nisole, S., Lynch, C., and Stoye, J.P. (2004). Trim5alpha protein restricts both HIV-1 and murine leukemia virus. Proc. Natl. Acad. Sci. U.S.A. *101*, 10786–10791.

Yap, M.W., Nisole, S., and Stoye, J.P. (2005). A single amino acid change in the SPRY domain of human Trim5alpha leads to HIV-1 restriction. Curr. Biol. *15*, 73–78.

Yap, M.W., Dodding, M.P., and Stoye, J.P. (2006). Trim-cyclophilin A fusion proteins can restrict human immunodeficiency virus type 1 infection at two distinct phases in the viral life cycle. J. Virol. *80*, 4061–4067.

Yap, M.W., Mortuza, G.B., Taylor, I.A., and Stoye, J.P. (2007). The design of artificial retroviral restriction factors. Virology *365*, 302–314.

Yap, M.W., Lindemann, D., Stanke, N., Reh, J., Westphal, D., Hanenberg, H., Ohkura, S., and Stoye, J.P. (2008). Restriction of foamy viruses by primate Trim5alpha. J. Virol. *82*, 5429–5439.

Yeung, M.L., Benkirane, M., and Jeang, K.T. (2007). Small non-coding RNAs, mammalian cells, and viruses: regulatory interactions? Retrovirology *4*, 74.

Yi, R., Qin, Y., Macara, I.G., and Cullen, B.R. (2003). Exportin-5 mediates the nuclear export of pre-microRNAs and short hairpin RNAs. Genes Dev. *17*, 3011–3016.

Ylinen, L.M., Keckesova, Z., Wilson, S.J., Ranasinghe, S., and Towers, G.J. (2005). Differential restriction of human immunodeficiency virus type 2 and simian immunodeficiency virus SIVmac by TRIM5alpha alleles. J. Virol. *79*, 11580–11587.

Ylinen, L.M., Keckesova, Z., Webb, B.L., Gifford, R.J., Smith, T.P., and Towers, G.J. (2006). Isolation of an active Lv1 gene from cattle indicates that tripartite motif protein-mediated innate immunity to retroviral infection is widespread among mammals. J. Virol. *80*, 7332–7338.

Ylinen, L.M., Schaller, T., Price, A., Fletcher, A.J., Noursadeghi, M., James, L.C., and Towers, G.J. (2009). Cyclophilin A levels dictate infection efficiency of human immunodeficiency virus type 1 capsid escape mutants A92E and G94D. J. Virol. *83*, 2044–2047.

Yoo, S., Myszka, D.G., Yeh, C., McMurray, M., Hill, C.P., and Sundquist, W.I. (1997). Molecular recognition in the HIV-1 capsid/cyclophilin A complex. J. Mol. Biol. *269*, 780–795.

Yu, Q., Chen, D., Konig, R., Mariani, R., Unutmaz, D., and Landau, N.R. (2004a). APOBEC3B and APOBEC3C are potent inhibitors of simian immunodeficiency virus replication. J. Biol. Chem. *279*, 53379–53386.

Yu, Q., Konig, R., Pillai, S., Chiles, K., Kearney, M., Palmer, S., Richman, D., Coffin, J.M., and Landau, N.R. (2004b). Single-strand specificity of APOBEC3G accounts for minus-strand deamination of the HIV genome. Nat. Struct. Mol. Biol. *11*, 435–442.

Yu, X., Yu, Y., Liu, B., Luo, K., Kong, W., Mao, P., and Yu, X.F. (2003). Induction of APOBEC3G ubiquitination and degradation by an HIV-1 Vif-Cul5-SCF complex. Science *302*, 1056–1060.

Yu, Y., Xiao, Z., Ehrlich, E.S., Yu, X., and Yu, X.F. (2004c). Selective assembly of HIV-1 Vif-Cul5-ElonginB-ElonginC E3 ubiquitin ligase complex through a novel SOCS box and upstream cysteines. Genes Dev. *18*, 2867–2872.

Zennou, V., Perez-Caballero, D., Gottlinger, H., and Bieniasz, P.D. (2004). APOBEC3G incorporation into human immunodeficiency virus type 1 particles. J. Virol. *78*, 12058–12061.

Zhang, F., Hatziioannou, T., Perez-Caballero, D., Derse, D., and Bieniasz, P.D. (2006). Antiretroviral potential of human tripartite motif-5 and related proteins. Virology *353*, 396–409.

Zhang, F., Perez-Caballero, D., Hatziioannou, T., and Bieniasz, P.D. (2008a). No effect of endogenous TRIM5alpha on HIV-1 production. Nat. Med. *14*, 235–236; author reply 236–238.

Zhang, H., Yang, B., Pomerantz, R.J., Zhang, C., Arunachalam, S.C., and Gao, L. (2003). The cytidine deaminase CEM15 induces hypermutation in newly synthesized HIV-1 DNA. Nature *424*, 94–98.

Zhang, L., Saadatmand, J., Li, X., Guo, F., Niu, M., Jiang, J., Kleiman, L., and Cen, S. (2008b). Function analysis of sequences in human APOBEC3G involved in Vif-mediated degradation. Virology *370*, 113–121.

Zhang, L., Li, X., Ma, J., Yu, L., Jiang, J., and Cen, S. (2008c). The incorporation of APOBEC3 proteins into murine leukemia viruses. Virology *378*, 69–78.

Zhang, W., Chen, G., Niewiadomska, A.M., Xu, R., and Yu, X.F. (2008d). Distinct determinants in HIV-1 Vif and human APOBEC3 proteins are required for the suppression of diverse host anti-viral proteins. PLoS ONE 3, e3963.

Zhang, W., Huang, M., Wang, T., Tan, L., Tian, C., Yu, X., Kong, W., and Yu, X.F. (2008e). Conserved and non-conserved features of HIV-1 and SIVagm Vif mediated suppression of APOBEC3 cytidine deaminases. Cell. Microbiol. *10*, 1662–1675.

Zhou, H., Xu, M., Huang, Q., Gates, A.T., Zhang, X.D., Castle, J.C., Stec, E., Ferrer, M., Strulovici, B., Hazuda, D.J., and Espeseth, A.S. (2008). Genome-scale RNAi screen for host factors required for HIV replication. Cell Host Microbe 4, 495–504.

Zhu, Y., and Gao, G. (2008). ZAP-mediated mRNA degradation. RNA Biol 5, 65–67.

# HIV-2 Tropism and Disease

5

Kelly Cheney and Áine McKnight

Abstract

Human immunodeficiency viruses (HIV-1 and HIV-2) have evolved from a reservoir of African non-human primate lentiviruses, the simian immunodeficiency viruses. In contrast to the epidemic nature of HIV-1 infections, HIV-2 is restricted in its worldwide distribution, with the lower viral loads established in asymptomatic infection a significant cause of its diminished transmission efficiency. HIV-2 is also much less pathogenic than HIV-1 and results in a reduced rate of progression to AIDS despite a substantial proviral burden. The majority of patients remain asymptomatic and die of causes unrelated to immunodeficiency.

The adoption of 'accessory genes' by HIV-2 and its more promiscuous pattern of coreceptor usage (including CD4-independence) may assist the virus in its adaptation to avoid innate restriction factors present in host cells. Adaptation to use normal cellular machinery to enable transmission and productive infection has also aided the establishment of HIV-2 replication in humans.

A survival strategy for any infectious agent is not to kill its host but ultimately to become a commensal organism. Having achieved a low pathogenicity, over time, variants more successful at transmission will be selected.

## Phylogenetics

Lentiviruses have been identified in at least 36 non-human primate species in sub-Saharan Africa (Bibollet-Ruche *et al.*, 2004); however, human immunodeficiency viruses HIV-1 and HIV-2 seem to have arisen from cross-species transmission events of only two of these simian immunodeficiency viruses (SIV). The three phylogenetic groups of HIV-1 – N, O and M – are believed to have evolved from relatively recent zoonotic transmissions of the strain $SIV_{cpz}$ from a particular subspecies of chimpanzee (*Pan troglodytes troglodytes*) (Fig. 5.1). Subsequent analysis of genome organization and sequence similarities has identified SIVs from this group of non-human primates as the natural precursors to HIV-1 infection in humans (Keele *et al.*, 2006). The M group of HIV-1 has radiated into the subtypes A–K (as well as many recombinant forms) and is responsible for the worldwide acquired immunodeficiency syndrome (AIDS) pandemic (Perrin *et al.*, 2003). HIV-2 is more closely related to $SIV_{sm}$ from sooty mangabeys (*Cercocebus atys*). This SIV strain was transferred both to humans in West Africa (on at least seven occasions giving rise to the HIV-2 groups A–H) and rhesus and cynomolgus macaques ($SIV_{mac}$) (Hirsch *et al.*, 1989). In contrast to the pandemic nature of HIV-1 infection, of the eight HIV-2 groups, only groups A and B are endemic in West Africa, with the other subtypes thought to be primary zoonotic infections (Gao *et al.*, 1994; Damond *et al.*, 2004). While phylogenetic studies can prove beyond reasonable doubt the relationship between the various SIV and HIV strains, no direct evidence of simian-to-human transmission events exist. This suggests that many transmissions occur but are not successful. It may be that a complex series of genetic adaptations are required to overcome species specific barriers to infection.

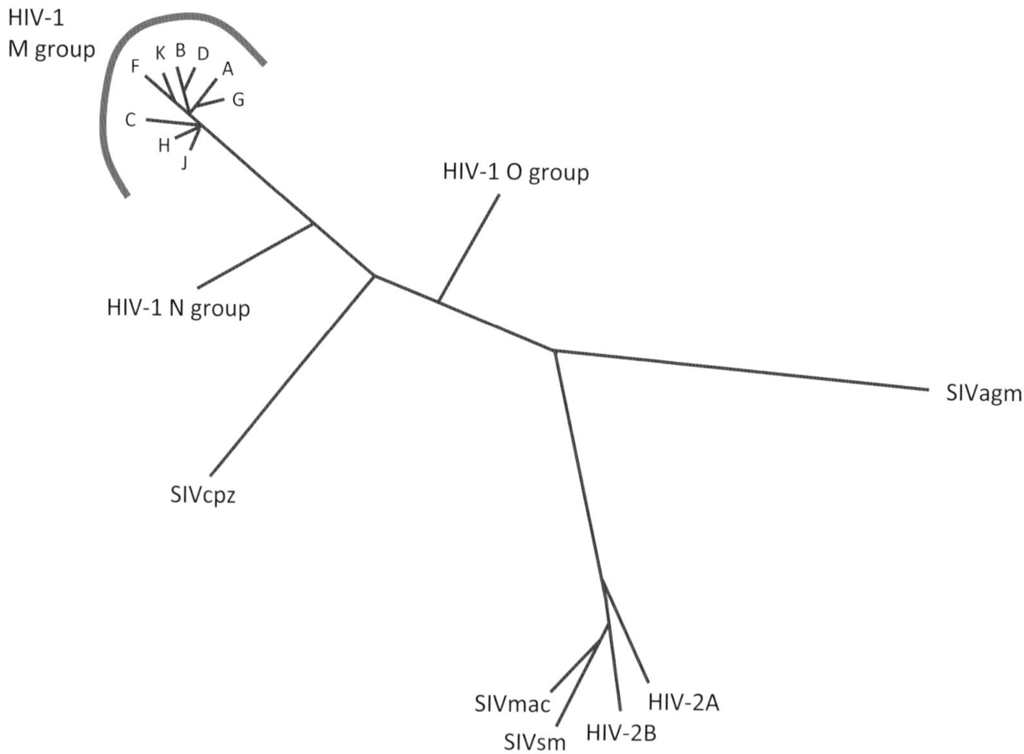


**Figure 5.1** Phylogenetic relationship of primate lentiviruses based on *pol* sequence alignments. $SIV_{cpz}$ gives rise to HIV-1 groups M, N and O while HIV-2 is more closely related to $SIV_{sm}$.

## Adaptation to establish infection in humans

Adaptation to allow use of a more effective route of transmission is likely to have been a factor in the ability of SIV strains to infect humans. While sexual transmission has been observed, the higher primates normally transmit SIV through biting and scratching. In comparison, the predominantly sexual route of HIV transmission in humans suggests that a stronger selection pressure may have initially advantaged zoonotic infection by viruses able to efficiently transgress and establish infection across mucosal surfaces rather than parenterally as would have been the case during aggressive primate interactions. The mucosal epithelium is approximately four cell layers thick. Just below this is found the greatest abundance of macrophages in the body (Lee *et al.*, 1985) which, in addition to many dendritic cells (DCs), are a primary target in the establishment of infection. The ability to interact with such cells may have been key in the adaptation of HIV to mucosal transmission. HIV establishes infection following entry through the mucosal tissues of either the genital or gastrointestinal tract. There, HIV encounters potential DC, lymphocyte and macrophage target cells in the lamina propria beneath the epithelium. The mechanism of viral transmission to these cells across the relatively impermeable epithelial barrier is poorly understood. This is compounded by inconsistent experimental data obtained from inadequate protocols including the use of immortalized cell lines. In spite of these inconsistencies, DCs have been established as having an important role in facilitating the transmission of HIVs to activated lymphocytes and other target cells such as macrophages in both the peripheral blood and tissues (reviewed in Wu and KewalRamani, 2006). Some work, including *ex vivo* experiments using isolated vaginal epithelium, suggests that HIV may disperse through the narrow gaps between the epithelial cells (Maher *et al.*, 2005; Hladik and McElrath, 2008), eventually contacting intraepithelial DCs and

CD4$^+$ T-cells (Fig. 5.2a). Additionally, these DCs may also capture free virus particles as part of a routine extension of processes into the lumen to sample contents and trap pathogens (Fig. 5.2b). Alternatively, HIV may come into contact with DCs and macrophages by crossing at breaches, lesions or points of thinning in the mucosa (Fig. 5.2c). Adaptation to binding to and/or infection of macrophages and DCs is likely to enhance the use of mucosal transmission as an option by a previously predominantly blood-borne virus.

Interaction of HIV with DCs occurs in a number of ways including binding to DC-specific ICAM-grabbing non-integrin (DC-SIGN). DC-SIGN is a C-type lectin and is expressed *in vivo* on DCs, macrophages, activated B-cells, the dermis of the skin, intestinal and genital mucosa and lymphoid tissues. DC-SIGN on the surface of cells captures HIV and may increase the efficiency of infection by aggregating virus particles on the cell surface, facilitating cross-infection of CD4$^+$ T-cells in the vicinity. Similarly, macrophages, which can express low levels of CD4, also have an abundance of heparan sulfate proteoglycans (HSPGs), especially syndecan. Like DC-SIGN, syndecan has been shown to allow the adsorption of HIV through binding to gp120. Adsorbed HIV retains its infectivity and can then be transmitted to CD4$^+$ T-cells in the surrounding environment (Bobardt *et al.*, 2003). The ability of syndecans and HSPGs to preserve the infectivity of HIV particles, together with the longevity of macrophages *in vivo* and their routine interactions with CD4$^+$ T-cells, may explain the gradual accomplishment of a stable mode of mucosal transmission for HIV among humans. While these routes are the same for both viruses, it is interesting to note that the majority of cases in HIV-2 endemic areas are the result of sexual (mucosal) rather than blood-borne encounters. Although HIV-2 has been introduced into the injecting drug using population (Singh *et al.*, 1995) unlike HIV-1 it has not established consistent transmission cycles among them.

## Accessory genes vpx/vpr: adaptation to host cells and divergent evolution

Primate hosts have evolved to counteract retroviral infections in a species-specific manner (discussed further in Chapter 4). The various orthologues of TRIM proteins, expressed in

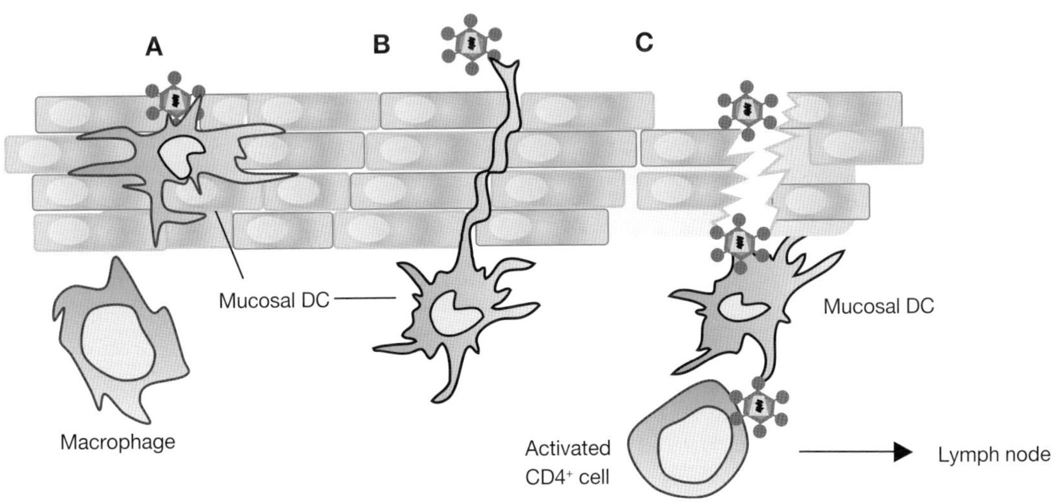


**Figure 5.2** DCs have an important role in facilitating the transmission of HIVs across mucosal surfaces to activated lymphocytes and other target cells such as macrophages. (A) HIV may disperse through the narrow gaps between the epithelial cells eventually contacting intraepithelial dendritic cells (DCs) and CD4$^+$ T-cells. (B) DCs may capture free virus particles as part of a routine extension of processes into the lumen to sample contents and trap pathogens. (C) HIV can also come into contact with DCs and macrophages by crossing at breaches, lesions or points of thinning in the mucosa.

host cells, provide protection against different viruses (reviewed in Ozato *et al.*, 2008), as do the APOBEC family of cellular cytidine deaminases. Some APOBEC proteins can provide immunity to infection by retroviruses, which is overcome in the case of HIV-1 through the action of its accessory protein Vif (reviewed in Goila-Gaur and Strebel, 2008). Thus, viruses must concomitantly adapt to both avoid innate restriction factors present in host cells and use normal cellular machinery to enable transmission and establish infection. Human lentiviruses have adopted 'accessory genes' to help accomplish this.

Of the 7 genera of retroviruses, the gammaretroviruses are the simplest, encoding only three groups of structural proteins, group-specific antigen (*gag*), polymerase (*pol*) and envelope (*env*). These retroviruses, such as the murine leukaemia viruses (MLVs), require cell division for efficient infection (Varmus *et al.*, 1977; Roe *et al.*, 1993), which may be related to a block in nuclear transport of the MLV genome in non-dividing cells (Lieber *et al.*, 2000). These constraints make such viruses unable to infect normally quiescent cells like macrophages, and may explain their use of primarily non-mucosal (i.e. blood-borne and vertical) routes of transmission. Additional genetic information is required to enable more complex interactions of a retrovirus with its host.

The genome of complex retroviruses such as HIV-1 and -2 includes several additional and overlapping open reading frames (ORFs) which encode 'accessory' proteins such as Vif, Nef and Vpr. The genomic organization of HIV-1 and -2 is extremely similar however HIV-1 contains an ORF for *vpu*, while members of the HIV-2/$SIV_{sm}$/$SIV_{mac}$ lineage have an additional gene encoding viral protein X (Vpx), originally derived from the African green monkey *vpr* gene (Fletcher *et al.*, 1996) (Fig. 5.3). In HIV-1 infected cells, Vpu overcomes the activity of tetherin to promote the release of virions from the cell surface (Neil *et al.*, 2008) while the HIV-2 Env protein provides this function in HIV-2 infection (Bour *et al.*, 1996; Bour and Strebel, 1996). Vpx is a paralogue of Vpr and although the amino acid sequences are closely related, the two proteins have different roles in the virus life cycle.

HIV-1 studies have shown that Vpr performs multiple functions; cell cycle arrest in the G2 phase, nuclear import of preintegration complexes, and induction of apoptosis in the infected cell. Vpr also enhances the infection of terminally differentiated macrophages and DCs (Eckstein *et al.*, 2001; Goujon *et al.*, 2007; Andersen *et al.*, 2008). For HIV-2/$SIV_{sm}$, these functions are divided between Vpr and Vpx, with Vpr only inducing cell cycle arrest while Vpx is shown to

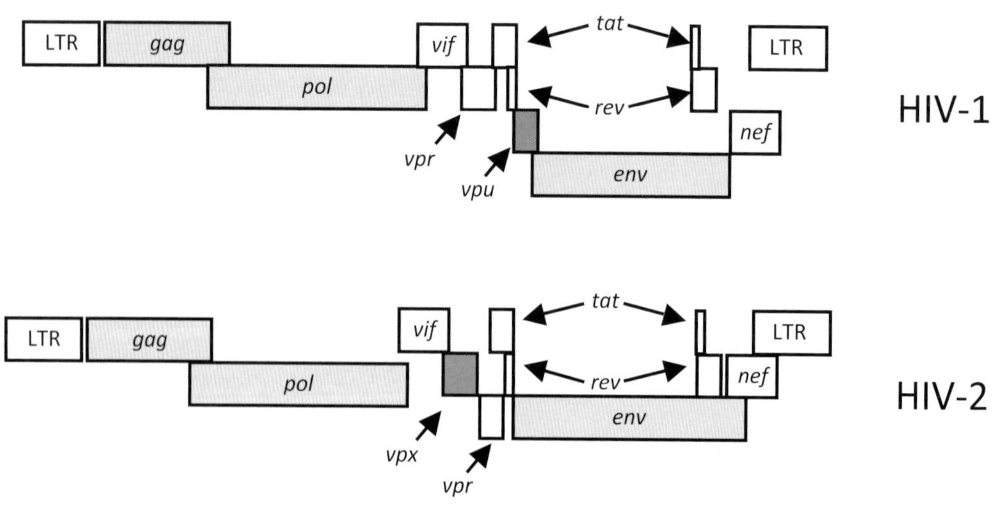


**Figure 5.3** Comparative genomic organization of HIV-1 and HIV-2 showing the location of accessory genes *vpu* and *vpx*.

be important for nuclear import and infection of primary macaque macrophages (Fletcher *et al.*, 1996; Sleigh *et al.*, 1998).

Vpx is packaged into viral particles through an interaction with the Gag p6 protein (Wu *et al.*, 1994) and thus is able to facilitate early viral replication events. Acquisition of Vpx may be a specific adaptation of HIV-2 to maximize infection of macrophages, and thus may be linked to the virus' adaptation to mucosal transmission and its ability to establish primary infection through these cells. It has recently been observed that HIV-2 and $SIV_{sm}$ Vpx proteins promote the accumulation of completed reverse transcribed genomes following infection of macrophages and DCs (Goujon *et al.*, 2007; Fujita *et al.*, 2008) overcoming an unidentified early post-entry restriction factor in these cells (Fig. 5.4). Damaged DNA binding protein 1 (DDB1) has been shown to be a cellular cofactor used in a process that overrides this restriction (Sharova *et al.*, 2008). DDB1 is a component of a ternary protein complex along with DET1 and DDB1-associated 1 (DDA1) which together with Vpr Binding Protein (VprBP, also termed DDB1 and CUL4-associated factor 1 (DCAF1)) form a putative substrate receptor for Cullin 4-based E3 ubiquitin ligase. The evidence suggests that VprBP is required to enable Vpx to overcome the block which occurs at the reverse transcription stage in macrophages and DCs, perhaps through inactivation of the cellular factor (Srivastava *et al.*, 2008). This may occur via a mechanism in which Vpx takes control of the Cullin 4 E3 ubiquitin ligase, binding to the DDB1-VprBP/DCAF1 module and recruiting the cellular restriction factor as a substrate for ubiquitination by the E3 ligase (Fig. 5.4). With the restriction factor polyubiquitinated and thus targeted for subsequent degradation, virus would be rescued from the early post-entry block to replication and efficient reverse transcription could proceed in these monocyte-derived cells. Manipulation of cellular degradation machinery by viral accessory proteins has previously been shown, with the targeting of signal transducer and activator of transcription (STAT) proteins by some RNA

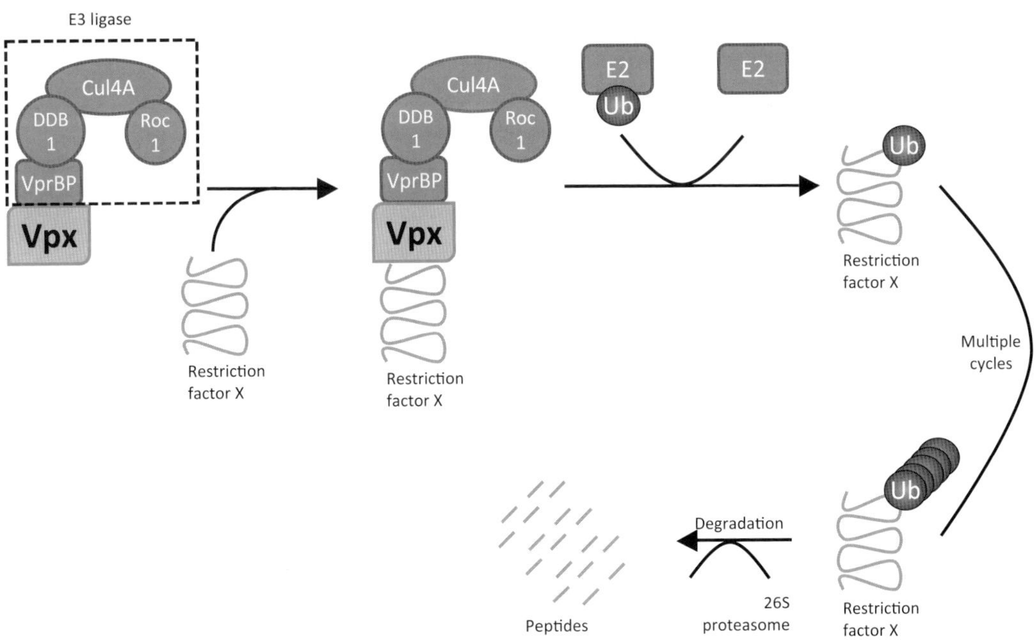


**Figure 5.4** Proposed model for interaction of Vpx with an E3 ubiquitin ligase and targeting of putative restriction factor for degradation. Vpx controls the Cullin 4 E3 ubiquitin ligase through binding to the DDB1-VprBP/DCAF1 module. The cellular restriction factor is then recruited as a substrate, is polyubiquitinated via E2 and E3 ubiquitin enzymes and targeted for subsequent degradation by the 26S proteasome.

virus members of the paramyxoviridae genus *Rubulavirus* (Ulane *et al.*, 2005; Barry and Fruh, 2006).

Thus the duplication and preservation of a separate gene (*vpx*) whose primary function is to rescue the virus from post-entry cellular restriction suggests that HIV-2 has specifically adapted to more efficient infection of macrophages.

## Towards commensalism: don't kill the host

A survival strategy for any infectious agent is not to kill its host and ultimately become a commensal organism. Chimpanzees and sooty mangabeys infected with their natural strains of SIV ($SIV_{cpz}$ and $SIV_{sm}$ respectively) do not develop AIDS (Kaur *et al.*, 1998; Rey-Cuille *et al.*, 1998; Silvestri *et al.*, 2003). Induction of symptomatic disease with these viruses is only associated with cross-species transmission from these SIV-infected populations to new, relatively unexposed naïve host species (for example the infection of $SIV_{mac}$ in rhesus macaques). Thus it may be that SIV in its native host has adapted over time not to kill the animal. Similarly, HIV-2 may have also evolved to become less pathogenic.

For HIV-1 too much lymphocyte activation can lead to immune exhaustion and ultimately to AIDS. Several studies suggest an association between high levels of immune activation and rapid disease progression/death in HIV-1 infected individuals (Giorgi *et al.*, 1993; Hazenberg *et al.*, 2000; Hazenberg *et al.*, 2003). In contrast to HIV-1, the global activation of lymphocytes does not appear to be a feature of HIV-2 infection. Although CD4-expressing lymphocytes are activated in both HIV-1 and -2 infections, $CD8^+$ cell activation is significantly lower in HIV-2 (Hanson *et al.*, 2005).

IFN-$\gamma$ is the hallmark cytokine of the Th1 immune response and is usually secreted by $CD4^+$ Th1 cells and $CD8^+$ CTLs. Of its many functions, it is able to increase the antigen presentation and lysosome activity of macrophages, ultimately determining the fate of infected cells through regulation of cell proliferation, differentiation and apoptosis causing generalized immune activation. In HIV-2 it is the $CD8^+$ rather than $CD4^+$ cells that mediate IFN-$\gamma$ stimulated cellular immune responses (Leligdowicz *et al.*, 2007). Moreover, in contrast to HIV-1, it appears that activated $CD4^+$ T-cells are associated with a reduction in proviral load in HIV-2 infection (Foxall *et al.*, 2008). The more specific immunity and avoidance of inappropriate activation may thereby contribute to the lower pathogenesis of HIV-2 infection.

In addition, HIV-2 patients differ from HIV-1 infected individuals in their ability to maintain the numbers of non-terminally differentiated HIV-specific $CD4^+$ T-cells which are able to proliferate and express multiple cytokines (such as IFN-$\gamma$ and IL-2 and to a lesser extent TNF-$\alpha$, MIP-1$\beta$ and CD107a) (Duvall *et al.*, 2006). A greater magnitude of HIV-2 specific $CD4^+$ T-cells responds to Gag peptides than is seen in HIV-1 infected individuals. It is possible that HIV-2, unlike HIV-1, does not preferentially infect or deplete HIV-specific $CD4^+$ T-cells, thus explaining why HIV-2 infected individuals have proviral loads equivalent to their HIV-1-infected counterparts, but much lower plasma virus (discussed below) (Berry *et al.*, 2002). Interestingly, polyfunctional IFN-$\gamma$- and IL-2-expressing HIV-specific $CD4^+$ T-cells have also been described in HIV-1 long-term non-progressors (LTNP) (Harari *et al.*, 2004) but were entirely absent in HIV-1 infected individuals with typical disease progression. Thus it is now well established that HIV-2 causes lower levels of immune activation and T-cell apoptosis and is usually less pathogenic than HIV-1 (Marlink *et al.*, 1994; Michel *et al.*, 2000; MacNeil *et al.*, 2007b).

Further evidence for a reduced level of immune activation in HIV-2 infection compared with HIV-1 is reported in a study showing that while antigen-specific $CD8^+$ T-cell responses can be mounted against all HIV-1 proteins (Addo *et al.*, 2003), immune responses to HIV-2 focus on a tiny proportion of Gag (4.5% of the total HIV-2 proteome), and in particular to a single peptide ($Gag_{298-315}$) which was recognized by over 30% of the cohort studied (Leligdowicz *et al.*, 2007). This peptide overlaps with the major homology region, located in an area equivalent to the HIV-1 capsid (CA) protein. This major homology region is highly conserved across all retroviruses and is essential for viral assembly, maturation and infectivity (Mammano *et al.*, 1994; Provitera *et al.*, 2001). Interestingly, the strongest IFN-$\gamma$ immune responses were most often from patients with undetectable plasma viral loads (<100 RNA copies/ml in this study), suggesting that conservation of the major homology region for its structural and functional properties may severely restrict the ability of HIV-2 to evade this CA-specific T and B-cell immune response, supporting a crucial role for the specific immune response in the control of HIV-2 replication.

The above observations indicate that a different balance between host and virus is reached in HIV-2 infection compared to HIV-1. This raises the possibility that the distinct focus of the host immune response mounted against HIV-2 infection, which seems aimed at avoiding global activation, may be at least partially responsible for the lower levels of viral pathogenesis in these individuals.

## Role of viral load in disease severity

Chimpanzees and sooty mangabeys infected with $SIV_{cpz}$ and $SIV_{sm}$ respectively do not develop AIDS. However unlike in HIV-2 infected individuals, these SIV infected animals carry high plasma viral burdens. This is surprising given the majority of studies on HIV-1 that conclude an individual's plasma load is the most important predictor of progression to AIDS (Popper *et al.*, 1999; Gottlieb *et al.*, 2002). It may be that in an effort to avoid killing its host, HIV-2 is far less pathogenic than HIV-1, but unlike SIV may have achieved this through lower plasma viral loads. While advantageous for persistence in a single host, this would seem to negatively impact its transmissibility, for which high levels of circulating virus are important.

HIV-1 and HIV-2 have a similar genomic organization and sequence homology of up to 60% (Guyader *et al.*, 1987). In the HIV-2 infected patients who eventually develop AIDS the clinical features are similar to HIV-1 infection. However HIV-2 infected individuals generally present with significantly lower levels of plasma RNA, regardless of similarities in age at infection and length of time infected (Popper *et al.*, 1999; Andersson *et al.*, 2000; Shanmugam *et al.*, 2000; Berry *et al.*, 2002). One study shows HIV-2 isolates are significantly less productive than HIV-1 with a reduced transmission capacity (Arien *et al.*, 2005). HIV-2 infection is associated with a reduced rate of progression to AIDS compared to HIV-1, with the majority of infected individuals remaining asymptomatic and dying of causes unrelated to immunodeficiency (Marlink *et al.*, 1994).

The course of HIV-1 infection is shown in Fig. 5.5. Approximately 3 weeks after initial infection, virus is detectable in the blood, often at high levels (>$10^7$ HIV-1 RNA copies/ml). After this point a partially successful immune response is mounted and is coincident with a 10–100 fold decrease in viral load. This relatively steady level of plasma virus is called the viral set-point and is typically reached within 6 months. Viral set-point is often used as a predictor of disease severity with high set-points correlated with higher infectivity, a poorer response to therapy and more rapid progression to AIDS (Mellors *et al.*, 1996, 1997; de Wolf *et al.*, 1997). The chronic phase of infection is characterized by a slow decline in $CD4^+$ T-cells for a time until the immune response can no longer control viral replication. This is marked by a rapid increase in plasma viral load and most individuals will progress to AIDS 8–10 years after initial infection. Whether HIV-2 reaches the same peak of replication in the acute phase of infection is unknown, but viral set-points for HIV-2 are much lower than for HIV-1, resulting in slower disease progression and thus a longer lifespan for the host. HIV-2 infected individuals observed for at least 8 years follow a clinical definition of LTNP. Long-term non-progression was initially described for individuals infected with HIV-1 for at least 7 years who maintain $CD4^+$ cell counts >500/μl and low viral loads, usually <$10^4$ RNA copies/ml in the absence of therapy (Pantaleo *et al.*, 1995; Cao *et al.*, 1996; Vesanen *et al.*, 1996). Therefore the stability of the HIV-2 plasma viral load and pattern of LTNP in these patients is consistent with the limited severity of disease in the majority of cases.

It is established for HIV-1 that the plasma viral load is the chief predictor for risk of infection following exposure, with no transmissions observed in a study of 51 patients with viral loads of <1500 HIV-1 copies/ml (Quinn *et al.*, 2000) and it is reasonable to assume this would be the case for HIV-2 also. As mentioned previously, long term persistence in its host has been achieved by HIV-2 through lower viral loads and lack of T-cell activation. However it would seem the price paid

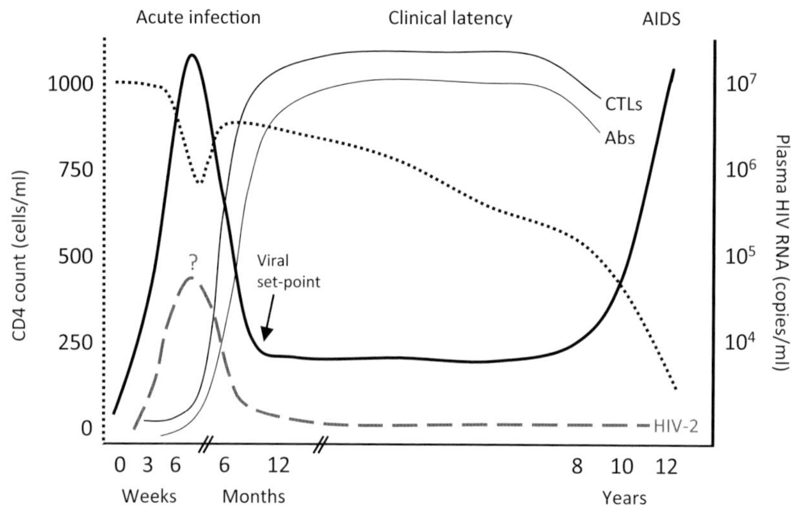


**Figure 5.5** Typical course of HIV-1 infection. Approximately 3 weeks after initial infection, virus is detectable in the blood, often to high levels (>$10^7$ HIV-1 RNA copies/ml). Antibody is detected soon after this (seroconversion). During acute infection a partially successful immune response is mounted and is coincident with a 10–100 fold decrease in viral load until set-point is reached. The period of clinical latency is generally asymptomatic and characterized by a slow decline in $CD4^+$ T-cells until the immune response can no longer control viral replication. This is marked by a rapid increase in plasma viral load and most individuals will progress to AIDS 8–10 years later. The extent of viral plasma RNA in acute HIV-2 infection is unknown, but viral set-points for HIV-2 are much lower than for HIV-1, resulting in slower disease progression and thus a longer lifespan for the host.

by HIV-2 for its lower pathogenicity is a diminished capacity for successful transmission. HIV-2 is transmitted by much the same routes as HIV-1, however for HIV-2 the frequency of transmission is reduced. For example, vertical transmissions of HIV-2 are rare (O'Donovan *et al.*, 2000), and heterosexual transmission rates are approximately 5–9 fold lower than for HIV-1 (Kanki *et al.*, 1994). Lower viral loads and reduced shedding of virus in semen in HIV-2 infected individuals are likely to contribute to its lower transmission rate (Gottlieb *et al.*, 2006).

## Nef and pathogenesis

The viral accessory protein negative regulatory factor (Nef) (encoded by *nef*), is present in all primate lentivirus genomes. In addition to its role in mediating the enhancement of viral entry into certain cell types, Nef has been characterized as a key component of pathogenesis, shown to enhance virion infectivity, stimulate viral replication and accelerate progression to AIDS. Previous studies have shown that many of the functions of Nef are conserved between HIV-1, -2 and $SIV_{mac}$ alleles; specifically, its ability to down-modulate the surface expression levels of CD4, CD28 and major histocompatibility complex (MHC) class I molecules (Schindler *et al.*, 2006; DeGottardi *et al.*, 2008). However, an interesting difference was observed between the various lentiviral lineages in the ability of Nef to down-modulate CD3, a key component of the T-cell receptor (TCR) complex. HIV-2 and most SIV Nefs down-modulate TCR-CD3 with high efficiency, while Nef proteins from HIV-1 were generally unable to affect surface TCR-CD3 expression levels. Down-modulation of TCR-CD3 acts to suppress the responsiveness of virally infected T-cells to stimulation and programmed cell death (Howe *et al.*, 1998). This difference between the actions of HIV-1 and -2 Nefs may be significant in explaining the lower pathogenicity of HIV-2.

Further data also shows that the frequency of defective or non-functional *nef* genes in proviral sequences from peripheral blood mononuclear cells (PBMCs) also seems to be higher in HIV-2 than HIV-1 infected individuals (Switzer *et al.*, 1998). It is reasonable to suggest that a lack of Nef function (i.e. no enhancement of viral infectivity) in these cases would be associated with lower viral loads and an attenuated clinical pattern of disease progression (Kestler *et al.*, 1991; Kirchhoff *et al.*, 1995).

In summary, the published evidence to date suggests that HIV-2 *nef* alleles are less effective than those of HIV-1 in enhancing viral infectivity and replication. However they are equally active in down-regulating MHC class I and even more potent in down-modulating the costimulatory factors of T-cell activation, CD28 and TCR-CD3 (Munch *et al.*, 2005; Munch *et al.*, 2007). So it would seem that overall, HIV-2 Nefs in contrast to those of HIV-1 are highly effective at functions involved in evading the host immune response (MHC down-regulation) and suppressing T-cell activation (decreased TCR-CD3 and CD28 levels) but less effective at those directly affecting the replicative capacity of the virion. The combination of these factors is likely to contribute to the lower observed virulence of HIV-2 compared with HIV-1.

## A role for latency in the survival of HIV-2

In order for a virus to be propagated within a population, it needs to persist in the host. Ideally, this would be manifested in a situation of high viral load to facilitate transmission and low pathogenicity to avoid killing the host, thereby ensuring persistence. While natural SIV infection seems to have attained this balance, HIV-1 is highly pathogenic and the viral loads of HIV-2 are usually too low to support infection of new hosts.

Latency is a form of infection utilized by some viruses to achieve persistence, characterized by periodic bursts of viral replication in between phases of dormancy. True latency is observed in herpes simplex virus type 1 (HSV-1) infection. HSV-1 establishes latent infection in sensory neurons as a circular episome associated with histones during which time the virus persists until reactivation, resulting in recurrent productive infection and transmission to new hosts (reviewed in Knipe and Cliffe, 200). This is distinct from 'clinical latency', referring to the period following acute HIV infection when an individual experiences the absence of symptoms but viral replication has never completely ceased, and 'proviral latency', the means by which HIV-1 and -2 achieve persistence by integrating their genome into the host chromosome. There is no direct evidence for complete transcriptional latency in HIV infection with basal levels of replication occurring throughout.

*In vitro* studies imply that there is no difference between the cytopathogenicity of HIV-1 and HIV-2 (Schramm *et al.*, 2000), but despite lower plasma HIV-2 loads the levels of proviral DNA in PBMCs are similar between patients infected with either HIV-1 or HIV-2 (Berry *et al.*, 1994; Sarr *et al.*, 1999). This suggests that HIV-2 is either latent or replicating in a different compartment to HIV-1 without contributing to the plasma viral load. Latency is consistent with findings from studies reporting RNA-to-DNA ratios from paired measurements of plasma viral RNA and proviral DNA. In HIV-1 infection this ratio is above 10.0 in the majority of individuals (Bagnarelli *et al.*, 1992; Verhofstede *et al.*, 1994), while quiescent infection (RNA to DNA ratio of 0.86) was observed in almost half the HIV-2 infected individuals studied (Popper *et al.*, 2000). A recent study further supports a latent stage for HIV-2. Various forms of HIV-2 replication intermediates (integrated proviral DNA and viral mRNA) were measured and show a lower replication rate for HIV-2 indicated by lower amounts of mRNA in HIV-2 infection despite similar levels of viral DNA intermediates (MacNeil *et al.*, 2007b).

The HIV long terminal repeat (LTR) functions like a eukaryotic transcriptional unit and controls proviral DNA expression through the binding of cellular activators such as NF-κB. Latency may therefore be due to differences in the cellular levels of such activators. The HIV-2 LTR is less responsive to tumour necrosis factor-alpha (TNF-α) (which induces transcription via the NF-κB pathway) than HIV-1 (Hannibal *et al.*, 1993). Additionally, it has been shown *in vitro* that recombinant HIV-2 Env protein is capable of stimulating higher levels of cytokines that inhibit viral replication, IFN-γ and interleukin (IL)-16, and lower levels of IL-4, which stimulates viral production (Sekigawa *et al.*, 1998). Transcriptional latency can also result following the integration of viral genome during productive infection (Adams *et al.*, 1994), with the location of the integrated provirus being a factor that influences basal transcription of HIV-1 (Jordan *et al.*, 2001). Proviral integration site selection could be involved in latency of HIV-2, with a higher proportion of genomes integrated in heterochromatin (associated with a block in viral transcription) in HIV-2 infected individuals compared with HIV-1 (MacNeil *et al.*, 2006).

Further suggestions of latency come from studies of sequence evolution of HIV-1 and -2. In HIV-1 infection intrapatient viral evolution has also been linked to disease progression (Mikhail *et al.*, 2005) and higher viral set-points (Mani *et al.*, 2002). If HIV-2 does in fact replicate at lower rates than HIV-1, one would expect a slower rate of sequence divergence and diversification in HIV-2 infected individuals. In a longitudinal study over approximately 10 years of intrapatient evolution of the *env* C2V3C3 region in HIV-2 infected individuals, there was no separation between sequences from early and late time points. This supports the premise that the turnover of viral populations occurs extremely slowly during asymptomatic HIV-2 infection (MacNeil *et al.*, 2007a). However, another study of HIV-2 patients who had progressed to AIDS showed a high turnover of HIV-2, indicating that rates of evolution may vary during different stages of disease (Brandin *et al.*, 2005). Interestingly, low sequence diversity of the *env* gene was reported in a recent study of HIV-1 LTNPs (Bailey *et al.*, 2006).

Nonetheless, the lack of phylogenetic separation and clustering of sequences from early and late time points in non-progressive HIV-2 infection in this and other studies supports the idea that the high proportion of HIV-2 proviral genomes observed in HIV-2 infection are present in a latent form within a long-lived cellular compartment *in vivo* (Esteves *et al.*, 2001). Macrophages, with their longevity and usually quiescent activation state as well as potential habitation of immune privileged sites (such as the resident microglial cells in the brain), represent a strong candidate for such a compartment (Marchant *et al.*, 2006).

*In vitro* studies support a latent state of HIV-2 in macrophages (Marchant *et al.*, 2006). This study found that stimulation of the NF-κB transcription factor following lipopolysaccharide (LPS) addition to infected monocyte-derived macrophage (MDM) cultures resulted in restimulation of HIV-2 replication. These findings suggest that MDMs are able to accommodate all steps of the HIV-2 viral life cycle as efficiently as HIV-1. Unlike HIV-1, HIV-2 has an initial burst of viral production followed by an apparent latent phase. An additional explanation for the higher than expected proviral loads may be a result of the preservation of $CD4^+$ T-cell (and thus infected cell) numbers in HIV-2 infection. This is in keeping with the proposal that HIV-2 is evolving not to kill its host.

## CD4 independence and tropism

In approximately 50% of HIV-1-infected individuals, progression to AIDS is associated with a broadening of coreceptor usage of the patients' virus from CCR5 to dual-tropic to CXCR4 only (Koot *et al.*, 1993; Richman and Bozzette, 1994). The general absence of symptomatic HIV-2 infections would predict a lack of coreceptor switch in HIV-2 patients. The development of coreceptor switch for HIV-2 is not known as sequential studies have not been done. However cross sectional studies initially suggested that counter intuitively, overall HIV-2 is capable of using a larger number of different coreceptors *in vitro* than HIV-1. Studies with lab adapted isolates show surprisingly broad coreceptor specificity including CXCR4, CCR5, CCR3, CCR8, CCR2b, GPR15 (BOB), CXCR6 (BONZO) and D6, all of which can be expressed on T-cells or macrophages (McKnight *et al.*, 1998; Clapham and McKnight, 2002; Neil *et al.*, 2005). A study of primary isolates (viruses not passaged in T-cell lines) showed them to be more restricted, using CCR5, BOB and BONZO only (Blaak *et al.*, 2005). Nevertheless although HIV-2 also has the potential for a broader cellular tropism based on its coreceptor use *in vitro* (Clapham *et al.*, 1991; McKnight *et al.*, 1994), this may not translate to *in vivo* tropism. This possibility is supported by *in situ* studies of HIV-2 in brain tissue where there was no evidence of infection in cells other than $CD4^+$ microglia (Morner *et al.*, 2002). The more promiscuous pattern of coreceptor usage and tropism of HIV-2 would seem to conflict with the lower pathogenicity of HIV-2, as broadening coreceptor tropism correlates with disease progression in HIV-1-infected individuals.

One striking difference between HIV-1 and HIV-2/SIV is the ability of some primary HIV-2 isolates to infect cells using a coreceptor in the absence of CD4 (Clapham *et al.*, 1992; Reeves and Schulz, 1997; Reeves *et al.*, 1999). In CD4-mediated infection, binding of viral envelope to CD4 induces conformational changes in the viral envelope that enable it to interact with the chemokine coreceptor (CXCR4 or CCR5). This eventually results in fusion of the viral and plasma membranes. In the case of CD4-independent infection presumably the coreceptor-binding site is partially exposed, enabling the viral Env to make direct contact with a coreceptor. Thus the conformational changes normally induced following interaction of HIV Env and CD4 are made redundant (Bhattacharya *et al.*, 2003). An analysis of Env determinants responsible for the CD4-independence of HIV-2 ROD-B, a variant of the CD4-dependent strain ROD-A, shows that only a limited number of amino acid mutations are required to confer efficient CD4-independent infection (Reeves *et al.*, 1997). Interestingly, CD4-independent HIV-2 strains that can normally use either CCR5 or CXCR4 (dual-tropic) to infect $CD4^+$ cells are only able to use one of these receptors in the absence of CD4, suggesting that infection via CD4 can occur through a wider range of seven-transmembrane (7TM) coreceptors (Reeves *et al.*, 1999). While many of the described primary HIV-2 isolates are CD4-independent *in vitro*, their infectivity is increased when CD4 is also expressed. Indeed, it seems that these isolates are unable to bypass the use of CD4 *in vivo*, according to the investigation of HIV-2 infection of brain astrocytes and microglia (Morner *et al.*, 2003). It may be that HIV-2 (and SIV) has acquired the ability to infect cells expressing low levels of CD4, thus reflecting an enhanced tropism for macrophages (Kazazi *et al.*, 1989). Indeed CD4-independence of SIV isolates has been shown to correlate with macrophage tropism (Puffer *et al.*, 2002) and it is reasonable to assume this is also the case for HIV-2.

Interestingly, CD4-independent variants are more susceptible to neutralizing antibodies than CD4-dependent viruses, with both significant autologous and heterologous neutralization observed across a panel of HIV-2 serum/viruses (Thomas *et al.*, 2003; Shi *et al.*, 2005). These results indicate that, unlike for HIV-1, neutralization escape is rare in HIV-2 infection. This suggests better immune control of HIV-2, which results in lower plasma virus levels than HIV-1 and the restricted pathogenicity and disease progression observed in the majority of patients. This may be because the coreceptor-binding site is already exposed on HIV-2 Env, presenting an ideal target for neutralizing antibodies. Thus, HIV may have evolved to use CD4 in part as a protective mechanism to shield conserved and essential Env epitopes from neutralizing antibodies.

## Summary

Given the similarities in genomic organization and sequence between HIV-1 and -2, HIV-2 provides a striking contrast to HIV-1 in most facets of transmission, pathogenicity and disease progression. In contrast to the epidemic nature of HIV-1 infections, HIV-2 is restricted in its worldwide distribution, with the lower viral loads established in asymptomatic infection a significant cause of its diminished transmission efficiency. HIV-2 is also much less pathogenic than HIV-1 and results in a reduced rate of progression to AIDS, with the majority of patients remaining asymptomatic and dying of causes unrelated to immunodeficiency.

However, the minority of individuals who eventually progress to AIDS are clinically indistinguishable from those infected with HIV-1 (Ariyoshi *et al.*, 2000), thus suggesting that HIV-2 is not just an attenuated virus and there may be different pathogenic forms. In fact, a recent publication reports the isolation of a newly emerging pathogenic strain of HIV-2 (Group F) from an individual from Sierra Leone with low CD4 counts (Smith *et al.*, 2008), indicating that virus groups previously thought to be non-epidemic are circulating, at least in West Africa, and capable of leading to immunosuppression.

For long term survival in a population, an intimate balance between plasma viral load, transmission and pathogenesis needs to evolve and ultimately be maintained. It could be that HIV-2 is still evolving in the human population and having achieved a low pathogenicity, it will adapt over time to being more successful at transmission, perhaps with high plasma viral loads.


### Acknowledgements

The authors thank Matthias Dittmar and Nidia Oliveira for critical reading of the manuscript and Andy Stagg for helpful discussions.

## References


Adams, M., Sharmeen, L., Kimpton, J., Romeo, J.M., Garcia, J.V., Peterlin, B.M., Groudine, M., and Emerman, M. (1994). Cellular latency in human immunodeficiency virus-infected individuals with high CD4 levels can be detected by the presence of promoter-proximal transcripts. Proc. Natl. Acad. Sci. U.S.A. *91*, 3862–3866.

Addo, M.M., Yu, X.G., Rathod, A., Cohen, D., Eldridge, R.L., Strick, D., Johnston, M.N., Corcoran, C., Wurcel, A.G., Fitzpatrick, C.A., *et al.* (2003). Comprehensive epitope analysis of human immunodeficiency virus type 1 (HIV-1)-specific T-cell responses directed against the entire expressed HIV-1 genome demonstrate broadly directed responses, but no correlation to viral load. J. Virol. *77*, 2081–2092.

Andersen, J.L., Le Rouzic, E., and Planelles, V. (2008). HIV-1 Vpr: mechanisms of G2 arrest and apoptosis. Exp. Mol. Pathol. *85*, 2–10.

Andersson, S., Norrgren, H., da Silva, Z., Biague, A., Bamba, S., Kwok, S., Christopherson, C., Biberfeld, G., and Albert, J. (2000). Plasma viral load in HIV-1 and HIV-2 singly and dually infected individuals in Guinea-Bissau, West Africa: significantly lower plasma virus set point in HIV-2 infection than in HIV-1 infection. Arch. Intern. Med. *160*, 3286–3293.

Arien, K.K., Abraha, A., Quinones-Mateu, M.E., Kestens, L., Vanham, G., and Arts, E.J. (2005). The replicative fitness of primary human immunodeficiency virus type 1 (HIV-1) group M, HIV-1 group O, and HIV-2 isolates. J. Virol. *79*, 8979–8990.

Ariyoshi, K., Jaffar, S., Alabi, A.S., Berry, N., Schim van der Loeff, M., Sabally, S., N'Gom, P.T., Corrah, T., Tedder, R., and Whittle, H. (2000). Plasma RNA viral load predicts the rate of CD4 T-cell decline and death in HIV-2-infected patients in West Africa. AIDS *14*, 339–344.

Bagnarelli, P., Menzo, S., Valenza, A., Manzin, A., Giacca, M., Ancarani, F., Scalise, G., Varaldo, P.E., and Clementi, M. (1992). Molecular profile of human immunodeficiency virus type 1 infection in symptomless patients and in patients with AIDS. J. Virol. *66*, 7328–7335.

Bailey, J.R., Lassen, K.G., Yang, H.C., Quinn, T.C., Ray, S.C., Blankson, J.N., and Siliciano, R.F. (2006). Neutralizing antibodies do not mediate suppression of human immunodeficiency virus type 1 in elite suppressors or selection of plasma virus variants in patients on highly active antiretroviral therapy. J. Virol. *80*, 4758–4770.

Barry, M., and Fruh, K. (2006). Viral modulators of cullin RING ubiquitin ligases: culling the host defense. Sci STKE *2006*, pe21.

Berry, N., Ariyoshi, K., Jobe, O., Ngum, P.T., Corrah, T., Wilkins, A., Whittle, H., and Tedder, R. (1994). HIV type 2 proviral load measured by quantitative polymerase chain reaction correlates with $CD4^+$ lymphopenia in HIV type 2-infected individuals. AIDS Res. Hum. Retroviruses *10*, 1031–1037.

Berry, N., Jaffar, S., Schim van der Loeff, M., Ariyoshi, K., Harding, E., N'Gom, P.T., Dias, F., Wilkins, A., Ricard, D., Aaby, P., *et al.* (2002). Low level viremia and high CD4% predict normal survival in a cohort of HIV type-2-infected villagers. AIDS Res. Hum. Retroviruses *18*, 1167–1173.

Bhattacharya, J., Peters, P.J., and Clapham, P.R. (2003). CD4-independent infection of HIV and SIV: implications for envelope conformation and cell tropism *in vivo*. AIDS *17 Suppl. 4*, S35–43.

Bibollet-Ruche, F., Bailes, E., Gao, F., Pourrut, X., Barlow, K.L., Clewley, J.P., Mwenda, J.M., Langat, D.K., Chege, G.K., McClure, H.M., *et al.* (2004). New simian immunodeficiency virus infecting De Brazza's monkeys (Cercopithecus neglectus): evidence for a cercopithecus monkey virus clade. J. Virol. *78*, 7748–7762.

Blaak, H., Boers, P.H., Gruters, R.A., Schuitemaker, H., van der Ende, M.E., and Osterhaus, A.D. (2005). CCR5, GPR*15*, and CXCR6 are major coreceptors of human immunodeficiency virus type 2 variants isolated from individuals with and without plasma viremia. J. Virol. *79*, 1686–1700.

Bobardt, M.D., Saphire, A.C., Hung, H.C., Yu, X., Van der Schueren, B., Zhang, Z., David, G., and Gallay, P.A. (2003). Syndecan captures, protects, and transmits HIV to T-lymphocytes. Immunity *18*, 27–39.

Bour, S., Schubert, U., Peden, K., and Strebel, K. (1996). The envelope glycoprotein of human immunodeficiency virus type 2 enhances viral particle release: a Vpu-like factor? J. Virol. *70*, 820–829.

Bour, S., and Strebel, K. (1996). The human immunodeficiency virus (HIV) type 2 envelope protein is a functional complement to HIV type 1 Vpu that enhances particle release of heterologous retroviruses. J. Virol. *70*, 8285–8300.

Brandin, E., Brostrom, C., Gille, E., Bonhoeffer, S., and Albert, J. (2005). Short communication: HIV type 2 dynamics. AIDS Res. Hum. Retroviruses *21*, 608–610.

Cao, Y., Qin, L., Zhang, L., Safrit, J., and Ho, D.D. (1996). Characterization of long-term survivors of human immunodeficiency virus type 1 infection. Immunol. Lett. *51*, 7–13.

Clapham, P.R., Blanc, D., and Weiss, R.A. (1991). Specific cell surface requirements for the infection of CD4-positive cells by human immunodeficiency virus types 1 and 2 and by Simian immunodeficiency virus. Virology *181*, 703–715.

Clapham, P.R., McKnight, A., and Weiss, R.A. (1992). Human immunodeficiency virus type 2 infection and fusion of CD4-negative human cell lines: induction and enhancement by soluble CD4. J. Virol. *66*, 3531–3537.

Clapham, P.R., and McKnight, A. (2002). Cell surface receptors, virus entry and tropism of primate lentiviruses. J. Gen. Virol. *83*, 1809–1829.

Damond, F., Worobey, M., Campa, P., Farfara, I., Colin, G., Matheron, S., Brun-Vezinet, F., Robertson, D.L., and Simon, F. (2004). Identification of a highly divergent HIV type 2 and proposal for a change in HIV type 2 classification. AIDS Res. Hum. Retroviruses *20*, 666–672.

de Wolf, F., Spijkerman, I., Schellekens, P.T., Langendam, M., Kuiken, C., Bakker, M., Roos, M., Coutinho, R., Miedema, F., and Goudsmit, J. (1997). AIDS prognosis based on HIV-1 RNA, $CD4^+$ T-cell count and function: markers with reciprocal predictive value over time after seroconversion. AIDS *11*, 1799–1806.

DeGottardi, M.Q., Specht, A., Metcalf, B., Kaur, A., Kirchhoff, F., and Evans, D.T. (2008). Selective downregulation of rhesus macaque and sooty mangabey major histocompatibility complex class I molecules by Nef alleles of simian immunodeficiency virus and human immunodeficiency virus type 2. J. Virol. *82*, 3139–3146.

Duvall, M.G., Jaye, A., Dong, T., Brenchley, J.M., Alabi, A.S., Jeffries, D.J., van der Sande, M., Togun, T.O., McConkey, S.J., Douek, D.C., *et al.* (2006). Maintenance of HIV-specific CD4$^+$ T-cell help distinguishes HIV-2 from HIV-1 infection. J. Immunol. *176*, 6973–6981.

Eckstein, D.A., Sherman, M.P., Penn, M.L., Chin, P.S., De Noronha, C.M., Greene, W.C., and Goldsmith, M.A. (2001). HIV-1 Vpr enhances viral burden by facilitating infection of tissue macrophages but not nondividing CD4$^+$ T-cells. J. Exp. Med. *194*, 1407–1419.

Esteves, A., Piedade, J., Santos, C., Venenno, T., Canas-Ferreira, W.F., and Parreira, R. (2001). Follow-up study of intrahost HIV type 2 variability reveals discontinuous evolution of C2V3 sequences. AIDS Res. Hum. Retroviruses *17*, 253–256.

Fletcher, T.M., 3rd, Brichacek, B., Sharova, N., Newman, M.A., Stivahtis, G., Sharp, P.M., Emerman, M., Hahn, B.H., and Stevenson, M. (1996). Nuclear import and cell cycle arrest functions of the HIV-1 Vpr protein are encoded by two separate genes in HIV-2/SIV(SM). EMBO J. *15*, 6155–6165.

Foxall, R.B., Cortesao, C.S., Albuquerque, A.S., Soares, R.S., Victorino, R.M., and Sousa, A.E. (2008). Gag-specific CD4$^+$ T-cell frequency is inversely correlated with proviral load and directly correlated with immune activation in infection with human immunodeficiency virus type 2 (HIV-2) but not HIV-1. J. Virol. *82*, 9795–9799.

Fujita, M., Otsuka, M., Miyoshi, M., Khamsri, B., Nomaguchi, M., and Adachi, A. (2008). Vpx is critical for reverse transcription of the human immunodeficiency virus type 2 genome in macrophages. J. Virol. *82*, 7752–7756.

Gao, F., Yue, L., Robertson, D.L., Hill, S.C., Hui, H., Biggar, R.J., Neequaye, A.E., Whelan, T.M., Ho, D.D., Shaw, G.M., and *et al.* (1994). Genetic diversity of human immunodeficiency virus type 2: evidence for distinct sequence subtypes with differences in virus biology. J. Virol. *68*, 7433–7447.

Giorgi, J.V., Liu, Z., Hultin, L.E., Cumberland, W.G., Hennessey, K., and Detels, R. (1993). Elevated levels of CD38+ CD8+ T-cells in HIV infection add to the prognostic value of low CD4$^+$ T-cell levels: results of 6 years of follow-up. The Los Angeles Center, Multicenter AIDS Cohort Study. J. Acquir. Immune Defic. Syndr. *6*, 904–912.

Goila-Gaur, R., and Strebel, K. (2008). HIV-1 Vif, APOBEC, and intrinsic immunity. Retrovirology *5*, 51.

Gottlieb, G.S., Sow, P.S., Hawes, S.E., Ndoye, I., Redman, M., Coll-Seck, A.M., Faye-Niang, M.A., Diop, A., Kuypers, J.M., Critchlow, C.W., *et al.* (2002). Equal plasma viral loads predict a similar rate of CD4$^+$ T-cell decline in human immunodeficiency virus (HIV) type 1- and HIV-2-infected individuals from Senegal, West Africa. J. Infect. Dis. *185*, 905–914.

Gottlieb, G.S., Hawes, S.E., Agne, H.D., Stern, J.E., Critchlow, C.W., Kiviat, N.B., and Sow, P.S. (2006). Lower levels of HIV RNA in semen in HIV-2 compared with HIV-1 infection: implications for differences in transmission. AIDS *20*, 895–900.

Goujon, C., Riviere, L., Jarrosson-Wuilleme, L., Bernaud, J., Rigal, D., Darlix, J.L., and Cimarelli, A. (2007). SIVSM/HIV-2 Vpx proteins promote retroviral escape from a proteasome-dependent restriction pathway present in human dendritic cells. Retrovirology *4*, 2.

Guyader, M., Emerman, M., Sonigo, P., Clavel, F., Montagnier, L., and Alizon, M. (1987). Genome organization and transactivation of the human immunodeficiency virus type 2. Nature *326*, 662–669.

Hannibal, M.C., Markovitz, D.M., Clark, N., and Nabel, G.J. (1993). Differential activation of human immunodeficiency virus type 1 and 2 transcription by specific T-cell activation signals. J. Virol. *67*, 5035–5040.

Hanson, A., Sarr, A.D., Shea, A., Jones, N., Mboup, S., Kanki, P., and Cao, H. (2005). Distinct profile of T-cell activation in HIV type 2 compared to HIV type 1 infection: differential mechanism for immunoprotection. AIDS Res. Hum. Retroviruses *21*, 791–798.

Harari, A., Petitpierre, S., Vallelian, F., and Pantaleo, G. (2004). Skewed representation of functionally distinct populations of virus-specific CD4 T-cells in HIV-1-infected subjects with progressive disease: changes after antiretroviral therapy. Blood *103*, 966–972.

Hazenberg, M.D., Otto, S.A., Cohen Stuart, J.W., Verschuren, M.C., Borleffs, J.C., Boucher, C.A., Coutinho, R.A., Lange, J.M., Rinke de Wit, T.F., Tsegaye, A., *et al.* (2000). Increased cell division but not thymic dysfunction rapidly affects the T-cell receptor excision circle content of the naive T-cell population in HIV-1 infection. Nat. Med. *6*, 1036–1042.

Hazenberg, M.D., Otto, S.A., van Benthem, B.H., Roos, M.T., Coutinho, R.A., Lange, J.M., Hamann, D., Prins, M., and Miedema, F. (2003). Persistent immune activation in HIV-1 infection is associated with progression to AIDS. AIDS *17*, 1881–1888.

Hirsch, V.M., Olmsted, R.A., Murphey-Corb, M., Purcell, R.H., and Johnson, P.R. (1989). An African primate lentivirus (SIVsm) closely related to HIV-2. Nature *339*, 389–392.

Hladik, F., and McElrath, M.J. (2008). Setting the stage: host invasion by HIV. Nat. Rev. Immunol. *8*, 447–457.

Howe, A.Y., Jung, J.U., and Desrosiers, R.C. (1998). Zeta chain of the T-cell receptor interacts with nef of simian immunodeficiency virus and human immunodeficiency virus type 2. J. Virol. *72*, 9827–9834.

Jordan, A., Defechereux, P., and Verdin, E. (2001). The site of HIV-1 integration in the human genome determines basal transcriptional activity and response to Tat transactivation. EMBO J. *20*, 1726–1738.

Kanki, P.J., Travers, K.U., Mboup, S., Hsieh, C.C., Marlink, R.G., Gueye, N.A., Siby, T., Thior, I., Hernandez-Avila, M., Sankale, J.L., and *et al.* (1994). Slower heterosexual spread of HIV-2 than HIV-1. Lancet *343*, 943–946.

Kaur, A., Grant, R.M., Means, R.E., McClure, H., Feinberg, M., and Johnson, R.P. (1998). Diverse host responses and outcomes following simian immunodeficiency virus SIVmac239 infection in sooty mangabeys and rhesus macaques. J. Virol. *72*, 9597–9611.

Kazazi, F., Mathijs, J.M., Foley, P., and Cunningham, A.L. (1989). Variations in CD4 expression by human monocytes and macrophages and their relationships to infection with the human immunodeficiency virus. J. Gen. Virol. *70 (Pt 10)*, 2661–2672.

Keele, B.F., Van Heuverswyn, F., Li, Y., Bailes, E., Takehisa, J., Santiago, M.L., Bibollet-Ruche, F., Chen, Y., Wain, L.V., Liegeois, F., *et al.* (2006). Chimpanzee reservoirs of pandemic and nonpandemic HIV-1. Science *313*, 523–526.

Kestler, H.W., 3rd, Ringler, D.J., Mori, K., Panicali, D.L., Sehgal, P.K., Daniel, M.D., and Desrosiers, R.C. (1991). Importance of the nef gene for maintenance of high virus loads and for development of AIDS. Cell *65*, 651–662.

Kirchhoff, F., Greenough, T.C., Brettler, D.B., Sullivan, J.L., and Desrosiers, R.C. (1995). Brief report: absence of intact nef sequences in a long-term survivor with nonprogressive HIV-1 infection. N. Engl. J. Med. *332*, 228–232.

Knipe, D.M., and Cliffe, A. (2008). Chromatin control of herpes simplex virus lytic and latent infection. Nat. Rev. Microbiol. *6*, 211–221.

Koot, M., Keet, I.P., Vos, A.H., de Goede, R.E., Roos, M.T., Coutinho, R.A., Miedema, F., Schellekens, P.T., and Tersmette, M. (1993). Prognostic value of HIV-1 syncytium-inducing phenotype for rate of $CD4^+$ cell depletion and progression to AIDS. Ann. Intern. Med. *118*, 681–688.

Lee, S.H., Starkey, P.M., and Gordon, S. (1985). Quantitative analysis of total macrophage content in adult mouse tissues. Immunochemical studies with monoclonal antibody F4/80. J. Exp. Med. *161*, 475–489.

Leligdowicz, A., Yindom, L.M., Onyango, C., Sarge-Njie, R., Alabi, A., Cotten, M., Vincent, T., da Costa, C., Aaby, P., Jaye, A., *et al.* (2007). Robust Gag-specific T-cell responses characterize viremia control in HIV-2 infection. J. Clin. Invest. *117*, 3067–3074.

Lieber, A., Kay, M.A., and Li, Z.Y. (2000). Nuclear import of moloney murine leukemia virus DNA mediated by adenovirus preterminal protein is not sufficient for efficient retroviral transduction in nondividing cells. J. Virol. *74*, 721–734.

MacNeil, A., Sankale, J.L., Meloni, S.T., Sarr, A.D., Mboup, S., and Kanki, P. (2006). Genomic sites of human immunodeficiency virus type 2 (HIV-2) integration: similarities to HIV-1 *in vitro* and possible differences *in vivo*. J. Virol. *80*, 7316–7321.

MacNeil, A., Sankale, J.L., Meloni, S.T., Sarr, A.D., Mboup, S., and Kanki, P. (2007a). Long-term intrapatient viral evolution during HIV-2 infection. J. Infect. Dis. *195*, 726–733.

MacNeil, A., Sarr, A.D., Sankale, J.L., Meloni, S.T., Mboup, S., and Kanki, P. (2007b). Direct evidence of lower viral replication rates *in vivo* in human immunodeficiency virus type 2 (HIV-2) infection than in HIV-1 infection. J. Virol. *81*, 5325–5330.

Maher, D., Wu, X., Schacker, T., Horbul, J., and Southern, P. (2005). HIV binding, penetration, and primary infection in human cervicovaginal tissue. Proc. Natl. Acad. Sci. U.S.A. *102*, 11504–11509.

Mammano, F., Ohagen, A., Hoglund, S., and Gottlinger, H.G. (1994). Role of the major homology region of human immunodeficiency virus type 1 in virion morphogenesis. J. Virol. *68*, 4927–4936.

Mani, I., Gilbert, P., Sankale, J.L., Eisen, G., Mboup, S., and Kanki, P.J. (2002). Intrapatient diversity and its correlation with viral setpoint in human immunodeficiency virus type 1 CRF02_A/G-IbNG infection. J. Virol. *76*, 10745–10755.

Marchant, D., Neil, S.J., and McKnight, A. (2006). Human immunodeficiency virus types 1 and 2 have different replication kinetics in human primary macrophage culture. J. Gen. Virol. *87*, 411–418.

Marlink, R., Kanki, P., Thior, I., Travers, K., Eisen, G., Siby, T., Traore, I., Hsieh, C.C., Dia, M.C., Gueye, E.H., and *et al.* (1994). Reduced rate of disease development after HIV-2 infection as compared to HIV-1. Science *265*, 1587–1590.

McKnight, A., Clapham, P.R., and Weiss, R.A. (1994). HIV-2 and SIV infection of nonprimate cell lines expressing human CD4: restrictions to replication at distinct stages. Virology *201*, 8–18.

McKnight, A., Dittmar, M.T., Moniz-Periera, J., Ariyoshi, K., Reeves, J.D., Hibbitts, S., Whitby, D., Aarons, E., Proudfoot, A.E., Whittle, H., and Clapham, P.R. (1998). A broad range of chemokine receptors are used by primary isolates of human immunodeficiency virus type 2 as coreceptors with CD4. J. Virol. *72*, 4065–4071.

Mellors, J.W., Rinaldo, C.R., Jr, Gupta, P., White, R.M., Todd, J.A., and Kingsley, L.A. (1996). Prognosis in HIV-1 infection predicted by the quantity of virus in plasma. Science *272*, 1167–1170.

Mellors, J.W., Munoz, A., Giorgi, J.V., Margolick, J.B., Tassoni, C.J., Gupta, P., Kingsley, L.A., Todd, J.A., Saah, A.J., Detels, R., *et al.* (1997). Plasma viral load and $CD4^+$ lymphocytes as prognostic markers of HIV-1 infection. Ann. Intern. Med. *126*, 946–954.

Michel, P., Balde, A.T., Roussilhon, C., Aribot, G., Sarthou, J.L., and Gougeon, M.L. (2000). Reduced immune activation and T-cell apoptosis in human immunodeficiency virus type 2 compared with type 1: correlation of T-cell apoptosis with beta2 microglobulin concentration and disease evolution. J. Infect. Dis. *181*, 64–75.

Mikhail, M., Wang, B., Lemey, P., Beckthold, B., Vandamme, A.M., Gill, M.J., and Saksena, N.K. (2005). Role of viral evolutionary rate in HIV-1 disease progression in a linked cohort. Retrovirology *2*, 41.

Morner, A., Bjorndal, A., Leandersson, A.C., Albert, J., Bjorling, E., and Jansson, M. (2002). CCR5 or CXCR4 is required for efficient infection of peripheral blood mononuclear cells by promiscuous human immunodeficiency virus type 2 primary isolates. AIDS Res. Hum. Retroviruses *18*, 193–200.

Morner, A., Thomas, J.A., Björling, E., Munson, P.J., Lucas, S.B., and McKnight, A. (2003). Productive HIV-2 infection in the brain is restricted to macrophages/microglia. AIDS *17*, 1451–1455.

Munch, J., Schindler, M., Wildum, S., Rücker, E., Bailer, N., Knoop, V., Novembre, F.J., and Kirchhoff, F. (2005). Primary sooty mangabey simian immunodeficiency virus and human immunodeficiency virus type 2 nef alleles modulate cell surface expression of various human receptors and enhance viral infectivity and replication. J. Virol. *79*, 10547–10560.

Munch, J., Rajan, D., Schindler, M., Specht, A., Rücker, E., Novembre, F.J., Nerrienet, E., Muller-Trutwin, M.C., Peeters, M., Hahn, B.H., and Kirchhoff, F. (2007). Nef-mediated enhancement of virion infectivity and stimulation of viral replication are fundamental properties of primate lentiviruses. J. Virol. *81*, 13852–13864.

Neil, S.J., Aasa-Chapman, M.M., Clapham, P.R., Nibbs, R.J., McKnight, A., and Weiss, R.A. (2005). The promiscuous CC chemokine receptor D6 is a functional coreceptor for primary isolates of human immunodeficiency virus type 1 (HIV-1) and HIV-2 on astrocytes. J. Virol. *79*, 9618–9624.

Neil, S.J., Zang, T., and Bieniasz, P.D. (2008). Tetherin inhibits retrovirus release and is antagonized by HIV-1 Vpu. Nature *451*, 425–430.

O'Donovan, D., Ariyoshi, K., Milligan, P., Ota, M., Yamuah, L., Sarge-Njie, R., and Whittle, H. (2000). Maternal plasma viral RNA levels determine marked differences in mother-to-child transmission rates of HIV-1 and HIV-2 in The Gambia. MRC/Gambia Government/University College London Medical School working group on mother-child transmission of HIV. AIDS *14*, 441–448.

Ozato, K., Shin, D.M., Chang, T.H., and Morse, H.C., 3rd (2008). TRIM family proteins and their emerging roles in innate immunity. Nat. Rev. Immunol. *8*, 849–860.

Pantaleo, G., Menzo, S., Vaccarezza, M., Graziosi, C., Cohen, O.J., Demarest, J.F., Montefiori, D., Orenstein, J.M., Fox, C., Schrager, L.K., and *et al.* (1995). Studies in subjects with long-term nonprogressive human immunodeficiency virus infection. N. Engl. J. Med. *332*, 209–216.

Perrin, L., Kaiser, L., and Yerly, S. (2003). Travel and the spread of HIV-1 genetic variants. Lancet Infect Dis 3, 22–27.

Popper, S.J., Sarr, A.D., Travers, K.U., Gueye-Ndiaye, A., Mboup, S., Essex, M.E., and Kanki, P.J. (1999). Lower human immunodeficiency virus (HIV) type 2 viral load reflects the difference in pathogenicity of HIV-1 and HIV-2. J. Infect. Dis. *180*, 1116–1121.

Popper, S.J., Sarr, A.D., Gueye-Ndiaye, A., Mboup, S., Essex, M.E., and Kanki, P.J. (2000). Low plasma human immunodeficiency virus type 2 viral load is independent of proviral load: low virus production *in vivo*. J. Virol. *74*, 1554–1557.

Provitera, P., Goff, A., Harenberg, A., Bouamr, F., Carter, C., and Scarlata, S. (2001). Role of the major homology region in assembly of HIV-1 Gag. Biochemistry *40*, 5565–5572.

Puffer, B.A., Pohlmann, S., Edinger, A.L., Carlin, D., Sanchez, M.D., Reitter, J., Watry, D.D., Fox, H.S., Desrosiers, R.C., and Doms, R.W. (2002). CD4 independence of simian immunodeficiency virus Envs is associated with macrophage tropism, neutralization sensitivity, and attenuated pathogenicity. J. Virol. *76*, 2595–2605.

Quinn, T.C., Wawer, M.J., Sewankambo, N., Serwadda, D., Li, C., Wabwire-Mangen, F., Meehan, M.O., Lutalo, T., and Gray, R.H. (2000). Viral load and heterosexual transmission of human immunodeficiency virus type 1. Rakai Project Study Group. N. Engl. J. Med. *342*, 921–929.

Reeves, J.D., and Schulz, T.F. (1997). The CD4-independent tropism of human immunodeficiency virus type 2 involves several regions of the envelope protein and correlates with a reduced activation threshold for envelope-mediated fusion. J. Virol. *71*, 1453–1465.

Reeves, J.D., McKnight, A., Potempa, S., Simmons, G., Gray, P.W., Power, C.A., Wells, T., Weiss, R.A., and Talbot, S.J. (1997). CD4-independent infection by HIV-2 (ROD/B): use of the 7-transmembrane receptors CXCR-4, CCR-3, and V28 for entry. Virology *231*, 130–134.

Reeves, J.D., Hibbitts, S., Simmons, G., McKnight, A., Azevedo-Pereira, J.M., Moniz-Pereira, J., and Clapham, P.R. (1999). Primary human immunodeficiency virus type 2 (HIV-2) isolates infect CD4-negative cells via CCR5 and CXCR4: comparison with HIV-1 and simian immunodeficiency virus and relevance to cell tropism *in vivo*. J. Virol. *73*, 7795–7804.

Rey-Cuille, M.A., Berthier, J.L., Bomsel-Demontoy, M.C., Chaduc, Y., Montagnier, L., Hovanessian, A.G., and Chakrabarti, L.A. (1998). Simian immunodeficiency virus replicates to high levels in sooty mangabeys without inducing disease. J. Virol. *72*, 3872–3886.

Richman, D.D., and Bozzette, S.A. (1994). The impact of the syncytium-inducing phenotype of human immunodeficiency virus on disease progression. J. Infect. Dis. *169*, 968–974.

Roe, T., Reynolds, T.C., Yu, G., and Brown, P.O. (1993). Integration of murine leukemia virus DNA depends on mitosis. EMBO J. *12*, 2099–2108.

Sarr, A.D., Popper, S., Thior, I., Hamel, D.J., Sankale, J.L., Siby, T., Marlink, R., Essex, M., Mboup, S., and Kanki, P. (1999). Relation between HIV-2 proviral load and $CD4^+$ lymphocyte count differs in monotypic and dual HIV infections. J Hum Virol 2, 45–51.

Schindler, M., Munch, J., Kutsch, O., Li, H., Santiago, M.L., Bibollet-Ruche, F., Muller-Trutwin, M.C., Novembre, F.J., Peeters, M., Courgnaud, V., *et al.* (2006). Nef-mediated suppression of T-cell activation was lost in a lentiviral lineage that gave rise to HIV-1. Cell *125*, 1055–1067.

Schramm, B., Penn, M.L., Palacios, E.H., Grant, R.M., Kirchhoff, F., and Goldsmith, M.A. (2000). Cytopathicity of human immunodeficiency virus type 2 (HIV-2) in human lymphoid tissue is coreceptor dependent and comparable to that of HIV-1. J. Virol. *74*, 9594–9600.

Sekigawa, I., Kaneko, H., Neoh, L.P., Takeda-Hirokawa, N., Akimoto, H., Hishikawa, T., Hashimoto, H., Hirose, S., Yamamoto, N., and Kaneko, Y. (1998). Differences of HIV envelope protein between HIV-1 and HIV-2: possible relation to the lower virulence of HIV-2. Viral Immunol. *11*, 1–8.

Shanmugam, V., Switzer, W.M., Nkengasong, J.N., Garcia-Lerma, G., Green, T.A., Ekpini, E., Sassan-Morokro, M., Antunes, F., Manshino, K., Soriano, V., *et al.* (2000). Lower HIV-2 plasma viral loads may explain differences between the natural histories of HIV-1 and HIV-2 infections. J. Acquir. Immune Defic. Syndr. *24*, 257–263.

Sharova, N., Wu, Y., Zhu, X., Stranska, R., Kaushik, R., Sharkey, M., and Stevenson, M. (2008). Primate lentiviral Vpx commandeers DDB1 to counteract a macrophage restriction. PLoS Pathog. *4*, e1000057.

Shi, Y., Brandin, E., Vincic, E., Jansson, M., Blaxhult, A., Gyllensten, K., Moberg, L., Brostrom, C., Fenyo, E.M., and Albert, J. (2005). Evolution of human immunodeficiency virus type 2 coreceptor usage, autologous neutralization, envelope sequence and glycosylation. J. Gen. Virol. *86*, 3385–3396.

Silvestri, G., Sodora, D.L., Koup, R.A., Paiardini, M., O'Neil, S.P., McClure, H.M., Staprans, S.I., and Feinberg, M.B. (2003). Nonpathogenic SIV infection of sooty mangabeys is characterized by limited bystander immunopathology despite chronic high-level viremia. Immunity *18*, 441–452.

Singh, N.B., Panda, S., Naik, T.N., Agarwal, A., Singh, H.L., Singh, Y.I., and Deb, B.C. (1995). HIV-2 strikes injecting drug users (IDUs) in India. J Infect *31*, 49–50.

Sleigh, R., Sharkey, M., Newman, M.A., Hahn, B., and Stevenson, M. (1998). Differential association of uracil DNA glycosylase with SIVSM Vpr and Vpx proteins. Virology *245*, 338–343.

Smith, S.M., Christian, D., de Lame, V., Shah, U., Austin, L., Gautam, R., Gautam, A., Apetrei, C., and Marx, P.A. (2008). Isolation of a new HIV-2 group in the US. Retrovirology 5, 103.

Srivastava, S., Swanson, S.K., Manel, N., Florens, L., Washburn, M.P., and Skowronski, J. (2008). Lentiviral Vpx accessory factor targets VprBP/DCAF1 substrate adaptor for cullin 4 E3 ubiquitin ligase to enable macrophage infection. PLoS Pathog. *4*, e1000059.

Switzer, W.M., Wiktor, S., Soriano, V., Silva-Graca, A., Mansinho, K., Coulibaly, I.M., Ekpini, E., Greenberg, A.E., Folks, T.M., and Heneine, W. (1998). Evidence of Nef truncation in human immunodeficiency virus type 2 infection. J. Infect. Dis. *177*, 65–71.

Thomas, E.R., Shotton, C., Weiss, R.A., Clapham, P.R., and McKnight, A. (2003). CD4-dependent and CD4-independent HIV-2: consequences for neutralization. AIDS *17*, 291–300.

Ulane, C.M., Kentsis, A., Cruz, C.D., Parisien, J.P., Schneider, K.L., and Horvath, C.M. (2005). Composition and assembly of STAT-targeting ubiquitin ligase complexes: paramyxovirus V protein carboxyl terminus is an oligomerization domain. J. Virol. *79*, 10180–10189.

Varmus, H.E., Padgett, T., Heasley, S., Simon, G., and Bishop, J.M. (1977). Cellular functions are required for the synthesis and integration of avian sarcoma virus-specific DNA. Cell *11*, 307–319.

Verhofstede, C., Reniers, S., Van Wanzeele, F., and Plum, J. (1994). Evaluation of proviral copy number and plasma RNA level as early indicators of progression in HIV-1 infection: correlation with virological and immunological markers of disease. AIDS *8*, 1421–1427.

Vesanen, M., Stevens, C.E., Taylor, P.E., Rubinstein, P., and Saksela, K. (1996). Stability in controlling viral replication identifies long-term nonprogressors as a distinct subgroup among human immunodeficiency virus type 1-infected persons. J. Virol. *70*, 9035–9040.

Wu, L., and KewalRamani, V.N. (2006). Dendritic-cell interactions with HIV: infection and viral dissemination. Nat. Rev. Immunol. *6*, 859–868.

Wu, X., Conway, J.A., Kim, J., and Kappes, J.C. (1994). Localization of the Vpx packaging signal within the C terminus of the human immunodeficiency virus type 2 Gag precursor protein. J. Virol. *68*, 6161–6169.

# Simian–Human Immunodeficiency Virus Model of Disease

6

Tatsuhiko Igarashi

## Abstract

Simian–human immunodeficiency virus (SHIV) was generated as an animal model for acquired immunodeficiency syndrome (AIDS) in order to overcome the narrow host range of human immunodeficiency virus (HIV-1). The first-generation SHIVs were non-pathogenic but evolved through animal-to-animal passage to become highly pathogenic viruses. In macaque monkeys, highly pathogenic SHIVs induce a distinct disease phenotype: a massive, systemic, and irreversible depletion of $CD4^+$ T-cells occurs within weeks of infection, followed by AIDS-like clinical manifestations. During the acute phase of infection, the virus predominantly infects and destroys $CD4^+$ T-cells. As a result, macrophages become the major virus-producing cell type. Virus isolated during the macrophage phase of infection exhibits macrophage tropism, a property not possessed by the inoculum SHIV. The V1/V2 region of the *env* gene was found to be responsible for the expanded cell tropism observed in the macrophage-tropic virus. The viral entry coreceptor, CXCR4, was maintained during the evolution of the virus. The vast majority of macrophage -tropic viruses are attenuated when inoculated into immune-competent monkeys.

## Generation of SHIV and rationale for use of SHIV

As with other viral infectious diseases, effective prophylactic or therapeutic measures against HIV-1 infection and AIDS cannot be devised without knowledge of the viral pathophysiology. To experimentally elucidate the pathophysiology of HIV, researchers require an animal infection model that is robust enough for a variety of manipulations. However, since the very early days of AIDS research, studies of HIV-1 have been hindered by the narrow host range of the virus. Although some studies have explored experimental HIV-1 infection of rabbit (Kulaga *et al.*, 1989), mouse (Morrow *et al.*, 1987), or gibbon ape (Lusso *et al.*, 1988), the chimpanzee is generally accepted to be the non-human species most susceptible to HIV-1 infection. Infected chimpanzees develop short-term viraemia and produce antibodies against HIV-1 proteins without exhibiting clinical manifestations of immune deficiency (Alter *et al.*, 1984; Desrosiers and Letvin, 1987). However, because they are an endangered species, the use of the chimpanzees in biomedical research is ethically unacceptable and they are therefore no longer used.

Efforts to develop other animal AIDS models have pursued three basic strategies. One strategy is to modify the host animal so that it becomes susceptible to HIV-1 infection, as in the SCID-hu mouse model (Namikawa *et al.*, 1988). Another strategy is to use a surrogate virus and its susceptible host species, as in the feline immunodeficiency virus/cat model (Pedersen *et al.*, 1987) or the simian immunodeficiency virus (SIV)/macaque monkey model (Murphey-Corb *et al.*, 1986). Finally, a third strategy is to manipulate HIV-1 so that it is infectious and pathogenic to a monkey species that can be used in biomedical research; this aim is accomplished through the generation of a chimeric SIV/HIV-1 virus (SHIV).

The first SHIV was developed in 1990 by Shibata *et al.* (1990), who made a series of $SIV_{agm}$/HIV-1 chimeras. Their experiments proved the feasibility of generating a chimera of two related but distinct primate lentiviruses. However, the authors did not assess the replicative capacity of their chimeric viruses in primary monkey cells. In 1991, Shibata *et al.* reported the development of another SHIV series; these viruses were generated by recombination between HIV-1 and $SIV_{mac}$ (Shibata *et al.*, 1991) and were the immediate predecessors to currently available SHIVs. One of their chimeric viruses, designated NM-3, carried the HIV-1 *tat, rev, vpu,* and *env* genes in an $SIV_{mac}$239 background. The genomic organizations of SHIV NM-3 and its parental HIV-1 and $SIV_{mac}$ are shown schematically in Fig. 6.1.

SHIV NM-3 has been shown to replicate in peripheral blood mononuclear cells (PBMCs) prepared from cynomolgus monkeys (crab-eating macaques) (Shibata *et al.*, 1991). This finding was momentous because it provided the first demonstration of SHIV replication in primary monkey cells, legitimizing further pursuit of an SHIV animal model. Furthermore, the experiment showed that the 5′ half of the HIV-1 genome is responsible for the inability of HIV-1 to replicate in monkey cells, ruling the entry step, which is mediated by the Env protein encoded in the 3′ half of the virus genome, out of the intrinsic species barrier.

The finding that a specific region of the HIV-1 genome defines species specificity, which was revealed during SHIV development efforts, was recently confirmed by the identification of host innate antiviral factors that contribute to the species specificity of HIV-1. These factors that are reviewed in Chapter 4 include TRIM5α (Stremlau *et al.*, 2004) and the proteins of the APOBEC3 family, including APOBEC3G (Sheehy *et al.*, 2002). TRIM5α targets the HIV-1 capsid protein Gag and restricts viral replication before reverse transcription. Gag apparently escapes restriction by human, but not simian, TRIM5α, which at least partially explains why HIV-1 does not replicate in monkey cells.

APOBEC3G is a DNA cytosine deaminase. An antiviral mechanism under current study suggests that the enzymatic activity of APOBEC3G is incorporated into virus particles, where it causes hypermutation of the virus genome during reverse transcription in newly infected cells and causes premature termination of the viral life cycle. The HIV-1 Vif protein may have evolved to counter the antiviral activity of APOBEC3G.

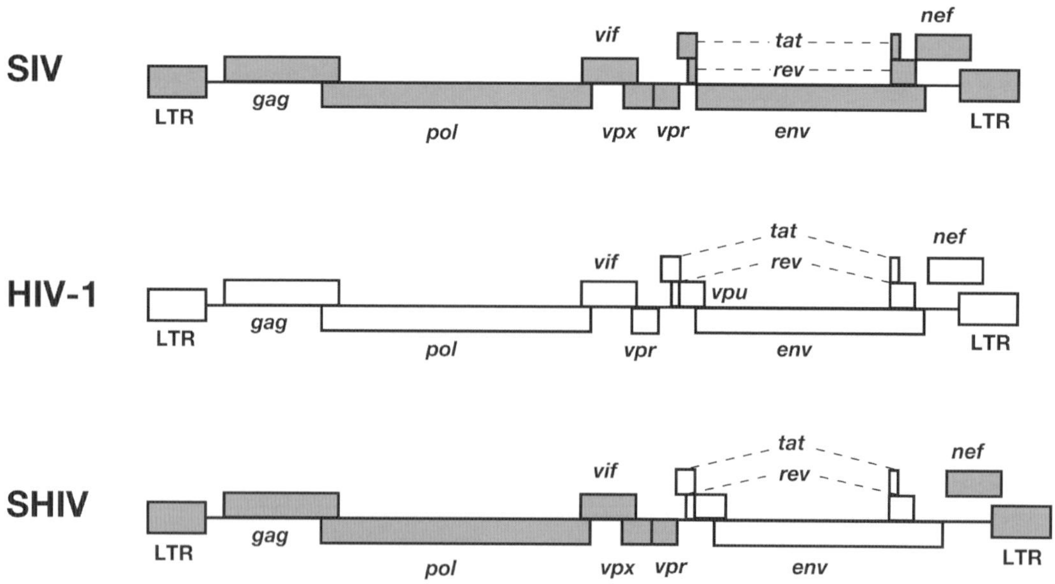


**Figure 6.1** Genomic organization of simian–human immunodeficiency virus (SHIV). Schematic depiction of the genomic organizations of prototypical SHIV and its parental HIV-1 and $SIV_{mac}$. Boxes indicate each viral gene and long terminal repeat (LTR). HIV-1 and SIV genes are shown as open and filled boxes, respectively.

In the cytoplasm of infected cells, Vif forms a complex with cellular proteins and functions as an E3 ubiquitin ligase, degrading APOBEC3G through the ubiquitin–proteasome pathway and preventing APOBEC3G from being incorporated into virus particles (Sheehy *et al.*, 2003; Yu *et al.*, 2003). However, HIV-1 Vif is unable to counteract monkey APOBEC3G (Mariani *et al.*, 2003). As would be expected, both the *gag* and *vif* genes are encoded in the 5′ half of the HIV-1 genome.

Since the acronym 'SHIV' applies to any chimeric simian-human immunodeficiency virus, some 'SHIVs' include combinations of viral genomes other than the prototypical NM-3-type combination. One of these 'non-NM-3-type' SHIVs is RT-SHIV, which carries the HIV-1 reverse transcriptase (RT) gene in an $SIV_{mac}239$ genomic background (Uberla *et al.*, 1995). The anti-HIV-1 compounds known as non-nucleoside RT inhibitors (NNRTIs) bind to the HIV-1 RT at a distinct pocket near its active site, inhibiting its enzymatic activity (Kohlstaedt *et al.*, 1992). Although the SIV and HIV-1 RT proteins are approximately 60% homologous in sequence (Isaka *et al.*, 1998), SIV RT is insensitive to NNRTIs that target HIV-1 RT. RT-SHIV was found to be sensitive to NNRTI, demonstrating the potential usefulness of RT-SHIV in studies of the mechanism of resistance and for testing therapeutic measures against NNRTI-resistant virus in experimental settings *in vivo* (Balzarini *et al.*, 1995).

The prototypical SHIV carries the HIV-1-derived Env protein, the principal target for virus neutralization. The Env proteins from HIV-1 and SIV are antigenically distinct, therefore, neutralizing antibodies directed against HIV-1 Env are unable to inhibit SIV infection. The SHIV/macaque model can be used to evaluate a neutralizing antibody-based strategy for prophylaxis or therapy against HIV-1 infection. SHIV NM-3, which carries the *tat, rev, vpu,* and *env* genes of HIV-1 NL4-3 in an SIV239 genomic background, is neutralized by the monoclonal antibody 0.5β, which also neutralizes the parental HIV-1 (Igarashi *et al.*, 1996). Monkeys infected with another SHIV, NM-3n, which carries the HIV-1 NL4–3 *tat, rev, vpu, env,* and *nef* genes in an SIV239 genomic background, raise antibodies capable of neutralizing both the parental HIV-1 NL4–3 and the chimeric virus itself (Igarashi *et al.*, 1994).

SHIV DH12 (formerly SHIV MD14YE), which was generated using the dual-tropic HIV-1 strain DH12 as the parental virus, was used to challenge monkeys passively immunized with IgG prepared from a neutralizing antibody-generating chimpanzee chronically infected with the parental HIV-1 DH12 (Shibata *et al.*, 1997). Animals harbouring a neutralizing antibody titre of <1:10 against the HIV-1 24 hours after passive transfer of the IgG preparation were protected from challenge with SHIV DH12. Macaques passively immunized with a cocktail of neutralizing monoclonal antibodies were also protected from challenge with the highly pathogenic SHIV strain 89.6PD (Mascola *et al.*, 2000). These studies demonstrate that neutralizing antibody-based prophylaxis is achievable in primate lentivirus infection.

## Derivation of highly pathogenic SHIVs

The first-generation SHIVs, SHIV NM-3 and NM-3n, were shown to be infectious in cynomolgus monkeys (Sakuragi *et al.*, 1992; Igarashi *et al.*, 1994); they caused low-level viraemia detectable by virus isolation from PBMCs of inoculated animals and established persistent infection. Immunoblot analysis showed that levels of antiviral antibodies increased over time in these animals, and infectious virus was recovered from the PBMCs of one animal infected with SHIV NM-3 3 years post infection (p.i.) (Igarashi *et al.*, 1996). However, these SHIVs did not cause any of the clinical manifestations observed in HIV-1-infected humans. Other first-generation SHIVs, including SHIV SF33, SF162 (Luciw *et al.*, 1995), and SHIV-4 (Li *et al.*, 1992) were non-pathogenic in macaque monkeys.

In 1996, two independent research groups reported the derivation of highly pathogenic SHIVs using the strategy of animal-to-animal passage. The same strategy was previously employed for SIV in a 1992 study (Sharma *et al.*, 1992) in which a neurovirulent variant of SIV239 was derived by the passage of bone marrow cells prepared from an infected animal to the brain of the recipient animal. Employing the same strategy, Joag *et al.* (1996) and Reimann *et al.* (1996a) derived two highly pathogenic SHIVs, SHIV KU-1 and SHIV 89.6P. Their reports made major contributions to the SHIV/macaque model by demonstrating animal-to-animal passage to be a powerful and possibly requisite step for converting a non-pathogenic SHIV into a virulent virus, and indicating that SHIV would be a vital animal model for AIDS studies.

The repeated passage of viruses in order to adapt them to new hosts is an established virological technique that has previously been applied to the development of live attenuated vaccine strains, including mumps (Hosai *et al.*, 1970), varicella (Takahashi, 1984), and yellow fever (Miller and Mitchell, 1986). Subsequent to the reports by Joag *et al.* (1996) and Reimann *et al.* (1996a), highly pathogenic SHIVs were derived from non-pathogenic prototypes by an essentially similar method (Igarashi *et al.*, 1999; Tan *et al.*, 1999; Chen *et al.*, 2000; Song *et al.*, 2006). However, the successful derivation of highly pathogenic SHIVs appears to depend largely on the individual animal and the particular HIV-1 *env* sequence used.

The mechanism by which pathogenicity is acquired through animal-to-animal passage is complex. A comparison of prototypical non-pathogenic SHIV with its highly pathogenic derivative reveals multiple genetic changes throughout the entire virus genome. In SHIV 89.6P, genomic changes responsible for the acquisition of virulence were mapped to the 3′ half of the viral genome (Karlsson *et al.*, 1997). Sadjadpour *et al.* (2004) isolated a pathogenic molecular clone (SHIV DH12R-CL-7) from a highly pathogenic SHIV DH12R swarm. Introduction of the *env* gene from SHIV DH12R-CL-7 into the parental, non-pathogenic SHIV DH12 molecular clone failed to confer the typical highly pathogenic disease phenotype, suggesting that genetic changes found outside *env* also contribute to pathogenicity.

Animals infected with either of the two highly pathogenic SHIVs, SHIV KU-1 or SHIV 89.6P, exhibit almost identical disease phenotypes. The viruses cause a massive depletion of circulating $CD4^+$ T-lymphocytes within weeks of infection. After this initial 'crash,' the $CD4^+$ T-cells do not recover until the infected animals develop AIDS-like symptoms, such as emaciation, diarrhoea refractory to treatment, and opportunistic infection. Other highly pathogenic SHIVs derived more recently induce similar disease states. Although the disease phenotype induced by the highly pathogenic SHIVs is not like that induced by SIV or HIV-1 infection, its rapid, drastic, and reproducible disease course appealed to researchers attempting to develop anti-HIV-1 vaccines. Thus, the 'second-generation' highly pathogenic SHIVs were used as challenge viruses in a variety of anti-HIV-1 vaccine studies (Barouch *et al.*, 2000; Amara *et al.*, 2001; Shiver *et al.*, 2002).

## Highly pathogenic SHIV-induced disease

The disease caused by highly pathogenic SHIV is distinct from the diseases induced by other primate lentiviruses. An example of the disease induced by highly pathogenic SHIV is that seen upon SHIV DH12R infection of rhesus macaque monkeys which can be divided into two phases (Igarashi *et al.*, 1999, 2001, 2002, 2003b). During the acute phase of infection, which lasts for about 3 weeks p.i., the plasma virus burden reaches its peak, and a massive depletion of the circulating $CD4^+$ T-lymphocytes occurs. The chronic phase of infection begins after the acute phase and lasts until the infected monkey requires euthanasia due to development of immune-deficiency-related clinical manifestations. Schematic representations of the plasma viral RNA burden and circulating $CD4^+$ T-lymphocyte level during these phases are shown in Fig. 6.2.

### *Acute phase*

As soon as 3 days after intravenous inoculation, viral replication can be detected in the plasma (about $1 \times 10^5$ RNA copies/ml), and provirus is found in the lymph nodes (about 1 copy/$10^5$ cells). The viral burden in the plasma increases exponentially and reaches its initial peak, ranging from $10^7$ to $10^8$ RNA copies/ml, between days 10 and 12 p.i. The initial detection and subsequent increase of provirus loads in the lymph nodes precede those in the PBMCs and thymus between days 3 and 10. By day 14 p.i., provirus loads in these compartments reach a comparable level of $>10^3$ copies/$10^5$ cells. Thus, a fair case can be made that replication of the highly pathogenic SHIV strain DH12R is nearly synchronous

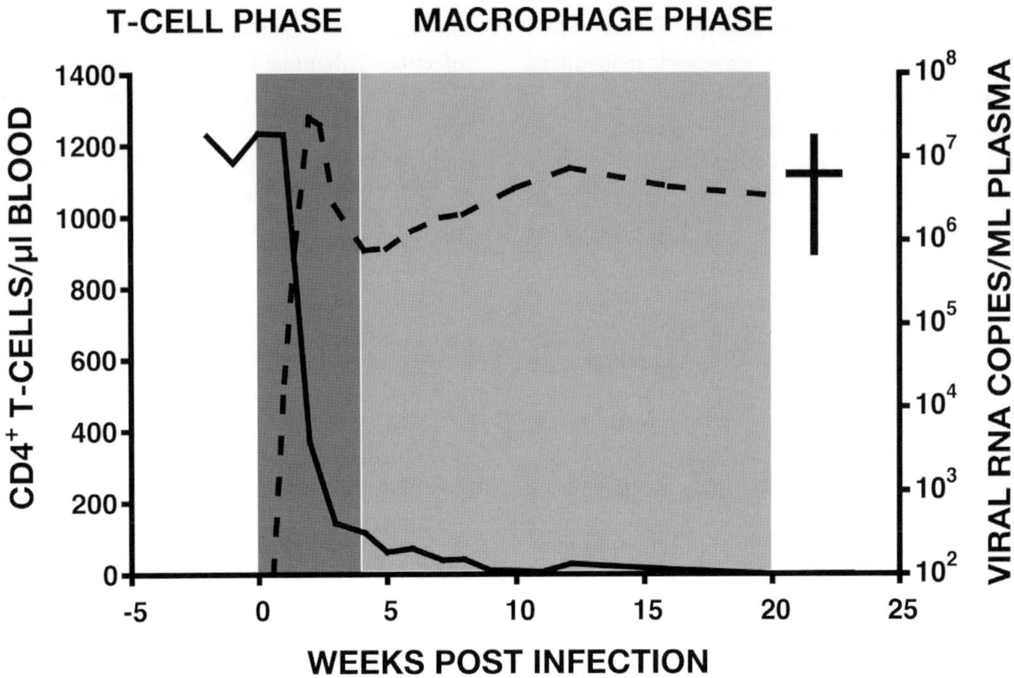


**Figure 6.2** Representative circulating CD4$^+$ T-lymphocytes and plasma viral burden in animals infected with highly pathogenic SHIV. Typical kinetics of the change in numbers of peripheral blood CD4$^+$ T-lymphocytes (per 1 µl blood; ———) and viral RNA copies (per 1 ml plasma; – – – –). Dark and light grey areas represent the acute (T-cell) and chronic (M) phases of infection, respectively. †, euthanization due to deteriorated clinical condition.

among the lymphatic tissues during the acute phase of infection.

Histopathological analysis of animals during the acute phase of SHIV DH12R infection reveals hyperplasia of the lymphatic tissues, including the lymph nodes and gut-associated lymphatic tissues. Unlike secondary lymphatic tissues, the thymus exhibits a mild cortical depletion. Immunohistochemical analysis reveals a remarkable depletion of CD4$^+$ cells in these tissues. By day 10, the levels of CD4$^+$ cells in the lymph nodes and thymus attain a level almost comparable to that in normal animals. On day 14, when a precipitous depletion of circulating CD4$^+$ cells occurs, the CD4$^+$ cells in these tissues are also markedly depleted (Igarashi *et al.*, 2002).

*Chronic phase*

After the initial peak of viraemia at about 2 weeks p.i., the plasma virus load decreases to about $10^5$–$10^7$ RNA copies/ml, and the viral burden remains constant or gradually increases until the infected animal develops AIDS-like symptoms. CD4$^+$ T-lymphocytes in the circulation remain depleted during this phase of infection.

The onset of clinical manifestations varies from animal to animal. Some macaques are moribund within 10 weeks p.i., while others survive up to 50 weeks p.i. The most commonly observed clinical manifestations in SHIV DH12R-infected monkeys are anorexia, weight loss, and diarrhoea. Histopathology of the lymphatic tissues from these animals at necropsy reveals a variety of changes. Although many monkeys develop lymphatic atrophy, some still exhibit hyperplasia of the lymph nodes. The monkeys commonly exhibit signs of opportunistic infections such as pneumonia induced by *Pneumocystis carinii*, cryptosporidiosis, oropharyngeal/oesophageal candidiasis, cytomegalovirus-induced enteritis, and mycobacterium infection.

Roughly 30% of SHIV DH12R-infected animals exhibit emaciation or idiopathic enteritis without recognizable opportunistic infection. Histopathologic analysis of the gastrointestinal tract in these cases often reveals lymphoplasmacytic or lymphohistiocytic enteritis. However, since these conditions are induced by a variety of agents, it is difficult to confirm that SHIV DH12R is the causal agent. Similarly, infiltration of histiocytes/lymphocytes in the intraepithelial space is observed in specimens collected from AIDS patients with idiopathic enteritis (Kotler *et al.*, 1984; Gillin *et al.*, 1985). Another major finding is the infiltration of multinucleated macrophages into a variety of tissues, including the intestinal lamina propria, pulmonary interstitial space, and lymph nodes. Unlike SIV-infected animals, SHIV DH12R-infected monkeys rarely develop neoplasm.

As described above, the disease induced by highly pathogenic SHIV is characterized by a massive, systemic, and irreversible depletion of $CD4^+$ T-lymphocytes followed 10–50 weeks after infection by the onset of AIDS-like symptoms. The disease course is much shorter than that induced by SIV or HIV-1, and neither SIV nor HIV-1 induces a comparable massive depletion of $CD4^+$ T-cells during the acute phase of infection.

## T-cell and M phases of highly pathogenic SHIV DH12R infection

During the first weeks of infection with highly pathogenic SHIV, viral replication represented by the initial peak in plasma viral RNA load occurs when the vast majority of $CD4^+$ T-cells are depleted. Viral RNA-positive cells are also positive for the T-cell marker CD3, as shown by combined *in situ* viral RNA hybridization/immunohistochemical analysis of a mesenteric lymph node collected from an SHIV DH12R-infected animal on day 10 p.i. (Igarashi *et al.*, 2003b). On day 10 p.i., as much as 88% of the circulating naïve subset of $CD4^+$ T-lymphocytes, which constitutes the vast majority of the peripheral $CD4^+$ T-cell population, are productively infected with SHIV DH12R (Nishimura *et al.*, 2005). Based on these observations, it is conceivable that the massive depletion of $CD4^+$ T-cells that occurs during the first weeks of SHIV DH12R infection is a result of cell killing by direct viral infection.

After near-complete systemic $CD4^+$ T-lymphocyte depletion, monkeys infected with SHIV DH12R survive for weeks to months with plasma viral RNA burdens as high as $10^7$ copies/ml. Viral RNA hybridization *in situ* reveals numerous viral RNA-positive cells in a variety of tissues (Igarashi *et al.*, 2001). These cells are larger than lymphocytes and resemble macrophages morphologically (Igarashi *et al.*, 2001). Combined *in situ* viral RNA hybridization/immunohistochemical analysis for CD3 molecules did not detect double-positive $CD3^+$/virus $RNA^+$ cells in the mesenteric lymph node collected at necropsy of an animal that was euthanized due to deteriorated clinical condition at 12 weeks p.i. (Igarashi *et al.*, 2001), indicating that the virus is not produced by T-lymphocytes during the end stage of SHIV DH12R infection.

Considering the near-complete depletion of circulating $CD4^+$ T-lymphocytes in infected animals, the remaining $CD3^+$ cells in the lymph nodes are most likely $CD8^+$ T-cells, which are not susceptible to SHIV infection. Based on the morphological features of the viral RNA-positive cells, another thin section taken from the same fixed tissue block used for CD3 staining was subjected to combined *in situ* hybridization/immunohistochemical analysis for the macrophage marker HAM56 (Dako); more than 95% of viral RNA-positive cells were also positive for HAM56, indicating that during the end stage of SHIV DH12R infection the vast majority of virus-producing cells are macrophages (Igarashi *et al.*, 2001).

To summarize the abovementioned findings, the acute phase of SHIV DH12R infection is characterized by the initial infection and resulting destruction of T-cells ('T-cell phase'), and the chronic phase is characterized by infection of macrophages ('macrophage phase' or 'M phase').

## Genetic changes accompanying the transition from the T-cell phase to the M phase in highly pathogenic SHIV DH12R infection

SHIV DH12R causes two distinct phases of infection in inoculated monkeys, the T-cell phase and the M phase. During the T-cell phase of infection, the viral RNA is predominantly detected in $CD3^+$ lymphocytes, and not in other cells, whereas during the subsequent M phase, the vast majority of virus-producing cells are macrophages. Since $CD8^+$ T-lymphocytes are not susceptible to primate lentivirus infection, the abovementioned virus RNA$^+$/CD3$^+$ cells are most likely to be the other population, namely, CD4$^+$ T-lymphocytes. It is conceivable that genetic alterations in the virus are behind this shift in target cell type. Since the viral *env* gene primarily governs host-cell tropism, the *env* genes of viruses collected from the plasma of seven SHIV DH12R-infected animals were analyzed to assess this possibility (Imamichi *et al.*, 2002).

There are two lineages of SHIV DH12R, the original SHIV DH12R and SHIV DH12R PS1. The original SHIV DH12R virus evolved in a single animal that was treated with anti-CD8 antibody three days prior to and four days following infection with the non-pathogenic SHIV strain DH12 (Matano *et al.*, 1998; Igarashi *et al.*, 1999). SHIV DH12R was recovered from a cell suspension of lymphatic tissue taken from the animal at 68 weeks p.i., when it was euthanized due to its deteriorated clinical condition. After successful recovery of the highly pathogenic SHIV DH12R virus, viral isolates from cryopreserved PBMC samples that had been serially collected from the same animal were analyzed to determine when the pathogenic phenotype had emerged. Infectious virus recovered from the plasma sample collected at 52 weeks p.i. was pathogenic when used to inoculate a monkey designated PS1. The viral stock prepared from PS1 was designated SHIV DH12R PS1. Sequence analysis of the *env* gene revealed that SHIV DH12R was highly heterogeneous, whereas SHIV DH12R PS1 was homogeneous.

Five animals infected with SHIV DH12R and two animals infected with SHIV DH12R PS1 exhibited typical highly pathogenic SHIV disease. Sequence analysis of multiple PCR-amplified *env* clones from plasma samples revealed that genetic changes were clustered in the V2 region (Fig. 6.3). Two weeks after infection, when the plasma virus burden reached its initial peak, particular genotypes were dominant in the circulation of the five SHIV DH12R-infected animals. Interestingly, the same genotypes were found to be dominant in multiple animals. SHIV DH12R PS1, which was homogeneous in nature, remained unchanged in the two SHIV DH12R PS1-infected animals. By the time they reached end stage, the M phase, the dominant *env* sequence had changed from that observed during the T-cell phase in all seven animals. In four of the five SHIV DH12R-infected animals, a genotype that had been present in the virus inoculum became dominant; this genotype was characterized by consecutive deletions of amino acid residues 186 and 187 of the Env protein.

The dominant Env genotype of the remaining SHIV DH12R-infected animal was characterized by deletions of consecutive amino acid residues 164 and 165 and two substitution mutations, I166K and N197D. The N197D change abolished a potential N-linked glycosylation site. This genotype was not found in the original viral stock, suggesting an outgrowth from a minor population or *de novo* evolution. Deletions at Env positions 164 and 165 were also found in the majority of genotypes from one of the SHIV DH12R PS1-infected animals; some of these genotypes also exhibited the N197D mutation. The other SHIV DH12R PS1-infected animal harboured a population with consecutive deletions of Env amino acid residues 169 and 170. Overall, the V2 loops in the virus particles circulating during M-phase SHIV DH12R or SHIV DH12R PS1 infection appeared to be shorter or less bulky than the V2 loops present during the T-cell phase.

Infectious viruses recovered from cell suspensions of lymph nodes collected from animals during M-phase SHIV DH12R infection lacked a 5-amino-acid sequence (L-K-N-G-T) in the Env V1 region, as well as residue 165 in the V2 loop, with or without simultaneous deletion of residue 164. Viewing V1 and V2 as a single unit, M-phase Env had a more compact V1/V2 region than that of Env of the inoculum virus and the virus present in the T-cell phase. The inoculum SHIV DH12R carries two copies of the L-K-N-G-T motif in the V1 loop (Igarashi *et al.*, 2003a; Igarashi *et al.*, 2007). The number of copies of this motif changed during the evolution of SHIV DH12R from non-pathogenic SHIV DH12 in a single animal (Igarashi *et al.*, 1999). It should be noted that the 'signature' changes (the five deletions in V1 and the two deletions in V2), emerged in multiple animals, even in an animal inoculated with a molecular clone of SHIV DH12R CL7 (Igarashi *et al.*, 2003a; Sadjadpour *et al.*, 2004; Igarashi *et al.*, 2007). These changes might be necessary for the transition from T-cell phase to M phase.

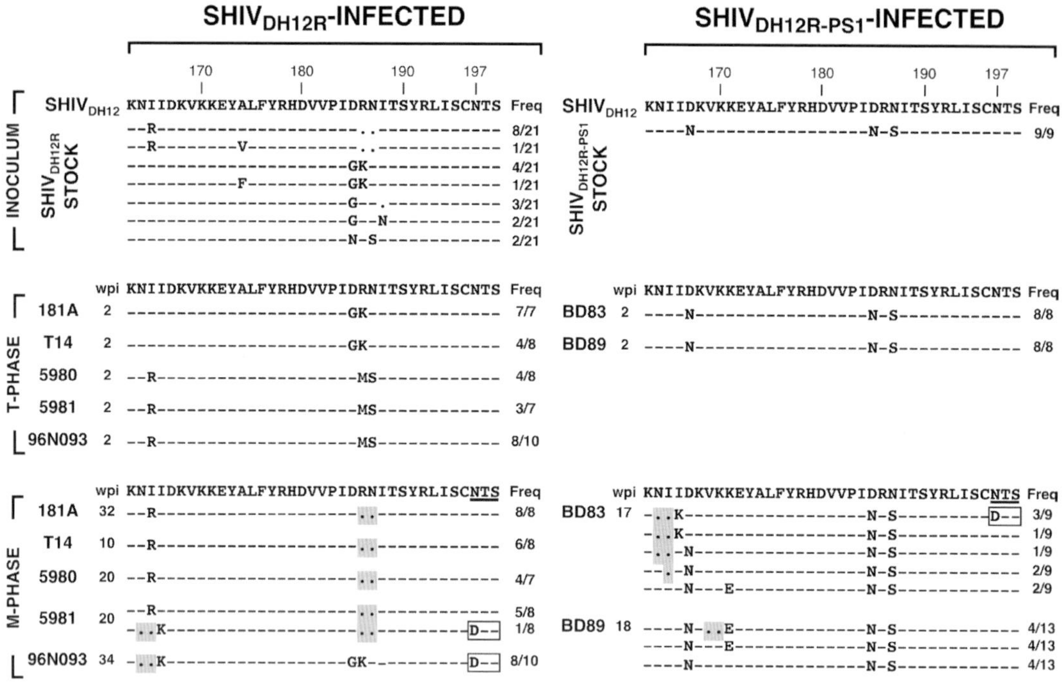


**Figure 6.3** V2 sequences of the *env* genes of the SHIV DH12R and SHIV DH12R CL7 inoculum and circulating viruses during T-cell and M phases. The deduced Env amino acid sequences from inoculum viruses and from viruses present in the plasma of five SHIV DH12R-infected (181A, T14, 5980, 5981 and 96N093) and two SHIV DH12R CL7-infected (BD83 and BD89) animals during acute (T-phase) and chronic (M-phase) phases of infection are aligned. –, amino acid identity with prototypical SHIV DH12 Env; •, amino acid deletion. Lost potential N-linked glycosylation sites are boxed. (From Imamichi *et al.* (2002) Proc. Natl. Acad. Sci. U.S.A. *99*, 13813–13818, with permission. Copyright (2002) National Academy of Sciences, U.S.A.)

## Genetic changes observed in the Env V2 region during M phase are sufficient to confer M-tropism on SHIV DH12R *in vitro*

Based on the histochemical stainings of the lymphatic tissues (Igarashi *et al.*, 2001), one might conclude that the M-phase virus has acquired M-tropism. To test this assumption, viruses emerging during M phase were assessed for their capacity to replicate in alveolar macrophages prepared from uninfected rhesus macaques (Imamichi *et al.*, 2002; Igarashi *et al.*, 2003a). Morphologically, alveolar macrophages are large mononuclear cells with abundant cytoplasm and projections (Fig. 6.4a and b). They are positive for non-specific esterase staining (Fig. 6.4b) and express CD206 (mannose receptor) and HLA-DR detectable by fluorescence-activated cell sorting (Fig. 6.4c and d). The cells were obtained by a bronchoalveolar lavage procedure (BAL) using a paediatric fibreoptic bronchoscope. After the lavage fluid had been passed through 70-μm mesh, the cells were sedimented by centrifugation and resuspended in Hanks' balanced salt solution (Imamichi *et al.*, 2002; Igarashi *et al.*, 2003a). BAL cell preparations typically contain bronchial epithelial cells, neutrophils, lymphocytes, and alveolar macrophages. BAL cells resuspended in serum-free Dulbecco's modified minimal essential medium (DMEM) were plated in 24-well culture plates and incubated at 37°C for 1 h to allow the macrophages to adhere to the well. After removal of non-adherent cells, the remaining cells were further cultured in DMEM supplemented with 10% human AB serum and 5% fetal bovine serum for 2 days.

The day after preparation, the alveolar macrophage preparations were inoculated with uncloned viruses isolated from M-phase lymph nodes or with SHIV DH12R CL7 mutant molecular clones carrying genetic changes in the V2

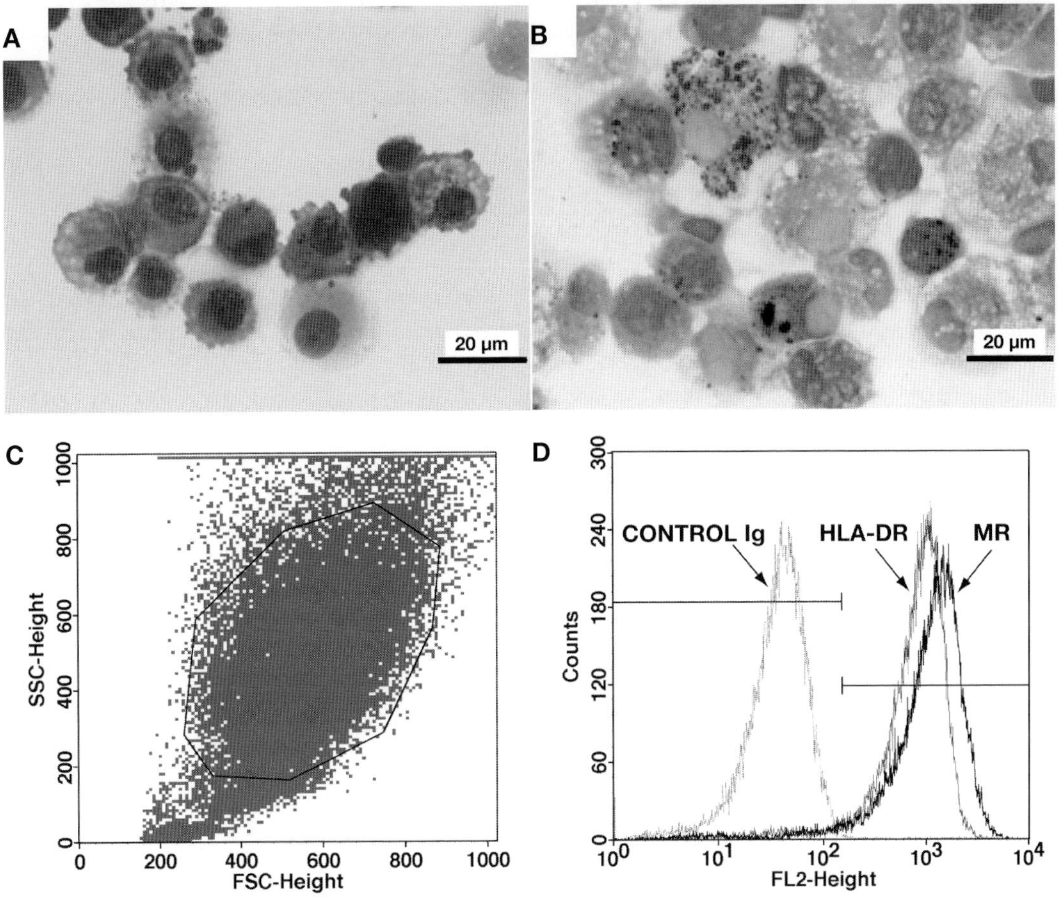


**Figure 6.4** Characterization of rhesus alveolar macrophages. (A) and (B) Staining of bronchoalveolar lavage (BAL) cells, which comprise several cell types. (A) Diff-Quik staining. The predominant cells are large mononuclear cells with abundant cytoplasm and multiple projections. (B) Non-specific esterase staining. Many of the BAL cells exhibit non-specific esterase activity in the cytoplasm. (C) and (D) Fluorescence-activated cell sorting of BAL cells. The vast majority of BAL cells express the mannose receptor CD206 (95.32% positive) and HLA-DR (95.84% positive) (d). (Contributed by M. Matsuyama.)

region. The viruses were normalized for infectious titre before inoculation. The T-cell-tropic strain SIV239 and its M-tropic variant, SIV316, which served as negative and positive controls, respectively, yielded the expected results in the viral replication assay (Igarashi *et al.*, 2003a). On the other hand, no progeny virus production was detected in the rhesus alveolar macrophages infected with either the inoculum SHIV DH12R or the molecularly cloned SHIV DH12R CL7; no RT activity was detected in the culture supernatant fraction (Igarashi *et al.*, 2003a). Lymph node viral isolates recovered from five different animals during M-phase SHIV DH12R infection replicated to high titres in the cells, with faster replication kinetics than that of SIV316 (Igarashi *et al.*, 2003a). Some, but not all, of the SHIV DH12R CL7 mutants carrying the V2 genetic changes observed in plasma virus also replicated in alveolar macrophages (Imamichi *et al.*, 2002). All the mutants that exhibited M-tropism carried Δ164 and Δ165 mutations, as well as the N197D mutation that abolishes a potential *N*-linked glycosylation site (Imamichi *et al.*, 2002). Thus, the conformational changes caused by these genetic alterations in Env appear to be necessary for DH12R to acquire a tropism for monkey macrophages.

It should be noted that the properties of Env protein assessed in monkey cells in the context of SHIV does not necessarily reproduce the properties identified in the context of parental HIV-1 assessed in human cells. HIV-1 DH12, the parental virus of the prototypical non-pathogenic SHIV DH12, replicates in human monocyte-derived macrophages (MDMs) (Shibata *et al.*, 1995), whereas SHIV DH12R does not replicate in monkey alveolar macrophages (Igarashi *et al.*, 2003a). Miller *et al.* (1998) reported that another highly pathogenic SHIV strain, 89.6PD, replicated in monkey MDMs, whereas SHIV 89.6 did not. SHIV 89.6PD evolved from the non-pathogenic SHIV 89.6, which carries Env derived from the dual-tropic HIV-1 strain HIV-1 89.6. This apparent inconsistency in results for human macrophages *vs.* their monkey counterparts might arise from differences in host cell factors, such as differences in the expression levels of CD4 and coreceptors or other unidentified factors.

These highly pathogenic SHIVs appear to have acquired M-tropism as a result of their passage through animals. In support of this conclusion, Liu *et al.* (1999) reported that the highly pathogenic SHIV KU-1 and its derivative, SHIV KU-2, replicated in monkey MDMs, whereas the prototypical SHIV-4, which carries Env derived from the T-cell line-adapted HIV-1 strain LAI (Li *et al.*, 1992), did not. However, contradictory results concerning the replication of SHIV-4 (also designated SHIV HXBc2) in monkey macrophages have been reported. Miller *et al.* reported that the virus replicated in MDMs, whereas others found that it failed to replicate in MDMs (Liu *et al.*, 1999) or in macrophages prepared from bone marrow cells (Reimann *et al.*, 1996b). Miller *et al.* and Liu *et al.* both confirmed that their preparations were predominantly positive for CD68, a macrophage marker. These groups differentiated macrophages in slightly dissimilar ways; Miller *et al.* added 200 units/ml of granulocyte–macrophage colony-stimulating factor to the medium, whereas Liu *et al.* used 15% conditioned L-cell medium, 5 units/ml of macrophage colony-stimulating factor, and 100 units/ml of granulocyte–macrophage colony-stimulating factor to the medium (Liu *et al.*, 1999). Different types of stimulation may alter the susceptibility of the differentiated cells to the virus. Variations in the donor animals may also come into play.

## 'Dual-tropic' HIV-1 Env utilizes CXCR4 as the entry coreceptor in monkey lymphocytes and macrophages

Primate lentiviruses, including HIV-1 and SIV, gain entry into target cells through a series of interactions between their Env proteins and cellular proteins. The relevant cellular proteins are CD4 (Klatzmann *et al.*, 1984; Maddon *et al.*, 1986) and the chemokine coreceptors, especially CXCR4 (Feng *et al.*, 1996) and also CCR5 (Alkhatib *et al.*, 1996; Choe *et al.*, 1996; Deng *et al.*, 1996). HIV-1 strains that use CXCR4 but not CCR5 for entry are known as X4 viruses. SIV and HIV-1 strains that use CCR5 but not CXCR4 are designated R5 viruses, and strains using both CXCR4 and CCR5 are known as R5X4 viruses. The observed relationship between coreceptor usage and cell tropism can be generally described as follows:

1. X4 and R5X4 strains replicate in permanent T-cell lines, such as Hut78, MT-2, and MT-4, and in peripheral blood mononuclear cells (PBMC). Although R5X4 strains replicate in MDMs, X4 viruses do not.
2. R5 viruses replicate in PBLs, but are unable to replicate in permanent T-cell lines. Many R5 viruses are replication-competent in MDMs.

The acquisition of changes in the V1/V2 region of SHIV DH12R and its resulting acquisition of M-tropism during M phase might indicate an alteration in its coreceptor usage. To assess this possibility, the inoculum SHIV DH12R virus and its M-tropic variants were assayed for coreceptor usage (Igarashi *et al.*, 2003a). Before inoculation, PBMC were pre-incubated with increasing amounts of a small-molecule inhibitor directed against either CCR5 or CXCR4. On day 5 p.i., the RT activity released in the culture supernatant fraction was determined as a measure of viral replication (Igarashi *et al.*, 2003a). Replication of SIV239 and its macrophage-tropic variant SIV316 were inhibited by the anti-CCR5 small-molecule inhibitors TAK779 (Baba *et al.*, 1999) and AD101 (Trkola *et al.*, 2002), but not by the anti-CXCR4 inhibitor AMD3100 (Donzella *et al.*, 1998), confirming that SIV239 uses CCR5, as previously reported (Zhang *et al.*, 2000). In this experiment, replication of the original SHIV DH12R and its prototype SHIV DH12 were not inhibited by TAK779 or AD101 (Igarashi *et al.*, 2003a). Instead, the infection was blocked by AMD3100, suggesting that SHIV DH12R predominantly uses CXCR4 as its coreceptor for PBMC entry (Igarashi *et al.*, 2003a). Of note, M-tropic variants of SHIV DH12R also maintained CXCR4 usage in these cells (Igarashi *et al.*, 2003a). When the M-tropic SHIVs were further assessed for their coreceptor preferences in the context of infection of monkey alveolar macrophages, the assay revealed that these viruses use CXCR4 as the coreceptor for alveolar macrophage entry (Fig. 6.5).

A few points bear mentioning here. First, SHIV DH12R carries Env derived from HIV-1 DH12, which was originally identified as an R5X4 strain (Trkola *et al.*, 1998) and behaved as an X4 virus in the context of monkey cell infection. SHIV 89.6PD, another highly pathogenic SHIV, was also found to be an X4 virus by the same assay (Zhang *et al.*, 2000); this virus evolved in monkeys from the non-pathogenic SHIV 89.6, which carries Env derived from the R5X4 HIV-1 strain 89.6 (Collman *et al.*, 1992; Simmons *et al.*, 1996).

Yi *et al.* (2005) reported that despite 'R5X4' label, R5X4 HIV-1 strains preferentially utilize CXCR4 to gain entry into human PBLs. They showed that both HIV-1 DH12 and HIV-1 89.6 use CXCR4 for infection of primary lymphocytes. Since these viruses were originally designated as R5X4 viruses based on the results obtained in coreceptor-transfected indicator cell lines, the CCR5 molecules expressed in these cell lines may not be identical to those expressed on primary lymphocytes (Yi *et al.*, 2005). It is also possible that these Env proteins cannot interact with monkey CCR5 molecules expressed on monkey cells to mediate viral entry.

The second point to be made is that infection of alveolar macrophages by SIV316 or SHIV DH12R M-tropic variants is mediated by CD4 and coreceptors, not through phagocytosis. Although observed only in one particular *ex vivo* experimental setting (Fig. 6.5), replication of these viruses in rhesus macaque alveolar

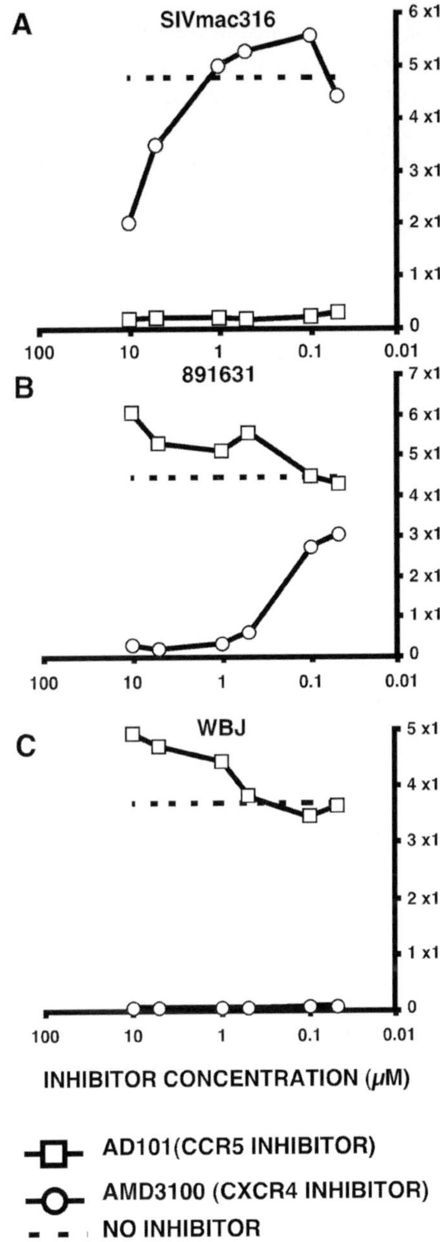


**Figure 6.5** Coreceptor usage of macrophage-tropic SHIVs in alveolar macrophages. Rhesus macaque alveolar macrophages were inoculated with M-tropic SIV316 (A) or two M-tropic SHIV isolates (B) and (C) in the presence of increasing amounts of a small-molecule inhibitor directed against either CXCR4 (AMD3100) or CCR5 (AD101). Viral replication was monitored by assaying RT activity released into the culture supernatant on day 5 p.i. Dashed lines represent viral replication in the absence of inhibitors. (From Igarashi *et al.* (2003) J. Virol. *77*, 13042–13052, with permission.)

macrophages was completely inhibited by small-molecule inhibitors (Igarashi *et al.*, 2003a). In fact, alveolar macrophages express CD4, CCR3, CCR5, and CXCR4 on their cell surface (Park *et al.*, 1999; Worgall *et al.*, 1999; Mori *et al.*, 2000), albeit at lower levels than observed in circulating lymphocytes and monocytes (Worgall *et al.*, 1999). Although this observation does not rule out entry of virus particles into macrophages through pathways other than the CD4/coreceptor pathway, these alternative modes of entry, if they exist, might not result in viral replication. Trujillo *et al.* (2007) have observed that entry of virus particles into macrophages *via* the mannose receptor does not lead to the infectious cycle.

The third and final point is that CXCR4 usage and the highly pathogenic disease phenotype appear to be tightly linked. SHIV 89.6PD and SHIV DH12R infect monkey lymphocytes using CXCR4. SHIV KU-2, which is derived from SHIV-4 carrying the X4 HIV-1 LAI-derived Env, maintains its CXCR4 usage (Liu *et al.*, 1999). Other highly pathogenic SHIVs, such as SHIV SBG (Dunn *et al.*, 1996) and SHIV MN229 (Fernandez *et al.*, 2005), also carry HIV-1 LAI-derived Env. Given the CXCR4 usage of SHIV KU-2, these highly pathogenic SHIVs are also likely to be X4 viruses.

## M-tropic SHIVs cause non-pathogenic infections in immunologically competent animals

M-tropic variants of SHIV DH12R were isolated during the end stage of highly pathogenic SHIV-induced disease. These viruses possess an expanded cell tropism, i.e. they are both M-tropic and lymphocyte-tropic. Thus, it seemed possible that M-tropic SHIV plays a major role in highly pathogenic SHIV disease, or that in addition to highly pathogenic SHIV disease, the virus induces distinct clinical manifestations based on its M-tropism, i.e. it might cause inflammatory disease, such as the pulmonary or neurological complications observed in sheep infected with visna–maedi virus, an ungulate lentivirus (Sigurdsson *et al.*, 1952,1957).

The availability of M-tropic SHIVs provided an opportunity to test this hypothesis. Six M-tropic virus isolates were inoculated independently into six pairs of rhesus macaques. Of the six virus isolates, three were uncloned viral swarms, one was a full-length molecular clone from a viral swarm isolate, and two were SHIV DH12R CL7 mutants carrying specific V2 changes known to confer M-tropism. Plasma virus burden, circulating lymphocyte subsets, and clinical manifestations were closely monitored for more than 4 years (Igarashi *et al.*, 2007). Five of the viruses failed to cause any detectable clinical manifestations; peak plasma virus burdens were 10–100 fold less than that induced by the parental SHIV DH12R and quickly fell to undetectable levels or remained at less than $10^4$ copies/ml. Circulating $CD4^+$ T-cell numbers changed only minimally, and the monkeys remained clinically healthy throughout the observation period.

Only one virus, SHIV M11, caused disease. This virus carries V2 changes in a SHIV DH12R CL7 background. The plasma virus load profile and observed disease phenotype (rapid and irreversible $CD4^+$ T-lymphocyte depletion within weeks of inoculation, followed by anorexia and a marked weight loss) induced by SHIV M11 were indistinguishable from those induced by the parental, T-cell tropic SHIV DH12R. No signs of a particular clinical condition related to macrophage infection or resulting activation (inflammation) were observed.

Since the concept of M-tropism arose from observations of viral replication in macrophages *in vitro*, an M-tropic virus may not necessarily replicate in macrophages *in vivo*. In fact, the M-tropic SIV316 virus does not replicate in macrophages *in vivo* (Borda *et al.*, 2004). To examine whether any M-tropic SHIVs actually replicate in macrophages *in vivo*, alveolar macrophages collected from animals infected with either of two M-tropic viruses were subjected to infectious centre assays. Cells from animals infected with SIV239, which is T-cell tropic *in vitro* but infects macrophages *in vivo* (Borda *et al.*, 2004), were also assayed. Serially diluted alveolar macrophages were co-cultivated with PBMCs prepared from uninfected animals, and the RT activity of the culture supernatant fraction was monitored for 3 weeks (Table 6.1). At 4 weeks p.i., fewer than 0.001% of cells prepared by BAL of SIV239-infected animals released virus, whereas 0.006% to 0.2% of those

**Table 6.1** Infectious centres in the pulmonary space of SHIV- or SIV-infected animals

| | | Number of virus producing cells/$10^6$ cells | |
|---|---|---|---|
| Animal ID | Virus inoculum | 4 wpi | 8 wpi |
| 95D139 | $SIV_{mac239}$ | 5.2 | <0.1 |
| H589 | $SIV_{mac239}$ | 2.3 | <0.1 |
| 4462 | $SHIV_{M3}$ | 243.0 | ND* |
| P12 | $SHIV_{BF37AM}$ | 60.9 | 243.0 |
| P25 | $SHIV_{BF37AM}$ | 2187.0 | 46.8 |

*ND, not done.

From Igarashi *et al.* (2007), with permission.

prepared from M-tropic SHIV-infected animals yielded infectious virus. At 8 weeks p.i., there was no detectable release of SIV239 in the lung space, but 0.004% to 0.024% of the cells from M-tropic SHIV-infected animals were still producing virus. These results demonstrate that M-tropic SHIVs preferentially infect macrophages *in vivo*.

Since the overwhelming majority (5/6) of the tested M-tropic SHIVs did not cause disease, it can be concluded that M-tropic SHIVs contribute little to the pathogenicity of high pathogenicity SHIV-induced disease, which is characterized by rapid, systemic, and irreversible $CD4^+$ T-cell depletion followed by AIDS-like clinical manifestations, such as emaciation, diarrhoea, and opportunistic infections. One of the M-tropic SHIVs, SHIV M11, induces disease indistinguishable from that caused by the inoculum SHIV DH12R (Igarashi *et al.*, 1999). Unlike ungulate lentiviruses, which are predominantly M-tropic (Gendelman *et al.*, 1985; Zink *et al.*, 1990; Sellon *et al.*, 1992; Brodie *et al.*, 1995), primate lentiviruses exhibit M-tropism that is inseparable from their lymphocyte-tropism. Considering the intrinsic tropism of primate lentiviruses, it is not surprising that the disease induced by SHIV M11 is directed against lymphocytes.

M-tropic SHIVs also do not cause any recognizable inflammatory conditions. Although it is possible that the viruses are too highly attenuated to induce any clinical signs, it is also possible that infrequent disease induction is inherent to M-tropic viruses. Not every animal infected with an M-tropic ungulate lentivirus develops clinical disease. Visna-maedi virus and its disease were not recognized until it was introduced into a herd of virus-naïve Icelandic sheep through the importation of chronically infected but clinically healthy Karakul sheep from Germany (Sigurdsson *et al.*, 1952, 1957; Sigurdsson and Palsson, 1958). Approximately 90% of horses experimentally infected with equine infectious anaemia virus survive the acute febrile phase of infection and become chronic carriers without any recognizable clinical manifestations (Sellon *et al.*, 1994). Compared with lymphocyte tropic HIV-1, these M-tropic lentiviruses cause disease less frequently. If the M-tropic SHIVs were inoculated into greater numbers of animals, inflammatory manifestations might be observed in some.

M-tropic SHIVs were probably forced to emerge because the original target cells, $CD4^+$ T-lymphocytes, were depleted. Presumably, the reduction in bulk of the V1/V2 region caused by the amino acid deletions and/or loss of potential glycosylation site observed in M-tropic SHIVs is advantageous for macrophage infection. If the V1/V2 region is a flap that shields part of the core region containing the binding sites for CD4 and coreceptors (Wyatt *et al.*, 1995), then the V1/V2 loop of the M-tropic SHIVs may expose neutralizing epitopes that are conserved to a greater extent among the M-tropic isolates. Indeed, M-tropic SHIVs were found to be more sensitive to a neutralizing antibody raised against the inoculum virus (Igarashi *et al.*, 2007).

## Implications for the roles of M-tropic viruses in the pathogenesis of HIV-1 infection and the potential research utility of SHIV

What roles do M-tropic viruses play in the pathogenesis of HIV-1 infection of humans? M-tropic HIV-1 is present during all phases of infection but becomes relatively abundant in individuals with impaired immunity (Appleman *et al.*, 1988; Sidtis and Price, 1990; Davis *et al.*, 1992; Orenstein *et al.*, 1997). As described above, the experimental infection of rhesus macaques with M-tropic SHIVs has indicated that the majority of M-tropic viruses are attenuated in nature. When the host immunity is debilitated, these viruses directly give rise to pathological conditions only in immune-privileged anatomical compartments, such as the central nervous system. During the asymptomatic phase of infection, however, these viruses infect macrophages and form a reservoir that undermines host immunity by continuously releasing virus particles capable of infecting and destroying CD4$^+$ T-cells.

The highly pathogenic SHIV/macaque system is a contrivance. Nonetheless, this system can provide valuable information for understanding the biology of primate lentiviruses in macrophages. During the M phase of infection *in vivo*, nearly all infected cells are macrophages. Considering the intrinsic tropism of primate lentiviruses, highly pathogenic SHIVs provide a rare opportunity. Highly active anti-retroviral therapy (HAART) was introduced as a therapeutic measure for HIV-1-infected individuals in the 1990s (Hammer *et al.*, 1996; Saravolatz *et al.*, 1996). Since then, the risk of disease progression among patients has been substantially reduced (Egger *et al.*, 2002). However, HAART does not eradicate the virus; the plasma virus burden quickly increases to pre-treatment levels when HAART is interrupted (Davey *et al.*, 1999; Garcia *et al.*, 1999; Harrigan *et al.*, 1999; Neumann *et al.*, 1999; Ortiz *et al.*, 1999), suggesting the existence of a virus reservoir. Indeed, studies of blood samples from patients undergoing HAART identified resting CD4$^+$ T-cells as the viral reservoir (Chun *et al.*, 1995; Chun *et al.*, 1997). However, other cell types have also been indicated to function as the virus reservoir (Chun *et al.*, 2000). Macrophages have been speculated to be the virus reservoir. There is no direct evidence for macrophages as the reservoir, however, possibly because studying virus-infected macrophages *in vivo* is technically difficult. By exploiting the M phase of highly pathogenic SHIV infection, it is possible to study the efficacy of HAART on the virus in macrophages and how the viral infection is maintained in macrophages *in vivo*.

In about 10–15% of infected macaques, known as rapid progressors, $SIV_{smm}$ or $SIV_{mac}$ infection induces a condition similar to that induced by highly pathogenic SHIVs (Brown *et al.*, 2007; Nishimura *et al.*, 2007). $SIV_{smm}$ and $SIV_{mac}$ use CCR5 as their entry coreceptor and target memory CD4$^+$ T-cells, which express CCR5. In rapid progressor animals, this subpopulation of CD4$^+$ lymphocytes is systemically depleted within weeks of infection through an unknown mechanism (Nishimura *et al.*, 2007), and the virus shifts its tropism to macrophages (Brown *et al.*, 2007). However, $SIV_{smm}$ and $SIV_{mac}$ are too limited by the low frequency of rapid progression to pursue them as experimental models.

## Summary

SHIV was generated as an animal model for AIDS in order to overcome the narrow host range of HIV-1. The first-generation SHIVs were non-pathogenic but evolved through animal-to-animal passage to become highly pathogenic viruses. Highly pathogenic SHIVs predominantly infect and destroy CD4$^+$ T-cells during the initial weeks of infection. After a near complete depletion of CD4$^+$ T-lymphocytes, macrophages become the major virus-producing cell type. Virus isolated during the macrophage phase of infection exhibits macrophage tropism, a property not possessed by the inoculum SHIV. The V1/V2 region of the *env* gene was found to be responsible for the expanded cell tropism observed in the macrophage-tropic virus. Although the macrophage-tropic variants emerged with expanded cell tropism in infected animals exhibiting AIDS-like clinical manifestations, the vast majority of macrophage-tropic viruses are attenuated when inoculated into immune-competent monkeys. Based on the abovementioned observations, it is concluded that macrophage-tropism contributes little to the highly pathogenic SHIV-induced disease phenotype. In HIV-1 infection, it may be also the case that macrophage-tropic strains do not play major roles for the onset of AIDS. However, the most relevant property of macrophage-tropic virus in primate lentivirus infection is that it forms a reservoir that can continuously release infectious virus particles that can infect and destroy both macrophages and CD4+ T-cells. Since HAART is unable to eradicate virus from infected individuals, HIV reservoir is an important therapeutic target. SHIV-induced macrophage phase of infection can serve as a valuable model to study this issue *in vivo*.

## References


Alkhatib, G., Combadiere, C., Broder, C.C., Feng, Y., Kennedy, P.E., Murphy, P.M., and Berger, E.A. (1996). CC CKR5: a RANTES, MIP-1alpha, MIP-1beta receptor as a fusion cofactor for macrophage-tropic HIV-1. Science *272*, 1955–1958.

Alter, H.J., Eichberg, J.W., Masur, H., Saxinger, W.C., Gallo, R., Macher, A.M., Lane, H.C., and Fauci, A.S. (1984). Transmission of HTLV-III infection from human plasma to chimpanzees: an animal model for AIDS. Science *226*, 549–552.

Amara, R.R., Villinger, F., Altman, J.D., Lydy, S.L., O'Neil, S.P., Staprans, S.I., Montefiori, D.C., Xu, Y., Herndon, J.G., Wyatt, L.S., *et al.* (2001). Control of a mucosal challenge and prevention of AIDS by a multiprotein DNA/MVA vaccine. Science *292*, 69–74.

Appleman, M.E., Marshall, D.W., Brey, R.L., Houk, R.W., Beatty, D.C., Winn, R.E., Melcher, G.P., Wise, M.G., Sumaya, C.V., and Boswell, R.N. (1988). Cerebrospinal fluid abnormalities in patients without AIDS who are seropositive for the human immunodeficiency virus. J. Infect. Dis. *158*, 193–199.

Baba, M., Nishimura, O., Kanzaki, N., Okamoto, M., Sawada, H., Iizawa, Y., Shiraishi, M., Aramaki, Y., Okonogi, K., Ogawa, Y., *et al.* (1999). A small-molecule, nonpeptide CCR5 antagonist with highly potent and selective anti-HIV-1 activity. Proc. Natl. Acad. Sci. U.S.A. *96*, 5698–5703.

Balzarini, J., Weeger, M., Camarasa, M.J., De Clercq, E., and Uberla, K. (1995). Sensitivity/resistance profile of a simian immunodeficiency virus containing the reverse transcriptase gene of human immunodeficiency virus type 1 (HIV-1) toward the HIV-1-specific non-nucleoside reverse transcriptase inhibitors. Biochem. Biophys. Res. Commun. *211*, 850–856.

Barouch, D.H., Santra, S., Schmitz, J.E., Kuroda, M.J., Fu, T.M., Wagner, W., Bilska, M., Craiu, A., Zheng, X.X., Krivulka, G.R., *et al.* (2000). Control of viremia and prevention of clinical AIDS in rhesus monkeys by cytokine-augmented DNA vaccination. Science *290*, 486–492.

Borda, J.T., Alvarez, X., Kondova, I., Aye, P., Simon, M.A., Desrosiers, R.C., and Lackner, A.A. (2004). Cell tropism of simian immunodeficiency virus in culture is not predictive of *in vivo* tropism or pathogenesis. Am. J. Pathol. *165*, 2111–2122.

Brodie, S.J., Pearson, L.D., Zink, M.C., Bickle, H.M., Anderson, B.C., Marcom, K.A., and DeMartini, J.C. (1995). Ovine lentivirus expression and disease. Virus replication, but not entry, is restricted to macrophages of specific tissues. Am. J. Pathol. *146*, 250–263.

Brown, C.R., Czapiga, M., Kabat, J., Dang, Q., Ourmanov, I., Nishimura, Y., Martin, M.A., and Hirsch, V.M. (2007). Unique pathology in simian immunodeficiency virus-infected rapid progressor macaques is consistent with a pathogenesis distinct from that of classical AIDS. J. Virol. *81*, 5594–5606.

Chen, Z., Huang, Y., Zhao, X., Skulsky, E., Lin, D., Ip, J., Gettie, A., and Ho, D.D. (2000). Enhanced infectivity of an R5-tropic simian/human immunodeficiency virus carrying human immunodeficiency virus type 1 subtype C envelope after serial passages in pig-tailed macaques (Macaca nemestrina). J. Virol. *74*, 6501–6510.

Choe, H., Farzan, M., Sun, Y., Sullivan, N., Rollins, B., Ponath, P.D., Wu, L., Mackay, C.R., LaRosa, G., Newman, W., *et al.* (1996). The beta-chemokine receptors CCR3 and CCR5 facilitate infection by primary HIV-1 isolates. Cell *85*, 1135–1148.

Chun, T.W., Finzi, D., Margolick, J., Chadwick, K., Schwartz, D., and Siliciano, R.F. (1995). *In vivo* fate of HIV-1-infected T-cells: quantitative analysis of the transition to stable latency. Nat. Med. *1*, 1284–1290.

Chun, T.W., Stuyver, L., Mizell, S.B., Ehler, L.A., Mican, J.A., Baseler, M., Lloyd, A.L., Nowak, M.A., and Fauci, A.S. (1997). Presence of an inducible HIV-1 latent reservoir during highly active antiretroviral therapy. Proc. Natl. Acad. Sci. U.S.A. *94*, 13193–13197.

Chun, T.W., Davey, R.T., Jr, Ostrowski, M., Shawn Justement, J., Engel, D., Mullins, J.I., and Fauci, A.S. (2000). Relationship between pre-existing viral reservoirs and the re-emergence of plasma viremia after discontinuation of highly active anti-retroviral therapy. Nat. Med. *6*, 757–761.

Collman, R., Balliet, J.W., Gregory, S.A., Friedman, H., Kolson, D.L., Nathanson, N., and Srinivasan, A. (1992). An infectious molecular clone of an unusual macrophage-tropic and highly cytopathic strain of human immunodeficiency virus type 1. J. Virol. *66*, 7517–7521.

Davey, R.T., Jr, Bhat, N., Yoder, C., Chun, T.W., Metcalf, J.A., Dewar, R., Natarajan, V., Lempicki, R.A., Adelsberger, J.W., Miller, K.D., *et al.* (1999). HIV-1 and T-cell dynamics after interruption of highly active antiretroviral therapy (HAART) in patients with a history of sustained viral suppression. Proc. Natl. Acad. Sci. U.S.A. *96*, 15109–15114.

Davis, L.E., Hjelle, B.L., Miller, V.E., Palmer, D.L., Llewellyn, A.L., Merlin, T.L., Young, S.A., Mills, R.G., Wachsman, W., and Wiley, C.A. (1992). Early viral brain invasion in iatrogenic human immunodeficiency virus infection. Neurology *42*, 1736–1739.

Deng, H., Liu, R., Ellmeier, W., Choe, S., Unutmaz, D., Burkhart, M., Di Marzio, P., Marmon, S., Sutton, R.E., Hill, C.M., *et al.* (1996). Identification of a major coreceptor for primary isolates of HIV-1. Nature *381*, 661–666.

Desrosiers, R.C., and Letvin, N.L. (1987). Animal models for acquired immunodeficiency syndrome. Rev Infect Dis *9*, 438–446.

Donzella, G.A., Schols, D., Lin, S.W., Este, J.A., Nagashima, K.A., Maddon, P.J., Allaway, G.P., Sakmar, T.P., Henson, G., De Clercq, E., and Moore, J.P. (1998). AMD3*100*, a small molecule inhibitor of HIV-1 entry via the CXCR4 co-receptor. Nat. Med. *4*, 72–77.

Dunn, C.S., Beyer, C., Kieny, M.P., Gloeckler, L., Schmitt, D., Gut, J.P., Kirn, A., and Aubertin, A.M. (1996). High viral load and CD4 lymphopenia in rhesus and cynomolgus macaques infected by a chimeric primate lentivirus constructed using the env, rev, tat, and vpu genes from HIV-1 Lai. Virology *223*, 351–361.

Egger, M., May, M., Chene, G., Phillips, A.N., Ledergerber, B., Dabis, F., Costagliola, D., D'Arminio Monforte, A., de Wolf, F., Reiss, P., *et al.* (2002). Prognosis of HIV-1-infected patients starting highly active antiretroviral therapy: a collaborative analysis of prospective studies. Lancet *360*, 119–129.

Feng, Y., Broder, C.C., Kennedy, P.E., and Berger, E.A. (1996). HIV-1 entry cofactor: functional cDNA cloning of a seven-transmembrane, G protein-coupled receptor. Science *272*, 872–877.

Fernandez, C.S., Stratov, I., De Rose, R., Walsh, K., Dale, C.J., Smith, M.Z., Agy, M.B., Hu, S.L., Krebs, K., Watkins, D.I., *et al.* (2005). Rapid viral escape at an immunodominant simian-human immunodeficiency virus cytotoxic T-lymphocyte epitope exacts a dramatic fitness cost. J. Virol. *79*, 5721–5731.

Garcia, F., Plana, M., Vidal, C., Cruceta, A., O'Brien, W.A., Pantaleo, G., Pumarola, T., Gallart, T., Miro, J.M., and Gatell, J.M. (1999). Dynamics of viral load rebound and immunological changes after stopping effective antiretroviral therapy. AIDS *13*, F79–86.

Gendelman, H.E., Narayan, O., Molineaux, S., Clements, J.E., and Ghotbi, Z. (1985). Slow, persistent replication of lentiviruses: role of tissue macrophages and macrophage precursors in bone marrow. Proc. Natl. Acad. Sci. U.S.A. *82*, 7086–7090.

Gillin, J.S., Shike, M., Alcock, N., Urmacher, C., Krown, S., Kurtz, R.C., Lightdale, C.J., and Winawer, S.J. (1985). Malabsorption and mucosal abnormalities of the small intestine in the acquired immunodeficiency syndrome. Ann. Intern. Med. *102*, 619–622.

Hammer, S.M., Katzenstein, D.A., Hughes, M.D., Gundacker, H., Schooley, R.T., Haubrich, R.H., Henry, W.K., Lederman, M.M., Phair, J.P., Niu, M., *et al.* (1996). A trial comparing nucleoside monotherapy with combination therapy in HIV-infected adults with CD4 cell counts from 200 to 500 per cubic millimeter. AIDS Clinical Trials Group Study 175 Study Team. N. Engl. J. Med. *335*, 1081–1090.

Harrigan, P.R., Whaley, M., and Montaner, J.S. (1999). Rate of HIV-1 RNA rebound upon stopping antiretroviral therapy. AIDS *13*, F59–62.

Hosai, H., Yamanishi, K., Ueda, S., Minekawa, Y., and Ogino, T. (1970). Studies on live attenuated mumps virus vaccine. 1. Attenuation of mumps virus by serial passage in the chorioallantoic cavity of developing chick embryos and field trials by the inhalation method. Biken J *13*, 121–126.

Igarashi, T., Shibata, R., Hasebe, F., Ami, Y., Shinohara, K., Komatsu, T., Stahl-Hennig, C., Petry, H., Hunsmann, G., Kuwata, T., and *et al.* (1994). Persistent infection with SIVmac chimeric virus having tat, rev, vpu, env and nef of HIV type 1 in macaque monkeys. AIDS Res. Hum. Retroviruses *10*, 1021–1029.

Igarashi, T., Kuwata, T., Takehisa, J., Ibuki, K., Shibata, R., Mukai, R., Komatsu, T., Adachi, A., Ido, E., and Hayami, M. (1996). Genomic and biological alteration of a human immunodeficiency virus type 1 (HIV-1)-simian immunodeficiency virus strain mac chimera, with HIV-1 Env, recovered from a long-term carrier monkey. J. Gen. Virol. *77 (Pt 8)*, 1649–1658.

Igarashi, T., Endo, Y., Englund, G., Sadjadpour, R., Matano, T., Buckler, C., Buckler-White, A., Plishka, R., Theodore, T., Shibata, R., and Martin, M. (1999). Emergence of a highly pathogenic simian/human immunodeficiency virus in a rhesus macaque treated with anti-CD8 mAb during a primary infection with a nonpathogenic virus. Proc. Natl. Acad. Sci. U.S.A. *96*, 14049–14054.

Igarashi, T., Brown, C.R., Endo, Y., Buckler-White, A., Plishka, R., Bischofberger, N., Hirsch, V., and Martin, M.A. (2001). Macrophage are the principal reservoir and sustain high virus loads in rhesus macaques after the depletion of CD4+ T-cells by a highly pathogenic simian immunodeficiency virus/HIV type 1 chimera (SHIV): Implications for HIV-1 infections of humans. Proc. Natl. Acad. Sci. U.S.A. *98*, 658–663.

Igarashi, T., Brown, C.R., Byrum, R.A., Nishimura, Y., Endo, Y., Plishka, R.J., Buckler, C., Buckler-White, A., Miller, G., Hirsch, V.M., and Martin, M.A. (2002). Rapid and irreversible CD4+ T-cell depletion induced by the highly pathogenic simian/human immunodeficiency virus SHIV(DH12R) is systemic and synchronous. J. Virol. *76*, 379–391.

Igarashi, T., Donau, O.K., Imamichi, H., Dumaurier, M.J., Sadjadpour, R., Plishka, R.J., Buckler-White, A., Buckler, C., Suffredini, A.F., Lane, H.C., *et al.* (2003a). Macrophage-tropic simian/human immunodeficiency virus chimeras use CXCR4, not CCR5, for infections of rhesus macaque peripheral blood mononuclear cells and alveolar macrophages. J. Virol. *77*, 13042–13052.

Igarashi, T., Imamichi, H., Brown, C.R., Hirsch, V.M., and Martin, M.A. (2003b). The emergence and characterization of macrophage-tropic SIV/HIV chimeric viruses (SHIVs) present in CD4+ T-cell-depleted rhesus monkeys. J. Leukoc. Biol. *74*, 772–780.

Igarashi, T., Donau, O.K., Imamichi, H., Nishimura, Y., Theodore, T.S., Iyengar, R., Erb, C., Buckler-White, A., Buckler, C.E., and Martin, M.A. (2007). Although macrophage-tropic simian/human immunodeficiency viruses can exhibit a range of pathogenic phenotypes, a majority of isolates induce no clinical disease in immunocompetent macaques. J. Virol. *81*, 10669–10679.

Imamichi, H., Igarashi, T., Imamichi, T., Donau, O.K., Endo, Y., Nishimura, Y., Willey, R.L., Suffredini, A.F., Lane, H.C., and Martin, M.A. (2002). Amino acid deletions are introduced into the V2 region of gp120 during independent pathogenic simian immunodeficiency virus/HIV chimeric virus (SHIV) infections of rhesus monkeys generating variants that are macrophage tropic. Proc. Natl. Acad. Sci. U.S.A. *99*, 13813–13818.

Isaka, Y., Sato, A., Kawauchi, S., Suyama, A., Miki, S., Hayami, M., and Fujiwara, T. (1998). Construction of the chimeric reverse transcriptase of simian immunodeficiency virus sensitive to nonnucleoside reverse transcriptase inhibitor. Microbiol Immunol *42*, 195–202.

Joag, S.V., Li, Z., Foresman, L., Stephens, E.B., Zhao, L.J., Adany, I., Pinson, D.M., McClure, H.M., and Narayan, O. (1996). Chimeric simian/human immunodeficiency virus that causes progressive loss of CD4+ T-cells and AIDS in pig-tailed macaques. J. Virol. *70*, 3189–3197.

Karlsson, G.B., Halloran, M., Li, J., Park, I.W., Gomila, R., Reimann, K.A., Axthelm, M.K., Iliff, S.A., Letvin, N.L., and Sodroski, J. (1997). Characterization of molecularly cloned simian-human immunodeficiency viruses causing rapid CD4+ lymphocyte depletion in rhesus monkeys. J. Virol. *71*, 4218–4225.

Klatzmann, D., Champagne, E., Chamaret, S., Gruest, J., Guetard, D., Hercend, T., Gluckman, J.C., and Montagnier, L. (1984). T-lymphocyte T4 molecule behaves as the receptor for human retrovirus LAV. Nature *312*, 767–768.

Kohlstaedt, L.A., Wang, J., Friedman, J.M., Rice, P.A., and Steitz, T.A. (1992). Crystal structure at 3.5 A resolution of HIV-1 reverse transcriptase complexed with an inhibitor. Science *256*, 1783–1790.

Kotler, D.P., Gaetz, H.P., Lange, M., Klein, E.B., and Holt, P.R. (1984). Enteropathy associated with the acquired immunodeficiency syndrome. Ann. Intern. Med. *101*, 421–428.

Kulaga, H., Folks, T., Rutledge, R., Truckenmiller, M.E., Gugel, E., and Kindt, T.J. (1989). Infection of rabbits with human immunodeficiency virus 1. A small animal model for acquired immunodeficiency syndrome. J. Exp. Med. *169*, 321–326.

Li, J., Lord, C.I., Haseltine, W., Letvin, N.L., and Sodroski, J. (1992). Infection of cynomolgus monkeys with a chimeric HIV-1/SIVmac virus that expresses the HIV-1 envelope glycoproteins. J. Acquir. Immune Defic. Syndr. *5*, 639–646.

Liu, Z.Q., Muhkerjee, S., Sahni, M., McCormick-Davis, C., Leung, K., Li, Z., Gattone, V.H., 2nd, Tian, C., Doms, R.W., Hoffman, T.L., *et al.* (1999). Derivation and biological characterization of a molecular clone of SHIV(KU-2) that causes AIDS, neurological disease, and renal disease in rhesus macaques. Virology *260*, 295–307.

Luciw, P.A., Pratt-Lowe, E., Shaw, K.E., Levy, J.A., and Cheng-Mayer, C. (1995). Persistent infection of rhesus macaques with T-cell-line-tropic and macrophage-tropic clones of simian/human immunodeficiency viruses (SHIV). Proc. Natl. Acad. Sci. U.S.A. 92, 7490–7494.

Lusso, P., Markham, P.D., Ranki, A., Earl, P., Moss, B., Dorner, F., Gallo, R.C., and Krohn, K.J. (1988). Cell-mediated immune response toward viral envelope and core antigens in gibbon apes (Hylobates lar) chronically infected with human immunodeficiency virus-1. J. Immunol. *141*, 2467–2473.

Maddon, P.J., Dalgleish, A.G., McDougal, J.S., Clapham, P.R., Weiss, R.A., and Axel, R. (1986). The T4 gene encodes the AIDS virus receptor and is expressed in the immune system and the brain. Cell *47*, 333–348.

Mariani, R., Chen, D., Schrofelbauer, B., Navarro, F., Konig, R., Bollman, B., Munk, C., Nymark-McMahon, H., and Landau, N.R. (2003). Species-specific exclusion of APOBEC3G from HIV-1 virions by Vif. Cell *114*, 21–31.

Mascola, J.R., Stiegler, G., VanCott, T.C., Katinger, H., Carpenter, C.B., Hanson, C.E., Beary, H., Hayes, D., Frankel, S.S., Birx, D.L., and Lewis, M.G. (2000). Protection of macaques against vaginal transmission of a pathogenic HIV-1/SIV chimeric virus by passive infusion of neutralizing antibodies. Nat. Med. *6*, 207–210.

Matano, T., Shibata, R., Siemon, C., Connors, M., Lane, H.C., and Martin, M.A. (1998). Administration of an anti-CD8 monoclonal antibody interferes with the clearance of chimeric simian/human immunodeficiency virus during primary infections of rhesus macaques. J. Virol. 72, 164–169.

Miller, B.R., and Mitchell, C.J. (1986). Passage of yellow fever virus: its effect on infection and transmission rates in Aedes aegypti. Am J Trop Med Hyg 35, 1302–1309.

Miller, C.J., Marthas, M., Greenier, J., Lu, D., Dailey, P.J., and Lu, Y. (1998). *In vivo* replication capacity rather than *in vitro* macrophage tropism predicts efficiency of vaginal transmission of simian immunodeficiency virus or simian/human immunodeficiency virus in rhesus macaques. J. Virol. 72, 3248–3258.

Mori, K., Rosenzweig, M., and Desrosiers, R.C. (2000). Mechanisms for adaptation of simian immunodeficiency virus to replication in alveolar macrophages. J. Virol. 74, 10852–10859.

Morrow, W.J., Wharton, M., Lau, D., and Levy, J.A. (1987). Small animals are not susceptible to human immunodeficiency virus infection. J. Gen. Virol. *68 (Pt 8)*, 2253–2257.

Murphey-Corb, M., Martin, L.N., Rangan, S.R., Baskin, G.B., Gormus, B.J., Wolf, R.H., Andes, W.A., West, M., and Montelaro, R.C. (1986). Isolation of an HTLV-III-related retrovirus from macaques with simian AIDS and its possible origin in asymptomatic mangabeys. Nature *321*, 435–437.

Namikawa, R., Kaneshima, H., Lieberman, M., Weissman, I.L., and McCune, J.M. (1988). Infection of the SCID-hu mouse by HIV-1. Science *242*, 1684–1686.

Neumann, A.U., Tubiana, R., Calvez, V., Robert, C., Li, T.S., Agut, H., Autran, B., and Katlama, C. (1999). HIV-1 rebound during interruption of highly active antiretroviral therapy has no deleterious effect on reinitiated treatment. Comet Study Group. AIDS *13*, 677–683.

Nishimura, Y., Brown, C.R., Mattapallil, J.J., Igarashi, T., Buckler-White, A., Lafont, B.A., Hirsch, V.M., Roederer, M., and Martin, M.A. (2005). Resting naive CD4+ T-cells are massively infected and eliminated by X4-tropic simian-human immunodeficiency viruses in macaques. Proc. Natl. Acad. Sci. U.S.A. *102*, 8000–8005.

Nishimura, Y., Igarashi, T., Buckler-White, A., Buckler, C., Imamichi, H., Goeken, R.M., Lee, W.R., Lafont, B.A., Byrum, R., Lane, H.C., *et al.* (2007). Loss of naive cells accompanies memory CD4+ T-cell depletion during long-term progression to AIDS in Simian immunodeficiency virus-infected macaques. J. Virol. *81*, 893–902.

Orenstein, J.M., Fox, C., and Wahl, S.M. (1997). Macrophages as a source of HIV during opportunistic infections. Science *276*, 1857–1861.

Ortiz, G.M., Nixon, D.F., Trkola, A., Binley, J., Jin, X., Bonhoeffer, S., Kuebler, P.J., Donahoe, S.M., Demoitie, M.A., Kakimoto, W.M., *et al.* (1999). HIV-1-specific immune responses in subjects who temporarily contain virus replication after discontinuation of highly active antiretroviral therapy. J. Clin. Invest. *104*, R13–18.

Park, I.W., Koziel, H., Hatch, W., Li, X., Du, B., and Groopman, J.E. (1999). CD4 receptor-dependent entry of human immunodeficiency virus type-1 env-pseudotypes into CCR5-, CCR3-, and CXCR4-expressing human alveolar macrophages is preferentially mediated by the CCR5 coreceptor. Am. J. Respir. Cell Mol. Biol. *20*, 864–871.

Pedersen, N.C., Ho, E.W., Brown, M.L., and Yamamoto, J.K. (1987). Isolation of a T-lymphotropic virus from domestic cats with an immunodeficiency-like syndrome. Science *235*, 790–793.

Reimann, K.A., Li, J.T., Veazey, R., Halloran, M., Park, I.W., Karlsson, G.B., Sodroski, J., and Letvin, N.L. (1996a). A chimeric simian/human immunodeficiency virus expressing a primary patient human immunodeficiency virus type 1 isolate env causes an AIDS-like disease after *in vivo* passage in rhesus monkeys. J. Virol. *70*, 6922–6928.

Reimann, K.A., Li, J.T., Voss, G., Lekutis, C., Tenner-Racz, K., Racz, P., Lin, W., Montefiori, D.C., Lee-Parritz, D.E., Lu, Y., *et al.* (1996b). An env gene derived from a primary human immunodeficiency virus type 1 isolate confers high *in vivo* replicative capacity to a chimeric simian/human immunodeficiency virus in rhesus monkeys. J. Virol. *70*, 3198–3206.

Sadjadpour, R., Theodore, T.S., Igarashi, T., Donau, O.K., Plishka, R.J., Buckler-White, A., and Martin, M.A. (2004). Induction of disease by a molecularly cloned highly pathogenic simian immunodeficiency virus/human immunodeficiency virus chimera is multigenic. J. Virol. *78*, 5513–5519.

Sakuragi, S., Shibata, R., Mukai, R., Komatsu, T., Fukasawa, M., Sakai, H., Sakuragi, J., Kawamura, M., Ibuki, K., Hayami, M., and *et al.* (1992). Infection of macaque monkeys with a chimeric human and simian immunodeficiency virus. J. Gen. Virol. 73 *(Pt 11)*, 2983–2987.

Saravolatz, L.D., Winslow, D.L., Collins, G., Hodges, J.S., Pettinelli, C., Stein, D.S., Markowitz, N., Reves, R., Loveless, M.O., Crane, L., *et al.* (1996). Zidovudine alone or in combination with didanosine or zalcitabine in HIV-infected patients with the acquired immunodeficiency syndrome or fewer than 200 CD4 cells per cubic millimeter. Investigators for the Terry Beirn Community Programs for Clinical Research on AIDS. N. Engl. J. Med. *335*, 1099–1106.

Sellon, D.C., Perry, S.T., Coggins, L., and Fuller, F.J. (1992). Wild-type equine infectious anemia virus replicates *in vivo* predominantly in tissue macrophages, not in peripheral blood monocytes. J. Virol. *66*, 5906–5913.

Sellon, D.C., Fuller, F.J., and McGuire, T.C. (1994). The immunopathogenesis of equine infectious anemia virus. Virus Res. *32*, 111–138.

Sharma, D.P., Zink, M.C., Anderson, M., Adams, R., Clements, J.E., Joag, S.V., and Narayan, O. (1992). Derivation of neurotropic simian immunodeficiency virus from exclusively lymphocytetropic parental virus: pathogenesis of infection in macaques. J. Virol. *66*, 3550–3556.

Sheehy, A.M., Gaddis, N.C., Choi, J.D., and Malim, M.H. (2002). Isolation of a human gene that inhibits HIV-1 infection and is suppressed by the viral Vif protein. Nature *418*, 646–650.

Sheehy, A.M., Gaddis, N.C., and Malim, M.H. (2003). The antiretroviral enzyme APOBEC3G is degraded by the proteasome in response to HIV-1 Vif. Nat. Med. *9*, 1404–1407.

Shibata, R., Sakai, H., Kiyomasu, T., Ishimoto, A., Hayami, M., and Adachi, A. (1990). Generation and characterization of infectious chimeric clones between human immunodeficiency virus type 1 and simian immunodeficiency virus from an African green monkey. J. Virol. *64*, 5861–5868.

Shibata, R., Kawamura, M., Sakai, H., Hayami, M., Ishimoto, A., and Adachi, A. (1991). Generation of a chimeric human and simian immunodeficiency virus infectious to monkey peripheral blood mononuclear cells. J. Virol. *65*, 3514–3520.

Shibata, R., Hoggan, M.D., Broscius, C., Englund, G., Theodore, T.S., Buckler-White, A., Arthur, L.O., Israel, Z., Schultz, A., Lane, H.C., and *et al.* (1995). Isolation and characterization of a syncytium-inducing, macrophage/T-cell line-tropic human immunodeficiency virus type 1 isolate that readily infects chimpanzee cells *in vitro* and *in vivo*. J. Virol. *69*, 4453–4462.

Shibata, R., Maldarelli, F., Siemon, C., Matano, T., Parta, M., Miller, G., Fredrickson, T., and Martin, M.A. (1997). Infection and pathogenicity of chimeric simian-human immunodeficiency viruses in macaques: determinants of high virus loads and CD4 cell killing. J. Infect. Dis. *176*, 362–373.

Shiver, J.W., Fu, T.M., Chen, L., Casimiro, D.R., Davies, M.E., Evans, R.K., Zhang, Z.Q., Simon, A.J., Trigona, W.L., Dubey, S.A., *et al.* (2002). Replication-incompetent adenoviral vaccine vector elicits effective anti-immunodeficiency-virus immunity. Nature *415*, 331–335.

Sidtis, J.J., and Price, R.W. (1990). Early HIV-1 infection and the AIDS dementia complex. Neurology *40*, 323–326.

Sigurdsson, B., Grimsson, H., and Palsson, P.A. (1952). maedi, a chronic, progressive infection of sheep's lungs. J. Infect. Dis. *90*, 233–241.

Sigurdsson, B., Palsson, P., and Grimsson, H. (1957). Visna, a demyelinating transmissible disease of sheep. J Neuropathol Exp Neurol *16*, 389–403.

Sigurdsson, B., and Palsson, P.A. (1958). Visna of sheep; a slow, demyelinating infection. Br. J. Exp. Pathol. *39*, 519–528.

Simmons, G., Wilkinson, D., Reeves, J.D., Dittmar, M.T., Beddows, S., Weber, J., Carnegie, G., Desselberger, U., Gray, P.W., Weiss, R.A., and Clapham, P.R. (1996). Primary, syncytium-inducing human immunodeficiency virus type 1 isolates are dual-tropic and most can use either Lestr or CCR5 as coreceptors for virus entry. J. Virol. *70*, 8355–8360.

Song, R.J., Chenine, A.L., Rasmussen, R.A., Ruprecht, C.R., Mirshahidi, S., Grisson, R.D., Xu, W., Whitney, J.B., Goins, L.M., Ong, H., *et al.* (2006). Molecularly cloned SHIV-1157ipd3N4: a highly replication- competent, mucosally transmissible R5 simian-human immunodeficiency virus encoding HIV clade C Env. J. Virol. *80*, 8729–8738.

Stremlau, M., Owens, C.M., Perron, M.J., Kiessling, M., Autissier, P., and Sodroski, J. (2004). The cytoplasmic body component TRIM5alpha restricts HIV-1 infection in Old World monkeys. Nature *427*, 848–853.

Takahashi, M. (1984). Development and characterization of a live varicella vaccine (Oka strain). Biken J. *27*, 31–36.

Tan, R.C., Harouse, J.M., Gettie, A., and Cheng-Mayer, C. (1999). *In vivo* adaptation of SHIV(SF162): chimeric virus expressing a NSI, CCR5-specific envelope protein. J. Med. Primatol. *28*, 164–168.

Trkola, A., Paxton, W.A., Monard, S.P., Hoxie, J.A., Siani, M.A., Thompson, D.A., Wu, L., Mackay, C.R., Horuk, R., and Moore, J.P. (1998). Genetic subtype-independent inhibition of human immunodeficiency virus type 1 replication by CC and CXC chemokines. J. Virol. *72*, 396–404.

Trkola, A., Kuhmann, S.E., Strizki, J.M., Maxwell, E., Ketas, T., Morgan, T., Pugach, P., Xu, S., Wojcik, L., Tagat, J., *et al.* (2002). HIV-1 escape from a small molecule, CCR5-specific entry inhibitor does not involve CXCR4 use. Proc. Natl. Acad. Sci. U.S.A. *99*, 395–400.

Trujillo, J.R., Rogers, R., Molina, R.M., Dangond, F., McLane, M.F., Essex, M., and Brain, J.D. (2007). Noninfectious entry of HIV-1 into peripheral and brain macrophages mediated by the mannose receptor. Proc. Natl. Acad. Sci. U.S.A. *104*, 5097–5102.

Uberla, K., Stahl-Hennig, C., Bottiger, D., Matz-Rensing, K., Kaup, F.J., Li, J., Haseltine, W.A., Fleckenstein, B., Hunsmann, G., Oberg, B., and *et al.* (1995). Animal model for the therapy of acquired immunodeficiency syndrome with reverse transcriptase inhibitors. Proc. Natl. Acad. Sci. U.S.A. *92*, 8210–8214.

Worgall, S., Connor, R., Kaner, R.J., Fenamore, E., Sheridan, K., Singh, R., and Crystal, R.G. (1999). Expression and use of human immunodeficiency virus type 1 coreceptors by human alveolar macrophages. J. Virol. *73*, 5865–5874.

Wyatt, R., Moore, J., Accola, M., Desjardin, E., Robinson, J., and Sodroski, J. (1995). Involvement of the V1/V2 variable loop structure in the exposure of human immunodeficiency virus type 1 gp120 epitopes induced by receptor binding. J. Virol. *69*, 5723–5733.

Yi, Y., Shaheen, F., and Collman, R.G. (2005). Preferential use of CXCR4 by R5X4 human immunodeficiency virus type 1 isolates for infection of primary lymphocytes. J. Virol. *79*, 1480–1486.

Yu, X., Yu, Y., Liu, B., Luo, K., Kong, W., Mao, P., and Yu, X.F. (2003). Induction of APOBEC3G ubiquitination and degradation by an HIV-1 Vif–Cul5–SCF complex. Science *302*, 1056–1060.

Zhang, Y., Lou, B., Lal, R.B., Gettie, A., Marx, P.A., and Moore, J.P. (2000). Use of inhibitors to evaluate coreceptor usage by simian and simian/human immunodeficiency viruses and human immunodeficiency virus type 2 in primary cells. J. Virol. *74*, 6893–6910.

Zink, M.C., Yager, J.A., and Myers, J.D. (1990). Pathogenesis of caprine arthritis encephalitis virus. Cellular localization of viral transcripts in tissues of infected goats. Am. J. Pathol. *136*, 843–854.

# SIV Pathogenic and Non-pathogenic Infections

7

Thaidra Gaufin, Ivona Pandrea and Cristian Apetrei

Abstract

Simian immunodeficiency viruses (SIVs) naturally infect African non-human primates. Cross-species transmissions of SIVs from naturally infected chimpanzees/gorillas and sooty mangabeys are at the origin of HIV-1 and HIV-2, respectively. Experimental or accidental transmission of $SIV_{smm}$ to different species of macaques resulted in the development of AIDS animal models. Differently from humans and macaques which, upon infection, invariably progress to AIDS, natural SIV hosts are generally spared of disease progression. Pathogenic and non-pathogenic SIV infections share some major features, such as a very active viral replication during both acute and chronic infection, a significant acute depletion of $CD4^+$ T-cells, which is more prominent at mucosal sites, and a partial control of the virus by both adaptive and innate immune responses. Although SIVs have the potential to infect macrophages, the bulk of viral replication *in vivo* is supported by $CD4^+$ T-cells in both pathogenic and non-pathogenic infections. The major differences between the two models are: maintenance of $CD4^+$ T-cell homeostasis in natural hosts, with rebounds to near preinfection levels of peripheral $CD4^+$ T-cells and significant recovery in the intestine; normal levels of T-cell immune activation, cell proliferation and apoptosis in natural infection, while increases in these three parameters are associated with disease progression in pathogenic infection. A better understanding of the mechanisms underlying the lack of disease in natural hosts for SIV infection will probably provide important clues as to the pathogenesis of AIDS in HIV-infected individuals.

## Introduction

More than 40 different SIV types naturally infect African non-human primates (NHPs) (VandeWoude and Apetrei, 2006). Cross-species transmissions of SIVs from naturally infected chimpanzees/gorillas and sooty mangabeys are at the origin of HIV-1 and HIV-2, respectively (Hahn *et al.*, 2000; Van Heuverswyn *et al.*, 2006; VandeWoude and Apetrei, 2006). Inadvertent cross-species transmission of $SIV_{smm}$ from naturally infected sooty mangabeys to different species of macaques resulted in severe immunodeficiency and subsequent development of the animal models for AIDS (Murphey-Corb *et al.*, 1986; Mansfield *et al.*, 1995; Apetrei *et al.*, 2005; Apetrei *et al.*, 2006). In marked contrast to HIV-infected humans and SIV infection of macaques who, if untreated, almost invariably develop progressive $CD4^+$ T-cell depletion and simian AIDS (Hirsch and Johnson, 1994), the African monkey species naturally infected with SIV are generally spared from any sign of disease (Pandrea *et al.*, 2008d).

By comparing pathogenic and non-pathogenic SIV infections in their natural hosts and in macaques, it becomes possible to compare and contrast the pathogenesis, immunological parameters and cellular host factors that are important in preventing progression to AIDS. Identifying the correlates of protection against disease progression may result in developing novel therapies for HIV-infected patients. This chapter will discuss both the immunological and cellular parameters in the non-pathogenic and pathogenic models of SIV infection.

## Animal models used to study HIV pathogenesis

Simian immunodeficiency virus (SIV) was discovered a few years after the discovery of HIV-1, in the Asian macaque colony at the New England Primate Research Centre (Daniel *et al.*, 1985; Letvin *et al.*, 1985). At first, SIV was thought to naturally infect macaques – rhesus and cynomolgus – hence the name $SIV_{mac}$ (Desrosiers *et al.*, 1989; Desrosiers, 1990). However, testing of macaques in Asia failed to reveal any evidence of SIV circulation in wild macaques (Lowenstine *et al.*, 1986; Ohta *et al.*, 1988). Moreover, it rapidly became evident that the natural host of $SIV_{mac}$-like viruses was the sooty mangabey, *Cercocebus atys* (Fultz *et al.*, 1986; Murphey-Corb *et al.*, 1986). The salient observations were that SIV probably had an African origin and that SIV-infected sooty mangabeys did not develop clinical signs of AIDS-like disease (Fultz *et al.*, 1986; Murphey-Corb *et al.*, 1986; Hirsch *et al.*, 1989). Therefore, this discovery is at the origin of both the pathogenic and non-pathogenic models for the study of HIV infection.

### Animal models for natural SIV infection

When discussing the biological properties of SIVs naturally infecting African NHPs, one should consider that most of these viruses are currently known only from sequences (VandeWoude and Apetrei, 2006). Moreover, numerous host species are heavily endangered or non-accessible to research facilities equipped to carry out detailed studies of SIV pathogenesis. Therefore, to date, only three models have been used for the study of natural SIV infections: the $SIV_{smm}$-infected sooty mangabey, the $SIV_{agm}$-infected African green monkey and the $SIV_{mnd}$-infected mandrill (Silvestri *et al.*, 2007; Pandrea *et al.*, 2008d; Paiardini *et al.*, 2009). However, there is limited access to sooty mangabeys and mandrills. Both these species are endangered, which precludes invasive studies aimed at investigating viral replication and immune responses in tissues. An animal model circumventing the use of endangered species as well as the difficulties of monkey importation from Africa was recently developed by infecting Caribbean African green monkeys with $SIV_{agm}$.sab92018 (Pandrea *et al.*, 2006b). During recent years, this model has been extensively used in pathogenesis studies aimed at understanding the mechanisms of the lack of disease progression in natural hosts of SIVs (Pandrea *et al.*, 2005, 2007a,b, 2008a,b; Gaufin *et al.*, 2009b,c; Mandell *et al.*, 2009).

### Animal models for pathogenic SIV infection

Accidental cross-species transmissions of SIVs to pigtailed macaques, crab-eating macaques, and rhesus macaques have resulted in progression to simian AIDS and the development of animal models for pathogenic SIV infection (Desrosiers, 1990; Apetrei *et al.*, 2005). Note that even prior to its discovery, SIV might have been adapted through serial passages in macaques (Gormus *et al.*, 2004; Apetrei *et al.*, 2006), which may account for the very high pathogenic potential in macaques of the currently available reference strains.

SIV infection in macaques closely mimics the pathogenesis, virology, immunology and pathology of HIV infection in humans and progression to AIDS, therefore being a very valuable animal model for HIV infection. Over the years, this model has been used for vaccine research (Desrosiers, 2004) and to demonstrate some of the key features of lentiviral infections showing that an infectious molecular clone can cause simian AIDS (Desrosiers, 1990) and that the mucosal tract is the major site of HIV replication and $CD4^+$ T-cell depletion (Smit-McBride *et al.*, 1998; Veazey *et al.*, 1998). Recently, studies on virus transmission and dissemination from the mucosal sites of entry were also carried out in macaques (Keele *et al.*, 2009a; Ma *et al.*, 2009). Therapeutic studies in macaques were used to develop several classes of antiretroviral drugs (Van Rompay *et al.*, 1998). Macaques infected with $SIV_{smm}$ also served as an animal model for transmission of virus during breastfeeding (Amedee *et al.*, 2004). Finally, macaque studies permit invasive approaches of selective depletion of specific components of the immune system to investigate the correlates of disease progression (Schmitz *et al.*, 1999b, 2003; Gaufin *et al.*, 2009a).

The major criticisms of the currently available macaque models of SIV infection are that (i) $SIV_{mac}$ infection in macaques is more pathogenic than HIV-1 in humans; (ii) SIV diversity does not cover the enormous diversity of HIV-1 and (iii) that $SIV_{mac}$ is significantly different from HIV-1 (Staprans and Feinberg, 2004). Approaches aimed at solving this third issue, especially for vaccine development, resulted in the generation of simian–human immunodeficiency viruses but these were recently shown to not entirely reproduce the properties of the parental strains (see Chapter 6).

## Virological aspects of SIV infection

### Coreceptor usage of SIVs

Both HIV and SIVs use the CD4 receptor, which is expressed on the surface of T-cells and macrophages, and chemokine receptor 5 (CCR5), which is the main coreceptor (Chen *et al.*, 1997; Schols and De Clercq, 1998; Owen *et al.*, 2000; Zhang *et al.*, 2000; Pandrea *et al.*, 2005). For HIV-1 replication *in vivo* another relevant coreceptor is CXCR4 (Moore *et al.*, 2004); in 50% of cases, progression to AIDS is characterized by a switch in viral tropism from CCR5 (so called 'macrophage'-tropic) to CXCR4 (so called 'lymphocyte'-tropic) viruses (Moore *et al.*, 2004). Note that this discrimination linking tropism with coreceptor usage may be confusing, as most of the SIVs use CCR5 exclusively as the coreceptor and infect mainly lymphocytes *in vivo*. Most of the SIVs which naturally infect African NHPs are reported to use the same system of receptors as HIV-1, but coreceptor switch does not generally occur during SIV infections (Moore *et al.*, 2004).

Moreover, in NHPs, no correlation can be established between the coreceptor usage and the pathogenesis *in vivo*. Thus, the highly pathogenic $SIV_{mac}$ is CCR5 tropic (Zhang *et al.*, 2000) in spite of being more virulent than HIV-1 (Hirsch and Johnson, 1994). $SIV_{mnd}$-1, $SIV_{agm}$.sab and some strains of $SIV_{smm}$ were reported to use CXCR4, but no clear pathogenic correlation has been described in these species (Schols and De Clercq, 1998; Owen *et al.*, 2000; Pandrea *et al.*, 2005). Experimental infection of sabaeus African green monkeys with CXCR4/CCR5 tropic $SIV_{agm}$.sab and of mandrills with $SIV_{mnd}$-1 does not show a different pattern of viral replication or disease progression compared to other SIV infections in natural NHP hosts (Onanga *et al.*, 2002, 2006; Pandrea *et al.*, 2006b, 2007b, 2008b,c); moreover, it was reported that replicative $SIV_{agm}$.sab uses only CCR5 as a major coreceptor *in vivo* (Kornfeld *et al.*, 2005; Pandrea *et al.*, 2005, 2007b), as does $SIV_{mnd}$-1 (Pandrea *et al.*, 2001). A recent study reported that expansion of coreceptor usage by $SIV_{smm}$ resulted in a massive $CD4^+$ T-cell depletion in sooty mangabeys, which was not associated with disease progression (Milush *et al.*, 2007).

*In vitro* studies identified another notable difference in coreceptor usage between $SIV_{agm}$.sab (that preferentially uses Bonzo/STRL33) and $SIV_{mac}$ (that preferentially uses Bob/GPR15). These two coreceptors were described for numerous strains *in vitro* but their *in vivo* biological significance seems to be negligible. This preferential usage of Bonzo and Bob correlates with the expression patterns of these two coreceptors on human T-cell lines. Thus, Bonzo is expressed on cells highly permissive for $SIV_{agm}$ (SupT1 and MOLT4 clone 8), whereas high levels of Bob, but not of Bonzo, are expressed on cells that preferentially replicate $SIV_{mac}$ (CEMx174 and HUT78) (Muller and Barre-Sinoussi, 2003) (I Pandrea, unpublished).

Finally, $SIV_{rcm}$, the virus that naturally infects red-capped mangabeys (*Cercocebus torquatus*) (Georges-Courbot *et al.*, 1998; Beer *et al.*, 2001; Bailes *et al.*, 2003) uses CCR2b instead of CCR5 or CXCR4 as the coreceptor for viral entry (Chen *et al.*, 1998; Beer *et al.*, 2001) due to a 24-base pair deletion in the *env* binding region of the CCR5, which occurs in the majority of red-capped mangabeys (Chen *et al.*, 1998). As such, $SIV_{rcm}$ is an example of a primate lentivirus that has adapted its coreceptor usage to allow productive replication and persistence in its natural host species.

Given the unique coreceptor usage, along with the observation that the CCR2 coreceptor is mainly expressed on macrophages and only at very low levels on $CD4^+$ T-cells, $SIV_{rcm}$ may be ideal for pathogenesis studies of macrophage-tropic viruses. To date, there is no available data on $SIV_{rcm}$ replication in its natural host, but it is believed that $SIV_{rcm}$ rarely causes disease in red-capped mangabeys, suggesting that it has been associated with its natural host for a long time. However, upon experimental infection of pigtailed macaques, the $SIV_{rcm}$ replication and memory $CD4^+$ T-cell depletion patterns were similar to those observed in lymphotropic SIV infections and it was demonstrated that following inoculation, $SIV_{rcm}$ expanded its coreceptor usage, becoming able to use CCR4 *in vivo* in addition to CCR2 (Gautam *et al.*, 2009). This result indicates that lentiviral adaptation to a new host may occur rapidly through strain selection and involve new pathogenic pathways, pointing to the need for *in vivo* studies to characterize the pathogenesis of primate lentiviruses. Similarly, studies have shown that the 'macrophage-tropic' $SIV_{mac}$ strains are actually predominantly infecting lymphocytes *in vivo* (Borda *et al.*, 2004).

The natural hosts of SIV infection, such as African green monkeys, sooty mangabeys, mandrills and chimpanzees, express lower levels of CCR5 on memory $CD4^+$ T-cells in PBMC and mucosal tissues, compared to disease-susceptible hosts such as macaques, baboons, and humans (Pandrea *et al.*, 2007a; Pandrea *et al.*, 2008c). Moreover, chimpanzees, which are considered a more recent host of SIV, show an intermediate level of $CD4^+CCR5^+$ expressing T-cells (Pandrea *et al.*, 2007a). A hallmark of pathogenic primate lentiviral infections is early and persistent depletion of mucosal memory/activated $CD4^+$ T-cells. Activated memory cells express high levels of CCR5, thus explaining this pathogenic feature in the context of CCR5 usage by the majority of SIV and HIV strains (Veazey *et al.*, 1998, 2000a,b; Brenchley *et al.*, 2004; Mehandru *et al.*, 2004, 2007; Li *et al.*, 2005; Mattapallil *et al.*, 2005). In this context, one would expect that African species that express lower levels of CCR5 on $CD4^+$ T-cells, especially at the mucosal sites, may be less susceptible to endemic naturally occurring SIVs because they have fewer receptor targets for infection (Pandrea *et al.*, 2007a). This hypothesis would provide an elegant evolutionary mechanism of 'peaceful coexistence' between SIV replication and natural host immune system function (Silvestri *et al.*, 2007).

While this possibility presents an attractive explanation for species differences in SIV disease expression, the fact that the levels of SIV replication in non-pathogenic infections are equivalent to those observed in pathogenic infections does not support a simple association between the number of susceptible target cells and disease. Moreover, a dramatic depletion of $CD4^+$ T-cells in the intestine has been reported in African green monkeys and sooty mangabeys during acute SIV infection (Gordon *et al.*, 2007; Pandrea *et al.*, 2007b; Favre *et al.*, 2009). Pathogenic infections are characterized by ongoing viral replication during chronic infection which is associated with continuous depletion of mucosal $CD4^+$ T-cells. In natural hosts there is a significant, albeit incomplete recovery of $CD4^+$ T-cells in the intestine during chronic infection, in spite of sustained viral replication, but in the context of normal levels of immune activation, T-cell proliferation and apoptosis. More studies of the *in vivo* dynamics of $SIV_{agm}$ and $SIV_{smm}$ replication are needed to investigate if the major target cells of these viruses differ from those of $SIV_{mac}$ and HIV.

## T-cell tropism vs. macrophage tropism

### *Cell and tissue tropism*

Upon infection, SIVs are disseminated to tissues by the blood. The major sites of SIV replication are the lymph nodes, gastrointestinal tract, spleen and other lymphoid tissues. The target cells are $CD4^+$ T-lymphocytes and macrophages, with lymphocytes vastly predominating in terms of numbers of infected cells. The preferential infection of T-lymphocytes by SIV may be related to expression of lower levels of CD4 molecule on the macrophage surface (Bannert *et al.*, 2000).

Studies have shown that, although $SIV_{mac}239$ does not replicate efficiently within macrophages in culture, it does infect macrophages *in vivo* (Veazey *et al.*, 1998; Wykrzykowska *et al.*, 1998; Williams *et al.*, 2001). Furthermore, 40% of rhesus macaques that die with simian AIDS after $SIV_{mac}239$ infection display lesions associated with infection of macrophages, such as encephalitis and giant cell disease (Sasseville and Lackner, 1997) (Fig. 7.1). Nine amino acid changes in the envelope protein were associated with the acquired ability of a $SIV_{mac}239$ variant (designated $SIV_{mac}239/316$) to replicate efficiently in monocytes/macrophages *in vitro* (Desrosiers *et al.*, 1991). Further work has shown that four out

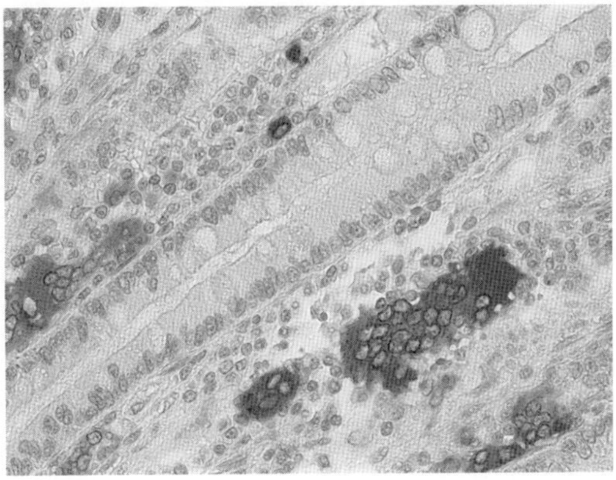

**Figure 7.1** Giant cell disease in simian AIDS. Numerous SIV-infected syncytial cells (dark grey, as detected by p27 Gag immunohistochemistry) can be observed in the lamina propria of the intestine of $SIV_{smm}$-infected rhesus macaques.

of these nine amino acid changes (V67M, K176E, G382R and K573T) are strongly associated with the ability to replicate well in cultured macrophages (Mori *et al.*, 1992; Mori *et al.*, 2000), a significant decrease in the requirement for CD4 before CCR5 binding and an increase in susceptibility to antibody neutralization (Edinger *et al.*, 1997a; Edinger *et al.*, 1997b; Means *et al.*, 2001; Puffer *et al.*, 2002). It was therefore speculated that $SIV_{mac}$239/316 may have a broader tropism than the parental $SIV_{mac}$239 with a particular preference for $CCR5^{high}$ $CD4^{low}$ populations such as alveolar macrophages, parenchymal microglia in brain and other cell types, particularly those in immunologically privileged sites such as the brain. However, *in vivo* studies failed to demonstrate a broadened cell tropism for $SIV_{mac}$239/316 compared to $SIV_{mac}$239 and reported that the only cell type infected *in vivo* by $SIV_{mac}$239/316 is T-cells (Borda *et al.*, 2004).

It was also suggested that the immune dysfunction caused by the highly pathogenic $SIV_{mac}$239 is due to increased cytokine production (Orandle *et al.*, 2001, 2002), up-regulation of adhesion molecules, and activation of tissue macrophages which thus become susceptible to infection by $SIV_{mac}$239/316. Studies that used $SIV_{mac}$239, $SIV_{mac}$239/316 and other macrophage-tropic viruses showed no evidence that expanded coreceptor usage was responsible for these alterations in cellular tropism (Puffer *et al.*, 2002).

Macrophage infection by $SIV_{mac}$239 has been consistently observed *in vivo* by 21 days after infection, before any evidence of SIV encephalitis or other pathology associated with macrophage infection (Veazey *et al.*, 1998; Wykrzykowska *et al.*, 1998; Williams *et al.*, 2001; Borda *et al.*, 2004). With regard to the mechanism of SIV encephalitis, studies have shown that infected macrophages in the brains of SIV-encephalitic macaques are predominantly distributed in perivascular areas, suggesting recent entry (Williams *et al.*, 2001). Therefore, development of SIV encephalitis may be the result of new virus entering the CNS. It is unclear whether these recently entered macrophages were infected before entering the brain or became infected upon entering the brain. Because disruption of the physical blood–brain barrier is a late event, cell-free plasma virus is unlikely to enter the CNS but rather virus probably enters the CNS within monocytic elements (Trojan horse theory) (Davis *et al.*, 1992). In this scenario, development of encephalitis may result from increased trafficking of SIV-infected monocytes (Chakrabarti *et al.*, 1991; Lane *et al.*, 1996). However, the number of blood monocyte-associated SIV DNA copies did not distinguish macaques that developed SIV encephalitis from those that did not. Paradoxically, macaques that did not develop encephalitis had more SIV-infected macrophages in the lungs and thymus than macaques with SIV encephalitis (Bissel *et al.*, 2006a,b). This suggests that there may be inherent differences in the ability of individual host monocytes to become productively infected or produce virus in animals that develop SIV encephalitis. Future studies are needed to establish whether monocytes from macaques that develop SIV encephalitis have greater susceptibility to be infected, produce virus, or traffic into the brain.

In natural SIV infections, studies thus far reported that the virus replicates mostly in lymphocytes (Pandrea *et al.*, 2007b, 2008b; Gordon *et al.*, 2008; Kaur *et al.*, 2008). Although high levels of viral replication were detected in the cerebrospinal fluid, no cases of encephalitis were reported in natural hosts (Broussard *et al.*, 2001). However, in the few cases that progressed to simian AIDS, giant cells harbouring SIV have been identified, suggesting that progression is associated with expanded cell tropism to macrophages

(Apetrei *et al.*, 2004; Ling *et al.*, 2004; Pandrea *et al.*, 2009).

*Viral dynamics studies of replication in short-lived versus long-lived cells*

Early mathematical modelling studies measuring the *in vivo* half-life of infected CD4$^+$ T-cells in HIV-1-infected patients using protease inhibitor treatments reported that the destruction to CD4$^+$ T-lymphocytes is of the order of $10^9$ cells per day. This observation also implies that around $10^9$ cells are replaced each day (Ho, 1995b). The clearance of free virions (virus half-life) occurs at a rate of 6 hours, as estimated in patients undergoing treatment with ritonavir. Some of the earlier studies have also predicted that the mean half-life for an infected CD4$^+$ T-cell is about 1.6 days (Perelson *et al.*, 1996). Other techniques such as plasma apheresis were used to measure virus clearance from the plasma and found that HIV had a half-life of approximately an hour (Perelson, 2002).

The basic reproductive number (Ro) of an HIV-1-infected cell (an approximation of the number of cells that will be infected by a single infected cell assuming that there are unlimited target cells available) was estimated at 7.1 days when using a mathematical formula that assumed no time delay in infecting other cells and production of new virions. However, when a fixed delay of 1 day between infection of a new cell and production of new virions (probably more biologically relevant) was included in the mathematical formula, a mean Ro value of 19.3 days was obtained. In these calculations, a half-life of infected CD4$^+$ T-cells of 2.4 days was established for HIV-infected patients without treatment while in patients on antiretroviral therapy the half-life of infected CD4$^+$ T-cells was 1.9 days, slightly lower than previously reported (Little *et al.*, 1999).

When more antiretroviral drugs were used in combination in HIV-1-infected patients, the mathematical models predicted a faster rate of clearance of HIV-1-infected CD4$^+$ T-cells. Thus, in chronic HIV-infected patients receiving quadruple antiretroviral therapy, the half-life for virus-infected cells was approximated to be 0.7 days, which also corresponds to a faster generation time of about 2 days for HIV-1 (Markowitz *et al.*, 2003). These values are probably more accurate because it can be assumed that the combined therapies provide approaching 100% effectiveness and that the half-lives of infected cells can be measured more precisely (Markowitz *et al.*, 2003).

A second, slower, decay profile was also reported for HIV-1-infected patients. This is probably caused by HIV infection in secondary sources, such as infected macrophages in the tissues, dendritic cells, and latent lymphocytes that have been activated (Perelson *et al.*, 1997). The rate of this secondary decay profile can be measured after the majority of virus has been eliminated from the short-lived cells, meaning that the secondary sources become the main producers of virus and was modelled in HIV patients taking a triple antiretroviral therapy (Perelson *et al.*, 1997). The long-lived cells do not contribute greatly to the production of HIV, as the authors maintain that 1–7% of the total virus is produced from long-lived cells. The half-life for long-lived cells infected with HIV was measured at 1–4 weeks with a mean of 14.1±7.5 days (Perelson *et al.*, 1997).

Mathematical modelling of $SIV_{mac}$ clearance was done after a bolus injection of virus particles into the femoral artery of one rhesus macaque and removal of blood from the other femoral artery to measure the disappearance of virus particles. The free virus decay rates in the plasma were measured at 3.3–4.6 min, which is a much faster rate than what has been determined in humans. Another method utilized intravenously injected $SIV_{mac}239$ virus particles at a constant rate and measured the plasma viral load (VL) before and after the virus particle injection. The rate of virus decay was measured at 1.3–4.3 min. Serial bolus injection of virus particles, with the first injection solely containing $SIV_{mac}239$ particles and the second one containing $SIV_{mac}239$ particles mixed with IgG anti-$SIV_{mac}239$ (to bind to all of the viral particles), showed a similar pattern of viral RNA increase followed by decrease in rhesus macaques, indicating that the half-lives of free viral particles are extremely short and that additional immunological components do not change this rate (Zhang *et al.*, 1999). The rate of infected CD4$^+$ T-cell clearance was estimated at 0.47 days for a first-phase decay of pathogenic $SIV_{mac}251$ in CD4$^+$ T-cells (Brandin *et al.*, 2006).

It has repeatedly been suggested that one of the reasons why natural hosts of SIVs do not progress to simian AIDS is because of lower viral cytopathicity compared to pathogenic infection, thus resulting in a longer life span of infected cells. However, the measurement of the lifespan of SIV-infected cells in both African green monkeys and sooty mangabeys during the chronic phase of infection when the animals had stable VLs actually showed that virus decay rates are much faster in natural hosts than has been described for HIV or pathogenic SIV infections in rhesus macaques (Gordon *et al.*, 2008; Pandrea *et al.*, 2008b). Thus, the *in vivo* lifespan for the $SIV_{agm}$ replication in infected cells was estimated at 4 to 9.5 hours (Pandrea *et al.*, 2008b). The half-life of $SIV_{smm}$ in infected $CD4^+$ T-cells in sooty mangabeys was slightly lower than for African green monkeys and was estimated at 1.1 days. At the end of treatment, the rebound of $SIV_{agm}$ replication was slower than that described for sooty mangabeys (Gordon *et al.*, 2008; Pandrea *et al.*, 2008b). In chronic $SIV_{agm}$ infection of African green monkeys it has been shown that the Ro value was 6.3±1.3 days, which is faster than what has been observed for HIV-infected humans, but also corresponds to the faster decay rate that was observed in these animals (Pandrea *et al.*, 2008b).

Altogether, these studies indicate the importance of short-lived cells for virus replication during the chronic phase of infection in both natural hosts of SIV and pathogenic SIV infections. Long-lived cells, such as macrophages, do not significantly contribute to the viral replication. These viral dynamics data were confirmed by *in situ* studies showing SIV colocalization with lymphocytes and not macrophages in both African green monkeys and sooty mangabeys. Only minor fluctuations in $CD4^+$ T-cells were observed in sooty mangabeys during and after antiretroviral therapy (Gordon *et al.*, 2008). Because there were no great changes in the levels of $CD4^+$ T-cells, it indicates that the recovery of $CD4^+$ T-cells during the steady-state of viral replication may be involved in maintaining T-cell homeostasis (Pandrea *et al.*, 2008b). Interestingly, following cessation of antiretroviral treatment, there was an increase in the proliferation of $CD4^+$ T-cells as compared to baseline levels (Gordon *et al.*, 2008; Pandrea *et al.*, 2008b).

In pathogenic SIV infections, there is a high turnover of $CD4^+$ T-cells, as suggested by labelling experiments with the nucleoside analogue bromodeoxyuridine (BrdU). The animals with the highest VL were reported to have the highest turnover of $CD4^+$ T-cells in the peripheral blood. The mean percentage for the rates of the daily replacement of $CD4^+$ T-cells was 0.9% for uninfected animals, 2.3% for the animals with low VL, and 2.7% for infected animals with high VLs (Mohri *et al.*, 1998). These rates indicate that there is a distinctive increase in the proliferation of $CD4^+$ T-cells in pathogenic SIV infections. However, this model was subsequently challenged by Grossman *et al.*, who suggested that the decrease in percentage of BrdU is caused by a burst of replication in unlabeled cells that are activated by new stimuli. These activated cells are excessively labelled in activated compartments but their death has little impact on $CD4^+$ depletion (Grossman *et al.*, 1999). Furthermore, it was also proposed that the signal is diluted in the activated cells and that labelled cells that are activated undergo a more rapid death, which is perhaps missed in the previous model (Grossman *et al.*, 1999). However, the observation that there is a steady increase in BrdU $CD4^+$ T-cells over a 3-week period rather than the occurrence of spikes and dips did not support this hypothesis (Grossman *et al.*, 1999). A simple explanation for all these discrepancies observed during the BrdU studies is that after the administration of HAART in HIV patients, there is a redistribution of $CD4^+$ T-cells from the lymph nodes into the blood along with a decrease in the levels of adhesion molecules in the lymph nodes, which controls the passage of lymphocytes between the blood and lymph nodes (Bucy *et al.*, 1999). This observation shows evidence in favour of a model in which activated cells are being replaced from the lymph nodes into the peripheral blood.

Despite the controversy surrounding the replacement of $CD4^+$ T-cells, it is quite apparent that short-lived activated $CD4^+$ T-cells are more important contributors to the VL in both pathogenic and non-pathogenic SIV infections in NHPs than are the long-lived cells, such as macrophages. In fact, it has recently been reported that experimental *in vivo* depletion of $CD4^+$ T-cells in $SIV_{smm}$-infected sooty mangabeys resulted in a significant decrease in the VL (of approximately 2 logs) in these animals (Klatt *et al.*, 2008). However, even though the majority of virus is produced in the short-lived cells, the underlying fact is that treatment should also target the minor producers of HIV and SIV in other secondary sources such as macrophages since these cells also contribute slightly to the VLs.

### Viral determinants of pathogenicity

Several viral factors were reported to be related to virulence in both natural and pathogenic SIV infections. $SIV_{mac}$ replicates poorly and has low pathogenicity in macaques when the nef gene is deleted (Kestler *et al.*, 1991). This observation was corroborated by the description of HIV-1-infected long-term progressor nef gene mutations (Mariani *et al.*, 1996). Studies have shown that some Nef functions include the ability to down-regulate CD4, CD28 and class I MHC (Wei *et al.*, 2003), resulting in virus immune evasion (Johnson and Desrosiers, 2002). Nef may also enhance the responsiveness of T-cells to activation (Fortin *et al.*, 2004; Fenard *et al.*, 2005), but this effect is not uniformly observed among primate lentiviruses (Bell *et al.*, 1998; Howe *et al.*, 1998; Munch *et al.*, 2005). However, all SIVs from African monkeys have open reading frames corresponding to a functional Nef, therefore nef structure may not entirely account for differences in pathogenicity. It was recently suggested that Nef from the great majority of primate lentiviruses, including HIV-2, down-modulates TCR-CD3 in infected T-cells in natural hosts, which subsequently blocks T-cell activation. In contrast, Nef derived from HIV-1 and a subset of closely related SIVs does not induce CD3 down-regulation, which may have predisposed the simian precursor of HIV-1 to greater pathogenicity in humans (Schindler *et al.*, 2006).

## Viral load (VL) as an indicator of pathogenicity

In HIV-infected patients and $SIV_{mac/smm}$-infected macaques levels of plasma VLs are the best predictor of disease progression (Hirsch *et al.*, 1996; Mellor's *et al.*, 1996; Lifson *et al.*, 1997; Nowak *et al.*, 1997; Watson *et al.*, 1997; Lyles *et al.*, 2000). Asymptomatic patients show low levels of VL, while progression of AIDS in patients resulting from failure of the immune system to control HIV replication or failure of antiretroviral treatments is always associated with significant increases in VLs (Lyles *et al.*, 2000). The corollary is that immune or therapeutic control of virus should result in low VLs (Ho, 1995a; Mellor's, 1998) and therefore, the study of viral replication is of interest in both pathogenic and non-pathogenic SIV and HIV infections.

In captive African NHPs naturally infected with SIVs like $SIV_{smm}$, $SIV_{agm}$, $SIV_{mnd\text{-}1}$ and $SIV_{mnd\text{-}2}$, VLs quantified during the chronic phase of infection are higher than those in HIV-1 chronically infected asymptomatic patients (Pandrea *et al.*, 2006a) (Fig. 7.2). The only investigation of SIV VLs in wild animals showed similar high levels of viral replication (Pandrea *et al.*, 2003). Longitudinal analyses of the dynamics of plasma viraemia in natural SIV infections suggest that the level of viral replication is relatively constant over time (Pandrea *et al.*, 2001; Pandrea *et al.*, 2003; Apetrei *et al.*, 2007). Interestingly, the $SIV_{mnd\text{-}1}$-infected mandrill and the $SIV_{smm}$-infected sooty mangabey progressed to simian AIDS when the set point level of viral replication was higher than average (Pandrea *et al.*, 2001; Pandrea *et al.*, 2003; Apetrei *et al.*, 2007).

Some species-specific differences could be observed in comparative studies (Pandrea *et al.*, 2006a). African green monkeys naturally infected with $SIV_{agm}$ display a considerably wider range of VLs than those observed in sooty mangabeys (Goldstein *et al.*, 2000; Broussard *et al.*, 2001; Silvestri *et al.*, 2003; Goldstein *et al.*, 2006; Apetrei *et al.*, 2007). Proviral load in lymph node mononuclear cells from African green monkeys are 100-fold lower than the viral DNA loads observed in mononuclear cells from naturally infected sooty mangabey or mandrill lymph nodes

**Figure 7.2 (opposite)** Comparison of major parameters of HIV and SIV nonpathogenic infections. (A) The levels of viral replication are similar between naturally-infected nonhuman primates and pathogenic HIV infection. The major difference occurs in the late stage of the chronic infection, when the steady state is maintained in natural SIV infection and viral loads increase in HIV-infected patients with disease progression.

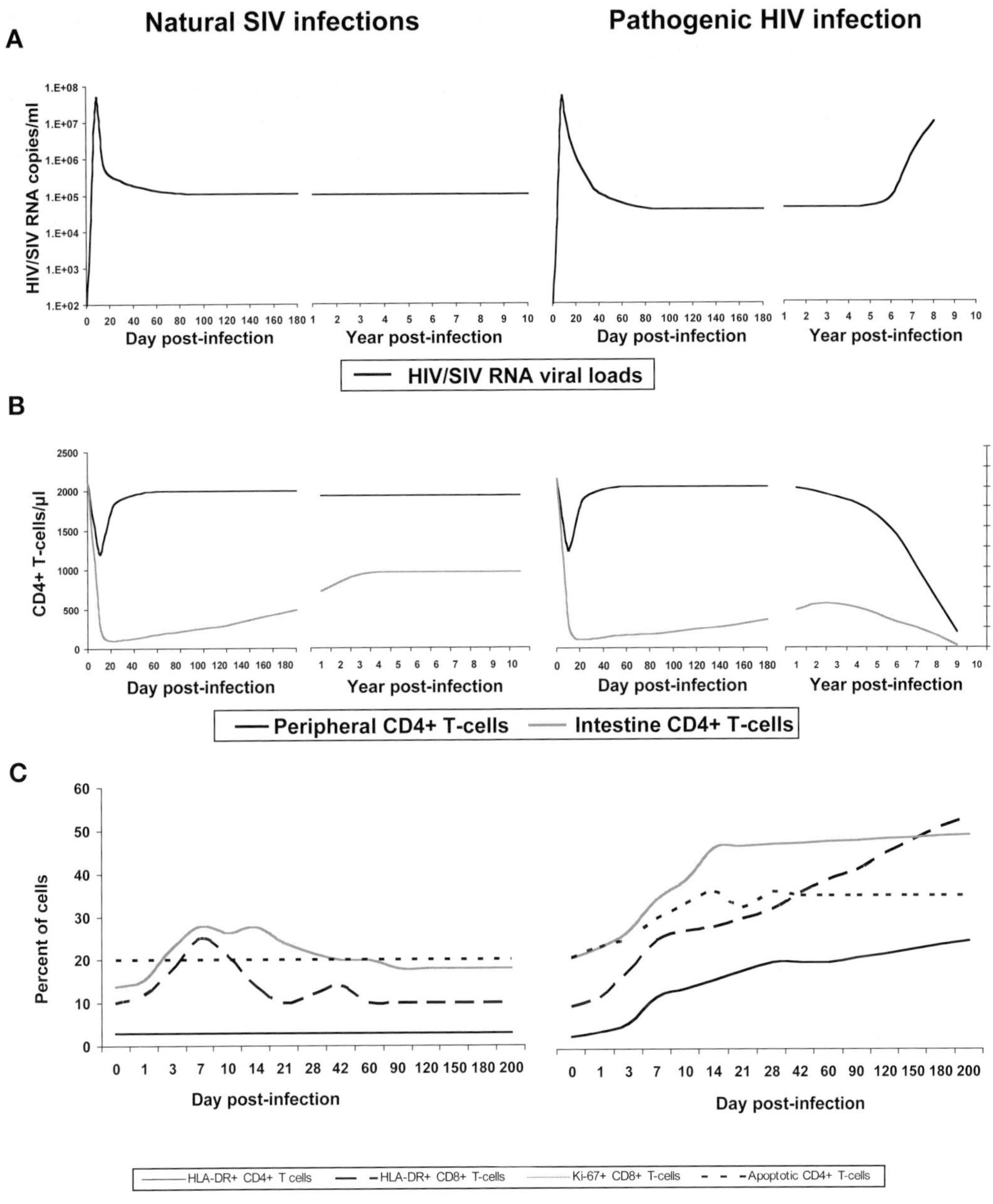


(B) The dynamics of peripheral CD4+ T-cells, as well as mucosal (intestinal) CD4+ T-cells is relatively similar during acute infection in both types of infection. During the chronic infection, there is a partial restoration in natural infections resulting in the stabilization of the mucosal CD4+ T-cell pool, while in pathogenic HIV infection, a partial restoration occurs only in the initial stages, and CD4+ T-cell depletion continues to exhaustion with disease progression. (C) Major differences are observed between natural SIV infections and pathogenic HIV infection, prevent disease progression in natural hosts and result in disease progression in HIV-infected patients. An anti-inflammatory milieu is rapidly installed in natural hosts after SIV infection. As a result, the levels of immune activation (illustrated here by the levels of –DR expression on both CD4+ and CD8+ T-cells), T-cell proliferation (illustrated by the levels of Ki-67+ CD8+ T-cells) and apoptosis of CD4+ T-cells in the intestine are relatively well controlled in natural infections. The integrity of the mucosal immunological barrier is maintained in natural hosts, which in turn prevents chronic immune activation and disease progression. In pathogenic HIV infection, a pro-inflammatory milieu results in excessive immune activation, apoptosis and T-cell proliferation, generating severe damage of the mucosal barrier and microbial translocation which maintains the chronic immune activation and results in disease progression. Cellular markers are presented as percentages of the major T-cell population. Pathogenic SIV infection of rhesus macaques display all the pathogenic features of HIV infection.

(Beer *et al.*, 1996; Onanga *et al.*, 2002; Onanga *et al.*, 2006). Altogether these data suggest that VL in $SIV_{agm}$-infected African green monkeys is generally lower than that observed in other African NHP species and that viral replication kinetics may differ among African NHP natural hosts of SIV without significant pathogenic consequences.

A series of studies has investigated the dynamics of early SIV replication by performing experimental SIV infections in natural hosts such as sooty mangabeys, African green monkeys, and mandrills (Dip *et al.*, 2000; Onanga *et al.*, 2002; Kornfeld *et al.*, 2005; Pandrea *et al.*, 2005, 2006a,b, 2008c; Silvestri *et al.*, 2005; Goldstein *et al.*, 2006; Onanga *et al.*, 2006; Gordon *et al.*, 2007). These studies showed a consistent pattern of SIV replication dynamics which consist of a peak of viraemia ($10^6$–$10^9$ copies/ml of plasma) occurring around days 9–11 post infection (Pandrea *et al.*, 2006a) (Fig. 7.2). Peak viraemia is followed by a sharp decline (1–2 logs) and attainment of a stable level of viral replication (set point), which is maintained during the chronic phase of infection (Pandrea *et al.*, 2006a). Therefore, it does not appear that the lack of disease in naturally infected monkeys is associated with effective host containment of viral replication in these species.

## Immunological aspects of SIV infection

One of the reasons why SIV is difficult to control is that it destroys important elements of the immune system and does not induce protective immune effectors. In this section we will review what is currently known about the immune response in both natural hosts and pathogenic SIV infections.

### Humoral immune response against SIV infection

In theory, neutralizing antibodies (Nabs) should protect against SIV infection. However, infusing SIV immunoglobulin into NHPs has resulted in mixed conclusions concerning the effects of passive immunization against SIV. It has been indicated that administering anti-SIV antibodies is effective and results in lower or undetectable VLs or better disease outcome (Plutonian *et al.*, 1991; Levy *et al.*, 1994; Conley *et al.*, 1996; Haigwood *et al.*, 1996; Yamamoto *et al.*, 2007). These studies have shown that sterilizing immunity is not reached, but that the early administration of immunoglobulins alters the disease course of the virus. It is important to note that passive immunization is not effective in all animals, with some experiencing sterilizing immunity while others become infected. Moreover, the levels of Nab titres correlate with disease progression, as rhesus macaques with the highest NAb titres were also demonstrated to have slower disease progression (Dykhuizen *et al.*, 1998). In contrast, administration of immunoglobulin during the chronic phase of infection has no impact on the VLs in primates (Binley *et al.*, 2000). Non-neutralizing antibodies appear two to three weeks after infection. Antibody-dependent cellular cytotoxicity (ADCC) is a mechanism by which the Fc portion of non-neutralizing antibodies binds to infected cells and is then engulfed by phagocytic cells such as macrophages, neutrophils, gamma-delta (gδ) T-cells, and natural killer (NK) cells. Higher ADCC levels in $SIV_{mac}$251-infected rhesus macaques correlate to lower VLs and thus, are thought to be involved in SIV control (Gomez-Roman *et al.*, 2005).

Depletion of CD20 B-cells have been performed in NHPs to determine the impact of ablating humoral immune responses on SIV replication and also upon disease progression. Depletion of CD20 B-cells is achieved using a chimeric human–mouse anti-CD20 monoclonal antibody, rituximab (Keystone, 2005). With $SIV_{mac}$251 infections of rhesus macaques, no difference in acute and chronic VLs was observed between rhesus macaques depleted of CD20 cells and undepleted controls (Johnson *et al.*, 2003; Schmitz *et al.*, 2003). A continuous administration of Rituximab in $SIV_{mac}$239-infected rhesus macaques resulted in no discernable changes in viral replication, but rather showed a significantly shorter duration of disease progression in $CD20^+$ B-cell-depleted rhesus macaques as compared to undepleted controls (Miller *et al.*, 2007). On the other hand, CD20 depletion studies on rhesus macaques using a neutralization-susceptible $SIV_{smm}$ (D215) strain demonstrated no significant change in viral replication or disease progression between CD20 depleted or undepleted rhesus macaques (Gaufin *et al.*, 2009a). The only study of CD20 depletions in African green monkeys infected with $SIV_{agm}$ also failed to identify any significant impact of humoral immune responses in controlling $SIV_{agm}$ replication (Gaufin *et al.*, 2009c).

Altogether, this series of experiments did not clearly establish a significant contribution of humoral immune responses in controlling SIV replication. Together with passive immunoglobulin infusion results that also failed to reveal any significant impact of antibodies on controlling viral replication during the chronic infection, these results suggest that antibodies play no major role in the control of SIV replication in natural hosts and pathogenic SIV infections.

## Cellular immune response during SIV infection

Together with humoral immune responses, cellular immune responses belong to the adaptive arm of the immune system. The cellular immune response comprises mainly $CD8^+$ T-cells, $CD4^+$ T-cells, Th17, and T regulatory cells (Tregs). $CD8^+$ T-cells exert cytotoxic T-lymphocyte (CTL) activity with the help of $CD4^+$ T helper cells (Hogan and Hammer, 2001) to destroy infected cells. Th17 cells are a subset of cells thought to control inflammation at tissue sites (Park *et al.*, 2005). Tregs suppress the activation of the immune system (Coleman *et al.*, 2007) and are described under the immune activation section. Due to the failure of an antibody-based vaccine trial (Flynn *et al.*, 2005) and other research showing the importance of $CD8^+$ T-cells in modulating HIV replication (Borrow *et al.*, 1994; Koup *et al.*, 1994; Jin *et al.*, 1999; Kuroda *et al.*, 1999; Schmitz *et al.*, 1999b), there has been a shift in the past ten years towards developing a vaccine based on CTL responses by $CD8^+$ cells only. CTLs have been proposed to be the most effective way to control viral replication, $CD4^+$ T-cell loss, and disease progression in HIV infection (McMichael, 2006; Watkins *et al.*, 2008). However, as shown recently, the first CTL-based vaccine trial carried out thus far failed to control HIV infection (Sekaly, 2008; Watkins *et al.*, 2008). In the following section, the roles of both $CD8^+$ and $CD4^+$ T-cells during SIV infection will be described.

### *$CD8^+$ T-cells*

In a typical $CD8^+$ T-cell response, viral peptides are presented as antigens to $CD8^+$ T-cells via HLA class I molecules (Hogan and Hammer, 2001). Upon CTL recognition of the foreign peptides presented by the antigen-presenting cells (APCs), the infected cells are destroyed (Hogan and Hammer, 2001). Through this recognition process, the naïve cells become activated and evolve to effector cells that eradicate cells infected with intracellular pathogens (Hogan and Hammer, 2001). Studies in NHPs have shown the importance of CTLs in controlling viral replication especially during the acute infection (Schmitz *et al.*, 1999b; Schmitz *et al.*, 1999a; Regoes *et al.*, 2004; Goulder and Watkins, 2008; Veazey *et al.*, 2008). This hypothesis is supported by several lines of evidence: (i) the post-peak decline of viraemia in acute HIV and SIV infections is coincidental with the expansion of HIV or SIV-specific CTLs, which suggests that CTLs are driving the initial contraction of VL (Borrow *et al.*, 1994; Koup *et al.*, 1994; Kuroda *et al.*, 1999); (ii) immunologic pressure exerted by CTLs results in viral escape mutations (Borrow *et al.*, 1994; Evans *et al.*, 1999; Goulder and Walker, 1999; Allen *et al.*, 2000; Evans and Desrosiers, 2001; Goulder and Watkins, 2008); (iii) strong T-cell responses in long-term non-progressors and elite controllers (Klein *et al.*, 1995; Rinaldo *et al.*, 1995; Rosenberg *et al.*, 1997; Propato *et al.*, 2001; Betts *et al.*, 2006) and (iv) experimental CD8 depletion in SIV-infected rhesus macaques, which resulted in increased VLs and rapid disease progression (Matano *et al.*, 1998; Jin *et al.*, 1999; Schmitz *et al.*, 1999b; Lifson *et al.*, 2000; Lifson *et al.*, 2001; Malkevitch *et al.*, 2006). There is, however, no clear inverse correlation between SIV-specific cell mediated immune responses and VL (Wang *et al.*, 2006).

Note that the anti-CD8 treatment may also induce depletion of the NK cells. However these cells probably play a minor role in loss of control of viral replication, as shown by a recent NK depletion study *in vivo* that demonstrated no significant impact of NKs on SIV replication (Choi *et al.*, 2007). It was also hypothesized that the increase in VL after anti-CD8 treatment may be partially related to increased activation and proliferation of $CD4^+$ T-cells through the proinflammatory signal resulting from the loss of large numbers of CD8+ T-cells and from an attempt of the T-lymphocyte compartment to restore the overall T-cell homeostasis in a blind fashion (Staprans and Feinberg, 2004). Such activated CD4+ T-cells would serve as highly fertile targets for SIV replication. However, a recent study has shown that simultaneous administration of an anti-IL-15 antibody together with the CD8-depleting antibody prevented CD4+ T-cell activation but not the increase in VL and the fast disease progression in $SIV_{mac}$-infected rhesus macaques (Okoye *et al.*, 2009).

In sooty mangabeys, *in vivo* depletion of CD8+ cells during the chronic phase of $SIV_{smm}$ infection was reported to have no significant impact on the control of viral replication (Barry *et al.*, 2007), and additional studies have also shown no correlation between cellular responses and the level of VLs during SIV infection of natural hosts (Dunham *et al.*, 2006; Zahn *et al.*, 2008). However, all of these studies were performed during the chronic stage of infection. CD8 depletion studies *in vivo* during acute infection of natural hosts have identified a potential role for cell-mediated immunity in the partial control of acute viral replication during natural SIV infections (Gaufin *et al.*, 2009b) (A. Kaur, unpublished and J.E. Schmitz, unpublished).

*CD4+ T-cells*

CD4+ T-cells become activated when they bind to regions of the virus antigen that are presented by MHC class II molecules and they secrete cytokines, such as IL-2, which help boost the function of CD8+ T-cell responses against intracellular viral pathogens and also aid the humoral immune response (Hogan and Hammer, 2001). During persistent viral infections like SIV, there is a lack of a CD4+ T helper response directed against the virus and it has thus been postulated that the lack of a response contributes to the dysregulation of the CD8+ T-cell response. Specific CD4+ T-cell responses directed against HIV or SIV are low during infection (Harari *et al.*, 2002; Iyasere *et al.*, 2003). However, studies have failed to establish a clear correlation between CD4+ T-cell responses against HIV and protection from AIDS, probably as a consequence of assessing the immune responses during the chronic infection (Jansen *et al.*, 2006). However, based on observations that the levels of CD4+ T-cell-specific responses against HIV increase when patients are receiving HAART (Rosenberg *et al.*, 2000) and are increased in long-term non-progressors (Boaz *et al.*, 2002), it is argued that specific CD4+ T-cell responses initiated early in the infection may be beneficial for the course of an HIV infection, although studies are not always in agreement (Jansen *et al.*, 2006). In contrast to the above described data, it has been shown that *in vivo* depletion of CD4+ cells results in a decrease of the VLs in $SIV_{smm}$-infected sooty mangabeys, indicating that CD4+ T-cells serve predominantly as a source of viral replication rather than to control viral replication (Klatt *et al.*, 2008).

CD4+ T-cell depletion is one of the hallmarks of HIV and SIV infections (Fig. 7.2). During progression to AIDS, there is a preferential depletion of the effector memory phenotype, although all CD4+ T-cell subsets are depleted (Jansen *et al.*, 2006). In SIV-infected rhesus macaques, there is a massive depletion of activated, memory CD4+ T-cells in the mucosal tissue of the gut (Smit-McBride *et al.*, 1998; Veazey *et al.*, 1998; Picker *et al.*, 2004; Li *et al.*, 2005; Mattapallil *et al.*, 2005) and to a lesser extent depletion of these same CD4+ T-cell subtypes in the peripheral blood and lymph nodes (Mattapallil *et al.*, 2005). The levels of activated memory cells are higher in the intestine, making it an ideal location for replication of SIV or HIV (Veazey *et al.*, 2001; Li *et al.*, 2005). With regard to the sites of CCR5+ CD4+ T-cell (target cells) depletion, at any stage of infection a more important depletion occurs in the gut as compared to the lymph nodes and peripheral blood (Brenchley *et al.*, 2004; Mehandru *et al.*, 2004) which is in agreement with higher levels of CCR5+ CD4+ T-cells at the mucosal sites (See above). The acute depletion in the gut is not predictive of disease outcome, as has been demonstrated in natural host species that do not progress to simian AIDS. A massive depletion of CD4+ T-cells occurs during the primary phase of infection, but is followed by a partial immune restoration during chronic infection in the presence of high levels of viral replication but normal levels of T-cell activation, proliferation and apoptosis (Gordon *et al.*, 2007; Pandrea *et al.*, 2007b).

CD4+ IL17+ cells have been described as effector cells that are involved in controlling inflammation in tissues, as well as affecting inflammatory chemokine production (Park *et al.*, 2005). An increasing amount of new research reported the preferential depletion of Th17 cells in chronic HIV or SIV infection, the levels of depletion being less significant in long-term non-progressors and patients receiving antiretroviral therapy (Macal *et al.*, 2008; Raffatellu *et al.*, 2008; Yue *et al.*, 2008; Favre *et al.*, 2009). One of these studies showed a significant difference in the depletion of CD4+ IL-17+ cells in the intestine of $SIV_{mac}$251-infected rhesus macaques coinfected with *Salmonella typhimurium* compared to SIV-uninfected controls. The bacteria were more readily detected in the mesenteric lymph nodes and the disease was more likely to be disseminated in the SIV-infected rhesus macaques compared to controls (Raffatellu *et al.*, 2008). Furthermore, NHP studies have shown that there is a depletion of Th17 cells that produces IL-17 in pathogenic SIV infections compared with natural hosts in the peripheral blood, lymph nodes, spleen, and colon (Raffatellu *et al.*, 2008; Favre *et al.*, 2009).

## Innate immunity during SIV infection

Innate immunity functions as a first-response to microbial pathogens and is mounted within hours after infection to limit the spread of the pathogen. It is thought that the function of the non-specific response allows more time for the adaptive immune response to react against an incoming pathogenic threat (Boasso and Shearer, 2008; Lehner *et al.*, 2008). At least four types of immune cells are involved in the innate immune response: dendritic cells (DCs), gδ T-cells, natural killer (NK) cells and monocytes. These cells are important during the initial innate response as they function as effector cells, but also later in infection when they act as regulators of the adaptive immune response (Lehner *et al.*, 2008). The cellular effectors of the innate immune system recognize invading organisms by pattern recognition receptors, distinguishing microorganisms and damaged cells from healthy cells (Boismenu and Havran, 1997).

### *Dendritic cells*

DCs function as antigen-presenting cells (APC) to naïve T-cells, promoting their maturation and differentiation (Banchereau *et al.*, 2000). DCs uptake antigen and travel to lymphoid tissues in the body such as lymph nodes, spleen, tonsils, adenoids, and the thymus (Banchereau *et al.*, 2000). In rhesus macaques infected with SIV through sexual routes, the first infected cell types are Langerhans DCs (Spira *et al.*, 1996). The virus can be rapidly identified in the draining lymph nodes, which suggests that once it infects mucosal tissue of the vaginal areas, it quickly spreads to the draining lymph nodes and is disseminated systemically throughout the host (Spira *et al.*, 1996).

Myeloid DCs (mDCs) and plasmacytoid (pDCs) or lymphoid DCs are the two major populations of DCs present in the peripheral blood. Peripheral mDCs and pDCs are immature and mature in the lymphoid tissues where they have further interaction with T-cells (Brown *et al.*, 2007). mDCs express CD11c and require granulocyte macrophage colony stimulating factor (GM-CSF) for growth (Pacanowski *et al.*, 2001). When mDCs are activated, they release IL-12, which is responsible for inducing cytotoxic T-cell functions and a type I response from T helper cells (Pacanowski *et al.*, 2001). In contrast, pDCs express CD123 (IL-3R), require IL-3 for growth and development, and secrete type I interferons (IFN) when activated in the presence of viruses and bacteria (Siegal *et al.*, 1999; Cella *et al.*, 2000).

It has been shown that DCs from humans and rhesus macaques can be infected *in vivo* with HIV and SIV, respectively, albeit at lower levels compared with CD4+ T-cells in lymphoid tissues (McIlroy *et al.*, 1995; Gautam *et al.*, 2007). pDCs are precursors to lymphoid DCs (Cella *et al.*, 2000; Colonna *et al.*, 2004; Boasso and Shearer, 2008) and have been observed to increase during the first week of infection of $SIV_{mac}$-infected rhesus macaques and then subsequently decrease with disease progression (Malleret *et al.*, 2008). Lower levels of pDCs and thus lower levels of type I IFNs have also been observed in the peripheral blood cells of infected HIV patients compared with uninfected subjects (Feldman *et al.*, 2001).

Lower levels of mDCs have also been reported to occur in HIV-infected patients during both acute and chronic stages of infection (Pacanowski *et al.*, 2001; Soumelis *et al.*, 2001; Chehimi *et al.*, 2002). Note, however, that even though mDCs are depleted during HIV infection, their ability to activate T-cells is maintained (Chehimi *et al.*, 2002). In macaques, however, conflicting results have been reported with regard to the dynamics of DCs: pDCs decreased, while mDCs increased or decreased during the chronic infection depending upon the marker used to measure levels of mDCs (CD11c *vs.* CD1c) in the peripheral blood of rhesus and cynomolgous macaques (Brown *et al.*, 2007; Malleret *et al.*, 2008).

In African green monkeys, it has been recently shown that during the acute phase of infection with $SIV_{agm}$, pDCs are slightly decreased in the peripheral blood, but slightly increased in the lymph nodes, and that there are small increases in IFN-α (Diop *et al.*, 2008). During the chronic phase, no changes are observed in either the peripheral blood, lymph nodes, or in IFN-α production (Diop *et al.*, 2008). In sooty mangabeys infected with $SIV_{smm}$, it was also shown that there are insignificant changes in the levels of pDCs in the blood, lymph nodes, and in the levels of IFN-α (Mandl *et al.*, 2008). In fact, pDCs in sooty mangabeys produce less IFN-α than rhesus macaques or humans, leading the authors to conclude that natural hosts that maintain this inherent ability to decrease production of IFN-α could correlate with the lower levels of immune activation and lack of disease progression in natural hosts (Mandl *et al.*, 2008).

### *g/δ T-cells*

g/δ T-cells are located in the vaginal and rectal epithelial tissue (Lehner *et al.*, 2008). In the lymphoid tissues these cells are found in smaller percentages compared to the α/β-positive $CD3^+$ T-cells (Boismenu and Havran, 1997; Lehner *et al.*, 2008). They differ from the α or β $CD3^+$ T-cells because they express a T-cell receptor that has a heterodimer of γ or δ-chains instead of a or - chains (Autran *et al.*, 1989). It is hypothesized that γ/δ T-cells recognize antigens through pattern recognition receptors as do macrophages (Boismenu and Havran, 1997).

The recombination of different segments leads to the production of both Vδ1 and Vδ2 cells in the peripheral blood (Casorati *et al.*, 1989). Vδ1 cells are more frequent in the intestine, whereas Vδ2 cells are more prevalent in the peripheral blood (Autran *et al.*, 1989; Parker *et al.*, 1990; Chowers *et al.*, 1994). In HIV infection, Vδ1 cells increase in the peripheral blood while Vδ2 cells decrease (Jitsukawa *et al.*, 1987; Autran *et al.*, 1989; Parker *et al.*, 1990; Kosub *et al.*, 2008). The level of Vδ1 cells in natural hosts like sooty mangabeys has been shown to remain stable throughout infection (Kosub *et al.*, 2008). The decrease in Vδ2 cells is associated also with a switch from a Th1 cytokine production (TNF-α and IFN-γ) to a Th2 cytokine production (IL-4 and IL-10) (Dobmeyer *et al.*, 2002). Vδ2 cells also decrease in HIV-infected patients when HAART is interrupted and viral levels begin to rise (Martini *et al.*, 2002). In contrast, sooty mangabeys maintain stable Vδ2 production levels of IFN-γ and TNF-α, although the TNF-α has been reported to be slightly increased (Kosub *et al.*, 2008).

It has been proposed that γ/δ cells are more effective at controlling acute infections than chronic infections, impacting the immune system negatively by contributing to disease pathology (Boismenu and Havran, 1997). In fact, it has been observed that protected rhesus macaques vaccinated with $SIV_{gp}120$ and p27 and challenged with $SIV_{mac}251$ J5 clone have higher numbers of γ/δ T-cells in the rectal mucosa and iliac lymph nodes (Lehner *et al.*, 2000).

### *Natural killer cells*

NKs are found in the blood and can migrate to the lymph nodes and intestine when they receive proinflammatory signals (Iannello *et al.*, 2008b). NKs are activated through toll-like receptors and once they are activated, they trigger the adaptive immune response, a critical function of NKs (Iannello *et al.*, 2008a). NK cells release cytokines to activate DCs or perforin and granzyme to destroy infected cells (Boyton and Altmann, 2007).

In $SIV_{mac}251$-infected rhesus macaques the levels of NK cytotoxicity reached a maximum at the same time as the virus peak, which is approximately 2 weeks post infection (Giavedoni *et al.*, 2000). During pathogenic SIV infection of rhesus macaques, levels of NKs secreting IFN-γ began to decrease three weeks after infection (LaBonte *et al.*, 2006). A comparative study has shown that functionality of cytotoxic NKs rose in uninfected sooty mangabeys compared to uninfected rhesus macaques despite the lower number of cytotoxic NKs present in sooty mangabeys (Pereira *et al.*, 2008). After SIV infection, in rhesus macaques with high VLs there is a decrease in cytotoxic function compared to uninfected rhesus macaques, while SIV-infected sooty mangabeys show an increase in NK cytotoxic function (Pereira *et al.*, 2008). Moreover, early increases in NK cells during the primary infection in both sooty mangabeys and rhesus macaques could potentially lower immune activation, resulting in a better pathogenic outcome (Pereira *et al.*, 2008).

*Monocytes/macrophages*

HIV infects macrophages, which function as a reservoir for the virus (Wu, 2008). The human tissues known to harbour HIV-infected macrophages are the brain (Koenig, 1996) and the lung (Gartner *et al.*, 1986; Lewin *et al.*, 1998; Worgall *et al.*, 1999). Infected macrophages can transfer virus to T-cells via the formation of 'infectious' synapses (Gousset *et al.*, 2008). Recent research has indicated that the turnover of peripheral blood monocytes in pathogenic rhesus macaque models is a predictor of progression to simian AIDS (Hasegawa *et al.*, 2009).

### T regulatory cells

Tregs function in down-regulating the immune response and are defined by expression of $CD4^{+}CD25^{+}FoxP3^{+}$ markers (Coleman *et al.*, 2007). Other markers expressed on the surface of Tregs include GITR and CTLA-4 (Coleman *et al.*, 2007). A consensus on the role of Tregs in HIV or SIV infection has yet to be reached. On one hand, it has been shown *in vitro* that the immune response from $CD4^{+}$ and $CD8^{+}$ cells is increased when Tregs are depleted from PBMC culture (Aandahl *et al.*, 2004). Yet, Tregs are increased in the lymph nodes, tonsils, and intestine of untreated HIV-infected patients, but decrease with the initiation of ART, indicating that Tregs may be more beneficial for the persistence of the virus, than they are to the host (Anderson *et al.*, 2005; Epple *et al.*, 2006; Nilsson *et al.*, 2006). It was also indicated that Tregs aid formation of collagen deposits in lymphatic tissues, causing further damage to the host (Estes *et al.*, 2007). In infant $SIV_{mac}251$-infected rhesus macaques, large numbers of Tregs may result in the suppression of $CD4^{+}$ anti-SIV responses (Hartigan-O'Connor *et al.*, 2007). Taken together, these studies have concluded that Tregs may be damaging to the host by down-regulating specific antiviral responses (Anderson *et al.*, 2005; Epple *et al.*, 2006; Estes *et al.*, 2007).

In contrast, a loss of Treg function was reported during progressive HIV infection (Weiss *et al.*, 2004; Tsunemi *et al.*, 2005). Other studies have found that loss of Tregs results in further viral persistence and increasing levels of immune activation (Oswald-Richter *et al.*, 2004; Eggena *et al.*, 2005; Baker *et al.*, 2007), as observed in $SIV_{mac}239$-infected rhesus and cynomolgus macaques infected with SIV/deltaB670 (Pereira *et al.*, 2008; Qin *et al.*, 2008). A decrease in the levels of TGFβ1 and IL-2 cytokines in the lymph nodes was also reported for the SIV-infected cynomolgus macaques, suggesting a decrease in the Treg population (Qin *et al.*, 2008). It has been shown that the peripheral levels of FoxP3 increased during $SIV_{agm}$ infection of African green monkeys, which has been correlated to the lower levels of immune activation observed in natural hosts during SIV infection (Kornfeld *et al.*, 2005). Furthermore, increased levels of Tregs were reported in the cord blood from neonates corresponding to lower levels of immune activation and better prognosis for the infant (Legrand *et al.*, 2006).

## Pathogenesis of SIV infection

### Clinical and pathological features

In contrast to macaques in which SIV infection constantly leads to an AIDS-like disease characterized by opportunistic infections and cancers, SIV infection in natural hosts generally does not show any clinical or pathological abnormalities (Pandrea *et al.*, 2008d). During progression to simian AIDS, SIV-infected rhesus macaques initially develop lymphadenopathy with confluent follicular hyperplasia, followed by lymph node atrophy due to lymphoid depletion and fibrosis (Hirsch, 2004). The destruction to the lymph node tissue architecture is considered an important factor in the progression to AIDS (Brenchley *et al.*, 2006a). There are no major changes in the tissue structure in the lymph nodes or intestines of African green monkeys or sooty mangabeys. Collagen deposition in secondary lymphoid tissue in the T-cell zones is observed to occur early on in the infection of rhesus macaques and then continues in the chronic phase, coincident with immune activation (Estes *et al.*, 2007; Estes *et al.*, 2008). However, the levels of collagen deposition are not observed to be modified at all during the course of infection in sooty mangabey lymphoid tissue, indicating that changes in the lymph node structure are minimal in natural host infections (Estes *et al.*, 2007; Estes *et al.*, 2008). No follicular trapping or $CD8^+$ T-cell infiltration of the germinal centres and no replacement of the normal lymph node architecture with connective tissue is observed in chronic SIV infection of natural hosts (Pandrea *et al.*, 2008d). Also, no thymic disinvolution, nodular lymphocytic infiltrates or giant cell disease are seen in SIV-infected natural hosts (Pandrea *et al.*, 2008d).

For 20 years, it was believed that natural SIV infections were non-pathogenic. This was a major paradox given the context of an active viral replication and high prevalence levels. However, occasionally, natural SIV infection of mandrills, African green monkeys and sooty mangabeys may eventually lead to the development of immunodeficiency (Pandrea *et al.*, 2009). Cases of progression to simian AIDS in African NHP hosts are rare, possibly because virus–host adaptation has occurred, resulting in a long-term persistent infection with an incubation period that exceeds the normal life span of the naturally infected animal (Pandrea *et al.*, 2001). In a recent report, increased mortality and AIDS-like immunopathology was described in wild chimpanzees infected with $SIV_{cpz}$ (Keele *et al.*, 2009b).

## Immunopathology of SIV infection

Acute SIV infection in the natural host induces massive mucosal $CD4^+$ T-cell depletion, of the same magnitude as in pathogenic HIV or SIV infection (Gordon *et al.*, 2007; Pandrea *et al.*, 2007b). However, in spite of persistent high viral replication, during chronic infection, there is a partial immune restoration of $CD4^+$ T-cells in natural hosts, probably as a consequence of the preservation of the mucosal immunological barrier and normal levels of immune activation (Pandrea *et al.*, 2007b). In marked contrast to HIV infection, in which a failure of the lymphoid regenerative capacity is an important factor in the pathogenesis of the immunodeficiency (Picker *et al.*, 2004), the regenerative capacity of the $CD4^+$ T-cell compartment is fully preserved in natural hosts and may play a key role in determining the lack of disease progression. Interleukin 7 (IL-7) has a critical role in preserving T-cell regeneration and in avoiding $CD4^+$ T-cell depletion and disease progression in natural infections (Muthukumar *et al.*, 2004; Sereti *et al.*, 2009).

## The role of immune activation

The immune system can control HIV infections, as illustrated by the cases of long-term non-progressors and elite controllers (Pacanowski *et al.*, 2001). One would expect that increased activation of the immune system would yield favourable results in terms of fighting an infection. However, in pathogenic HIV and SIV infections, the immune system is chronically activated, which is not beneficial to the host and is in fact the root cause of deficient immune responses that eventually lead to AIDS (Sodora and Silvestri, 2008). In humans, increased levels of immune activation on both $CD4^+$ and $CD8^+$ T-cells correlate with increased disease progression rates (Giorgi *et al.*, 1996; Giorgi *et al.*, 1999). B-cells are also activated during the course of HIV infection and become functionally dysregulated by spontaneously secreting antibodies and having decreased responses to antigens presented by T-cells (Martinez-Maza *et al.*, 1987). Furthermore, Martinez-Maza *et al.* reported an increase in numbers of immature B-cells in the peripheral blood during HIV infection (Martinez-Maza *et al.*, 1987). Likewise, pathogenic SIV infections display increased levels of immune activation and these up-regulated levels are highly predictive of disease progression (Brenchley *et al.*, 2006a; Brenchley *et al.*, 2006b; Grossman *et al.*, 2006).

One of the most important arguments for immune activation being the driving force behind HIV and SIV disease progression is that in natural hosts of SIV infection, which generally do not progress to simian AIDS, the major factor associated with the lack of disease progression is the low level of T-cell activation (Silvestri *et al.*, 2003; Kornfeld *et al.*, 2005; Onanga *et al.*, 2006; Pandrea *et al.*, 2007b; Pandrea *et al.*, 2008a). Moreover, we recently demonstrated that experimentally induced increases in immune activation in African green monkeys result in increases in viral replication and $CD4^+$ T-cell depletion (Pandrea *et al.*, 2008a). In a recent review, Sodora and Silvestri found six different mechanisms that may lead to immune activation in HIV infection (Sodora and Silvestri, 2008) including (i) direct activation of the innate immune system by the virus and subsequent activation of both the cellular and humoral arms of the adaptive immune response through innate immunity; (ii) direct activation of T-cells via viral peptides; (iii) microbial translocation from the intestine to the peripheral blood; (iv) other pathogenic coinfections; (v) induction of non-specific immune activation by proinflammatory cytokines and (vi) dysfunction of Tregs (Sodora and Silvestri, 2008).

The question is why does HIV-induced chronic immune activation result in progression to AIDS. It is known that general $CD4^+$ T-cell depletion is not enough to cause the pathogenesis that leads to AIDS (Milush *et al.*, 2007). In natural hosts, even though there is a significant $CD4^+$ T-cell depletion in the gut, residual $CD4^+$ T-cells still maintain low levels of immune activation, demonstrating that $CD4^+$ T-cell depletion alone is not sufficient to cause chronic immune activation (Grossman *et al.*, 2006; Gordon *et al.*, 2007; Milush *et al.*, 2007; Pandrea *et al.*, 2007b). HIV uses immune activation to its advantage because it infects cells that are actively replicating (Sodora and Silvestri, 2008). $CD4^+$ T-cell counts can result from excessive immune activation that induces depletion of the $CD4^+$ T central memory (TCM) cells and the ability to regenerate the $CD4^+$ T effector memory (TEM) population (Grossman *et al.*, 2006; Okoye *et al.*, 2007). In this scenario, the continuous depletion of $CD4^+$ TCM cells results in an inability to restore the $CD4^+$ TEM population in sites such as the intestine. When the $CD4^+$ TEM population is reduced below a threshold level, progression to simian AIDS occurs (Okoye *et al.*, 2007). It is proposed that the $CD4^+$ TCM population is depleted as a result of immune activation, since this causes the $CD4^+$ TCM population to become more activated and to proliferate, increasing the possibility that these cells become infected and lose their ability to regenerate the effector memory cells (Okoye *et al.*, 2007). Another mechanism by which excessive immune activation can cause disease progression is by speeding up the ageing process of the immune system, which is characterized by a decreased regenerative capacity of T-cells (measured by shorter telomeres). Finally, the increased levels of proinflammatory cytokines observed in both HIV patients and elderly non-infected HIV patients may explain why both groups of patients show signs of osteoporosis, atherosclerosis, and overall frailty (Appay and Sauce, 2008).

It is unknown how natural hosts control immune activation or why the pathogenic hosts lose control of immune activation. One contributing factor is microbial translocation, determined by measuring levels of lipopolysaccharide (LPS) from Gram-negative bacteria found in the plasma of SIV-infected animals to ascertain the amount of bacterial translocation from the lumen of the gut to the peripheral blood (Brenchley *et al.*, 2006b). LPS levels during the acute phase for both pathogenic and natural SIV hosts are increased (Gordon *et al.*, 2007; Pandrea *et al.*, 2007b). However, it is during the chronic phase that levels of LPS are elevated in pathogenic infections as compared to natural hosts, which display little or no evidence of bacterial translocation during this time (Brenchley *et al.*, 2006b). It is thus possible that microbial translocation fuels additional immune activation, especially in the local environment of the gut, where it further perpetuates the damage (Silvestri *et al.*, 2007; Douek *et al.*, 2009; Pandrea *et al.*, 2008d).

The convergence point of the studies carried out in recent years is that immune activation is an important distinction between natural hosts and pathogenic SIV infections in the progression to simian AIDS; natural hosts have the ability to control immune activation that is observed during acute infection, while the pathogenic hosts lack this ability and contain elevated levels of immune activation during their course of infection. Because immune activation is an important determinant for AIDS progression, it will be essential to incorporate analysis of immune activation during vaccine trials and also for future drug candidates.

## Apoptosis and progression to AIDS

Humans infected with HIV-1 and macaques infected with $SIV_{mac}$ experience a progressive decline in CD4+ T-cell numbers. The mechanism of CD4+ T-cell loss during infection is the result of a combination of factors: impaired cell production (Picker *et al.*, 2004), direct viral lysis of the infected cells that ensure virus propagation to other cells (Veazey *et al.*, 2000a; Mattapallil *et al.*, 2005) and apoptosis of the non-infected cells (Li *et al.*, 2005). As a limited number of cells are infected with HIV or SIV and the regenerative bone marrow capacity is severely affected by the lentiviral infection mainly in the late stages, it becomes clear that increased apoptosis of non-infected inflammatory or immune cells represents the major cause of CD4+ T-cell destruction during HIV or SIV infection (Gougeon *et al.*, 1993; Estaquier *et al.*, 1994; Cumont *et al.*, 2008; Elbim *et al.*, 2008). This bystander apoptosis represents the viral strategy to escape the host's innate and adaptive defence systems. The chronic systemic immune activation described in HIV patients and SIV-infected macaques is responsible for triggering both the extrinsic and the intrinsic cell death pathways (Gougeon and Piacentini, 2009). There are several evidences that increased levels of immune activation and increased susceptibility to apoptosis of T-lymphocytes are markers of HIV and SIV disease progression: (i) the magnitude of immune activation and apoptosis correlates positively with the stage of disease and inversely with CD4+ T-cell depletion; (ii) in HIV-2 infection, which is characterized by a slower progression, higher CD4+ T-cell counts and low levels of plasma viraemia, there are also lower levels of immune activation and CD4+ T-cell apoptosis; (iii) the levels of immune activation and associated cell apoptosis are lower in macaques infected with live attenuated $SIV_{mac}$239 delta nef, where a slower progression to simian AIDS is also observed; (iv) apoptosis in blood and tissues decreases after the HAART treatment, preceded by a decrease in VL (Gougeon and Piacentini, 2009). Finally, levels of CD4+ T-cell immune activation, proliferation and apoptosis are the critical factors that distinguish SIV pathogenic infection of macaques from that of natural African NHP hosts (Estaquier *et al.*, 1994; Li *et al.*, 2005; Pandrea *et al.*, 2007b).

There are several mechanisms that trigger or block apoptosis of the non-infected immune cells in HIV or SIV infection: (i) activation-induced cell death; (ii) death by neglect due to loss of mitochondrial homeostasis; (iii) the levels of pro- and anti-inflammatory cytokines; (iv) autologous infected cell-mediated killing; (v) cell cycle dysregulation; (vi) Gp120-induced apoptosis and (vii) apoptosis induced by other HIV proteins (Gougeon and Piacentini, 2009).

The lower levels of apoptosis in natural hosts such as African green monkeys, mandrills and sooty mangabeys may be explained by the lower levels of immune activation-induced cell death. However, interaction between the envelope glycoprotein and CD4/coreceptors may also play a role in diminishing the levels of bystander killing in these species, as has been recently shown in natural hosts that have a lower expression of CCR5 on their CD4+ T-cells.

## Concluding remarks

Comparative assessment of the main epidemiological, virological, and immunological features of pathogenic and non-pathogenic SIV infections may address some crucial but still unanswered question of SIV pathogenesis: why do SIV- infected natural hosts avoid AIDS while most HIV-infected humans do not? How can we use these features in order to improve the clinical management of HIV infection? At this time, unfortunately, a simple and straightforward answer to this question is still missing. However, extensive research carried out during recent years identified some essential pieces of a puzzle that are now fitting together. While the viral factors (i.e., reduced intrinsic cytopathicity, restricted tropism for CD4+ T-cells and/or Nef-induced CD3 down-regulation) are far from being ruled out, most of the available studies suggest that host response to infection play a key role in the control of SIV disease progression in natural hosts. These features appear to be crucial, not in terms of controlling viral replication per se (a task that is obviously not accomplished in natural SIV hosts), but by avoiding the generation of a state of chronic hyperactivation of the immune system that is ultimately more harmful than beneficial to the host. The molecular and cellular mechanisms responsible for the attenuated immune response observed in naturally SIV-infected monkeys in the face of chronically high viral replication are currently under investigation and are likely to be complex. While additional evolutionary mechanisms may play an important role in avoiding disease progression in natural hosts (i.e., lower CCR5 expression, more effective CD4$^+$ T-cell regeneration) it is now clear that the co-evolution of SIV and its natural host monkey species has led to a state of disease-resistance that does not involve immunity to viral infection or viral replication, but rather reflects an evolutionary adaptation that allows for a specific co-existence of host and virus. Further comparative studies are needed to improve our understanding of the pathogenesis of HIV infection in humans. These conceptual advances will eventually translate into substantial improvements in the clinical management of HIV-infected patients.

## References


Aandahl, E.M., Michaelsson, J., Moretto, W.J., Hecht, F.M., and Nixon, D.F. (2004). Human CD4$^+$ CD25+ regulatory T-cells control T-cell responses to human immunodeficiency virus and cytomegalovirus antigens. J. Virol. *78*, 2454–2459.

Allen, T.M., O'Connor, D.H., Jing, P., Dzuris, J.L., Mothe, B.R., Vogel, T.U., Dunphy, E., Liebl, M.E., Emerson, C., Wilson, N., *et al.* (2000). Tat-specific cytotoxic T-lymphocytes select for SIV escape variants during resolution of primary viraemia. Nature *407*, 386–390.

Amedee, A.M., Rychert, J., Lacour, N., Fresh, L., and Ratterree, M. (2004). Viral and immunological factors associated with breast milk transmission of SIV in rhesus macaques. Retrovirology *1*, 17.

Anderson, P.O., Sundstedt, A., Yazici, Z., Minaee, S., O'Neill, E.J., Woolf, R., Nicolson, K., Whitley, N., Li, L., Li, S., *et al.* (2005). IL-2 overcomes the unresponsiveness but fails to reverse the regulatory function of antigen-induced T regulatory cells. J. Immunol. *174*, 310–319.

Apetrei, C., Gormus, B., Pandrea, I., Metzger, M., ten Haaft, P., Martin, L.N., Bohm, R., Alvarez, X., Koopman, G., Murphey-Corb, M., *et al.* (2004). Direct inoculation of simian immunodeficiency virus from sooty mangabeys in black mangabeys (Lophocebus aterrimus): first evidence of AIDS in a heterologous African species and different pathologic outcomes of experimental infection. J. Virol. *78*, 11506–11518.

Apetrei, C., Kaur, A., Lerche, N.W., Metzger, M., Pandrea, I., Hardcastle, J., Fakelstein, S., Bohm, R., Kohler, J., Traina-Dorge, V., *et al.* (2005). Molecular epidemiology of SIVsm in US Primate Centers unravels the origin of SIVmac and SIVstm. J. Virol. *79*, 8991–9005.

Apetrei, C., Lerche, N.W., Pandrea, I., Gormus, B., Metzger, M., Silvestri, G., Kaur, A., Bohm, R., Robertson, D.L., Hardcastle, J., *et al.* (2006). Kuru experiments triggered the emergence of pathogenic SIVmac. AIDS *20*, 317–321.

Apetrei, C., Gautam, R., Sumpter, B., Carter, A.C., Gaufin, T., Staprans, S.I., Else, J., Barnes, M., Cao, R., Jr, Garg, S., *et al.* (2007). Virus-subtype specific features of natural SIVsmm infection in sooty mangabeys. J. Virol. *81*, 7913–7923.

Appay, V., and Sauce, D. (2008). Immune activation and inflammation in HIV-1 infection: causes and consequences. J. Pathol. *214*, 231–241.

Autran, B., Triebel, F., Katlama, C., Rozenbaum, W., Hercend, T., and Debre, P. (1989). T-cell receptor gamma/delta+ lymphocyte subsets during HIV infection. Clin. Exp. Immunol. *75*, 206–210.

Bailes, E., Gao, F., Bibollet-Ruche, F., Courgnaud, V., Peeters, M., Marx, P.A., Hahn, B.H., and Sharp, P.M. (2003). Hybrid origin of SIV in chimpanzees. Science *300*, 1713.

Baker, C.A., Clark, R., Ventura, F., Jones, N.G., Guzman, D., Bangsberg, D.R., and Cao, H. (2007). Peripheral CD4 loss of regulatory T-cells is associated with persistent viraemia in chronic HIV infection. Clin. Exp. Immunol. *147*, 533–539.

Banchereau, J., Briere, F., Caux, C., Davoust, J., Lebecque, S., Liu, Y.J., Pulendran, B., and Palucka, K. (2000). Immunobiology of dendritic cells. Annu. Rev. Immunol. *18*, 767–811.

Bannert, N., Schenten, D., Craig, S., and Sodroski, J. (2000). The level of CD4 expression limits infection of primary rhesus monkey macrophages by a T-tropic simian immunodeficiency virus and macrophage-tropic human immunodeficiency viruses. J. Virol. *74*, 10984–10993.

Barry, A.P., Silvestri, G., Safrit, J.T., Sumpter, B., Kozyr, N., McClure, H.M., Staprans, S.I., and Feinberg, M.B. (2007). Depletion of CD8+ cells in sooty mangabey monkeys naturally infected with simian immunodeficiency virus reveals limited role for immune control of virus replication in a natural host species. J. Immunol. *178*, 8002–8012.

Beer, B., Scherer, J., zur Megede, J., Norley, S., Baier, M., and Kurth, R. (1996). Lack of dichotomy between virus load of peripheral blood and lymph nodes during long-term simian immunodeficiency virus infection of African green monkeys. Virology *219*, 367–375.

Beer, B.E., Foley, B.T., Kuiken, C.L., Tooze, Z., Goeken, R.M., Brown, C.R., Hu, J., St Claire, M., Korber, B.T., and Hirsch, V.M. (2001). Characterization of novel simian immunodeficiency viruses from red-capped mangabeys from Nigeria (SIVrcmNG409 and -NG411). J. Virol. *75*, 12014–12027.

Bell, I., Ashman, C., Maughan, J., Hooker, E., Cook, F., and Reinhart, T.A. (1998). Association of simian immunodeficiency virus Nef with the T-cell receptor (TCR) zeta chain leads to TCR down-modulation. J. Gen. Virol. *79 (Pt 11)*, 2717–2727.

Betts, M.R., Nason, M.C., West, S.M., De Rosa, S.C., Migueles, S.A., Abraham, J., Lederman, M.M., Benito, J.M., Goepfert, P.A., Connors, M., *et al.* (2006).

HIV nonprogressors preferentially maintain highly functional HIV-specific CD8+ T-cells. Blood *107*, 4781–4789.

Binley, J.M., Clas, B., Gettie, A., Vesanen, M., Montefiori, D.C., Sawyer, L., Booth, J., Lewis, M., Marx, P.A., Bonhoeffer, S., and Moore, J.P. (2000). Passive infusion of immune serum into simian immunodeficiency virus-infected rhesus macaques undergoing a rapid disease course has minimal effect on plasma viremia. Virology *270*, 237–249.

Bissel, S.J., Wang, G., Trichel, A.M., Murphey-Corb, M., and Wiley, C.A. (2006a). Longitudinal analysis of monocyte/macrophage infection in simian immunodeficiency virus-infected, CD8+ T-cell-depleted macaques that develop lentiviral encephalitis. Am. J. Pathol. *168*, 1553–1569.

Bissel, S.J., Wang, G., Trichel, A.M., Murphey-Corb, M., and Wiley, C.A. (2006b). Longitudinal analysis of activation markers on monocyte subsets during the development of simian immunodeficiency virus encephalitis. J. Neuroimmunol. *177*, 85–98.

Boasso, A., and Shearer, G.M. (2008). Chronic innate immune activation as a cause of HIV-1 immunopathogenesis. Clin. Immunol. *126*, 235–242.

Boaz, M.J., Waters, A., Murad, S., Easterbrook, P.J., and Vyakarnam, A. (2002). Presence of HIV-1 Gag-specific IFN-γamma+IL-2+ and CD28+IL-2+ CD4 T-cell responses is associated with nonprogression in HIV-1 infection. J. Immunol. *169*, 6376–6385.

Boismenu, R., and Havran, W.L. (1997). An innate view of gamma delta T-cells. Curr. Opin. Immunol. *9*, 57–63.

Borda, J.T., Alvarez, X., Kondova, I., Aye, P., Simon, M.A., Desrosiers, R.C., and Lackner, A.A. (2004). Cell tropism of simian immunodeficiency virus in culture is not predictive of *in vivo* tropism or pathogenesis. Am. J. Pathol. *165*, 2111–2122.

Borrow, P., Lewicki, H., Hahn, B.H., Shaw, G.M., and Oldstone, M.B. (1994). Virus-specific CD8+ cytotoxic T-lymphocyte activity associated with control of viremia in primary human immunodeficiency virus type 1 infection. J. Virol. *68*, 6103–6110.

Boyton, R.J., and Altmann, D.M. (2007). Natural killer cells, killer immunoglobulin-like receptors and human leucocyte antigen class I in disease. Clin. Exp. Immunol. *149*, 1–8.

Brandin, E., Thorstensson, R., Bonhoeffer, S., and Albert, J. (2006). Rapid viral decay in simian immunodeficiency virus-infected macaques receiving quadruple antiretroviral therapy. J. Virol. *80*, 9861–9864.

Brenchley, J.M., Schacker, T.W., Ruff, L.E., Price, D.A., Taylor, J.H., Beilman, G.J., Nguyen, P.L., Khoruts, A., Larson, M., Haase, A.T., and Douek, D.C. (2004). $CD4^+$ T-cell depletion during all stages of HIV disease occurs predominantly in the gastrointestinal tract. J. Exp. Med. *200*, 749–759.

Brenchley, J.M., Price, D.A., and Douek, D.C. (2006a). HIV disease: fallout from a mucosal catastrophe? Nat. Immunol. *7*, 235–239.

Brenchley, J.M., Price, D.A., Schacker, T.W., Asher, T.E., Silvestri, G., Rao, S., Kazzaz, Z., Lambotte, O., Altmann, D., Blazar, B.R., *et al.* (2006b). Microbial translocation is a cause of systemic immune activation in chronic HIV infection. Nat. Med. *12*, 1365–1371.

Broussard, S.R., Staprans, S.I., White, R., Whitehead, E.M., Feinberg, M.B., and Allan, J.S. (2001). Simian immunodeficiency virus replicates to high levels in naturally infected African green monkeys without inducing immunologic or neurologic disease. J. Virol. *75*, 2262–2275.

Brown, K.N., Trichel, A., and Barratt-Boyes, S.M. (2007). Parallel loss of myeloid and plasmacytoid dendritic cells from blood and lymphoid tissue in simian AIDS. J. Immunol. *178*, 6958–6967.

Bucy, R.P., Hockett, R.D., Derdeyn, C.A., Saag, M.S., Squires, K., Sillers, M., Mitsuyasu, R.T., and Kilby, J.M. (1999). Initial increase in blood CD4(+) lymphocytes after HIV antiretroviral therapy reflects redistribution from lymphoid tissues. J. Clin. Invest. *103*, 1391–1398.

Casorati, G., De Libero, G., Lanzavecchia, A., and Migone, N. (1989). Molecular analysis of human gamma/delta+ clones from thymus and peripheral blood. J. Exp. Med. *170*, 1521–1535.

Cella, M., Facchetti, F., Lanzavecchia, A., and Colonna, M. (2000). Plasmacytoid dendritic cells activated by influenza virus and CD40L drive a potent TH1 polarization. Nat. Immunol. *1*, 305–310.

Chakrabarti, L., Hurtrel, M., Maire, M.A., Vazeux, R., Dormont, D., Montagnier, L., and Hurtrel, B. (1991). Early viral replication in the brain of SIV-infected rhesus monkeys. Am. J. Pathol. *139*, 1273–1280.

Chehimi, J., Campbell, D.E., Azzoni, L., Bacheller, D., Papasavvas, E., Jerandi, G., Mounzer, K., Kostman, J., Trinchieri, G., and Montaner, L.J. (2002). Persistent decreases in blood plasmacytoid dendritic cell number and function despite effective highly active antiretroviral therapy and increased blood myeloid dendritic cells in HIV-infected individuals. J. Immunol. *168*, 4796–4801.

Chen, Z., Zhou, P., Ho, D.D., Landau, N.R., and Marx, P.A. (1997). Genetically divergent strains of simian immunodeficiency virus use CCR5 as a coreceptor for entry. J. Virol. *71*, 2705–2714.

Chen, Z., Kwon, D., Jin, Z., Monard, S., Telfer, P., Jones, M.S., Lu, C.Y., Aguilar, R.F., Ho, D.D., and Marx, P.A. (1998). Natural infection of a homozygous delta24 CCR5 red-capped mangabey with an R2b-tropic simian immunodeficiency virus. J. Exp. Med. *188*, 2057–2065.

Choi, E.I., Reimann, K.A., and Letvin, N.L. (2007). *In vivo* natural killer cell depletion during primary SIV infection in rhesus monkeys J. Virol. *82*, 6758–6761.

Chowers, Y., Holtmeier, W., Harwood, J., Morzycka-Wroblewska, E., and Kagnoff, M.F. (1994). The V delta 1 T-cell receptor repertoire in human small intestine and colon. J. Exp. Med. *180*, 183–190.

Coleman, C.A., Muller-Trutwin, M.C., Apetrei, C., and Pandrea, I. (2007). T regulatory cells: aid or hindrance in the clearance of disease? J Cell Mol Med *11*, 1291–1325.

Colonna, M., Trinchieri, G., and Liu, Y.J. (2004). Plasmacytoid dendritic cells in immunity. Nat. Immunol. *5*, 1219–1226.

Conley, A.J., Kessler, J.A., II, Boots, L.J., McKenna, P.M., Schleif, W.A., Emini, E.A., Mark, G.E., III, Katinger, H., Cobb, E.K., Lunceford, S.M., *et al.* (1996). The consequence of passive administration of an anti-human immunodeficiency virus type 1 neutralizing monoclonal antibody before challenge of chimpanzees with a primary virus isolate. J. Virol. *70*, 6751–6758.

Cumont, M.C., Diop, O., Vaslin, B., Elbim, C., Viollet, L., Monceaux, V., Lay, S., Silvestri, G., Le Grand, R., Muller-Trutwin, M., *et al.* (2008). Early divergence in lymphoid tissue apoptosis between pathogenic and nonpathogenic simian immunodeficiency virus infections of nonhuman primates. J. Virol. *82*, 1175–1184.

Daniel, M.D., Letvin, N.L., King, N.W., Kannagi, M., Sehgal, P.K., Hunt, R.D., Kanki, P.J., Essex, M., and Desrosiers, R.C. (1985). Isolation of T-cell tropic HTLV-III-like retrovirus from macaques. Science *228*, 1201–1204.

Davis, L.E., Hjelle, B.L., Miller, V.E., Palmer, D.L., Llewellyn, A.L., Merlin, T.L., Young, S.A., Mills, R.G., Wachsman, W., and Wiley, C.A. (1992). Early viral brain invasion in iatrogenic human immunodeficiency virus infection. Neurology *42*, 1736–1739.

Desrosiers, R.C., Daniel, M.D., and Li, Y. (1989). HIV-related lentiviruses of nonhuman primates. AIDS Res. Hum. Retroviruses *5*, 465–473.

Desrosiers, R.C. (1990). The simian immunodeficiency viruses. Annu. Rev. Immunol. *8*, 557–578.

Desrosiers, R.C., Hansen-Moosa, A., Mori, K., Bouvier, D.P., King, N.W., Daniel, M.D., and Ringler, D.J. (1991). Macrophage-tropic variants of SIV are associated with specific AIDS-related lesions but are not essential for the development of AIDS. Am. J. Pathol. *139*, 29–35.

Desrosiers, R.C. (2004). Prospects for an AIDS vaccine. Nat. Med. *10*, 221–223.

Diop, O.M., Gueye, A., Dias-Tavares, M., Kornfeld, C., Faye, A., Ave, P., Huerre, M., Corbet, S., Barre-Sinoussi, F., and Muller-Trutwin, M.C. (2000). High levels of viral replication during primary simian immunodeficiency virus SIVagm infection are rapidly and strongly controlled in African green monkeys. J. Virol. *74*, 7538–7547.

Diop, O.M., Ploquin, M.J.-Y., Mortara, L., Faye, A., Jacquelin, B., Kunkel, D., Lebon, P., Butor, C., Hosmalin, A., Barré-Sinoussi, F., and Müller-Trutwin, M.C. (2008). Plasmacytoid dendritic cell dynamics and alpha interferon production during simian immunodeficiency virus infection with a nonpathogenic outcome. J. Virol. *82*, 5145–5152.

Dobmeyer, T.S., Dobmeyer, R., Wesch, D., Helm, E.B., Hoelzer, D., and Kabelitz, D. (2002). Reciprocal alterations of Th1/Th2 function in gammadelta T-cell subsets of human immunodeficiency virus-1-infected patients. Br J Haematol *118*, 282–288.

Douek, D.C., Roederer, M., and Koup, R.A. (2009). Emerging Concepts in the Immunopathogenesis of AIDS. Annu. Rev. Med. *60*, 471–484.

Dunham, R., Pagliardini, P., Gordon, S., Sumpter, B., Engram, J., Moanna, A., Lawson, B., McClure, H.M., Xian-Xu, H., Ibegbu, C., *et al.* (2006). The AIDS-resistance of naturally SIV-infected sooty mangabeys is independent of cellular immunity to the virus. Blood *108*, 209–217.

Dykhuizen, M., Mitchen, J.L., Montefiori, D.C., Thomson, J., Acker, L., Lardy, H., and Pauza, C.D. (1998). Determinants of disease in the simian immunodeficiency virus-infected rhesus macaque: characterizing animals with low antibody responses and rapid progression. J. Gen. Virol. *79 (Pt 10)*, 2461–2467.

Edinger, A.L., Amedee, A., Miller, K., Doranz, B.J., Endres, M., Sharron, M., Samson, M., Lu, Z.H., Clements, J.E., Murphey-Corb, M., *et al.* (1997a). Differential utilization of CCR5 by macrophage and T-cell tropic simian immunodeficiency virus strains. Proc. Natl. Acad. Sci. U.S.A. *94*, 4005–4010.

Edinger, A.L., Mankowski, J.L., Doranz, B.J., Margulies, B.J., Lee, B., Rucker, J., Sharron, M., Hoffman, T.L., Berson, J.F., Zink, M.C., *et al.* (1997b). CD4-independent, CCR5-dependent infection of brain capillary endothelial cells by a neurovirulent simian immunodeficiency virus strain. Proc. Natl. Acad. Sci. U.S.A. *94*, 14742–14747.

Eggena, M.P., Barugahare, B., Jones, N., Okello, M., Mutalya, S., Kityo, C., Mugyenyi, P., and Cao, H. (2005). Depletion of regulatory T-cells in HIV infection is associated with immune activation. J. Immunol. *174*, 4407–4414.

Elbim, C., Monceaux, V., Mueller, Y.M., Lewis, M.G., Francois, S., Diop, O., Akarid, K., Hurtrel, B., Gougerot-Pocidalo, M.A., Levy, Y., *et al.* (2008). Early divergence in neutrophil apoptosis between pathogenic and nonpathogenic simian immunodeficiency virus infections of nonhuman primates. J. Immunol. *181*, 8613–8623.

Epple, H.J., Loddenkemper, C., Kunkel, D., Troger, H., Maul, J., Moos, V., Berg, E., Ullrich, R., Schulzke, J.D., Stein, H., *et al.* (2006). Mucosal but not peripheral FOXP3+ regulatory T-cells are highly increased in untreated HIV infection and normalize after suppressive HAART. Blood *108*, 3072–3078.

Estaquier, J., Idziorek, T., de Bels, F., Barre-Sinoussi, F., Hurtrel, B., Aubertin, A.M., Venet, A., Mehtali, M., Muchmore, E., Michel, P., and *et al.* (1994). Programmed cell death and AIDS: significance of T-cell apoptosis in pathogenic and nonpathogenic primate lentiviral infections. Proc. Natl. Acad. Sci. U.S.A. *91*, 9431–9435.

Estes, J.D., Wietgrefe, S., Schacker, T., Southern, P., Beilman, G., Reilly, C., Milush, J.M., Lifson, J.D., Sodora, D.L., Carlis, J.V., and Haase, A.T. (2007). Simian immunodeficiency virus-induced lymphatic tissue fibrosis is mediated by transforming growth factor beta 1-positive regulatory T-cells and begins in early infection. J. Infect. Dis. *195*, 551–561.

Estes, J.D., Gordon, S.N., Zeng, M., Chahroudi, A.M., Dunham, R.M., Staprans, S.I., Reilly, C.S., Silvestri, G., and Haase, A.T. (2008). Early resolution of acute immune activation and induction of PD-1 in SIV-infected sooty mangabeys distinguishes nonpathogenic from pathogenic infection in rhesus macaques. J. Immunol. *180*, 6798–6807.

Evans, D.T., O'Connor, D.H., Jing, P., Dzuris, J.L., Sidney, J., da Silva, J., Allen, T.M., Horton, H., Venham, J.E.,

Rudersdorf, R.A., *et al.* (1999). Virus-specific cytotoxic T-lymphocyte responses select for amino-acid variation in simian immunodeficiency virus Env and Nef. Nat. Med. *5*, 1270–1276.

Evans, D.T., and Desrosiers, R.C. (2001). Immune evasion strategies of the primate lentiviruses. Immunol Rev *183*, 141–158.

Favre, D., Lederer, S., Kanwar, B., Ma, Z.M., Proll, S., Kasakow, Z., Mold, J., Swainson, L., Barbour, J.D., Baskin, C.R., *et al.* (2009). Critical loss of the balance between Th17 and T regulatory cell populations in pathogenic SIV infection. Plos Pathogens *5*, e1000295.

Feldman, S., Stein, D., Amrute, S., Denny, T., Garcia, Z., Kloser, P., Sun, Y., Megjugorac, N., and Fitzgerald-Bocarsly, P. (2001). Decreased interferon-alpha production in HIV-infected patients correlates with numerical and functional deficiencies in circulating type 2 dendritic cell precursors. Clin. Immunol. *101*, 201–210.

Fenard, D., Yonemoto, W., de Noronha, C., Cavrois, M., Williams, S.A., and Greene, W.C. (2005). Nef is physically recruited into the immunological synapse and potentiates T-cell activation early after TCR engagement. J. Immunol. *175*, 6050–6057.

Flynn, N.M., Forthal, D.N., Harro, C.D., Judson, F.N., Mayer, K.H., and Para, M.F. (2005). Placebo-controlled phase 3 trial of a recombinant glycoprotein 120 vaccine to prevent HIV-1 infection. J. Infect. Dis. *191*, 654–665.

Fortin, J.F., Barat, C., Beausejour, Y., Barbeau, B., and Tremblay, M.J. (2004). Hyper-responsiveness to stimulation of human immunodeficiency virus-infected CD4+ T-cells requires Nef and Tat virus gene products and results from higher NFAT, NF-kappaB, and AP-1 induction. J. Biol. Chem. *279*, 39520–39531.

Fultz, P.N., McClure, H.M., Anderson, D.C., Swenson, R.B., Anand, R., and Srinivasan, A. (1986). Isolation of a T-lymphotropic retrovirus from naturally infected sooty mangabey monkeys (Cercocebus atys). Proc. Natl. Acad. Sci. U.S.A. *83*, 5286–5290.

Gartner, S., Markovits, P., Markovitz, D.M., Kaplan, M.H., Gallo, R.C., and Popovic, M. (1986). The role of mononuclear phagocytes in HTLV-III/LAV infection. Science *233*, 215–219.

Gaufin, T., Gautam, R., Kasheta, M., Ribeiro, R.M., Ribka, E., Barnes, M., Pattison, M., Tatum, C., MacFarland, J., Montefiori, D., *et al.* (2009a). Limited ability of humoral immune responses in control of viremia during infection with SIVsmmD215 strain. Blood *113*, 4250–4261.

Gaufin, T., Ribeiro, R.M., Gautam, R., Dufour, J., Pattison, M., MacFarland, J., Tatum, C., Stoulig, C., Apetrei, C., and Pandrea, I. (2009b). Experimental depletion of CD8+ cells in acutely SIVagm-infected African green monkeys results in increase in viral replication. Submitted.

Gaufin, T., Pattison, M., Stoulig, C., Gautam, R., Dufour, J., Barnes, M., MacFarland, J., Monjure, C., Mandell, D., Tatum, C., *et al.* (2009c). Effect of B-cell depletion on viral replication and clinical outcome of SIV infection in a natural host. J. Virol. 83, 10347–10357.

Gautam, R., Carter, A.C., Katz, N., Butler, I.F., Barnes, M., Hasegawa, A., Ratterree, M., Silvestri, G., Marx, P.A., Hirsch, V.M., *et al.* (2007). *In vitro* characterization of primary SIVsmm isolates belonging to different lineages. *In vitro* growth on rhesus macaque cells is not predictive for *in vivo* replication in rhesus macaques. Virology *362*, 257–270.

Gautam, R., Gaufin, T., Butler, I., Gautam, A., Barnes, M., Mandell, D., Pattison, M., Tatum, C., Macfarland, J., Monjure, C., *et al.* (2009). SIVrcm, a unique CCR2-tropic virus, selectively depletes memory CD4+ T-cells in pigtailed macaques through rapid coreceptor expansion *in vivo*. J. Virol. *83*, 7894–7908.

Georges-Courbot, M.C., Lu, C.Y., Makuwa, M., Telfer, P., Onanga, R., Dubreuil, G., Chen, Z., Smith, S.M., Georges, A., Gao, F., *et al.* (1998). Natural infection of a household pet red-capped mangabey (Cercocebus torquatus torquatus) with a new simian immunodeficiency virus. J. Virol. *72*, 600–608.

Giavedoni, L.D., Velasquillo, M.C., Parodi, L.M., Hubbard, G.B., and Hodara, V.L. (2000). Cytokine expression, natural killer cell activation, and phenotypic changes in lymphoid cells from rhesus macaques during acute infection with pathogenic simian immunodeficiency virus. J. Virol. *74*, 1648–1657.

Giorgi, J.V., Hultin, L.E., and Desrosiers, R.C. (1996). The immunopathogenesis of retroviral diseases: no immunophenotypic alterations in T, B, and NK cell subsets in SIVmac239-challenged rhesus macaques protected by SIV delta nef vaccination. J. Med. Primatol. *25*, 186–191.

Giorgi, J.V., Hultin, L.E., McKeating, J.A., Johnson, T.D., Owens, B., Jacobson, L.P., Shih, R., Lewis, J., Wiley, D.J., Phair, J.P., *et al.* (1999). Shorter survival in advanced human immunodeficiency virus type 1 infection is more closely associated with T-lymphocyte activation than with plasma virus burden or virus chemokine coreceptor usage. J. Infect. Dis. *179*, 859–870.

Goldstein, S., Ourmanov, I., Brown, C.R., Beer, B.E., Elkins, W.R., Plishka, R., Buckler-White, A., and Hirsch, V.M. (2000). Wide range of viral load in healthy african green monkeys naturally infected with simian immunodeficiency virus. J. Virol. *74*, 11744–11753.

Goldstein, S., Brown, C.R., Ourmanov, I., Pandrea, I., Buckler-White, A., Erb, C., Nandi, J.S., Foster, G.J., Autissier, P., Schmitz, J.E., and Hirsch, V.M. (2006). Comparison of simian immunodeficiency virus SIVagmVer replication and CD4+ T-cell dynamics in vervet and sabaeus African green monkeys. J. Virol. *80*, 4868–4877.

Gomez-Roman, V.R., Patterson, L.J., Venzon, D., Liewehr, D., Aldrich, K., Florese, R., and Robert-Guroff, M. (2005). Vaccine-elicited antibodies mediate antibody-dependent cellular cytotoxicity correlated with significantly reduced acute viremia in rhesus macaques challenged with SIVmac251. J. Immunol. *174*, 2185–2189.

Gordon, S., Klatt, N.R., Milush, J.M., Engram, J., Dunham, R.M., Paiardini, M., Strobert, E.A., Apetrei, C., Pandrea, I., Staprans, S., *et al.* (2007). Severe depletion of mucosal CD4+ T-cells in AIDS-free SIV-infected sooty mangabeys. J. Immunol. *179*, 3026–3034.

Gordon, S., Dunham, R.M., Engram, J.C., Estes, J., Klatt, N.R., Pandrea, I., Apetrei, C., Sodora, D.L., Lee, H.Y., Haase, A.T., *et al.* (2008). Short-lived infected cells support the bulk of virus replication in naturally SIV-infected sooty mangabeys: implications for AIDS pathogenesis. J. Virol. *82*, 3725–3735.

Gormus, B.J., Martin, L.N., and Baskin, G.B. (2004). A brief history of the discovery of natural simian immunodeficiency virus (SIV) infections in captive sooty mangabey monkeys. Front Biosci *9*, 216–224.

Gougeon, M.L., Garcia, S., Heeney, J., Tschopp, R., Lecoeur, H., Guetard, D., Rame, V., Dauguet, C., and Montagnier, L. (1993). Programmed cell death in AIDS-related HIV and SIV infections. AIDS Res. Hum. Retroviruses *9*, 553–563.

Gougeon, M.L., and Piacentini, M. (2009). New insights on the role of apoptosis and autophagy in HIV pathogenesis. Apoptosis *14*, 501–508.

Goulder, P.J., and Walker, B.D. (1999). The great escape – AIDS viruses and immune control. Nat. Med. *5*, 1233–1235.

Goulder, P.J., and Watkins, D.I. (2008). Impact of MHC class I diversity on immune control of immunodeficiency virus replication. Nat. Rev. Immunol. *8*, 619–630.

Gousset, K., Ablan, S.D., Coren, L.V., Ono, A., Soheilian, F., Nagashima, K., Ott, D.E., and Freed, E.O. (2008). Real-time visualization of HIV-1 GAG trafficking in infected macrophages. PLoS Pathog. *4*, e1000015.

Grossman, Z., Herberman, R.B., Dimitrov, D.S., Rouzine, I., M;, C., Perelson, A.J., Bonhoeffer, S., Mohri, H., and Ho, D.D. (1999). T Cell Turnover in SIV Infection Science *248*, 555.

Grossman, Z., Meier-Schellersheim, M., Paul, W.E., and Picker, L.J. (2006). Pathogenesis of HIV infection: what the virus spares is as important as what it destroys. Nat. Med. *12*, 289–295.

Hahn, B.H., Shaw, G.M., De Cock, K.M., and Sharp, P.M. (2000). AIDS as a zoonosis: scientific and public health implications. Science *287*, 607–614.

Haigwood, N.L., Watson, A., Sutton, W.F., McClure, J., Lewis, A., Ranchalis, J., Travis, B., Voss, G., Letvin, N.L., Hu, S.L., *et al.* (1996). Passive immune globulin therapy in the SIV/macaque model: early intervention can alter disease profile. Immunol. Lett. *51*, 107–114.

Harari, A., Rizzardi, G.P., Ellefsen, K., Ciuffreda, D., Champagne, P., Bart, P.A., Kaufmann, D., Telenti, A., Sahli, R., Tambussi, G., *et al.* (2002). Analysis of HIV-1- and CMV-specific memory CD4 T-cell responses during primary and chronic infection. Blood *100*, 1381–1387.

Hartigan-O'Connor, D.J., Abel, K., and McCune, J.M. (2007). Suppression of SIV-specific CD4+ T-cells by infant but not adult macaque regulatory T-cells: implications for SIV disease progression. J. Exp. Med. *204*, 2679–2692.

Hasegawa, A., Liu, H., Ling, B., Borda, J.T., Alvarez, X., Sugimoto, C., Vinet-Oliphant, H., Kim, W.K., Williams, K.C., Ribeiro, R.M., *et al.* (2009). The level of monocyte turnover predicts disease progression in the macaque model of AIDS. Blood *114*, 2917–2925.

Hirsch, V.M., Dapolito, G., McGann, C., Olmsted, R.A., Purcell, R.H., and Johnson, P.R. (1989). Molecular cloning of SIV from sooty mangabey monkeys. J. Med. Primatol. *18*, 279–285.

Hirsch, V.M., and Johnson, P.R. (1994). Pathogenic diversity of simian immunodeficiency viruses. Virus Res. *32*, 183–203.

Hirsch, V.M., Fuerst, T.R., Sutter, G., Carroll, M.W., Yang, L.C., Goldstein, S., Piatak, M., Jr, Elkins, W.R., Alvord, W.G., Montefiori, D.C., *et al.* (1996). Patterns of viral replication correlate with outcome in simian immunodeficiency virus (SIV)-infected macaques: effect of prior immunization with a trivalent SIV vaccine in modified vaccinia virus Ankara. J. Virol. *70*, 3741–3752.

Hirsch, V.M. (2004). What can natural infection of African monkeys with simian immunodeficiency virus tell us about the pathogenesis of AIDS? AIDS Rev *6*, 40–53.

Ho, D.D. (1995a). Time to hit HIV, early and hard. N. Engl. J. Med. *333*, 450–451.

Ho, D.D. (1995b). HIV-1 dynamics *in vivo.* J Biol Regul Homeost Agents *9*, 76–77.

Hogan, C.M., and Hammer, S.M. (2001). Host determinants in HIV infection and disease. Part 1: cellular and humoral immune responses. Ann. Intern. Med. *134*, 761–776.

Howe, A.Y., Jung, J.U., and Desrosiers, R.C. (1998). Zeta chain of the T-cell receptor interacts with nef of simian immunodeficiency virus and human immunodeficiency virus type 2. J. Virol. *72*, 9827–9834.

Iannello, A., Debbeche, O., Samarani, S., and Ahmad, A. (2008a). Antiviral NK cell responses in HIV infection: II. viral strategies for evasion and lessons for immunotherapy and vaccination. J. Leukoc. Biol. *84*, 27–49.

Iannello, A., Debbeche, O., Samarani, S., and Ahmad, A. (2008b). Antiviral NK cell responses in HIV infection: I. NK cell receptor genes as determinants of HIV resistance and progression to AIDS. J. Leukoc. Biol. *84*, 1–26.

Iyasere, C., Tilton, J.C., Johnson, A.J., Younes, S., Yassine-Diab, B., Sekaly, R.P., Kwok, W.W., Migueles, S.A., Laborico, A.C., Shupert, W.L., *et al.* (2003). Diminished proliferation of human immunodeficiency virus-specific CD4+ T-cells is associated with diminished interleukin-2 (IL-2) production and is recovered by exogenous IL-2. J. Virol. *77*, 10900–10909.

Jansen, C.A., De Cuyper, I.M., Hooibrink, B., van der Bij, A.K., van Baarle, D., and Miedema, F. (2006). Prognostic value of HIV-1 Gag-specific CD4+ T-cell responses for progression to AIDS analyzed in a prospective cohort study. Blood *107*, 1427–1433.

Jin, X., Bauer, D.E., Tuttleton, S.E., Lewin, S., Gettie, A., Blanchard, J., Irwin, C.E., Safrit, J.T., Mittler, J., Weinberger, L., *et al.* (1999). Dramatic rise in plasma viremia after CD8+ T-cell depletion in simian immunodeficiency virus-infected macaques. J. Exp. Med. *189*, 991–998.

Jitsukawa, S., Faure, F., Lipinski, M., Triebel, F., and Hercend, T. (1987). A novel subset of human lymphocytes with a T-cell receptor–gamma complex. J. Exp. Med. *166*, 1192–1197.

Johnson, W.E., and Desrosiers, R.C. (2002). Viral persistance: HIV's strategies of immune system evasion. Annu Rev Med *53*, 499–518.

Johnson, W.E., Lifson, J.D., Lang, S.M., Johnson, R.P., and Desrosiers, R.C. (2003). Importance of B-cell responses for immunological control of variant strains of simian immunodeficiency virus. J. Virol. *77*, 375–381.

Kaur, A., Di Mascio, M., Barabasz, A., Rosenzweig, M., McClure, H.M., Perelson, A.S., Ribeiro, R.M., and Johnson, R.P. (2008). Dynamics of T- and B-lymphocyte turnover in a natural host of simian immunodeficiency virus. J. Virol. *82*, 1084–1093.

Keele, B.F., Li, H., Learn, G.H., Hraber, P., Giorgi, E.E., Grayson, T., Sun, C., Chen, Y., Yeh, W.W., Letvin, N.L., *et al.* (2009a). Low dose rectal inoculation of rhesus macaques by SIVsmE660 or SIVmac251 recapitulates human mucosal infection by HIV-1. J. Exp. Med. *206*, 1117–1134.

Keele, B.F., Holland Jones, J., Terio, K., Estes, J.D., Rudicell, R.S., Wilson, M.L., Li, Y., Learn, G.H., Beasley, M., Schumacher-Stankey, J., *et al.* (2009b). Increased mortality and AIDS-like immunopathology in wild chimpanzees infected with SIVcpz. Nature *460*, 515–519.

Kestler, H.W., 3rd, Ringler, D.J., Mori, K., Panicali, D.L., Sehgal, P.K., Daniel, M.D., and Desrosiers, R.C. (1991). Importance of the nef gene for maintenance of high virus loads and for development of AIDS. Cell *65*, 651–662.

Keystone, E.C. (2005). B-cells in rheumatoid arthritis: from hypothesis to the clinic. Rheumatology (Oxford) *44 Suppl. 2*, ii8-ii12.

Klatt, N.R., Villinger, F., Bostik, P., Gordon, S.N., Pereira, L., Engram, J.C., Mayne, A., Dunham, R.M., Lawson, B., Ratcliffe, S.J., *et al.* (2008). Availability of activated $CD4^+$ T-cells dictates the level of viremia in naturally SIV-infected sooty mangabeys. J. Clin. Invest. *118*, 2039–2049.

Klein, M.R., van Baalen, C.A., Holwerda, A.M., Kerkhof Garde, S.R., Bende, R.J., Keet, I.P., Eeftinck-Schattenkerk, J.K., Osterhaus, A.D., Schuitemaker, H., and Miedema, F. (1995). Kinetics of Gag-specific cytotoxic T-lymphocyte responses during the clinical course of HIV-1 infection: a longitudinal analysis of rapid progressors and long-term asymptomatics. J. Exp. Med. *181*, 1365–1372.

Koenig, R. (1996). Koch keeps new watch on infections. Science 272, 1412–1414.

Kornfeld, C., Ploquin, M.J., Pandrea, I., Faye, A., Onanga, R., Apetrei, C., Poaty-Mavoungou, V., Rouquet, P., Estaquier, J., Mortara, L., *et al.* (2005). Anti-inflammatory profiles during primary SIV infection in African green monkeys are associated with protection against AIDS. J. Clin. Invest. *115*, 1082–1091.

Kosub, D.A., Lehrman, G., Milush, J.M., Zhou, D., Chacko, E., Leone, A., Gordon, S., Silvestri, G., Else, J.G., Keiser, P., *et al.* (2008). Gamma/Delta T-cell functional responses differ after pathogenic human immunodeficiency virus and nonpathogenic simian immunodeficiency virus infections. J. Virol. *82*, 1155–1165.

Koup, R.A., Safrit, J.T., Cao, Y., Andrews, C.A., McLeod, G., Borkowsky, W., Farthing, C., and Ho, D.D. (1994). Temporal association of cellular immune responses with the initial control of viremia in primary human immunodeficiency virus type 1 syndrome. J. Virol. *68*, 4650–4655.

Kuroda, M.J., Schmitz, J.E., Charini, W.A., Nickerson, C.E., Lifton, M.A., Lord, C.I., Forman, M.A., and Letvin, N.L. (1999). Emergence of CTL coincides with clearance of virus during primary simian immunodeficiency virus infection in rhesus monkeys. J. Immunol. *162*, 5127–5133.

LaBonte, M.L., McKay, P.F., and Letvin, N.L. (2006). Evidence of NK cell dysfunction in SIV-infected rhesus monkeys: impairment of cytokine secretion and NKG2C/C2 expression. Eur J. Immunol. *36*, 2424–2433.

Lane, J.H., Sasseville, V.G., Smith, M.O., Vogel, P., Pauley, D.R., Heyes, M.P., and Lackner, A.A. (1996). Neuroinvasion by simian immunodeficiency virus coincides with increased numbers of perivascular macrophages/microglia and intrathecal immune activation. J Neurovirol *2*, 423–432.

Legrand, F.A., Nixon, D.F., Loo, C.P., Ono, E., Chapman, J.M., Miyamoto, M., Diaz, R.S., Santos, A.M., Succi, R.C., Abadi, J., *et al.* (2006). Strong HIV-1-specific T-cell responses in HIV-1-exposed uninfected infants and neonates revealed after regulatory T-cell removal. PLoS ONE *1*, e102.

Lehner, T., Mitchell, E., Bergmeier, L., Singh, M., Spallek, R., Cranage, M., Hall, G., Dennis, M., Villinger, F., and Wang, Y. (2000). The role of gammadelta T-cells in generating antiviral factors and beta-chemokines in protection against mucosal simian immunodeficiency virus infection. Eur J. Immunol. *30*, 2245–2256.

Lehner, T., Wang, Y., Pido-Lopez, J., Whittall, T., Bergmeier, L.A., and Babaahmady, K. (2008). The emerging role of innate immunity in protection against HIV-1 infection. Vaccine *26*, 2997–3001.

Letvin, N.L., Daniel, M.D., Sehgal, P.K., Desrosiers, R.C., Hunt, R.D., Waldron, L.M., MacKey, J.J., Schmidt, D.K., Chalifoux, L.V., and King, N.W. (1985). Induction of AIDS-like disease in macaque monkeys with T-cell tropic retrovirus STLV-III. Science *230*, 71–73.

Levy, J., Youvan, T., and Lee, M.L. (1994). Passive hyperimmune plasma therapy in the treatment of acquired immunodeficiency syndrome: results of a 12-month multicenter double-blind controlled trial. The Passive Hyperimmune Therapy Study Group. Blood *84*, 2130–2135.

Lewin, S.R., Kirihara, J., Sonza, S., Irving, L., Mills, J., and Crowe, S.M. (1998). HIV-1 DNA and mRNA concentrations are similar in peripheral blood monocytes and alveolar macrophages in HIV-1-infected individuals. Aids *12*, 719–727.

Li, Q., Duan, L., Estes, J.D., Ma, Z.M., Rourke, T., Wang, Y., Reilly, C., Carlis, J., Miller, C.J., and Haase, A.T. (2005). Peak SIV replication in resting memory $CD4^+$ T-cells depletes gut lamina propria $CD4^+$ T-cells. Nature *434*, 1148–1152.

Lifson, J.D., Nowak, M.A., Goldstein, S., Rossio, J.L., Kinter, A., Vasquez, G., Wiltrout, T.A., Brown, C.,

Schneider, D., Wahl, L., *et al.* (1997). The extent of early viral replication is a critical determinant of the natural history of simian immunodeficiency virus infection. J. Virol. *71*, 9508–9514.

Lifson, J.D., Rossio, J.L., Arnaout, R., Li, L., Parks, T.L., Schneider, D.K., Kiser, R.F., Coalter, V.J., Walsh, G., Imming, R.J., *et al.* (2000). Containment of simian immunodeficiency virus infection: cellular immune responses and protection from rechallenge following transient postinoculation antiretroviral treatment. J. Virol. *74*, 2584–2593.

Lifson, J.D., Rossio, J.L., Piatak, M., Jr, Parks, T., Li, L., Kiser, R., Coalter, V., Fisher, B., Flynn, B.M., Czajak, S., *et al.* (2001). Role of CD8(+) lymphocytes in control of simian immunodeficiency virus infection and resistance to rechallenge after transient early antiretroviral treatment. J. Virol. *75*, 10187–10199.

Ling, B., Apetrei, C., Pandrea, I., Veazey, R.S., Lackner, A.A., Gormus, B., and Marx, P.A. (2004). Classic AIDS in a sooty mangabey after an 18-year natural infection. J. Virol. *78*, 8902–8908.

Little, S.J., McLean, A.R., Spina, C.A., Richman, D.D., and Havlir, D.V. (1999). Viral dynamics of acute HIV-1 infection. J. Exp. Med. *190*, 841–850.

Lowenstine, L.J., Pedersen, N.C., Higgins, J., Pallis, K.C., Uyeda, A., Marx, P., Lerche, N.W., Munn, R.J., and Gardner, M.B. (1986). Seroepidemiologic survey of captive Old-World primates for antibodies to human and simian retroviruses, and isolation of a lentivirus from sooty mangabeys (Cercocebus atys). Int. J. Cancer *38*, 563–574.

Lyles, R.H., Munoz, A., Yamashita, T.E., Bazmi, H., Detels, R., Rinaldo, C.R., Margolick, J.B., Phair, J.P., and Mellors, J.W. (2000). Natural history of human immunodeficiency virus type 1 viremia after seroconversion and proximal to AIDS in a large cohort of homosexual men. Multicenter AIDS Cohort Study. J. Infect. Dis. *181*, 872–880.

Ma, Z.M., Stone, M., Piatak, M., Jr, Schweighardt, B., Haigwood, N.L., Montefiori, D., Lifson, J.D., Busch, M.P., and Miller, C.J. (2009). High specific infectivity of plasma virus from the pre-ramp-up and ramp-up stages of acute simian immunodeficiency virus infection. J. Virol. *83*, 3288–3297.

Macal, M., Sankaran, S., Chun, T.-W., Reay, E., Flamm, J., Prindiville, T.J., and Dandekar, S. (2008). Effective $CD4^+$ T-cell restoration in gut-associated lymphoid tissue of HIV-infected patients is associated with enhanced Th17 cells and polyfunctional HIV-specific T-cell responses. Mucosal Immunol *1*, 475–488.

Malkevitch, N.V., Patterson, L.J., Aldrich, M.K., Wu, Y., Venzon, D., Florese, R.H., Kalyanaraman, V.S., Pal, R., Lee, E.M., Zhao, J., *et al.* (2006). Durable protection of rhesus macaques immunized with a replicating adenovirus-SIV multigene prime/protein boost vaccine regimen against a second SIVmac251 rectal challenge: role of SIV-specific CD8+ T-cell responses. Virology 353, 83–98.

Malleret, B., Karlsson, I., Maneglier, B., Brochard, P., Delache, B., Andrieu, T., Muller-Trutwin, M., Beaumont, T., McCune, J.M., Banchereau, J., *et al.* (2008). Effect of SIVmac infection on plasmacytoid and CD1c+ myeloid dendritic cells in cynomolgus macaques. Immunology *124*, 223–233.

Mandell, D., Gaufin, T., Gautam, R., Sandler, N., Pattison, M., Dufour, J., Tatum, C., MacFarland, J., Douek, D.C., Apetrei, C., and Pandrea, I. (2009). A single serial passage of SIVagm.sab in pigtailed macaques results in a dramatic increase in pathogenicity of infection. Submitted.

Mandl, J.N., Barry, A.P., Vanderford, T.H., Kozyr, N., Chavan, R., Klucking, S., Barrat, F.J., Coffman, R.L., Staprans, S.I., and Feinberg, M.B. (2008). Divergent TLR7 and TLR9 signaling and type I interferon production distinguish pathogenic and nonpathogenic AIDS virus infections. Nat. Med. *14*, 1077–1087.

Mansfield, K.G., Lerche, N.W., Gardner, M.B., and Lackner, A.A. (1995). Origins of simian immunodeficiency virus infection in macaques at the New England Regional Primate Research Center. J. Med. Primatol. *24*, 116–122.

Mariani, R., Kirchhoff, F., Greenough, T.C., Sullivan, J.L., Desrosiers, R.C., and Skowronski, J. (1996). High frequency of defective nef alleles in a long-term survivor with nonprogressive human immunodeficiency virus type 1 infection. J. Virol. *70*, 7752–7764.

Markowitz, M., Louie, M., Hurley, A., Sun, E., Di Mascio, M., Perelson, A.S., and Ho, D.D. (2003). A novel antiviral intervention results in more accurate assessment of human immunodeficiency virus type 1 replication dynamics and T-cell decay *in vivo*. J. Virol. *77*, 5037–5038.

Martinez-Maza, O., Crabb, E., Mitsuyasu, R.T., Fahey, J.L., and Giorgi, J.V. (1987). Infection with the human immunodeficiency virus (HIV) is associated with an *in vivo* increase in B-lymphocyte activation and immaturity. J. Immunol. *138*, 3720–3724.

Martini, F., Poccia, F., Goletti, D., Carrara, S., Vincenti, D., D'Offizi, G., Agrati, C., Ippolito, G., Colizzi, V., Pucillo, L.P., and Montesano, C. (2002). Acute human immunodeficiency virus replication causes a rapid and persistent impairment of Vgamma9Vdelta2 T-cells in chronically infected patients undergoing structured treatment interruption. J. Infect. Dis. *186*, 847–850.

Matano, T., Shibata, R., Siemon, C., Connors, M., Lane, H.C., and Martin, M.A. (1998). Administration of an anti-CD8 monoclonal antibody interferes with the clearance of chimeric simian/human immunodeficiency virus during primary infections of rhesus macaques. J. Virol. *72*, 164–169.

Mattapallil, J.J., Douek, D.C., Hill, B., Nishimura, Y., Martin, M., and Roederer, M. (2005). Massive infection and loss of memory $CD4^+$ T-cells in multiple tissues during acute SIV infection. Nature *434*, 1093–1097.

McIlroy, D., Autran, B., Cheynier, R., Wain-Hobson, S., Clauvel, J.P., Oksenhendler, E., Debre, P., and Hosmalin, A. (1995). Infection frequency of dendritic cells and $CD4^+$ T-lymphocytes in spleens of human immunodeficiency virus-positive patients. J. Virol. *69*, 4737–4745.

McMichael, A.J. (2006). HIV vaccines. Annu. Rev. Immunol. *24*, 227–255.

Means, R.E., Matthews, T., Hoxie, J.A., Malim, M.H., Kodama, T., and Desrosiers, R.C. (2001). Ability of the V3 loop of simian immunodeficiency virus to serve as a target for antibody-mediated neutralization: correlation of neutralization sensitivity, growth in macrophages, and decreased dependence on CD4. J. Virol. 75, 3903–3915.

Mehandru, S., Poles, M.A., Tenner-Racz, K., Horowitz, A., Hurley, A., Hogan, C., Boden, D., Racz, P., and Markowitz, M. (2004). Primary HIV-1 infection is associated with preferential depletion of CD4+ T-lymphocytes from effector sites in the gastrointestinal tract. J. Exp. Med. *200*, 761–770.

Mehandru, S., Poles, M.A., Tenner-Racz, K., Manuelli, V., Jean-Pierre, P., Lopez, P., Shet, A., Low, A., Mohri, H., Boden, D., *et al.* (2007). Mechanisms of gastrointestinal CD4+ T-cell depletion during acute and early human immunodeficiency virus type 1 infection. J. Virol. *81*, 599–612.

Mellors, J.W., Rinaldo, C.R., Jr, Gupta, P., White, R.M., Todd, J.A., and Kingsley, L.A. (1996). Prognosis in HIV-1 infection predicted by the quantity of virus in plasma. Science *272*, 1167–1170.

Mellors, J.W. (1998). Viral-load tests provide valuable answers. Sci. Am. *279*, 90–93.

Miller, C.J., Genesca, M., Abel, K., Montefiori, D., Forthal, D., Bost, K., Li, J., Favre, D., and McCune, J.M. (2007). Antiviral antibodies are necessary for control of simian immunodeficiency virus replication. J. Virol. *81*, 5024–5035.

Milush, J.M., Reeves, J.D., Gordon, S., Zhou, D., Muthukumar, A., Kosub, D., Chacko, E., Giavedoni, L., Ibegbu, C., Stefano-Cole, K., *et al.* (2007). Virally-induced CD4+ T-cell depletion is not sufficient to induce AIDS in a natural host. J. Immunol. *179*, 3047–3056.

Mohri, H., Bonhoeffer, S., Monard, S., Perelson, A.S., and Ho, D.D. (1998). Rapid turnover of T-lymphocytes in SIV-infected rhesus macaques. Science *279*, 1223–1227.

Moore, J.P., Kitchen, S.G., Pugach, P., and Zack, J.A. (2004). The CCR5 and CXCR4 coreceptors--central to understanding the transmission and pathogenesis of human immunodeficiency virus type 1 infection. AIDS Res. Hum. Retroviruses *20*, 111–126.

Mori, K., Ringler, D.J., Kodama, T., and Desrosiers, R.C. (1992). Complex determinants of macrophage tropism in env of simian immunodeficiency virus. J. Virol. *66*, 2067–2075.

Mori, K., Rosenzweig, M., and Desrosiers, R.C. (2000). Mechanisms for adaptation of simian immunodeficiency virus to replication in alveolar macrophages. J. Virol. *74*, 10852–10859.

Muller, M.C., and Barre-Sinoussi, F. (2003). SIVagm: genetic and biological features associated with replication. Front. Biosci. *8*, D1170–1185.

Munch, J., Schindler, M., Wildum, S., Rucker, E., Bailer, N., Knoop, V., Novembre, F.J., and Kirchhoff, F. (2005). Primary sooty mangabey simian immunodeficiency virus and human immunodeficiency virus type 2 nef alleles modulate cell surface expression of various human receptors and enhance viral infectivity and replication. J. Virol. *79*, 10547–10560.

Murphey-Corb, M., Martin, L.N., Rangan, S.R., Baskin, G.B., Gormus, B.J., Wolf, R.H., Andes, W.A., West, M., and Montelaro, R.C. (1986). Isolation of an HTLV-III-related retrovirus from macaques with simian AIDS and its possible origin in asymptomatic mangabeys. Nature *321*, 435–437.

Muthukumar, A., Wozniakowski, A., Gauduin, M.C., Paiardini, M., McClure, H.M., Johnson, R.P., Silvestri, G., and Sodora, D.L. (2004). Elevated interleukin-7 levels not sufficient to maintain T-cell homeostasis during simian immunodeficiency virus-induced disease progression. Blood *103*, 973–979.

Nilsson, J., Boasso, A., Velilla, P.A., Zhang, R., Vaccari, M., Franchini, G., Shearer, G.M., Andersson, J., and Chougnet, C. (2006). HIV-1-driven regulatory T-cell accumulation in lymphoid tissues is associated with disease progression in HIV/AIDS. Blood *108*, 3808–3817.

Nowak, M.A., Lloyd, A.L., Vasquez, G.M., Wiltrout, T.A., Wahl, L.M., Bischofberger, N., Williams, J., Kinter, A., Fauci, A.S., Hirsch, V.M., and Lifson, J.D. (1997). Viral dynamics of primary viremia and antiretroviral therapy in simian immunodeficiency virus infection. J. Virol. *71*, 7518–7525.

Ohta, Y., Masuda, T., Tsujimoto, H., Ishikawa, K., Kodama, T., Morikawa, S., Nakai, M., Honjo, S., and Hayami, M. (1988). Isolation of simian immunodeficiency virus from African green monkeys and seroepidemiologic survey of the virus in various non-human primates. Int. J. Cancer *41*, 115–122.

Okoye, A., Meier-Schellersheim, M., Brenchley, J.M., Hagen, S.I., Walker, J.M., Rohankhedkar, M., Lum, R., Edgar, J.B., Planer, S.L., Legasse, A., *et al.* (2007). Progressive CD4+ central memory T-cell decline results in CD4+ effector memory insufficiency and overt disease in chronic SIV infection. J. Exp. Med. *204*, 2171–2185.

Okoye A., Park, H., Rohankhedkar, M., Coyne-Johnson L., Lum, R., Walker, J.M., Planer, S.L., Legasse, A.W., Sylwester, A.W., Piatak, M. Jr, *et al.* (2009). Profound CD4+/CCR5+ T cell expansion is induced by CD8+ lymphocyte depletion but does not account for accelerated SIV pathogenesis. J. Exp. Med. *206*, 1575–1588.

Onanga, R., Kornfeld, C., Pandrea, I., Estaquier, J., Souquiere, S., Rouquet, P., Mavoungou, V.P., Bourry, O., M'Boup, S., Barre-Sinoussi, F., *et al.* (2002). High levels of viral replication contrast with only transient changes in CD4+ and CD8+ cell numbers during the early phase of experimental infection with simian immunodeficiency virus SIVmnd-1 in *Mandrillus sphinx*. J. Virol. *76*, 10256–10263.

Onanga, R., Souquiere, S., Makuwa, M., Mouinga-Ondeme, A., Simon, F., Apetrei, C., and Roques, P. (2006). Primary simian immunodeficiency virus SIVmnd-2 infection in mandrills (*Mandrillus sphinx*). J. Virol. *80*, 3303–3309.

Orandle, M.S., Williams, K.C., MacLean, A.G., Westmoreland, S.V., and Lackner, A.A. (2001). Macaques with rapid disease progression and simian immunodeficiency virus encephalitis have a unique cytokine profile in peripheral lymphoid tissues. J. Virol. 75, 4448–4452.

Orandle, M.S., MacLean, A.G., Sasseville, V.G., Alvarez, X., and Lackner, A.A. (2002). Enhanced expression of proinflammatory cytokines in the central nervous system is associated with neuroinvasion by simian immunodeficiency virus and the development of encephalitis. J. Virol. *76*, 5797–5802.

Oswald-Richter, K., Grill, S.M., Shariat, N., Leelawong, M., Sundrud, M.S., Haas, D.W., and D., U. (2004). HIV Infection of Naturally Occurring and Genetically Reprogrammed Human Regulatory T-cells. PLoS Biol. *2*, E198.

Owen, S.M., Masciotra, S., Novembre, F., Yee, J., Switzer, W.M., Ostyula, M., and Lal, R.B. (2000). Simian immunodeficiency viruses of diverse origin can use CXCR4 as a coreceptor for entry into human cells. J. Virol. *74*, 5702–5708.

Pacanowski, J., Kahi, S., Baillet, M., Lebon, P., Deveau, C., Goujard, C., Meyer, L., Oksenhendler, E., Sinet, M., and Hosmalin, A. (2001). Reduced blood CD123+ (lymphoid) and CD11c+ (myeloid) dendritic cell numbers in primary HIV-1 infection. Blood *98*, 3016–3021.

Paiardini, M., Pandrea, I., Apetrei, C., and Silvestri, G. (2009). Lessons learned from the natural hosts of HIV-related viruses. Annu Rev Med *60*, 485–495.

Pandrea, I., Onanga, R., Rouquet, P., Bourry, O., Ngari, P., Wickings, E.J., Roques, P., and Apetrei, C. (2001). Chronic SIV infection ultimately causes immunodeficiency in African non-human primates. AIDS *15*, 2461–2462.

Pandrea, I., Onanga, R., Kornfeld, C., Rouquet, P., Bourry, O., Clifford, S., Telfer, P.T., Abernethy, K., White, L.T., Ngari, P., *et al.* (2003). High levels of SIVmnd-1 replication in chronically infected *Mandrillus sphinx*. Virology *317*, 119–127.

Pandrea, I., Kornfeld, C., Ploquin, M.J.-I., Apetrei, C., Faye, A., Rouquet, P., Roques, P., Simon, F., Barré-Sinoussi, F., Müller-Trutwin, M.C., and Diop, O.M. (2005). Impact of viral factors on very early *in vivo* replication profiles in SIVagm-infected African green monkeys. J. Virol. *79*, 6249–6259.

Pandrea, I., Silvestri, G., Onanga, R., Veazey, R.S., Marx, P.A., Hirsch, V.M., and Apetrei, C. (2006a). Simian immunodeficiency viruses replication dynamics in African non-human primate hosts: common patterns and species-specific differences. J. Med. Primatol. *35*, 194–201.

Pandrea, I., Apetrei, C., Dufour, J., Dillon, N., Barbercheck, J., Metzger, M., Jacquelin, B., Bohm, R., Marx, P.A., Barre-Sinoussi, F., *et al.* (2006b). Simian immunodeficiency virus (SIV) SIVagm.sab infection of Caribbean African green monkeys: New model of the study of SIV pathogenesis in natural hosts J. Virol. *80*, 4858–4867.

Pandrea, I., Apetrei, C., Gordon, S., Barbercheck, J., Dufour, J., Bohm, R., Sumpter, B., Roques, P., Marx, P.A., Hirsch, V.M., *et al.* (2007a). Paucity of $CD4^+$CCR5+ T-cells is a typical feature of natural SIV hosts. Blood *109*, 1069–1076.

Pandrea, I., Gautam, R., Ribeiro, R., Brenchley, J.M., Butler, I.F., Pattison, M., Rasmussen, T., Marx, P.A., Silvestri, G., Lackner, A.A., *et al.* (2007b). Acute loss of intestinal $CD4^+$ T-cells is not predictive of SIV virulence. J. Immunol. *179*, 3035–3046.

Pandrea, I., Gaufin, T., Brenchley, J.M., Gautam, R., Monjure, C., Gautam, A., Coleman, C., Lackner, A.A., Ribeiro, R., Douek, D.C., and Apetrei, C. (2008a). Experimentally-induced immune activation in natural hosts of SIV induces significant increases in viral replication and $CD4^+$ T-cell depletion. J. Immunol. *181*, 6687–6691.

Pandrea, I., Ribeiro, R.M., Gautam, R., Gaufin, T., Pattison, M., Barnes, M., Monjure, C., Stoulig, C., Silvestri, G., Miller, M., *et al.* (2008b). Simian immunodeficiency virus SIVagm dynamics in African green monkeys. J. Virol. *82*, 3713–3724.

Pandrea, I., Onanga, R., Souquiere, S., Mouinga-Ondéme, A., Bourry, O., Makuwa, M., Rouquet, P., Silvestri, G., Simon, F., Roques, P., and Apetrei, C. (2008c). Paucity of $CD4^+$CCR5+ T-cells may prevent breastfeeding transmission of SIV in natural non-human primate hosts J. Virol. *82*, 5501–5509.

Pandrea, I., Sodora, D.L., Silvestri, G., and Apetrei, C. (2008d). Into the wild: simian immunodeficiency virus (SIV) infection in natural hosts. Trends Immunol *29*, 419–428.

Pandrea, I., Silvestri, G., and Apetrei, C. (2009). AIDS in African nonhuman primate hosts of SIVs: A new paradigm of SIV infection. Curr HIV Res. *6*, 57–72.

Park, H., Li, Z., Yang, X.O., Chang, S.H., Nurieva, R., Wang, Y.H., Wang, Y., Hood, L., Zhu, Z., Tian, Q., and Dong, C. (2005). A distinct lineage of CD4 T-cells regulates tissue inflammation by producing interleukin 17. Nat. Immunol. *6*, 1133–1141.

Parker, C.M., Groh, V., Band, H., Porcelli, S.A., Morita, C., Fabbi, M., Glass, D., Strominger, J.L., and Brenner, M.B. (1990). Evidence for extrathymic changes in the T-cell receptor gamma/delta repertoire. J. Exp. Med. *171*, 1597–1612.

Pereira, L.E., Johnson, R.P., and Ansari, A.A. (2008). Sooty mangabeys and rhesus macaques exhibit significant divergent natural killer cell responses during both acute and chronic phases of SIV infection. Cell Immunol *254*, 10–19.

Perelson, A.S., Neumann, A.U., Markowitz, M., Leonard, J.M., and Ho, D.D. (1996). HIV-1 dynamics *in vivo*: virion clearance rate, infected cell life-span, and viral generation time. Science *271*, 1582–1586.

Perelson, A.S., Essunger, P., Cao, Y., Vesanen, M., Hurley, A., Saksela, K., Markowitz, M., and Ho, D.D. (1997). Decay characteristics of HIV-1-infected compartments during combination therapy. Nature *387*, 188–191.

Perelson, A.S. (2002). Modelling viral and immune system dynamics. Nat. Rev. Immunol. *2*, 28–36.

Picker, L.J., Hagen, S.I., Lum, R., Reed-Inderbitzin, E.F., Daly, L.M., Sylwester, A.W., Walker, J.M., Siess, D.C., Piatak, M., Jr, Wang, C., *et al.* (2004). Insufficient production and tissue delivery of $CD4^+$ memory T-cells in rapidly progressive simian immunodeficiency virus infection. J. Exp. Med. *200*, 1299–1314.

Propato, A., Schiaffella, E., Vicenzi, E., Francavilla, V., Baloni, L., Paroli, M., Finocchi, L., Tanigaki, N., Ghezzi, S., Ferrara, R., *et al.* (2001). Spreading of

HIV-specific CD8+ T-cell repertoire in long-term nonprogressors and its role in the control of viral load and disease activity. Hum Immunol 62, 561–576.

Puffer, B.A., Pohlmann, S., Edinger, A.L., Carlin, D., Sanchez, M.D., Reitter, J., Watry, D.D., Fox, H.S., Desrosiers, R.C., and Doms, R.W. (2002). CD4 independence of simian immunodeficiency virus Envs is associated with macrophage tropism, neutralization sensitivity, and attenuated pathogenicity. J. Virol. 76, 2595–2605.

Putkonen, P., Thorstensson, R., Ghavamzadeh, L., Albert, J., Hild, K., Biberfeld, G., and Norrby, E. (1991). Prevention of HIV-2 and SIVsm infection by passive immunization in cynomolgus monkeys. Nature 352, 436–438.

Qin, S., Sui, Y., Soloff, A.C., Junecko, B.A., Kirschner, D.E., Murphey-Corb, M.A., Watkins, S.C., Tarwater, P.M., Pease, J.E., Barratt-Boyes, S.M., and Reinhart, T.A. (2008). Chemokine and cytokine mediated loss of regulatory T-cells in lymph nodes during pathogenic simian immunodeficiency virus infection. J. Immunol. *180*, 5530–5536.

Raffatellu, M., Santos, R.L., Verhoeven, D.E., George, M.D., Wilson, R.P., Winter, S.E., Godinez, I., Sankaran, S., Paixao, T.A., Gordon, M.A., *et al.* (2008). Simian immunodeficiency virus-induced mucosal interleukin-17 deficiency promotes Salmonella dissemination from the gut. Nat. Med. *14*, 421–428.

Regoes, R.R., Antia, R., Garber, D.A., Silvestri, G., Feinberg, M.B., and Staprans, S.I. (2004). Roles of target cells and virus-specific cellular immunity in primary simian immunodeficiency virus infection. J. Virol. 78, 4866–4875.

Rinaldo, C., Huang, X.L., Fan, Z.F., Ding, M., Beltz, L., Logar, A., Panicali, D., Mazzara, G., Liebmann, J., Cottrill, M., and *et al.* (1995). High levels of anti-human immunodeficiency virus type 1 (HIV-1) memory cytotoxic T-lymphocyte activity and low viral load are associated with lack of disease in HIV-1-infected long-term nonprogressors. J. Virol. 69, 5838–5842.

Rosenberg, E.S., Billingsley, J.M., Caliendo, A.M., Boswell, S.L., Sax, P.E., Kalams, S.A., and Walker, B.D. (1997). Vigorous HIV-1-specific $CD4^+$ T-cell responses associated with control of viremia. Science *278*, 1447–1450.

Rosenberg, E.S., Altfeld, M., Poon, S.H., Phillips, M.N., Wilkes, B.M., Eldridge, R.L., Robbins, G.K., D'Aquila, R.T., Goulder, P.J., and Walker, B.D. (2000). Immune control of HIV-1 after early treatment of acute infection. Nature *407*, 523–526.

Sasseville, V.G., and Lackner, A.A. (1997). Neuropathogenesis of simian immunodeficiency virus infection in macaque monkeys. J Neurovirol 3, 1–9.

Schindler, M., Munch, J., Kutsch, O., Li, H., Santiago, M.L., Bibollet-Ruche, F., Muller-Trutwin, M.C., Novembre, F.J., Peeters, M., Courgnaud, V., *et al.* (2006). Nef-mediated suppression of T-cell activation was lost in a lentiviral lineage that gave rise to HIV-1. Cell *125*, 1055–1067.

Schmitz, J.E., Simon, M.A., Kuroda, M.J., Lifton, M.A., Ollert, M.W., Vogel, C.W., Racz, P., Tenner-Racz, K., Scallon, B.J., Dalesandro, M., *et al.* (1999a). A nonhuman primate model for the selective elimination of CD8+ lymphocytes using a mouse-human chimeric monoclonal antibody. Am. J. Pathol. *154*, 1923–1932.

Schmitz, J.E., Kuroda, M.J., Santra, S., Sasseville, V.G., Simon, M.A., Lifton, M.A., Racz, P., Tenner-Racz, K., Dalesandro, M., Scallon, B.J., *et al.* (1999b). Control of viremia in simian immunodeficiency virus infection by CD8+ lymphocytes. Science *283*, 857–860.

Schmitz, J.E., Kuroda, M.J., Santra, S., Simon, M.A., Lifton, M.A., Lin, W., Khunkhun, R., Piatak, M., Lifson, J.D., Grosschupff, G., *et al.* (2003). Effect of humoral immune responses on controlling viremia during primary infection of rhesus monkeys with simian immunodeficiency virus. J. Virol. *77*, 2165–2173.

Schols, D., and De Clercq, E. (1998). The simian immunodeficiency virus mnd(GB-1) strain uses CXCR4, not CCR5, as coreceptor for entry in human cells. J. Gen. Virol. *79 (Pt 9)*, 2203–2205.

Sekaly, R.P. (2008). The failed HIV Merck vaccine study: a step back or a launching point for future vaccine development? J. Exp. Med. *205*, 7–12.

Sereti, I., Dunham, R.M., Spritzler, J., Aga, E., Proschan, M.A., Medvik, K., Battaglia, C.A., Landay, A.L., Pahwa, S., Fischl, M.A., *et al.* (2009). IL-7 administration drives T-cell cycle entry and expansion in HIV-1 infection. Blood *113*, 6304–6314.

Siegal, F.P., Kadowaki, N., Shodell, M., Fitzgerald-Bocarsly, P.A., Shah, K., Ho, S., Antonenko, S., and Liu, Y.J. (1999). The nature of the principal type 1 interferon-producing cells in human blood. Science *284*, 1835–1837.

Silvestri, G., Sodora, D.L., Koup, R.A., Paiardini, M., O'Neil, S.P., McClure, H.M., Staprans, S.I., and Feinberg, M.B. (2003). Nonpathogenic SIV infection of sooty mangabeys is characterized by limited bystander immunopathology despite chronic high-level viremia. Immunity *18*, 441–452.

Silvestri, G., Fedanov, A., Germon, S., Kozyr, N., Kaiser, W.J., Garber, D.A., McClure, H., Feinberg, M.B., and Staprans, S.I. (2005). Divergent host responses during primary simian immunodeficiency virus SIVsm infection of natural sooty mangabey and nonnatural rhesus macaque hosts. J. Virol. *79*, 4043–4054.

Silvestri, G., Paiardini, M., Pandrea, I., Lederman, M.M., and Sodora, D.L. (2007). Understanding the benign nature of SIV infection in natural hosts. J. Clin. Invest. *117*, 3148–3154.

Smit-McBride, Z., Mattapallil, J.J., Villinger, F., Ansari, A.A., and Dandekar, S. (1998). Intracellular cytokine expression in the $CD4^+$ and CD8+ T-cells from intestinal mucosa of simian immunodeficiency virus infected macaques. J. Med. Primatol. *27*, 129–140.

Sodora, D.L., and Silvestri, G. (2008). Immune activation and AIDS pathogenesis. Aids *22*, 439–446.

Soumelis, V., Scott, I., Gheyas, F., Bouhour, D., Cozon, G., Cotte, L., Huang, L., Levy, J.A., and Liu, Y.J. (2001). Depletion of circulating natural type 1 interferon-producing cells in HIV-infected AIDS patients. Blood *98*, 906–912.

Spira, A.I., Marx, P.A., Patterson, B.K., Mahoney, J., Koup, R.A., Wolinsky, S.M., and Ho, D.D. (1996). Cellular targets of infection and route of viral dissemination after an intravaginal inoculation of simian immunodeficiency virus into rhesus macaques. J. Exp. Med. *183*, 215–225.

Staprans, S.I., and Feinberg, M.B. (2004). The roles of nonhuman primates in the preclinical evaluation of candidate AIDS vaccines. Expert Rev Vaccines *3*, 5–32.

Tsunemi, S., Iwasaki, T., Imado, T., Higasa, S., Kakishita, E., Shirasaka, T., and Sano, H. (2005). Relationship of CD4+CD25+ regulatory T-cells to immune status in HIV-infected patients. Aids *19*, 879–886.

Van Heuverswyn, F., Li, Y., Neel, C., Bailes, E., Keele, B.F., Liu, W., Loul, S., Butel, C., Liegeois, F., Bienvenue, Y., *et al.* (2006). Human immunodeficiency viruses: SIV infection in wild gorillas. Nature *444*, 164.

Van Rompay, K.K., Marthas, M.L., Lifson, J.D., Berardi, C.J., Vasquez, G.M., Agatep, E., Dehqanzada, Z.A., Cundy, K.C., Bischofberger, N., and Pedersen, N.C. (1998). Administration of 9-[2-(phosphonomethoxy)propyl]adenine (PMPA) for prevention of perinatal simian immunodeficiency virus infection in rhesus macaques. AIDS Res. Hum. Retroviruses *14*, 761–773.

VandeWoude, S., and Apetrei, C. (2006). Going wild: Lessons from T-lymphotropic naturally occurring lentiviruses. Clin. Microbiol. Rev. *19*, 728–762.

Veazey, R.S., DeMaria, M., Chalifoux, L.V., Shvetz, D.E., Pauley, D.R., Knight, H.L., Rosenzweig, M., Johnson, R.P., Desrosiers, R.C., and Lackner, A.A. (1998). Gastrointestinal tract as a major site of CD4+ T-cell depletion and viral replication in SIV infection. Science *280*, 427–431.

Veazey, R.S., Tham, I.C., Mansfield, K.G., DeMaria, M., Forand, A.E., Shvetz, D.E., Chalifoux, L.V., Sehgal, P.K., and Lackner, A.A. (2000a). Identifying the target cell in primary simian immunodeficiency virus (SIV) infection: highly activated memory CD4+ T-cells are rapidly eliminated in early SIV infection *in vivo*. J. Virol. *74*, 57–64.

Veazey, R.S., Mansfield, K.G., Tham, I.C., Carville, A.C., Shvetz, D.E., Forand, A.E., and Lackner, A.A. (2000b). Dynamics of CCR5 expression by CD4(+) T-cells in lymphoid tissues during simian immunodeficiency virus infection. J. Virol. *74*, 11001–11007.

Veazey, R.S., Marx, P.A., and Lackner, A.A. (2001). The mucosal immune system: primary target for HIV infection and AIDS. Trends Immunol *22*, 626–633.

Veazey, R.S., Acierno, P.M., McEvers, K.J., Baumeister, S.H., Foster, G.J., Rett, M.D., Newberg, M.H., Kuroda, M.J., Williams, K., Kim, E.Y., *et al.* (2008). Increased loss of CCR5+ CD45RA- CD4+ T-cells in CD8+ lymphocyte-depleted Simian immunodeficiency virus-infected rhesus monkeys. J. Virol. *82*, 5618–5630.

Wang, Z., Metcalf, B., Ribeiro, R.M., McClure, H., and Kaur, A. (2006). Th-1-type cytotoxic CD8+ T-lymphocyte responses to simian immunodeficiency virus (SIV) are a consistent feature of natural SIV infection in sooty mangabeys. J. Virol. *80*, 2771–2783.

Watkins, D.I., Burton, D.R., Kallas, E.G., Moore, J.P., and Koff, W.C. (2008). Nonhuman primate models and the failure of the Merck HIV-1 vaccine in humans. Nat. Med. *14*, 617–621.

Watson, A., Ranchalis, J., Travis, B., McClure, J., Sutton, W., Johnson, P.R., Hu, S.L., and Haigwood, N.L. (1997). Plasma viremia in macaques infected with simian immunodeficiency virus: plasma viral load early in infection predicts survival. J. Virol. *71*, 284–290.

Wei, X., Decker, J.M., Wang, S., Hui, H., Kappes, J.C., Wu, X., Salazar-Gonzalez, J.F., Salazar, M.G., Kilby, J.M., Saag, M.S., *et al.* (2003). Antibody neutralization and escape by HIV-1. Nature *422*, 307–312.

Weiss, L., Donkova-Petrini, V., Caccavelli, L., Balbo, M., Carbonneil, C., and Levy, Y. (2004). Human immunodeficiency virus-driven expansion of CD4+CD25+ regulatory T-cells, which suppress HIV-specific CD4 T-cell responses in HIV-infected patients. Blood *104*, 3249–3256.

Williams, K.C., Corey, S., Westmoreland, S.V., Pauley, D., Knight, H., deBakker, C., Alvarez, X., and Lackner, A.A. (2001). Perivascular macrophages are the primary cell type productively infected by simian immunodeficiency virus in the brains of macaques: implications for the neuropathogenesis of AIDS. J. Exp. Med. *193*, 905–915.

Worgall, S., Connor, R., Kaner, R.J., Fenamore, E., Sheridan, K., Singh, R., and Crystal, R.G. (1999). Expression and use of human immunodeficiency virus type 1 coreceptors by human alveolar macrophages. J. Virol. *73*, 5865–5874.

Wu, L. (2008). Biology of HIV mucosal transmission. Curr Opin HIV AIDS *3*, 534–540.

Wykrzykowska, J.J., Rosenzweig, M., Veazey, R.S., Simon, M.A., Halvorsen, K., Desrosiers, R.C., Johnson, R.P., and Lackner, A.A. (1998). Early regeneration of thymic progenitors in rhesus macaques infected with simian immunodeficiency virus. J. Exp. Med. *187*, 1767–1778.

Yamamoto, H., Kawada, M., Takeda, A., Igarashi, H., and Matano, T. (2007). Post-infection immunodeficiency virus control by neutralizing antibodies. PLoS ONE *2*, e540.

Yue, F.Y., Merchant, A., Kovacs, C.M., Loutfy, M., Persad, D., and Ostrowski, M.A. (2008). Virus-specific interleukin-17-producing CD4+ T-cells are detectable in early human immunodeficiency virus type 1 infection. J. Virol. *82*, 6767–6771.

Zahn, R.C., Rett, M.D., Korioth-Schmitz, B., Sun, Y., Buzby, A.P., Goldstein, S., Brown, C.R., Byrum, R.A., Freeman, G.J., Letvin, N.L., *et al.* (2008). Simian Immunodeficiency Virus (SIV)-specific CD8+ T-cell responses in chronically SIVagm-infected vervet African green monkeys. J. Virol. *82*, 11577–11588.

Zhang, L., Dailey, P.J., He, T., Gettie, A., Bonhoeffer, S., Perelson, A.S., and Ho, D.D. (1999). Rapid clearance of simian immunodeficiency virus particles from plasma of rhesus macaques. J. Virol. *73*, 855–860.

Zhang, Y., Lou, B., Lal, R.B., Gettie, A., Marx, P.A., and Moore, J.P. (2000). Use of inhibitors to evaluate coreceptor usage by simian and simian/human immunodeficiency viruses and human immunodeficiency virus type 2 in primary cells. J. Virol. *74*, 6893–6910.

# The Felid Immunodeficiency Viruses: Viral Cell Tropism and the Pathogenesis of Feline AIDS

8

Brian J. Willett and Margaret J. Hosie

Abstract

Lentiviruses are widespread pathogens of primates, ungulates and felids. While the ungulate lentiviruses induce a disease state typical of a chronic inflammatory condition, the felid and primate lentiviruses induce an immunodeficiency characterized by a progressive depletion of $CD4^+$ T-helper cells. FIV infection of the domestic cat may lead to a spectrum of diseases, ranging from a rapid, acute-onset immunodeficiency to a chronic wasting disease with concomitant neuropathology and persistent recurring opportunistic infections. Here, we examine the host and viral determinants of FIV cell tropism and pathogenicity. The virus targets activated $CD4^+$ T-cells selectively by interactions with its primary receptor, CD134 and co-receptor, CXCR4. We discuss the impact of the virus–receptor interaction on the normal cellular function of the viral receptors and the cells on which they are expressed and assess how this contributes to pathogenicity, immunity to infection and the prospects for the development of FIV vaccines.

## Introduction

Lentiviruses are widespread pathogens of the Felidae, infecting both domestic cats and non-domestic felids (Brown *et al.*, 1993; Brown *et al.*, 1994; Carpenter *et al.*, 1996; Hofmann-Lehmann *et al.*, 1996; Troyer *et al.*, 2004; Troyer *et al.*, 2005). Within the domestic cat population, feline immunodeficiency virus (FIV) is a significant cause of disease, with infection resulting in an immunodeficiency syndrome characterized by recurrent gingivitis–stomatitis, cachexia, wasting, neuropathology and an increased risk of tumour development (Pedersen *et al.*, 1987; Hosie *et al.*, 1989; Sparger *et al.*, 1989; Yamamoto *et al.*, 1989; Ackley *et al.*, 1990; Torten *et al.*, 1991; Callanan *et al.*, 1992; Pedersen, 1993; Callanan *et al.*, 1996). The association between lentiviral infection and disease development in non-domestic felids is less clear, however lentivirus-infected African lions show a dramatic decline in $CD4^+$ T-lymphocytes (Bull *et al.*, 2003; Roelke *et al.*, 2006) and the expansion of an activated $CD8\alpha/\beta^+$ lymphocyte sub-population, analogous to observations in FIV (Willett *et al.*, 1993; Shimojima *et al.*, 1998) and HIV (Schmitz *et al.*, 1998) infections. Moreover, captive lions have succumbed to lentiviral-associated diseases after prolonged infection (Poli *et al.*, 1995; Brennan *et al.*, 2006) and specific pathological changes including lymphadenopathy, gingivitis, tongue papillomas and cachexia have been described recently in free-ranging lions (Roelke *et al.*, 2008).

In this chapter, we explore the role of cell tropism in the pathogenesis of feline AIDS, examining the link between the expression of the viral receptors CD134 and CXCR4, cell tropism and pathogenesis. Are there truly 'macrophage-tropic' strains of FIV and which determinants in the viral genome contribute to this biological phenotype?

## The cell tropism of FIV

Infection with the lentivirus of the domestic cat (*Felis catus*), $FIV_{Fca}$ leads to the development of an immune dysfunction characterized by a progressive depletion of $CD4^+$ T-lymphocytes (Ackley *et al.*, 1990; Novotney *et al.*, 1990; Hoffmann-Fezer *et al.*, 1992; Willett *et al.*, 1993). These data led to speculation that FIV and HIV shared usage of CD4 as a primary high-affinity binding receptor. However, in the cat, CD4 expression is restricted to helper T-cells and their thymic precursors. This is in contrast to primates where CD4 is also expressed on cells of the monocyte/macrophage lineage. While productive infection of helper T-cells with $FIV_{Fca}$ leads to a down-regulation of CD4, $FIV_{Fca}$ also infects CD4-negative cell lines (Willett *et al.*, 1991). Moreover, monoclonal antibodies against feline CD4 do not inhibit infection with $FIV_{Fca}$ (Hosie *et al.*, 1993). *In vivo*, $FIV_{Fca}$ targets not only $CD4^+$ T-cells, but also $CD8^+$ T-cells and B-cells (English *et al.*, 1993; Dean *et al.*, 1996), providing further compelling evidence for the existence of a non-CD4 receptor.

The identity of the primary receptor for FIV was revealed in 2004, when researchers in Tokyo, London and Glasgow identified an Env binding protein from a feline helper T-cell-derived cDNA expression library (Shimojima *et al.*, 2004). The cDNA encoded the feline homologue of CD134 (OX40, ACT35) and ectopic expression of CD134 was sufficient to render cells permissive for infection with diverse strains of $FIV_{Fca}$ (Shimojima *et al.*, 2004). As CD134 was initially characterized as OX40, an antigen of activated rat $CD4^+$ helper T-cells, the restricted expression of pattern of CD134 would be consistent with targeting of helper T-cells by FIV. However, the cell tropism of $FIV_{Fca}$ is complex; while $CD4^+$ T-cells are the predominant target in early/acute infection, in late/chronic infection FIV targets a broader range of cell types including $CD8^+$ T-cells and B-cells (English *et al.*, 1993; Dean *et al.*, 1996) and FIV replicates in cells of the macrophage (Brunner and Pedersen, 1989; Dow *et al.*, 1992) and neuronal lineages (Dow *et al.*, 1990). The key to understanding the complex cell tropism of $FIV_{Fca}$ may lie in the expression pattern of the viral receptor(s), strain-specific differences in the way the virus interacts with its receptor(s), and the expression of minor open reading frames (ORFs).

## The expression of the FIV receptors

CD134 is the primary receptor for FIV, however CD134 expression alone is insufficient to confer susceptibility to infection upon non-permissive cells. Viral entry requires expression of a co-receptor, CXCR4, a seven transmembrane domain, G-protein-coupled (7TM) receptor. CXCR4 was identified as an essential co-factor for infection with T-cell line-adapted strains of HIV in 1996 (Feng *et al.*, 1996). Subsequent studies revealed that the related 7TM receptors CCR5, and (to a lesser extent) CCR3 served similar roles in infection with primary, macrophage-tropic strains of HIV (Alkhatib *et al.*, 1996; Choe *et al.*, 1996; Deng *et al.*, 1996; Doranz *et al.*, 1996; Dragic *et al.*, 1996). The finding that cell culture (T-cell line) adapted strains of HIV used CXCR4 as a co-receptor for infection shed light on the mechanism of infection with FIV and in 1997 feline CXCR4 was identified as a necessary co-factor for infection with cell culture-adapted strains of FIV (Willett *et al.*, 1997a; Willett *et al.*, 1997c). However, the development of selective CXCR4 antagonists such as AMD3100, or SDF-1 (stromal-derived factor-1/CXCL12) the natural ligand for CXCR4, demonstrated that cell culture-adapted and primary isolates of FIV-alike were dependent on CXCR4 for infection (Willett *et al.*, 1997b; Hosie *et al.*, 1998; Egberink *et al.*, 1999; Richardson *et al.*, 1999). Following the discovery of the primary viral receptor CD134, it was then possible to confirm that CXCR4 acted as a 'co-receptor' for FIV infection (Shimojima *et al.*, 2004), where viral entry mediated by CD134 is dependent upon the co-expression of CXCR4. For infection to occur, the virus must interact sequentially with CD134 followed by CXCR4 (de Parseval *et al.*, 2006). This interaction triggers a conformational change in Env that exposes the co-receptor binding site and facilitates attachment to CXCR4, insertion of the fusion peptide into the target cell membrane and ultimately, fusion of the viral and cellular membranes (Fig. 8.1).

The expression pattern of the viral receptor and co-receptor determines cell tropism and ultimately pathogenesis. In humans, CD4 expression is not restricted to helper T-cells, lower levels of CD4 are expressed on cells of the monocyte/macrophage (Maddon *et al.*, 1986), including monocyte/macrophage-lineage cells within the central nervous system such as microglia (Jordan *et al.*, 1991). FIV targets $CD4^+$ T-lymphocytes by utilizing CD134 (OX40) as a primary attachment

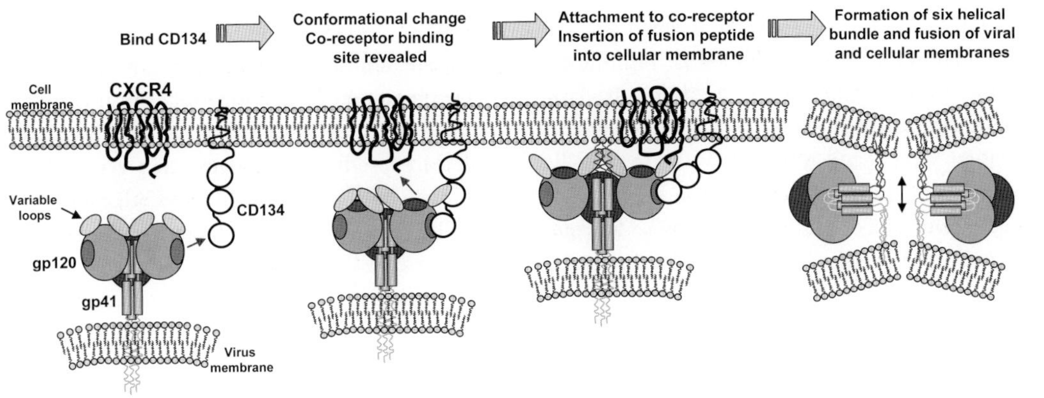


**Figure 8.1** Infection with FIV is mediated by sequential interactions between the FIV Env and CD134 followed by CXCR4. In this model, the initial high affinity engagement between Env and CD134 induces a conformational change in Env that reveals the CXCR4 binding site. Attachment to CXCR4 triggers exposure of the fusion peptide and initiates the fusion of the viral and cellular membranes.

receptor (Shimojima *et al.*, 2004) and CXCR4 as a co-receptor (Willett *et al.*, 1997a; Willett *et al.*, 1997c). Surface expression of CD134 in the cat is restricted predominantly to activated CD4$^+$ T-lymphocytes, although a minority of activated CD8$^+$ T-cells, and B-cells also express the antigen (Willett *et al.*, 2007). CD134 can also be detected on feline macrophages, consistent with the reported macrophage-tropism of FIV (Brunner and Pedersen, 1989; Dow *et al.*, 1992). As CD134 expression is up-regulated following stimulation of macrophages with bacterial lipopolysaccharide (LPS) (Willett *et al.*, 2007), the immunological activation status of the host may impact negatively upon susceptibility to infection and disease progression by modulating the expression of the viral receptor and co-receptor on key target cell populations. CD134 is expressed at a higher level on CD4$^+$CD25$^+$ Treg cells than on CD4$^+$CD25- cells (Joshi *et al.*, 2005), a population of cells shown previously to support FIV replication preferentially (Joshi *et al.*, 2004).

In the cat, CXCR4 is a T-cell activation antigen; resting T-cells express low levels of CXCR4 but this is up-regulated within 48 hours of mitogenic stimulation with concanavalin A (ConA) (Richardson *et al.*, 1999) and established IL2-dependent T-cell lines such as MYA-1 are rich in CXCR4 mRNA (Willett *et al.*, 1997c). Indeed, overnight culture of feline lymphoid cell suspensions in the absence of mitogen is sufficient to induce a marked up-regulation of CXCR4 expression (Troth *et al.*, 2008). Flow cytometric analyses have revealed high levels of surface CXCR4 expression on (B220$^+$) B-cells and (CD14$^+$) monocytes (Willett *et al.*, 2003). However, in direct comparison with human T-cells, where the majority of peripheral cells are CXCR4 positive, surface CXCR4 expression is not detectable on feline peripheral blood T-cells (Willett *et al.*, 2003). As the widespread expression on human T-cells of CXCR4 is thought to contribute to the cytopathicity of X4 strains of HIV in the symptomatic stage of disease, the restricted expression of feline CXCR4 to activated T-cells may explain why FIV produces a similar disease in cats to HIV in humans; in effect, T-cell expression of CXCR4 in the cat resembles T-cell expression of CCR5 in humans. All strains of FIV tested to date use CXCR4 as a co-receptor and thus target activated T-cells selectively. The specificity of the targeting would be determined by level of CD134 dependency for entry and the affinity of the virus receptor/co-receptor interaction rather than a switch in co-receptor usage from CCR5 to CXCR4 as is observed in HIV infection (Connor *et al.*, 1997). The expression of CXCR4 *in vivo* in lymphoid tissues has been examined by a combination of fluorescence-activated cell sorting (FACS) and real-time PCR (Troth *et al.*, 2008). CXCR4 mRNA was abundant in lymphoid tissues, consistent with the detection of surface CXCR4 expression by flow cytometry (Troth *et al.*, 2008). CXCR4 expression has also been examined on lymph node-derived CD4+/CD25+ Treg cells, where preferential binding to, and infection of CD4+/CD25+ Treg cells correlated with surface expression of CD134 and CXCR4 (Joshi *et al.*, 2005). CD4+/CD25- cells responded to mitogenic stimulation with ConA by up-regulating CXCR4 and in doing so became more susceptible to infection with FIV, a positive correlation between CXCR4 expression and proviral copy number being noted (Joshi *et al.*, 2005).

## Distinct modes of interaction between FIV Env and CD134

CD134 is a member of the tumour necrosis factor receptor (TNFR) family of molecules and consists of three cysteine-rich domains (CRDs). The binding of soluble monomeric SU-Fc fusion protein from $FIV_{PPR}$, a CD134-dependent strain of FIV, to CD134 requires determinants in the CRD1; mutation of aspartate residues D60 and D62 ablates SU-Fc binding and renders the molecule non-functional as a viral receptor while incorporation of the mutation SN60,62DD into human CD134 restores binding of PPR SU (de Parseval *et al.*, 2004a). However, although a chimaeric CD134 molecule comprising CRD1 of feline CD134 in the context of human CD134 binds soluble FIV SU-Fc, it is insufficient to confer susceptibility to infection with the majority of primary isolates of FIV (Willett *et al.*, 2006b) indicating that for viral entry to proceed, additional determinants outwith CRD1 are required. By comparing the amino acid sequence of human and feline CD134 CRD2, and then mutating in turn each of the residues in human CRD2 to the corresponding sequence of feline CD134 CRD2, a critical determinant was identified (NYE) that was required to restore receptor function upon the feline/human CD134 chimera (Willett *et al.*, 2006a). The identification of the NYE motif in feline CD134 was intriguing as the predicted three dimensional structure of feline CD134 placed the motif on the opposite face of the molecule from the SU-Fc binding site (Fig. 8.2).

Moreover, the motif formed a critical contact point between CD134 and CD134-ligand (CD134L), since it was shown that mutation of the motif ablated binding of CD134L to CD134. Why then is a motif in CRD2 critical for receptor function? To support viral entry, the virus may require a complex high affinity interaction and determinants in CRD2 may be critical for maintaining the correct conformation of the receptor. Structure predictions suggest that tyr-79 forms hydrogen bonds with tyr-37 on an adjacent strand of CRD1 which in turn interacts with cys-57 on the face of CD134 which is predicted to be the Env binding face. Mutation of tyr-79 may modulate the conformation of the Env binding face and as tyr-79 is critical for the CD134–CD134L interaction, the interaction of CD134L with CD134 may also alter the conformation for the Env binding face of the molecule. Alternatively, as the NYE motif forms a critical point of interaction between CD134 and CD134L, it is possible that engagement of CD134L by CD134 promotes the formation a trimeric form of CD134 that is preferred by primary FIV isolates (trimerization of CD134 requires complex formation with CD134L). A role for the CD134L–CD134 interaction in viral entry was revealed when it was observed that infection could be inhibited by soluble CD134L. As CD134L binds primarily to CRD2 on the face of CD134 opposite to the FIV SU-Fc binding site on CRD1, the modulation of FIV entry by the soluble ligand is most likely to be mediated by an allosteric effect, similar to the modulation of herpes simplex virus glycoprotein D (HSV gD) to its receptor HveA (Sarrias *et al.*, 2000). However, it is notable that unlike the interaction between TNF and TNFR, the receptor–ligand interface extends beyond CRD2 of CD134 to encompass residues within CRD1 (Compaan and Hymowitz, 2006) and therefore it is conceivable that CD134L may directly affect the interaction between FIV Env and CD134 (Willett *et al.*, 2006a).

## Significance of the CD134 interaction to pathogenesis

In early infection, FIV targets CD4+ T-cells preferentially while in later infection the virus disseminates into additional cellular compartments including CD8+ T-cells and B-cells. As the virus–receptor interaction is a primary determinant of cell tropism, these data suggest that there may be a fundamental difference in the biological properties of early and late viruses. In HIV infection, the dominant viral species in early infection uses CCR5 as a co-receptor (R5

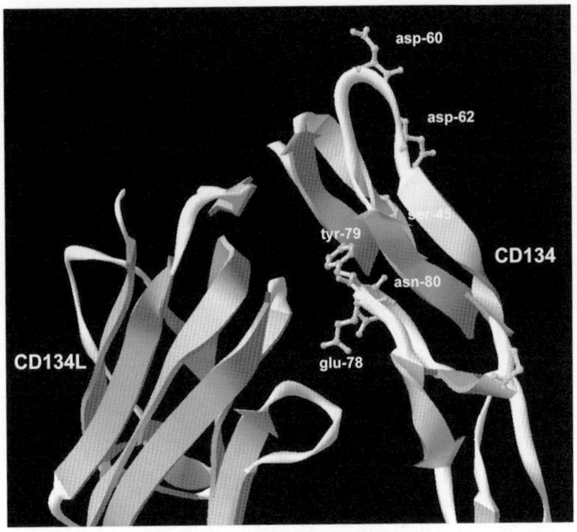


**Figure 8.2** Predicted 3D-structure of feline CD134 and CD134L. Structures were generated using the SWISS-MODEL modelling server in 'First Approach' mode and are based on the solved structures of human CD134 complexed with human and murine CD134L (Compaan and Hymowitz, 2006). Asp-60, asp-62 and ser-45 residues important for binding of soluble FIV-PPR SU-Fc (de Parseval *et al.*, 2004a) and the glu-78, tyr-79 and asn-80 residues critical for infection with GL8 and CPG41 (Willett *et al.*, 2006a) are shown.

viruses) (Connor *et al.*, 1997), possibly the result of selective 'filtering' of viruses using CXCR4 (X4 viruses) during the process of transmission, the 'gatekeeper' hypothesis (Margolis and Shattock, 2006). Disease progression in HIV infection is often associated with the emergence of X4 viruses (Connor *et al.*, 1997) (or dual-tropic viruses capable of using both CXCR4 and CCR5, 'X4R5' viruses), leading to speculation on a causal relationship between CXCR4 usage and disease progression. However, X4 or X4R5 viruses are absent in 50–60% of HIV-1-infected individuals who progress to symptomatic disease/AIDS (Jansson *et al.*, 1999; Li *et al.*, 1999; Karlsson *et al.*, 2004). In spite of the expanded cell tropism of late versus early strains of FIV, all strains of FIV tested to date use CXCR4 as a coreceptor. Why then does the tropism of FIV expand with disease progression? Two scenarios can be envisaged:

1 The nature of the virus–receptor interaction changes with disease progression.
2 The expression pattern of the viral receptors in the host alters with disease progression.

Macrophage-tropic R5 strains of HIV from late infection require lower levels of CD4 and CCR5 expression for entry and are less sensitive to entry inhibitors (Gray *et al.*, 2005). Moreover, the nature of the interaction between Env and CCR5 alters with disease progression with a 'narrow' highly constrained interaction, sensitive to modulation by RANTES dominating early in infection. As the disease progresses the mode of interaction 'broadens' such that the virus is less constrained/more flexible in its interaction, becoming less sensitive to modulation by RANTES and more prone to mutating to permit usage of CXCR4 (Karlsson *et al.*, 2004). Thus the cell tropism of a virus may alter with disease progression, manifesting as an enhanced ability to utilize lower levels of primary receptor and co-receptor. If FIV were to employ a similar mechanism for expanding its cell tropism, then this hypothesis would predict that late isolates of FIV would display a propensity for infection of cells expressing reduced levels of CD134 and CXCR4. Further, the hypothesis would suggest that the interactions with CD134 may alter such that the virus has a less constrained interaction and a propensity to evolve towards CD134-independent infection. In order to detect a broadening of the interaction between HIV and CCR5, chimaeric co-receptors were generated between CCR5 and CXCR4 (Karlsson *et al.*, 2004); a 'narrow' specificity being defined as a virus reliant on contacts with multiple determinants within CCR5 for infection while a 'broadening' of the interaction was revealed by an acquisition of the ability to use chimaeric CCR5/CXCR4 molecules for entry. In this context, the finding that FIV strains may be differentiated on the basis of whether they displayed an absolute requirement for an interaction with feline CD134 for infection, or whether a chimaeric feline CD134 CRD1 in the context of human CD134 could substitute as a viral receptor (see above) would suggest a commonality of pathogenetic mechanisms between the primate and felid lentiviruses. An initial analysis of 15 isolates of FIV (Fig. 8.3) emphasizes the range of specificities in usage of CD134 across FIV isolates.

Whether the emergence of FIV strains that have a less stringent interaction with CD134 correlates with disease progression remains to be established. Further important questions to

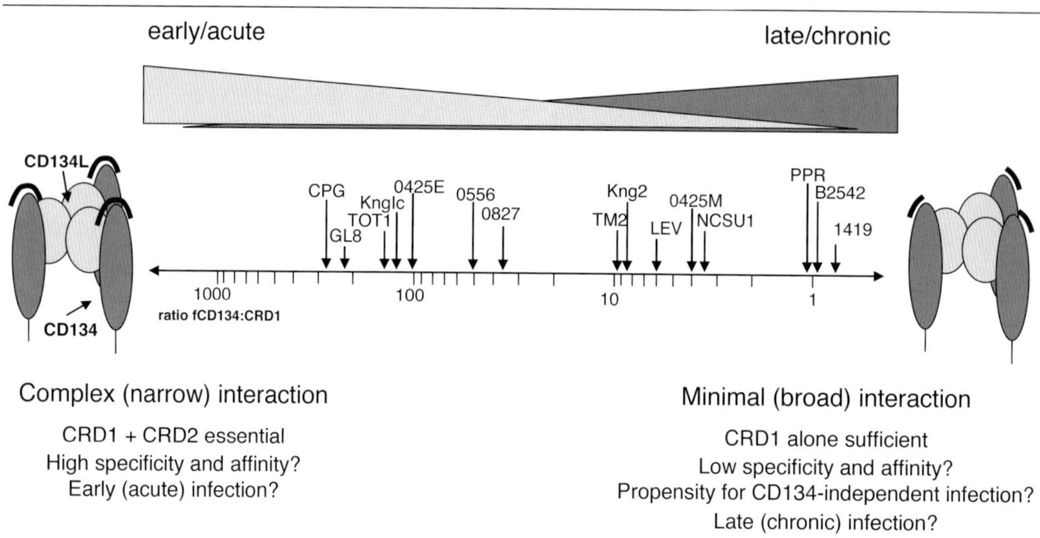


**Figure 8.3** Model for the differential utilization of CD134 by diverse isolates of FIV. HIV(FIV) luciferase pseudotypes bearing Envs from 15 FIVs were plated onto MCC cells expressing native feline CD134 or a feline CD134 CRD1 x human CD134 chimaera (FFHH). Luciferase activity was monitored three days post infection and the ratio of fCD134:CRD1 derived. Isolates such as CPG and GL8 had an absolute requirement for both CRD1 and CRD2 of feline CD134 while at the other extreme, PPR and B2542 infected efficiently using the chimaera expressing the feline CRD1 only. The model predicts that viruses with a complex interaction will dominate in early infection whereas those with a minimal interaction will be more abundant in late infection.

be resolved include the association between the specificity of interaction and the affinity of interaction, and the link between CD134-independent infection and affinity for CXCR4. While it has been long-established that cell culture-adapted strains of FIV are able to infect cells using CXCR4 alone (Willett *et al.*, 1997a; Willett *et al.*, 1997c), a recent study demonstrated that the PPR strain will infect cells that have been engineered to over express CXCR4 alone (de Parseval *et al.*, 2004c). That the PPR strain is also capable of infecting cells via an interaction with CD134 CRD1 alone (see Fig. 8.3) is consistent with a model whereby a reduced requirement for a complex interaction with CD134 manifests as a propensity for CD134-independent infection. Accordingly, the CD134-dependency of the PPR strain can be circumvented by increasing levels of surface CXCR4; primary isolates such as GL8 and CPG41 remain unable to infect cells expressing enhanced levels of CXCR4 (Willett *et al.*, 2008a).

## Escape from virus neutralizing antibody

The V3 loop of the FIV and HIV Env proteins is both a primary determinant of cell tropism and a target for the humoral immune response. This was illustrated clearly *in vitro* when HIV-1 was selected for the ability to grow in brain-derived cells and viral variants emerged that had lost their ability to replicate in macrophages and which were sensitive to neutralization by a monoclonal antibody targeting the V3 loop (McKnight *et al.*, 1995). Conversely, selection of the brain-tropic variants for resistance to neutralization by the anti-V3 antibody resulted in the loss of their tropism for brain cells and a re-acquisition of a macrophage-tropic phenotype (McKnight *et al.*, 1995). The HIV-1 V3 loop contributes to the interaction between Env and the viral co-receptor(s), principally CCR5 or CXCR4, and mutations in V3 are sufficient to switch co-receptor usage (reviewed in Hartley *et al.*, 2005; Huang *et al.*, 2005). Structural studies of HIV gp120 complexed with CD4 and the anti-gp120 antibody X5 indicated that the V3 loop would protrude downwards towards the host cell membrane, facilitating an interaction with the co-receptor but exposing the V3 loop to neutralizing antibody (Huang *et al.*, 2005). Analogous to the role of the HIV V3 loop in cell tropism and virus neutralization, the V3 loop homologue of FIV is a primary determinant of the cell tropism of FIV. Initially, it was found that mutations at the crown of the FIV V3 loop that increased the net charge of the loop, were acquired during the process of adaptation for growth in CrFK (Crandell feline kidney) cells (Siebelink *et al.*, 1995b; Verschoor *et al.*, 1995). As described above, the molecular basis for this phenomenon was found subsequently to result from an enhanced ability to interact directly with CXCR4, the FIV co-receptor (Willett *et al.*, 1997a; Willett *et al.*, 1997c). Previous studies examining the sensitivity of FIV to virus neutralizing antibody used viruses that had been propagated in CrFK cells (Tozzini *et al.*, 1992; Lombardi *et al.*, 1993; de Ronde *et al.*, 1993; Osborne *et al.*, 1994), and accordingly, these studies revealed an important role for the V3 loop of FIV in virus neutralization. However, when the substrate for the assays was switched to IL2-dependent T-cell lines, viruses were refractory to neutralization by sera that had formerly given high neutralizing titres on CrFK cells (Baldinotti *et al.*, 1994). Taken together, these studies indicate that assay systems in which the V3 loop is the primary dependent of tropism, mediating a direct interaction between V3 and CXCR4, reveal high titres of anti-V3 antibodies in a range of cat sera. Accordingly, should such viruses emerge *in vivo*, it is likely that CD134-independent infection routes would be suppressed by the potent anti-V3 humoral immune response. As recent studies have demonstrated that when a CD134-dependent strain of FIV (PPR) is pre-incubated with soluble CD134, the CD134-pretreated virus becomes susceptible to neutralization by monoclonal antibodies targeting V3. These antibodies failed to neutralize infection of CD134-expressing cells with the untreated virus (de Parseval *et al.*, 2006). These data are consistent with a model whereby the emergence of CD134-independent strains of FIV would be suppressed *in vivo* by anti-V3 antibodies. When primary virus was grown *in vitro* in the presence of antibody from an infected cat, the virus evaded neutralization by acquiring mutations in V4 and V5 (Siebelink *et al.*, 1993; Siebelink *et al.*, 1995a; Siebelink *et al.*, 1996). These findings were recapitulated independently *in vivo*, with evasion of the host immune response being accompanied by mutations in both V4 and V5 (Bendinelli *et al.*, 2001). Assuming that virus targets activated T-cells *in vivo*, as indicated by the restricted expression of the primary receptor for the virus (de Parseval *et al.*, 2004b; Willett *et al.*, 2007), then *in vitro* neutralization assays that are based on IL2-dependent T-cells are likely to be most informative about virus neutralization and the control of viral replication *in vivo*. It is likely then that V4 and V5 may represent significant targets for neutralizing antibody *in vivo*. Accordingly, we have found that following infection of cats with an uncloned biological isolate of FIV GL8, the animals raised a potent neutralizing antibody response that is targeted primarily to V5.

## Vaccination and cell tropism

Experimental vaccine induced protection from infection with FIV in domestic cats has been achieved using diverse immunogens, including whole inactivated virus/infected cells (Yamamoto *et al.*, 1991b, 1993; Hosie *et al.*, 1995; Matteucci *et al.*, 1997, 2000; Pu *et al.*, 2005), canarypoxvirus-based vectors (Tellier *et al.*, 1998) and nucleic acid-based vaccines (Hosie *et al.*, 2000; Lockridge *et al.*, 2000; reviewed in Hosie and Beatty, 2007). To date only one vaccine platform has been launched commercially, Fel-O-Vax® FIV (Fort Dodge Animal Health, USA) which has been licensed for use by veterinarians in the USA, Australia, New Zealand and Japan. The Fel-O-Vax® FIV vaccine is a combination whole inactivated virus/fixed cell vaccine and has shown variable efficacy against infection with heterologous strains of virus (Pu *et al.*, 2001, 2005; Dunham *et al.*, 2006). Although the utility of this vaccine remains equivocal, some experimental studies have demonstrated clear evidence of virus-specific immunity (Pu *et al.*, 2001) and understanding the mechanism whereby vaccine-induced protection from infection is afforded may provide a valuable insight into the development of the next generation of lentiviral vaccines. Fel-O-Vax® FIV was generated from the cell line FL4 (Yamamoto *et al.*, 1991a), a cell line that is persistently infected with a cell-culture adapted variant (FL4) of the prototypic Petaluma strain of FIV. Although immunity to infection with homologous virus induced by FL4-derived vaccines was correlated with virus neutralizing antibody titre (Hosie *et al.*, 1995), purified Env from the FL4 strain of FIV could not recapitulate these findings, inducing at best, a suppression of virus post-challenge viral burden; failing to induce protection in a significant number of vaccinates (Hosie *et al.*, 1996). These studies indicated that the conformation of the viral Env protein on fixed, inactivated viral particles may be of importance to the induction of virus neutralizing antibodies. A close examination of the FL4 immunogen reveals the presence of novel features acquired during the process of adaptation to cell culture that may have contributed to its reported enhanced immunogenicity. We have identified two potential sites for N-linked glycosylation in the FIV Env that were highly conserved among field isolates of FIV but which were absent from the FL4 Env protein (Willett *et al.*, 2008b). By introducing similar mutations into the GL8 and CPG41 primary isolates of FIV, we demonstrated that N-linked glycosylation at these sites impacted upon the virus–receptor interaction (Willett *et al.*, 2008b). Thus the FL4 virus may be uniquely immunogenic due to selective deglycosylation of Env and the exposure of cryptic epitopes for virus neutralization. The predicted N-linked glycosylation sites ablated in FL4 (T271 and N342Y) lie in a region of FIV Env equivalent to the stem of the HIV V1V2 loop and are strikingly similar to previously described N-linked glycosylation sites in the Envs of SIV and HIV thought to contribute to enhanced immunogenicity of Env-based immunogens and an altered receptor usage *in vitro* (Stamatatos and Cheng-Mayer, 1998; Kolchinsky *et al.*, 2001; Puffer *et al.*, 2002; Li *et al.*, 2008). Removal of a site for N-linked glycosylation at the base of the V1V2 stem of HIV-1 ADA (N197) facilitated CD4-independent infection of canine cells expressing human CCR5 (Kolchinsky *et al.*, 2001). How might selective deglycosylation of this region affect receptor usage? It has been postulated that deglycosylation repositions the V1V2 loop and in doing so exposes the chemokine receptor binding site. Indeed, additional studies have shown that mutations in this region can modulate both primary receptor (CD4) and co-receptor (CXCR4 and CCR5) usage (Ogert *et al.*, 2001; Pastore *et al.*, 2006). Consistent with this hypothesis, the FL4 strain of FIV infects CD134-negative cells efficiently through a direct interaction with CXCR4. It is likely that the process of adaptation for growth in the CD134-negative FL4 cells resulted in the acquisition of mutations in V3 and at the stem of the V1V2 loop that enhanced the ability of the virus to interact directly with CXCR4. Future studies will address the contribution of the unique features of the FL4 Env to the induction of neutralizing antibodies and whether similar mutations may be introduced into representative Envs from primary strains of FIV to induce high titre broadly acting neutralizing antibodies.

## Enhancement of infection following vaccination

Post-vaccinal outcomes after challenge may be efficacious, neutral or in extreme circumstances deleterious. Vaccine-associated enhancement of virus replication has been described for experimental lentiviral vaccines targeting HIV, SIV, EIAV (equine infectious anaemia virus), CAEV (caprine arthritis encephalitis virus) and FIV (Hosie *et al.*, 1992; Russo *et al.*, 1993; Wang *et al.*, 1994; Siebelink *et al.*, 1995c; Richardson *et al.*, 1997; Hammond *et al.*, 1999; Raabe *et al.*, 1999; Giannecchini *et al.*, 2002; Staprans *et al.*, 2004). Thus, understanding enhancement of infection following vaccination is of pivotal importance to the development of successful lentiviral vaccines. Enhancement following vaccination against FIV has been observed with a range of vaccine formulations (Hosie *et al.*, 1992; Siebelink *et al.*, 1995c; Osterhaus *et al.*, 1996; Richardson *et al.*, 1997; Karlas *et al.*, 1998, 1999; Giannecchini *et al.*, 2002), manifesting as an accelerated viraemia post-challenge, a higher viral load during acute infection and an increased likelihood of becoming infected following experimental challenge. Antibody-dependent enhancement of infection is well-documented in HIV and EIAV infections (Kostrikis *et al.*, 1996; Hammond *et al.*, 1999) and a comparison of X4 and R5 variants of HIV showed that mAbs against gp120 which either neutralized or had a neutral effect against X4 variants, enhanced infection with R5 variants (Schutten *et al.*, 1994, 1997; Guillon *et al.*, 2002). The enhancement of infectivity was mediated by improved utilization of CCR5 by the viruses, reducing the sensitivity of the virus to inhibition by anti-CCR5 antibody (Guillon *et al.*, 2002). Thus, the same humoral immune response may protect against an X4 strain of virus but enhance infectivity with an R5 strain of virus. The role of antibody in vaccine-mediated enhancement of FIV infection is unclear (Karlas *et al.*, 1998; Huisman *et al.*, 2004). Exchange of plasma between animals transfers enhancement of infection (Siebelink *et al.*, 1995c) indicating that it can be a contributory factor. Recently, we have uncovered evidence to suggest that plasma from animals infected with primary virulent strains of FIV contains antibodies that neutralize homologous virus efficiently, but have a neutral effect or show enhancing activity against heterologous virus (Fig. 8.4).

The second confounding factor to the induction of protective immunity is an expansion of the number of susceptible cells accessible to the incoming virus following immunization. HIV infection itself drives the expansion of an HIV-specific CD4⁺ T-cell subset that is highly susceptible to infection (Douek *et al.*, 2002). Pre-existing immunity to human Adenovirus was thought to be a major contributing factor to the enhanced risk of HIV infection in some of the participants of the Merck STEP trial following vaccination with a recombinant adenovirus expressing HIV antigens (Cohen, 2007; Ledford, 2007). The STEP trial was terminated prematurely when it was discovered that there was a strong correlation between the level of pre-existing humoral immunity to adenovirus 5 and the likelihood of becoming infected. Prior FIV studies in the domestic cat had suggested that immune activation may predispose the vaccinate to an enhanced risk of infection and with the discovery that the primary viral receptor was CD134, a surface antigen restricted predominantly to activated CD4⁺ T-cells, a cellular basis for post-vaccination enhancement of infection was proposed (Shimojima *et al.*, 2004). Indeed, as the expression of both CXCR4 and CD134 in the cat are restricted to activated T-cells, then immunological activation may drive up-regulation of both the primary viral receptor and sole co-receptor on target cells. Indeed, a vaccine trial using recombinant FIV ORF-A noted that in the absence of protective immunity, there was an association between CD134 expression *in vivo* and post-vaccination enhancement of infection (Pistello *et al.*, 2006). Subsequent to the termination of the STEP trial, it was shown that adenovirus 5 immune complexes stimulated dendritic cells and promoted HIV infection of dendritic cells/T-cell co-cultures, indicating that the humoral immune response to the adenovirus 5 vaccine vector may have created a permissive environment for HIV infection in vaccine recipients (Perreau *et al.*, 2008). Addressing the phenomenon of enhancement following vaccination is crucial to the development of reliable and efficacious lentiviral vaccines and a critical evaluation of the role of the FIV–CD134–CXCR4 interaction in enhancement will be pivotal.

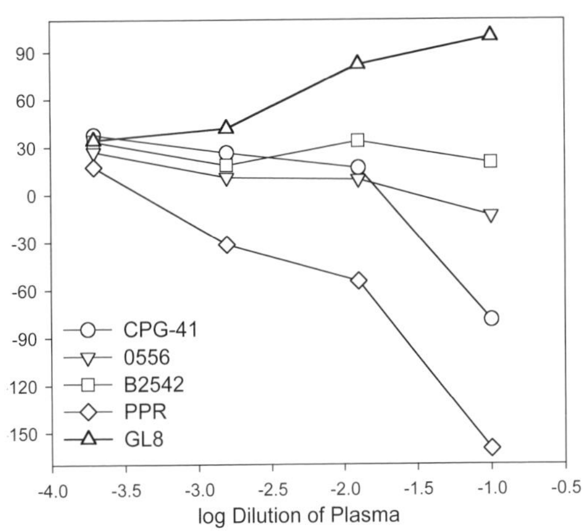


**Figure 8.4** Antibody-mediated enhancement of infection with FIV. HIV (FIV) luciferase pseudotypes bearing Envs from the GL8, PPR, B2542, 0556 and CPG41 strains of FIV were incubated with dilutions of plasma from a GL8-infected cat and plated onto indicator cells. While neutralizing activity was detected against the homologous GL8 strain, no activity was detected against heterologous strains 0556 and B2542 while infection with CPG41 and PPR was enhanced markedly.

## Receptor usage of lentiviruses in other felids

Lentiviruses are widespread pathogens of the Felidae, infecting both domestic cats and non-domestic felids (Brown *et al.*, 1993; Brown *et al.*, 1994; Carpenter *et al.*, 1996; Hofmann-Lehmann *et al.*, 1996; Troyer *et al.*, 2004; Troyer *et al.*, 2005). Among free-ranging felids, lentiviruses are endemic in Africa (in lions, leopards and cheetahs), South America (Geoffroy's cat, jaguar, margay, ocelot, puma and tigrina) and Asia (Pallas' cat and the Asian leopard cat). In Africa, lentiviruses have also been detected in one Hyaenidae species (the spotted hyena, *Crocuta crocuta*) (Troyer *et al.*, 2005). Although there are relatively few isolates of non-domestic cat lentiviruses available for study, preliminary studies on lion ($FIV_{Ple}$) and puma ($FIV_{Pco}$) lentiviruses have indicated that the viruses may not use the same receptor molecule(s) as domestic cat lentiviruses ($FIV_{Fca}$) (Smirnova *et al.*, 2005; Willett and Hosie, 2008a). While infection with diverse strains of $FIV_{Fca}$ *in vitro* is blocked reliably by the CXCR4 antagonist AMD3100 (Egberink *et al.*, 1998; Richardson *et al.*, 1999), similar studies using $FIV_{Ple}$ or $FIV_{Pco}$ found no evidence of sensitivity to this CXCR4 antagonist (Smirnova *et al.*, 2005). Further, $FIV_{Ple}$ and $FIV_{Pco}$ display evidence of an expanded cell tropism relative to $FIV_{Fca}$ (Smirnova *et al.*, 2005), consistent with differences in receptor utilization. However not all isolates of $FIV_{Ple}$ have the same origin; recent phylogenetic data have indicated that although subtype B and subtype E $FIV_{Ple}$ are monophyletic across the majority of the genome, the *env* ORFs provide evidence of distinct origins of the two viruses. The *env* of $FIV_{Ple}$-B clusters phylogenetically with the Pallas' cat lentivirus FIV-Oma while the *env* of $FIV_{Ple}$-E is more closely related to the virus of the domestic cat. As the early studies on $FIV_{Ple}$ tropism focussed on isolates of $FIV_{Ple}$-B, the tropism and receptor usage of $FIV_{Ple}$-E was compared with that of the $FIV_{Fca}$. The subtype E virus infected cells in a CXCR4 and CD134-dependent manner and was sensitive to antagonism by both AMD3100 and the CD134 antagonist Fc-TNC-CD134L (McEwan *et al.*, 2008). While ectopic expression of CD134 on the canine cell line CLL was required for infection with $FIV_{Fca}$ and $FIV_{Ple}$-E, the parent cell line was susceptible to infection with $FIV_{Ple}$-B in the absence of CD134 suggesting divergence of receptor usage (McEwan *et al.*, 2008). The distinct receptor usages of $FIV_{Ple}$-B and $FIV_{Ple}$-E raise the possibility that these two subtypes may differ in their pathogenicities *in vivo*.

## Macrophage-tropic strains of FIV?

FIV infects cells of diverse lineages, including monocytes and macrophages (Brunner and Pedersen, 1989; Dow *et al.*, 1992; Beebe *et al.*, 1994; Dow *et al.*, 1999). Macrophages isolated by peritoneal lavage from healthy cats were susceptible to infection with FIV, marked by syncytium formation, *in vitro* (Brunner and Pedersen, 1989). Moreover, virus could be recovered from peritoneal macrophages isolated from FIV-infected cats although, in general, virus isolation required stimulation with phorbol ester followed by co-cultivation with mitogen-stimulated PBMC (Brunner and Pedersen, 1989). Peripheral blood monocytes from cats infected with FIV B2542 were found to express increasing levels of viral capsid protein P24 (detected by immunostaining) with time in culture *ex vivo* indicating productive viral replication (Dow *et al.*, 1999). Similarly, bone marrow-derived monocyte/macrophages expressed FIV antigens after 72 hours in culture suggesting a latent *in vivo* infection that could be reactivated following culture *in vitro* (Dow *et al.*, 1999). Importantly, subtype A isolate 2531 and B isolate 2542 were identified as being macrophage-tropic in comparison with A isolate 2546 and B isolates 2561 and 2560 (Dow *et al.*, 1999), suggesting that the degree of macrophage tropism may be strain-specific. Beebe *et al.* suggested that monocyte/macrophage infection is more noticeable in the later stages of infection *in vivo* (Beebe *et al.*, 1994). As isolate B 2542 interacts with a minimal determinant on CD134 (Willett *et al.*, 2006a; Willett *et al.*, 2006b), it is possible that it represents a typical 'late stage' virus, able to infect and replicate in macrophages.

A primary determinant of macrophage tropism maps to *env*, (Vahlenkamp *et al.*, 1999), suggesting that the virus–receptor interaction is a critical determinant. When a series of *envs* was amplified from a cat infected with UT-48 (derived from a naturally infected cat displaying signs of neuropathology) and cloned into a derivative of the Petaluma F14 molecular clone, the resulting chimaeric viruses could be differentiated by their ability to replicate in macrophages (Vahlenkamp *et al.*, 1999). UT-48 clone 27 grew well in bone marrow-derived macrophages while clone 8 displayed markedly delayed replication kinetics. By generating chimeric *envs* between clones 8 and 27, the determinant of macrophage tropism was mapped to the V3–V4 region of *env*. Although the precise residues involved in conferring the ability to replicate in macrophages upon clone 27 were not identified, differential glycosylation patterns between the V3–V4 regions of clones 8 and 27 were noted and implicated (Vahlenkamp *et al.*, 1999).

## Accessory genes and the viral LTRs in cell tropism

The presence of a small open reading frame (ORF) in the region 3′ of vif and 5′ of rev (exon 1) was noted in the first molecular clones of FIV (ORF-A; Talbott *et al.*, 1989) and ORF-2 (Olmsted *et al.*, 1989a,b). 'ORF-A', as it is now known, is defective in the 34TF10 molecular clone of the Petaluma strain of FIV and this impaired the ability of the 34TF10 virus to grow in mitogen-activated peripheral blood mononuclear cells (PBMC) or IL2-dependent T-cell line MCH5–4 (Phillips *et al.*, 1990; Waters *et al.*, 1996). The defect in the 34TF10 molecular clone was a premature stop codon and when this was repaired the virus was able to grow in both PBMC and MCH5–4 cells, indicating an important role for ORF-A in viral growth. In these early studies, the molecular clones being compared were 34TF10 and PPR; it has since been established that 34TF10 also bears mutations in *env* that facilitate CD134-independent entry (Shimojima *et al.*, 2004) and therefore the broad tropism of 34TF10 was in fact mediated primarily by its ability to infect through a direct interaction with CXCR4. However, the failure of the parent 34TF10 clone to replicate in PBMC and MCH5–4 could be attributed directly to the defect in ORF-A, raising the question of the role of ORF-A in viral replication. *In vivo* studies have recapitulated the critical role for ORF-A in FIV growth (Inoshima *et al.*, 1996; Dean *et al.*, 1999; Pistello *et al.*, 2002; Novak *et al.*, 2007). Initial suggestions were that ORF-A may be a functional homologue of HIV Tat, however, although ORF-A increases FIV LTR-driven gene expression (de Parseval and Elder, 1999; Chatterji *et al.*, 2002), the FIV LTR lacks a homologue of the HIV TAR element and the basal level of transcription is significantly higher than that driven by the HIV LTR (Sparger *et al.*, 1992). When the ORF-A gene of the PPR (CD134-dependent) molecular clone of FIV was mutated, the restricted replicative capacity of the virus in mitogen-stimulated PBMC and IL2-dependent T-cell lines was reproduced (Gemeniano *et al.*, 2003). Analyses of these mutant viruses revealed a minimal effect of ORF-A mutation on LTR-driven gene expression, although a marked reduction in virus release and virion infectivity was observed (Gemeniano *et al.*, 2003). Combined with the observed nuclear localization of ORF-A and its ability to induce cell cycle arrest, these data are more consistent with an ORF-A function similar to HIV-1 Vpr. The HIV-1 Vpr protein is thought to permit infection and replication of HIV in monocyte/macrophage lineage cells (Connor *et al.*, 1995; Eckstein *et al.*, 2001; Sherman *et al.*, 2003); mutant R5 viruses defective in Vpr replicate efficiently in CD4$^+$ T-cell blasts but are defective for growth in tissue macrophages (Eckstein *et al.*, 2001). It is thought that the ability of HIV to grow in non-dividing cells is due to concerted actions of the matrix (MA) (Bukrinsky *et al.*, 1993) and integrase (IN) (Gallay *et al.*, 1995, 1997) proteins, the central DNA flap (Zennou *et al.*, 2000), and the actions of Vpr. Simple retroviruses require the breakdown of the nuclear envelope that associates normal cellular division for productive infection, in contrast, the HIV preintegration complex (PIC) is targeted to the nucleus by nuclear localization signals in MA and IN and the highly nucleophilic Vpr protein. ORF-A may serve a similar function in FIV as Vpr does to HIV, facilitating nuclear localization of the FIV PIC in monocyte/macrophage-lineage cells. Further, as HIV Vpr manipulates the cycle to favour viral replication (HIV LTR expression is optimal during the G2 phase of the cell cycle) (Goh *et al.*, 1998), then a similar function attributed to FIV ORF-A (Gemeniano *et al.*, 2004) may facilitate FIV LTR expression in non-dividing monocyte/macrophage-lineage cells.

The deoxyuridine-triphosphatase (dUTPase)-encoding DU gene is present in the non-primate lentiviruses FIV, EIAV, CAEV and MVV (Elder *et al.*, 1992; Wagaman *et al.*, 1993). By catalysing the interconversion of dUTP to dUMP, dUTPase provides a pool of dUMP for dTTP biosynthesis whilst maintaining the low dUTP/dTTP ratio that is essential to avoid misincorporation of uracil into DNA. A DU mutant of FIV was generated from the PPR strain of FIV and was found to replicate efficiently in the CrFK cell line however it appeared to be replication defective in macrophages (Wagaman *et al.*, 1993). Macrophages have low levels of endogenous dUTPase and similar studies on EIAV have indicated that the dUTPase facilitates replication in these cells (Threadgill *et al.*, 1993). *In vivo* studies with DU$^-$ FIV in macrophages revealed that although the virus replicated with a similar efficiency to wild-type virus, G→A mutations accumulated consistent with deoxyuridine misincorporation (Lerner *et al.*, 1995). The FIV *vif* gene is similar in size and genomic location to the *vif* gene of HIV-1 (Tomonaga *et al.*, 1992). Vif functions to overcome the antiviral activity of APOBEC3G, a cytidine deaminase that targets the nascent viral (–) strand RNA genome at a late stage of viral replication, deaminating cytosine (C) to uracil (U) and thus inducing hypermutation of guanine (G) to adenine (A) on the (+) strand proviral DNA genome (Sheehy *et al.*, 2002; Harris *et al.*, 2003; Mangeat *et al.*, 2003). The feline genome contains one APOBEC3H gene, three APOBEC3C genes (3Ca,3Cb and 3Cc) and one APOBEC3CH gene (Munk *et al.*, 2008). Feline immunodeficiency virus and feline leukaemia virus are targeted by feline APOBECS 3H and 3CH while feline foamy virus is inactivated by APOBECs 3Ca, 3Cb and 3Cc primarily. The FIV Vif counteracts the antiviral activity of APOBECs 3H and 3CH in feline cells and as such has a similar function to HIV-1 Vif (Munk *et al.*, 2008). FIV Vif mutants are defective for replication in both peripheral blood mononuclear cells and monocyte-derived macrophages *in vitro* (Lockridge *et al.*, 1999) and their replication is severely disabled *in vivo* (Gupta *et al.*, 2006). Studies with maedi-visna virus (MVV) have identified a natural isolate of MVV that has defective replication in macrophages *in vitro* and which is apathogenic (Gudmundsson *et al.*, 2005), and this restricted growth was attributed to point mutations in CA and Vif (neither mutation alone was sufficient to reproduce the phenotype; Gudmundsson *et al.*, 2005). However, to date there are no reports of mutations in FIV Vif that affect cell tropism selectively.

The FIV LTR controls viral gene expression in infected cells. There is a high degree of basal LTR activity in a range of cell types (Sparger *et al.*, 1992; Thompson *et al.*, 1994), critical elements within the LTR for gene expression per se being the AP-1, ATF and C/EBP sites (Kawaguchi *et al.*, 1992; Sparger *et al.*, 1992; Thompson *et al.*, 1994). The role of the LTR in lineage-specific gene expression is less clear. While deletion of AP-1 and AP-4 sites from FIV reduced basal promoter activity in classical reporter assays, the mutant provirus replicated efficiently in feline T-lymphoblastoid lines (Miyazawa *et al.*, 1993). However, growth of AP-1-deleted virus was impaired *in vivo* suggesting that *in vitro* assays do not recapitulate fully the host cells of the virus in natural infection (Inoshima *et al.*, 1996). In contrast, deletion of the C/EBP binding site reduced both basal promoter activity and viral replication in T-cells (Kawaguchi *et al.*, 1995). To date, little is known about LTR function in feline macrophages. However it is interesting to note that the LTRs of the puma lentiviruses (but not the domestic cat viruses) contain additional putative binding sites for interferon response factors, IRF-1 and IRF-3 (Poss and Ross, 2008), and thus replication of these viruses may be enhanced in cells responsive to interferons such as macrophages.

## Concluding remarks

The cell tropism of FIV is governed by factors acting at several stages of the viral replicative cycle. In order to target cells of the monocyte/macrophage lineage, the virus must utilize efficiently the low levels of CD134 that are expressed on these cells for attachment. As the viral co-receptor CXCR4 is expressed at a higher level on macrophages, strains that are able to infect independently of CD134 may be able to target macrophages more efficiently *in vitro*. Such viruses may be suppressed *in vivo* by the humoral immune response targeted at the V3 loop, only emerging late in infection when a functional immunodeficiency develops. A shift towards CD134-independence may accompany disease progression and this shift may manifest as a reduction in the stringency of the interaction between Env and CD134. Whether such viral variants contribute to dissemination within the chronically infected host remains to be established. Are such viruses less pathogenic or less likely to be transmitted? If they are transmitted, are they more likely to establish an inapparent infection?

Further fine-tuning of replication in macrophages may have occurred through the acquisition of a dUTPase activity, enabling the virus to replicate in cell types such as macrophages where endogenous dTTP pools are low. In addition, the *orfA* gene product may provide a function analogous to HIV-1 Vpr, facilitating nuclear localization of the pre-integration complex in non-dividing cells. Little is known about the contribution of the viral LTR to cell type-specific targeting; however, the presence of putative binding sites for interferon response factors, IRF-1 and IRF-3 in the LTRs of the puma (but not domestic cat) lentiviruses may indicate adaptation of the non-domestic cat lentiviruses for growth in cells such as macrophages that are known to respond to interferons. As the experimental techniques with which we may study the feline lentiviruses improve each year, there has never been a better time to revisit the role of cell tropism in the pathogenesis of the non-primate lentiviruses.

## Acknowledgements

Work in the authors' laboratory is supported by the National Institute of Allergy and Infectious Diseases (Public Health Service grant AI049765), the Biotechnology and Biosciences Research Council and The Wellcome Trust.

## References

Ackley, C.D., Yamamoto, J.K., Levy, N., Pedersen, N.C., and Cooper, M.D. (1990). Immunologic abnormalities in pathogen-free cats experimentally infected with feline immunodeficiency virus. J. Virol. *64*, 5652–5655.

Alkhatib, G., Combadiere, C., Broder, C.C., Feng, Y., Kennedy, P.E., Murphy, P.M., and Berger, E.A. (1996). CC CKR5: A RANTES, MIP-1α, MIP-1β receptor as a fusion cofactor for macrophage-tropic HIV-1. Science *272*, 1955–1958.

Baldinotti, F., Matteucci, D., Mazzetti, P., Giannelli, C., Bandecchi, P., Tozzini, F., and Bendinelli, M. (1994). Serum neutralization of feline immunodeficiency virus is markedly dependent on passage history of the virus and host system. J. Virol. *68*, 4572–4579.

Beebe, A.M., Dua, N., Faith, T.G., Moore, P.F., Pedersen, N.C., and Dandekar, S. (1994). Primary stage of feline immunodeficiency virus infection: viral dissemination and cellular targets. J. Virol. *68*, 3080–3091.

Bendinelli, M., Pistello, M., Del Mauro, D., Cammarota, G., Maggi, F., Leonildi, A., Giannecchini, S., Bergamini, C., and Matteucci, D. (2001). During readaptation *in vivo*, a tissue culture-adapted strain of feline immunodeficiency virus reverts to broad neutralization resistance at different times in individual hosts but through changes at the same position of the surface glycoprotein. J. Virol. *75*, 4584–4593.

Brennan, G., Podell, M.D., Wack, R., Kraft, S., Troyer, J.L., Bielefeldt-Ohmann, H., and VandeWoude, S. (2006). Neurologic Disease in Captive Lions (Panthera leo) with Low-Titer Lion Lentivirus Infection. J. Clin. Microbiol. *44*, 4345–4352.

Brown, E.W., Miththapala, S., and OBrien, S.J. (1993). Prevalence of exposure to feline immunodeficiency virus in exotic felid species. J. Zoo Wildlife Med. *24*, 357–364.

Brown, E.W., Yuhki, N., Packer, C., and OBrien, S.J. (1994). A lion lentivirus related to feline immunodeficiency virus – epidemiologic and phylogenetic aspects. J. Virol. *68*, 5953–5968.

Brunner, D., and Pedersen, N.C. (1989). Infection of peritoneal macrophages *in vitro* and *in vivo* with with feline immunodeficiency virus. J. Virol. *63*, 5483–5488.

Bukrinsky, M.I., Haggerty, S., Dempsey, M.P., Sharova, N., Adzhubel, A., Spitz, L., Lewis, P., Goldfarb, D., Emerman, M., and Stevenson, M. (1993). A nuclear localization signal within HIV-1 matrix protein that governs infection of non-dividing cells. Nature *365*, 666–669.

Bull, M.E., Kennedy-Stoskopf, S., Levine, J.F., Loomis, M., Gebhard, D.G., and Tompkins, W.A. (2003). Evaluation of T-lymphocytes in captive african lions (Panthera leo) infected with feline immunodeficiency virus. Am. J. Vet. Res. *64*, 1293–1300.

Callanan, J.J., Thompson, H., Toth, S.R., O'Neil, B., Lawrence, C.E., Willett, B.J., and Jarrett, O. (1992). Clinical and pathological findings in feline immunodeficiency virus experimental infection. Vet. Immunol. Immunopathol. *35*, 3–13.

Callanan, J.J., Jones, B.A., Irvine, J., Willett, B.J., McCandlish, I.A.P., and Jarrett, O. (1996). Histological classification and immunophenotype of lymphosarcomas in cats with naturally and experimentally acquired feline immunodeficiency virus infections. Vet. Pathol. *33*, 264–272.

Carpenter, M.A., Brown, E.W., Culver, M., Johnson, W.E., Pecon-Slattery, J., Brousset, D., and O'Brien, S.J. (1996). Genetic and phylogenetic divergence of feline immunodeficiency virus in the puma (*Puma concolor*). J. Virol. *70*, 6682–6693.

Chatterji, U., de Parseval, A., and Elder, J.H. (2002). Feline immunodeficiency virus OrfA is distinct from other lentivirus transactivators. J. Virol. *76*, 9624–9634.

Choe, H., Farzan, M., Sun, Y., Sullivan, N., Rollins, B., Ponath, P.D., Wu, L., Mackay, C.R., LaRosa, G., Newman, W., Gerard, N., Gerard, C., and Sodroski, J. (1996). The β-chemokine receptors CCR3 and CCR5 facilitate infection by primary HIV-1 isolates. Cell *85*, 1135–1148.

Cohen, J. (2007). AIDS research. Did Merck's failed HIV vaccine cause harm? Science *318*, 1048–1049.

Compaan, D.M., and Hymowitz, S.G. (2006). The crystal structure of the costimulatory OX40–OX40L complex. Structure. *14*, 1321–1330.

Connor, R.I., Chen, B.K., Choe, S., and Landau, N.R. (1995). Vpr is required for efficient replication of human immunodeficiency virus type-1 in mononuclear phagocytes. Virology *206*, 935–944.

Connor, R.I., Sheridan, K.E., Ceradini, D., Choe, S., and Landau, N.R. (1997). Change in coreceptor use correlates with disease progression in HIV- 1-infected individuals. J. Exp. Med. *185*, 621–628.

de Parseval, A., and Elder, J.H. (1999). Demonstration that orf2 encodes the feline immunodeficiency virus transactivating (Tat) protein and characterization of a unique gene product with partial rev activity. J. Virol. *73*, 608–617.

de Parseval, A., Chatterji, U., Morris, G., Sun, P., Olson, A.J., and Elder, J.H. (2004a). Structural mapping of CD134 residues critical for interaction with feline immunodeficiency virus. Nat. Struct. Mol. Biol. *12*, 60–66.

de Parseval, A., Chatterji, U., Sun, P., and Elder, J.H. (2004b). Feline immunodeficiency virus targets activated CD4+ T-cells by using CD134 as a binding receptor. Proc. Natl. Acad. Sci. U.S.A. *101*, 13044–13049.

de Parseval, A., Ngo, S., Sun, P., and Elder, J.H. (2004c). Factors that increase the effective concentration of CXCR4 dictate feline immunodeficiency virus tropism and kinetics of replication. J. Virol. *78*, 9132–9143.

de Parseval, A., Grant, C.K., Sastry, K.J., and Elder, J.H. (2006). Sequential CD134–CXCR4 interactions in feline immunodeficiency virus (FIV): soluble CD134 activates FIV Env for CXCR4-dependent entry and reveals a cryptic neutralization epitope. J. Virol. *80*, 3088–3091.

de Ronde, A., Stam, J.G., Boers, P., Langedijk, H., Meloen, R., Hesselink, W., Keldermans, L.C.E.J.M., van Vliet, A., Verschoor, E., Horzinek, M.C., and Egberink, H.F. (1993). Antibody response in cats to the envelope proteins of feline immunodeficiency virus: identification of an immunodominant neutralization domain. Virology *198*.

Dean, G.A., Reubel, G.H., Moore, P.F., and Pedersen, N.C. (1996). Proviral burden and infection kinetics of feline immunodeficiency virus in lymphocyte subsets of blood and lymph node. J. Virol. *70*, 5165–5169.

Dean, G.A., Himathongkham, S., and Sparger, E.E. (1999). Differential cell tropism of feline immunodeficiency virus molecular clones *in vivo*. J. Virol. *73*, 2596–2603.

Deng, H., Liu, R., Ellmeir, W., Choe, S., Unutmaz, D., Burkhart, M., Di Marzio, P., Marmon, S., Sutton, R.E., Hill, C.M., Davis, C.B., Peiper, S.C., Schall, T.J., Littman, D.R., and Landau, N.R. (1996). Identification of a major co-receptor for primary isolates of HIV-1. Nature *381*, 661–666.

Doranz, B.J., Rucker, J., Yi, Y., Smyth, R.J., Samson, M., Peiper, S.C., Parmentier, M., Collman, R.G., and Doms, R.W. (1996). A dual-tropic primary HIV-1 isolate that uses fusin and the β-chemokine receptors CKR-5, CKR-3, and CKR-2b as fusion co-factors. Cell *85*, 1149–1158.

Douek, D.C., Brenchley, J.M., Betts, M.R., Ambrozak, D.R., Hill, B.J., Okamoto, Y., Casazza, J.P., Kuruppu, J., Kunstman, K., Wolinsky, S., Grossman, Z., Dybul, M., Oxenius, A., Price, D.A., Connors, M., and Koup, R.A. (2002). HIV preferentially infects HIV-specific CD4+ T-cells. Nature *417*, 95–98.

Dow, S.W., Poss, M.L., and Hoover, E.A. (1990). Feline immunodeficiency virus: a neurotropic lentivirus. J. AIDS 3, 658–668.

Dow, S.W., Dreitz, M.J., and Hoover, E.A. (1992). Feline immunodeficiency virus neurotropism: evidence that astrocytes and microglia are the primary target cells. Vet. Immunol. Immunopathol. *35*, 23–35.

Dow, S.W., Mathiason, C.K., and Hoover, E.A. (1999). *In vivo* monocyte tropism of pathogenic feline immunodeficiency viruses. J. Virol. *73*, 6852–6861.

Dragic, T., Litwin, T., Allaway, G.P., Martin, S.R., Huang, Y., Nagashima, K.A., Cayanan, C., Maddon, P.J., Koup, R.A., Moore, J.P., and Paxton, W.A. (1996). HIV-1 entry into CD4+ cells is mediated by the chemokine receptor CC-CKR5. Nature *381*, 667–673.

Dunham, S.P., Bruce, J., MacKay, S., Golder, M., Jarrett, O., and Neil, J.C. (2006). Limited efficacy of an inactivated feline immunodeficiency virus vaccine. Vet. Rec. *158*, 561–562.

Eckstein, D.A., Sherman, M.P., Penn, M.L., Chin, P.S., De Noronha, C.M., Greene, W.C., and Goldsmith, M.A. (2001). HIV-1 Vpr enhances viral burden by facilitating infection of tissue macrophages but not nondividing CD4+ T-cells. J. Exp. Med. *194*, 1407–1419.

Egberink, H.F., De Clercq, E., Henson, G., Balzarini, J., Horzinek, M.C., and Schols, D. (1998). The bicyclam AMD3*100*, a selective antagonist of the human chemokine receptor CXCR4, inhibits feline immunodeficiency virus (FIV) replication *in vitro*. Fourth International Feline Retrovirus Research Symposium *A3.3*, 26.

Egberink, H.F., De Clerq, E., Van Vliet, A.L., Balzarini, J., Bridger, G.J., Henson, G., Horzinek, M.C., and Schols, D. (1999). Bicyclams, selective antagonists of the human chemokine receptor CXCR4, potently inhibit feline immunodeficiency virus replication. J. Virol. *73*, 6346–6352.

Elder, J.H., Lerner, D.L., Hasselkus-Light, C.S., Fontenot, D.J., Hunter, E., Luciw, P.A., Montelaro, R.C., and Phillips, T.R. (1992). Distinct subsets of retroviruses encode dUTPase. J. Virol. *66*, 1791–1794.

English, R.V., Johnson, C.M., Gebhard, D.H., and Tompkins, M.B. (1993). *In vivo* lymphocyte tropism of feline immunodeficiency virus. J. Virol. *67*, 5175–5186.

Feng, Y., Broder, C.C., Kennedy, P.E., and Berger, E.A. (1996). HIV-1 entry co-factor: functional cDNA cloning of a seven-transmembrane, G protein-coupled receptor. Science 272, 872–877.

Gallay, P., Swingler, S., Song, J., Bushman, F., and Trono, D. (1995). HIV nuclear import is governed by the phosphotyrosine-mediated binding of matrix to the core domain of integrase. Cell *83*, 569–576.

Gallay, P., Hope, T., Chin, D., and Trono, D. (1997). HIV-1 infection of nondividing cells through the recognition of integrase by the importin/karyopherin pathway. Proc. Natl. Acad. Sci. U.S.A. *94*, 9825–9830.

Gemeniano, M.C., Sawai, E.T., Leutenegger, C.M., and Sparger, E.E. (2003). Feline immunodeficiency virus ORF-Ais required for virus particle formation and virus infectivity. J. Virol. *77*, 8819–8830.

Gemeniano, M.C., Sawai, E.T., and Sparger, E.E. (2004). Feline immunodeficiency virus Orf-A localizes to the nucleus and induces cell cycle arrest. Virology *325*, 167–174.

Giannecchini, S., Isola, P., Sichi, O., Matteucci, D., Pistello, M., Zaccaro, L., Del Mauro, D., and Bendinelli, M.

(2002). AIDS vaccination studies using an *ex vivo* feline immunodeficiency virus model: failure to protect and possible enhancement of challenge infection by four cell-based vaccines prepared with autologous lymphoblasts. J. Virol. *76*, 6882–6892.

Goh, W.C., Rogel, M.E., Kinsey, C.M., Michael, S.F., Fultz, P.N., Nowak, M.A., Hahn, B.H., and Emerman, M. (1998). HIV-1 Vpr increases viral expression by manipulation of the cell cycle: a mechanism for selection of Vpr *in vivo*. Nat. Med. *4*, 65–71.

Gray, L., Sterjovski, J., Churchill, M., Ellery, P., Nasr, N., Lewin, S.R., Crowe, S.M., Wesselingh, S.L., Cunningham, A.L., and Gorry, P.R. (2005). Uncoupling coreceptor usage of human immunodeficiency virus type 1 (HIV-1) from macrophage tropism reveals biological properties of CCR5-restricted HIV-1 isolates from patients with acquired immunodeficiency syndrome. Virology *337*, 384–398.

Gudmundsson, B., Jonsson, S.R., Olafsson, O., Agnarsdottir, G., Matthiasdottir, S., Georgsson, G., Torsteinsdottir, S., Svansson, V., Kristbjornsdottir, H.B., Franzdottir, S.R., Andresson, O.S., and Andresdottir, V. (2005). Simultaneous mutations in CA and Vif of maedi-Visna virus cause attenuated replication in macrophages and reduced infectivity *in vivo*. J. Virol. *79*, 15038–15042.

Guillon, C., Schutten, M., Boers, P.H., Gruters, R.A., and Osterhaus, A.D. (2002). Antibody-mediated enhancement of human immunodeficiency virus type 1 infectivity is determined by the structure of gp120 and depends on modulation of the gp120–CCR5 interaction. J. Virol. *76*, 2827–2834.

Gupta, S., Leutenegger, C., Dean, G., and Sparger, E. (2006). Construction and characterization of feline immunodeficiency virus proviral mutants that coexpress interferon gamma and green fluorescent protein. Aids Res. Hum. Retroviruses *22*, 342–349.

Hammond, S.A., Raabe, M.L., Issel, C.J., and Montelaro, R.C. (1999). Evaluation of antibody parameters as potential correlates of protection or enhancement by experimental vaccines to equine infectious anemia virus. Virology *262*, 416–430.

Harris, R.S., Bishop, K.N., Sheehy, A.M., Craig, H.M., Petersen-Mahrt, S.K., Watt, I.N., Neuberger, M.S., and Malim, M.H. (2003). DNA deamination mediates innate immunity to retroviral infection. Cell *113*, 803–809.

Hartley, O., Klasse, P.J., Sattentau, Q.J., and Moore, J.P. (2005). V3: HIV's switch-hitter. Aids Res. Hum. Retroviruses *21*, 171–189.

Hoffmann-Fezer, G., Thum, J., Ackley, C., Herbold, M., Mysliwietz, J., Thefeld, S., Hartmann, K., and Kraft, W. (1992). Decline in CD4$^+$ cell numbers in cats with naturally acquired feline immunodeficiency virus infection. J. Virol. *66*, 1484–1488.

Hofmann-Lehmann, R., Fehr, D., Grob, M., Elgizoli, M., Packer, C., Martenson, J.S., O'Brien, S.J., and Lutz, H. (1996). Prevalence of antibodies to feline parvovirus, calicivirus, herpesvirus, coronavirus, and immunodeficiency virus and of feline leukemia virus antigen and the interrelationship of these viral infections in free-ranging lions in east Africa. Clin. Diagn. Lab Immunol. *3*, 554–562.

Hosie, M.J., Robertson, C., and Jarrett, O. (1989). Prevalence of feline leukaemia virus and antibodies to feline immunodeficiency virus in cats in the United Kingdom. Vet. Rec. *125*, 293–297.

Hosie, M.J., Osborne, R., Reid, G., Neil, J.C., and Jarrett, O. (1992). Enhancement after feline immunodeficiency virus vaccination. Vet. Immunol. Immunopathol. *35*, 191–198.

Hosie, M.J., Willett, B.J., Dunsford, T.H., Jarrett, O., and Neil, J.C. (1993). A monoclonal antibody which blocks infection with feline immunodeficiency virus identifies a possible non-CD4 receptor. J. Virol. *67*, 1667–1671.

Hosie, M., Osborne, R., Yamamoto, J.K., Neil, J.C., and Jarrett, O. (1995). Protection against homologous but not heterologous challenge induced by inactivated feline immunodeficiency virus vaccines. J. Virol. *69*, 1253–1255.

Hosie, M.J., Dunsford, T.D., de Ronde, A., Willett, B.J., Cannon, C.A., Neil, J.C., and Jarrett, O. (1996). Suppression of virus burden by immunisation with feline immunodeficiency virus Env protein. Vaccine *14*, 405–411.

Hosie, M.J., Broere, N., Hesselgesser, J., Turner, J.D., Hoxie, J.A., Neil, J.C., and Willett, B.J. (1998). Modulation of feline immunodeficiency virus infection by stromal cell-derived factor (SDF-1). J. Virol. *72*, 2097–2104.

Hosie, M.J., Dunsford, T., Klein, D., Willett, B.J., Cannon, C., Osborne, R., Macdonald, J., Spibey, N., Mackay, N., Jarrett, O., and Neil, J.C. (2000). Vaccination with inactivated virus but not viral DNA reduces virus load following challenge with a heterologous and virulent isolate of feline immunodeficiency virus. J. Virol. *74*, 9403–9411.

Hosie, M.J., and Beatty, J.A. (2007). Vaccine protection against feline immunodeficiency virus: setting the challenge. Aust. Vet. J. *85*, 5–12.

Huang, C.C., Tang, M., Zhang, M.Y., Majeed, S., Montabana, E., Stanfield, R.L., Dimitrov, D.S., Korber, B., Sodroski, J., Wilson, I.A., Wyatt, R., and Kwong, P.D. (2005). Structure of a V3-containing HIV-1 gp120 core. Science *310*, 1025–1028.

Huisman, W., Schrauwen, E.J., Pas, S.D., Karlas, J.A., Rimmelzwaan, G.F., and Osterhaus, A.D. (2004). Antibodies specific for hypervariable regions 3 to 5 of the feline immunodeficiency virus envelope glycoprotein are not solely responsible for vaccine-induced acceleration of challenge infection in cats. J. Gen. Virol. *85*, 1833–1841.

Inoshima, Y., Kohmoto, M., Ikeda, Y., Yamada, H., Kawaguchi, Y., Tomonaga, K., Miyazawa, T., Kai, C., Umemura, T., and Mikami, T. (1996). Roles of the auxiliary genes and AP-1 binding site in the long terminal repeat of feline immunodeficiency virus in the early stage of infection in cats. J. Virol. *70*, 8518–8526.

Jansson, M., Backstrom, E., Bjorndal, A., Holmberg, V., Rossi, P., Fenyo, E.M., Popovic, M., Albert, J., and Wigzell, H. (1999). Coreceptor usage and RANTES sensitivity of non-syncytium-inducing HIV-1 isolates obtained from patients with AIDS. J. Hum. Virol. *2*, 325–338.

Jordan, C.A., Watkins, B.A., Kufta, C., and Dubois-Dalq, M. (1991). Infection of brain microglial cells by human immunodeficiency virus type 1 is CD4 dependent. J. Virol. *65*, 736–742.

Joshi, A., Vahlenkamp, T.W., Garg, H., Tompkins, W.A., and Tompkins, M.B. (2004). Preferential replication of FIV in activated CD4(+)CD25(+)T-cells independent of cellular proliferation. Virology *321*, 307–322.

Joshi, A., Garg, H., Tompkins, M.B., and Tompkins, W.A. (2005). Preferential Feline Immunodeficiency Virus (FIV) Infection of CD4$^{+}$ CD25+ T-Regulatory Cells Correlates both with Surface Expression of CXCR4 and Activation of FIV Long Terminal Repeat Binding Cellular Transcriptional Factors. J. Virol. *79*, 4965–4976.

Karlas, J.A., Siebelink, K.H., Peer, M.A., Huisman, W., Rimmelzwaan, G.F., and Osterhaus, A.D. (1998). Accelerated viraemia in cats vaccinated with fixed autologous FIV-infected cells. Vet. Immunol. Immunopathol. *65*, 353–365.

Karlas, J.A., Siebelink, K.H., Peer, M.A., Huisman, W., Cuisinier, A.M., Rimmelzwaan, G.F., and Osterhaus, A.D. (1999). Vaccination with experimental feline immunodeficiency virus vaccines, based on autologous infected cells, elicits enhancement of homologous challenge infection. J. Gen. Virol. *80 (Pt 3)*, 761–765.

Karlsson, I., Antonsson, L., Shi, Y., Oberg, M., Karlsson, A., Albert, J., Olde, B., Owman, C., Jansson, M., and Fenyo, E.M. (2004). Coevolution of RANTES sensitivity and mode of CCR5 receptor use by human immunodeficiency virus type 1 of the R5 phenotype. J. Virol. *78*, 11807–11815.

Kawaguchi, Y., Norimine, J., Miyazawa, T., Kai, C., and Mikami, T. (1992). Sequences within the feline immunodeficiency virus long terminal repeat that regulate gene expression and respond to activation by feline herpesvirus type 1. Virology *190*, 465–468.

Kawaguchi, Y., Tomonaga, K., Maeda, K., Ono, M., Miyazawa, T., Kohmoto, M., Tohya, Y., and Mikami, T. (1995). The C/EBP site in the feline immunodeficiency virus (FIV) long terminal repeat (LTR) is necessary for its efficient replication and is also involved in the inhibition of FIV LTR-directed gene expression by pseudorabies virus ICP4. Virology *208*, 492–499.

Kolchinsky, P., Kiprilov, E., Bartley, P., Rubinstein, R., and Sodroski, J. (2001). Loss of a single N-linked glycan allows CD4-independent human immunodeficiency virus type 1 infection by altering the position of the gp120 V1/V2 variable loops. J. Virol. *75*, 3435–3443.

Kostrikis, L.G., Cao, Y., Ngai, H., Moore, J.P., and Ho, D.D. (1996). Quantitative analysis of serum neutralization of human immunodeficiency virus type 1 from subtypes A, B, C, D, E, F, and I: lack of direct correlation between neutralization serotypes and genetic subtypes and evidence for prevalent serum-dependent infectivity enhancement. J. Virol. *70*, 445–458.

Ledford, H. (2007). HIV vaccine may raise risk. Nature *450*, 325.

Lerner, D.L., Wagaman, P.C., Phillips, T.R., Prospero-Garcia, O., Henriksen, S.J., Fox, H.S., Bloom, F.E., and Elder, J.H. (1995). Increased mutation frequency of feline immunodeficiency virus lacking functional deoxyuridine-triphosphatase. Proc. Natl. Acad. Sci. U.S.A. *92*, 7480–7484.

Li, S., Juarez, J., Alali, M., Dwyer, D., Collman, R., Cunningham, A., and Naif, H.M. (1999). Persistent CCR5 utilization and enhanced macrophage tropism by primary blood human immunodeficiency virus type 1 isolates from advanced stages of disease and comparison to tissue-derived isolates. J. Virol. *73*, 9741–9755.

Li, Y., Cleveland, B., Klots, I., Travis, B., Richardson, B.A., Anderson, D., Montefiori, D., Polacino, P., and Hu, S.L. (2008). Removal of a single N-linked glycan in human immunodeficiency virus type 1 gp120 results in an enhanced ability to induce neutralizing antibody responses. J. Virol. *82*, 638–651.

Lockridge, K.M., Himathongkham, S., Sawai, E.T., Chienand, M., and Sparger, E.E. (1999). The feline immunodeficiency virus vif gene is required for productive infection of feline peripheral blood mononuclear cells and monocyte-derived macrophages. Virology *261*, 25–30.

Lockridge, K.M., Chien, M., Dean, G.A., Stefano, C.K., Montelaro, R.C., Luciw, P.A., and Sparger, E.E. (2000). Protective immunity against feline immunodeficiency virus induced by inoculation with vif-deleted proviral DNA. Virology *273*, 67–79.

Lombardi, S., Garzelli, C., La Rosa, C., Zaccaro, L., Specter, S., Malvaldi, G., Tozzini, F., Esposito, F., and Bendinelli, M. (1993). Identification of a linear neutralization site within the third variable region of the feline immunodeficiency virus envelope. J. Virol. *67*, 4742–4749.

Maddon, P.J., Dalgleish, A.G., McDougal, J.S., Clapham, P.R., Weiss, R.A., and Axel, R. (1986). The T4 gene encodes the AIDS virus receptor and is expressed in the immune system and the brain. Cell *47*, 333–348.

Mangeat, B., Turelli, P., Caron, G., Friedli, M., Perrin, L., and Trono, D. (2003). Broad antiretroviral defence by human APOBEC3G through lethal editing of nascent reverse transcripts. Nature *424*, 99–103.

Margolis, L., and Shattock, R. (2006). Selective transmission of CCR5-utilizing HIV-1: the 'gatekeeper' problem resolved? Nat. Rev. Microbiol. *4*, 312–317.

Matteucci, D., Pistello, M., Mazzetti, P., Giannecchini, S., Del Mauro, D., Lonetti, I., Zaccaro, L., Pollera, C., Specter, S., and Bendinelli, M. (1997). Studies of AIDS vaccination using an *ex vivo* feline immunodeficiency virus model: protection conferred by a fixed-cell vaccine against cell-free and cell-associated challenge differs in duration and is not easily boosted. J. Virol. *71*, 8368–8376.

Matteucci, D., Poli, A., Mazzetti, P., Sozzi, S., Bonci, F., Isola, P., Zaccaro, L., Giannecchini, S., Calandrella, M., Pistello, M., Specter, S., and Bendinelli, M. (2000). Immunogenicity of an anti-clade B feline immunodeficiency fixed-cell virus vaccine in field cats. J. Virol. *74*, 10911–10919.

McEwan, W.A., McMonagle, E.L., logan, N., Serra, R.C., Kat, P., VandeWoude, S., Hosie, M.J., and Willett, B.J. (2008). Genetically divergent strains of feline immunodeficiency virus from the domestic cat (Felis catus) and the African lion (Panthera leo) share usage of CD134 and CXCR4 as entry receptors. J. Virol. *82*, 10953–10958.

McKnight, A., Weiss, R.A., Shotton, C., Takeuchi, Y., Hoshino, H., and Clapham, P.R. (1995). Change in tropism upon immune escape by human immunodeficiency virus. J. Virol. *69*, 3167–3170.

Miyazawa, T., Kohmoto, M., Kawaguchi, Y., Tomonaga, K., Toyosaki, T., Ikuta, K., Adachi, A., and Mikami, T. (1993). The AP-1 binding site in the feline immunodeficiency virus long terminal repeat is not required for virus replication in feline T-lymphocytes. J. Gen. Virol. *74*, 1573–1580.

Munk, C., Beck, T., Zielonka, J., Hotz-Wagenblatt, A., Chareza, S., Battenberg, M., Thielebein, J., Cichutek, K., Bravo, I.G., O'Brien, S.J., Lochelt, M., and Yuhki, N. (2008). Functions, structure, and read-through alternative splicing of feline APOBEC3 genes. Genome Biol. *9*, R48.

Novak, J.M., Crawford, P.C., Kolenda-Roberts, H.M., Johnson, C.M., and Mergia, A. (2007). Viral gene expression and provirus load of Orf-A defective FIV in lymphoid tissues and lymphocyte subpopulations of neonatal cats during acute and chronic infections. Virus Res. *130*, 110–120.

Novotney, C., English, R.V., Housman, J., Davidson, M.G., Nasisse, M.P., Jeng, C.R., Davis, W.C., and Tompkins, M.B. (1990). Lymphocyte population changes in cats naturally infected with feline immunodeficiency virus. AIDS *4*, 1213–1218.

Ogert, R.A., Lee, M.K., Ross, W., Buckler-White, A., Martin, M.A., and Cho, M.W. (2001). N-linked glycosylation sites adjacent to and within the V1/V2 and the V3 loops of dualtropic human immunodeficiency virus type 1 isolate DH12 gp120 affect coreceptor usage and cellular tropism. J. Virol. *75*, 5998–6006.

Olmsted, R.A., Barnes, A.K., Yamamoto, J.K., Hirsch, V.M., Purcell, R.H., and Johnson, P.R. (1989a). Molecular cloning of feline immunodeficiency virus. Proc. Natl. Acad. Sci. U.S.A. *86*, 2448–2452.

Olmsted, R.A., Hirsch, V.M., Purcell, R.H., and Johnson, P.R. (1989b). Nucleotide-sequence analysis of feline immunodeficiency virus – genome organization and relationship to other lentiviruses. Proceedings Of The National Academy Of Sciences Of The United States Of America *86*, 8088–8092.

Osborne, R., Rigby, M., Siebelink, K., Neil, J.C., and Jarrett, O. (1994). Virus neutralization reveals antigenic variation among feline immunodeficiency virus isolates. J. Gen. Virol. *75*, 3641–3645.

Osterhaus, A.D., Tijhaar, E., Huisman, R.C., Huisman, W., Darby, I.H., Francis, M.J., Rimmelzwaan, G.F., and Siebelink, K.H. (1996). Accelerated viremia in cats vaccinated with recombinant vaccinia virus expressing envelope glycoprotein of feline immunodeficiency virus. Aids Res. Hum. Retroviruses *12*, 437–441.

Pastore, C., Nedellec, R., Ramos, A., Pontow, S., Ratner, L., and Mosier, D.E. (2006). Human immunodeficiency virus type 1 coreceptor switching: V1/V2 gain-of-fitness mutations compensate for V3 loss-of-fitness mutations. J. Virol. *80*, 750–758.

Pedersen, N.C., Ho, E.W., Brown, M.L., and Yamamoto, J.K. (1987). Isolation of a T-lymphotropic virus from domestic cats with an immunodeficiency syndrome. Science *235*, 790–793.

Pedersen,N.C. (1993). The feline immunodeficiency virus. In The Retroviridae, J.A.Levy, ed. (New York: Plenum Press), pp. 181–228.

Perreau, M., Pantaleo, G., and Kremer, E.J. (2008). Activation of a dendritic cell-T-cell axis by Ad5 immune complexes creates an improved environment for replication of HIV in T-cells. J. Exp. Med. *205*, 2717–2725.

Phillips, T.R., Talbott, R.L., Lamont, C., Muir, S., Lovelace, K., and Elder, J.H. (1990). Comparison of two host cell range variants of feline immunodeficiency virus. J. Virol. *64*, 4605–4613.

Pistello, M., Moscardini, M., Mazzetti, P., Bonci, F., Zaccaro, L., Isola, P., Freer, G., Specter, S., Matteucci, D., and Bendinelli, M. (2002). Development of feline immunodeficiency virus ORF-A (tat) mutants: *in vitro* and *in vivo* characterization. Virology *298*, 84–95.

Pistello, M., Bonci, F., Flynn, J.N., Mazzetti, P., Isola, P., Zabogli, E., Camerini, V., Matteucci, D., Freer, G., Pelosi, P., and Bendinelli, M. (2006). AIDS vaccination studies with an *ex vivo* feline immunodeficiency virus model: analysis of the accessory ORF-A protein and DNA as protective immunogens. J. Virol. *80*, 8856–8868.

Poli, A., Abramo, F., Cavicchio, P., Bandecchi, P., Ghelardi, E., and Pistello, M. (1995). Lentivirus infection in an African lion: a clinical, pathologic and virologic study. J. Wildl. Dis. *31*, 70–74.

Poss, M., and Ross, H. (2008). Evolution of the long terminal repeat and accessory genes of feline immunodeficiency virus genomes from naturally infected cougars. Virology *370*, 55–62.

Pu, R., Coleman, J., Omori, M., Arai, M., Hohdatsu, T., Huang, C., Tanabe, T., and Yamamoto, J.K. (2001). Dual-subtype FIV vaccine protects cats against *in vivo* swarms of both homologous and heterologous subtype FIV isolates. AIDS *15*, 1225–1237.

Pu, R., Coleman, J., Coisman, J., Sato, E., Tanabe, T., Arai, M., and Yamamoto, J.K. (2005). Dual-subtype FIV vaccine (Fel-O-Vax FIV) protection against a heterologous subtype B FIV isolate. J. Feline. Med. Surg. *7*, 65–70.

Puffer, B.A., Pohlmann, S., Edinger, A.L., Carlin, D., Sanchez, M.D., Reitter, J., Watry, D.D., Fox, H.S., Desrosiers, R.C., and Doms, R.W. (2002). CD4 independence of simian immunodeficiency virus Envs is associated with macrophage tropism, neutralization sensitivity, and attenuated pathogenicity. J. Virol. *76*, 2595–2605.

Raabe, M.L., Issel, C.J., and Montelaro, R.C. (1999). *In vitro* antibody-dependent enhancement assays are insensitive indicators of *in vivo* vaccine enhancement of equine infectious anemia virus. Virology *259*, 416–427.

Richardson, J., Moraillon, A., Baud, S., Cuisinier, A.M., Sonigo, P., and Pancino, G. (1997). Enhancement of feline immunodeficiency virus (FIV) infection after DNA vaccination with the FIV envelope. J. Virol. *71*, 9640–9649.

Richardson, J., Pancino, G., Leste-Lasserre, T., Schneider-Mergener, J., Alizon, M., Sonigo, P., and Heveker,

N. (1999). Shared usage of the chemokine receptor CXCR4 by primary and laboratory-adapted strains of feline immunodeficiency virus. J. Virol. 73, 3661–3671.

Roelke, M.E., Pecon-Slattery, J., Taylor, S., Citino, S., Brown, E., Packer, C., VandeWoude, S., and O'Brien, S.J. (2006). T-lymphocyte profiles in FIV-infected wild lions and pumas reveal CD4 depletion. J. Wildl. Dis. 42, 234–248.

Roelke, M. E., Brown, M., Winterbach, H., Winterbach, C., Hemson, G., Troyer, J. L., Smith, D., Johnson, R. C., Pecon-Slattery, J., Roca, A. L., Alexander, K., Klein, L., Martinelli, P., and O'Brien, S. J. Pathological Manifestations of Feline Immunodeficiency Virus Infection in Wild Lions. 9th International Feline Retrovirus Research Symposium. 2008.

Russo, P., Vitu, C., Fontaine, J.J., and Vignoni, M. (1993). [Caprine arthritis-encephalitis: trial of an adjuvant vaccine preparation. I. Clinical and virological study]. Comp Immunol. Microbiol. Infect. Dis. *16*, 131–136.

Sarrias, M.R., Whitbeck, J.C., Rooney, I., Ware, C.F., Eisenberg, R.J., Cohen, G.H., and Lambris, J.D. (2000). The three HveA receptor ligands, gD, LT-alpha and LIGHT bind to distinct sites on HveA. Mol. Immunol. *37*, 665–673.

Schmitz, J.E., Forman, M.A., Lifton, M.A., Concepcion, O., Reimann, K.A.J., Crumpacker, C.S., Daley, J.F., Gelman, R.S., and Letvin, N.L. (1998). Expression of the CD8alpha beta-heterodimer on CD8(+) T-lymphocytes in peripheral blood lymphocytes of human immunodeficiency virus- and human immunodeficiency virus+ individuals. Blood *92*, 198–206.

Schutten, M., Andeweg, A.C., Bosch, M.L., and Osterhaus, A.D.M.E. (1994). Enhancement of infectivity of a non-syncytium inducing HIV-1 by sCD4 and by human antibodies that neutralise syncytium inducing HIV-1. Scand. J. Immunol. *41*, 18–22.

Schutten, M., Andeweg, A.C., Rimmelzwaan, G.F., and Osterhaus, A.D. (1997). Modulation of primary human immunodeficiency virus type 1 envelope glycoprotein-mediated entry by human antibodies. J. Gen. Virol. *78 (Pt 5)*, 999–1006.

Sheehy, A.M., Gaddis, N.C., Choi, J.D., and Malim, M.H. (2002). Isolation of a human gene that inhibits HIV-1 infection and is suppressed by the viral Vif protein. Nature *418*, 646–650.

Sherman, M.P., De Noronha, C.M., Eckstein, L.A., Hataye, J., Mundt, P., Williams, S.A., Neidleman, J.A., Goldsmith, M.A., and Greene, W.C. (2003). Nuclear export of Vpr is required for efficient replication of human immunodeficiency virus type 1 in tissue macrophages. J. Virol. *77*, 7582–7589.

Shimojima, M., Miyazawa, T., Kohmoto, M., Ikeda, Y., Nishimura, Y., Maeda, K., Tohya, Y., and Mikami, T. (1998). Expansion of CD8alpha+beta- cells in cats infected with feline immunodeficiency virus. J. Gen. Virol. *79*, 91–94.

Shimojima, M., Miyazawa, T., Ikeda, Y., McMonagle, E.L., Haining, H., Akashi, H., Takeuchi, Y., Hosie, M.J., and Willett, B.J. (2004). Use of CD134 as a primary receptor by the feline immunodeficiency virus. Science *303*, 1192–1195.

Siebelink, K.H.J., Rimmelzwaan, G.F., Bosch, M.L., Meloen, R.H., and Osterhaus, A.D.M.E. (1993). A single amino acid substitution in hypervariable region 5 of the envelope protein of feline immunodeficiency virus allows escape from virus neutralisation. J. Virol. *67*, 2202–2208.

Siebelink, K.H., Bosch, M.L., Rimmelzwaan, G.F., Meloen, R.H., and Osterhaus, A.D. (1995a). Two different mutations in the envelope protein of feline immunodeficiency virus allow the virus to escape from neutralization by feline serum antibodies. Vet. Immunol. Immunopathol. *46*, 51–59.

Siebelink, K.H.J., Karlas, J.A., Rimmelzwaan, G.F., Osterhaus, A.D.M.E., and Bosch, M.L. (1995b). A determinant of feline immunodeficiency virus involved in CrFK tropism. Vet. Immunol. Immunopathol. *46*, 61–69.

Siebelink, K.H.J., Tijhaar, E., Huisman, R.C., Huisman, W., de Ronde, A., Darby, I.H., Francis, M.J., Rimmelzwaan, G.F., and Osterhaus, A.D.M.E. (1995c). Enhancement of feline immunodeficiency virus infection after immunisation with envelope glycoprotein subunit vaccines. J. Virol. *69*, 3704–3711.

Siebelink, K.H.J., Huisman, W., Karlas, J.A., Rimmelzwaan, G.F., Bosch, M.L., and Osterhaus, A.D.M.E. (1996). Neutralization of feline immunodeficiency virus by polyclonal feline antibody: simultaneous involvement of hypervariable regions 4 and 5 of the surface glycoprotein. J. Virol. *69*, 5124–5127.

Smirnova, N., Troyer, J.L., Schissler, J., TerWee, J., Poss, M., and VandeWoude, S. (2005). Feline lentiviruses demonstrate differences in receptor repertoire and envelope structural elements. Virology *342*, 60–76.

Sparger, E.E., Luciw, P.A., Elder, J.H., Yamamoto, J.K., Lowenstine, L.J., and Pedersen, N.C. (1989). Feline immunodeficiency virus is a lentivirus associated with an AIDS-like disease in cats. AIDS *3 (suppl-1)*, S43-S49.

Sparger, E.E., Shacklett, B.L., Renshaw-Gegg, L., Barry, P.A., Pedersen, N.C., Elder, J.H., and Luciw, P.A. (1992). Regulation of gene expression directed by the long terminal repeat of the feline immunodeficiency virus. Virology *187*, 165–177.

Stamatatos, L., and Cheng-Mayer, C. (1998). An envelope modification that renders a primary, neutralization-resistant clade B human immunodeficiency virus type 1 isolate highly susceptible to neutralization by sera from other clades. J. Virol. 72, 7840–7845.

Staprans, S.I., Barry, A.P., Silvestri, G., Safrit, J.T., Kozyr, N., Sumpter, B., Nguyen, H., McClure, H., Montefiori, D., Cohen, J.I., and Feinberg, M.B. (2004). Enhanced SIV replication and accelerated progression to AIDS in macaques primed to mount a CD4 T-cell response to the SIV envelope protein. Proc. Natl. Acad. Sci. U.S.A. *101*, 13026–13031.

Talbott, R.L., Sparger, E.E., Lovelace, K.M., Fitch, W.M., Pedersen, N.C., Luciw, P.A., and Elder, J.H. (1989). Nucleotide-sequence and genomic organization of feline immunodeficiency virus. Proc. Natl. Acad. Sci. U.S.A. *86*, 5743–5747.

Tellier, M.C., Pu, R., Pollock, D., Vitsky, A., Tartaglia, J., Paoletti, E., and Yamamoto, J.K. (1998). Efficacy evaluation of prime-boost protocol: canarypoxvirus-based feline immunodeficiency virus (FIV) vaccine and inactivated FIV-infected cell vaccine against heterologous FIV challenge in cats. AIDS *12*, 11–18.

Thompson, F.J., Elder, J., and Neil, J.C. (1994). Cis- and trans-regulation of feline immunodeficiency virus: identification of functional binding sites in the long terminal repeat. J. Gen. Virol. *75 (Pt 3)*, 545–554.

Threadgill, D.S., Steagall, W.K., Flaherty, M.T., Fuller, F.J., Perry, S.T., Rushlow, K.E., Le Grice, S.F., and Payne, S.L. (1993). Characterization of equine infectious anemia virus dUTPase: growth properties of a dUTPase-deficient mutant. J. Virol. *67*, 2592–2600.

Tomonaga, K., Norimine, J., Shin, Y.S., Fukasawa, M., Miyazawa, T., Adachi, A., Toyosaki, T., Kawaguchi, Y., Kai, C., and Mikami, T. (1992). Identification of a feline immunodeficiency virus gene which is essential for cell-free virus infectivity. J. Virol. *66*, 6181–6185.

Torten, M., Franchini, M., Barlough, J.E., George, J.W., Mozes, E., Lutz, H., and Pedersen, N.C. (1991). Progressive immune dysfunction in cats experimentally infected with feline immunodeficiency virus. J. Virol. *65*, 2225–2230.

Tozzini, F., Matteucci, D., Bandecchi, P., Baldinotti, F., Poli, A., Pistello, M., Siebelink, K.H.J., Ceccherini-Nelli, L., and Bendinelli, M. (1992). Simple *in vitro* methods for titrating feline immunodeficiency virus (FIV) and FIV neutralising antibodies. J. Virol. Meth. *37*, 241–252.

Troth, S.P., Dean, A.D., and Hoover, E.A. (2008). *In vivo* CXCR4 expression, lymphoid cell phenotype, and feline immunodeficiency virus infection. Vet. Immunol. Immunopathol. *123*, 97–105.

Troyer, J.L., Pecon-Slattery, J., Roelke, M.E., Black, L., Packer, C., and O'Brien, S.J. (2004). Patterns of feline immunodeficiency virus multiple infection and genome divergence in a free-ranging population of African lions. J. Virol. *78*, 3777–3791.

Troyer, J.L., Pecon-Slattery, J., Roelke, M.E., Johnson, W., VandeWoude, S., Vazquez-Salat, N., Brown, M., Frank, L., Woodroffe, R., Winterbach, C., Winterbach, H., Hemson, G., Bush, M., Alexander, K.A., Revilla, E., and O'Brien, S.J. (2005). Seroprevalence and genomic divergence of circulating strains of feline immunodeficiency virus among Felidae and Hyaenidae species. J. Virol. *79*, 8282–8294.

Vahlenkamp, T.W., de Ronde, A., Schuurman, N.N., Van Vliet, A.L., van Drunen, J., Horzinek, M.C., and Egberink, H.F. (1999). Envelope gene sequences encoding variable regions 3 and 4 are involved in macrophage tropism of feline immunodeficiency virus. J. Gen. Virol. *80 (Pt 10)*, 2639–2646.

Verschoor, E.J., Boven, L.A., Blaak, H., van Vliet, A.R.W., Horzinek, M.C., and de Ronde, A. (1995). A single mutation within the V3 envelope neutralization domain of feline immunodeficiency virus determines its tropism for CRFK cells. J. Virol. *69*, 4752–4757.

Wagaman, P.C., Hasselkus-Light, C.S., Henson, M., Lerner, D.L., Phillips, T.R., and Elder, J.H. (1993). Molecular cloning and characterization of deoxyuridine triphosphatase from feline immunodeficiency virus (FIV). Virology *196*, 451–457.

Wang, S.Z., Rushlow, K.E., Issel, C.J., Cook, R.F., Cook, S.J., Raabe, M.L., Chong, Y.H., Costa, L., and Montelaro, R.C. (1994). Enhancement of EIAV replication and disease by immunization with a baculovirus-expressed recombinant envelope surface glycoprotein. Virology *199*, 247–251.

Waters, A.K., De Parseval, A.P., Lerner, D.L., Neil, J.C., Thompson, F.J., and Elder, J.H. (1996). Influence of ORF2 on host cell tropism of feline immunodeficiency virus. Virology *215*, 10–16.

Willett, B.J., Hosie, M.J., Dunsford, T.H., Neil, J.C., and Jarrett, O. (1991). Productive infection of T-helper lymphocytes with feline immunodeficiency virus is accompanied by reduced expression of CD4. AIDS *5*, 1469–1475.

Willett, B.J., Hosie, M.J., Callanan, J.J., Neil, J.C., and Jarrett, O. (1993). Infection with feline immunodeficiency virus is followed by the rapid expansion of a CD8+ lymphocyte subset. Immunology *78*, 1–6.

Willett, B.J., Hosie, M.J., Neil, J.C., Turner, J.D., and Hoxie, J.A. (1997a). Common mechanism of infection by lentiviruses. Nature *385*, 587.

Willett, B.J., Hosie, M.J., Shaw, A.R.E., and Neil, J.C. (1997b). Inhibition of feline immunodeficiency virus infection by CD9 antibody operates after viral entry and is independent of virus tropism. J. Gen. Virol. *78*, 611–618.

Willett, B.J., Picard, L., Hosie, M.J., Turner, J.D., Adema, K., and Clapham, P.R. (1997c). Shared usage of the chemokine receptor CXCR4 by the feline and human immunodeficiency viruses. J. Virol. *71*, 6407–6415.

Willett, B.J., Cannon, C.A., and Hosie, M.J. (2003). Expression of CXCR4 on feline peripheral blood mononuclear cells: effect of feline immunodeficiency virus infection. J. Virol. *77*, 709–712.

Willett, B.J., McMonagle, E.L., Bonci, F., Pistello, M., and Hosie, M.J. (2006a). Mapping the domains of CD134 as a functional receptor for feline immunodeficiency virus. J. Virol. *80*, 7744–7747.

Willett, B.J., McMonagle, E.L., Ridha, S., and Hosie, M.J. (2006b). Differential utilization of CD134 as a functional receptor by diverse strains of feline immunodeficiency virus (FIV). J. Virol. *80*, 3386–3394.

Willett, B.J., McMonagle, E.L., logan, N., Spiller, O.B., Schneider, P., and Hosie, M.J. (2007). Probing the interaction between the feline immunodeficiency virus and CD134 using a novel monoclonal antibody 7D6 and CD134L (OX40L). J. Virol. *81*, 9665–9679.

Willett, B.J., and Hosie, M.J. (2008a). Chemokine receptors and co-stimulatory molecules: Unravelling feline immunodeficiency virus infection. Vet. Immunol. Immunopathol. *123*, 56–64.

Willett, B.J., McMonagle, E.L., logan, N., Samman, A., and Hosie, M.J. (2008b). A single site for N-linked glycosylation in the envelope glycoprotein of feline immunodeficiency virus modulates the virus–receptor interaction. Retrovirology. *5*, 77.

Yamamoto, J.K., Hansen, H., Ho, E.W., Morishita, T.Y., Okuda, T., Saura, T.R., Nakamura, R.M., and Pedersen, N.C. (1989). Epidemiologic and clinical aspects of FIV infection in cats from the continental

U.S. and Canada and possible mode of transmission. J. Am. Vet. Med. Assoc. *194*, 213–220.

Yamamoto, J.K., Ackley, C.D., Zochlinski, H., Louie, H., Pembroke, E., Torten, M., Hansen, H., Munn, R., and Okuda, T. (1991a). Development of IL-2-independent feline lymphoid cell lines chronically infected with feline immunodeficiency viruses: importance for diagnostic reagents and vaccines. Intervirology *32*, 361–375.

Yamamoto, J.K., Okuda, T., Ackley, C.D., Louie, H., Pembroke, E., Zochlinski, H., Munn, R.J., and Gardner, M.B. (1991b). Experimental vaccine protection against feline immunodeficiency virus. Aids Res. Hum. Retroviruses *7*, 911–921.

Yamamoto, J.K., Hohdatsu, T., Olmsted, R.A., Pu, R., Louie, H., Zochlinski, H.A., Acevedo, V., Johnson, H.M., Soulds, G.A., and Gardner, M.B. (1993). Experimental vaccine protection against homologous and heterologous strains of feline immunodeficiency virus. J. Virol. *67*, 601–605.

Zennou, V., Petit, C., Guetard, D., Nerhbass, U., Montagnier, L., and Charneau, P. (2000). HIV-1 genome nuclear import is mediated by a central DNA flap. Cell *101*, 173–185.

# Equine Infectious Anaemia Virus Pathogenesis and Replication

9

Wendy Maury and J. Lindsay Oaks

Abstract

Equine infectious anaemia virus (EIAV) is an ungulate lentivirus related to human immunodeficiency virus (HIV). Much of the understanding of lentiviral infection of macrophages comes from HIV studies that have provided insights into molecular regulation of all lentiviruses. However, numerous aspects of the life cycle of each lentivirus are unique and associated with specific pathological consequences. *In vivo* EIAV is primarily if not exclusively a macrophage-tropic virus. As a consequence of this targeted tropism, EIAV causes an acute and sometimes fulminant disease associated with high-titred viraemia with no associated immunodeficiency. Investigations have only begun to unravel the molecular mechanisms leading to cell-specific replication of EIAV. This chapter provides an overview of the molecular biology of EIAV replication and the pathogenesis associated with infection of macrophages.

## EIAV pathogenesis

### Clinical disease and immunological control

Equine infectious anaemia virus (EIAV) is a lentivirus that persistently infects horses and other equids. The resulting disease, called equine infectious anaemia (EIA), is primarily an intermittent febrile disease that may or may not be noticed clinically. EIA can be accurately diagnosed by detection of anti-EIAV antibody in serum by an immunoagar diffusion test (Coggins test). However, many natural EIAV infections go unnoticed or are not clinically suspected, and thus are not tested for. Also, antibody testing for surveillance and control purposes is performed only on selected horse populations. Consequently, EIA remains endemic in many developed and developing countries worldwide.

From the vantage point of understanding lentiviral replication and pathogenesis within the context of macrophage biology, horses infected with EIAV provide a unique *in vivo* opportunity. Dissecting the role of macrophages during both acute infection and virus persistence is possible since individual horses have alternating episodes of disease and subclinical infection. Also, the monocyte–macrophage tropism of EIAV provides an important advantage since there is not the confounding effect of infected lymphocyte compartments that may serve as alternative or additional sites of persistence.

Although there can be considerable variation in the severity and forms of disease, the clinical course typically includes an initial acute phase, a chronic phase and eventually a long-term inapparent phase (Kono, 1973; Kono *et al.*, 1973a; Sellon *et al.*, 1994; Belshan *et al.*, 2001). In contrast to other pathogenic lentiviruses, clinical disease is not usually progressive, and most infected horses clinically recover from EIA. Plasma viraemia is associated with clinical signs of disease (McGuire *et al.*, 1990).

The acute phase is characterized by high fever (up to 42°C), depression and anorexia that may last 1–2 weeks. Contrary to its name, anaemia is not a consistent feature of acute disease, and the most salient clinicopathological and often the most clinically important feature of acute

EIA is thrombocytopenia (Clabough *et al.*, 1991; Crawford *et al.*, 1996; Oaks *et al.*, 1998). Fatalities rarely occur during acute infection and, when it does, is usually due to complications of severe thrombocytopenia and rarely to anaemia. The resolution of acute disease and control of viraemia is coincident with the appearance of antiviral antibodies and cell-mediated immune responses (Carpenter *et al.*, 1987; O'Rourke *et al.*, 1989; McGuire *et al.*, 1994; Hammond *et al.*, 1997; Mealey *et al.*, 2001). The requirement for adaptive immune responses in the control of viraemia and disease has been demonstrated by the inability of B- and T-lymphocyte deficient foals to control viraemia (Perryman *et al.*, 1988) and the ability to recrudesce virus in carrier horses with lymphotoxic drugs (Kono *et al.*, 1976; Tumas *et al.*, 1994).

The chronic phase is characterized by recurrent episodes of viraemia with accompanying clinical disease. These episodes usually last a few days in duration and are characterized by fever, thrombocytopenia and sometimes anaemia. This intermittent disease phase may last as long as a year, during which time the disease episodes tend to decrease in both frequency and intensity. These recurrent episodes are associated with antigenic variants that escape both neutralizing antibody (Kono *et al.*, 1973b; Montelaro *et al.*, 1984) as well as cytotoxic T-lymphocytes (Mealey *et al.*, 2003). Variation associated with immune escape has been documented in the SU protein (Zheng *et al.*, 1997a,b; Howe *et al.*, 2002; Mealey *et al.*, 2003), and has been detected in cytotoxic T-lymphocyte epitopes in the capsid protein (Zhang *et al.*, 1999).

Finally, most infected horses become clinically quiescent. They have normal platelet and erythrocyte counts, and only rare and mild episodes of disease. Long term control of viraemia is most likely the result of the maturation of both antibody and cell-mediated immune responses (Hammond *et al.*, 1997). However, as is typical of lentiviruses, EIAV-infected horses remain infected, and infectious, for life (Coggins, 1984).

Notable variations on this general disease theme may range from never showing overt clinical disease, to a progressive fatal form with persistent high-level viraemia, fever, thrombocytopenia, anaemia and immune-mediated renal disease (Henson and McGuire, 1971; Henson and McGuire, 1974; Issel and Coggins, 1979; Sellon *et al.*, 1994). Because of the unrelenting disease course and chronic renal disease, these horses often develop severe cachexia, and are referred to colloquially as 'swampers'. Another uncommon variant and fatal form of disease is severe and progressive leukoencephalitis that presents as neurologic disease, and often occurs independently of any other clinical disease manifestation (Oaks *et al.*, 2004).

One of the most unique features of EIAV is that it is a lentivirus that is naturally transmitted by insect vectors. The transmission occurs mechanically by contaminating the mouth parts of biting horseflies in the Tabanidae family (Foil *et al.*, 1987). This mode of transmission is presumably facilitated by the relatively large volume of blood carried on the mouth parts (up to 5 nl), the interrupted feeding behaviour of these flies, and unusually high titres of cell-free viraemia of up to $10^8$ horse infectious doses per ml of plasma in infected horses (Oaks *et al.*, 1998).

## Sites of EIAV replication *in vivo* during clinical disease

The primary target cell for EIAV *in vivo* is the tissue macrophage, and most of the known biology of EIA can be attributed to infection and replication in these cells. Unlike immunodeficiency-causing lentiviruses, EIAV does not replicate in lymphocytes (Sellon *et al.*, 1992)(Oaks, unpublished data), and, while not formally proven, there is no evidence that lymphocytes are infected. Transient suppression of cellular proliferative responses can be demonstrated *in vitro* (Newman *et al.*, 1991) and decreased numbers of $CD4^+$ and $CD8^+$ T-lymphocytes are observed *in vivo* during acute disease (Murakami *et al.*, 1999); however, immunosuppressive diseases and associated opportunistic infections are not characteristic of EIAV infection. The clinical signs of EIA are for the most part directly related to active viral replication in tissue macrophages, as indicated by the direct correlation shown between fever, thrombocytopenia, and viral burden assessed by measuring cell-free viraemia, cell-associated viraemia, viral DNA or RNA loads in tissues, and numbers of infected cells observed by *in situ* techniques (Rice *et al.*, 1989; Sellon *et al.*, 1992; Crawford *et al.*, 1996; Oaks *et al.*, 1998; Harrold *et al.*, 2000). Replication in tissue macrophages during acute clinical disease was first demonstrated by McGuire by immunofluorescent imaging of viral antigen expression in cells with the location and morphology consistent with macrophages (McGuire *et al.*, 1971). These observations were subsequently confirmed by Sellon and Oaks (Sellon *et al.*, 1992; Oaks *et al.*, 1998) using double labelling methods to demonstrate viral *gag* RNA in cells that expressed macrophage markers (Fig. 9.1). In horses with acute disease and high-titred viraemia, large numbers of RNA- and antigen-expressing cells can be readily observed in virtually all tissues, suggesting that these cells are producing large amounts of virus and are the primary source of viraemia (McGuire *et al.*, 1971; Sellon *et al.*, 1992; Oaks *et al.*, 1998).

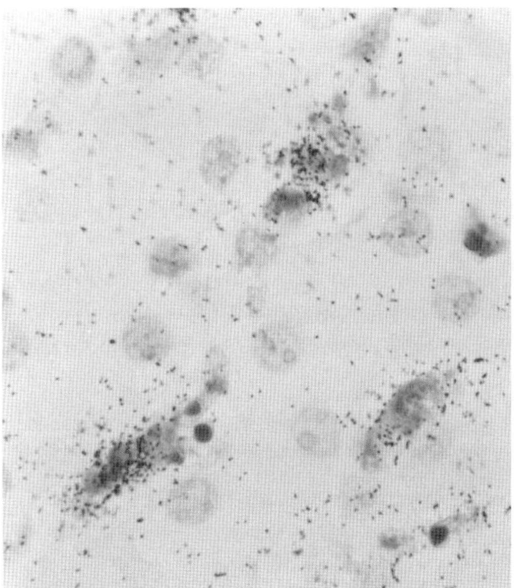

**Figure 9.1** Photomicrograph of a section of spleen from a horse acutely infected with the virulent Wyoming strain, showing EIAV-infected macrophages. The dark dots are silver grains that represent EIAV *gag* RNA detected by *in situ* hybridization with a radiolabelled riboprobe. The dark cells are stained with DAB chromogen, and are the result of immunohistochemistry for lysozyme, a macrophage marker. Co-localization of viral RNA and lysozyme confirms infection of macrophages. Details of this procedure can be found in (Oaks *et al.*, 1998).

Endothelial cells have also been shown to support virus replication during acute disease (Oaks *et al.*, 1999) and *in vitro* (Maury *et al.*, 1998). This was first observed *in vivo* through *in situ* hybridization studies for viral RNA, where *gag* RNA was observed in cells lining the lumen of blood vessels in numerous tissues. The identity of these infected cells was subsequently confirmed by double-labelling these cells by *in situ* hybridization for viral RNA and by immunohistochemistry for the endothelial cell marker Von Willebrand's factor (Oaks *et al.*, 1999). However, necrotizing vasculitis is not observed and the overall significance of these infected cells to either the pathogenesis and/or persistence of EIAV is not known.

Infection of peripheral blood mononuclear cells (PBMCs) can also be readily demonstrated *in vivo* during acute clinical disease by PCR (O'Rourke *et al.*, 1991; Sellon *et al.*, 1992; Oaks *et al.*, 1998; Harrold *et al.*, 2000). However, cell-associated RNA is not present or present in very limited amounts, suggesting that these cells are mainly latently infected. The infected mononuclear cell type is inferred to be the monocyte, based on the affinity of EIAV for tissue macrophages, lack of viral replication in lymphocytes, and the ability to recover infectious virus from monocyte-derived macrophage cultures derived from infected horses (Sellon *et al.*, 1996a). However, due to the lack of viral replication in these cells, they presumably do not play a direct role in the pathogenesis of acute disease. It has been suggested that monocytes may serve as viral reservoirs that maintain virus by passive replication of viral DNA along with host DNA in the bone marrow monocytic stem cells, and serve to seed tissues with infected macrophages in the 'Trojan horse' manner described for ovine visna virus (Peluso *et al.*, 1985). Alternatively, it is possible that monocyte infection may be a consequence of active viral replication rather than a source of it. Consistent with this possibility is the observation that monocytes are susceptible to infection *in vitro* (Maury, 1994). Also high titred cell-free viraemia and virus load during acute infection correspond with levels of proviral DNA in PBMCs (Rice *et al.*, 1989; Oaks *et al.*, 1998; Harrold *et al.*, 2000).

## Pathogenesis of clinical EIA

Thrombocytopenia is one of the most salient clinical features of EIA, and is the result of both increased platelet consumption in the peripheral circulation as well as deficits in platelet production in the bone marrow. One of the primary causes of platelet consumption is believed to be immune-mediated. Platelet-associated immunoglobulins can be detected in acutely infected horses at the onset of the drop in platelet numbers (Clabough *et al.*, 1991). However, it is not certain that these antibodies are antiviral as viral particle binding to platelets was not detected, and a drop in platelet numbers is evident before antiviral antibodies can be demonstrated. Platelet-bound fibrinogen indicative of platelet activation has also been demonstrated in acute infection (Russell *et al.*, 1999), which may indicate another mechanism by which circulating platelets are consumed. Another possible mechanism of early platelet consumption is endothelial cell infection which may cause platelet activation and aggregation in response to either overt vascular damage or to activated endothelial cells (Oaks *et al.*, 1999), although infected horses do not show evidence of disseminated intravascular coagulation associated with decreased platelets (Crawford *et al.*, 1996). Although the actual mechanisms for loss of circulating platelets are still poorly defined, decreased platelet survival in acute infection was demonstrated with 51Cr labelling studies (Crawford *et al.*, 1996). Interestingly, decreased platelet lifespan occurred with equal magnitude in severe combined immunodeficient foals (although these foals did receive passively transferred antibody), suggesting that acquired immune responses and/or anti-EIAV antibody is not required for consumptive thrombocytopenia.

Suppressed bone marrow production of platelets has also been demonstrated by decreased metabolic uptake of radioactive selenomethionine (Crawford *et al.*, 1996). This was evident in both immunocompetent as well as combined immunodeficient foals with acute EIA, whose platelet production dropped by 35–89%. Several morphometric studies also support the finding of bone marrow suppression. In foals with acute EIA, there are no significant changes in the total numbers, morphology or size of the megakaryocytes (Crawford *et al.*, 1996; Wardrop *et al.*, 1996), changes that would be expected with a compensatory response by normal bone marrow to loss of circulating platelets. However, another study found some evidence of megakaryocytopoiesis as indicated by increased megakaryocyte size, nuclear area, and numbers of circulating immature platelets (Russell *et al.*, 1999), suggesting the effect of EIA on the bone marrow is variable and/or incomplete. The cause of decreased platelet production does not appear to be due to direct cytolytic effects of virus on the megakaryocytes, as there is no evidence of EIAV replication in these cells (Clabough *et al.*, 1991; Crawford *et al.*, 1996)(Oaks, unpublished data). Bone marrow suppression is most likely the result of cytokines released from infected macrophages in the bone marrow acting in a paracrine fashion to suppress megakaryocytopoiesis. Viral antigen and RNA are readily detected in bone marrow macrophages by *in situ* methods (McGuire *et al.*, 1971; Oaks *et al.*, 1998). Tornquist *et al.* (1997) found increases in three cytokines secreted by macrophages which potentially suppress megakaryocytopoiesis, including tumour necrosis factor (TNF)-α, transforming growth factor (TGF)-β, and interferon (IFN)-α, in the sera and/or bone marrow of infected foals at the onset of thrombocytopenia. Serum from the immediately pre-thrombocytopenic foals was also found to suppress megakaryocytes *in vitro*, and this effect could be partially reversed by neutralization of TNF-α and TGF-β (Tornquist and Crawford, 1997).

The pathogenesis of anaemia appears to be similar to that for thrombocytopenia, involving both immune-mediated destruction of circulating erythrocytes and suppression of bone marrow erythropoiesis. Erythrocytes from infected horses have been shown to be coated with complement (C′3), be osmotically fragile, have decreased half-lives, and exhibit both intravascular and extravascular haemolysis (McGuire and Henson, 1969; McGuire *et al.*, 1969a; McGuire *et al.*, 1969c; Sentsui and Kono, 1988). Interestingly, unlike platelets, erythrocytes from infected horses do not appear to have surface immunoglobulins (McGuire *et al.*, 1969a; Sentsui and Kono, 1988).

The role of EIAV-specific mechanisms in the suppression of erythrocyte production has not been explored in as much detail as the effect on platelets, but if present also probably represents paracrine cytokine effects of macrophages on erythropoiesis. Consistent with an indirect cause of anaemia by EIAV, no evidence of viral replication has been noted in haematopoietic cells *in vivo* by *in situ* hybridization for viral RNA (Oaks, unpublished data). Bone marrow suppression was demonstrated with ferrokinetic studies and showed a lack of compensatory increases in bone marrow erythroid precursors (McGuire *et al.*, 1969b), although it is likely that this mechanism is not EIA-specific and it most likely represents chronic-disorder type anaemia (Swardson *et al.*, 1992). Bone marrow suppression is also suggested by the observation that severe combined immunodeficient foals develop anaemia and do not have erythrocyte-associated complement (Perryman *et al.*, 1988), and the ability of EIAV to selectively suppress erythroid precursors *in vitro* (Swardson *et al.*, 1992).

The pathogenesis of leukoencephalitis is also not known, and is unique in that it tends to occur independently of the other clinical manifestations of EIAV (McIlwraith and Kitchen, 1978; McClure *et al.*, 1982; Oaks *et al.*, 2004). The one study that examined virological activity by *in situ* hybridization for viral RNA (Oaks *et al.*, 2004) found that the lesion was directly associated with viral replication, and that this was occurring preferentially within the lesion. The infected cell type was most likely a macrophage or microglial cell, but this could not be confirmed by double-labelling. In contrast, brains from other EIAV-infected horses with overt clinical EIA, but no central nervous signs or histopathological lesions, did not have detectable RNA-expressing cells (Oaks *et al.*, 2004).

Macrophages and their response to infection play a pivotal role in EIA. Because EIAV destruction of macrophages does not lead directly to organ dysfunction per se, macrophage-mediated disease appears to be primarily cytokine related. In EIAV-infected horses, the most important effects are related to those in the bone marrow. Substantiating this concept of a macrophage/cytokine mediated disease is the work of Lim *et al.* (2005) showing that macrophage cultures *in vitro* infected with virulent strains of EIAV produced significant increases in interleukins IL-1α, IL-1β, IL-6, IL-10 and TNF-α, while avirulent EIAV did not increase expression of these cytokines. Production of IL-1α, IL-1β, IL-6 and TNF-α can account for EIAV-associated febrile responses. In addition to the suppressive effects of TNF-α, IFN-α and TGF-β on thrombopoiesis, secretion of IL-1 and IL-6 have been suggested as mechanism of activating platelets and contributing to consumptive thrombocytopenias. IL-6 and IL-10 expression has been suggested to enhance the B-cell activation and hypergammaglobulinaemia involved in immune complex glomerulonephritis, immune-mediated thrombocytopenia and anaemia, and possibly antibody-mediated enhancement of infection (Lim *et al.*, 2005).

### Subclinical EIAV infections and virus persistence

While there has been progress in understanding the pathogenesis of clinical EIA, far less is known about the sites and mechanisms of viral persistence in the subclinically infected horse. Compared to clinical disease, subclinical EIAV infections are characterized by dramatically decreased viraemia and tissue viral loads (Rice *et al.*, 1989; Sellon *et al.*, 1992; Kim and Casey, 1994; Crawford *et al.*, 1996; Langemeier *et al.*, 1996; Oaks *et al.*, 1998; Harrold *et al.*, 2000). The adaptive immune system is clearly very important in control of viral replication and disease based on the inability of immunodeficient foals to control viral replication (Perryman *et al.*, 1988), the ability of immunosuppressive drugs to recrudesce virus (Kono *et al.*, 1976; Tumas *et al.*, 1994), and the correlation between control of disease and appearance of cellular and/or neutralizing antibody responses (Carpenter *et al.*, 1987; O'Rourke *et al.*, 1989; McGuire *et al.*, 1994; Hammond *et al.*, 1997; Mealey *et al.*, 2003). However, it is also very likely that viral factors, viral–macrophage interactions and/or innate immune responses also play a role in viral persistence by restricting replication and avoiding host immune responses. Although virus can be reduced to very low levels in the subclinical horse, as with other lentiviruses viral replication never ceases entirely as indicated by the ability to detect viral unspliced or singly spliced RNA transcripts in tissues if sensitive enough methods such as nested RT-PCR are used (Langemeier *et al.*, 1996; Oaks *et al.*, 1998; Harrold *et al.*, 2000). Additionally, Leroux *et al.* (2001) have demonstrated that *env* gene evolution during persistence is independent of plasma virus levels suggesting that ongoing virus replication occurs. It is unknown if true viral latency occurs *in vivo* with EIAV, since PCR done on tissues cannot determine if the viral RNA is coming from some or all of the infected cells. Monocytes *in vivo* are generally regarded as latently infected since in some studies DNA, but not cell-associated RNA, could be detected in PBMC (Sellon *et al.*, 1992; Maury, 1994; Sellon *et al.*, 1996a)(Oaks, unpublished). A more recent study however detected low levels of RNA in PBMC (Harrold *et al.*, 2000), questioning whether true viral latency actually occurs in these cells.

Evidence indicates that the cellular site of persistence in subclinical EIA is the tissue macrophage (Oaks *et al.*, 1998). This finding was somewhat enigmatic since it would appear that the same cell type that is highly permissive during acute disease is also able to either maintain latent infections and/or restricted replication. *In vivo* data indicate that restricted replication occurs at the level of proviral transcription. In tissues from horses with subclinical EIA, *in situ* PCR for viral DNA was able to detect infected cells, but viral RNA could not be detected in these same tissues by *in situ* hybridization, even though viral RNA could be detected by RT-PCR (Oaks *et al.*, 1998). This suggested that no individual cells were expressing viral RNA at the level seen during clinical disease. Another study also found decreased DNA:RNA ratios during subclinical infection, suggesting decreased transcription levels (Harrold *et al.*, 2000). The actual mechanism of restricted replication is unknown. One study identified strain variation in the composition of the EIAV long terminal repeat (LTR) enhancer (Maury *et al.*, 1997), however longitudinal studies have shown that the LTR enhancer does not change within infected horses between periods of high and low viral loads or replication (Reis *et al.*, 2003; Maury *et al.*, 2005). Another possible mechanism is decreased function of Rev, a viral protein that regulates splicing and nuclear export of viral mRNAs necessary for expression of structural proteins. Belshan *et al.* (2001) found variation in *rev* mRNA transcripts throughout the course of infection, and that Rev proteins associated with clinical disease had higher activities than those associated with subclinical infection. In support of this observation, we have also found that in horses with subclinical infections, there are selective decreases in Rev-dependent transcripts (Oaks, unpublished data).

One interesting possible mechanism of restricted replication is a change in the phenotype of the infected macrophage. It has been shown that differentiation from monocytes to macrophages *in vitro* up-regulates EIAV replication (Maury, 1994; Sellon *et al.*, 1996a). *In vitro Escherichia coli* lipopolysaccharide or phorbol ester activation of EIAV-infected macrophages suppresses levels of supernatant reverse transcriptase activity (Sellon *et al.*, 1996b) and virus infectivity (Smith *et al.*, 1998). The study by Smith *et al.* (1998) found that the block to EIAV replication was due to defects in Rev-mediated nuclear export and/or splicing of mRNA based on the accumulation of Rev-independent, multiply spliced mRNA transcripts in the activated macrophages. We have also shown that in lymph nodes there is an inverse relationship between disease status, virus load and TNF-α mRNA, and selective decreases in Rev-dependent transcripts associated with higher levels of TNF-α expression (Oaks, unpublished data). Thus, macrophage activation during or following episodes of plasma viraemia may result in reduced virus replication. This effect may help to explain the transient, episodic nature of viraemia during EIAV infection. Collectively, however, the available evidence to date suggests that multiple viral and viral–host interactions may regulate EIAV replication *in vivo*, and further studies are necessary to establish their relative contributions to EIAV persistence.

## Molecular regulation of EIAV replication

### Genomic structure and viral transcripts

Like other lentiviruses, EIAV contains a complex genome (Fig. 9.2). The 5′ and 3′ long terminal repeats (LTRs) are direct repeat sequences that flank the EIAV genome. The EIAV LTR is smaller than most other retroviral LTRs containing about 320 nucleotides. It is functionally and

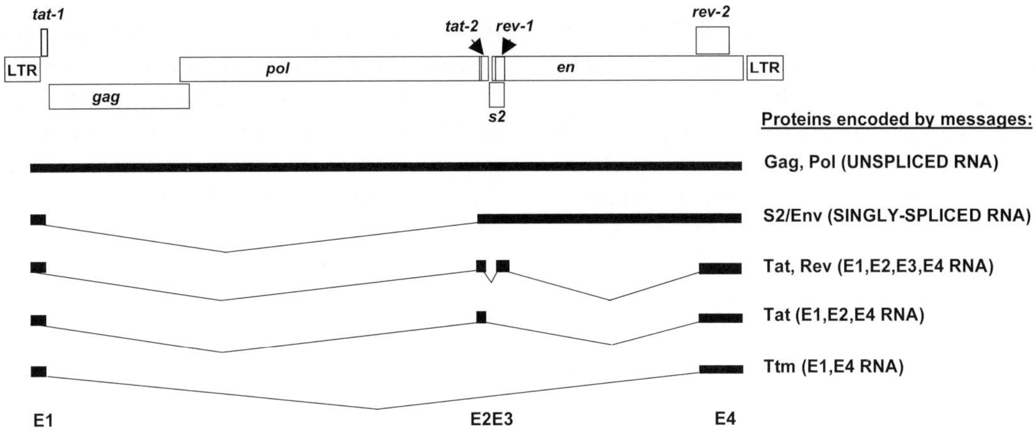


**Figure 9.2** Schematic of the EIAV genome and most commonly produced transcripts. The approximate size and location of each open reading frame of the genome is shown. The bolded bars in the lower part of the figure represent transcripts produced by the virus and the genes encoded by each of the transcripts are noted to the right of the transcript. E1 through E4 are the most common exons found in the multiply spliced transcripts.

structurally divided into three regions: U3, R and U5 that contain cis acting sequences that are responsible for RNA polymerase II recruitment to the viral promoter, transcriptional initiation, transcriptional transactivation and polyadenylation of the viral RNA. A small exon (E1 in Fig. 9.2) that is discussed in more detail below is encoded by sequences immediately downstream from the LTR. Overlapping with these sequences is a poorly defined packaging sequence composed of approximately 750 bp (Olsen, 1998). The packaging or ψ sequence in viral RNA interacts with the Gag polypeptides recruiting the EIAV genome into nascently forming viral particles. The large *gag* gene encodes the viral structural proteins. Overlapping with the 3′ end of *gag* is the *pol* gene responsible for encoding the enzymatic functions of virus. Immediately downstream from *pol* is the second exon of *tat*, which encodes the viral transcriptional transactivator protein. S2 that is located after the termination codon of *tat* is a conserved gene of unknown function and overlaps the 5′ end of *env*. The *env* gene encodes the viral glycoprotein. As with all retroviral glycoproteins, *env* encodes a single polypeptide that is cleaved within the Golgi by the cellular protease furin to yield the mature, surface (SU gp90) and transmembrane (TM gp45) glycoproteins. In the case of EIAV, these glycoproteins interact non-covalently. Interestingly, in addition to encoding both S2 and SU, two other important functions are located in this same region including the first exon of rev (Stephens *et al.*, 1990) and the Rev responsive element (RRE) (Belshan *et al.*, 2000; Harris *et al.*, 1998). At the 3′ end of the genome, the second exon of rev overlaps with the *env* gene. The Rev protein binds to RRE sequences facilitating export of RRE-containing transcripts out of the nucleus making these transcripts available for translation (Saman *et al.*, 1990; Stephens *et al.*, 1990; Martarano *et al.*, 1994).

While all transcripts initiate at the U3-R border within the LTR, differential splicing is responsible for the production of three different size classes of messages: multiply spliced, singly spliced and unspliced (Fig. 9.2). A number of small, multiply spliced transcripts have been described (Stephens *et al.*, 1990; Carroll *et al.*, 1991; Noiman *et al.*, 1991; Schiltz *et al.*, 1992; Rosin-Arbesfeld *et al.*, 1993). One such transcript was described in a single study that identified an RNA that encodes a 27-kDa chimeric transmembrane protein Ttm (Beisel *et al.*, 1993). Ttm was demonstrated to be composed of a portion of the second exon of Tat and much of the carboxy terminal domain of Env TM. The role of Ttm remains elusive.

Another multiply spliced EIAV mRNA transcript is a unique four exon bicistronic message. This primarily encodes the regulatory proteins

Tat and Rev with Tat encoded by the first two exons and Rev encoded by the last two (Carroll and Derse, 1993). The first exon of Tat is present at the 5′ end of all multiply spliced messages. Interestingly, this 16-residue exon does not contain a methionine start site; translation initiation occurs within this exon at a leucine codon (Derse *et al.*, 1990; Stephens *et al.*, 1990). Inefficient translation initiation within this exon is thought to lead to ribosomal scanning to the downstream methionine that initiates Rev in the four exon transcript (Carroll and Derse, 1993). The third exon of this four exon transcript contains the purine-rich sequences that compose a portion of the RRE as well as serving as an exonic splicing enhancer (ESE) (Belshan *et al.*, 1998; Gontarek and Derse, 1996; Lee *et al.*, 2008). This ESE binds to two different sets of proteins: the cellular splicing factor SF2/ASF and EIAV Rev (Gontarek and Derse, 1996; Harris *et al.*, 1998; Belshan *et al.*, 2000; Chung and Derse, 2001; Liao *et al.*, 2004). Inclusion of exon 3 within the multiply spliced transcript occurs depending upon which set of proteins is bound to the exon with binding of Rev to exon 3 resulting in the exclusion of exon 3 from the message as well as export of singly spliced and unspliced messages from the nucleus (Gontarek and Derse, 1996; Belshan *et al.*, 2000; Chung and Derse, 2001; Liao *et al.*, 2004). The alternative splicing of exon 3 is believed to be a mechanism to control the cellular level of Rev protein and, thus, maintain a balance between the production and translation of spliced and unspliced messages (Belshan *et al.*, 2000; Liao *et al.*, 2004).

Mid-sized messages encode the viral receptor protein SU gp90 and a small accessory protein S2. The singly spliced, mid-sized message can also potentially code for Tat (Schiltz *et al.*, 1992); however, whether Tat is synthesized from this message during an infection is not known. Both S2 and Env messages are transcribed from the singly spliced transcript with *S2* located slightly 5′ and in a +1 reading frame relative to *env*. The ribosome binding site that initiates translation of S2 is not optimal and ribosomes presumably scan through the start of S2 frequently to initiate translation at the consensus Kozak site found at the beginning of *env*. Another transcript that has been detected both *in vitro* and *in vivo* is a variation of the singly spliced transcript in which the 5′ splice donor site at the end of the first exon of *tat* splices to a splice acceptor site that is 77 nucleotides into the second exon of *tat* (Oaks, unpublished data). The function of this transcript is unknown. It may decrease the total amounts of functional Tat produced, and/or may affect the translation of SU gp90/TM gp45 and/or S2. However, its consistent presence *in vivo* suggests that it is relevant to the biology of EIAV.

Full length messages are responsible for the synthesis of the Gag and Pol polyproteins that are processed within extracellular virions by the virally encoded protease to produce the structural and enzymatic proteins of the virus, respectively. In addition, the full length transcripts are packaged into virions to produce infectious particles.

## Entry of EIAV

A cellular receptor for EIAV has been recently identified to be a tumour necrosis factor receptor superfamily member (Zhang *et al.*, 2005). This plasma membrane protein, equine lentivirus receptor 1 (ELR1), is the equine orthologue of HveA or Tumour Necrosis Factor Receptor Superfamily, Member 14 (TNFSF14) that serves as a cellular receptor for herpes simplex virus (Montgomery *et al.*, 1996). While it has not been directly shown that EIAV entry into macrophages is dependent upon ELR1, the necessity of ELR1 expression for EIAV uptake into a number of other permissive cells such as equine dermal cells (ED) has been demonstrated (Zhang *et al.*, 2005; Brindley, 2008; Zhang and Montelaro, 2009).

Two indirect lines of evidence suggest that ELR1 may be utilized by EIAV for productive virus entry into macrophages. Equine macrophages express abundant amounts of ELR1 mRNA (Zhang *et al.*, 2005) and modest levels of ELR1 are present on the surface of monocyte-derived macrophages (MDM) (Fidalgo-Carvalho *et al.*, 2008). In addition, ELR1-dependent entry of EIAV in ED cells is mediated through a endosomal, low pH-dependent pathway (Brindley and Maury, 2005; Jin *et al.*, 2005b) and EIAV entry into both equine macrophages and ED cells is inhibited by similar concentrations of the lysosomal blocking agent ammonium chloride (Brindley and Maury, 2005). While neither of these observations definitively demonstrates that ELR1 is required for EIAV entry into macrophages, they are consistent with the possibility.

Nonetheless, a number of questions remain unanswered about the role of ELR1 in EIAV infection *in vivo*. For instance, tissue culture strains of EIAV such as $EIAV_{MA-1}$ that have clearly been shown to use ELR1 for entry in ED cells (Brindley *et al.*, 2008) do not replicate or replicate poorly in MDMs and are unable to replicate *in vivo* (Carpenter and Chesebro, 1989). Whether this restriction is due to the inability of this strain to enter macrophages or results from a block at later steps within the viral life cycle has not been investigated. In addition, the tissue distribution of the human orthologue of ELR1 would suggest that ELR1 is present on a variety of cells that are not permissive for EIAV. These cell populations include lymphocytes, cardiomyocytes, adipose cells and keratinocytes (Harrop *et al.*, 1998; Hung *et al.*, 2002; Fradette *et al.*, 2005; Dahl *et al.*, 2008). If ELR1 serves as the receptor *in vivo*, the absence of a broader cell tropism of the virus remains puzzling.

The molecular interactions of SU gp90 and ELR1 that are needed for productive virus entry have only begun to be explored. SU gp90 deletion and substitution studies suggest that residues important for ELR1 binding reside in the C-terminal two-thirds of the surface glycoprotein gp90 (Sun *et al.*, 2008). Complementary studies that explored residues of ELR1 required for EIAV entry demonstrated that L70 within the cysteine rich domain-1 of ELR1 is important for SU gp90 binding and optimal virus infectivity (Zhang *et al.*, 2008).

EIAV entry into cells requires virus uptake into endosomes rather than the direct viral fusion at the cell surface that occurs with primate lentiviruses. Upon SU gp90/ELR1 interactions, EIAV enters endosomal vesicles via receptor-mediated endocytosis (Brindley and Maury, 2008). Virus entry is blocked by inhibitors of clathrin-dependent endocytosis indicating that clathrin coated pits serve as an important endocytic pathway for productive EIAV infection (Brindley and Maury, 2008). Upon vesicle endocytosis, low pH dependent events are required for productive infection (Brindley and Maury, 2005; Jin *et al.*, 2005b). However, low pH events by themselves are insufficient to mediate EIAV fusion (Brindley and Maury, 2005) suggesting that a two-step mechanism involving ELR1 binding followed by low pH dependent events is needed for EIAV entry. Receptor binding followed by low pH events have previously been described for uptake of a number of retroviruses as well as filoviruses and coronaviruses (Matsuyama *et al.*, 2004; Chandran *et al.*, 2005; Delos *et al.*, 2005; Simmons *et al.*, 2005; Schornberg *et al.*, 2006; Kumar *et al.*, 2007). As low pH-dependent proteases have recently been appreciated to be required for entry of murine leukaemia virus, Ebola virus and the SARS coronavirus (Chandran *et al.*, 2005; Simmons *et al.*, 2005; Schornberg *et al.*, 2006; Kumar *et al.*, 2007), the role of cathepsins in EIAV entry was explored (Brindley and Maury, 2008). However, these cellular proteases are not involved in the low pH dependent events that are required for EIAV uptake. This suggests that the low pH events needed for EIAV entry following ELR1 interactions may mediate SU gp90 conformational changes that lead to TM gp45 fusion events in a manner similar to that found with avian retroviruses (Matsuyama *et al.*, 2004; Delos *et al.*, 2005;).

## Uncoating

Uncoating of EIAV particles in macrophages has not been studied; however, studies in other cell populations have begun to provide insights into EIAV uncoating and have identified a cellular restriction mechanism that interferes with uncoating. Transduction of EIAV pseudovirions composed of the EIAV Gag proteins was shown to be restricted by the same cellular factor that restricts primate lentiviruses in many cell types (Hatziioannou *et al.*, 2003). Evidence of species specificity of this factor was provided with the rhesus macaque factor able to restrict transduction of both EIAV and HIV pseudovirions whereas the human factor restricted transduction of EIAV but not HIV (Hatziioannou *et al.*, 2004; Saenz *et al.*, 2005). Subsequent studies identified this factor as TRIM5α (Stremlau *et al.*, 2004). TRIM5α has been shown to bind to the retroviral capsid resulting in proteasome-dependent capsid/TRIM5α degradation (Anderson *et al.*, 2006; Wu *et al.*, 2006; Campbell *et al.*, 2008; Rold and Aiken, 2008). Species restriction results from species-specific amino acid changes in TRIM5α that control the ability of the protein to interact with the capsid of retroviruses and interfere with productive uncoating. In general, retroviruses have evolved to successfully avoid interactions with TRIM5α made by their targeted species, but not those of other species (see Chapter 4).

Although cellular, monomeric cyclophilin A is known to be associated with HIV capsids and helps to mediate uncoating of the capsid structure in a timely and productive manner (Braaten *et al.*, 1996), there is no evidence that cyclophilin A is found in EIAV particles.

## Reverse transcription

EIAV reverse transcriptase (RT) is error prone having a high and similar rate of nucleotide misincorporation to that of other lentiviruses (Bakhanashvili and Hizi, 1993). The lacking of proof-reading activity of lentiviral RT is thought to lead to at least one nucleotide misincorporation in each DNA genome produced during reverse transcription (Preston *et al.*, 1988; Roberts *et al.*, 1988).

dNTP pools in terminally differentiated macrophages may be as much as 20-fold lower than that those found in an actively dividing cell (Traut, 1994; Diamond *et al.*, 2004). Lentiviruses have developed strategies for efficient DNA strand elongation under low dNTP environments (Diamond *et al.*, 2004). For instance, the RTs of HIV and SIV, but not murine leukaemia virus reverse transcriptase, are highly processive when nucleotide pools are at concentrations as low as 40 nM – concentrations that approximate levels observed in macrophages (Diamond *et al.*, 2004). While these studies have not been performed with EIAV reverse transcriptase, a similar processivity under low dNTP conditions might be anticipated.

In addition to the potential for EIAV reverse transcriptase to be highly processive under low dNTP conditions, EIAV encodes a dUTPase that boosts TTP levels within the macrophages (Threadgill *et al.*, 1993; Payne and Elder, 2001). Endogenous levels of dUTPase activity in terminally differentiated cells is low, whereas expression of this enzyme is elevated in actively dividing cells (Hokari *et al.*, 1995). dUTPase converts dUTP to dUMP and PPi providing a source of TTP precursor (Threadgill *et al.*, 1993). Concomitantly this enzyme reduces cellular dUTP levels altering the TTP to dUTP ratio within the cell and diminishing dUTP incorporation into elongating DNA (Steagall *et al.*, 1995). In the non-dividing macrophage, infection of a dUTPase-deficient EIAV results in dUTP incorporation into EIAV DNA by reverse transcriptase and reduction in viral RNA transcription from the dUTP-containing DNA (Steagall *et al.*, 1995). Loss of EIAV dUTPase activity prevents virus replication in macrophages, but does not affect virus replication in actively dividing cell cultures indicating that expression of this gene serves as a virulence factor during *in vivo* infection (Threadgill *et al.*, 1993; Steagall *et al.*, 1995; Payne and Elder, 2001).

An intriguing question that has recently been addressed is how EIAV copes with the inhibitory activity of equine APOBEC3 proteins. Human APOBEC3 proteins that contain two tandem cytosine deaminase (CDA) domains such as ABOBEC3G inhibit primate lentivirus replication (Gooch and Cullen, 2008). Human APOBEC3G binds to HIV nucleocapsid and is incorporated into budding virions (Sheehy *et al.*, 2002; Mariani *et al.*, 2003; Cen *et al.*, 2004). Following infection of a cell, virion-associated APOBEC3G interferes with productive replication via several mechanisms that include deoxycytidine deamination (Harris *et al.*, 2003) and additional mechanisms that block reverse transcriptase activity (Bishop *et al.*, 2008). The HIV *vif* gene prevents APOBEC3G incorporation into virions. Equine APOBEC3s that contain two CDA domains, eAPOBEC3F1 and eAPOBEC3F2, bind to HIV and EIAV nucleocapsids, incorporate into virions and eAPOBEC3F2 blocks HIV infectivity (Bogerd *et al.*, 2008). Interestingly, these equine APOBEC3 proteins only modestly inhibit EIAV replication. This modest inhibition by eAPOBECs may explain the absence of a *vif*-like gene within the EIAV genome. In addition, EIAV may also encode other means of inhibiting the effects of this family of anti-retroviral proteins. As cytidine deamination leads to AT rich genomes, it is perhaps not surprising that the EIAV genome is highly AT rich with 61.5% of nucleotides of the *env* gene either adenine or thymidine (Cook *et al.*, 2005).

## Nuclear import and viral DNA integration

Following reverse transcription, the newly synthesized viral DNA is complexed with viral integrase and cellular proteins forming the pre-integration complex (PIC). PICs migrate into the nucleus of the cell where the viral DNA integrates into the genome of the host cell. EIAV is readily able to infect non-dividing cells such as tissue macrophages and, as these cells do not undergo mitosis and concomitant dissolution of the nuclear membrane, EIAV PICs must contain one or more active mechanisms that mediate PIC nuclear entry. Trafficking of other lentiviral PICs require PIC-associated proteins containing nuclear localization signals (NLS). NLS interactions with the nuclear pore complex are thought to mediate active transport of the complex across the nuclear membrane in a RanGTP-dependent manner (Anderson and Hope, 2005); however, the protein(s) responsible for nuclear import of the EIAV PIC has not been identified. Extensive studies with HIV have suggested a number of viral and cellular proteins may be involved in nuclear import. Early studies indicated that the viral proteins Vpr, integrase and/or matrix can mediate nuclear import of the HIV PIC (von Schwedler *et al.*, 1994; Connor *et al.*, 1995; Gallay *et al.*, 1997). More recently, the central polypurine tract has also been implicated (Arhel *et al.*, 2007). Cellular proteins associated with the PIC may also mediate import. Transportin-SR2 may mediate nuclear import as it was recently shown to interact with HIV integrase and reduction of this cellular protein by siRNA reduced HIV PIC entry into the nucleus (Christ *et al.*, 2008). In addition, LEDGF, Nup98, Nup358, Nup153 and importin 7 interact with the HIV PIC and may be involved in nuclear import of HIV (Cherepanov *et al.*, 2003; Fassati *et al.*, 2003; Ebina *et al.*, 2004; Busschots *et al.*, 2005; Brass *et al.*, 2008; Konig *et al.*, 2008; Zaitseva *et al.*, 2009). Studies have demonstrated that importin 7 does not serve a similar nuclear import role for EIAV (Zaitseva *et al.*, 2009), but the ability of the other cellular or viral proteins to mediate EIAV transport across the nuclear membrane is not known. One intriguing candidate that may be involved in EIAV PIC entrance into the nucleus and viral DNA integration is EIAV Gag late protein (p9). Jin *et al.* (2005a) have demonstrated that in ED cells p9 truncation results in a defect that is consistent with a problem with nuclear import. The authors demonstrate that the defect in virus replication occurs following completion of reverse transcription but prior to or during nuclear entry. Thus, it is tempting to speculate that p9 may play a role in one or more of these important steps.

The integration pattern of a EIAV vector was recently investigated and compared with the integration pattern observed for HIV (Hacker *et al.*, 2006). Like other lentiviruses, EIAV has a tendency to integrate into the bodies of actively transcribed genes rather than in promoter/enhancer regions or intergenic regions. However, unlike HIV, EIAV was found to integrate in AT rich regions. The nuclear protein LEDGF (lens epithelium-derived growth factor) interacts with EIAV integrase and enhances integration providing additional evidence for the importance of this cellular protein for successful integration of lentiviral DNA into the host's genome.

## Initiation of EIAV transcription and enhancer variation

Upon integration of the EIAV genome into the host cell DNA, the virus utilizes cellular transcriptional machinery to generate viral transcripts. This machinery interacts with sequences within the LTR. The EIAV LTR is composed of three regions: U3, R and U5 (Fig. 9.3). The known cis-acting sequences that interact with cellular transcriptional activator proteins and lead to transcription initiation reside within the LTR U3 sequences (Derse *et al.*, 1987; Dorn and Derse, 1988; Sherman *et al.*, 1989; Carvalho and Derse, 1993a; Carvalho and Derse, 1993b; Carvalho *et al.*, 1993; Maury, 1994; Maury *et al.*, 1997; Maury *et al.*, 2000; Hines *et al.*, 2004) These include an invariant TATA box that recruits basal transcriptional machinery and a number of transcription factor (TF) binding motifs that sit within the U3 enhancer region.

These TF binding motifs are well documented to evolve during cell tropism adaptation of the virus and are known to impact viral virulence (Carpenter *et al.*, 1991; Maury *et al.*, 2000; Payne *et al.*, 2004; Maury *et al.*, 2005). The

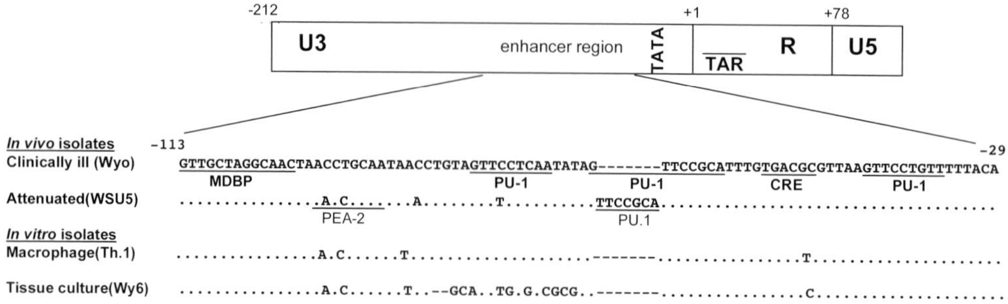


**Figure 9.3** EIAV long terminal repeat (LTR). The EIAV LTR composed of U3, R and U5 sequences contains about 320 nucleotides. Except for the enhancer region that is located immediately upstream from the TATA box, the EIAV LTR is relatively conserved between isolates. Within the enhancer region, genetic variation is notable consisting of point mutations, deletion and insertions. Shown here are examples of enhancer regions of LTRs from four different groups of EIAV: virulent strains (represented by Wyoming (Wyo)), attenuated strains that causes mild disease in horses and can be readily passaged in tissue culture (represented by WSU5), *in vitro* passaged strains that grows well in monocyte-derived macrophages, but not tissue culture cell lines (Th.1) and strains that grows well in fibroblastic cell lines (λ6 from Carvalho and Derse, 1993). LTR enhancers from virulent strains of virus such as Wyoming appear to be conserved as a group. The LTR enhancers from the other strains tend to have large amounts of enhancer diversity both within and between the groups. Noted on the nucleotides are transcription factor binding motifs that have been identified in the enhancer. Dots represent nucleotides that are identical to that found in Wyoming. Gaps in the sequence alignment are shown as dashes. Additional motifs not shown here have been identified in other isolates.

enhancer region of the EIAV LTR from different strains is found to be variable with loss and gain of TF binding sites. In one study that compared the LTRs found in horses acutely ill with EIA to those found in healthy, long term EIAV-infected horses, 45% of the nucleotide positions within the enhancer region from seven horses were variable due to point mutations, whereas only 5% variation was found in the remainder of the LTR between the *in vivo* isolates (Maury *et al.*, 1997). Similar high levels of variation within the enhancer, but not the rest of the LTR, are also found between individual tissue culture-derived isolates and between LTRs from virulent strains of virus and tissue culture-derived strains of virus (Alexandersen and Carpenter, 1991; Carvalho and Derse, 1993a; Payne *et al.*, 1994). For example, in a study that demonstrated that the LTR is important for virulence, there were a total of 16 nucleotide differences within the enhancer sequence of the virulent pSP17 strain when compared to the avirulent pSP19 strain (Payne *et al.*, 2004). Enhancer variation does appear limited in highly virulent strains (Maury *et al.*, 1997) suggesting that there are specific TF binding motifs that play a role in EIAV virulence. Enhancer region variation primarily due to single point mutations is unique for EIAV (Maury *et al.*, 1997). Even in Type C retroviruses where extensive LTR variation has been described, the variation tends to involve sequence duplication rather than individual point mutations (Li *et al.*, 1987; Short *et al.*, 1987; Golemis *et al.*, 1989). In total, these studies suggest that strong selective pressures are exerted on the enhancer region of EIAV leading to high levels of genetic variation.

Studies to identify TF binding motifs within the EIAV enhancer which are important for expression have been performed using a number of different cell types. This review will focus on studies that identify the TFs important for expression in primary equine macrophages and myelomonocytic cell lines. These studies have primarily used a combination of protein/DNA interaction studies, such as DNase footprinting and electrophoretic mobility shift assays (EMSAs), and reporter gene assays to measure the transcriptional activity of the LTR. Because of the nucleotide variation and concomitant changes in TF binding motifs present in different EIAV LTRs, the enhancer sequences analyzed in each study may be quite different. In macrophages, *ets* sites that bind to the cellular TF PU.1 are required for optimal LTR-driven reporter expression in macrophages (Maury, 1994; Hines *et al.*, 2004). Three *ets* sites are routinely found in virulent strains of EIAV, but one or two sites that are distal to the promoter are frequently lost in attenuated strains (Maury *et al.*, 1997). The three *ets* sites present in virulent strains have slight sequence variation, but all bind to PU.1 with similar affinity (Hines *et al.*, 2004). In an LTR construct that only contained *ets* sites, loss of the promoter-proximal site dramatically decreased Tat-dependent transcription (Hines *et al.*, 2004). This study further demonstrated that the position of this *ets* site within the LTR rather than the specific sequence of the site was important for maximal LTR activity.

PU.1 is a TF that is important for both myeloid and B-cell development (Fisher and Scott, 1998; Lloberas *et al.*, 1999; Oikawa *et al.*, 1999). Interestingly, the level of PU.1 expression in progenitor cells controls the development of haematopoietic lineages with high levels of PU.1 expression generating myeloid precursors (Singh *et al.*, 1999; DeKoter and Singh, 2000). PU.1 levels remain high in mature macrophages. Thus, transcriptional initiation of the macrophage-tropic lentivirus EIAV is controlled by a TF that is specifically expressed at high levels in macrophages and is responsible for the generation of the myeloid lineage. This developmentally controlled transcriptional regulation of EIAV stands in contrast to that observed with HIV. HIV expression is principally controlled by NF-κB, a cellular TF whose activity is stimulated by detection of microbial 'danger' signals. Consequently, HIV transcription is activated and virus is produced as NF-κB stimulates expression of genes in immune cells to fight HIV and other infections (Nabel and Baltimore, 1987; Leonard *et al.*, 1989).

As demonstrated with both reporter gene studies and virus infectivity studies, the three *ets* sites by themselves are insufficient for robust EIAV LTR activity and detectable virus replication in macrophages (Hines *et al.*, 2004). Additional TF motifs within the EIAV LTR needed for virus replication in macrophages have yet to be definitively identified although binding motifs for other TFs have been characterized in the enhancer region of virulent strains of virus (Maury, 1994). TF motifs that are usually found in virulent strains include an almost ubiquitously present methylation-dependent binding site (MDBP) that flanks the 5′ end of the enhancer and a CRE binding site that is located between the middle and promoter-proximal *ets* sites. Loss of the MDBP site delays, but does not abrogate, virus replication in macrophages (Hines *et al.*, 2004) and other permissive cells (W. Maury, unpublished). Primary equine macrophage nuclear protein extracts do not appear to interact with the CRE site in EMSAs suggesting this site, while present, may not be important for LTR expression in macrophages (Maury, 1994). Alternatively, as PU.1 binding to *ets* sites in cellular enhancers has been shown to synergize with adjacent TFs/motif interactions, it is possible that unappreciated synergies are needed for maximal transcriptional activity. Early studies demonstrated that a PEA-2 site present in most tissue culture strains of EIAV binds nuclear extracts from the monocytic line THP-1 (Carvalho and Derse, 1993b; Carvalho *et al.*, 1993), but the PEA-2 site is not present in most virulent strains of virus (Maury *et al.*, 1997).

## Tat transactivation of EIAV LTR expression

In a manner similar to that found with primate lentiviral Tat/TAR interactions, EIAV-encoded transcriptional transactivator protein, Tat, dramatically enhances EIAV LTR expression through interactions with RNA stem-loop sequences called TAR that are present at the 5′ end of all nascent EIAV transcripts. Early studies localized the EIAV transactivating activity to the S1 ORF making it clear that S1 encoded Tat (Dorn *et al.*, 1990; Stephens *et al.*, 1990). EIAV Tat expression is required for detection of viral antigens in cells transfected with full length molecular clones of EIAV (Maury *et al.*, 1994). As with HIV, the cellular kinase positive transcription elongation factor-b (P-TEFb) interacts with the EIAV Tat and TAR to stimulate EIAV transcript elongation (Bieniasz *et al.*, 1999; Albrecht *et al.*, 2000; Taube *et al.*, 2000). The active P-TEFb complex that stimulates EIAV transcription is composed of Cdk9 and cyclin T1. P-TEFb has been shown to phosphorylate several negative transcription elongation factors that bind to nascent transcripts and block transcript elongation (Wada *et al.*,

1998; Fujinaga *et al.*, 2004; Cheng and Price, 2007). P-TEFb-dependent phosphorylation causes release of these factors. In addition, P-TEFb phosphorylates serine 2 of the C-terminal domain of RNA polymerase II enhancing processivity of the polymerase (Marshall *et al.*, 1996). A number of residues in the cyclin domain of Cyclin T1 interact with EIAV Tat residues (Bieniasz *et al.*, 1999; Albrecht *et al.*, 2000; Taube *et al.*, 2000; Anand *et al.*, 2007, 2008) as well as bind directly to the upper loop of TAR (Anand *et al.*, 2008). A highly basic region of Tat also interacts with the TAR loop and upper stem nucleotides (Anand *et al.*, 2008). Recruitment of P-TEFb produces localized enhanced P-TEFb activity at nascent EIAV transcripts resulting in viral RNA elongation.

### Nuclear export of EIAV transcripts

Completely spliced messages are synthesized early in infection (Stephens *et al.*, 1990; Martarano *et al.*, 1994) and studies with HIV have shown that this population of messages is exported from the nucleus via export pathways used by cellular mRNAs (Cullen, 2003). Singly spliced and unspliced EIAV messages require binding of and nuclear export facilitated by Rev (Fridell *et al.*, 1993; Mancuso *et al.*, 1994; Martarano *et al.*, 1994). For export of Rev-dependent messages, Rev must bind to the Rev responsive element (RRE) that has recently been shown to be composed of two RNA stem–loop domains (Lee *et al.*, 2008). EIAV Rev multimerizes on the RRE and transports RRE-containing messages out of the nucleus through a chromosome region maintenance 1-dependent pathway, thus making these messages available for translation. While each EIAV RRE sequence can function independently, reporter assays for Rev activity suggest that the two RREs are synergistic and the combination results in significantly greater nuclear export than either element by itself (Martarano *et al.*, 1994; Belshan *et al.*, 1998; Lee *et al.*, 2008). Two short arginine-rich domains (ARMs) within EIAV Rev are required for RRE binding and the Rev bipartite binding site is unique for EIAV (Lee *et al.*, 2006). It is tempting to speculate that the two Rev ARMs may interact with the two RRE that have been identified thereby leading to message export from the nucleus.

Nucleotide variation within the second exon of *rev* has been characterized in multiple episodes of viraemia of an EIAV infected horse (Belshan *et al.*, 1998; Baccam *et al.*, 2003; Sparks *et al.*, 2008). Nucleotide changes in *rev* result in Rev amino acid alterations that modify Rev activity (Belshan *et al.*, 1998). Variants that increased Rev activity were found to be associated with recurrent fever episodes suggesting that *rev* alterations during the course of infection impact the clinical status of the infected horse.

### Packaging and budding of new virions

Recently, good progress has been made in understanding how retroviruses are packaged. Retroviral assembly is driven by the multimerization of the Gag polyprotein. Gag is thought to localize to cellular membranes, recruiting the full length EIAV genome to the location through nucleocapsid interactions with φ sequences present in the genomic RNA. Two genomes are packaged as a result of viral RNA dimeric interactions. The cellular location of lentiviral packaging and budding remains controversial, especially in macrophages. Budding of HIV from T-cells appears to occur at the plasma membrane (Jouvenet *et al.*, 2008). In contrast, vacuoles within infected macrophages and some cell lines used to produce virus are filled with HIV particles suggesting that the lentivirus is budding into intracellular membranes in these cells (Gendelman *et al.*, 1988; Orenstein *et al.*, 1988; Grigorov *et al.*, 2006). Thus, the location of lentiviral budding may be dependent upon the infected cell, with virions budding into vesicles of macrophages whereas plasma membrane-associated budding may occur in most other cell populations. It has been recently demonstrated that the membranes of HIV-containing vacuoles from macrophages contain plasma membrane-associated proteins suggesting that these vacuoles may be derived from plasma membranes perhaps through exocytosis (Deneka *et al.*, 2007). Additional studies by Welsch *et al.* (2007) have proposed that these membranous structures are not vacuoles, but invaginations of the plasma membrane with direct contact with the extracellular milieu allowing potential direct release of viral particles. This issue is central to EIAV replication as macrophages are the primary target of EIAV *in vivo*; however, no studies with EIAV directly address this question. As EIAV is lytic to macrophages, release of EIAV from a potentially intracellular compartment may not pose an issue for the virus.

The small carboxy terminal Gag protein, termed the Late Protein or p9, is important for efficient EIAV egress from cells. In the case of EIAV, the Late Protein contains a YPDL motif that interacts with the endosomal sorting factor ALIX/AIP1 leading to efficient budding of viral particles (Strack *et al.*, 2003). Ubiquitin or other retroviral sorting motifs such as PT/SAP or PPPY can functionally substitute for the YPDL motif (Li *et al.*, 2002; Shehu-Xhilaga *et al.*, 2004; Joshi *et al.*, 2008). The Gag p9 YPDL motif also interacts with the μ2 subunit of adaptor protein AP2 and this interaction also appears to facilitate budding (Puffer *et al.*, 1998; Chen *et al.*, 2001; Chen *et al.*, 2005). Overexpression of either the μ2 subunit or the p9 binding region of ALIX leads to budding inhibition and the combination of both inhibitors provides an additive effect (Chen *et al.*, 2001).

Once the virion buds from the cell, Gag and Pol processing occurs producing a mature, infectious viral particle. This alters the morphology of the virion into its characteristic bullet shape. Evidence suggests that the cytoplasmic tail of the TM protein may also be processed within the virion. Rice *et al.* (1990) have shown that Env gp45 is processed into a glycosylated gp32 and a carboxy terminus, non-glycosylated protein, p20 within the virion. The role of this further processing in EIAV pathogenesis is not known, but may be required to generate a fusion-ready viral glycoprotein that upon ELR1 interactions leads to productive entry of the infectious virion.

### Role of S2 in pathogenesis

S2 is not required for EIAV replication in tissue culture, although it has been demonstrated that in both MDM and ED cells EIAV replication can be delayed by the introduction of stop codons in the S2 ORF (Li *et al.*, 1998; Fagerness *et al.*, 2006). The S2 gene is conserved across all EIAV isolates and has been shown to significantly contribute to pathogenesis (Li *et al.*, 2000; Fagerness *et al.*, 2006). As a consequence, one EIAV vaccine strategy being developed is deletion of the S2 gene (Li *et al.*, 2003; Craigo *et al.*, 2005, 2007a,b). S2 has no significant homology to other known viral or cellular proteins providing no insight into the mechanism mediating enhanced pathogenesis. Motifs that are present in EIAV S2 proteins include two myristoylation sites, a casein kinase 2 phosphorylation site, a SH3 binding site and a hydrophobic domain that would like interact with membranes (Li *et al.*, 1998; Fagerness *et al.*, 2006). Which, if any, of these motifs/domains is important for S2 expression, S2 function, and EIAV pathogenesis remains to be determined.

## EIAV strains: adaptation and attenuation

Field strains of EIAV are macrophage-tropic *in vitro*, replicating only in primary equine MDM cultures (Kobayashi and Kono, 1967a,b; Carpenter and Chesebro, 1989). Field strains can be adapted *in vitro* to replicate in a variety of other cell types including equine dermal (ED) (Malmquist *et al.*, 1973; Carpenter and Chesebro, 1989; Maury *et al.*, 2005) and equine kidney cell cultures (Kono and Yokomizo, 1968). This adaptation was a significant milestone in EIAV research, as it allowed propagation and studies of EIAV in cell types readily generated and maintained in tissue culture, in contrast to the laborious and difficult isolation and maintenance of equine MDM cultures. Subsequent studies found that cell lines of canine and feline origin, including the canine fetal thymus cell line Cf2Th (Benton *et al.*, 1981), the canine malignant histiocytosis macrophage-like cell line DH82 (Hines and Maury, 2001), and the feline fibroblast cell line FEA (Benton *et al.*, 1981) could also be infected by tissue-culture adapted EIAV strains. Although the ED cell-adapted Malmquist strain is derived from the highly virulent Wyoming strain, this process significantly attenuated its virulence (Gutekunst and Becvar, 1979; Orrego *et al.*, 1982). In an attempt to restore virulence for *in vivo* pathogenesis studies, the Malmquist virus was serially back passaged through horses by two different research groups (Orrego *et al.*, 1982; O'Rourke *et al.*, 1989) to give rise to the WSU5 and PV strains. This approach was successful, although the level of virulence was reduced from that of the original virus. Interestingly, the ability to replicate in non-macrophage origin cells was retained by these adapted strains. Much of the research on EIAV has thus been based on work with the Wyoming strain of EIAV, or its derivative the Malmquist strain and its respective derivatives. Many of the studies from Japan are also based on the Wyoming strain or its derivatives V26 and V70 (Kono and Kobayashi, 1970). In addition, most of the available infectious molecular clones of EIAV, such as pCL22 (Whetter *et al.*, 1990), pSP19 and its recombinant variants (Payne *et al.*, 1994; Payne *et al.*, 1998), and pUK (Cook *et al.*, 1998) are derived from the Malmquist strain. Although the tissue culture adapted strains of EIAV have been enormously useful, it should be kept in mind that findings using these somewhat modified and *in vitro* manipulated virus strains may not be truly representative of field isolates. Similarly, the highly virulent Wyoming strain, which is often referred to as a 'field' or 'wild type' strain and has been used for many *in vivo* studies, has been maintained by repeated serial passage in horses, with virus stocks harvested during acute disease. Thus, it may be that this virus, in particular its extreme virulence, may also be somewhat artificial.

EIAV genomic changes that are important for adaptation to non-macrophage cell types and attenuation of virulence have been explored. Mapping studies indicate that both LTR and *env* sequence changes play a role in virus adaptation to ED cells (Perry *et al.*, 1992; Zheng *et al.*, 2000b; Maury *et al.*, 2005). LTR changes are primarily localized to the enhancer region, with both the addition and deletion of TF binding motifs (Carpenter and Chesebro, 1989; Maury, 1994; Maury *et al.*, 2000, 2005; Zheng *et al.*, 2000a). SU changes that are important for tissue culture adaptation are poorly characterized. Perry *et al.* (1992) demonstrated that cell tropism of a molecular clone derived from a tissue culture isolate was restricted in Cf2Th cells by substitution of 3′ SU gp90 sequences from the virulent Wyoming strain. Interestingly, Sun *et al.* (2008) have recently shown that this same region of SU gp90 is important for binding to EIAV's cognate cellular receptor ELR1.

As loss of viral virulence occurs concurrently with cell tropism adaptation, it is perhaps not surprising that changes in LTR and SU gp90 sequences have also been identified to contribute to virulence (Cook *et al.*, 1998; Payne *et al.*, 2004; Liang *et al.*, 2006; Shen *et al.*, 2006). However, whether the same sequence alterations are responsible for both tropism and virulence changes has not been studied and identification of specific sequences responsible for virulence awaits investigation.

The observation that Env surface protein SU gp90 sequences play a role in both cell tropism and virulence suggests that changes in SU gp90/cellular receptor interactions may be important in these viral characteristics. This highlights the need for additional studies to understand SU gp90 sequence heterogeneity, sequence requirements for ELR1 interaction and the possibility that other plasma membrane-associated proteins may serve as receptors for EIAV. An alternative possible explanation for loss of virulence mapping to SU gp90 may be due to changes in cytokine production and endotoxin-like activity elicited by SU gp90 from virulent versus avirulent strains of virus (Ball *et al.*, 2005; Lim *et al.*, 2005).

One possible explanation for virulence and tropism changes mapping to SU gp90 sequences is that SU gp90 may interact with additional cellular receptors *in vivo*. The evolution of EIAV receptor usage has been recently documented in tissue culture (Brindley *et al.*, 2008), and these data suggest that utilization of multiple receptors *in vivo* is possible. A variant of the tissue culture adapted strain MA-1 arose during passage in ED cells. This variant, vMA-1c, is a superinfecting strain of virus that rapidly kills ED cells and elicits uncharacteristic large syncytia cells in the infected culture (Maury *et al.*, 2003). While vMA-1c can enter ED cells through interactions with ELR1, uptake and superinfection results from vMA-1c interactions with a different, currently uncharacterized cellular receptor that leads to virus uptake by direct fusion with the plasma membrane in a manner similar to that used by HIV-1 (Brindley and Maury, 2005, 2008; Brindley *et al.*, 2008).

## Vaccine development

Interestingly, a tissue culture adapted strain of EIAV with highly attenuated virulence was developed in China in the 1970s and was successfully used as a vaccine to control clinical EIA problems in China. This virus was created by passaging a field strain 133 times through donkeys (which enhanced virulence), followed by attenuation by 120 passages in primary donkey macrophages cultures (Shen *et al.*, 2006). This vaccine apparently offered protection against homologous and heterologous challenges, and has shown no evidence of reversion to virulence. However, similar attempts in Japan found protection to be incomplete when challenged with virulent viruses (Kono *et al.*, 1970).

Recently, a more systematic approach has been taken to develop a vaccine against EIAV and in parallel provides insights into vaccine development against other lentiviruses such as HIV. Initial vaccine studies with formalin-fixed viral particles demonstrated protection against homologous challenge, but not against strains that were antigenically divergent from the vaccine strain (Issel *et al.*, 1992). These same studies also used purified SU gp90 protein as a vaccine, but this immunogen actually exacerbated disease in some vaccinated animals upon challenge with a virulent strain of virus. A baculovirus-produced SU vaccine also exacerbated disease following challenge, and in addition this recombinant protein did not generate anamnestic antibody responses in EIAV-challenged horses suggesting that the baculovirus-produced protein differed antigenically from the native SU gp90 (Grund *et al.*, 1994; Wang *et al.*, 1994). Vaccination with EIAV Gag and Pol peptides known to contain CD4 and/or CD8-specific epitopes also did not provide protection upon virus challenge (Fraser *et al.*, 2005). Immunization with codon optimized *env* DNA enhanced production of SU, but did not result in protection upon challenge with a virulent strain of EIAV (Cook *et al.*, 2005). These studies provide evidence that recombinant vaccine approaches or killed virus do not provide sufficient or appropriate anti-EIAV antigenic stimulation to generate protective immunity against challenge with virulent EIAV. These disappointing findings are not dissimilar to those reported for recombinant vaccines against primate lentiviruses.

More recent studies have focused on the development of a live attenuated strain of EIAV as a vaccine. Vaccination with virus produced from an infectious clone that contained an inactivated *s2* gene elicited significant anti-EIAV antibodies and lymphoproliferative response that protected horses against challenge with a related, virulent virus (Li *et al.*, 2003). Subsequent studies indicated that this vaccine resulted in sterilizing immunity in 50% of the vaccinated horses (Craigo *et al.*, 2007a). Most recently, studies demonstrate that protection by the vaccine was reduced as SU gp90 sequences present in the challenge strain diverged from the vaccine strain (Craigo *et al.*, 2007b; Tagmyer *et al.*, 2008). While the use of a live attenuated, multivalent vaccine may prove to be an effective domestic animal vaccine, a similar approach for the development of a HIV vaccine will not be possible as integration of viral DNA may lead to insertional mutagenesis events.

## Concluding comments

While EIA was first described in 1843 in France, shown to be contagious in 1859 and the virus was one of the first filterable agents characterized (Dreguss, 1954), many molecular and clinical aspects of EIAV remain unaddressed today. Much of the molecular biology of the virus has been gleamed from studying other lentiviruses. Yet as this chapter highlights, EIAV and the disease it causes are unique. Furthermore, it is critical to understand the role of macrophages in lentiviral pathogenesis and, as a macrophage-tropic virus, EIAV is able to provide information to help unravel details of lentiviral/macrophage interactions. Hopefully, future studies will provide insights into the many unanswered questions about this interesting and important virus.

### References


Albrecht, T.R., Lund, L.H., and Garcia-Blanco, M.A. (2000). Canine cyclin T1 rescues equine infectious anemia virus tat trans-activation in human cells. Virology *268*, 7–11.

Alexandersen, S., and Carpenter, S. (1991). Characterization of variable regions in the envelope and S3 open reading frame of equine infectious anemia virus. J. Virol. *65*, 4255–4262.

Anand, K., Schulte, A., Fujinaga, K., Scheffzek, K., and Geyer, M. (2007). Cyclin box structure of the P-TEFb subunit cyclin T1 derived from a fusion complex with EIAV tat. J. Mol. Biol. *370*, 826–836.

Anand, K., Schulte, A., Vogel-Bachmayr, K., Scheffzek, K., and Geyer, M. (2008). Structural insights into the cyclin T1-Tat-TAR RNA transcription activation complex from EIAV. Nat. Struct. Mol. Biol. *15*, 1287–1292.

Anderson, J.L., and Hope, T.J. (2005). Intracellular trafficking of retroviral vectors: obstacles and advances. Gene Ther. *12*, 1667–1678.

Anderson, J.L., Campbell, E.M., Wu, X., Vandegraaff, N., Engelman, A., and Hope, T.J. (2006). Proteasome inhibition reveals that a functional preintegration complex intermediate can be generated during restriction by diverse TRIM5 proteins. J. Virol. *80*, 9754–9760.

Arhel, N.J., Souquere-Besse, S., Munier, S., Souque, P., Guadagnini, S., Rutherford, S., Prevost, M.C., Allen, T.D., and Charneau, P. (2007). HIV-1 DNA Flap formation promotes uncoating of the pre-integration complex at the nuclear pore. Embo J. *26*, 3025–3037.

Baccam, P., Thompson, R.J., Li, Y., Sparks, W.O., Belshan, M., Dorman, K.S., Wannemuehler, Y., Oaks, J.L., Cornette, J.L., and Carpenter, S. (2003). Subpopulations of equine infectious anemia virus Rev coexist *in vivo* and differ in phenotype. J. Virol. *77*, 12122–12131.

Bakhanashvili, M., and Hizi, A. (1993). Fidelity of DNA synthesis exhibited *in vitro* by the reverse transcriptase of the lentivirus equine infectious anemia virus. Biochemistry *32*, 7559–7567.

Ball, J.M., Swaggerty, C.L., Pei, X., Lim, W.S., Xu, X., Cox, V.C., and Payne, S.L. (2005). SU proteins from virulent and avirulent EIAV demonstrate distinct biological properties. Virology *333*, 132–144.

Beisel, C.E., Edwards, J.F., Dunn, L.L., and Rice, N.R. (1993). Analysis of multiple mRNAs from pathogenic equine infectious anemia virus (EIAV) in an acutely infected horse reveals a novel protein, Ttm, derived from the carboxy terminus of the EIAV transmembrane protein. J. Virol. *67*, 832–842.

Belshan, M., Harris, M.E., Shoemaker, A.E., Hope, T.J., and Carpenter, S. (1998). Biological characterization of Rev variation in equine infectious anemia virus. J. Virol. *72*, 4421–4426.

Belshan, M., Park, G.S., Bilodeau, P., Stoltzfus, C.M., and Carpenter, S. (2000). Binding of equine infectious anemia virus rev to an exon splicing enhancer mediates alternative splicing and nuclear export of viral mRNAs. Mol. Cell Biol. *20*, 3550–3557.

Belshan, M., Baccam, P., Oaks, J.L., Sponseller, B.A., Murphy, S.C., Cornette, J., and Carpenter, S. (2001). Genetic and biological variation in equine infectious anemia virus Rev correlates with variable stages of clinical disease in an experimentally infected pony. Virology *279*, 185–200.

Benton, C.V., Brown, B.L., Harshman, J.S., and Gilden, R.V. (1981). *In vitro* host range of equine infectious anemia virus. Intervirology *16*, 225–232.

Bieniasz, P.D., Grdina, T.A., Bogerd, H.P., and Cullen, B.R. (1999). Highly divergent lentiviral Tat proteins activate viral gene expression by a common mechanism. Mol. Cell Biol. *19*, 4592–4599.

Bishop, K.N., Verma, M., Kim, E.Y., Wolinsky, S.M., and Malim, M.H. (2008). APOBEC3G inhibits elongation of HIV-1 reverse transcripts. PLoS Pathog. *4*, e1000231.

Bogerd, H.P., Tallmadge, R.L., Oaks, J.L., Carpenter, S., and Cullen, B.R. (2008). Equine infectious anemia virus resists the antiretroviral activity of equine APOBEC3 proteins through a packaging-independent mechanism. J. Virol. *82*, 11889–11901.

Braaten, D., Franke, E.K., and Luban, J. (1996). Cyclophilin A is required for an early step in the life cycle of human immunodeficiency virus type 1 before the initiation of reverse transcription. J. Virol. *70*, 3551–3560.

Brass, A.L., Dykxhoorn, D.M., Benita, Y., Yan, N., Engelman, A., Xavier, R.J., Lieberman, J., and Elledge, S.J. (2008). Identification of host proteins required for HIV infection through a functional genomic screen. Science *319*, 921–926.

Brindley, M.A., and Maury, W. (2005). Endocytosis and a low-pH step are required for productive entry of equine infectious anemia virus. J. Virol. *79*, 14482–14488.

Brindley, M.A., and Maury, W. (2008). Equine infectious anemia virus entry occurs through clathrin-mediated endocytosis. J. Virol. *82*, 1628–1637.

Brindley, M.A., Zhang, B., Montelaro, R.C., and Maury, W. (2008). An equine infectious anemia virus variant superinfects cells through novel receptor interactions. J. Virol. *82*, 9425–9432.

Busschots, K., Vercammen, J., Emiliani, S., Benarous, R., Engelborghs, Y., Christ, F., and Debyser, Z. (2005). The interaction of LEDGF/p75 with integrase is lentivirus-specific and promotes DNA binding. J. Biol. Chem. *280*, 17841–17847.

Campbell, E.M., Perez, O., Anderson, J.L., and Hope, T.J. (2008). Visualization of a proteasome-independent intermediate during restriction of HIV-1 by rhesus TRIM5alpha. J. Cell Biol. *180*, 549–561.

Carpenter, S., Evans, L.H., Sevoian, M., and Chesebro, B. (1987). Role of the host immune response in selection of equine infectious anemia virus variants. J. Virol. *61*, 3783–3789.

Carpenter, S., and Chesebro, B. (1989). Change in host cell tropism associated with *in vitro* replication of equine infectious anemia virus. J. Virol. *63*, 2492–2496.

Carpenter, S., Alexandersen, S., Long, M.J., Perryman, S., and Chesebro, B. (1991). Identification of a hypervariable region in the long terminal repeat of equine infectious anemia virus. J. Virol. *65*, 1605–1610.

Carroll, R., Martarano, L., and Derse, D. (1991). Identification of lentivirus tat functional domains through generation of equine infectious anemia virus/human immunodeficiency virus type 1 tat gene chimeras. J. Virol. *65*, 3460–3467.

Carroll, R., and Derse, D. (1993). Translation of equine infectious anemia virus bicistronic tat-rev mRNA requires leaky ribosome scanning of the tat CTG initiation codon. J. Virol. *67*, 1433–1440.

Carvalho, M., Kirkland, M., and Derse, D. (1993). Protein interactions with DNA elements in variant equine infectious anemia virus enhancers and their impact on transcriptional activity. J. Virol. *67*, 6586–6595.

Carvalho, M., and Derse, D. (1993a). Physical and functional characterization of transcriptional control elements in the equine infectious anemia virus promoter. J. Virol. *67*, 2064–2074.

Carvalho, M., and Derse, D. (1993b). The PU.1/Spi-1 proto-oncogene is a transcriptional regulator of a lentivirus promoter. J. Virol. *67*, 3885–3890.

Cen, S., Guo, F., Niu, M., Saadatmand, J., Deflassieux, J., and Kleiman, L. (2004). The interaction between HIV-1 Gag and APOBEC3G. J. Biol. Chem. *279*, 33177–33184.

Chandran, K., Sullivan, N.J., Felbor, U., Whelan, S.P., and Cunningham, J.M. (2005). Endosomal proteolysis of the Ebola virus glycoprotein is necessary for infection. Science *308*, 1643–1645.

Chen, C., Li, F., and Montelaro, R.C. (2001). Functional roles of equine infectious anemia virus Gag p9 in viral budding and infection. J. Virol. *75*, 9762–9770.

Chen, C., Vincent, O., Jin, J., Weisz, O.A., and Montelaro, R.C. (2005). Functions of early (AP-2) and late (AIP1/ALIX) endocytic proteins in equine infectious anemia virus budding. J. Biol. Chem. *280*, 40474–40480.

Cheng, B., and Price, D.H. (2007). Properties of RNA polymerase II elongation complexes before and after the P-TEFb-mediated transition into productive elongation. J. Biol. Chem. *282*, 21901–21912.

Cherepanov, P., Maertens, G., Proost, P., Devreese, B., Van Beeumen, J., Engelborghs, Y., De Clercq, E., and Debyser, Z. (2003). HIV-1 integrase forms stable tetramers and associates with LEDGF/p75 protein in human cells. J. Biol. Chem. *278*, 372–381.

Christ, F., Thys, W., De Rijck, J., Gijsbers, R., Albanese, A., Arosio, D., Emiliani, S., Rain, J.C., Benarous, R., Cereseto, A., and Debyser, Z. (2008). Transportin-SR2 imports HIV into the nucleus. Curr. Biol. *18*, 1192–1202.

Chung, H., and Derse, D. (2001). Binding sites for Rev and ASF/SF2 map to a 55-nucleotide purine-rich exonic element in equine infectious anemia virus RNA. J. Biol. Chem. *276*, 18960–18967.

Clabough, D.L., Gebhard, D., Flaherty, M.T., Whetter, L.E., Perry, S.T., Coggins, L., and Fuller, F.J. (1991). Immune-mediated thrombocytopenia in horses infected with equine infectious anemia virus. J. Virol. *65*, 6242–6251.

Coggins, L. (1984). Carriers of equine infectious anemia virus. J. Am. Vet. Med. Assoc. *184*, 279–281.

Connor, R.I., Chen, B.K., Choe, S., and Landau, N.R. (1995). Vpr is required for efficient replication of human immunodeficiency virus type-1 in mononuclear phagocytes. Virology *206*, 935–944.

Cook, R.F., Leroux, C., Cook, S.J., Berger, S.L., Lichtenstein, D.L., Ghabrial, N.N., Montelaro, R.C., and Issel, C.J. (1998). Development and characterization of an *in vivo* pathogenic molecular clone of equine infectious anemia virus. J. Virol. *72*, 1383–1393.

Cook, R.F., Cook, S.J., Bolin, P.S., Howe, L.J., Zhou, W., Montelaro, R.C., and Issel, C.J. (2005). Genetic immunization with codon-optimized equine infectious anemia virus (EIAV) surface unit (SU) envelope protein gene sequences stimulates immune responses in ponies. Vet. Microbiol. *108*, 23–37.

Craigo, J.K., Li, F., Steckbeck, J.D., Durkin, S., Howe, L., Cook, S.J., Issel, C., and Montelaro, R.C. (2005). Discerning an effective balance between equine infectious anemia virus attenuation and vaccine efficacy. J. Virol. *79*, 2666–2677.

Craigo, J.K., Durkin, S., Sturgeon, T.J., Tagmyer, T., Cook, S.J., Issel, C.J., and Montelaro, R.C. (2007a). Immune suppression of challenged vaccinates as a rigorous assessment of sterile protection by lentiviral vaccines. Vaccine *25*, 834–845.

Craigo, J.K., Zhang, B., Barnes, S., Tagmyer, T.L., Cook, S.J., Issel, C.J., and Montelaro, R.C. (2007b). Envelope variation as a primary determinant of lentiviral vaccine efficacy. Proc. Natl. Acad. Sci. U.S.A. *104*, 15105–15110.

Crawford, T.B., Wardrop, K.J., Tornquist, S.J., Reilich, E., Meyers, K.M., and McGuire, T.C. (1996). A primary production deficit in the thrombocytopenia of equine infectious anemia. J. Virol. *70*, 7842–7850.

Cullen, B.R. (2003). Nuclear mRNA export: insights from virology. Trends Biochem. Sci *28*, 419–424.

Dahl, C.P., Gullestad, L., Fevang, B., Holm, A.M., Landro, L., Vinge, L.E., Fiane, A.E., Sandberg, W.J., Otterdal, K., Froland, S.S., *et al.* (2008). Increased expression of LIGHT/TNFSF14 and its receptors in experimental and clinical heart failure. Eur. J. Heart Fail *10*, 352–359.

DeKoter, R P., and Singh, H. (2000). Regulation of B-lymphocyte and macrophage development by graded expression of PU.1. Science *288*, 1439–1441.

Delos, S.E., Godby, J.A., and White, J.M. (2005). Receptor-induced conformational changes in the SU subunit of the avian sarcoma/leukosis virus A envelope protein: implications for fusion activation. J. Virol. *79*, 3488–3499.

Deneka, M., Pelchen-Matthews, A., Byland, R., Ruiz-Mateos, E., and Marsh, M. (2007). In macrophages, HIV-1 assembles into an intracellular plasma membrane domain containing the tetraspanins CD*81*, CD9, and CD53. J. Cell Biol. *177*, 329–341.

Derse, D., Dorn, P.L., Levy, L., Stephens, R.M., Rice, N.R., and Casey, J.W. (1987). Characterization of equine infectious anemia virus long terminal repeat. J. Virol. *61*, 743–747.

Derse, D., Dorn, P., DaSilva, L., and Martarano, L. (1990). Structure and expression of the equine infectious anemia virus transcriptional trans-activator (tat). Dev. Biol. Stand. *72*, 39–48.

Diamond, T.L., Roshal, M., Jamburuthugoda, V.K., Reynolds, H.M., Merriam, A.R., Lee, K.Y., Balakrishnan, M., Bambara, R.A., Planelles, V., Dewhurst, S., and Kim, B. (2004). Macrophage tropism of HIV-1 depends on efficient cellular dNTP utilization by reverse transcriptase. J. Biol. Chem. *279*, 51545–51553.

Dorn, P.L., and Derse, D. (1988). cis- and trans-acting regulation of gene expression of equine infectious anemia virus. J. Virol. *62*, 3522–3526.

Dorn, P., DaSilva, L., Martarano, L., and Derse, D. (1990). Equine infectious anemia virus tat: insights into the structure, function, and evolution of lentivirus trans-activator proteins. J. Virol. *64*, 1616–1624.

Dreguss, M.N. (1954). Experimental Studies in Equine Infectious Anemia (Philadelphia: University of Pennsylvania Press).

Ebina, H., Aoki, J., Hatta, S., Yoshida, T., and Koyanagi, Y. (2004). Role of Nup98 in nuclear entry of human immunodeficiency virus type 1 cDNA. Microbes Infect. *6*, 715–724.

Fagerness, A.J., Flaherty, M.T., Perry, S.T., Jia, B., Payne, S.L., and Fuller, F.J. (2006). The S2 accessory gene of equine infectious anemia virus is essential for expression of disease in ponies. Virology *349*, 22–30.

Fassati, A., Gorlich, D., Harrison, I., Zaytseva, L., and Mingot, J.M. (2003). Nuclear import of HIV-1 intracellular reverse transcription complexes is mediated by importin 7. Embo J. *22*, 3675–3685.

Fidalgo-Carvalho, I., Craigo, J.K., Barnes, S., Costa-Ramos, C., and Montelaro, R.C. (2008). Characterization of an equine macrophage cell line: Application to studies of EIAV infection. Vet. Microbiol. *136*, 8–19.

Fisher, R.C., and Scott, E.W. (1998). Role of PU.1 in hematopoiesis. Stem Cells *16*, 25–37.

Foil, L.D., Adams, W.V., McManus, J.M., and Issel, C.J. (1987). Bloodmeal residues on mouthparts of Tabanus fuscicostatus (Diptera: Tabanidae) and the potential for mechanical transmission of pathogens. J. Med. Entomol. *24*, 613–616.

Fradette, J., Wolfe, D., Goins, W.F., Huang, S., Flanigan, R M., and Glorioso, J.C. (2005). HSV vector-mediated transduction and GDNF secretion from adipose cells. Gene Ther. *12*, 48–58.

Fraser, D.G., Leib, S.R., Zhang, B.S., Mealey, R.H., Brown, W.C., and McGuire, T.C. (2005). Lymphocyte proliferation responses induced to broadly reactive Th peptides did not protect against equine infectious anemia virus challenge. Clin. Diagn. Lab. Immunol. *12*, 983–993.

Fridell, R.A., Partin, K.M., Carpenter, S., and Cullen, B.R. (1993). Identification of the activation domain of equine infectious anemia virus rev. J. Virol. *67*, 7317–7323.

Fujinaga, K., Irwin, D., Huang, Y., Taube, R., Kurosu, T., and Peterlin, B.M. (2004). Dynamics of human immunodeficiency virus transcription: P-TEFb phosphorylates RD and dissociates negative effectors from the transactivation response element. Mol. Cell Biol. *24*, 787–795.

Gallay, P., Hope, T., Chin, D., and Trono, D. (1997). HIV-1 infection of nondividing cells through the recognition of integrase by the importin/karyopherin pathway. Proc. Natl. Acad. Sci. U.S.A. *94*, 9825–9830.

Gendelman, H.E., Orenstein, J.M., Martin, M.A., Ferrua, C., Mitra, R., Phipps, T., Wahl, L.A., Lane, H.C., Fauci, A.S., Burke, D.S., and *et al.* (1988). Efficient isolation and propagation of human immunodeficiency virus on recombinant colony-stimulating factor 1-treated monocytes. J. Exp. Med. *167*, 1428–1441.

Golemis, E., Li, Y., Fredrickson, T.N., Hartley, J.W., and Hopkins, N. (1989). Distinct segments within the enhancer region collaborate to specify the type of leukemia induced by nondefective Friend and Moloney viruses. J. Virol. *63*, 328–337.

Gontarek, R.R., and Derse, D. (1996). Interactions among SR proteins, an exonic splicing enhancer, and a lentivirus Rev protein regulate alternative splicing. Mol. Cell Biol. *16*, 2325–2331.

Gooch, B.D., and Cullen, B.R. (2008). Functional domain organization of human APOBEC3G. Virology *379*, 118–124.

Grigorov, B., Arcanger, F., Roingeard, P., Darlix, J.L., and Muriaux, D. (2006). Assembly of infectious HIV-1 in human epithelial and T-lymphoblastic cell lines. J. Mol. Biol. *359*, 848–862.

Grund, C.H., Lechman, E.R., Issel, C.J., Montelaro, R.C., and Rushlow, K.E. (1994). Lentivirus cross-reactive determinants present in the capsid protein of equine infectious anaemia virus. J. Gen. Virol. *75*, 657–662.

Gutekunst, D.E., and Becvar, C.S. (1979). Responses in horses infected with equine infectious anemia virus adapted to tissue culture. Am. J. Vet. Res. *40*, 974–977.

Hacker, C.V., Vink, C.A., Wardell, T.W., Lee, S., Treasure, P., Kingsman, S.M., Mitrophanous, K.A., and Miskin, J.E. (2006). The integration profile of EIAV-based vectors. Mol. Ther. *14*, 536–545.

Hammond, S.A., Cook, S.J., Lichtenstein, D.L., Issel, C.J., and Montelaro, R.C. (1997). Maturation of the cellular and humoral immune responses to persistent infection in horses by equine infectious anemia virus is a complex and lengthy process. J. Virol. *71*, 3840–3852.

Harris, M.E., Gontarek, R.R., Derse, D., and Hope, T.J. (1998). Differential requirements for alternative splicing and nuclear export functions of equine infectious anemia virus Rev protein. Mol. Cell Biol. *18*, 3889–3899.

Harris, R.S., Bishop, K.N., Sheehy, A.M., Craig, H.M., Petersen-Mahrt, S.K., Watt, I.N., Neuberger, M.S., and Malim, M.H. (2003). DNA deamination mediates innate immunity to retroviral infection. Cell *113*, 803–809.

Harrold, S.M., Cook, S.J., Cook, R.F., Rushlow, K.E., Issel, C.J., and Montelaro, R.C. (2000). Tissue sites of persistent infection and active replication of equine infectious anemia virus during acute disease and asymptomatic infection in experimentally infected equids. J. Virol. *74*, 3112–3121.

Harrop, J.A., Reddy, M., Dede, K., Brigham-Burke, M., Lyn, S., Tan, K.B., Silverman, C., Eichman, C., DiPrinzio, R., Spampanato, J., *et al.* (1998). Antibodies to TR2 (herpesvirus entry mediator), a new member of the TNF receptor superfamily, block T-cell proliferation, expression of activation markers, and production of cytokines. J. Immunol. *161*, 1786–1794.

Hatziioannou, T., Cowan, S., Goff, S.P., Bieniasz, P.D., and Towers, G.J. (2003). Restriction of multiple divergent retroviruses by Lv1 and Ref1. Embo J. *22*, 385–394.

Hatziioannou, T., Cowan, S., Von Schwedler, U.K., Sundquist, W.I., and Bieniasz, P.D. (2004). Species-specific tropism determinants in the human immunodeficiency virus type 1 capsid. J. Virol. *78*, 6005–6012.

Henson, J.B., and McGuire, T.C. (1971). Immunopathology of equine infectious anemia. Am J. Clin. Pathol. *56*, 306–313.

Henson, J.B., and McGuire, T.C. (1974). Equine infectious anemia. Prog Med Virol *18*, 143–159.

Hines, R., and Maury, W. (2001). DH82 cells: a macrophage cell line for the replication and study of equine infectious anemia virus. J. Virol. Methods *95*, 47–56.

Hines, R., Sorensen, B.R., Shea, M.A., and Maury, W. (2004). PU.1 binding to ets motifs within the equine infectious anemia virus long terminal repeat (LTR) enhancer: regulation of LTR activity and virus replication in macrophages. J. Virol. *78*, 3407–3418.

Hokari, S., Horikawa, S., Tsukada, K., and Sakagishi, Y. (1995). Expression of deoxyuridine triphosphatase during liver regeneration in rat. Biochem. Mol. Biol. Int. *37*, 583–590.

Howe, L., Leroux, C., Issel, C.J., and Montelaro, R.C. (2002). Equine infectious anemia virus envelope evolution *in vivo* during persistent infection progressively increases resistance to *in vitro* serum antibody neutralization as a dominant phenotype. J. Virol. *76*, 10588–10597.

Hung, S.L., Cheng, Y.Y., Wang, Y.H., Chang, K.W., and Chen, Y.T. (2002). Expression and roles of herpesvirus entry mediators A and C in cells of oral origin. Oral Microbiol Immunol *17*, 215–223.

Issel, C.J., and Coggins, L. (1979). Equine infectious anemia: current knowledge. J. Am. Vet. Med. Assoc. *174*, 727–733.

Issel, C.J., Horohov, D.W., Lea, D.F., Adams, W.V., Jr, Hagius, S.D., McManus, J.M., Allison, A.C., and Montelaro, R.C. (1992). Efficacy of inactivated whole-virus and subunit vaccines in preventing infection and disease caused by equine infectious anemia virus. J. Virol. *66*, 3398–3408.

Jin, S., Chen, C., and Montelaro, R.C. (2005a). Equine infectious anemia virus Gag p9 function in early steps of virus infection and provirus production. J. Virol. *79*, 8793–8801.

Jin, S., Zhang, B., Weisz, O.A., and Montelaro, R.C. (2005b). Receptor-mediated entry by equine infectious anemia virus utilizes a pH-dependent endocytic pathway. J. Virol. *79*, 14489–14497.

Joshi, A., Munshi, U., Ablan, S.D., Nagashima, K., and Freed, E.O. (2008). Functional replacement of a retroviral late domain by ubiquitin fusion. Traffic *9*, 1972–1983.

Jouvenet, N., Bieniasz, P.D., and Simon, S.M. (2008). Imaging the biogenesis of individual HIV-1 virions in live cells. Nature *454*, 236–240.

Kim, C.H., and Casey, J.W. (1994). *In vivo* replicative status and envelope heterogeneity of equine infectious anemia virus in an inapparent carrier. J. Virol. *68*, 2777–2780.

Kobayashi, K., and Kono, Y. (1967a). Propagation and titration of equine infectious anemia virus in horse leukocyte culture. Natl. Inst. Anim Health Q. (Tokyo) 7, 8–20.

Kobayashi, K., and Kono, Y. (1967b). Serial passages of equine infectious anemia virus in horse leukocyte culture. Natl. Inst. Anim Health Q. (Tokyo) 7, 1–7.

Konig, R., Zhou, Y., Elleder, D., Diamond, T.L., Bonamy, G.M., Irelan, J.T., Chiang, C.Y., Tu, B.P., De Jesus, P.D., Lilley, C.E., *et al.* (2008). Global analysis of host–pathogen interactions that regulate early-stage HIV-1 replication. Cell *135*, 49–60.

Kono, Y., and Yokomizo, Y. (1968). Attempts to cultivate the equine infectious anemia virus in various types of cells. Natl Inst Anim Health Q (Tokyo) 8, 182–186.

Kono, Y., and Kobayashi, K. (1970). Changes in pathogenicity of equine infectious anemia virus during passages in horse leukocyte cultures. Natl. Inst. Anim. Health Q. (Tokyo) *10*, 106–112.

Kono, Y., Kobayashi, K., and Fukunaga, Y. (1970). Immunization of horses against equine infectious anemia (EIA) with an attenuated EIA virus. Natl. Inst. Anim. Health Q. (Tokyo) *10*, 113–122.

Kono, Y. (1973). Equine infectious anemia--recent researches and prospect of the study (author's transl.). Uirusu *23*, 1–12.

Kono, Y., Fukunaga, Y., and Kobayashi, K. (1973a). Excretion of equine infectious anemia virus from horses infected with the virus. Natl. Inst. Anim Health Q. (Tokyo) *13*, 182–186.

Kono, Y., Kobayashi, K., and Fukunaga, Y. (1973b). Antigenic drift of equine infectious anemia virus in chronically infected horses. Arch. Gesamte Virusforsch *41*, 1–10.

Kono, Y., Hirasawa, K., Fukunaga, Y., and Taniguchi, T. (1976). Recrudescence of equine infectious anemia by treatment with immunosuppressive drugs. Natl. Inst .Anim Health Q (Tokyo) *16*, 8–15.

Kumar, P., Nachagari, D., Fields, C., Franks, J., and Albritton, L.M. (2007). Host cell cathepsins potentiate Moloney murine leukemia virus infection. J. Virol. *81*, 10506–10514.

Langemeier, J.L., Cook, S.J., Cook, R.F., Rushlow, K.E., Montelaro, R.C., and Issel, C.J. (1996). Detection of equine infectious anemia viral RNA in plasma samples from recently infected and long-term inapparent carrier animals by PCR. J. Clin. Microbiol. *34*, 1481–1487.

Lee, J.H., Murphy, S.C., Belshan, M., Sparks, W.O., Wannemuehler, Y., Liu, S., Hope, T.J., Dobbs, D., and Carpenter, S. (2006). Characterization of functional domains of equine infectious anemia virus Rev suggests a bipartite RNA-binding domain. J. Virol. *80*, 3844–3852.

Lee, J.H., Culver, G., Carpenter, S., and Dobbs, D. (2008). Analysis of the EIAV Rev-responsive element (RRE) reveals a conserved RNA motif required for high affinity Rev binding in both HIV-1 and EIAV. PLoS ONE *3*, e2272.

Leonard, J., Parrott, C., Buckler-White, A.J., Turner, W., Ross, E.K., Martin, M.A., and Rabson, A.B. (1989). The NF-kappa B binding sites in the human immunodeficiency virus type 1 long terminal repeat are not required for virus infectivity. J. Virol. *63*, 4919–4924.

Leroux, C., Craigo, J.K., Issel, C.J., and Montelaro, R.C. (2001). Equine infectious anemia virus genomic evolution in progressor and nonprogressor ponies. J. Virol. *75*, 4570–4583.

Li, F., Puffer, B.A., and Montelaro, R.C. (1998). The S2 gene of equine infectious anemia virus is dispensable for viral replication *in vitro*. J. Virol. *72*, 8344–8348.

Li, F., Leroux, C., Craigo, J.K., Cook, S.J., Issel, C.J., and Montelaro, R.C. (2000). The S2 gene of equine infectious anemia virus is a highly conserved determinant of viral replication and virulence properties in experimentally infected ponies. J. Virol. *74*, 573–579.

Li, F., Chen, C., Puffer, B.A., and Montelaro, R.C. (2002). Functional replacement and positional dependence of homologous and heterologous L domains in equine infectious anemia virus replication. J. Virol. *76*, 1569–1577.

Li, F., Craigo, J.K., Howe, L., Steckbeck, J.D., Cook, S., Issel, C., and Montelaro, R.C. (2003). A live attenuated equine infectious anemia virus proviral vaccine with a modified S2 gene provides protection from detectable infection by intravenous virulent virus challenge of experimentally inoculated horses. J. Virol. *77*, 7244–7253.

Li, Y., Golemis, E., Hartley, J.W., and Hopkins, N. (1987). Disease specificity of nondefective Friend and Moloney murine leukemia viruses is controlled by a small number of nucleotides. J. Virol. *61*, 693–700.

Liang, H., He, X., Shen, R.X., Shen, T., Tong, X., Ma, Y., Xiang, W.H., Zhang, X.Y., and Shao, Y.M. (2006). Combined amino acid mutations occurring in the envelope closely correlate with pathogenicity of EIAV. Arch. Virol. *151*, 1387–1403.

Liao, H.J., Baker, C.C., Princler, G.L., and Derse, D. (2004). cis-Acting and trans-acting modulation of equine infectious anemia virus alternative RNA splicing. Virology *323*, 131–140.

Lim, W.S., Payne, S.L., Edwards, J.F., Kim, I., and Ball, J.M. (2005). Differential effects of virulent and avirulent equine infectious anemia virus on macrophage cytokine expression. Virology *332*, 295–306.

Lloberas, J., Soler, C., and Celada, A. (1999). The key role of PU.1/SPI-1 in B-cells, myeloid cells and macrophages. Immunol. Today *20*, 184–189.

Malmquist, W.A., Barnett, D., and Becvar, C.S. (1973). Production of equine infectious anemia antigen in a persistently infected cell line. Arch. Gesamte Virusforsch *42*, 361–370.

Mancuso, V.A., Hope, T.J., Zhu, L., Derse, D., Phillips, T., and Parslow, T.G. (1994). Posttranscriptional effector domains in the Rev proteins of feline immunodeficiency virus and equine infectious anemia virus. J. Virol. *68*, 1998–2001.

Mariani, R., Chen, D., Schrofelbauer, B., Navarro, F., Konig, R., Bollman, B., Munk, C., Nymark-McMahon, H., and Landau, N.R. (2003). Species-specific exclusion of APOBEC3G from HIV-1 virions by Vif. Cell *114*, 21–31.

Marshall, N.F., Peng, J., Xie, Z., and Price, D.H. (1996). Control of RNA polymerase II elongation potential by a novel carboxyl-terminal domain kinase. J. Biol. Chem. *271*, 27176–27183.

Martarano, L., Stephens, R., Rice, N., and Derse, D. (1994). Equine infectious anemia virus transregulatory protein Rev controls viral mRNA stability, accumulation, and alternative splicing. J. Virol. *68*, 3102–3111.

Matsuyama, S., Delos, S.E., and White, J.M. (2004). Sequential roles of receptor binding and low pH in forming prehairpin and hairpin conformations of a retroviral envelope glycoprotein. J. Virol. *78*, 8201–8209.

Maury, W. (1994). Monocyte maturation controls expression of equine infectious anemia virus. J. Virol. *68*, 6270–6279.

Maury, W.J., Carpenter, S., Graves, K., and Chesebro, B. (1994). Cellular and viral specificity of equine infectious anemia virus Tat transactivation. Virology *200*, 632–642.

Maury, W., Perryman, S., Oaks, J.L., Seid, B.K., Crawford, T., McGuire, T., and Carpenter, S. (1997). Localized sequence heterogeneity in the long terminal repeats of *in vivo* isolates of equine infectious anemia virus. J. Virol. *71*, 4929–4937.

Maury, W., Oaks, J.L., and Bradley, S. (1998). Equine endothelial cells support productive infection of equine infectious anemia virus. J. Virol. *72*, 9291–9297.

Maury, W., Bradley, S., Wright, B., and Hines, R. (2000). Cell specificity of the transcription-factor repertoire used by a lentivirus: motifs important for expression of equine infectious anemia virus in nonmonocytic cells. Virology *267*, 267–278.

Maury, W., Wright, P.J., and Bradley, S. (2003). Characterization of a cytolytic strain of equine infectious anemia virus. J. Virol. *77*, 2385–2399.

Maury, W., Thompson, R.J., Jones, Q., Bradley, S., Denke, T., Baccam, P., Smazik, M., and Oaks, J.L. (2005). Evolution of the Equine Infectious Anemia Virus Long Terminal Repeat during the Alteration of Cell Tropism. J. Virol. *79*, 5653–5664.

McClure, J.J., Lindsay, W.A., Taylor, W., Ochoa, R., Issel, C.J., and Coulter, S.J. (1982). Ataxia in four horses with equine infectious anemia. J. Am. Vet. Med. Assoc. *180*, 279–283.

McGuire, T.C., and Henson, J.B. (1969). The detection of intravascular haemolysis in the horse. Br Vet. J. *125*, v-vi.

McGuire, T.C., Henson, J.B., and Burger, D. (1969a). Complement (C'3)-coated red blood cells following infection with the virus of equine infectious anemia. J. Immunol. *103*, 293–299.

McGuire, T.C., Henson, J.B., and Quist, S.E. (1969b). Impaired bone marrow response in equine infectious anemia. Am. J. Vet. Res. *30*, 2099–2104.

McGuire, T.C., Henson, J.B., and Quist, S.E. (1969c). Viral-induced hemolysis in equine infectious anemia. Am. J. Vet. Res. *30*, 2091–2097.

McGuire, T.C., Crawford, T.B., and Henson, J.B. (1971). Immunofluorescent localization of equine infectious anemia virus in tissue. Am. J. Pathol. *62*, 283–294.

McGuire, T.C., O'Rourke, K.I., and Perryman, L.E. (1990). Immunopathogenesis of equine infectious anemia lentivirus disease. Dev. Biol. Stand. *72*, 31–37.

McGuire, T.C., Tumas, D.B., Byrne, K.M., Hines, M.T., Leib, S.R., Brassfield, A.L., O'Rourke, K.I., and Perryman, L.E. (1994). Major histocompatibility complex-restricted CD8+ cytotoxic T-lymphocytes from horses with equine infectious anemia virus recognize Env and Gag/PR proteins. J. Virol. *68*, 1459–1467.

McIlwraith, C.W., and Kitchen, D.N. (1978). Neurologic signs and neuropathology associated with a case of equine infectious anemia. Cornell Vet. *68*, 238–249.

Mealey, R.H., Fraser, D.G., Oaks, J.L., Cantor, G.H., and McGuire, T.C. (2001). Immune reconstitution prevents continuous equine infectious anemia virus replication in an Arabian foal with severe combined immunodeficiency: lessons for control of lentiviruses. Clin. Immunol. *101*, 237–247.

Mealey, R.H., Zhang, B., Leib, S.R., Littke, M.H., and McGuire, T.C. (2003). Epitope specificity is critical for high and moderate avidity cytotoxic T-lymphocytes associated with control of viral load and clinical disease in horses with equine infectious anemia virus. Virology *313*, 537–552.

Montelaro, R.C., Parekh, B., Orrego, A., and Issel, C.J. (1984). Antigenic variation during persistent infection by equine infectious anemia virus, a retrovirus. J. Biol. Chem. *259*, 10539–10544.

Montgomery, R.I., Warner, M.S., Lum, B.J., and Spear, P.G. (1996). Herpes simplex virus-1 entry into cells mediated by a novel member of the TNF/NGF receptor family. Cell *87*, 427–436.

Murakami, K., Sentsui, H., Shibahara, T., and Yokoyama, T. (1999). Reduction of CD4+ and CD8+ T-lymphocytes during febrile periods in horses experimentally infected with equine infectious anemia virus. Vet. Immunol. Immunopathol. *67*, 131–140.

Nabel, G., and Baltimore, D. (1987). An inducible transcription factor activates expression of human immunodeficiency virus in T-cells. Nature *326*, 711–713.

Newman, M.J., Issel, C.J., Truax, R.E., Powell, M.D., Horohov, D.W., and Montelaro, R.C. (1991). Transient suppression of equine immune responses by equine infectious anemia virus (EIAV). Virology *184*, 55–66.

Noiman, S., Yaniv, A., Tsach, T., Miki, T., Tronick, S.R., and Gazit, A. (1991). The Tat protein of equine infectious anemia virus is encoded by at least three types of transcripts. Virology *184*, 521–530.

O'Rourke, K.I., Perryman, L.E., and McGuire, T.C. (1989). Cross-neutralizing and subclass characteristics of antibody from horses with equine infectious anemia virus. Vet. Immunol. Immunopathol. *23*, 41–49.

O'Rourke, K.I., Besola, M.L., and McGuire, T.C. (1991). Proviral sequences detected by polymerase chain reaction in peripheral blood cells of horses with equine infectious anemia lentivirus. Arch. Virol. *117*, 109–119.

Oaks, J.L., McGuire, T.C., Ulibarri, C., and Crawford, T.B. (1998). Equine infectious anemia virus is found in tissue macrophages during subclinical infection. J. Virol. *72*, 7263–7269.

Oaks, J.L., Ulibarri, C., and Crawford, T.B. (1999). Endothelial cell infection *in vivo* by equine infectious anaemia virus. J. Gen. Virol. *80 (Pt 9)*, 2393–2397.

Oaks, J.L., Long, M.T., and Baszler, T.V. (2004). Leukoencephalitis associated with selective viral replication in the brain of a pony with experimental chronic equine infectious anemia virus infection. Vet. Pathol. *41*, 527–532.

Oikawa, T., Yamada, T., Kihara-Negishi, F., Yamamoto, H., Kondoh, N., Hitomi, Y., and Hashimoto, Y. (1999). The role of Ets family transcription factor PU.1 in hematopoietic cell differentiation, proliferation and apoptosis. Cell Death Differ. *6*, 599–608.

Olsen, J.C. (1998). Gene transfer vectors derived from equine infectious anemia virus. Gene Ther. *5*, 1481–1487.

Orenstein, J.M., Meltzer, M.S., Phipps, T., and Gendelman, H.E. (1988). Cytoplasmic assembly and accumulation of human immunodeficiency virus types 1 and 2 in recombinant human colony-stimulating factor-1-treated human monocytes: an ultrastructural study. J. Virol. *62*, 2578–2586.

Orrego, A., Issel, C.J., Montelaro, R.C., and Adams, W.V., Jr (1982). Virulence and *in vitro* growth of a cell-adapted strain of equine infectious anemia virus after serial passage in ponies. Am. J. Vet. Res. *43*, 1556–1560.

Payne, S.L., Rausch, J., Rushlow, K., Montelaro, R.C., Issel, C., Flaherty, M., Perry, S., Sellon, D., and Fuller, F. (1994). Characterization of infectious molecular clones of equine infectious anaemia virus. J. Gen. Virol. *75*, 425–429.

Payne, S.L., Qi, X.M., Shao, H., Dwyer, A., and Fuller, F.J. (1998). Disease induction by virus derived from molecular clones of equine infectious anemia virus. J. Virol. *72*, 483–487.

Payne, S.L., and Elder, J.H. (2001). The role of retroviral dUTPases in replication and virulence. Curr Protein Pept. Sci. *2*, 381–388.

Payne, S.L., Pei, X.F., Jia, B., Fagerness, A., and Fuller, F.J. (2004). Influence of long terminal repeat and env on the virulence phenotype of equine infectious anemia virus. J. Virol. *78*, 2478–2485.

Peluso, R., Haase, A., Stowring, L., Edwards, M., and Ventura, P. (1985). A Trojan Horse mechanism for the spread of visna virus in monocytes. Virology *147*, 231–236.

Perry, S.T., Flaherty, M.T., Kelley, M.J., Clabough, D.L., Tronick, S.R., Coggins, L., Whetter, L., Lengel, C.R., and Fuller, F. (1992). The surface envelope protein gene region of equine infectious anemia virus is not an important determinant of tropism *in vitro*. J. Virol. *66*, 4085–4097.

Perryman, L.E., O'Rourke, K.I., and McGuire, T.C. (1988). Immune responses are required to terminate viremia in equine infectious anemia lentivirus infection. J. Virol. *62*, 3073–3076.

Preston, B.D., Poiesz, B.J., and Loeb, L.A. (1988). Fidelity of HIV-1 reverse transcriptase. Science *242*, 1168–1171.

Puffer, B.A., Watkins, S.C., and Montelaro, R.C. (1998). Equine infectious anemia virus Gag polyprotein late domain specifically recruits cellular AP-2 adapter protein complexes during virion assembly. J. Virol. *72*, 10218–10221.

Reis, J.K., Craigo, J.K., Cook, S.J., Issel, C.J., and Montelaro, R.C. (2003). Characterization of EIAV LTR variability and compartmentalization in various reservoir tissues of long-term inapparent carrier ponies. Virology *311*, 169–180.

Rice, N.R., Lequarre, A.S., Casey, J.W., Lahn, S., Stephens, R.M., and Edwards, J. (1989). Viral DNA in horses infected with equine infectious anemia virus. J. Virol. *63*, 5194–5200.

Rice, N.R., Henderson, L.E., Sowder, R.C., Copeland, T.D., Oroszlan, S., and Edwards, J.F. (1990). Synthesis and processing of the transmembrane envelope protein of equine infectious anemia virus. J. Virol. *64*, 3770–3778.

Roberts, J.D., Bebenek, K., and Kunkel, T.A. (1988). The accuracy of reverse transcriptase from HIV-1. Science *242*, 1171–1173.

Rold, C.J., and Aiken, C. (2008). Proteasomal degradation of TRIM5alpha during retrovirus restriction. PLoS Pathog. *4*, e1000074.

Rosin-Arbesfeld, R., Rivlin, M., Noiman, S., Mashiah, P., Yaniv, A., Miki, T., Tronick, S.R., and Gazit, A. (1993). Structural and functional characterization of rev-like transcripts of equine infectious anemia virus. J. Virol. *67*, 5640–5646.

Russell, K.E., Perkins, P.C., Hoffman, M.R., Miller, R.T., Walker, K.M., Fuller, F.J., and Sellon, D.C. (1999). Platelets from thrombocytopenic ponies acutely infected with equine infectious anemia virus are activated *in vivo* and hypofunctional. Virology *259*, 7–19.

Saenz, D.T., Teo, W., Olsen, J.C., and Poeschla, E.M. (2005). Restriction of feline immunodeficiency virus by Ref1, Lv1, and primate TRIM5alpha proteins. J. Virol. *79*, 15175–15188.

Saman, E., Breugelmans, K., Heyndricks, L., and Merregaert, J. (1990). The open reading frame ORF S3 of equine infectious anemia virus is expressed during the viral life cycle. J. Virol. *64*, 6319–6324.

Schiltz, R.L., Shih, D.S., Rasty, S., Montelaro, R.C., and Rushlow, K.E. (1992). Equine infectious anemia virus gene expression: characterization of the RNA splicing pattern and the protein products encoded by open reading frames S1 and S2. J. Virol. *66*, 3455–3465.

Schornberg, K., Matsuyama, S., Kabsch, K., Delos, S., Bouton, A., and White, J. (2006). Role of endosomal cathepsins in entry mediated by the Ebola virus glycoprotein. J. Virol. *80*, 4174–4178.

Sellon, D.C., Perry, S.T., Coggins, L., and Fuller, F.J. (1992). Wild-type equine infectious anemia virus replicates *in vivo* predominantly in tissue macrophages, not in peripheral blood monocytes. J. Virol. *66*, 5906–5913.

Sellon, D.C., Fuller, F.J., and McGuire, T.C. (1994). The immunopathogenesis of equine infectious anemia virus. Virus Res. *32*, 111–138.

Sellon, D.C., Walker, K.M., Russell, K.E., Perry, S.T., Covington, P., and Fuller, F.J. (1996a). Equine infectious anemia virus replication is up-regulated during differentiation of blood monocytes from acutely infected horses. J. Virol. *70*, 590–594.

Sellon, D.C., Walker, K.M., Russell, K.E., Perry, S.T., and Fuller, F.J. (1996b). Phorbol ester stimulation of equine macrophage cultures alters expression of equine infectious anemia virus. Vet. Microbiol. *52*, 209–221.

Sentsui, H., and Kono, Y. (1988). Studies on viral-induced anemia in horses infected with equine infectious anemia virus. Nippon Juigaku Zasshi *50*, 303–311.

Sheehy, A.M., Gaddis, N.C., Choi, J.D., and Malim, M.H. (2002). Isolation of a human gene that inhibits HIV-1 infection and is suppressed by the viral Vif protein. Nature *418*, 646–650.

Shehu-Xhilaga, M., Ablan, S., Demirov, D.G., Chen, C., Montelaro, R.C., and Freed, E.O. (2004). Late domain-dependent inhibition of equine infectious anemia virus budding. J. Virol. *78*, 724–732.

Shen, T., Liang, H., Tong, X., Fan, X., He, X., Ma, Y., Xiang, W., Shen, R., Zhang, X., and Shao, Y. (2006). Amino acid mutations of the infectious clone from Chinese EIAV attenuated vaccine resulted in reversion of virulence. Vaccine *24*, 738–749.

Sherman, L., Yaniv, A., Lichtman-Pleban, H., Tronick, S.R., and Gazit, A. (1989). Analysis of regulatory elements of the equine infectious anemia virus and caprine arthritis-encephalitis virus long terminal repeats. J. Virol. *63*, 4925–4931.

Short, M.K., Okenquist, S.A., and Lenz, J. (1987). Correlation of leukemogenic potential of murine retroviruses with transcriptional tissue preference of the viral long terminal repeats. J. Virol. *61*, 1067–1072.

Simmons, G., Gosalia, D.N., Rennekamp, A.J., Reeves, J.D., Diamond, S.L., and Bates, P. (2005). Inhibitors of cathepsin L prevent severe acute respiratory syndrome coronavirus entry. Proc. Natl. Acad. Sci. U.S.A. *102*, 11876–11881.

Singh, H., DeKoter, R.P., and Walsh, J.C. (1999). PU.1, a shared transcriptional regulator of lymphoid and myeloid cell fates. Cold Spring Harb Symp Quant Biol *64*, 13–20.

Smith, T.A., Davis, E., and Carpenter, S. (1998). Endotoxin treatment of equine infectious anaemia virus-infected horse macrophage cultures decreases production of infectious virus. J. Gen. Virol. *79*, 747–755.

Sparks, W.O., Dorman, K.S., Liu, S., and Carpenter, S. (2008). Naturally arising point mutations in non-essential domains of equine infectious anemia virus Rev alter Rev-dependent nuclear-export activity. J. Gen. Virol. *89*, 1043–1048.

Steagall, W.K., Robek, M.D., Perry, S.T., Fuller, F.J., and Payne, S.L. (1995). Incorporation of uracil into viral DNA correlates with reduced replication of EIAV in macrophages. Virology *210*, 302–313.

Stephens, R.M., Derse, D., and Rice, N.R. (1990). Cloning and characterization of cDNAs encoding equine infectious anemia virus tat and putative Rev proteins. J. Virol. *64*, 3716–3725.

Strack, B., Calistri, A., Craig, S., Popova, E., and Gottlinger, H.G. (2003). AIP1/ALIX is a binding partner for HIV-1, p6 and EIAV p9 functioning in virus budding. Cell *114*, 689–699.

Stremlau, M., Owens, C.M., Perron, M.J., Kiessling, M., Autissier, P., and Sodroski, J. (2004). The cytoplasmic body component TRIM5alpha restricts HIV-1 infection in Old World monkeys. Nature *427*, 848–853.

Sun, C., Zhang, B., Jin, J., and Montelaro, R.C. (2008). Binding of equine infectious anemia virus to the equine lentivirus receptor-1 is mediated by complex discontinuous sequences in the viral envelope gp90 protein. J. Gen. Virol. *89*, 2011–2019.

Swardson, C.J., Kociba, G.J., and Perryman, L.E. (1992). Effects of equine infectious anemia virus on hematopoietic progenitors *in vitro*. Am. J. Vet. Res. *53*, 1176–1179.

Tagmyer, T.L., Craigo, J.K., Cook, S.J., Even, D.L., Issel, C.J., and Montelaro, R.C. (2008). Envelope determinants of equine infectious anemia virus vaccine protection and the effects of sequence variation on immune recognition. J. Virol. *82*, 4052–4063.

Taube, R., Fujinaga, K., Irwin, D., Wimmer, J., Geyer, M., and Peterlin, B.M. (2000). Interactions between equine cyclin T1, Tat, and TAR are disrupted by a leucine-to-valine substitution found in human cyclin T1. J. Virol. *74*, 892–898.

Threadgill, D.S., Steagall, W.K., Flaherty, M.T., Fuller, F.J., Perry, S.T., Rushlow, K.E., Le Grice, S.F., and Payne, S.L. (1993). Characterization of equine infectious anemia virus dUTPase: growth properties of a dUTPase-deficient mutant. J. Virol. *67*, 2592–2600.

Tornquist, S.J., and Crawford, T.B. (1997). Suppression of megakaryocyte colony growth by plasma from foals infected with equine infectious anemia virus. Blood *90*, 2357–2363.

Tornquist, S.J., Oaks, J.L., and Crawford, T.B. (1997). Elevation of cytokines associated with the thrombocytopenia of equine infectious anaemia. J. Gen. Virol. *78 (Pt 10)*, 2541–2548.

Traut, T.W. (1994). Physiological concentrations of purines and pyrimidines. Mol. Cell Biochem *140*, 1–22.

Tumas, D.B., Hines, M.T., Perryman, L.E., Davis, W.C., and McGuire, T.C. (1994). Corticosteroid immunosuppression and monoclonal antibody-mediated CD5+ T-lymphocyte depletion in normal and equine infectious anaemia virus-carrier horses. J. Gen. Virol. *75*, 959–968.

von Schwedler, U., Kornbluth, R.S., and Trono, D. (1994). The nuclear localization signal of the matrix protein of human immunodeficiency virus type 1 allows the establishment of infection in macrophages and quiescent T-lymphocytes. Proc. Natl. Acad. Sci. U.S.A. *91*, 6992–6996.

Wada, T., Takagi, T., Yamaguchi, Y., Watanabe, D., and Handa, H. (1998). Evidence that P-TEFb alleviates the negative effect of DSIF on RNA polymerase II-dependent transcription *in vitro*. Embo J. *17*, 7395–7403.

Wang, S.Z., Rushlow, K.E., Issel, C.J., Cook, R.F., Cook, S.J., Raabe, M.L., Chong, Y.H., Costa, L., and Montelaro, R.C. (1994). Enhancement of EIAV replication and disease by immunization with a baculovirus-expressed recombinant envelope surface glycoprotein. Virology *199*, 247–251.

Wardrop, K.J., Baszler, T.V., Reilich, E., and Crawford, T.B. (1996). A morphometric study of bone marrow megakaryocytes in foals infected with equine infectious anemia virus. Vet Pathol 33, 222–227.

Welsch, S., Keppler, O.T., Habermann, A., Allespach, I., Krijnse-Locker, J., and Krausslich, H.G. (2007). HIV-1 buds predominantly at the plasma membrane of primary human macrophages. PLoS Pathog. 3, e36.

Whetter, L., Archambault, D., Perry, S., Gazit, A., Coggins, L., Yaniv, A., Clabough, D., Dahlberg, J., Fuller, F., and Tronick, S. (1990). Equine infectious anemia virus derived from a molecular clone persistently infects horses. J. Virol. *64*, 5750–5756.

Wu, X., Anderson, J.L., Campbell, E.M., Joseph, A.M., and Hope, T.J. (2006). Proteasome inhibitors uncouple rhesus TRIM5alpha restriction of HIV-1 reverse transcription and infection. Proc. Natl. Acad. Sci. U.S.A. *103*, 7465–7470.

Zaitseva, L., Cherepanov, P., Leyens, L., Wilson, S.J., Rasaiyaah, J., and Fassati, A. (2009). HIV-1 exploits importin 7 to maximize nuclear import of its DNA genome. Retrovirology *6*, 11.

Zhang, B., Jin, S., Jin, J., Li, F., and Montelaro, R.C. (2005). A tumor necrosis factor receptor family protein serves as a cellular receptor for the macrophage-tropic equine lentivirus. Proc. Natl. Acad. Sci. U.S.A. *102*, 9918–9923.

Zhang, B., Sun, C., Jin, S., Cascio, M., and Montelaro, R.C. (2008). Mapping of equine lentivirus receptor 1 residues critical for equine infectious anemia virus envelope binding. J. Virol. *82*, 1204–1213.

Zhang, B., and Montelaro, R.C. (2009). Replication of equine infectious anemia virus in engineered mouse NIH 3T3 cells. J. Virol. *83*, 2034–2037.

Zhang, W., Auyong, D.B., Oaks, J.L., and McGuire, T.C. (1999). Natural variation of equine infectious anemia virus Gag protein cytotoxic T-lymphocyte epitopes. Virology *261*, 242–252.

Zheng, Y.H., Nakaya, T., Sentsui, H., Kameoka, M., Kishi, M., Hagiwara, K., Takahashi, H., Kono, Y., and Ikuta, K. (1997a). Insertions, duplications and substitutions in restricted gp90 regions of equine infectious anaemia virus during febrile episodes in an experimentally infected horse. J. Gen. Virol. *78*, 807–820.

Zheng, Y.H., Sentsui, H., Nakaya, T., Kono, Y., and Ikuta, K. (1997b). *In vivo* dynamics of equine infectious anemia viruses emerging during febrile episodes: insertions/duplications at the principal neutralizing domain. J. Virol. *71*, 5031–5039.

Zheng, Y.H., Sentsui, H., Kono, Y., and Ikuta, K. (2000a). Mutations occurring during serial passage of Japanese equine infectious anemia virus in primary horse macrophages. Virus Res. *68*, 93–98.

Zheng, Y.H., Sentsui, H., Sugita, M., Nakaya, T., Kishi, M., Hagiwara, K., Inoshima, Y., Ishihara, C., Kono, Y., Lu, J.L., and Ikuta, K. (2000b). Replication ability *in vitro* and *in vivo* of equine infectious anemia virus avirulent Japanese strain. Virology *266*, 129–139.

# Small Ruminant Lentiviruses and Cross-species Transmission

10

Giuseppe Bertoni and Barbara Blacklaws

Abstract

The visna–maedi virus (VMV) and the caprine arthritis encephalitis virus (CAEV) were considered to be specific pathogens of sheep and goats, respectively.

The finding that these lentiviruses frequently cross the species barrier between sheep and goats, and vice versa, has changed our view of the epidemiology of these viruses and they are now referred to as small ruminant lentiviruses (SRLV). A brief review of the molecular epidemiology of these lentiviruses will illustrate the diffusion and intermixing of these viruses in the two target species and documented cases of double infection and recombination between VMV and CAEV will be discussed.

Monocytes–macrophages and dendritic cells are the main target cells of SLRV. Monocytes carrying the lentiviral provirus in their genome show little or no viral transcription. These latently infected cells are believed to function as 'Trojan horses', capable of spreading the virus to different organs, while eluding the host immune response. The terminal maturation of monocytes to macrophages activates the expression of the transcription factors that, by interacting with the control elements in the viral LTRs, promote the production of infectious virus. The LTR sequences of different SRLV isolates are quite heterogeneous and may control the tissue-specific replication of these viruses and their virulence. SRLV replicate unrestrictedly and to high titres in differentiated macrophages *in vitro*, whereas *in vivo* virus replication is tightly controlled by mechanisms involving innate immunity and the adaptive immune system, and the intrinsic resistance of cells to retrovirus replication.

SRLV manipulate the expression of different cytokines in infected cells and modulate the cytokine response of these cells to stimulation of the various receptors involved in recognizing pathogen-associated molecular patterns (PAMPS).

The genetic background of animals influences the clinical outcome of SRLV infection which, in contrast to HIV infected humans, is mainly benign in the majority of infected animals. Most animals therefore fulfil the criteria defining long-term non-progressors.

Finally, the various strategies adopted by SRLV to manipulate the aforementioned immunological and non-immunological antiviral mechanisms directly influence the efficacy of vaccination strategies as documented by the paradoxical effects induced by experimental vaccines on viral load and pathological manifestations in vaccinated and challenged animals.

## Introductory note

The aim of this review is to summarize the wealth of information on the interactions between the small ruminant lentiviruses, VMV and CAEV, and their principal target cells, the monocyte/macrophages and dendritic cells. The review is focused on the two principal target species i.e. sheep and goats, even though other ungulates such as ibex, mouflon and cattle have been shown to be more or less permissive host species (Guiguen *et al.*, 2000; Morin *et al.*, 2003; Erhouma *et al.*, 2008). Monocyte/macrophages and dendritic cells are the principal cells providing an innate immunological response to SRLV infection and are indispensable for the ensuing adaptive immune response, mediated mainly by T- and B- cells. The consequences of SRLV infection on the activity of both arms of the immune system are reviewed in separate sections. The contribution of T-cells and B-cells/antiviral antibody in defining and controlling the viral load of the infected animals will also be reviewed in this chapter. The vaccine section is the obvious corollary to these sections reviewing the more or less frustrating attempts to protect the vaccinated hosts by manipulating the immune response to SRLV with different antigen preparations and immunization routes. Several aspects of SRLV infections such as clinical consequences, detailed pathological manifestations or diagnostic procedures could not be dealt in detail in this review and readers are referred to specialized reviews for more information on these subjects (Pepin *et al.*, 1998; Peterhans *et al.*, 2004; de Andres *et al.*, 2005; Torsteinsdottir *et al.*, 2007a; Bertoni, 2007)

## Molecular epidemiology/cross-species transmission

The visna–maedi virus (VMV) is the prototypic lentivirus and was isolated and characterized by Sigurdsson and colleagues in 1960 (Sigurdsson *et al.*, 1960; Sigurdardottir and Thormar, 1964). These authors had previously described the two main pathological manifestations of VMV infection, i.e. '*maedi*', Icelandic for dyspnoea, and '*visna*', an Icelandic term meaning wasting (Sigurdsson *et al.*, 1952; Sigurdsson *et al.*, 1957). The virus was introduced into Iceland in 1933 by the importation of apparently healthy Karakul sheep from Germany. Due to the slow, progressive course of the disease the epidemic remained undetected for years. This atypical course of virus-induced disease prompted Sigurdsson to introduce the still valid concept of 'slow disease' that describes pathological changes induced by infectious agents, such as lentiviruses or prions, months or even years after the primary infection (Pepin *et al.*, 1998; Torsteinsdottir *et al.*, 2007a). The term visna–maedi virus is used in this chapter according to the official taxonomic index of the International Committee on Taxonomy of Viruses (http://www.ictvonline.org/index.asp).

Caprine arthritis and encephalitis were first recognized as distinct clinical diseases of goats with an unknown aetiology (Cork *et al.*, 1974; Cork and Narayan, 1980). The virus was isolated and characterized in 1980 and the first experimental infections demonstrated the common aetiology of the two clinical manifestations (Crawford *et al.*, 1980; Narayan and Cork, 1985).

The first complete sequences of 4 VMV and one CAEV molecular clones confirmed the apparent connection between the target species specificity of these viruses and their phylogenetic relationship (Sonigo *et al.*, 1985; Pyper *et al.*, 1986; Querat *et al.*, 1990; Saltarelli *et al.*, 1990; Sargan *et al.*, 1991; Andresson *et al.*, 1993). The published VMV sequences were more divergent than other species specific groups of lentiviruses but still clearly more closely related to each other than to the CAEV sequence. For a long time viruses isolated from goats were classified as CAEV- and those from sheep as VMV strains. Experimental infection of sheep with CAEV and goats with VMV were described but cross-species infections were considered unlikely to occur under field conditions (Oliver *et al.*, 1982; Banks *et al.*, 1983; Smith *et al.*, 1985; Dickson and Ellis, 1989). The first pieces of evidence indicating that strict species specificity does not apply to all strains of CAEV or VMV came from molecular epidemiological studies in France. Leroux and colleagues investigated the genomic heterogeneity in the *pol* region of lentiviruses isolated from sheep in southern France, and to their surprise, found that in both nucleotide and deduced amino acid sequences these viruses were more closely related to prototypic CAEV than VMV sequences (Leroux *et al.*, 1995). These results were confirmed by the analysis of newly obtained *env* sequences and additional *pol* sequences (Leroux *et al.*, 1997). Finally, the authors drew the conclusion that these viruses have little or no host specificity and proposed to substitute the terms VMV and CAEV with the more general term small ruminant lentiviruses (SRLV), which has become widely accepted. These original observations were confirmed in several studies from different countries (Castro *et al.*, 1999; Grego *et al.*, 2002; Rolland *et al.*, 2002; Shah *et al.*, 2004a,b; Pisoni *et al.*, 2005; Germain and Valas, 2006; Gjerset *et al.*, 2007; Grego *et al.*, 2007). The transmission of different VMV or CAEV genotypes between sheep and goats, and vice versa, clearly indicates that these lentiviruses frequently cross the species barrier under field conditions.

Zanoni in 1998 first proposed a phylogenetic classification of SRLV into six different clades with no clear separation between strains derived from goats and sheep (Zanoni, 1998). Based on numerous additional sequences obtained from viruses isolated from sheep and goats in Switzerland, Shah and colleagues extended this classification and proposed the existence of four principal sequence groups, A to D (Shah *et al.*, 2004a). In analogy to the phylogenetic classification of HIV groups and subtypes, sequences were assigned to the different groups according to pairwise DNA distance analysis (cut-off was set at 25% difference). Additionally, group A and B were subdivided into 7 and 2 subtypes, respectively, differing by more than 15%. Molecular epidemiological studies showed that some SRLV genotypes such as A3, A4, A6, B2 and C circulated in sheep as well as goat flocks. Other subtypes such as A1 and B1, however, appeared to be strictly species specific for sheep and goats, respectively. The prototypic B1 lentivirus CAEV Cork (CAEV-CO) was also shown to replicate very efficiently in goat but not in sheep cells *in vitro,* confirming the apparent species specificity of this subtype. This target cell specificity was shown to rely on the amino acid sequence of the envelope glycoprotein that influences the post-translational cleavage of Env (Chebloune *et al.*, 1996). The Env glycoprotein is also considered to be the principal barrier preventing the infection of human cells by CAEV (Mselli-Lakhal *et al.*, 2000). Additional field studies, however, showed that the A1 and B1 subtypes do indeed cross the species barrier under natural conditions (Pisoni *et al.*, 2005; Grego *et al.*, 2007). Recently, a novel group E was identified in different flocks of a local Italian goat breed (Grego *et al.*, 2007). A full genome analysis of one representative of this novel SRLV genotype revealed two large deletions eliminating almost the entire dUTPase subunit of the *pol* and the complete *tat* (*vpr*-like) accessory genes (Reina *et al.*, 2009). Additionally, this virus has also lost a 70-bp repeat of the U3 region in the long terminal repeat (LTR). Deletions in these regions of the viral genome were previously shown to confer an attenuated phenotype to SRLV, suggesting the existence of host-adapted, low-pathogenicity SRLV strains (Harmache *et al.*, 1996b; Turelli *et al.*, 1997; Barros *et al.*, 2004; Angelopoulou *et al.*, 2008). The absence of SRLV induced clinical signs in flocks infected exclusively with genotype E SRLV supports this conclusion.

The extensive genomic polymorphism of SRLV and their efficient circulation in and across their target species suggested that, in analogy to HIV, some individuals might acquire multiple infections, setting the scene for potential recombination events between different SRLV strains (Germain *et al.*, 2008). Indeed, animals coinfected with SRLV from the A and B genotypes were detected and from one of these goats a recombinant virus was isolated (Pisoni *et al.*, 2007a). The recombination event generated a chimeric *env* gene including fragments from both A and B genotypes. The *env* gene was previously described as a privileged target of genome recombination events in experimentally infected sheep (Andresdottir, 2003). Recombination events, however, do not appear to occur at high frequency in flocks infected with different SRLV genotypes, possibly due to the low viral load and frequency of infected cells (Grego *et al.*, 2007). Extensive molecular epidemiology studies will be required to determine the importance of recombination in the epidemiology and evolution of SRLV.

## SRLV/macrophage (DC) interactions – persistence by the 'Trojan horse' strategy

### Monocyte/macrophage infection

Monocyte macrophages and dendritic cells are the main target cells of SRLV both *in vivo* and *in vitro* (Narayan *et al.*, 1982; Gendelman *et al.*, 1985; Zink *et al.*, 1990; Gorrell *et al.*, 1992; Ryan *et al.*, 2000). Virus can enter, reverse transcribe and integrate provirus into monocytes such that these cells carry the lentiviral provirus in their genome but they show little or no viral transcription. This is called 'restricted' replication: viral RNA is expressed to low copy number and no viral protein is detectable (Haase *et al.*, 1977;

Narayan *et al.*, 1983; Geballe *et al.*, 1985). It is terminal differentiation and maturation/activation of monocytes to macrophages that allows viral replication to become fully productive (high copy numbers of viral RNA and expression of protein). This occurs in the tissues, especially those where virus induced pathology is seen e.g. lung, CNS, mammary gland or synovium. There is evidence that not all types of tissue macrophages express SRLV RNA or protein e.g. Kupffer cells of the liver (Gendelman *et al.*, 1985; Kennedy *et al.*, 1989; Brodie *et al.*, 1995; Brellou *et al.*, 2007). Other cell types *in vivo* show the presence of viral nucleic acid and/or protein: cells of importance during pathogenesis may be microglia and endothelial cells in the CNS (Stowring *et al.*, 1985; Georgsson *et al.*, 1989) and epithelial cells in the lung, gut and mammary glands may play a role in pathogenesis and transmission (Zink *et al.*, 1990; Staskus *et al.*, 1991; Brodie *et al.*, 1995; Lerondelle *et al.*, 1999; Carrozza *et al.*, 2003; Preziuso *et al.*, 2004; Angelopoulou *et al.*, 2006; Bolea *et al.*, 2006). Endothelial cells and epithelial cells in the renal tubules and thyroid follicles have also been shown to carry viral nucleic acid (Zink *et al.*, 1990). Whether these cells are truly infected i.e. contain integrated provirus and produce infectious virions is a matter of debate. Other explanations for viral positive cells could be uptake of viral nucleic acid from inocula, phagocytosis of viral antigen or transcytosis of virus from milk and colostrum. The SRLV show a wider cell tropism *in vitro* with macrophages and fibroblastic cells including sheep choroid plexus (SCP), goat synovial membrane (GSM) and skin cells being the usual cell types in which to culture virus. The cell type infected as well as virus strain may impact on the cytopathic effect seen: *in vitro*, macrophage infection is often non-cytopathic and persistent whilst the fibroblast infection is cytopathic (Narayan *et al.*, 1982; Lairmore *et al.*, 1987).

Unlike HIV, SRLV cell tropism is not dependent on which cells and species express the viral receptor, despite the fact that productive infection *in vivo* is limited to small ruminants and few cell types. The receptor for SRLV has not been identified but is known to be expressed widely on different cell types and in different species (Lyall *et al.*, 2000; Hötzel and Cheevers, 2002a). However the cellular receptor for CAEV and some VMV strains has a more restricted pattern of expression in different species and the cellular receptor for the two viruses (CAEV and some VMV strains) appears to be different when interference and inhibition assays are used (Jolly *et al.*, 1989a; Hötzel and Cheevers, 2001; Hötzel and Cheevers, 2002b). This may therefore help to cause a species restriction for some of the SRLV strains (Mselli-Lakhal *et al.*, 2000).

A major factor determining in which tissues or cells the virus will replicate comes from the activity of the proviral LTR (Small *et al.*, 1989; Agnarsdottir *et al.*, 2000). It is the LTR that controls levels of viral gene expression in monocytes versus macrophages. Terminal differentiation and maturation/activation of monocytes to macrophages in tissues leads to the expression of transcription factors used to regulate macrophage specific genes. These transcription factors can then interact with control elements in the LTR to promote production of viral RNA (Small *et al.*, 1989; Clements and Payne, 1994; Clements and Zink, 1996). Indeed replication of VMV has been shown to need activation of the MAPK (Barber *et al.*, 2002) or protein kinase C pathway (Small *et al.*, 1989). Like all retroviruses the LTR of the SRLV are direct repeat regions at the termini of the provirus which are formed during reverse transcription of the viral RNA into DNA (see Fig. 10.1). There are 3 major regions, U3, R and U5 with the U3 region containing the major promoter/enhancer regions and the transcriptional start site of the virus. The LTRs of VMV and CAEV show sequence homology especially over certain regions of U3 including the TATA box and polyadenylation sequences and in the U5 region (Hess *et al.*, 1986). It has been shown that there is variation in the LTR between strains of VMV and this can lead to strains with distinct transcriptional activities such as the slow/low or rapid/high phenotypes (Sargan *et al.*, 1995; Barros *et al.*, 2004). Increased levels of virus expression *in vivo* are linked to increased pathology in infected animals (Brodie *et al.*, 1992; Zhang *et al.*, 2000; Ravazzolo *et al.*, 2006) and the efficiency of replication of viral strains in macrophages *in vitro* correlates with their *in vivo* pathogenicity (Lairmore *et al.*, 1987; Barros *et al.*, 2005). Therefore the activity of the LTR may profoundly affect the

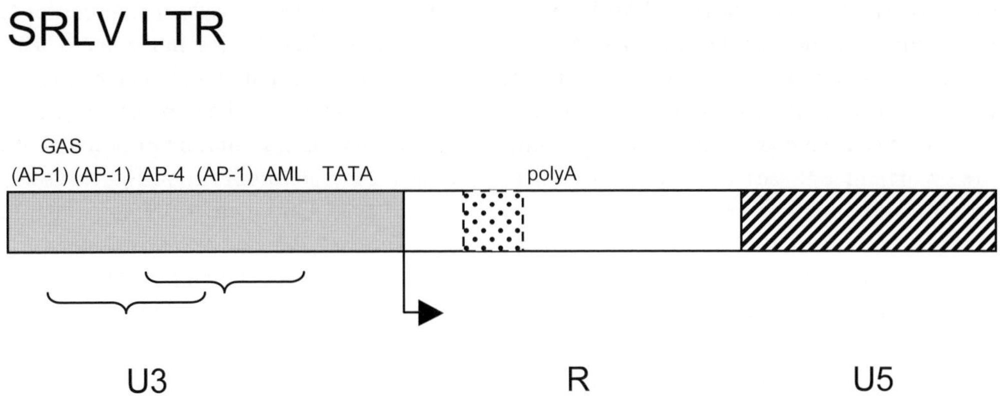


**Figure 10.1** Diagram of the SRLV LTR. The U3 region contains promoter elements including (degenerate) AP-1, AP-4, AML (*vis*)/PEBP2/CBF and in CAEV GAS sites. The TATA box, transcriptional start site and polyadenylation signal are indicated. The transcriptional start site is usually 17 bp downstream of the TATA box at the 3′ end of U3 in CAEV and 26 bp downstream in VMV at the 5′ end of the R region. Many SRLV isolates contain repeats within the U3 region which cause duplication of transcription factor binding sites. These vary between isolates as indicated by the curly brackets({}). The deletion in R defined by Angelopoulou *et al.* (2008) is indicated (⁝⁝).

disease caused after infection by SRLV. Indeed changes within the LTR can be directly linked to pathogenic phenotype and neurovirulence (Andresdottir *et al.*, 1998; Oskarsson *et al.*, 2007).

The LTR and the viral transcript produced show no evidence of a *trans*-activating response region (TAR), a stable stem loop structure formed in R region RNA, which is present in HIV and EIAV (Gdovin and Clements, 1992). This is confirmed by the fact that the SRLV LTRs have a high basal level of transcriptional activity in transient transfection assays, with LTR-marker gene fusion constructs exhibiting between 16 to 40 fold more activity in comparison to HIV LTR (Mercier *et al.*, 1994; Villet *et al.*, 2003b; Barros *et al.*, 2005). The SRLV LTR often contain repeat motifs [37, 43, or 53 bp in VMV (Hess *et al.*, 1989; Sargan *et al.*, 1995; Campbell and Avery, 1996; Agnarsdottir *et al.*, 2000; Barros *et al.*, 2004; Barros *et al.*, 2005) and 71 bp in CAEV (Hess *et al.*, 1986)], which allow repetition of transcription factor binding sites including activator protein (AP)–1, AP-4, CANNTG E-box and AML (*vis*)/PEBP2/CBF (PEA-2) sites. It is the combinations of varying numbers and orientations of these motifs in cells undergoing different stimuli that provide the variation in viral replication abilities that can be seen.

Early work on the VMV LTR used the first sequenced neurovirulent VMV strain, K1514. With this LTR, an AP-4 site is required for basal transcriptional activity but not the serum responsiveness of the LTR (Hess *et al.*, 1989). This AP-4 site is conserved between isolates and may be the most important transcription factor binding site for basal LTR activity (Gdovin and Clements, 1992; Campbell and Avery, 1996). However the TATA box (TATATAA) and U5 regions are also important for basal transcriptional activity and enhancement (Hess *et al.*, 1989; Campbell and Avery, 1996). Phorbol ester activation of macrophages transfected with K1514 LTR-marker gene fusions showed that up-regulation of transcription was dependent on the AP-1 site proximal to the TATA box in this LTR (Hess *et al.*, 1989; Shih *et al.*, 1992). It binds AP-1/jun and fos complexes and is involved in up-regulation of transcription in the presence of Ha-*ras* (Mercier *et al.*, 1994). However in other strains of VMV, the AP-1 site is mutated and so may not be as important for replication as first thought (Gabuzda *et al.*, 1989; Querat *et al.*, 1990; Sargan *et al.*, 1991).

The majority of VMV strains contain conserved AML (*vis*)/PEBP2/CBF (PEA-2) site(s) that may be more important for replication of the virus (Hess *et al.*, 1986; Gabuzda *et al.*, 1989; Gdovin and Clements, 1992; Campbell and Avery, 1996; Sutton *et al.*, 1997). The EV1 strain AML (*vis*) site binds AML-1, -2, and -3 in macrophage extracts (Sutton *et al.*, 1997). In the LTR of K1514 the proximal AP-1 site and AML (*vis*) site are very close and AP-1 and AML bind to a K1514 oligonucleotide competitively (Sutton *et al.*, 1997) suggesting that both sites could be involved in transcriptional regulation of this virus strain. Variants with different numbers of repeats of the AML (*vis*) site show different abilities to grow in SCP cells with a link between increased number of sites and increased replication (Barros *et al.*, 2005).

A consensus E-box site (CAAATG) between the AML (*vis*) and proximal AP-1 site is part of a 54 bp repeat in VMV that has been linked to neurovirulence. In strains which cause visna and grow well in SCP cells there are two copies of the repeat and an E-box site whilst strains causing maedi or which grow poorly in SCP have only one copy of the repeat and E-box site. When this consensus E-box site is engineered into a proviral clone it extends the tropism of the virus in tissue culture although the transcriptional activity of the LTR is not increased (Oskarsson *et al.*, 2007).

The R and U5 regions of the LTR are 3′ of the transcription start site but may also be involved in LTR activity. They have been implicated in transcriptional activation of an American strain of VMV (CU1) but not K1514 (Campbell and Avery, 1996; Craig *et al.*, 1997). Similarly a natural variant with a 13–14 bp deletion in the R region is also associated with decreased antigen expression and pathology in the lung but not the brain or mammary gland (Angelopoulou *et al.*, 2008). This deletion is also present in the Portuguese P1OLV isolate which has been characterized as a slow/low persistent infection *in vitro*. However this strain also has deletions in the AML binding sites (Barros *et al.*, 2004) and as this region is incorporated into RNA it may contribute to virus gene expression by allowing extension of mRNA transcripts by the polymerase or by stabilization of the RNA itself.

Although a major determinant of viral tropism, the LTR is active in cells from different species and transfection of human or mouse cell lines is often used to study LTR transcriptional activity. U937 human promonocytic cells have been used to show that transcription from SRLV LTR is activated by phorbol ester and serum stimulation of macrophages (Gabuzda *et al.*, 1989; Hess *et al.*, 1989; Tong-Starksen *et al.*, 1996). Phorbol ester treatment causes the differentiation of U937 cells into macrophages and induces expression of c-jun and c-fos, especially of large multimeric forms of the proteins in the treated cells (Shih *et al.*, 1992). Thus these results help to explain the restricted replication of SRLV in monocytes but productive replication in differentiated macrophages. In mice transgenic for a VMV LTR-CAT construct, CAT is expressed in macrophages (Small *et al.*, 1989). There are also low levels of CAT expression in bone marrow, brain and lymphocytes. To express CAT to high levels, macrophages from these mice must be activated, either using thioglycolate *in vivo* or phorbol ester *in vitro*. However CAT activity in the transgenic lymphocytes is not up-regulated by stimulation of the lymphocytes with different agents including a lymphocyte mitogen. The expression characteristics of this transgenic construct, especially in the macrophage cell lineage, mirror that seen in natural infection of small ruminants: limited expression of provirus in immature macrophage precursors but high levels of expression in mature activated macrophages confirming the importance of the viral LTR in controlling virus expression after natural infection.

The S open reading frame of VMV and later CAEV was originally thought to encode a Tat protein similar to that of HIV or EIAV. However, as stated above, there is no TAR element in the viral RNA R region, the usual site of interaction of Tat and viral RNA (Clements and Payne, 1994; Clements and Zink, 1996) and unlike HIV-1 Tat, CAEV Tat is not always necessary for replication of the virus *in vitro* or *in vivo* (Saltarelli *et al.*, 1993; Harmache *et al.*, 1995). This suggested that the product of this open reading frame may not function in the same way as HIV Tat. However co-expression of VMV Tat does increase transcription from the VMV and CAEV LTRs by about 2 to 46 fold (Hess *et al.*, 1986; Gourdou *et al.*, 1989; Hess *et al.*, 1989; Gdovin and Clements, 1992; Campbell and Avery, 1996; Carruth *et al.*, 1996; Villet *et al.*, 2003a; Villet *et al.*, 2003b; Barros *et al.*, 2005). It has been more difficult showing *trans*-activation of transcription using CAEV Tat: the increase in transcription is either none or only 2–10 fold using this Tat (Hess *et al.*, 1986; Sherman *et al.*, 1989; Saltarelli *et al.*, 1990; Jackson *et al.*, 1991; Kalinski *et al.*, 1994; Villet *et al.*, 2003b). Some experiments may not be optimal when studying the ability of Tat to *trans*-activate the LTR. Its activity is cell type dependent (Hess *et al.*, 1986; Kalinski *et al.*, 1994) and to see maximal activation, LTR fusion constructs should include the R region (Hess *et al.*, 1985). However, Tat is known to increase the transcriptional activity of the LTR (see below) as well as causing post-transcriptional effects that are less well characterized but may include stabilizing the RNA or increasing the processivity of the polymerase (Gdovin and Clements, 1992). Tat may therefore interact with the RNA produced, perhaps via the R region, although not in the same way as HIV Tat through a TAR element. With the LTR of K1514, *trans*-activation occurs through the AP-1 and AP-4 sites in the U3 region of the LTR (Hess *et al.*, 1985, 1986; Neuveut *et al.*, 1993). Tat itself does not bind directly to DNA but with cellular proteins (Carruth *et al.*, 1994). It interacts with fos and jun via the bZIP regions of a fos/jun dimer and the central leucine rich region of Tat (Gabuzda *et al.*, 1989; Carruth *et al.*, 1996; Morse *et al.*, 1999). Via its N-terminal region, it will also bind TATA box binding protein (Morse *et al.*, 1999). SRLV Tat *trans*-activates cellular genes including c-jun (but not c-fos) (Hess *et al.*, 1985; Neuveut *et al.*, 1993) and increased availability of these proteins may account for some of the LTR *trans*-activation seen in the presence of Tat. It is now known that Tat is not secreted from the cell but is incorporated into virions, induces G2 arrest (Villet *et al.*, 2003a), and can cause apoptosis of cells (Duval *et al.*, 2002a; Duval *et al.*, 2002b). From yeast two-hybrid systems CAEV Tat also interacts with human Notch2 EGF-like repeats and the epithelin/Granulin precursor (Shoham *et al.*, 2003). However the functional significance of this interaction is not clear given that SRLV Tat is not secreted, the EGF-like repeats are extracellular and epithelin is secreted. The C terminus of the Tat protein contains conserved cysteine residues that may be important in dimerization and function (Villet *et al.*, 2003a). From its primary sequence, the lack of strong *trans*-activation activity as well as the other activities given above, it is now thought that SRLV Tat is more similar to HIV-1 Vpr than Tat.

The tight control of viral gene expression by the LTR led to the 'Trojan horse' theory for the maintenance of virus within the persistently infected host (Haase *et al.*, 1977; Peluso *et al.*, 1985; Haase, 1986). In this theory virus infection is maintained in immature macrophages, probably myeloid stem cells in the bone marrow (Gendelman *et al.*, 1985), by tightly controlling its replication in a 'restricted' state. The infected precursor cells do not express antigens that the immune system can recognize to target their removal. These then provide a constant source of infected cells that seed into the periphery and tissues. In the tissues where terminal macrophage differentiation occurs, there is up-regulation of the viral LTR and high levels of expression of viral mRNA and protein, and productive infection. The immune response to this viral antigen then initiates the lymphoproliferative pathology that typifies SRLV infection. It is the constant cycle of tissue damage attracting more monocytes that then mature into macrophages and express viral antigen that causes the gradual worsening of pathology and leads to the slow nature of SRLV disease (Sigurdsson, 1954a; Sigurdsson, 1954b) (Fig. 10.2). Whilst this theory is valid today for the way in which pathology develops, the site of viral persistence has been challenged recently using PCR based methods in CAEV infected goats. Ravazzolo *et al.* (2006) found very few viral positive bone marrow samples in persistently infected goats suggesting that bone marrow stem cells may not be a major reservoir of infection. However, stromal cells within the bone marrow are viral antigen positive and may provide a continuous source of virus for infection of monocyte precursors (Grossi *et al.*, 2005).

Pathology is caused by an immune response to SRLV antigen with only a minor component attributed to the virus itself, although Tat has been implicated in cell death, CNS pathology and lymphoproliferation (Philippon *et al.*, 1994; Vellutini *et al.*, 1994; Strijbos *et al.*, 1995; Starling *et al.*, 1999; Rea-Boutrois *et al.*, 2009). Evidence to support this comes from several studies that show a correlation between levels of viral antigen or nucleic acid detected and pathology seen (Brodie *et al.*, 1992; Zhang *et al.*, 2000; Ravazzolo *et al.*, 2006). The direct action of the immune system in development of pathology is taken from the fact that immunosuppression reduces the number of viral isolations and severity of CNS

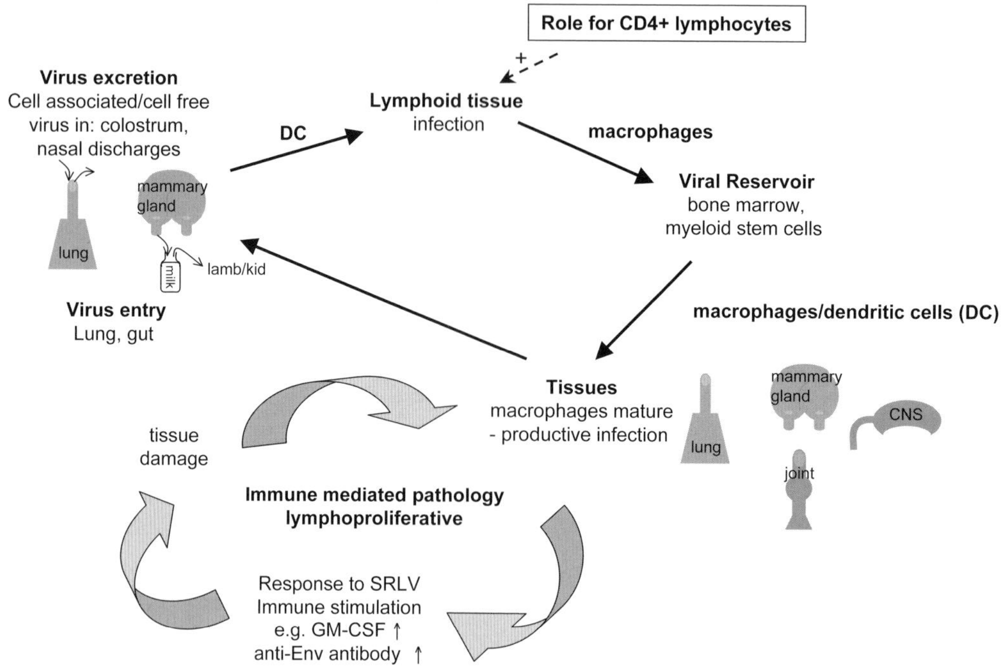


**Figure 10.2** SRLV immunopathogenesis model. The major route of virus entry to the host is via the respiratory tract or the gut. Virus is taken up by dendritic cells and transferred to the lymphoid tissue where infection is transferred to macrophages. Infected macrophages may then carry the virus to a viral reservoir (bone marrow has been suggested) where precursor cells may maintain the virus in a restricted replication state. When precursors leave the bone marrow and circulate to tissue, they mature and express viral antigen. The immune response to this antigen creates inflammatory tissue damage which attracts more macrophages to establish a cycle of pathogenesis that gradually worsens. As lesions become severe, there are enough macrophages expressing virus that cell free virus is detected. The free virus or virus-infected cells then transmit the virus horizontally by respiratory aerosol to other sheep/goats or via colostrum/milk to lambs/kids. The infection and pathology seen in the joints and CNS is therefore a dead end for the virus as it does not help in transmission.

lesions (Nathanson *et al.*, 1976) whilst immune stimulation especially after SRLV vaccination (i.e. increasing antigen specific responses) increases levels of virus expression and may accelerate lesion formation (Nathanson *et al.*, 1981; McGuire *et al.*, 1986). The acceleration of tissue damage by virus specific immune responses suggests that not only does tissue damage create conditions which attract potentially infected macrophages to the lesion but that there may be a close link between viral gene expression in these newly recruited macrophages and immune cell mediators e.g. cytokine stimulation. It is difficult *in vivo* to obtain direct evidence of which cytokines are important in activating virus replication versus bystander effects of inflammation and increased cell recruitment. Several studies have addressed this *in vitro*. The CAEV LTR is weakly activated (3-fold) by TNF-α or GM-CSF early (4 h) after stimulation of U937 cells via sites in the U3 70 bp repeat region. This is AP-1 independent (Murphy *et al.*, 2006). The transcriptional activity of the LTR reduces to basal levels if activity is measured 24 h after stimulation with TNF-α (or LPS) but at this time IFN-γ stimulation of U937 cells leads to increased transcription from the LTR (Tong-Starksen *et al.*, 1996). This is mediated by binding of STAT1 (signal transducer and activators of transcription-1) via a consensus IFN-γ activated sequence (GAS) element proximal to the AP-1 binding site within the 70 bp repeat and is a different site to that required for phorbol ester responsiveness (Tong-Starksen *et al.*, 1996; Sepp and Tong-Starksen, 1997). The activation of the LTR by TNF-α is supported by the increase in virus replication after treatment of SRLV infected macrophages with bovine TNF-α (Ellis *et al.*, 1994). Similarly IL-1β and GM-CSF treatment of ovine macrophages increases VMV replication (Ellis *et al.*, 1994; Zhang *et al.*, 2002). A link to IL-8 stimulation has also been put forward (Legastelois *et al.*, 1996, 1998). Treatment of infected cells *in vitro* with type I interferon (IFN), either bovine IFN-α, human IFN-α or ovine IFN-τ (a pregnancy recognition signal secreted by ruminant trophoblasts) decreases replication in this system (Canivet *et al.*, 1989; Ellis *et al.*, 1994; Juste *et al.*, 1996) and similarly *in vivo* IFN-τ treated lambs show reduced infected cell loads early after infection (Juste *et al.*, 2000). This inhibition of viral replication may not be a direct effect on the viral LTR but an effect of the anti-viral pathways induced in cells after type I IFN treatment. Therefore cytokines produced during inflammatory responses do help to activate virus replication within the lesion but may be in competition with anti-viral cytokines.

## SRLV-macrophage (DC) interaction – manipulation of innate immunity

As described in the previous section monocyte/macrophages and dendritic cells are the principal target cells of SRLV. These professional antigen presenting cells (APC) play a pivotal role at the interface between innate and adaptive immunity. Therefore, the potential modulatory activity of SRLV infection on APC has been the focus of several investigations *in vitro* and *in vivo*.

Several pieces of evidence indicate that SRLV infections *in vivo* promote the induction of a type 2 T-helper cell immune response, characterized by a relative dominance of IL-4 expression over IFN-γ (Perry *et al.*, 1995; Cheevers *et al.*, 1997). The humoral immune response to SRLV antigens is robust and restricted (in sheep) or heavily skewed (in goats) to the IgG1 antibody isotype, supporting the concept of a type 2 immune response (Bird *et al.*, 1995; Trujillo *et al.*, 2004a). These results, however, should be placed in the context of the ruminant immune system that, in contrast to the human and murine immune system, shows a less clear cut polarization into type 1 and 2 responses (Estes and Brown, 2002). *In situ* hybridization experiments show that IFN-γ and viral RNA are co-expressed in the joints of goats with a high viral load and a histological picture of severe arthritis, pointing to the capacity of the immune system to mount a robust type 1 response (Lechner *et al.*, 1997b). Different areas of the same joint, however, are characterized by infiltrations with plasma cells and the formation of lymphoid-like follicles indicative of a type 2-centred humoral immune response. These contrasting observations support the concept of a compartmentalized immune response with both type 1 and 2 responses present in the same affected organs but in different locations. The detection of both IFN-γ and IL-4 mRNA in the lung tissue of VMV infected sheep with severe virus induced lymphoid interstitial pneumonia supports this model (Woodall *et al.*, 1997). Several cytokines and chemokines are expressed in the arthritic joints or in the associated draining lymph nodes of CAEV infected goats. MCP-1 mRNA is abundantly expressed at all stages of disease, predominantly at sites of virus replication, while TNF-α and IL-6 expressing cells are not consistently co-localized with virus mRNA expression. IL-10 expression is present in draining lymph nodes showing a high viral load and IFN-γ expression (Lechner *et al.*, 1996; Lechner *et al.*, 1997b). High IL-8 expression was reported in the lungs of VMV infected sheep but is not detected in the arthritic joint of CAEV infected arthritic goats (Legastelois *et al.*, 1996; Lechner *et al.*, 1997b; Legastelois *et al.*, 1998). In sheep affected by VMV induced lymphoid interstitial pneumonia, IL-1β and above all GM-CSF were shown to be highly expressed in the affected tissues, suggesting a direct or indirect modulatory effect of these cytokines on virus replication (Woodall *et al.*, 1997). The analysis of different cytokines expressed in freshly isolated macrophages obtained from naturally VMV infected animals confirmed these observations and pointed to GM-CSF as a key cytokine in stimulating virus replication *in vitro* and potentially *in vivo*, thus promoting the formation of lung lesions (Zhang *et al.*, 2000). GM-CSF expression is up-regulated in alveolar macrophages following a VMV experimental infection via lower lung instillation (McNeilly *et al.*, 2008). The authors showed that lung instillation with free VMV or *ex vivo* infected alveolar macrophages efficiently mediated systemic infections. The transfer of virus from the infected lung to the draining lymph node did not involve a direct migration of VMV infected macrophages, suggesting that intermediate cells, such as respiratory tract dendritic cells, whose antigen presenting activity is stimulated by GM-CSF, may perform this task (McNeilly *et al.*, 2008). Interestingly, goats immunized with a particular CAEV Gag peptide and subsequently challenged with CAEV-CO showed enhanced virus replication at 55 days post infection. The vaccine induced enhancement of provirus load correlated with the expression GM-CSF mRNA in peripheral mononuclear cells stimulated *in vitro* with the Gag peptide or virus antigen, again pointing to GM-CSF as an important cytokine promoting virus replication (Nenci *et al.*, 2007).

Importantly, all these studies reinforced the concept of a highly restricted expression of viral mRNA *in vivo*, accompanied by a good correlation between virus mRNA expression and severity of lesions. Interestingly, the *in situ* hybridization studies in joints of goats affected by CAEV induced arthritis revealed a peculiar pattern of mRNA expression in cells resembling macrophages. In these cells the hybridization signal was more abundant in the nucleus than in the cytoplasm indicating a possible mechanism of virus restriction, possibly based, as proposed by Ravazzolo and colleagues, on the regulation of *rev* expression (Ravazzolo *et al.*, 2006).

Contrasting with a tightly restricted SRLV mRNA expression *in vivo*, macrophages, GSM cells and SCP cells infected *in vitro* are fully permissive for SRLV replication that reaches high titres in these cells. This dramatic difference in permissiveness for virus replication of macrophages *in vitro* versus *in vivo* should be kept in mind while interpreting the modulatory activity of SRLV infection on the expression of cytokines *in vitro*.

*In vitro*, CAEV infected macrophages show an increased expression of IL-8 and MCP-1 mRNA at the peak of virus replication at day 7 post infection, whereas the constitutive expression of TGF-β1 was inhibited at this time point (Lechner *et al.*, 1997a). Enhanced IL-8 expression was also observed in VMV infected pulmonary macrophages but, in contrast to CAEV infected cells, this activity was detected in cells stimulated with UV inactivated virus as well (Legastelois *et al.*, 1996; Legastelois *et al.*, 1998). The response of macrophages to stimulation with bacterial lipopolysaccharide (LPS), a TLR-4 agonist, was considerably affected by CAEV infection (Lechner *et al.*, 1997a). Macrophages infected for 7 days with CAEV responded to LPS stimulation with a decreased expression of TNF-α, IL-1β, IL-6 and IL-12 p40. In contrast, the expression of GM-CSF, mentioned above as an SRLV promoting cytokine, was increased in CAEV infected macrophages. These modulatory effects, however, are observed only in cells where CAEV is replicating at maximal titres that may never be reached *in vivo*.

Virus infection is a potent activator of the type I interferon (IFN) system via stimulation of a variety of extra- and intracellular receptors designed to bind particular molecular patterns associated with virus infections such as double stranded viral RNA of different sizes, detected by TLR-3 (30–80 bp), MDA5 (>2–4 kbp) and RIG-I (<2 kbp), or single-stranded viral RNA, sensed by TLR-7/8 and RIG-I (Haller *et al.*, 2006; Takeuchi and Akira, 2009). Retroviruses, however, are considered to be poor inducers of type I IFN expression in infected cells and, in this respect, SRLV are no exception. SCP fibroblasts or peripheral blood derived sheep macrophages infected with the prototypic Icelandic VMV strain K1514 do not produce any detectable bioactive IFN. The same is true for GSM and goat macrophages infected with the CAEV-CO strain (Narayan *et al.*, 1985). We repeated these experiments using Mx gene expression as a sensitive marker of IFN induction (von Wussow *et al.*, 1990). The molecular clone CAEV-CO and the VMV strain K1514 replicated to titres approaching $10^5$ $TCID_{50}$/ml in GSM and lamb synovial membrane cells, respectively, without inducing an Mx response in the infected cells. Mx expression was seen only late in infection, when the virus titres climbed above this threshold value (L. Hüsser, unpublished). SRLV-infected macrophages were even more refractory to type I IFN induction and Mx was detected only sporadically and in cultures replicating the viruses to very high titres (C. Reinhard, unpublished).

Interestingly, although SRLV infected macrophages do not produce IFN, they are capable of inducing a robust IFN response in lymphocytes making contact with these infected cells. T-cells appear to be the principal source of IFN in this type of response, which, remarkably, is not MHC restricted (Narayan *et al.*, 1985). The lentivirus-induced IFN (LV-IFN) detected in the aforementioned co-cultures enhances the expression of MHC II in spleen and lymph node cells treated *in vitro* (Kennedy *et al.*, 1985). In view of the T-cell origin of LV-IFN and its biological activity, the conclusion may be drawn that this factor is IFN-$\gamma$ (or type II IFN). The physicochemical properties of LV-IFN, however, such as a low pH and high temperature resistance, classify this activity as type I rather than type II IFN. LV-IFN is spontaneously produced *in vitro* by cultures of pulmonary lymph node lymphocytes obtained from sheep infected with ovine lentivirus (Lairmore *et al.*, 1988). The LV-IFN activity detected in the supernatant of these cultures correlated with the severity of the lymphoid interstitial pneumonia detected at necropsy in these animals, suggesting that LV-IFN may contribute to the immunopathology of SRLV induced disease (Lairmore *et al.*, 1988). We have repeated the experiments reported by Narayan *et al.* using the Mx response as a marker of IFN activity and detected a strong Mx response following the addition of PBMC to SRLV infected macrophages (D. Stenger, unpublished).

The sensitivity of SRLV to the antiviral effects of type I IFN remains a controversial issue. Using sheep IFN produced by polyriboinosinic-polyribocytidylic acid (poly(I:C)) stimulation in fetal lambs, Carroll and colleagues reported that VMV replication was unaffected by doses that completely inhibited the replication of IFN sensitive viruses such as herpes simplex virus, vesicular stomatitis virus, and vaccinia virus in sheep choroid plexus cells (Carroll *et al.*, 1978). In contrast, high doses of recombinant bovine IFN-$\alpha$ (2.5 mg/ml) were shown to inhibit the replication of a biologically cloned strain of ovine lentivirus (OvLV 85-34) in pulmonary alveolar macrophages (Ellis *et al.*, 1994). Recombinant human IFN-$\alpha$ (1000 IU/ml) strongly and specifically inhibited the replication of the prototypic Icelandic visna strain K1514 in ovine choroid plexus fibroblasts, possibly by blocking the synthesis of the major viral proteins, thereby reducing the virus induced cytopathic effect on these cells (Canivet *et al.*, 1989). IFN-$\tau$ has a similar antiviral activity as type I IFN but lower cytotoxicity (Spencer *et al.*, 2007). Juste and colleagues showed that this particular type I interferon had a moderate inhibitory effect on the replication of ovine lentivirus in GSM cells (82% titre reduction) (Juste *et al.*, 1996). Immortalized macrophages treated with LV-IFN and infected with a particular strain of VMV showed a 3 log reduction in virus titres (Kennedy *et al.*, 1985). In contrast, GSM cells repeatedly exposed to LV-IFN showed only a small reduction in virus titre compared to control cells (Lairmore *et al.*, 1988). Finally, clear cut results were obtained by treating freshly isolated monocytes with LV-IFN prior to infection with VMV. Virus titres were reduced by more than 99% in LV-IFN treated monocyte cultures, possibly because IFN interferes with the proliferation and differentiation of monocytes into macrophages, which, as opposed to monocytes, are fully permissive for virus replication (Zink and Narayan, 1989).

In an attempt to clarify these controversial results, we tested the effect of recombinant bovine IFN-$\alpha$ on CAEV-CO replication in GSM cells and macrophages (C. Reinhard, unpublished). The cells were treated with up to 100 ng/ml of rbIFN-$\alpha$ 18 hours before infection and with the same dosage every third day after infection. This regimen completely blocked the replication of the bovine viral diarrhoea virus (BVDV) strain SD1 and vesicular stomatitis virus (VSV) in the IFN treated cells. In contrast, the CAEV titres were reduced by less than 1 log (<90% inhibition), demonstrating that the CAEV-CO molecular clone can replicate to high titres in GSM cells (>$10^6$ $TCID_{50}$/ml) and macrophages (>$10^5$ $TCID_{50}$/ml), showing a very active antiviral state, as demonstrated by the complete suppression of BVDV and VSV replication. However, the same IFN regimen applied to freshly isolated monocytes had a profound effect on CAEV replication, reducing the virus titre measured 7 days post infection by more than 4 logs (<99.99% inhibition). This confirms the results reported *in vitro* by Zink and Narayan with LV-IFN and suggests a possible mechanism of action to explain the reduced viral load and protection from lentivirus induced lung lesions observed in lambs treated with IFN-$\tau$ immediately after an experimental infection with OvLV 85-34 (Zink and Narayan, 1989; Juste *et al.*, 2000).

In addition to the effect on the differentiation of monocytes to macrophages, IFN may activate the expression of the Apobec3 family proteins, such as Apobec3G that is known to interfere with HIV replication in macrophages and dendritic cells (Peng *et al.*, 2006; Hou *et al.*, 2009; Trapp *et al.*, 2009). This may represent a direct link between innate immunity and several intrinsic resistance mechanisms known to protect mammalian cells from retrovirus infections, such as the above mentioned Apobec family proteins or the TRIM family proteins whose expression is enhanced in IFN treated cells as well (Harris and Liddament, 2004; Asaoka *et al.*, 2005; Luban, 2007; Sakuma *et al.*, 2007; Carthagena *et al.*, 2008). Strong indirect evidence supporting the involvement of Apobec3G in intrinsic resistance to SRLV infections was provided by the phenotypic analysis of *vif* deletion mutants of CAEV and VMV (Harmache *et al.*, 1996a; Harmache *et al.*, 1996b; Kristbjornsdottir *et al.*, 2004). These *vif* mutants are highly attenuated *in vitro*, particularly in macrophages, and *in vivo*, where they are unable to sustain a persistent infection. Interestingly, a VMV *vif* mutant replicating in semi-permissive cells demonstrated higher mutation rates in proviral DNA, possibly as a result of the cytosine deaminase activity of Apobec3G (Kristbjornsdottir *et al.*, 2004). Additional evidence supporting the importance of different intrinsic mechanisms of resistance to SRLV infection was provided by the analysis of the attenuated VMV isolate KS1 (Gudmundsson *et al.*, 2005). This particular virus does not replicate in macrophages and shows an attenuated phenotype *in vivo*. The molecular lesions leading to this attenuated phenotype were mapped to two amino acid mutations, one in the capsid protein of Gag and the other in Vif, yet again suggesting the involvement of Apobec and possibly TRIM family members as potential intrinsic barriers against SRLV infections (Gudmundsson *et al.*, 2005).

## Role of the adaptive immune response in controlling virus replication – manipulation of responses

After infection with a virus it is assumed that the host will mount an immune response that clears the pathogen. For many acute infections such as influenza this is true. Where the immune response fails to clear the virus, a persistent infection may be established. In this situation the immune system may control the level of replication of the virus for many years such that the infection is sub-clinical and indeed, this may be seen for many years after HIV infection in long-term non-progressors. After infection by SRLV, there is an early burst of virus replication with a wave of infected cells leaving the site of infection and draining lymph node and entering the peripheral blood (Bird *et al.*, 1993; Blacklaws *et al.*, 1995b). The frequency of infected cells then decreases to very low levels ($< 1$ in $10^5$ infected mononuclear cells is not uncommon) that are maintained for long periods of time. This occurs when viral specific immune responses are induced but proving if the immune response is important in reducing this initial burst of replication and then in continuing to suppress replication to minimal levels has been more problematic.

Both antibody and T-cell responses are induced after SRLV infection. Antibody is first detectable in serum within 1 to 3 months of experimental infection (dose and route of virus administration affect time to detection) (Griffin *et al.*, 1978; Adams *et al.*, 1980; Sihvonen, 1981; Larsen *et al.*, 1982a; Larsen *et al.*, 1982b; Sihvonen, 1984;). After natural infection it may be 6 months to 2 years before antibody is apparent in the infected animal (Houwers and van der Molen, 1987). This makes detection of infection by serology insensitive so that repeated testing is necessary. The antibody response may be biphasic, with the first peak following the control of the primary infection and a second response consequent to the establishment of persistent infection. However these peaks probably correlate with IgM induction then later class switching to IgG (Bird *et al.*, 1993; Cheevers *et al.*, 2003; Mordasini *et al.*, 2006). After SRLV infection in sheep, IgG1, a complement fixing isotype, is detected (Mehta and Thormar, 1974; Bird *et al.*, 1995) whilst goats induce both IgG1 and 2 isotypes (Trujillo *et al.*, 2004a). The lack of IgG2 antibody production in sheep is not seen after immunization with viral Gag antigen (Mehta and Thormar, 1974; Pétursson *et al.*, 1985; Bird *et al.*, 1995) and may be due to a dysfunction of infected antigen presenting cells leading to aberrant $CD4^+$ T-lymphocyte activation such that the cytokines required for an IgG2 class switch are not produced (in ruminants IFN-$\gamma$ is required; Estes *et al.*, 1994). In some animals antibody responses may disappear after initial seroconversion although the reason for this effect is not known (Houwers and van der Molen, 1987). Neutralizing antibodies are induced in sheep after SRLV infection and are found in both the IgM and IgG1 antibody classes (Mehta and Thormar, 1974). In goats, even when both IgG antibody isotypes are produced the neutralizing antibody response is present in the IgG1 fraction of serum; however, the lack of activity in the IgG2 fraction may be due to low concentrations of immunoglobulin used in the assays (Narayan *et al.*, 1984). Antibody responses to all the viral gene products may eventually be detected but of most interest are those induced to the structural proteins of the virion which have roles in control of virus infection and replication, and in detection of infected animals. An antibody response to both Gag and Env are induced (Lacerenza *et al.*, 2006) and the affinity of the response increases with time (Bertoni *et al.*, 1994, 2000; Trujillo *et al.*, 2004a; Mordasini *et al.*, 2006). There is debate in the literature as to which anti-viral protein response is seen first and it may be the assay used which is important in determining this, with Gag p25 responses being seen more easily when ELISA or immunoblot methods are used, and anti-Env responses when immunoprecipitation is the method of choice (Johnson *et al.*, 1983; Bertoni *et al.*, 1994). This may reflect the ability of the antigens to maintain their conformation in the different assay systems, the targeting of antibody responses to linear versus conformational epitopes, or the strain of virus used for infection (Kajikawa *et al.*, 1990). Bearing this in mind, after VMV infection of sheep the anti-Gag response is usually detected first with the anti-Env antibody response becoming apparent and dominating later after infection (Houwers and Nauta, 1989; Kajikawa *et al.*, 1990). In CAEV infected goats, anti-Env and Gag responses are detectable by 1–3 weeks post infection and the anti-Env response is dominant 6 months after infection (Johnson *et al.*, 1983; Bertoni *et al.*, 1994).

Antibodies have several effector functions against viruses that may be used to protect the host from virus infection. These include neutralization of cell free virus, antibody-dependent cellular cytotoxicity (ADCC) and complement-mediated cytotoxicity. The last two functions require the Fc portion of the antibody first, to interact with Fc receptors on effector cells, or second to interact with complement. Determining whether antibody activities *in vitro* correspond with protection from viral disease *in vivo* is more difficult but it is commonly thought that neutralization of virus is an important function, especially after vaccination (Law and Hangartner, 2008). Neutralizing antibody activity occurs against virion surface components and this is also true with SRLV. Neutralizing activity against SRLV is found in the anti-Env antibodies but the ability to show induction of a neutralizing antibody response varies between experimental studies. This may be due to the strain of virus, cell type used for assay, as well as host genetic factors (Cheevers *et al.*, 1999). Neutralizing antibody responses are seen 2–5 months after infection when they are usually specific for the virus strain used to infect the animal (type-specific). This early response broadens later after infection so that different strains of SRLV may then be neutralized (Narayan *et al.*, 1981; Andresdottir *et al.*, 2002). Early studies suggested that neutralizing antibody was not induced by CAEV infection, as goats infected with CAEV did not induce a neutralizing antibody response but did to VMV and conversely sheep infected with CAEV did not induce a neutralizing antibody response (Klevjer-Anderson and McGuire, 1982; Narayan *et al.*, 1984). This was probably due to the strain of CAEV used but even when neutralizing antibody is induced against CAEV infection, it is of low titre whilst the neutralizing response induced by VMV (K1514) is much better (Narayan *et al.*, 1981; McGuire *et al.*, 1988; Bertoni *et al.*, 2000; Kemp *et al.*, 2000).

The structure of SRLV Env has not been elucidated but it is assumed that the Env protein is processed and functions in a very similar way to that of HIV-1 (Freed and Martin, 2007). Env is synthesized as a large precursor protein that traffics to the cell membrane. As it is processed and put onto the cell surface the precursor molecule is cleaved by a cellular protease to form two closely associated proteins called SU (surface subunit) and TM (transmembrane subunit). This dimer associates with other Env molecules to form trimers functional for cell receptor binding and membrane fusion. The fusion peptide is the amino terminal peptide of the TM subunit (Crane *et al.*, 1991). Virions bud through Env rich membranes and are released from the cell to propagate infection. After cell receptor binding there is a conformational change in Env that exposes the TM fusion domain to the cell membrane where it initiates membrane fusion. Therefore there are several sites on Env and steps in the infection process where antibodies could interact with virions to cause neutralization of the virus: inhibition of cell receptor binding, conformational change and fusion being three. However Env is heavily glycosylated and highly variable suggesting it has and is evolving to evade antibody responses.

To determine which regions of SRLV Env are immunodominant several studies have looked at variability of the *env* gene and antibody reactivity to different regions of the protein (see Fig. 10.3). Variability in the CAEV *env* gene has been mapped in different isolates to five major regions (V1–5) (Valas *et al.*, 2000) with two hypervariable sites seen in neutralization escape mutants arising during persistent infection (Hötzel and Cheevers, 2002a). One of these hypervariable sites is within V4 (Hötzel and Cheevers, 2002a) and similarly a neutralization epitope in VMV has also been mapped to V4 (Skraban *et al.*, 1999). Although the linear B-cell epitopes in one strain of CAEV have been defined, less is known about neutralizing epitopes more commonly thought to be contained in conformationally sensitive regions. CAEV-CO has at least six linear B-cell epitopes in TM and 5 in SU (Bertoni *et al.*, 1994; Bertoni *et al.*, 2000; Valas *et al.*, 2000) whilst VMV is known to have at least five overlapping epitopes in SU (Stanley *et al.*, 1987). SU3 and SU5 are the immunodominant epitopes inducing early immune responses but are also highly variable (Bertoni *et al.*, 2000; Mordasini *et al.*, 2006) however this variability is not linked to neutralization. In contrast the majority of mutations arising with time during persistent infection of sheep are located in the immunodominant neutralization epitope of VMV's V4 region that contains a conformationally sensitive epitope, a possible loop, as well as linear epitopes (Skraban *et al.*, 1999; Andresdottir *et al.*, 2002). Mutations in potential glycosylation sites and possible

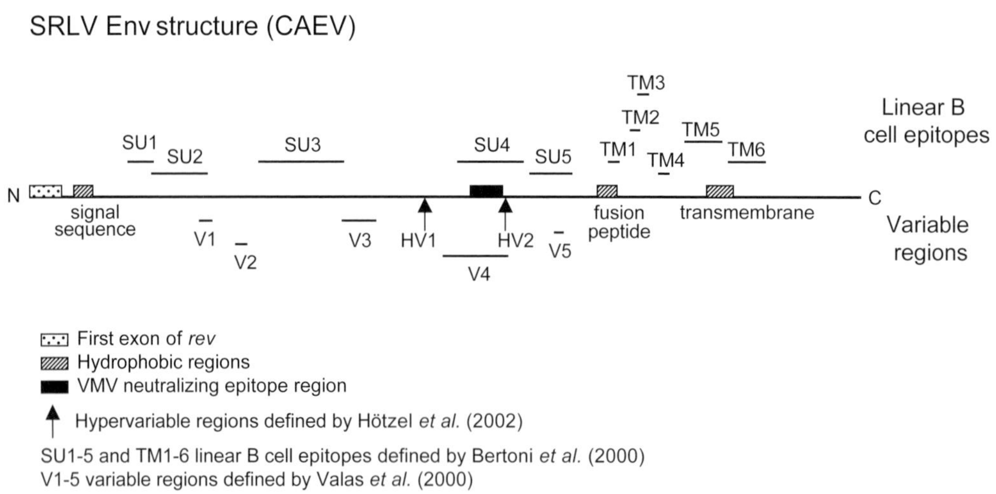


**Figure 10.3** Diagram of CAEV Env structure. The variable (V1–5) and hypervariable (HV1–2) domains defined by sequence variation in CAEV isolates by Valas *et al.* (2000) and Hötzel *et al.* (2002) are shown below the line. The linear B-cell epitopes defined using an Env expression library by Bertoni *et al.* (2000) are shown above the line. The first exon of *rev* has been indicated (▫) (*rev* and *env* share a common ATG) as has the hydrophobic regions encoding the signal sequence, fusion peptide and transmembrane region of the protein. The position of a neutralizing epitope defined in VMV by Skraban *et al.* (1999) is also shown (■). The precursor Env protein is cleaved by a host protease at the amino terminal of the fusion peptide (proposed recognition sequence RKKR/(GXG)) to form the surface subunit, SU, and the transmembrane subunit, TM.

structurally important cysteine residues around this loop alter antibody recognition of this site (Haflidadottir *et al.*, 2008). Loss of glycosylation as well as other mutations may alter the structure of epitopes such that they will no longer be recognized by neutralizing antibody (Clements *et al.*, 1988; Lichtensteiger *et al.*, 1991; Andresdottir *et al.*, 2002). Similarly masking linear antigenic sites (in SU4 of CAEV) by addition of glycosylation sites can redirect the antibody response to give a stronger neutralizing response although whether this leads to protection is not known (Trujillo *et al.*, 2004b).

Factors suggesting that neutralizing antibody activity is of little importance against SRLV infection are as follows. Firstly, SRLV are strongly cell associated with free virus rarely found during persistent infection in blood, draining lymphatics or CSF (Griffin *et al.*, 1978; Bird *et al.*, 1993; Ryan *et al.*, 2000) although virus may be isolated from diseased tissue such as synovial fluid from arthritic joints (Adams *et al.*, 1980), bronchiolar lavage fluid from lungs with late stage pathology (McNeilly *et al.*, 2007) or colostrum/milk produced from mammary glands with mastitis (Pisoni *et al.*, 2007b). Therefore, within the host during persistent infection, there is little free virus to inhibit. Secondly, although neutralization of virus is detectable in culture, it may not function *in vivo*. This comes from two strands of evidence: assays using macrophages and affinity of the virus for antibody. The ability of antibody to neutralize virus varies with the cell type used in assays. If Fc receptor negative cells such as fibroblasts are used, then neutralization is possible with both whole antibody or $F(ab)_2$ fragments blocking infectivity at the cell surface. If Fc receptor positive, phagocytic cells such as macrophages are used then there may be enhancement of viral entry by antibody coated virions. This need not lead to infection and certainly there is an apparent inhibition of reverse transcription. However this is delayed not absolute inhibition of virus replication and with longer incubation times (> 5 days) the antibody–virion complexes may dissociate to allow productive replication (Kennedy-Stoskopf and Narayan, 1986; Jolly *et al.*, 1989a; Jolly *et al.*, 1989b). In addition the interaction of neutralizing antibody with SRLV is slow implying low affinity, indeed it is much slower than the rate at which virus binds its cellular receptor which leads to very inefficient inhibition of the virus (Kennedy-Stoskopf and Narayan, 1986). The last piece of evidence against a role for neutralizing antibody in protection is that viral variants survive in the persistently infected host for many years in the face of neutralizing antibody responses (Thormar *et al.*, 1983; Cheevers *et al.*, 1993).

However there is also evidence that neutralizing antibody does play a role in control of infection. The early appearance of strong neutralizing antibody responses has been associated with reduced CNS pathology caused by VMV (Georgsson *et al.*, 1993) and there is selection of neutralization escape mutants (Gudnadottir, 1974; Narayan *et al.*, 1977, 1978, 1981; Clements *et al.*, 1982, 1988; Ellis *et al.*, 1987; McGuire *et al.*, 1988; Cheevers *et al.*, 1991). The development of mutations that escape early type-specific antibody responses suggests a selective advantage especially as some have been shown to grow more slowly in macrophages than the original strain (Andresdottir *et al.*, 2002). However, these variants are often neutralized by later antibody responses and it can be shown that the original neutralization serotype is maintained in the presence of the variants and antibody throughout infection. Antigenic variants may also arise in the absence of neutralizing antibody. Therefore antigenic drift is not essential for viral persistence or indeed for pathogenesis of lesions (Lutley *et al.*, 1983; Thormar *et al.*, 1983; Cheevers *et al.*, 1993, 1999; Andresdottir *et al.*, 2002). This led Cheevers *et al.* (1993) to postulate that antigenic variants were not selected to avoid neutralization but perhaps to interact with non-neutralizing antibodies such that infectivity for macrophages could be enhanced i.e. positive selection is occurring. However other functions of anti-viral antibody may be of more importance in selection of mutants. Guinea pigs immunized with VMV Env induce neutralizing and fusion-blocking antibodies (Huso, 1997). It may be the fusion blocking function that is important *in vivo*: as the virus is cell-associated this activity may inhibit spread of the virus from cell to cell (Crane *et al.*, 1988). Certainly in the presence of antibody the budding of virions from the surface of cells is altered with the appearance of virion aggregates on the cell surface and possible phagocytosis into the cell (Dubois-Dalcq *et al.*, 1979). Also recent data from the immunodeficiency virus field suggests ADCC is an important effector mechanism *in vivo* in control of SHIV replication (Hessell *et al.*, 2007). ADCC, not neutralization, may be more important in control of SRLV and help to drive variability. It is therefore interesting to note that after SRLV infection of sheep the antibody isotype which mediates ADCC is absent (Singh *et al.*, 2006b) although present in most goats (Trujillo *et al.*, 2004a).

Overall, the presence of a silent infected cell reservoir with the ability of the virus to disseminate by cell to cell spread would reduce the efficiency of the antibody response in controlling virus replication once infection is established.

Antibody responses are correlated with development of pathology after SRLV infection (see Fig. 10.2). In goats, one-third of infected animals may develop clinical arthritis. This is associated with depressed CD4$^+$ T-lymphocyte count and increased antibody responses to immunodominant epitopes in Env TM and SU regions but not Gag epitopes (Knowles *et al.*, 1990; McGuire *et al.*, 1992; Bertoni *et al.*, 1994; Perry *et al.*, 1995). Similarly the anti-Env response is correlated with the development of interstitial pneumonitis in sheep (Kajikawa *et al.*, 1990; Juste *et al.*, 2000). The link to development of pathology in goats is also related to isotype as both IgG1 and 2 isotypes are induced by viral infection. Production of large amounts of IgG1 anti-SU with little IgG2 is a marker for development of arthritis (Trujillo *et al.*, 2004a). This is a result of a Th2-biased CD4$^+$ T-cell response (Cheevers *et al.*, 1997). When more IgG2 is induced, e.g. after vaccination, then the appearance of arthritis may be delayed.

Both CD4$^+$ and CD8$^+$ T-cell responses are induced in response to SRLV infection. Early experiments to analyse the presence of VMV specific T-cells by lymphoproliferation showed transient and inconsistent responses in blood from 1 week to 3 months post infection and then throughout persistent infection (Griffin *et al.*, 1978; Sihvonen, 1981, 1984; Larsen *et al.*, 1982a,b; Torsteinsdottir *et al.*, 1992). Early T-cell responses in goats show a similar trend with inconsistent responses by 3 weeks post infection with CAEV but later responses (>4 months) being detectable more consistently (Adams *et al.*, 1980). When the viral antigen used to stimulate PBMC is concentrated sufficiently or recombinant antigen is used then T-cell proliferation is detected much more consistently during persistent infection (Adams *et al.*, 1980; Reyburn *et al.*, 1992). SRLV reactive T-cells can also be shown in the CSF and lymphoid tissues (Griffin *et al.*, 1978; Blacklaws *et al.*, 1995b). Lymphoproliferation usually measures CD4$^+$ T-cell responses (Perry *et al.*, 1995) but CD8$^+$ T-lymphocytes may contribute to responses to whole virus antigen (Reyburn *et al.*, 1992). Both VMV Gag and Env induce lymphoproliferative responses (Reyburn *et al.*, 1992; Rafnar *et al.*, 1998; Singh *et al.*, 2006a) and peptides can also be used to show reactivity to Gag after CAEV infection (Fluri *et al.*, 2006). It is the lack of purified antigen that restricts the study of the other viral gene products as possible CD4$^+$ T-cell targets. Infected macrophages present Gag antigen to Gag specific-CD4$^+$ T-cells and so may be involved in lesion formation in tissues (Singh *et al.*, 2006a). As animals become clinically ill, their anti-SRLV CD4$^+$ T-cell response declines (Perry *et al.*, 1995; Reina *et al.*, 2007) and this has been linked to reduced expression of CD80 co-stimulatory molecules in PBMC (Reina *et al.*, 2007) suggesting a subtle loss of antigen presenting cell function with time after infection.

CD8$^+$ cytotoxic T-lymphocytes (CTL) are also induced within 3 weeks to 3 months in blood after infection by SRLV and in a small proportion of sheep can kill infected targets directly *ex vivo* suggesting a high proportion of viral specific CD8$^+$ T-cells in these animals (Lichtensteiger *et al.*, 1993; Blacklaws *et al.*, 1994). VMV infected macrophages are killed by VMV specific CTL and can also act as antigen presenting cells to stimulate VMV specific CTL *in vitro* (Blacklaws *et al.*, 1994; Lee *et al.*, 1994). All the gene products of VMV contain epitopes recognized by VMV specific CTL except for *tat* (Wu *et al.*, 2008). This is probably due to the limited number of animals and therefore MHC haplotypes used in this study. There is no apparent reason why Tat should not contain epitopes as is seen in other lentiviral infections. There is also evidence that there are antigenic variants in CTL epitopes as different strains of virus are recognized differentially (Lichtensteiger *et al.*, 1993; Blacklaws *et al.*, 1994). CD8$^+$ T-lymphocytes are often present in high proportions in severe lesions induced by SRLV infections (Watt *et al.*, 1992; Lujan *et al.*, 1995; von Bodungen *et al.*, 1998) but their role in lesion development or control of replication is unknown.

It has not been possible to show a role for either $CD4^+$ or $CD8^+$ T-lymphocytes in protection of the host from VMV infection using *in vivo* depletion of lymphocytes with monoclonal antibodies. Using this methodology, good depletion of $CD8^+$ cells is achieved in the blood and lymph for up to 2 weeks but in tissues there is only a 50% drop in cell numbers (Eriksson *et al.*, 1999b). This may not be sufficient to cause complete loss of function as CTL were evident when $CD8^+$ T-lymphocyte numbers recovered. With similar efficiencies of $CD8^+$ cell depletion in SIV/HIV models of infection, there is clear evidence of loss of control of virus replication with increased virus load in plasma (Matano *et al.*, 1998; Jin *et al.*, 1999; Schmitz *et al.*, 1999). Therefore there may be differences in the control mechanisms needed for macrophage-tropic virus versus $CD4^+$ T-lymphocyte-tropic lentiviral infections. *In vivo* depletion of $CD4^+$ cells gave surprising results: in the absence of $CD4^+$ lymphocytes there is very inefficient infection of sheep by VMV. Although VMV does not require $CD4^+$ T-lymphocytes for its transfer in dendritic cells from the periphery to the lymph node, it does require $CD4^+$ T-lymphocytes to establish an efficient infection in macrophages once it reaches the node (Eriksson *et al.*, 1999a). We may not understand fully whether the $CD4^+$ lymphocyte is affecting cell migration in the node, or replication of virus within dendritic cells as well as macrophages but it is quite clear that VMV uses the T-cell response to help it to infect and replicate in macrophages even in the very early stages of infection (see Fig. 10.2).

The immune response to the SRLV was originally thought to be slow and inconsistent (see above) and this was considered to be one of the reasons the virus was persistent. However, the infectious dose may affect the time at which responses are seen and similarly the route of infection could delay the responses seen in the periphery/blood, the usual compartment sampled. Therefore an acute infection model of VMV using cannulation of the lymphatics draining the site of infection (skin) has been established in sheep. This allows the sensitive detection of viral specific responses before they are diluted by the systemic circulation (Bird *et al.*, 1993; Blacklaws *et al.*, 1995a). Antibody specific for Gag in this system precedes the anti-Env response and is detected by 4 days post infection whilst anti-Env is detected by day 12 post infection at a similar time to the appearance of neutralizing antibody. The response is biphasic with a clear switch from IgM to IgG1 production and the appearance of neutralizing antibody coincided with the appearance of IgG1. T-lymphocyte proliferative responses were detectable consistently from 2 weeks post infection and were specific for both whole viral antigen and Gag. Directly active CTL responses were also detectable in one sheep between day 10 and 14 post infection in the efferent lymphatics. $CD8^+$ CTL precursors were induced consistently within the first 2 weeks of infection. Infected cells are present in the draining lymph node by day 4 post infection and their frequency is maximal approximately 12–14 days post infection. This coincides with their exit into the efferent lymphatics in a wave that peaks at approximately day 9–14 post infection. Detection of virus infected cells in the lymph node precedes the induction of the immune response and the cells are not cleared at this early stage of infection. However as the strength of the immune response increases, the frequency of virus infected cells decreases to very low levels. Therefore, immune responses to SRLV infection are induced rapidly after infection and may help to control virus replication but they are induced after the virus has exited the lymphoid tissue draining the site of infection and disseminated to other sites in the body. This allows the virus time to establish reservoirs in other tissues.

## Immunosuppression

There have been publications reporting immunosuppression in SRLV infected animals but these are often countered by others showing no immunosuppressive effects. These differences may be due to the stage of disease at which samples are taken. Early reports of systemic alterations in the $CD4^+/CD8^+$ T-cell ratio have not been substantiated (Kennedy-Stoskopf *et al.*, 1989; Perry *et al.*, 1995; Grezel *et al.*, 1997). However there is evidence of a decreased delayed-type hypersensitivity (DTH) response in the skin which may be caused by direct effects of infection on macrophages (e.g. chemotaxis) that are important antigen presenting cells in DTH lesions (Larsen *et al.*, 1982b; Myer *et al.*, 1988; Pyrah and Watt, 1996). There is also a reduction in the expression of mRNA for the co-stimulatory factor CD80 in PBMC from sheep clinically affected by VMV infection that leads to reduced lymphocyte responses to VMV and previously encountered antigens (Reina *et al.*, 2007). There has also been the suggestion that lymphocytes from lymph nodes of sheep infected with VMV are unable to produce IL-2 like factors as well as having reduced responses to high concentrations of mitogen, however this was not consistent across all lymph nodes (Ellis and DeMartini, 1985) or all studies (Perry *et al.*, 1995). These effects with the known cytokine dysregulation that SRLV infection of macrophages causes (Lechner *et al.*, 1997a) may lead to a mild immunosuppression allowing secondary infection although very few studies substantiate this (Myer *et al.*, 1988).

### Autoimmunity

Several studies have investigated the development of autoimmunity after infection by SRLV. The pathology caused in the tissues, especially arthritis and visna are very similar to diseases of man known to be mediated by autoimmunity. One report has shown the development of autoreactive antibodies after VMV infection of sheep. These are specific for IgM or IgG and rheumatoid factor (sheep Fc region) as well as ssDNA, cardiolipin or histones (Harkiss *et al.*, 1993) but there is no evidence that these are involved in disease pathogenesis. Studies to show reactivity to CNS antigens in animals suffering from visna were unsuccessful (Panitch *et al.*, 1976; Davies *et al.*, 1996).

## Role of host genetics in SRLV infection and disease

It is clear from the VMV epidemic in Iceland that host genetics play a major role in determining the course of lentiviral infection and lesion development. The Icelandic epidemic was started by the importation of 20 Karakul sheep onto the island to improve the local sheep breed. The imported sheep never showed clinical signs of SRLV infection and the parental flock from which they came also did not show signs of disease. Unfortunately Icelandic sheep are particularly susceptible to disease, especially visna, hence the epidemic. Where purebred and crossbred sheep have been infected with VMV there are clear differences in susceptibility to lesion development, however all the sheep do become infected (reviewed in Thormar, 2005). Also the seroprevalence of CAEV is apparently affected by goat breed (Grewal *et al.*, 1986). Host genetics are also important in determining which target organ is the site of lesion development: in sheep, arthritis (Anderson *et al.*, 1994) is a rare sequelae to SRLV infection whilst one third of infected goats present with this disease (Perry *et al.*, 1995). The goat leucocyte antigen system (major histocompatibility system) has been linked to CAEV induced arthritis with CLA Be7 being a protective MHC class I allele (Ruff and Lazary, 1988; Ruff *et al.*, 1993). It has not been determined if this protection is due to presentation of peptides to $CD8^+$ CTL. The CLA system has also been shown to influence the response of immunized goats to CAEV antigen (Fluri *et al.*, 2006). Whether it is peptide presentation on MHC that controls the inability of goats to initiate an IFNγ positive type 1 T-lymphocyte response and so the IgG2 anti-viral antibody that correlates with protection from arthritis is not known (Cheevers *et al.*, 1997).

## Vaccination

Early attempts to vaccinate against the SRLV used whole virus antigen preparations administered in adjuvant. The outcome after challenge was poor, with either no evidence of protection (Cutlip *et al.*, 1987) or the induction of worse pathology (McGuire *et al.*, 1986; Russo *et al.*, 1993). Since then various attempts at using subunit vaccines, different delivery methods and addition of immune stimulator molecules have been attempted. None have succeeded at inducing sterile immunity.

Attenuated viruses have been used as vaccines against both VMV and CAEV. These attenuated vaccines themselves cause persistent infection and do not induce sterile immunity to a challenge strain of virus. After immunization with a *tat* negative CAEV strain, immunized goats showed less wild type virus in the blood and inoculated joint by RT-PCR and also less pathology in the joints after challenge (Harmache *et al.*, 1996b; Harmache *et al.*, 1998). However the *tat* negative virus does itself induce some arthritis. The use of a *vif* negative CAEV virus preparation did not inhibit virus infection by wild-type virus but did reduce lesion severity (this was in two animals). Vif is essential for replication of SRLV and *vif* negative virus does not establish a persistent infection itself (Harmache *et al.*, 1996a; Harmache *et al.*, 1996b; Kristbjornsdottir *et al.*, 2004). Intratracheal inoculation with an attenuated VMV clone induced partial protection to intratracheal challenge with pathogenic VMV; this was only seen late after challenge (> 40 weeks) when there was a reduced frequency of viral isolations from blood and lung (Petursson *et al.*, 2005).

Subunit vaccines in a variety of different vector/delivery systems have also been tried. In goats the link between progression to pathology and a type 2 $CD4^+$ T helper cell IL-4 driven IgG1 dominated response to the Env protein of the virus (Cheevers *et al.*, 1997; Trujillo *et al.*, 2004a) means that both protection from infection and from progression to arthritis may be investigated. When vaccinia virus expressing CAEV *env*, or Env SU preparations administered in Freund's adjuvant were given to goats there was no protection from challenge with the homologous virus or alterations in the severity of the arthritis which developed subsequently (Cheevers *et al.*, 1994). Plasmid administration of *env* induces an IgG2 biased anti-Env response whilst that from vaccinia virus was IgG1 biased. This was maintained after boosting with SU protein (Beyer *et al.*, 2001). The vaccinia vector induced IgG1 bias could not be changed by co-administration of a vaccinia vector expressing caprine IL-12, a cytokine known to aid the induction of IFN-γ in other species (Cheevers *et al.*, 2000). Although after plasmid injection the T-lymphocyte response to SU antigen was dominated by IFN-γ production, this was not boosted by co-injection of plasmid encoding IFN-γ and interestingly in the presence of Tat, IFN-γ decreased the antibody response to Env (Cheevers *et al.*, 2001). A combination vaccine with a DNA prime using full length *env* and viral SU boost showed reduced viral replication after challenge and reduced joint pathology (Cheevers *et al.*, 2003). These studies also point out the importance of the vector used for delivery of the immunogen. The use of a T-cell-priming Gag peptide to immunize goats caused increased virus replication early after challenge (Nenci *et al.*, 2007). This was linked to increased GM-CSF levels in the vaccinated animals although in chronic infection this was not confirmed.

The development of pathology in VMV is not clearly linked to specific anti-viral responses. Early studies in Icelandic sheep with the neuropathogenic strains of virus investigated the CNS as the site of pathology. However more recent research uses the lung as the major site of pathology as this is the more common tissue in which pathology develops in non-Icelandic breeds of sheep. But this can be a difficult organ in which to detect and quantify lesion severity. To quantify viral replication, usually viral load is determined in the blood and lung in vaccine studies in sheep. Subunit vaccines using plasmid expressing *env* have been performed. Gene gun administration of an *env* expression plasmid into a mucosal site using a boost in the presence of an IFN-γ expression plasmid induced early protection from intratracheal challenge with VMV as measured by reduced viral load in the blood. This was in the absence of neutralizing antibody (Gonzalez *et al.*, 2005). Comparison of mucosal immunization with *gag* and/or *env* using a DNA prime/vaccinia virus boost strategy showed that although *env* caused a transient reduction in proviral load in blood it did not protect against the formation of lesions in the lung. However the use of *gag* alone did reduce the severity of lung lesions albeit in the presence of increased proviral load in the lung. In a similar way to that seen with CAEV (above), the presence of IFN-γ at priming did not alter these results (Reina *et al.*, 2008). When the same expression constructs were used to immunize sheep systemically, slightly different results were seen. Here the use of *gag* as an immunogen caused a reduction in provirus load in blood and lymphoid tissue whilst *env* did not alter provirus load but did increase the severity of lesions in the lung (Niesalla *et al.*, 2009). A study using a systemic DNA prime/protein boost with VMV *gag* also failed to induce sterilising immunity against intratracheal challenge and may even have increased the infectivity of the challenge virus as virus was isolated earlier from vaccinated sheep than from controls (Torsteinsdottir *et al.*, 2007b). Therefore a protective immunogen has not yet been identified in the SRLV although the link between Env and pathology is becoming clearer.

## Conclusion

We have presented data to show current information about the interaction of SRLV and monocyte/macrophages and dendritic cells and the immune system of the infected sheep or goat. This helps to clarify the use the virus makes of the immune system in maintaining a persistent infection and in causing pathology, and so our lack of success to date in developing good vaccines to control the virus. These include natural immune processes such as monocyte to macrophage maturation to allow virus replication, the mobility of dendritic cells and macrophages to exit sites of infection or lymphoid tissue and disseminate infection before the immune response is induced, and the movement of macrophages into tissues. The virus may alter cytokine expression to aid its gene expression and certainly makes use of the host's own cytokine production by the immune response to stimulate virus replication. This may have deleterious consequences for the host as pathology caused by the immune response to viral antigen develops and tissue function is lost. But even this may aid virus transmission to new hosts as more macrophages are recruited to sites of tissue damage and higher levels of virus replication are found. Finally, SRLV are superb models to study virus-host coevolution and lentivirus pathogenesis in a natural model characterized by the absence of lymphocyte infection and ensuing immunodeficiency.

### References


Adams, D.S., Crawford, T.B., Banks, K.L., McGuire, T.C., and Perryman, L.E. (1980). Immune responses of goats persistently infected with caprine arthritis- encephalitis virus. Infect. Immun. *28*, 421–427.

Agnarsdottir, G., Thorsteinsdottir, H., Oskarsson, T., Matthiasdottir, S., St Haflidadottir, B., Andresson, O.S., and Andresdottir, V. (2000). The long terminal repeat is a determinant of cell tropism of maedi-visna virus. J. Gen. Virol. *81*, 1901–1905.

Anderson, A.A., Harkiss, G.D., and Watt, N.J. (1994). Quantitative analysis of immunohistological changes in the synovial membrane of sheep infected with maedi-Visna virus. Clin. Immunol. Immunopathol. *72*, 21–29.

Andresdottir, V., Tang, X., Agnarsdottir, G., Andresson, O.S., Georgsson, G., Skraban, R., Torsteinsdottir, S., Rafnar, B., Benediktsdottir, E., Matthiasdottir, S., Arnadottir, S., Hognadottir, S., Palsson, P.A., and Petursson, G. (1998). Biological and genetic differences between lung- and brain-derived isolates of maedi-visna virus. Virus Genes. *16*, 281–293.

Andresdottir, V., Skraban, R., Matthiasdottir, S., Lutley, R., Agnarsdottir, G., and Thorsteinsdottir, H. (2002). Selection of antigenic variants in maedi-visna virus infection. J. Gen. Virol. *83*, 2543–2551.

Andresdottir, V. (2003). Evidence for recombination in the envelope gene of maedi-visna virus. Virus Genes. *27*, 5–9.

Andresson, O.S., Elser, J.E., Tobin, G.J., Greenwood, J.D., Gonda, M.A., Georgsson, G., Andresdottir, V., Benediktsdottir, E., Carlsdottir, H.M., Maentylae, E.O., Rafnar, B., Palsson, P.A., Casey, J.W., and Petursson, G. (1993). Nucleotide sequence and biological properties of a pathogenic proviral molecular clone of neurovirulent visna virus. Virology *193*, 89–105.

Angelopoulou, K., Brellou, G.D., and Vlemmas, I. (2006). Detection of maedi-visna virus in the kidneys of naturally infected sheep. J. Comp Pathol. *134*, 329–335.

Angelopoulou, K., Poutahidis, T., Brellou, G.D., Greenland, T., and Vlemmas, I. (2008). A deletion in the R region of long terminal repeats in small ruminant lentiviruses is associated with decreased pathology in the lung. Vet. J. *175*, 346–355.

Asaoka, K., Ikeda, K., Hishinuma, T., Horie-Inoue, K., Takeda, S., and Inoue, S. (2005). A retrovirus restriction factor TRIM5alpha is transcriptionally regulated by interferons. Biochem. Biophys. Res. Commun. *338*, 1950–1956.

Banks, K.L., Adams, D.S., McGuire, T.C., and Carlson, J. (1983). Experimental infection of sheep by caprine arthritis- encephalitis virus and goats by progressive pneumonia virus. Am. J. Vet. Res. *44*, 2307–2311.

Barber, S.A., Bruett, L., Douglass, B.R., Herbst, D.S., Zink, M.C., and Clements, J.E. (2002). Visna virus-induced activation of MAPK is required for virus replication and correlates with virus-induced neuropathology. J. Virol. *76*, 817–828.

Barros, S.C., Ramos, F., Duarte, M., Fagulha, T., Cruz, B., and Fevereiro, M. (2004). Genomic characterization of a slow/low maedi visna virus. Virus Genes *29*, 199–210.

Barros, S.C., Andresdottir, V., and Fevereiro, M. (2005). Cellular specificity and replication rate of maedi Visna virus *in vitro* can be controlled by LTR sequences. Arch. Virol. *150*, 201–213.

Bertoni, G., Zahno, M.-L., Zanoni, R., Vogt, H.-R., Peterhans, E., Ruff, G., Cheevers, W.P., Sonigo, P., and Pancino, G. (1994). Antibody reactivity to the immunodominant epitopes of the caprine arthritis-encephalitis virus gp38 transmembrane protein associates with the development of arthritis. J. Virol. *68*, 7139–7147.

Bertoni, G., Hertig, C., Zahno, M.L., Vogt, H.-R., Dufour, S., Cordano, P., Peterhans, E., Cheevers, W.P., Sonigo, P., and Pancino, G. (2000). B-cell epitopes of the envelope glycoprotein of caprine arthritis-encephalitis virus and antibody response in infected goats. J. Gen. Virol. *81*, 2929–2940.

Bertoni, G. (2007). Caprine Arthritis Encephalitis Complex. In Recent Advances in Goat Diseases, M. Tempesta, ed. (Ithaca, NY, USA: International Veterinary Information Service (www.ivis.org)).

Beyer, J.C., Chebloune, Y., Mselli-Lakhal, L., Hötzel, I., Kumpula-McWhirter, N., and Cheevers, W.P. (2001). Immunization with plasmid DNA expressing the caprine arthritis-encephalitis virus envelope gene: quantitative and qualitative aspects of antibody response to viral surface glycoprotein. Vaccine *19*, 1643–1651.

Bird, P., Blacklaws, B.A., Reyburn, H.T., Allen, D., Hopkins, J., Sargan, D., and McConnell, I. (1993). Early events in immune evasion by the lentivirus maedi-visna occurring within infected lymphoid tissue. J. Virol. *67*, 5187–5197.

Bird, P., Reyburn, H.T., Blacklaws, B.A., Allen, D., Nettleton, P., Yirrell, D.L., Watt, N., Sargan, D., and McConnell, I. (1995). The restricted IgG1 antibody response to maedi visna virus is seen following infection but not following immunization with recombinant gag protein. Clin. Exp. Immunol. *102*, 274–280.

Blacklaws, B.A., Bird, P., Allen, D., and McConnell, I. (1994). Circulating cytotoxic T-lymphocyte precursors in maedi-visna virus-infected sheep. J. Gen. Virol. *75*, 1589–1596.

Blacklaws, B.A., Bird, P., Allen, D., Roy, D.J., MacLennan, I.C.M., Hopkins, J., Sargan, D.R., and McConnell, I. (1995a). Initial lentivirus–host interactions within lymph nodes: A study of maedi-visna virus infection in sheep. J. Virol. *69*, 1400–1407.

Blacklaws, B.A., Bird, P., and McConnell, I. (1995b). Early events in infection of lymphoid tissue by a lentivirus, maedi- visna. Trends. Microbiol. *3*, 434–440.

Bolea, R., Monleon, E., Carrasco, L., Vargas, A., de Andres, D., Amorena, B., Badiola, J.J., and Lujan, L. (2006). maedi-visna virus infection of ovine mammary epithelial cells. Vet. Res. *37*, 133–144.

Brellou, G.D., Angelopoulou, K., Poutahidis, T., and Vlemmas, I. (2007). Detection of maedi-visna virus in the liver and heart of naturally infected sheep. J. Comp Pathol. *136*, 27–35.

Brodie, S.J., Marcom, K.A., Pearson, L.D., Anderson, B.C., Concha-Bermejillo, A., Ellis, J.A., and DeMartini, J.C. (1992). Effects of virus load in the pathogenesis of lentivirus-induced lymphoid interstitial pneumonia. J Infect. Dis. *166*, 531–541.

Brodie, S.J., Bickle, H.M., and DeMartini, J.C. (1995). Virological markers in cerebrospinal fluid are predictive of ovine lentivirus-associated subclinical encephalomyelitis. Clin. Immunol. Immunopathol. *77*, 14–18.

Campbell, B.J., and Avery, R.J. (1996). Sequence analysis and transcriptional activity of the LTR of OLV-CU1, a North American ovine lentivirus. J. Gen. Virol. *77 (Pt 12)*, 2999–3004.

Canivet, M., Hurpin, C., Rozain, F., Tanaka, A., Lasneret, J., Debons-Guillemin, M.C., and Peries, J. (1989). Effect of recombinant human alpha A interferon on Visna virus multiplication in acutely infected ovine cells. J. Biol. Regul. Homeost. Agents. *3*, 122–127.

Carroll, D., Ventura, P., Haase, A., Rinaldo, C.R., Jr, Overall, J.C., Jr, and Glasgow, L.A. (1978). Resistance of visna virus to interferon. J. Infect. Dis. *138*, 614–617.

Carrozza, M.L., Mazzei, M., Bandecchi, P., Arispici, M., and Tolari, F. (2003). In situ PCR-associated immunohistochemistry identifies cell types harbouring the maedi-Visna virus genome in tissue sections of sheep infected naturally. J Virol. Methods *107*, 121–127.

Carruth, L.M., Hardwick, J.M., Morse, B.A., and Clements, J.E. (1994). Visna virus Tat protein: A potent transcription factor with both activator and suppressor domains. J. Virol. *68*, 6137–6146.

Carruth, L.M., Morse, B.A., and Clements, J.E. (1996). The leucine domain of the visna virus Tat protein mediates targeting to an AP-1 site in the viral long terminal repeat. J. Virol. *70*, 4338–4344.

Carthagena, L., Parise, M.C., Ringeard, M., Chelbi-Alix, M.K., Hazan, U., and Nisole, S. (2008). Implication of TRIM alpha and TRIMCyp in interferon-induced anti-retroviral restriction activities. Retrovirology. *5*, 59.

Castro, R.S., Greenland, T., Leite, R.C., Gouveia, A., Mornex, J.F., and Cordier, G. (1999). Conserved sequence motifs involving the tat reading frame of Brazilian caprine lentiviruses indicate affiliations to both caprine arthritis-encephalitis virus and visna–maedi virus. J. Gen. Virol. *80 (Pt 7)*, 1583–1589.

Chebloune, Y., Sheffer, D., Karr, B.M., Stephens, E., and Narayan, O. (1996). Restrictive type of replication of ovine/caprine lentiviruses in ovine fibroblast cell cultures. Virology *222*, 21–30.

Cheevers, W.P., Knowles, D.P., Jr, and Norton, L.K. (1991). Neutralization-resistant antigenic variants of caprine arthritis- encephalitis lentivirus associated with progressive arthritis. J. Infect. Dis. *164*, 679–685.

Cheevers, W.P., McGuire, T.C., Norton, L.K., Cordery-Cotter, R., and Knowles, D.P. (1993). Failure of neutralizing antibody to regulate CAE lentivirus expression *in vivo*. Virology *196*, 835–839.

Cheevers, W.P., Knowles, D.P., McGuire, T.C., Baszler, T.V., and Hullinger, G.A. (1994). Caprine arthritis-encephalitis lentivirus (CAEV) challenge of goats immunized with recombinant vaccinia virus expressing CAEV surface and transmembrane envelope glycoproteins. Vet. Immunol. Immunopathol. *42*, 237–251.

Cheevers, W.P., Beyer, J.C., and Knowles, D.P. (1997). Type 1 and type 2 cytokine gene expression by viral gp135 surface protein-activated T-lymphocytes in caprine arthritis-encephalitis lentivirus infection. J. Virol. *71*, 6259–6263.

Cheevers, W.P., Cordery-Cotter, R., McGuire, T.C., and DeMartini, J.C. (1999). Neutralizing antibody responses and evolution of antigenic variants in monozygotic twin lambs infected with phenotypically distinct ovine lentiviruses. Virology *258*, 382–388.

Cheevers, W.P., Hötzel, I., Beyer, J.C., and Kumpula-McWhirter, N. (2000). Immune response to caprine arthritis-encephalitis virus surface protein induced by coimmunization with recombinant vaccinia viruses expressing the caprine arthritis-encephalitis virus envelope gene and caprine interleukin-12. Vaccine *18*, 2494–2503.

Cheevers, W.P., Beyer, J.C., and Hötzel, I. (2001). Plasmid DNA encoding caprine interferon gamma inhibits antibody response to caprine arthritis-encephalitis virus (CAEV) surface protein encoded by a co-administered plasmid expressing CAEV env and tat genes. Vaccine *19*, 3209–3215.

Cheevers, W.P., Snekvik, K.R., Trujillo, J.D., Kumpula-McWhirter, N.M., Pretty On Top K.J., and Knowles, D.P. (2003). Prime-boost vaccination with plasmid DNA encoding caprine-arthritis encephalitis lentivirus env and viral SU suppresses challenge virus and development of arthritis. Virology *306*, 116–125.

Clements, J.E., D'Antonio, N., and Narayan, O. (1982). Genomic changes associated with antigenic variation of visna virus. II. Common nucleotide sequence changes detected in variants from independent isolations. J. Mol. Biol. *158*, 415–434.

Clements, J.E., Gdovin, S.L., Montelaro, R.C., and Narayan, O. (1988). Antigenic variation in lentiviral diseases. Ann. Rev. Immunol. *6*, 139–159.

Clements, J.E., and Payne, S.L. (1994). Molecular basis of the pathobiology of lentiviruses. Virus Res. *32*, 97–109.

Clements, J.E., and Zink, M.C. (1996). Molecular biology and pathogenesis of animal lentivirus infections. Clin. Microbiol. Rev. *9*, 100–117.

Cork, L.C., Hadlow, W.J., Crawford, T.B., Gorham, J.R., and Piper, R.C. (1974). Infectious leukoencephalomyelitis of young goats. J. Infect. Dis. *129*, 134–141.

Cork, L.C., and Narayan, O. (1980). The pathogenesis of viral leukoencephalomyelitis-arthritis of goats. I. Persistent viral infection with progressive pathologic changes. Lab. Invest. *42*, 596–602.

Craig, L.E., Sheffer, D., Meyer, A.L., Hauer, D., Lechner, F., Peterhans, E., Adams, R.J., Clements, J.E., Narayan, O., and Zink, M.C. (1997). Pathogenesis of ovine lentiviral encephalitis: derivation of a neurovirulent strain by *in vivo* passage. J. Neurovirol. 3, 417–427.

Crane, S.E., Clements, J.E., and Narayan, O. (1988). Separate epitopes in the envelope of visna virus are responsible for fusion and neutralization: biological implications for anti- fusion antibodies in limiting virus replication. J. Virol. *62*, 2680–2685.

Crane, S.E., Kanda, P., and Clements, J.E. (1991). Identification of the fusion domain in the visna virus transmembrane protein. Virology *185*, 488–492.

Crawford, T.B., Adams, D.S., Cheevers, W.P., and Cork, L.C. (1980). Chronic arthritis in goats caused by a retrovirus. Science *207*, 997–999.

Cutlip, R.C., Lehmkuhl, H.D., Brogden, K.A., and Schmerr, M.J.F. (1987). Failure of experimental vaccines to protect against infection with ovine progressive pneumonia (maedi-visna) virus. Vet. Microbiol. *13*, 201–204.

Davies, J.M., Watt, N.J., Torsteinsdottir, S., and Carnegie, P.R. (1996). Mimicry of a 21.5 kDa myelin basic protein peptide by a maedi Visna virus polymerase peptide does not contribute to the pathogenesis of encephalitis in sheep. Vet. Immunol. Immunopathol. *55*, 127–139.

de Andres, D., Klein, D., Watt, N.J., Berriatua, E., Torsteinsdottir, S., Blacklaws, B.A., and Harkiss, G.D. (2005). Diagnostic tests for small ruminant lentiviruses. Vet. Microbiol. *107*, 49–62.

Dickson, J., and Ellis, T.M. (1989). Experimental Caprine Retrovirus Infection in Sheep. Vet. Rec. *125*, 649.

Dubois-Dalcq, M., Narayan, O., and Griffin, D.E. (1979). Cell surface changes associated with mutation of visna virus in antibody-treated cell cultures. Virology *92*, 353–366.

Duval, R., Bellet, V., Delebassee, S., and Bosgiraud, C. (2002a). Implication of caspases during maedi-visna virus-induced apoptosis. J. Gen. Virol. *83*, 3153–3161.

Duval, R., Delebassee, S., Cardot, P.J., and Bosgiraud, C. (2002b). Visna virus-induced cytopathic effect *in vitro* is caused by apoptosis. Arch. Virol. *147*, 943–959.

Ellis, J.A., and DeMartini, J.C. (1985). Ovine interleukin-2: partial purification and assay in normal sheep and sheep with ovine progressive pneumonia. Vet. Immunol. Immunopathol. *8*, 15–25.

Ellis, T.M., Wilcox, G.E., and Robinson, W.F. (1987). Antigenic variation of caprine arthritis-encephalitis virus during persistent infection of goats. J. Gen. Virol. *68*, 3145–3152.

Ellis, J.A., Russell, H.I., and Du, C.W. (1994). Effect of selected cytokines on the replication of *Corynebacterium pseudotuberculosis* and ovine lentiviruses in pulmonary macrophages. Vet. Immunol. Immunopathol. *40*, 31–47.

Erhouma, E., Guiguen, F., Chebloune, Y., Gauthier, D., Lakhal, L.M., Greenland, T., Mornex, J.F., Leroux, C., and Alogninouwa, T. (2008). Small ruminant lentivirus proviral sequences from wild ibexes in contact with domestic goats. J. Gen. Virol. *89*, 1478–1484.

Eriksson, K., McInnes, E., Ryan, S., Tonks, P., McConnell, I., and Blacklaws, B.A. (1999a). CD4(+) T-cells are required for the establishment of maedi-visna virus infection in macrophages but not dendritic cells *in vivo*. Virology *258*, 355–364.

Eriksson, K., McInnes, E., Ryan, S., Tonks, P., McConnell, I., and Blacklaws, B.A. (1999b). *In vivo* depletion of CD8+ cells does not affect primary maedi visna virus infection in sheep. Vet. Immunol. Immunopathol. *70*, 173–187.

Estes, D.M., Closser, N.M., and Allen, G.K. (1994). IFN-γamma stimulates IgG2 production from bovine B-cells costimulated with anti-mu and mitogen. Cell Immunol. *154*, 287–295.

Estes, D.M., and Brown, W.C. (2002). Type 1 and type 2 responses in regulation of Ig isotype expression in cattle. Vet. Immunol. Immunopathol. *90*, 1–10.

Fluri, A., Nenci, C., Zahno, M.L., Vogt, H.-R., Charan, S., Busato, A., Pancino, G., Peterhans, E., Obexer-Ruff, G., and Bertoni, G. (2006). The MHC-haplotype influences primary, but not memory, immune responses to an immunodominant peptide containing T- and B-cell epitopes of the caprine arthritis encephalitis virus Gag protein. Vaccine *24*, 597–606.

Freed, E.O., and Martin, M.A. (2007). HIVs and their replication. In Fields Virology, D.M. Knipe and P.M.

Howley, eds. (Philadelphia: Lippincott Williams and Wilkins), pp. 2107–2185.

Gabuzda, D.H., Hess, J.L., Small, J.A., and Clements, J.E. (1989). Regulation of the visna virus long terminal repeat in macrophages involves cellular factors that bind sequences containing AP-1 sites. Mol. Cell Biol. *9*, 2728–2733.

Gdovin, S.L., and Clements, J.E. (1992). Molecular mechanisms of visna virus Tat: Identification of the targets for transcriptional activation and evidence for a post-transcriptional effect. Virology *188*, 438–450.

Geballe, A.P., Ventura, P., Stowring, L., and Haase, A.T. (1985). Quantitative analysis of visna virus replication *in vivo*. Virology *141*, 148–154.

Gendelman, H.E., Narayan, O., Molineaux, S., Clements, J.E., and Ghotbi, Z. (1985). Slow, persistent replication of lentiviruses: role of tissue macrophages and macrophage precursors in bone marrow. Proc. Natl. Acad. Sci. U.S.A. *82*, 7086–7090.

Georgsson, G., Houwers, D.J., Palsson, P.A., and Petursson, G. (1989). Expression of viral antigens in the central nervous system of visna-infected sheep: an immunohistochemical study on experimental visna induced by virus strains of increased neurovirulence. Acta Neuropathol. 77, 299–306.

Georgsson, G., Torsteinsdóttir, S., Pétursson, G., Pálsson, P.A., and Andrésson, Ó.S. (1993). Role of the immune response in visna, a lentiviral central nervous system disease of sheep. In Animal models of HIV and other retroviral infections, P. Racz, N.L. Letvin, and J.C. Gluckman, eds. (Basel: Karger), pp. 183–195.

Germain, K., and Valas, S. (2006). Distribution and heterogeneity of small ruminant lentivirus envelope subtypes in naturally infected French sheep. Virus Res. *120*, 156–162.

Germain, K., Croise, B., and Valas, S. (2008). Field evaluation of a gag/env heteroduplex mobility assay for genetic subtyping of small-ruminant lentiviruses. J. Gen. Virol. *89*, 2020–2028.

Gjerset, B., Jonassen, C.M., and Rimstad, E. (2007). Natural transmission and comparative analysis of small ruminant lentiviruses in the Norwegian sheep and goat populations. Virus Res. *125*, 153–161.

Gonzalez, B., Reina, R., Garcia, I., Andres, S., Glaria, I., Alzueta, M., Mora, M.I., Jugo, B.M., Arrieta-Aguirre, I., de la Lastra, J.M., Rodriguez, D., Rodriguez, J.R., Esteban, M., Grillo, M.J., Blacklaws, B.A., Harkiss, G.D., Chebloune, Y., Lujan, L., de Andres, D., and Amorena, B. (2005). Mucosal immunization of sheep with a maedi-Visna virus (MVV) env DNA vaccine protects against early MVV productive infection. Vaccine *23*, 4342–4352.

Gorrell, M.D., Brandon, M.R., Sheffer, D., Adams, R.J., and Narayan, O. (1992). Ovine lentivirus is macrophagetropic and does not replicate productively in T-lymphocytes. J. Virol. *66*, 2679–2688.

Gourdou, I., Mazarin, V., Querat, G., Sauze, N., and Vigne, R. (1989). The open reading frame S of visna virus genome is a trans-activating gene. Virology *171*, 170–178.

Grego, E., Profiti, M., Giammarioli, M., Giannino, L., Rutili, D., Woodall, C., and Rosati, S. (2002). Genetic heterogeneity of small ruminant lentiviruses involves immunodominant epitope of capsid antigen and affects sensitivity of single-strain-based immunoassay. Clin. Diagn. Lab Immunol. *9*, 828–832.

Grego, E., Bertolotti, L., Quasso, A., Profiti, M., Lacerenza, D., Muz, D., and Rosati, S. (2007). Genetic characterization of small ruminant lentivirus in Italian mixed flocks: evidence for a novel genotype circulating in a local goat population. J. Gen. Virol. *88*, 3423–3427.

Grewal, A.S., Greenwood, P.E., Burton, R.W., Smith, J.E., Batty, E.M., and North, R. (1986). Caprine retrovirus infection in New South Wales: virus isolations, clinical and histopathological findings and prevalence of antibody. Aust. Vet. J. *63*, 245–248.

Grezel, D., Forestier, J., Guiguen, F., and Mornex, J.F. (1997). T-lymphocyte populations in the blood of caprine arthritis encephalitis virus-infected goats. Vet. Immunol. Immunopathol. *57*, 99–104.

Griffin, D.E., Narayan, O., and Adams, R.J. (1978). Early immune responses in visna, a slow viral disease of sheep. J. Infect. Dis. *138*, 340–350.

Grossi, P., Giudice, C., Bertoletti, I., Cioccarelli, G., Brocchi, E., Cammarata, G., and Gelmetti, D. (2005). Immunohistochemical Detection of the p27 Capsid Protein of Caprine Arthritis-Encephalitis Virus (CAEV) in Bone-marrow Cells of Seropositive Goats. J. Comp. Pathol. *133*, 197–200.

Gudmundsson, B., Jonsson, S.R., Olafsson, O., Agnarsdottir, G., Matthiasdottir, S., Georgsson, G., Torsteinsdottir, S., Svansson, V., Kristbjornsdottir, H.B., Franzdottir, S.R., Andresson, O.S., and Andresdottir, V. (2005). Simultaneous mutations in CA and Vif of maedi-Visna virus cause attenuated replication in macrophages and reduced infectivity *in vivo*. J. Virol. *79*, 15038–15042.

Gudnadottir, M. (1974). Visna-maedi in sheep. Prog. Med. Virol. *18*, 336–349.

Guiguen, F., Mselli-Lakhal, L., Durand, J., Du, J., Favier, C., Fornazero, C., Grezel, D., Balleydier, S., Hausmann, E., and Chebloune, Y. (2000). Experimental infection of Mouflon-domestic sheep hybrids with caprine arthritis-encephalitis virus. Am. J. Vet. Res. *61*, 456–461.

Haase, A.T., Stowring, L., Narayan, P., Griffin, D., and Price, D. (1977). Slow persistent infection caused by visna virus: role of host restriction. Science *195*, 175–177.

Haase, A.T. (1986). Pathogenesis of lentivirus infections. Nature *322*, 130–136.

Haflidadottir, B.S., Matthiasdottir, S., Agnarsdottir, G., Torsteinsdottir, S., Petursson, G., Andresson, O.S., and Andresdottir, V. (2008). Mutational analysis of a principal neutralization domain of visna/maedi virus envelope glycoprotein. J. Gen. Virol. *89*, 716–721.

Haller, O., Kochs, G., and Weber, F. (2006). The interferon response circuit: Induction and suppression by pathogenic viruses. Virology *344*, 119–130.

Harkiss, G.D., Veitch, D., Dickson, L., and Watt, N.J. (1993). Autoimmune reactivity in sheep induced by the visna retrovirus. J. Autoimmun. *6*, 63–75.

Harmache, A., Vitu, C., Russo, P., Bouyac, M., Hieblot, C., Peveri, P., Vigne, R., and Suzan, M. (1995). The caprine arthritis encephalitis virus tat gene is dispensable for efficient viral replication *in vitro* and *in vivo*. J. Virol. *69*, 5445–5454.

Harmache, A., Russo, P., Guiguen, F., Vitu, C., Vignon, M., Bouyac, M., Hieblot, C., Pepin, M., Vigne, R., and Suzan, M. (1996a). Requirement of caprine arthritis encephalitis virus vif gene for *in vivo* replication. Virology *224*, 246–255.

Harmache, A., Russo, P., Vitu, C., Guiguen, F., Mornex, J.F., Pepin, M., Vigne, R., and Suzan, M. (1996b). Replication in goats *in vivo* of caprine arthritis-encephalitis virus deleted in vif or tat genes: possible use of these deletion mutants as live vaccines. AIDS Res. Hum. Retroviruses *12*, 409–411.

Harmache, A., Vitu, C., Guiguen, F., Russo, P., Bertoni, G., Pepin, M., Vigne, R., and Suzan, M. (1998). Priming with tat-deleted caprine arthritis encephalitis virus (CAEV) proviral DNA or live virus protects goats from challenge with pathogenic CAEV. J. Virol. *72*, 6796–6804.

Harris, R.S., and Liddament, M.T. (2004). Retroviral restriction by APOBEC proteins. Nat. Rev. Immunol. *4*, 868–877.

Hess, J.L., Clements, J.E., and Narayan, O. (1985). cis- and trans-acting transcriptional regulation of visna virus. Science *229*, 482–485.

Hess, J.L., Pyper, J.M., and Clements, J.E. (1986). Nucleotide sequence and transcriptional activity of the caprine arthritis-encephalitis virus long terminal repeat. J. Virol. *60*, 385–393.

Hess, J.L., Small, J.A., and Clements, J.E. (1989). Sequences in the visna virus long terminal repeat that control transcriptional activity and respond to viral trans-activation: Involvement of AP-1 sites in basal activity and trans- activation. J. Virol. *63*, 3001–3015.

Hessell, A.J., Hangartner, L., Hunter, M., Havenith, C.E., Beurskens, F.J., Bakker, J.M., Lanigan, C.M., Landucci, G., Forthal, D.N., Parren, P.W., Marx, P.A., and Burton, D.R. (2007). Fc receptor but not complement binding is important in antibody protection against HIV. Nature *449*, 101–104.

Hötzel, I., and Cheevers, W.P. (2001). Host range of small-ruminant lentivirus cytopathic variants determined with a selectable caprine arthritis- encephalitis virus pseudotype system. J. Virol. 75, 7384–7391.

Hötzel, I., and Cheevers, W.P. (2002a). A maedi-visna virus strain K1514 receptor gene is located in sheep chromosome 3p and the syntenic region of human chromosome 2. J. Gen. Virol. *83*, 1759–1764.

Hötzel, I., and Cheevers, W.P. (2002b). Differential receptor usage of small ruminant lentiviruses in ovine and caprine cells: host range but not cytopathic phenotype is determined by receptor usage. Virology *301*, 21–31.

Hötzel, I., Kumpula-McWhirter, N., and Cheevers, W.P. (2002). Rapid evolution of two discrete regions of the caprine arthritis-encephalitis virus envelope surface glycoprotein during persistent infection. Virus Res. 84: 17-25.

Hou, W., Wang, X., Ye, L., Zhou, L., Yang, Z.Q., Riedel, E., and Ho, W.Z. (2009). Lambda interferon inhibits human immunodeficiency virus type 1 infection of macrophages. J. Virol. *83*, 3834–3842.

Houwers, D.J., and van der Molen, E.J. (1987). A five-year serological study of natural transmission of maedi-visna virus in a flock of sheep, completed with post mortem investigation. Zentralbl. Veterinarmed. B *34*, 421–431.

Houwers, D.J., and Nauta, I.M. (1989). Immunoblot analysis of the antibody response to ovine lentivirus infections. Vet. Microbiol. *19*, 127–139.

Huso, D.L. (1997). Neutralizing and fusion-blocking antibodies induced by recombinant visna virus envelope glycoprotein. Viral. Immunol. *10*, 15–20.

Jackson, M.K., Knowles, D.P., Stem, T.A., Harwood, W.G., Robinson, M.M., and Cheevers, W.P. (1991). Genetic structure of the *pol-env* region of the caprine arthritis-encephalitis lentivirus genome. Virology *180*, 389–394.

Jin, X., Bauer, D.E., Tuttleton, S.E., Lewin, S., Gettie, A., Blanchard, J., Irwin, C.E., Safrit, J.T., Mittler, J., Weinberger, L., Kostrikis, L.G., Zhang, L., Perelson, A.S., and Ho, D.D. (1999). Dramatic rise in plasma viremia after CD8(+) T-cell depletion in simian immunodeficiency virus-infected macaques. J. Exp. Med. *189*, 991–998.

Johnson, G.C., Barbet, A.F., Klevjer-Anderson, P., and McGuire, T.C. (1983). Preferential Immune Response to virion surface glycoproteins by caprine arthritis-encephalitis virus-infected goats. Infect. Immun. *41*, 657–665.

Jolly, P.E., Huso, D., Hart, G., and Narayan, O. (1989a). Modulation of lentivirus replication by antibodies. Non- neutralizing antibodies to caprine arthritis-encephalitis virus enhance early stages of infection in macrophages, but do not cause increased production of virions. J. Gen. Virol. *70*, 2221–2226.

Jolly, P.E., Huso, D.L., Sheffer, D., and Narayan, O. (1989b). Modulation of lentivirus replication by antibodies: Fc portion of immunoglobulin molecule is essential for enhancement of binding, internalization, and neutralization of visna virus in macrophages. J. Virol. *63*, 1811–1813.

Juste, R.A., Ott, T.L., Kwang, J., Bazer, F.W., and De la Concha-Bermejillo, A. (1996). Effects of recombinant interferon-tau on ovine lentivirus replication. J. Interferon. Cytokine. Res. *16*, 989–994.

Juste, R.A., Ott, T.L., Kwang, J., Bazer, F.W., and Concha-Bermejillo, A. (2000). Effects of recombinant ovine interferon-tau on ovine lentivirus replication and progression of disease. J. Gen. Virol. *81*, 525–532.

Kajikawa, O., Lairmore, M.D., and DeMartini, J.C. (1990). Analysis of Antibody Responses to Phenotypically Distinct Lentiviruses. J. Clin. Microbiol. *28*, 764–770.

Kalinski, H., Mashiah, P., Rotem, D., Orzech, Y., Sherman, L., Miki, T., Yaniv, A., Gazit, A., and Tronick, S.R. (1994). Characterization of cDNAs species encoding the Tat protein of caprine arthritis encephalitis virus. Virology *204*, 828–834.

Kemp, R.K., Knowles, D.P., Perry, L.L., McGuire, T.C., Besser, T.E., and Cheevers, W.P. (2000). Crossreactive neutralizing antibodies induced by immunization with caprine arthritis-encephalitis virus surface glycoprotein. Vaccine *18*, 1282–1287.

Kennedy, P.G., Narayan, O., Ghotbi, Z., Hopkins, J., Gendelman, H.E., and Clements, J.E. (1985). Persistent expression of Ia antigen and viral genome in visna–maedi virus-induced inflammatory cells. Possible role of lentivirus-induced interferon. J. Exp. Med. *162*, 1970–1982.

Kennedy, P.G., Narayan, O., Zink, M.C., Hess, J., Clements, J.E., and Adams, R.J. (1989). The pathogenesis of visna, a lentivirus induced immunopathologic disease of the central nervous system. In Clinical and Molecular Aspects of Neurotropic Virus Infection, D.H. Gilden and H.L. Lipton, eds. (New York: Kluwer Academic Publishers), pp. 393–421.

Kennedy-Stoskopf, S., and Narayan, O. (1986). Neutralizing antibodies to visna lentivirus: mechanism of action and possible role in virus persistence. J. Virol. *59*, 37–44.

Kennedy-Stoskopf, S., Zink, C., and Narayan, O. (1989). Pathogenesis of ovine lentivirus-induced arthritis: Phenotypic evaluation of T-lymphocytes in synovial fluid, synovium, and peripheral circulation. Clin. Immunol. Immunopathol. *52*, 323–330.

Klevjer-Anderson, P., and McGuire, T.C. (1982). Neutralizing Antibody Response of Rabbits and Goats to Caprine Arthritis-Encephalitis Virus. Infect. Immun. *38*, 455–461.

Knowles, D.P., Cheevers, W.P., McGuire, T.C., Stem, T.A., and Gorham, J.R. (1990). Severity of Arthritis Is Predicted by Antibody Response to Gp135 in Chronic Infection with Caprine Arthritis- Encephalitis Virus. J. Virol. *64*, 2396–2398.

Kristbjornsdottir, H.B., Andresdottir, V., Svansson, V., Torsteinsdottir, S., Matthiasdottir, S., and Andresson, O.S. (2004). The vif gene of maedi-visna virus is essential for infectivity *in vivo* and *in vitro*. Virology *318*, 350–359.

Lacerenza, D., Giammarioli, M., Grego, E., Marini, C., Profiti, M., Rutili, D., and Rosati, S. (2006). Antibody response in sheep experimentally infected with different small ruminant lentivirus genotypes. Vet. Immunol. Immunopathol. *112*, 264–271.

Lairmore, M.D., Akita, G.Y., Russell, H.I., and DeMartini, J.C. (1987). Replication and cytopathic effects of ovine lentivirus strains in alveolar macrophages correlate with *in vivo* pathogenicity. J. Virol. *61*, 4038–4042.

Lairmore, M.D., Butera, S.T., Callahan, G.N., and DeMartini, J.C. (1988). Spontaneous interferon production by pulmonary leukocytes is associated with lentivirus-induced lymphoid interstitial pneumonia. J. Immunol. *140*, 779–785.

Larsen, H.J., Hyllseth, B., and Krogsrud, J. (1982a). Experimental maedi virus infection in sheep: cellular and humoral immune response during three years following intranasal inoculation. Am. J. Vet. Res. *43*, 384–389.

Larsen, H.J., Hyllseth, B., and Krogsrud, J. (1982b). Experimental maedi virus infection in sheep: early cellular and humoral immune response following parenteral inoculation. Am. J. Vet. Res. *43*, 379–383.

Law, M., and Hangartner, L. (2008). Antibodies against viruses: passive and active immunization. Curr. Opin. Immunol. *20*, 486–492.

Lechner, F., Vogt, H.-R., Seow, H.F., von Bodungen, U., Bertoni, G., Zurbriggen, A., and Peterhans, E. (1996). Expression of TNF alpha in arthritis caused by caprine arthritis encephalitis virus. Vet. Immunol. Immunopathol. *54*, 281–289.

Lechner, F., Machado, J., Bertoni, G., Seow, H.F., Dobbelaere, D.A., and Peterhans, E. (1997a). Caprine arthritis encephalitis virus dysregulates the expression of cytokines in macrophages. J. Virol. *71*, 7488–7497.

Lechner, F., Vogt, H.-R., Seow, H.F., Bertoni, G., Cheevers, W.P., von Bodungen, U., Zurbriggen, A., and Peterhans, E. (1997b). Expression of cytokine mRNA in lentivirus-induced arthritis. Am. J. Pathol. *151*, 1053–1065.

Lee, W.C., McConnell, I., and Blacklaws, B.A. (1994). Cytotoxic activity against maedi-visna virus-infected macrophages. J. Virol. *68*, 8331–8338.

Legastelois, I., Cordier, G., Cozon, G., Cadore, J.L., Guiguen, F., Greenland, T., and Mornex, J.F. (1996). Visna-maedi virus-induced expression of interleukin-8 gene in sheep alveolar cells following experimental *in vitro* and *in vivo* infection. Res. Virol. *147*, 191–197.

Legastelois, I., Levrey, H., Greenland, T., Mornex, J.F., and Cordier, G. (1998). Visna-maedi virus induces interleukin-8 in sheep alveolar macrophages through a tyrosine-kinase signaling pathway. Am. J. Respir. Cell Mol. Biol. *18*, 532–537.

Lerondelle, C., Godet, M., and Mornex, J.F. (1999). Infection of primary cultures of mammary epithelial cells by small ruminant lentiviruses. Vet. Res. *30*, 467–474.

Leroux, C., Vuillermoz, S., Mornex, J.-F., and Greenland, T. (1995). Genomic heterogeneity in the pol region of ovine lentiviruses obtained from bronchoalveolar cells of infected sheep from France. J. Gen. Virol. *76*, 1533–1537.

Leroux, C., Chastang, J., Greenland, T., and Mornex, J.F. (1997). Genomic heterogeneity of small ruminant lentiviruses: existence of heterogeneous populations in sheep and of the same lentiviral genotypes in sheep and goats. Arch. Virol. *142*, 1125–1137.

Lichtensteiger, C.A., Knowles, D.P., Jr, McGuire, T.C., and Cheevers, W.P. (1991). Recombinant gp135 envelope glycoproteins of caprine arthritis- encephalitis lentivirus variants inhibit homologous and heterologous variant-specific neutralizing antibodies. Virology *185*, 2–9.

Lichtensteiger, C.A., Cheevers, W.P., and Davis, W.C. (1993). $CD8^+$ cytotoxic T-lymphocytes against antigenic variants of caprine arthritis-encephalitis virus. J. Gen. Virol. *74*, 2111–2116.

Luban, J. (2007). Cyclophilin A, TRIM5, and resistance to human immunodeficiency virus type 1 infection. J. Virol. *81*, 1054–1061.

Lujan, L., Begara, I., Collie, D.D.S., and Watt, N.J. (1995). CD8+ lymphocytes in bronchoalveolar lavage and blood: *In vivo* indicators of lung pathology caused by maedi-visna virus. Vet. Immunol. Immunopathol. *49*, 89–100.

Lutley, R., Petursson, G., Palsson, P.A., Georgsson, G., Klein, J., and Nathanson, N. (1983). Antigenic drift in visna: virus variation during long-term infection of Icelandic sheep. J. Gen. Virol. *64 (Pt 7)*, 1433–1440.

Lyall, J.W., Solanky, N., and Tiley, L.S. (2000). Restricted species tropism of maedi-visna virus strain EV-1 is not due to limited receptor distribution. J. Gen. Virol. *81*, 2919–2927.

Matano, T., Shibata, R., Siemon, C., Connors, M., Lane, H.C., and Martin, M.A. (1998). Administration of an anti-CD8 monoclonal antibody interferes with the clearance of chimeric simian/human immunodeficiency virus during primary infections of rhesus macaques. J. Virol. *72*, 164–169.

McGuire, T.C., Adams, D.S., Johnson, G.C., Klevjer-Anderson, P., Barbee, D.D., and Gorham, J.R. (1986). Acute arthritis in caprine arthritis-encephalitis virus challenge exposure of vaccinated or persistently infected goats. Am. J. Vet. Res. *47*, 537–540.

McGuire, T.C., Norton, L.K., O'Rourke, K.I., and Cheevers, W.P. (1988). Antigenic variation of neutralization-sensitive epitopes of caprine arthritis-encephalitis lentivirus during persistent arthritis. J. Virol. *62*, 3488–3492.

McGuire, T.C., Knowles, D.P., Jr, Davis, W.C., Brassfield, A.L., Stem, T.A., and Cheevers, W.P. (1992). Transmembrane protein oligomers of caprine arthritis-encephalitis lentivirus are immunodominant in goats with progressive arthritis. J. Virol. *66*, 3247–3250.

McNeilly, T.N., Tennant, P., Lujan, L., Perez, M., and Harkiss, G.D. (2007). Differential infection efficiencies of peripheral lung and tracheal tissues in sheep infected with Visna/maedi virus via the respiratory tract. J. Gen. Virol. *88*, 670–679.

McNeilly, T.N., Baker, A., Brown, J.K., Collie, D., Maclachlan, G., Rhind, S.M., and Harkiss, G.D. (2008). Role of alveolar macrophages in respiratory transmission of visna/maedi virus. J. Virol. *82*, 1526–1536.

Mehta, P.D., and Thormar, H. (1974). Neutralizing activity in isolated serum antibody fractions from visna-infected sheep. Infect. Immun. *10*, 678–680.

Mercier, G., Galien, R., and Emanoil-Ravier, R. (1994). Differential effects of ras and jun family members on complex retrovirus promoter activities. Res. Virol. *145*, 361–367.

Mordasini, F., Vogt, H.R., Zahno, M.L., Maeschli, A., Nenci, C., Zanoni, R., Peterhans, E., and Bertoni, G. (2006). Analysis of the antibody response to an immunodominant epitope of the envelope glycoprotein of a lentivirus and its diagnostic potential. J. Clin. Microbiol. *44*, 981–991.

Morin, T., Guiguen, F., Bouzar, B.A., Villet, S., Greenland, T., Grezel, D., Gounel, F., Gallay, K., Garnier, C., Durand, J., Alogninouwa, T., Mselli-Lakhal, L., Mornex, J.F., and Chebloune, Y. (2003). Clearance of a productive lentivirus infection in calves experimentally inoculated with caprine arthritis-encephalitis virus. J. Virol. *77*, 6430–6437.

Morse, B.A., Carruth, L.M., and Clements, J.E. (1999). Targeting of the visna virus tat protein to AP-1 sites: interactions with the bZIP domains of fos and jun *in vitro* and *in vivo*. J. Virol. *73*, 37–45.

Mselli-Lakhal, L., Favier, C., Leung, K., Guiguen, F., Grezel, D., Miossec, P., Mornex, J.F., Narayan, O., Querat, G., and Chebloune, Y. (2000). Lack of functional receptors is the only barrier that prevents caprine arthritis-encephalitis virus from infecting human cells. J. Virol. *74*, 8343–8348.

Murphy, B.G., Hötzel, I., Jasmer, D.P., Davis, W.C., and Knowles, D. (2006). TNFαlpha and GM-CSF-induced activation of the CAEV promoter is independent of AP-1. Virology *352*, 188–199.

Myer, M.S., Huchzermeyer, H.F., York, D.F., Hunter, P., Verwoerd, D.W., and Garnett, H.M. (1988). The possible involvement of immunosuppression caused by a lentivirus in the aetiology of jaagsiekte and pasteurellosis in sheep. Onderstepoort J. Vet. Res. *55*, 127–133.

Narayan, O., Griffin, D.E., and Chase, J. (1977). Antigenic shift of visna virus in persistently infected sheep. Science *197*, 376–378.

Narayan, O., Griffin, D.E., and Clements, J.E. (1978). Virus mutation during 'slow infection': temporal development and characterization of mutants of visna virus recovered from sheep. J. Gen. Virol. *41*, 343–352.

Narayan, O., Clements, J.E., Griffin, D.E., and Wolinsky, J.S. (1981). Neutralizing antibody spectrum determines the antigenic profiles of emerging mutants of visna virus. Infect Immun. *32*, 1045–1050.

Narayan, O., Wolinsky, J.S., Clements, J.E., Strandberg, J.D., Griffin, D.E., and Cork, L.C. (1982). Slow virus replication: the role of macrophages in the persistence and expression of visna viruses of sheep and goats. J. Gen. Virol. *59*, 345–356.

Narayan, O., Kennedy-Stoskopf, S., Sheffer, D., Griffin, D.E., and Clements, J.E. (1983). Activation of caprine arthritis-encephalitis virus expression during maturation of monocytes to macrophages. Infect. Immun. *41*, 67–73.

Narayan, O., Sheffer, D., Griffin, D.E., Clements, J., and Hess, J. (1984). Lack of neutralizing antibodies to caprine arthritis-encephalitis lentivirus in persistently infected goats can be overcome by immunization with inactivated Mycobacterium tuberculosis. J. Virol. *49*, 349–355.

Narayan, O., and Cork, L.C. (1985). Lentiviral diseases of sheep and goats: chronic pneumonia leukoencephalomyelitis and arthritis. Rev. Infect. Dis. *7*, 89–98.

Narayan, O., Sheffer, D., Clements, J.E., and Tenekoon, G. (1985). Restricted replication of lentiviruses: visna viruses induce a unique interferon during interaction between lymphocytes and infected macrophages. J. Exp. Med. *162*, 1954–1969.

Nathanson, N., Panitch, H., Palsson, P.A., Petursson, G., and Georgsson, G. (1976). Pathogenesis of visna. II. Effect of immunosuppression upon early central nervous system lesions. Lab. Invest. 35, 444–451.

Nathanson, N., Martin, J.R., Georgsson, G., Palsson, P.A., Lutley, R.E., and Petursson, G. (1981). The effect of post-infection immunization on the severity of experimental visna. J Comp Pathol. *91*, 185–191.

Nenci, C., Zahno, M.L., Vogt, H.-R., Obexer-Ruff, G., Doherr, M.G., Zanoni, R., Peterhans, E., and Bertoni, G. (2007). Vaccination with a T-cell-priming Gag peptide of caprine arthritis encephalitis virus enhances virus replication transiently *in vivo*. J. Gen. Virol. *88*, 1589–1593.

Neuveut, C., Vigne, R., Clements, J.E., and Sire, J. (1993). The visna transcriptional activator TAT: Effects on the viral LTR and on cellular genes. Virology *197*, 236–244.

Niesalla, H., de Andrés, X., Barbezange, C., Fraisier, C., Reina, R., Arnarson, H., Biescas, E., Mazzei, M., McNeilly, T.N., Liu, C., Watkins, C., Perez, M., Carrozza, M.L., Bandecchi, P., Solano, C., Crespo, H., Glaria, I., Huard, C., Shaw, D.J., de Blas, I., de Andres, D., Tolari, F., Rosati, S., Suzan-Monti, M., Andresdottir, V., Torsteinsdottir, S., Petursson, G., Badiola, J., Lujan, L., Pepin, M., Amorena, B., Blacklaws, B., and Harkiss, G.D. (2009). Systemic DNA immunization against ovine lentivirus using particle-mediated epidermal delivery and modified vaccinia Ankara encoding the gag and/or env genes. Vaccine *27*, 260–269.

Oliver, R.E., McNiven, R.A., Julian, A.F., and Poole, W.S. (1982). Experimental infection of sheep and goats with caprine arthritis-encephalitis virus. N. Z. Vet. J. *30*, 158–159.

Oskarsson, T., Hreggvidsdottir, H.S., Agnarsdottir, G., Matthiasdottir, S., Ogmundsdottir, M.H., Jonsson, S.R., Georgsson, G., Ingvarsson, S., Andresson, O.S., and Andresdottir, V. (2007). Duplicated sequence motif in the long terminal repeat of maedi-visna virus extends cell tropism and is associated with neurovirulence. J. Virol. *81*, 4052–4057.

Panitch, H., Petursson, G., Georgsson, G., Palsson, P.A., and Nathanson, N. (1976). Pathogenesis of visna. III. Immune responses to central nervous system antigens in experimental allergic encephalomyelitis and visna. Lab. Invest. *35*, 452–460.

Peluso, R., Haase, A., Stowring, L., Edwards, M., and Ventura, P. (1985). A Trojan Horse mechanism for the spread of visna virus in monocytes. Virology *147*, 231–236.

Peng, G., Lei, K.J., Jin, W., Greenwell-Wild, T., and Wahl, S.M. (2006). Induction of APOBEC3 family proteins, a defensive maneuver underlying interferon-induced anti-HIV-1 activity. J. Exp. Med. *203*, 41–46.

Pepin, M., Vitu, C., Russo, P., Mornex, J.F., and Peterhans, E. (1998). maedi-visna virus infection in sheep: a review. Vet. Res. *29*, 341–367.

Perry, L.L., Wilkerson, M.J., Hullinger, G.A., and Cheevers, W.P. (1995). Depressed $CD4^+$ T-lymphocyte proliferative response and enhanced antibody response to viral antigen in chronic lentivirus- induced arthritis. J. Infect. Dis. *171*, 328–334.

Peterhans, E., Greenland, T., Badiola, J., Harkiss, G., Bertoni, G., Amorena, B., Eliaszewicz, M., Juste, R.A., Krassnig, R., Lafont, J.P., Lenihan, P., Petursson, G., Pritchard, G., Thorley, J., Vitu, C., Mornex, J.F., and Pepin, M. (2004). Routes of transmission and consequences of small ruminant lentiviruses (SRLVs) infection and eradication schemes. Vet. Res. *35*, 257–274.

Pétursson, G.,, D.B.M., and Lutley, R. (1985). Immunoglobulin subclass distribution and restriction of antibody response in visna. (Brussels: Commission of teh European Communities, Directorate General for Agriculture Coordination of Agricultural Research).

Petursson, G., Matthiasdottir, S., Svansson, V., Andresdottir, V., Georgsson, G., Martin, A.H., Agnarsdottir, G., Gisladottir, E., Arnadottir, S., Hognadottir, S., Jonsson, S.R., Andresson, O.S., and Torsteinsdottir, S. (2005). Mucosal vaccination with an attenuated maedi-visna virus clone. Vaccine *23*, 3223–3228.

Philippon, V., Vellutini, C., Gambarelli, D., Harkiss, G., Arbuthnott, G., Metzger, D., Roubin, R., and Filippi, P. (1994). The basic domain of the lentiviral Tat protein is responsible for damages in mouse brain: involvement of cytokines. Virology *205*, 519–529.

Pisoni, G., Quasso, A., and Moroni, P. (2005). Phylogenetic analysis of small-ruminant lentivirus subtype B1 in mixed flocks: Evidence for natural transmission from goats to sheep. Virology *339*, 147–152.

Pisoni, G., Bertoni, G., Puricelli, M., Maccalli, M., and Moroni, P. (2007a). Demonstration of coinfection with and recombination by caprine arthritis-encephalitis virus and maedi-visna virus in naturally infected goats. J. Virol. *81*, 4948–4955.

Pisoni, G., Moroni, P., Turin, L., and Bertoni, G. (2007b). Compartmentalization of small ruminant lentivirus between blood and colostrum in infected goats. Virology *369*, 119–130.

Preziuso, S., Renzoni, G., Allen, T.E., Taccini, E., Rossi, G., DeMartini, J.C., and Braca, G. (2004). Colostral transmission of maedi visna virus: sites of viral entry in lambs born from experimentally infected ewes. Vet. Microbiol. *104*, 157–164.

Pyper, J.M., Clements, J.E., Gonda, M.A., and Narayan, O. (1986). Sequence homology between cloned caprine arthritis encephalitis virus and visna virus, two neurotropic lentiviruses. J. Virol. *58*, 665–700.

Pyrah, I.T.G., and Watt, N.J. (1996). Immunohistological study of the depressed cutaneous DTH response in sheep naturally infected with an ovine lentivirus (maedi-Visna virus). Clin. Exp. Immunol. *104*, 32–36.

Querat, G., Audoly, G., Sonigo, P., and Vigne, R. (1990). Nucleotide sequence analysis of SA-OMVV, a visna-related ovine lentivirus: phylogenetic history of lentiviruses. Virology *175*, 434–447.

Rafnar, B., Tobin, G.J., Nagashima, K., Gonda, M.A., Gunnarsson, E., Andresson, O.S., Georgsson, G., and Torsteinsdottir, S. (1998). Immune response to recombinant visna virus Gag and Env precursor proteins synthesized in insect cells. Virus Res. *53*, 107–120.

Ravazzolo, A.P., Nenci, C., Vogt, H.-R., Waldvogel, A., Obexer-Ruff, G., Peterhans, E., and Bertoni, G. (2006). Viral load, organ distribution, histopathological lesions, and cytokine mRNA expression in goats infected with a molecular clone of the caprine arthritis encephalitis virus. Virology *350*, 116–127.

Rea-Boutrois, A., Villet, S., Greenland, T., Mehlen, P., Chebloune, Y., Verdier, G., and Legras-Lachuer, C. (2009). Small ruminant lentivirus Tat protein induces apoptosis in caprine cells *in vitro* by the intrinsic pathway. Virology *383*, 93–102.

Reina, R., Glaria, I., Benavides, J., de Andrés, X., Crespo, H., Solano, C., Perez, V., Lujan, L., Perez, M.M., de la Lastra, J.M., Rosati, S., Blacklaws, B., Harkiss, G., de Andres, D., and Amorena, B. (2007). Association of CD80 and CD86 Expression Levels with Disease Status of Visna/maedi Virus Infected Sheep. Viral Immunol. *20*, 609–622.

Reina, R., Barbezange, C., Niesalla, H., de Andrés, X., Arnarson, H., Biescas, E., Mazzei, M., Fraisier, C., McNeilly, T.N., Liu, C., Perez, M., Carrozza, M.L., Bandecchi, P., Solano, C., Crespo, H., Glaria, I., Huard, C., Shaw, D.J., de Blas, I., de Andres, D., Tolari, F., Rosati, S., Suzan-Monti, M., Andresdottir, V., Torsteinsdottir, S., Petursson, G., Lujan, L., Pepin, M., Amorena, B., Blacklaws, B., and Harkiss, G.D. (2008). Mucosal immunization against ovine lentivirus using PEI–DNA complexes and modified vaccinia Ankara encoding the gag and/or env genes. Vaccine *26*, 4494–4505.

Reina, R., Grego, E., Bertolotti, L., De Meneghi, D., and Rosati, S. (2009). Genome analysis of small-ruminant lentivirus genotype E: a caprine lentivirus with natural deletions of the dUTPase subunit, vpr-like accessory gene, and 70-base-pair repeat of the U3 region. J. Virol. *83*, 1152–1155.

Reyburn, H.T., Roy, D.J., Blacklaws, B.A., Sargan, D.R., Watt, N.J., and McConnell, I. (1992). Characteristics of the T-cell-mediated immune response to maedi-visna virus. Virology *191*, 1009–1012.

Rolland, M., Mooney, J., Valas, S., Perrin, G., and Mamoun, R.Z. (2002). Characterisation of an Irish caprine lentivirus strain--SRLV phylogeny revisited. Virus Res. *85*, 29–39.

Ruff, G., and Lazary, S. (1988). Evidence for linkage between the caprine leucocyte antigen (CLA) system and susceptibility to CAE virus-induced arthritis in goats. Immunogenetics 28, 303–309.

Ruff, G., Regli, J.G., and Lazary, S. (1993). Occurrence of caprine leucocyte class I and II antigens in Saanen goats affected by caprine arthritis (CAE). Eur. J. Immunogenet. *20*, 285–288.

Russo, P., Vitu, C., Fontaine, J.J., and Vignoni, M. (1993). (Caprine arthritis-encephalitis: trial of an adjuvant vaccine preparation. I. Clinical and virological study) Arthrite- encephalite caprine: essai d'une preparation vaccinale adjuvee-- I. Etude clinique et virologique. Comp. Immunol. Microbiol. Infect. Dis. *16*, 131–136.

Ryan, S., Tiley, L., McConnell, I., and Blacklaws, B.A. (2000). Infection of dendritic cells by the maedi-Visna lentivirus. J Virol. *74*, 10096–10103.

Sakuma, R., Mael, A.A., and Ikeda, Y. (2007). Alpha interferon enhances TRIM5alpha-mediated antiviral activities in human and rhesus monkey cells. J. Virol. *81*, 10201–10206.

Saltarelli, M., Querat, G., Konings, D.A.M., Vigne, R., and Clements, J.E. (1990). Nucleotide sequence and transcriptional analysis of molecular clones of CAEV which generate infectious virus. Virology *179*, 347–364.

Saltarelli, M.J., Schoborg, R., Gdovin, S.L., and Clements, J.E. (1993). The CAEV *tat* gene *trans*-activates the viral LTR and is necessary for efficient viral replication. Virology *197*, 35–44.

Sargan, D.R., Bennet, I.D., Cousens, C., Roy, D.J., Blacklaws, B.A., Dalziel, R.G., Watt, N.J., and McConnell, I. (1991). Nucleotide sequence of EV1, a British isolate of maedi-visna virus. J. Gen. Virol. *72*, 1893–1903.

Sargan, D.R., Sutton, K.A., Bennet, I.D., McConnell, I., and Harkiss, G.D. (1995). Sequence and repeat structure variants in the long terminal repeat of maedi-visna virus EV1. Virology *208*, 343–348.

Schmitz, J.E., Simon, M.A., Kuroda, M.J., Lifton, M.A., Ollert, M.W., Vogel, C.W., Racz, P., Tenner-Racz, K., Scallon, B.J., Dalesandro, M., Ghrayeb, J., Rieber, E.P., Sasseville, V.G., and Reimann, K.A. (1999). A nonhuman primate model for the selective elimination of CD8+ lymphocytes using a mouse-human chimeric monoclonal antibody. Am. J. Pathol. *154*, 1923–1932.

Sepp, T., and Tong-Starksen, S.E. (1997). STAT1 pathway is involved in activation of caprine arthritis- encephalitis virus long terminal repeat in monocytes. J. Virol. *71*, 771–777.

Shah, C., Boni, J., Huder, J.B., Vogt, H.-R., Muhlherr, J., Zanoni, R., Miserez, R., Lutz, H., and Schupbach, J. (2004a). Phylogenetic analysis and reclassification of caprine and ovine lentiviruses based on 104 new isolates: evidence for regular sheep-to-goat transmission and worldwide propagation through livestock trade. Virology *319*, 12–26.

Shah, C., Huder, J.B., Boni, J., Schonmann, M., Muhlherr, J., Lutz, H., and Schupbach, J. (2004b). Direct evidence for natural transmission of small-ruminant lentiviruses of subtype A4 from goats to sheep and vice versa. J. Virol. *78*, 7518–7522.

Sherman, L., Yaniv, A., Lichtman-Pleban, H., Tronick, S.R., and Gazit, A. (1989). Analysis of regulatory elements of the equine infectious anemia virus and caprine arthritis-encephalitis virus long terminal repeats. J. Virol. *63*, 4925–4931.

Shih, D.S., Carruth, L.M., Anderson, M., and Clements, J.E. (1992). Involvement of FOS and JUN in the activation of visna virus gene expression in macrophages through an AP-1 site in the viral LTR. Virology *190*, 84–91.

Shoham, N., Cohen, L., Gazit, A., and Yaniv, A. (2003). The Tat protein of the caprine arthritis encephalitis virus interacts with the Notch2 EGF-like repeats and the epithelin/granulin precursor. Intervirology *46*, 239–244.

Sigurdardottir, B., and Thormar, H. (1964). Isolation of a viral agent from the lungs of sheep affected with maedi. J. Infect. Dis. *114*, 55–60.

Sigurdsson, B., Grimsson, H., and Palsson, P.A. (1952). maedi, a chronic, progressive infection of sheep's lungs. J. Infect. Dis. *90*, 233–241.

Sigurdsson, B. (1954a). maedi, a slow progressive pneumonia of sheep: an epizoological and a pathological study. Brit. Vet. J. *110*, 255–270.

Sigurdsson, B. (1954b). Rida, a chronic encephalitis of sheep: With general remarks on infections which develop slowly and some of their special characteristics. Brit. Vet. J. *110*, 341–354.

Sigurdsson, B., Palsson, P., and Grimsson, H. (1957). Visna, a demyelinating transmissible disease of sheep. J. Neuropathol. Exp. Neurol. *16*, 389–403.

Sigurdsson, B., Thormar, H., and Palsson, P. A. (1960). Cultivation of visna virus in tissue culture. Arch. Ges. Virusforsch. *10*, 368–380.

Sihvonen, L. (1981). Early immune responses in experimental maedi. Res. Vet. Sci. *30*, 217–222.

Sihvonen, L. (1984). Late immune responses in experimental maedi. Vet. Microbiol. *9*, 205–213.

Singh, I., McConnell, I., and Blacklaws, B. (2006a). Immune response to individual maedi-visna virus gag antigens. J. Virol. *80*, 912–919.

Singh, I., McConnell, I., Dalziel, R., and Blacklaws, B.A. (2006b). Serum containing ovine IgG2 antibody specific for maedi visna virus envelope glycoprotein mediates antibody dependent cellular cytotoxicity. Vet. Immunol. Immunopathol. *113*, 357–366.

Skraban, R., Matthiasdottir, S., Torsteinsdottir, S., Agnarsdottir, G., Gudmundsson, B., Georgsson, G., Meloen, R.H., Andresson, O.S., Staskus, K.A., Thormar, H., and Andresdottir, V. (1999). Naturally occurring mutations within 39 amino acids in the envelope glycoprotein of maedi-visna virus alter the neutralization phenotype. J Virol. *73*, 8064–8072.

Small, J.A., Bieberich, C., Ghotbi, Z., Hess, J., Scangos, G.A., and Clements, J.E. (1989). The visna virus long terminal repeat directs expression of a reporter gene in activated macrophages, lymphocytes, and the central nervous systems of transgenic mice. J. Virol. *63*, 1891–1896.

Smith, V.W., Dickson, J., Coackley, W., and Carman, H. (1985). Response of merino sheep to inoculation with a caprine retrovirus. Vet. Rec. *117*, 61–63.

Sonigo, P., Alizon, M., Staskus, K., Klatzmann, D., Cole, S., Danos, O., Retzel, E., Tiollais, P., Haase, A., and Wain-Hobson, S. (1985). Nucleotide sequence of the visna lentivirus: relationship to the AIDS virus. Cell *42*, 369–382.

Spencer, T.E., Johnson, G.A., Bazer, F.W., Burghardt, R.C., and Palmarini, M. (2007). Pregnancy recognition and conceptus implantation in domestic ruminants: roles of progesterone, interferons and endogenous retroviruses. Reprod. Fertil. Dev. *19*, 65–78.

Stanley, J., Bhaduri, L.M., Narayan, O., and Clements, J.E. (1987). Topographical rearrangements of visna virus envelope glycoprotein during antigenic drift. J. Virol. *61*, 1019–1028.

Starling, I., Wright, A., Arbuthnott, G., and Harkiss, G. (1999). Acute *in vivo* neurotoxicity of peptides from maedi Visna virus transactivating protein Tat. Brain Res. *830*, 285–291.

Staskus, K.A., Couch, L., Bitterman, P., Retzel, E.F., Zupancic, M., List, J., and Haase, A.T. (1991). *In situ* amplification of visna virus DNA in tissue sections reveals a reservoir of latently infected cells. Microb. Pathog. *11*, 67–76.

Stowring, L., Haase, A.T., Petursson, G., Georgsson, G., Palsson, P., Lutley, R., Roos, R., and Szuchet, S. (1985). Detection of visna virus antigens and RNA in glial cells in foci of demyelination. Virology *141*, 311–318.

Strijbos, P.J.L.M., Zamani, M.R., Rothwell, N.J., Arbuthnott, G., and Harkiss, G. (1995). Neurotoxic mechanisms of transactivating protein tat of maedi-Visna virus. Neurosci. Lett. *197*, 215–218.

Sutton, K.A., Lin, C.T., Harkiss, G.D., McConnell, I., and Sargan, D.R. (1997). Regulation of the long terminal repeat in visna virus by a transcription factor related to the AML/PEBP2/CBF superfamily. Virology *229*, 240–250.

Takeuchi, O., and Akira, S. (2009). Innate immunity to virus infection. Immunol. Rev. *227*, 75–86.

Thormar, H., Barshatzky, M.R., Arnesen, K., and Kozlowski, P.B. (1983). The emergence of antigenic variants is a rare event in long-term visna virus infection *in vivo*. J. Gen. Virol. *64 (Pt 7)*, 1427–1432.

Thormar, H. (2005). maedi-visna virus and its relationship to human immunodeficiency virus. AIDS Rev. *7*, 233–245.

Tong-Starksen, S.E., Sepp, T., and Pagtakhan, A.S. (1996). Activation of caprine arthritis-encephalitis virus long terminal repeat by gamma interferon. J. Virol. *70*, 595–599.

Torsteinsdottir, S., Georgsson, G., Gisladottir, E., Rafnar, B., Palsson, P.A., and Petursson, G. (1992). Pathogenesis of central nervous system lesions in visna: cell-mediated immunity and lymphocyte subsets in blood, brain and cerebrospinal fluid. J Neuroimmunol. *41*, 149–158.

Torsteinsdottir, S., Andresdottir, V., Arnarson, H., and Petursson, G. (2007a). Immune response to maedi-visna virus. Front Biosci. *12*, 1532–1543.

Torsteinsdottir, S., Carlsdottir, H.M., Svansson, V., Matthiasdottir, S., Martin, A.H., and Petursson, G. (2007b). Vaccination of sheep with maedi-visna virus gag gene and protein, beneficial or harmful? Vaccine *25*, 6713–6720.

Trapp, S., Derby, N.R., Singer, R., Shaw, A., Williams, V.G., Turville, S.G., Bess, J.W., Jr, Lifson, J.D., and Robbiani, M. (2009). Double-stranded RNA analog poly(I:C) inhibits human immunodeficiency virus amplification in dendritic cells via type I interferon-mediated activation of APOBEC3G. J. Virol. *83*, 884–895.

Trujillo, J.D., Hötzel, K.J., Snekvik, K.R., and Cheevers, W.P. (2004a). Antibody response to the surface envelope of caprine arthritis-encephalitis lentivirus: disease status is predicted by SU antibody isotype. Virology *325*, 129–136.

Trujillo, J.D., Kumpula-McWhirter, N.M., Hötzel, K.J., Gonzalez, M., and Cheevers, W.P. (2004b). Glycosylation of immunodominant linear epitopes in the carboxy-terminal region of the caprine arthritis-encephalitis virus surface envelope enhances vaccine-induced type-specific and cross-reactive neutralizing antibody responses. J. Virol. *78*, 9190–9202.

Turelli, P., Guiguen, F., Mornex, J.F., Vigne, R., and Querat, G. (1997). dUTPase-minus caprine arthritis-encephalitis virus is attenuated for pathogenesis and accumulates G-to-A substitutions. J. Virol. *71*, 4522–4530.

Valas, S., Benoit, C., Baudry, G., Perrin, G., and Mamoun, R.Z. (2000). Variability and immunogenicity of caprine arthritis-encephalitis virus surface glycoprotein. J. Virol. *74*, 6178–6185.

Vellutini, C., Philippon, V., Gambarelli, D., Horschowski, N., Nave, K.-A., Navarro, J.M., Auphan, M., Courcoul, M.-A., and Filippi, P. (1994). The maedi-Visna virus Tat protein induces multiorgan lymphoid hyperplasia in transgenic mice. J. Virol. *68*, 4955–4962.

Villet, S., Bouzar, B.A., Morin, T., Verdier, G., Legras, C., and Chebloune, Y. (2003a). maedi-visna virus and caprine arthritis encephalitis virus genomes encode a Vpr-like but no Tat protein. J. Virol. *77*, 9632–9638.

Villet, S., Faure, C., Bouzar, B.A., Morin, T., Verdier, G., Chebloune, Y., and Legras, C. (2003b). Lack of trans-activation function for maedi Visna virus and Caprine arthritis encephalitis virus Tat proteins. Virology *307*, 317–327.

von Bodungen, U., Lechner, F., Pfister, H., Vogt, H.-R., Cheevers, W.P., Bertoni, G., Jungi, T.W., and Peterhans, E. (1998). Immunohistology of the early course of lentivirus-induced arthritis. Clin. Exp. Immunol. *111*, 384–390.

von Wussow, P., Jakschies, D., Hochkeppel, H.K., Fibich, C., Penner, L., and Deicher, H. (1990). The human intracellular Mx-homologous protein is specifically induced by type I interferons. Eur. J. Immunol. *20*, 2015–2019.

Watt, N.J., MacIntyre, N., Collie, D., Sargan, D., and McConnell, I. (1992). Phenotypic analysis of lymphocyte populations in the lungs and regional lymphoid tissue of sheep naturally infected with maedi visna virus. Clin. Exp. Immunol. *90*, 204–208.

Woodall, C.J., Maclaren, L.J., and Watt, N.J. (1997). Differential levels of mRNAs for cytokines, the interleukin-2 receptor and class II DR/DQ genes in ovine interstitial pneumonia induced by maedi visna virus infection. Vet. Pathol. *34*, 204–211.

Wu, C., Barbezange, C., McConnell, I., and Blacklaws, B.A. (2008). Mapping and characterization of visna/maedi virus cytotoxic T-lymphocyte epitopes. J. Gen. Virol. *89*, 2586–2596.

Zanoni, R.G. (1998). Phylogenetic analysis of small ruminant lentiviruses. J. Gen. Virol. *79*, 1951–1961.

Zhang, Z., Watt, N.J., Hopkins, J., Harkiss, G., and Woodall, C.J. (2000). Quantitative analysis of maedi-visna virus DNA load in peripheral blood monocytes and alveolar macrophages. J. Virol. Methods *86*, 13–20.

Zhang, Z., Harkiss, G.D., Hopkins, J., and Woodall, C.J. (2002). Granulocyte macrophage colony stimulating factor is elevated in alveolar macrophages from sheep naturally infected with maedi-visna virus and stimulates maedi-visna virus replication in macrophages *in vitro*. Clin. Exp. Immunol. *129*, 240–246.

Zink, M.C., and Narayan, O. (1989). Lentivirus-induced interferon inhibits maturation and proliferation of monocytes and restricts the replication of caprine arthritis- encephalitis virus. J. Virol. *63*, 2578–2584.

Zink, M.C., Yager, J.A., and Myers, J.D. (1990). Pathogenesis of caprine arthritis encephalitis virus – cellular localization of viral transcripts in tissues of infected goats. Am. J. Pathol. *136*, 843–854.

# The Bovine Lentiviruses: Pathogenesis and Cell Tropism

# 11

Moira Desport and Sandy McLachlan

## Abstract

Infections with the bovine lentiviruses, bovine immunodeficiency virus (BIV) or Jembrana disease virus (JDV) represent the extremes of lentivirus induced disease. BIV has a broad cell tropism and causes a mild lymphoproliferative disorder with low viral titres and no reproducible disease sequelae. JDV has a more restricted cell tropism than BIV and infects Bali cattle in Indonesia, replicating to high titres during an acute disease period characterized by lymph node enlargement, leucopenia and high rectal temperatures. The similarities and differences between these two genetically and antigenically closely related viruses and between other lentiviruses will be reviewed in this chapter with particular regard to their cell tropism, pathogenesis and genetic composition.

## Introduction

Bovine immunodeficiency virus (BIV) and Jembrana disease virus (JDV) are the two genetically and antigenically related members of the bovine lentivirus family. BIV was first discovered in the late 1960s and is widely believed to represent a lentivirus that has co-evolved with its host to the point that it is not associated with a particular disease syndrome. It can replicate in a variety of hosts including rabbits, goats, buffalo and cattle and has a worldwide distribution in the latter species. JDV was also first described in the 1960s and is presumed to have arisen from BIV via an unknown transmission event and causes an acute, occasionally lethal, infection in Bali cattle in Indonesia. This chapter will compare and contrast the cell tropism, disease pathogenesis and host responses of these two bovine lentiviruses.

## Virion characteristics

A syncytium-inducing virus with the morphology of a lentivirus was first isolated from a cow with persistent lymphocytosis and found to be similar to visna–maedi virus (VMV) in sheep (Van der Maaten *et al.*, 1972). The virus was found to encode a reverse transcriptase with $Mg^{2+}$ cation preference and have a morphology most similar to HIV and was thus named bovine immunodeficiency virus (BIV) (Gonda *et al.*, 1987). Early electron microscope studies identified 3 distinct morphological identities of the virion and raised the question of its role in infected cell cultures. Intracytoplasmic particles ranging from 98 to 116 nm in diameter were never observed budding through any of the cell membranes indicating that release or infectivity must rely on destruction of the host cell. The budding form of 120–130 nm and an abundant extracellular form 80–130 nm in diameter were found to contain an eccentrically located electron dense core of 30–43 nm (Boothe and Van der Maaten, 1974).

Jembrana disease was first described after outbreaks of severe disease occurred in Bali cattle in Indonesia in 1964 but the causative agent, JDV, was not identified as a lentivirus with an antigenic relationship to BIV until 1993 (Kertayadnya *et al.*, 1993). It is described as 'not cell-associated' due to the large numbers of spherical nucleoid-containing viruses of 75–130 nm diameter that are found in sections of pelleted plasma by electron microscopy (Wilcox *et al.*, 1992). Nucleoids of 30–44 nm diameter are present in over 50% of particles and are located centrally and eccentrically. The high titre of infectious particles in the plasma of cattle during the acute stage of the disease suggests that the viral particles in plasma are of a mature form, although the higher viral loads detected in plasma by indirect methods, e.g. qRT-PCR, indicate the presence of immature or defective particles as well (Stewart *et al.*, 2005). Extracellular viruses are also found in spleen and lymph nodes both as particles and in membrane-bound cytoplasmic vacuoles. Assembly and maturation of JDV occurs via C-type budding through either intracytoplasmic membranes or the plasma membrane. Immature particles with relatively clear centres are formed which mature into virus particles with an electron dense core (Kertayadnya *et al.*, 1993). A bar-shaped nucleoid that is described for BIV and other lentiviruses is not seen with JDV but eccentric nucleoids have been described as in HIV-1 studies (Palmer *et al.*, 1985)

## Pathogenesis and replication dynamics

BIV infection in cattle has been associated with a variety of clinical signs including lymphadenopathy, lymphocytosis, CNS lesions, progressive weakness and emaciation (Carpenter *et al.*, 1992; Munro *et al.*, 1998; Snider *et al.*, 2003). In the original experimental infections infectious material from cow R29 was inoculated into calves and leukaemia was consistently observed (Van der Maaten *et al.*, 1972). More recent field isolates obtained from a seropositive herd in Florida (FL112 and FL491) exhibit different replication characteristics *in vitro* from the original R29 strain including high co-cultivation rates and fewer nuclei per syncytium. The differences are assumed to be due, at least in part, to attenuation of R29 during adaptation to cell culture passage. Flaming *et al.* concluded that this was the case after poor virus recovery and seroconversion rates to BIV infection with R29 strain. They observed that BIV infection does cause alterations in lymphocyte and neutrophil function by 4–5 months after infection but that the effects of BIV may become more marked with time (Flaming *et al.*, 1993). Experimental infection of calves with the Florida strains results in a transient leucocytosis consisting mainly of mononuclear cells and enlargement of lymph nodes around the site of inoculation (Suarez *et al.*, 1993). The major gross manifestation of BIV strain FL112 infection was lymphadenopathy with a considerable increase in the size, number and distribution of haemal lymph nodes (Munro *et al.*, 1998).

In natural infections with BIV a high level of co-infection with bovine leukaemia virus (BLV) is often reported. Under field conditions BIV is observed to have a different effect on the host than under experimental conditions (Snider *et al.*, 1996); however, co-infection with BIV and BLV does not consistently cause a different response than sole infection with either virus (Flaming *et al.*, 1997). Natural BIV infection consistently resulted in enlargement of haemal nodes and hyperplasia of lymphoid follicles. Splenic changes were not consistent, ranging from normal to four-fold enlargement of the organ. Overall, there was expansion of both lymphoid and vascular tissue compartments with retention of a relatively normal architectural pattern. Clinical abnormalities include progressive weight loss, immunodeficiency and behavioural changes such as torpidity, stupor and dullness (Snider *et al.*, 2003). Perturbations of the immune system are observed in experimental infection with BIV in the first 7 weeks after infection. An increased *in vitro* proliferative response to gag protein together with a decreased CD4/CD8 ratio is suggestive of an immune dysfunction. Delayed and significantly lower antibody responses are found after exposure to infectious agents or vaccines during this early stage of infection (Zhang *et al.*, 1997b).

Many of the clinical and pathological features of Jembrana disease in Bali cattle (*Bos javanicus*) differ from those described for BIV infection of *Bos taurus* (Carpenter *et al.*, 1992) or indeed other animal species (Clements and Zink, 1996). The lesions in Bali cattle with Jembrana disease are dominated by the intense lymphoproliferative response, but once this acute stage of disease has resolved the majority of cattle recover. In a series of recent experimental infections with 2 different strains of JDV, 73% of Bali cattle developed clinical Jembrana disease after an incubation period of 5–12 days. Febrile responses developed within 3–10 days of infection, coinciding with a leucopenia, which both returned to normal levels between 13 and 17 days post infection (Fig. 11.1). The groups that were infected with strain $JDV_{PUL/01}$ developed peak rectal temperatures on average 0.7 days before those infected with strain $JDV_{TAB/87}$. Four animals did not develop a typical febrile response but instead had transient mild fever for 1–2 days and lower levels of detectable circulating virus than the typical responders (Table 11.1). In addition, three cattle died after infection with $JDV_{PUL/01}$. Failure to clear the virus was a common feature of these animals, since all had high plasma viral loads at death (Desport *et al.*, 2009a).

We have recently determined that monitoring the viral load (VL) in plasma by quantifying the levels of viral RNA using qRT-PCR reveals the benefits of vaccination (see below) and provides evidence of varying replication dynamics between the two virus strains used in this study (Desport *et al.*, 2009a; Ditcham *et al.*, 2009). Plasma VL $\geq 10^6$ genomes/ml was calculated as this represents 99% of the VL during the acute stage of the disease and is a conservative estimate of the VL required for successful transmission of JDV during an outbreak. The mean plasma VL on the first day of the febrile response in the cattle infected with $JDV_{PUL/01}$ is significantly lower than in cattle infected with $JDV_{TAB/87}$. Similarly, the mean peak VL is significantly lower in $JDV_{PUL/01}$ infected animals (Table 11.1). During the acute stage of Jembrana disease $JDV_{TAB/87}$ infected cattle are found to have a mean $\log_{10}$ plasma VL $\geq 10^6$ genomes/ml of 11.5 versus 10.4 for $JDV_{PUL/01}$ infected cattle. There is a significant difference of 3 days in the mean number of days when plasma VL is above the threshold of $\geq 10^6$ genome copies/ml between the 2 groups of animals.

It is clear from these *in vivo* infection studies in host cattle species that a spectrum of responses is seen after infection with JDV or BIV. Although closely related at the genome level, these viruses replicate at very different rates *in vivo* which may be related to genomic differences between strains of virus, virus attenuation during *in vitro* passage (BIV only) or more fundamental differences related to the cell types that these viruses can infect.

## Cell tropism *in vitro*

The formation of syncytia *in vitro* results from the interaction of viral gene products with host cell membranes and is linked to the spread of virus from infected to uninfected cells. BIV can

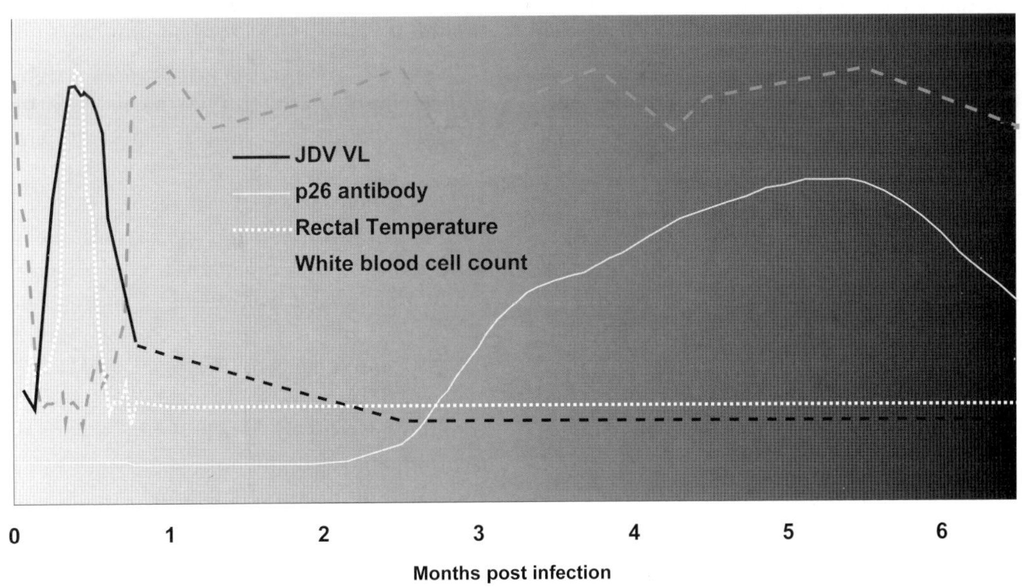


**Figure 11.1 Infection** of Bali cattle with JDV is characterized by an acute transient plasma viraemia with infectious viral loads of up to $10^8$ $ID_{50}$/ml coinciding with elevated rectal temperatures (≥39.3°C) and a leucopenia. Cattle that survive the acute stage of Jembrana disease do not suffer any apparent relapses and develop humoral antibody responses within 5–15 weeks of infection.

**Table 11.1** Summary of observed mean data for groups of cattle infected with $JDV_{TAB/87}$, $JDV_{PUL/01}$, atypical responders, fatalities and vaccinated with JDVacc prior to infection with $JDV_{TAB/87}$

| | No. of days post challenge | | Duration (days) | | Viral load ($log_{10}$) | | |
|---|---|---|---|---|---|---|---|
| Group of cattle | Peak VL | Peak °C | ≥39.3°C | AUC ≥$10^6$ genome copies/ml | First day ≥39.3°C | Peak | AUC ≥$10^6$ genome copies/ml |
| $JDV_{TAB/87}$ ($n=10$) | 11.1 | 11.6 | 7.2 | 12.6 | 9.8 | 11.0 | 11.5 |
| $JDV_{PUL/01}$ ($n=9$) | 11.0 | 10.9 | 6.0 | 9.6 | 8.7 | 10.0 | 10.4 |
| Atypicals ($n=4$) | 13.3 | 11.3 | 1.3 | 7.1 | 6.7 | 8.8 | 9.1 |
| Fatalities ($n=3$) | 11.7 | 11.3 | 8.0 | 9.9 | 8.7 | 10.7 | 11.2 |
| Vaccinated ($n=4$) | 13.3 | 13.8 | 4.3 | 6.8 | 7.8 | 9.9 | 10.2 |

be propagated in a variety of primary cell lines including bovine leucocyte adherent (BLAC-20) cells derived from long term culture of purified PBMC. These cultures contain adherent cells with the morphology of monocytes/macrophages and productive infection of these cells with BIV is determined by measurement of RT activity in the culture supernatant and cytopathic effect (Pifat *et al.*, 1992). As well as being infected by BIV, *in vitro* monocyte function, including enhanced monocyte chemotaxis and random migration is affected by an active component of the BIV Gag precursor located near or within the p26 capsid protein. A second uncharacterized factor also appears to be present in BIV supernatants as monocyte antibody-dependent cell-mediated cytotoxicity is suppressed by BIV supernatants but not by BIV Gag proteins. The mechanism by which the BIV Gag proteins alter monocyte function is unknown but is likely to be a receptor-mediated event and include cytokine induction from monocytes by Gag or cause a direct alteration in signal transduction leading to changes in the microenvironment of the monocyte. During *in vivo* infections with BIV, circulating monocytes do not appear to be greatly affected and virus levels are low but Gag protein expression at local sites may be sufficiently high to interfere with the normal function of tissue macrophages (Rovid and Roth, 1997). It is possible that BIV p26 may be involved in viral modulation of immune function *in vivo* including a role in antigen presentation defects and the loss of reactivity to p26 by infected cattle (Isaacson *et al.*, 1995).

Very little is known about the cell receptor/s that BIV uses but identification of a bovine CCR5 homologue on the surface of foetal bovine lung cells together with evidence that syncytium formation *in vitro* is decreased after pre-treatment with a cocktail of β-chemokines provides preliminary evidence that BIV may utilize the CCR5 receptor (Wright *et al.*, 2002). BIV can become transcriptionally active *in vitro* after superinfection with other bovine viruses that commonly cause disease in cattle under natural conditions. Bovine herpes virus 1 (BHV-1) can induce cellular factors which bind to the NF-κB-like sequences on the BIV LTR (Geng *et al.*, 1992).

Attempts to culture JDV *in vitro* have included continuous and primary Bali cattle derived cells and have so far been uniformly unsuccessful. Uninfected mononuclear cell cultures do not display any cytopathic effect after inoculation with plasma from infected cattle but clinical signs of Jembrana disease are reproduced when cultures maintained for 7–28 days are subinoculated into susceptible Bali cattle (Wilcox *et al.*, 1992).

## Cell tropism and *in vivo* pathogenesis

Several investigations have been carried out in order to determine the *in vivo* cell-tropism of BIV. Early pathogenic effects following experimental infection with BIV include mild proliferative changes during the first 6 weeks which manifest as follicular hyperplasia in the lymph nodes, haemal nodes and spleen as well as an increased number of lymphocytes in peripheral blood which persist for more than a year. Virus can be successfully isolated from PBMC after 3–4 weeks but becomes increasingly difficult thereafter requiring 2–3 passages in cell culture. Reisolation of virus from cattle inoculated with a cell free supernatant from cells containing the molecularly cloned BIV106 isolate was not possible. This may indicate that this clone replicates poorly *in vivo* or that, similar to what has been observed *in vitro*, BIV transmission occurs more readily if the inoculum contains infected cells rather than cell-free virus (Carpenter *et al.*, 1992).

BIV infection of blood monocytes, T-cells and B-cells from a BIV-infected animal between 4 and 7 weeks post infection was assessed after sequential purification of each cell subset to between 82–93% using plastic adherent methods. Infection was determined by monitoring the development of syncytia *in vitro* and the monocyte fraction was consistently found to be infected with approximately 9% of this fraction found to be antigen positive. B-cells were found to contain BIV, albeit with fewer syncytia/well compared to the monocyte fraction. No evidence of infection in the T-cell subset was found. The phagocytic activity of the monocytes was significantly suppressed (Onuma *et al.*, 1992). In longer term infection studies with the same strain, BIV was consistently recovered from unfractionated mononuclear cells over a period of 2 years post infection. Virus isolation from monocytes was only achieved from PBMC of one of four infected cattle at 4–8 months post infection but by 18–24 months BIV could be isolated from monocytes from the three remaining cattle. The virus isolate obtained from monocytes at 4–8 months was inoculated into another calf and virus was then consistently isolated from monocytes as early as 2 weeks post infection and for at least the first year. This suggests that this isolate retained its ability to productively infect monocytes and is the first *in vivo* evidence that BIV strains may have differing cell tropisms or that during persistent infections BIV may alter its tropism. Whilst monocyte dysfunction *in vitro* has been observed after infection with BIV, no significant effect of infection on monocyte function was observed between the BIV and control groups. However, when uninfected monocyte cultures were treated with virus containing supernatants *in vitro*, all monocyte functions were affected. Low viral loads have been suggested as a factor in this observation (Rovid *et al.*, 1995).

The lymphocytosis that generally occurs following experimental infections with BIV was identified as being of B-cell origin with other lymphocyte subtypes being relatively unaffected. The proliferation could occur as a direct result of infection of B-cells by BIV or may be related to B-cell stimulatory activity which has been attributed to a region in the carboxyl terminus of gp41 in HIV-1 (Chirmule *et al.*, 1990; Whetstone *et al.*, 1997).

The major inconsistency between results obtained from different experimental infection studies is that T-cells appear to be infected when PCR based techniques are used whereas when *in vitro* culture is used, the tropism appeared to be restricted to monocytes and B-cells. The pantropic nature of BIV infections *in vivo* was confirmed using the original R29 isolate of BIV and different cell surface markers to separate mononuclear cell populations. Heaton *et al.* used both PCR and cell culture techniques on sorted cell populations and found that $CD21^+$ B-cells, $CD14^+$ monocytes, αβT-cells and γδ T-cells are productively infected as well as harbouring BIV proviral DNA (Heaton *et al.*, 1998). An interesting observation worthy of further investigation is the reduced binding ability of monoclonal antibodies against WC1 and CD21 surface markers on BIV-infected but not uninfected cells indicating a possible down-regulation of cell surface markers in infected cells. More recent studies have identified both BIV genome and mRNA transcripts in $CD3^+$ T-cells, B-cells and CD14 monocytes indicating that virus is transcriptionally active in all of these cell types, confirming the pantropic nature of BIV (Wu *et al.*, 2003).

*In situ* amplification was used to determine the tropism of BIV in tissues at one year post infection. Proviral DNA is present predominantly in neural tissues and some lymphoid tissues and is found in neurons, microglial cells, lymphocytes, septal macrophages, smooth muscle cells and endothelial cells, confirming the broad nature of the cell tropism *in vivo* (Zhang *et al.*, 1997a).

The major differences between the *env* sequences of JDV and BIV, the very different pathogenesis of the diseases that they induce, and the absence of a cell culture system for JDV all suggest that the cell tropism of JDV may not be as broad as that of BIV. In the first week after infection with JDV a predominantly non-follicular lymphoproliferative effect has been consistently reported in lymphoid organs with a similar infiltrative process in the liver, kidney and other non-lymphoid tissues. The response in the spleen is similar to the lymph nodes with the peri-arterial areas and the interfollicular red pulp showing active proliferation of large pleomorphic mononuclear cells (Dharma *et al.*, 1991). In lymph nodes from two necropsied animals during the acute stage of disease only a few CD4 and CD8 cells are found in the parafollicular areas where a prominent lymphoproliferative response is observed. IgG-positive cells, in particular, are prominent in the medullary cords of the lymph nodes and non-follicular compartments of the spleen where they are presumed to be undergoing maturation (Dharma *et al.*, 1994). ISH studies revealed that the spleen contains the most infected cells during the early phase of infection and that this possibly acts as a reservoir for further dissemination of virus as the disease progresses so that infected cells are found in lymph node, lung, liver, kidneys, bone marrow, etc. In lung sections infected cells are predominantly within alveolar septae and vessels. Bone marrow contains high levels of infected cells despite appearing normal in routine histopathology and is suggested to be an important source of infected cells in the circulation (Chadwick *et al.*, 1998). Infected cells have the appearance of large lymphocytes or haematopoietic precursor cells. Many infected cells are found within blood vessels, suggesting that significant numbers of infected cells are present in circulation. From these initial tropism studies it was concluded that infected cells were most probably of lymphoid origin or of the monocyte/macrophage lineage.

Pleomorphic basophilic intracytoplasmic inclusions were reported to occur in the original study in 1991 but were not observed in a study undertaken more recently (I.W.M. Tenaya, unpublished). We observed that the pleomorphic cells have variable amounts of intensely purple cytoplasm, marked nuclear pleomorphism and anisokaryosis with bizarre coarsely clumped chromatin, thus resembling centroblasts. Early investigators originally suggested the phenotype of the lymphocytes proliferating during the acute stage of the disease is T-cells. It had been widely assumed although not proven, that the high level viraemia which occurs during the acute phase of infection results from replication of virus in the population of proliferating cells. Follicular architecture was found to be obliterated by proliferating cells by 8 days after infection and marked follicular lymphoid reactions and plasma cell formation were only observed from the fifth week after infection (Dharma *et al.*, 1991).

The distribution of infected cells in post-mortem tissues taken at the acute stage of disease is predominantly in the parafollicular areas of the spleen and lymph node with little or no signal in the follicles (Chadwick *et al.*, 1998). The germinal centres are depleted and the remains of the mantle zone are surrounded by a wide zone of pleomorphic centroblast-like cells in densely packed sheets. A loss of IgG-containing cells and a decrease in CD4:CD8 lymphocyte numbers during the first 2 weeks of infection was reported in earlier studies (Dharma *et al.*, 1991; Dharma *et al.*, 1994). A recent study using immunohistochemical techniques identified viral capsid antigen in the cytoplasm of some large pleomorphic centroblast-like cells that were scattered throughout the mantle zone and red pulp of the spleen, paracortex and medullary cords of lymph nodes and in circulation. A proportion of JDV capsid containing cells were identified as B-lineage by fluorescent double immunolabelling with anti-JDV capsid and anti-CD79a MAb. During the early stages of acute infection sheets of centroblast-like cells surrounded germinal centres and coalesced to efface the paracortex of lymph nodes and tonsils, and expanded the marginal zone of splenic lymphoid follicles. A subset of plasma cells, identified by the presence of cytoplasmic IgG, also contained JDV-capsid protein (Desport *et al.*, 2009b).

Bovine monocyte/macrophage-lineage cells are identified in fixed tissues using two different MAbs. Recently blood-derived monocytes and neutrophils are identified using MAb MAC387 whilst MAb EBM11, which recognizes CD68, identifies mature tissue macrophages (Bielefeldt-Ohmann *et al.*, 1988; Greywoode *et al.*, 1990). The distribution of EBM11 positive cells compared to MAC387 positive cells in infected lymph nodes and lung is different confirming that these antibodies identify different subsets of cells of the mononuclear phagocyte system. Despite similarities in the size, morphology and distribution of MAC387 positive cells and JDV-infected cells, there is no co-localization of staining of monocytes or neutrophils and JDV-infected cells. The distribution of mature tissue macrophages identified using EBM11 in the medullary sinuses of the lymph node is different to that of the JDV-positive cells which occupy the medullary cords (Desport *et al.*, 2009b).

EBM11 positive cells are present in the alveolar septae of the lung and JDV capsid antigen is occasionally observed in large vacuolated macrophage-like cells together with remnants of erythrocytes (Fig. 11.2). JDV capsid-immunopositive cells are scattered throughout the alveolar septae and, because double immunostaining is not possible with this combination of MAbs, it is unclear whether the positive staining in macrophage-like cells was due to productive infection or phagocytosis of infected cells. It is possible that at least some alveolar macrophages are infected (Desport *et al.*, 2009b).

Many JDV capsid-containing cells are large, pleomorphic, have abundant cytoplasm and coarsely clumped nuclear chromatin (Fig. 11.3). The infected cells often occur in small clusters in lymphoid organs (Fig. 11.3 inset) and in mononuclear cell infiltrates in non-lymphoid organs such as liver and kidney. Whilst proliferative responses involving $CD3^+$ T-cells occur during and after the acute stage of infections with JDV, T-cells are not infected with JDV. This is supported by the absence of any positive ISH staining in mitotic cells (Chadwick *et al.*, 1998). The apparent tropism of JDV for cells of a non-T-cell lineage is perhaps not surprising as there are marked differences in the disease pathogenesis and genetic variability of JDV compared with most of the other lentiviruses, particularly those that are T-cell tropic (Soesanto *et al.*, 1990; Desport *et al.*, 2007). The detection of JDV capsid protein in a subpopulation of B-cells and in terminally differentiated plasma cells indicates that JDV can replicate in cells of this lineage. The location of the infected cells in the mantle zone, outside the germinal centres, suggests that productive infection occurs in mature rather than proliferating immature B-cells. Interestingly, B-cells are infected at a low frequency by $SIV_{smmPBj14}$, an acutely lethal lentivirus infection in pigtail macaques with a similar pathogenesis to JDV (O'Neil *et al.*, 1999). In addition, B-cells, particularly IgG-containing cells, are a major reservoir for FIV in chronically infected cats (English *et al.*, 1993).

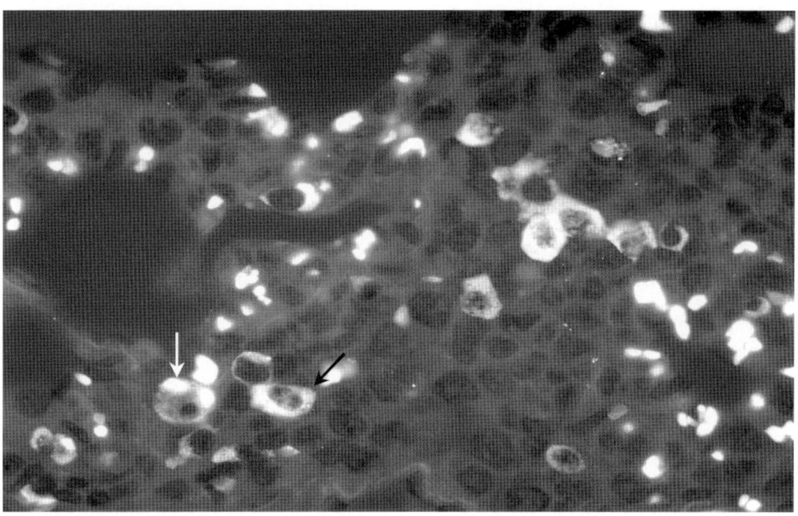

**Figure 11.2** In the lungs, the JDV capsid positive cells (light grey) were mainly detected in the alveolar septa and occasionally in the lumen of blood vessels. Capsid antigen was observed in large vacuolated (black arrow) macrophage-like cells sometimes together with remnants of erythrocytes (white arrow).

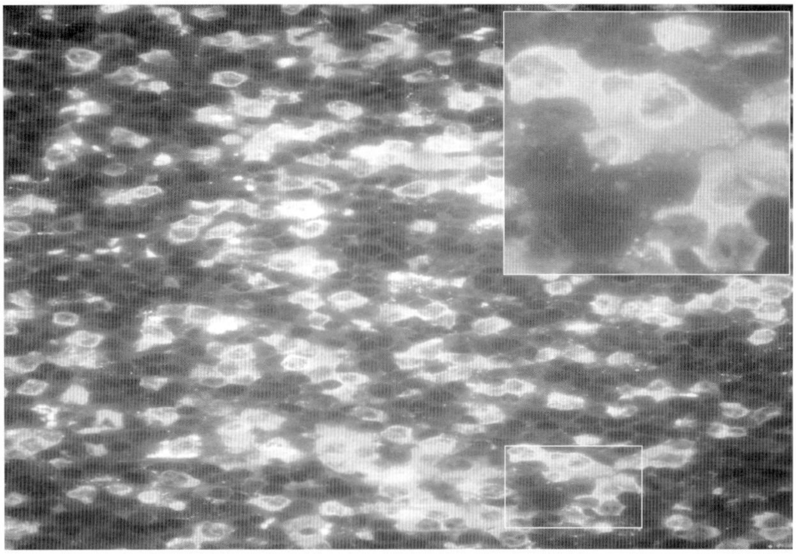

**Figure 11.3** Viral antigen detected using a monoclonal antibody specific for the JDV capsid protein was most prominently observed in the spleen during acute Jembrana disease. Numerous large pleomorphic cells with strong cytoplasmic staining (light grey) were disseminated throughout the red pulp, often in small clusters, largely sparing lymphoid follicles and periarteriolar lymphoid sheaths.

## Are macrophages a target cell?

BIV exhibits a variably broad cell tropism *in vivo* which includes T-cells, B-cells and monocyte/macrophage cells (Whetstone *et al.*, 1997; Heaton *et al.*, 1998). Other experimental infection studies in cattle have reported that BIV could only be isolated from monocytes but not from T-cells (Onuma *et al.*, 1992) and when inoculated into rabbits, BIV antigen was only detected in atypical blastic mononuclear cells which were presumed to be of the macrophage lineage in the red pulp of the spleen (Pifat *et al.*, 1992). A consistent finding, then, has been that BIV infects monocyte/macrophage cells both *in vitro* and *in vivo*.

Whilst there are similarities in the acute diseases induced by the M-tropic EIAV in horses and JDV in Bali cattle, macrophages have not been conclusively identified as a target cell population for infection by JDV. MAb MAC387 recognizes leucocyte protein L1 in neutrophils and blood-derived monocytes and is lost during maturation into tissue macrophages (Brandtzaeg, 1988; Poston and Hussain, 1993). Using double immunofluorescent staining methods it was clear that MAC387$^+$ cells were not infected by the virus (Desport *et al.*, 2009b). CD68 is an intracytoplasmic marker of mature tissue macrophages and is commonly used to identify macrophages infected by primate and other lentiviruses (Chakrabarti *et al.*, 1991; Fischer-Smith *et al.*, 2004). The distribution of CD68 (EBM11) positive cells in lymphoid tissues from JDV-infected cattle was different to the cells containing JDV capsid although there were similarities in the size and morphology of the infected cells (Bielefeldt-Ohmann *et al.*, 1988; Ackermann *et al.*, 1994). Unlike many M-tropic lentiviruses including BIV, JDV has never been detected in brain tissues nor associated with any neurological lesions (Soesanto *et al.*, 1990; Andresdottir *et al.*, 1998; Chadwick *et al.*, 1998; Smit *et al.*, 2001). The genomes of the bovine lentiviruses differ from other non-primate lentiviruses by lacking a functioning dUTPase (McGeoch, 1990; Chadwick *et al.*, 1995) although a non-functioning dUTPase related segment has been identified more recently (Avidan and Hizi, 2008). Retroviral dUTPases have a central role in productive viral replication in non-dividing cells, such as macrophages, where cellular dUTPases and the pool of available deoxynucleotides are at low levels (Terai and Carson, 1991; Whetstone *et al.*, 1997; Chen *et al.*, 2002). The accumulation of G-to-A substitutions in CAEV has been shown to be prevented by the virally encoded dUTPase (Turelli *et al.*, 1997) and replication in macrophages without a mechanism for preventing these substitutions leads to a drift of the genome towards poly(A). The genomes of the bovine lentiviruses exhibit dramatic differences in their genome compositions in their 5′ compared with their 3′ halves. Lentivirus genomes are characteristically A-rich along their entire length, but both BIV and JDV genomes were found to be A-rich only in their 5′ halves (Foley, 2000). The significance of this observation with regard to the cell tropism of JDV remains to be determined.

The identification of cells containing JDV capsid and RNA in the circulation as well as in tissues indicates that it is a circulating cell type that is infected, not tissue macrophages. Circulating monocytes are only rarely found to be infected in other lentiviral infections such as EIAV, and usually it is only after maturation to tissue macrophages that productive viral replication occurs (Sellon *et al.*, 1992). One of the hallmarks of Jembrana disease is the delay in seroconversion to viral antigens until 5–15 weeks post infection (Fig. 11.1) (Hartaningsih *et al.*, 1994; Desport *et al.*, 2009a). This suggests that during the acute phase of the disease the normal process of antigen presentation and antibody production is disrupted. The presence of virus in plasma cells might be anticipated to affect the production of IgG and the decline in the number of these cells during the acute phase of the disease indicates that they do not survive being hijacked by JDV (Dharma *et al.*, 1994).

## Animal models and host range

Sheep appear to be relatively resistant to infection with BIV, compared to cattle, but continued presence of antibody in the absence of positive virus isolation has been interpreted as indicative of persistent infection in sheep and cattle (Whetstone *et al.*, 1991; Jacobs *et al.*, 1994). Isolation of BIV is difficult and even in persistently infected cattle the frequency of infected cells in peripheral blood is extremely low (Carpenter *et al.*, 1992). An increased cellularity involving lymphocytes, macrophages and plasma cells in both B and T-cell areas of the lymphoid organs is observed in the early stages of infection but appears to be transient (Jacobs *et al.*, 1994). Lymphoid hyperplasia is thought to be due to factors other than local production of viral antigen e.g. viral interference with cytokine regulatory pathways. Experimental infection of sheep results in transiently and persistently seropositive and PCR- positive animals. Although most sheep seroconvert to BIV antigens within a few weeks of infection, only 1/5 infected sheep remained seropositive 1 year after infection with BIV (Jacobs *et al.*, 1996). This transient seropositivity is observed in both experimentally and naturally infected cattle and is proposed to be unique to ruminants (Whetstone *et al.*, 1990; Isaacson *et al.*, 1995). BIV has been isolated from buffaloes in Pakistan and draught animals in Cambodia and deduced amino acid sequences were not so variable between isolates from different breeds of cattle and buffalo suggesting that this is not a factor in susceptibility to infection with BIV (Meas *et al.*, 2001).

Rabbits can be infected with BIV with rather more success than sheep and in one study were found to be persistently infected and to demonstrate a rapid and sustained humoral response to BIV antigens. All rabbits were clinically normal with no changes observed in the differential and total leucocyte counts. BIV was rescued from lymph nodes, PBMC and spleens of approx 90% of the infected rabbits and antigen-containing mononuclear cells appeared to be of the macrophage-lineage and were atypical and blastic prompting the hypothesis that BIV replicates in monocyte/macrophage cells *in vivo* (Pifat *et al.*, 1992). In a long-term infection study in rabbits, PBMC were proposed as the first cellular targets of BIV in early infection with detection of proviral DNA in lymph nodes and spleen 3–6 weeks after inoculation. Brain, spinal cord and mucosal surfaces of the gastro-intestinal tract were also infected and splenomegaly and lymphadenopathy were consistent histopathological findings (Walder *et al.*, 2001).

BIV is considered to have co-evolved over time with its bovine host and it was expected that an increase in pathogenicity might have been observed when it was used to infect different hosts, e.g. sheep or goats, as has occurred when other lentiviruses have crossed species boundaries. Whilst this was not observed in sheep under experimental conditions (Jacobs *et al.*, 1994), or in calves infected with CAEV (Morin *et al.*, 2003), the emergence of JDV in a different bovine species almost certainly represents such an event. Clinical signs of Jembrana disease have only been described in Bali cattle, but a range of related bovine species and other ungulates have been successfully infected with the virus. Ongole cattle (*Bos indicus*), Friesian cattle (*Bos taurus*), buffaloes (*Bubalus bubalis*) and pigs all respond to JDV infection by developing a transient febrile response, although no response is detected in sheep. Persistent infections can be detected in Ongole cattle for 3 months but not after 6 months, in Friesian cattle for 1 month but not after 4 months, in buffaloes for 9 months and in sheep for 4 months after inoculation. Buffalo are a common draught animal in Indonesia and may be capable of maintaining JDV in populations of animals in areas with few Bali cattle (Soeharsono *et al.*, 1990). When cross-breeds of Bali cattle were experimentally infected with JDV, only 80% of Madura (*Bos javanicus* x *Bos indicus*) and 60% of Rambon (first generation *Bos javanicus* × *Bos indicus*) cattle developed clinical or haematological signs of disease; however, an antibody response developed in all except one of the Rambon cattle (Soeharsono *et al.*, 1995b).

## Viral transmission

BIV can be transmitted *in utero* and is found in semen, milk-derived leucocytes and milk suggesting that horizontal transmission through natural or artificial insemination as well as iatrogenic or blood-sucking insects may occur (StCyr Coats *et al.*, 1994; Nash *et al.*, 1995a; Nash *et al.*, 1995b; Meas *et al.*, 2000; Meas *et al.*, 2002).

The transmission of JDV is believed to occur via biting arthropods or by the use of contaminated needles during mass vaccination procedures which are commonly practiced in Bali for the control of haemorrhagic septicaemia. The incidence of Jembrana disease is greater during the wet season in Bali which would be consistent with high numbers of biting arthropods. Close contact can be a route of transmission with detection of virus in saliva, milk and urine during the febrile phase of the disease. Oral, intranasal and conjunctival routes of infection have all successfully lead to the reproduction of Jembrana disease in susceptible Bali cattle. All of these routes of virus transmission are only successful when there are high titres of infectious virus present in acutely infected cattle. During the recovery period when viral titres drop to 10 $ID_{50}$/ml plasma the likelihood of mechanical or iatrogenic transmission is low and the role of recovered cattle in transmission is unclear (Soeharsono *et al.*, 1995a).

## Strain variation and virus origin

BIV and JDV differ from other lentiviruses by being genetically stable over time with apparently lower viral mutation rates than the other lentivirus genomes. This observation may be related to their tropism for cells with a long life span (Carpenter *et al.*, 2000; Desport *et al.*, 2007, 2009b). Phylogenetic analysis of 90 amino acids from the RT domain of pol suggest that BIV and EIAV may have diverged from a lentivirus predecessor before the other members of the lentivirus family (Garvey *et al.*, 1990). JDV and BIV sequences cluster tightly irrespective of whether pol or gag sequences are used in phylogenetic analysis (Foley, 2000). BIV (and by association, JDV) is therefore classified as a unique and distinct member of the lentiviruses.

BIV demonstrates a 10% nucleotide substitution rate in *pol* sequences derived from naturally infected cattle (Cooper *et al.*, 1999). As with the other members of the lentivirus family, BIV *env* exhibits the most sequence diversity. When compared to SU sequences from American BIV isolates, Asian isolates are smaller with several base substitutions in the V1 region and deletions in the V2 region, forming a phylogenetically separate branch from the American isolates (Meas *et al.*, 2001). Sequence analysis of American field isolates of BIV, including the strains from Florida, Oklahoma and derivatives of the original R29 strain, reveal both hypervariable and conserved regions with SU genes showing up to 50% amino acid sequence divergence. This contrasts with an 11% deduced sequence divergence in a conserved region of the reverse transcriptase gene. Size variations in the V2 region of the SU genes vary by as much as 104 amino acids. Four field isolates of BIV have SU genes larger than the largest R29 derived strain (Suarez and Whetstone, 1997). The R29 isolates with the smallest SU gene appear to be the easiest to recover from experimentally infected cattle and are non-pathogenic. The Florida isolates with an intermediate recovery rate cause mononuclear cell proliferation early in infection whilst the Oklahoma strains, with the largest SU genes, remain unculturable. This indicates that variations in this region of the genome may confer a selective advantage *in vivo*. Larger SU protein variants containing duplications/insertions predominate in samples taken 1 year after infection (Suarez *et al.*, 1995). This type of sequence variation is thought to occur through template switching/strand transfer during reverse transcription possibly due to *cis*-acting elements (Li and Carpenter, 2001).

The high levels of variation in nucleotide sequences derived from different field isolates of BIV contrast with the antigenic and genetic stability of BIV R29 observed during a 4-year study. Nucleotide diversity of *env* sequences from the original inoculum and virus recovered 4 years after infection ranged from 0.4% to 1.74% with consequent amino acid variation of 1–5%. No disease progression or overt clinical disease was seen in any of the five cattle in this experiment and it was hypothesized that this was due to low levels of viral replication and the development of virus-neutralizing (VN) antibody. The antigenic and genetic stability of the viruses isolated at 4 years post infection suggests that VN antibody does not play a major role in controlling virus replication *in vivo* and it was surprising that the host was unable to clear the infection (Carpenter *et al.*, 2000).

The genome of JDV is approximately 7732 bp, 750 bp smaller than BIV127, and is the smallest of the lentivirus genomes. When comparing the two genomes it is without precedent that two such closely related lentiviruses should have so many insertions and deletions relative to each other and it may reflect a basic functional difference in the reverse transcriptase encoded by these viruses compared to other lentiviruses (Chadwick *et al.*, 1995). Sequence analysis of regions of *gag, pol* and the entire *env* of JDV strains from different regions of Bali taken over a period of nearly 20 years are highly conserved with less than 5% amino acid variation compared to the $JDV_{TAB/87}$ strain. In support of the hypothesis that JDV originated from BIV, potential recombination hotspots have been identified by comparing genome sequences. A deletion in the SU region of strain $JDV_{PUL/01}$ occurs at exactly the same position as a 468-bp insertion in BIV127 SU. Similarly, an insertion of 7bp in the U3 region of $JDV_{PUL/01}$ relative to $JDV_{TAB/87}$ occurs at the same position in BIV127 U3 as an insertion of 157 bp of unrelated sequence (Desport *et al.*, 2007).

Quantification of the magnitude and duration of VL $\geq 10^6$ genome copies/ml has revealed the benefits of using JDVacc, a whole virus inactivated vaccine that is currently used in Indonesia, and highlighted problems associated with correlating febrile responses with vaccine efficacy (Ditcham *et al.*, 2009). Despite the low sequence variation, experimental infection with different strains of JDV results in a broad spectrum of responses, ranging from an absence of any clinical signs of disease in 15% of cattle to fatalities in 11.5% of cattle. We have identified differences in the magnitude and duration of plasma VL generated after infection with $JDV_{PUL/01}$ compared to $JDV_{TAB/87}$ which could be associated with the phylogenetic separation of $JDV_{PUL/01}$ from other Bali strains based on *env* sequences. Animals infected with $JDV_{TAB/87}$ exhibit higher and longer plasma VL compared to those infected with $JDV_{PUL/01}$ which could be due to both host and virus factors (Table 11.1). Differences in LTR sequences correlate with increased replication capacity of EIAV *in vitro* (Madden and Shih, 1996) and it is possible that a 7 nucleotide insertion in the U3 region of the LTR of $JDV_{PUL/01}$ may account for at least some of the differences in the viral replication dynamics (Desport *et al.*, 2007; Desport *et al.*, 2009a).

Expression and regulation of mRNA transcripts can influence both lentiviral replication and disease pathogenesis (Unger *et al.*, 1991; Schiltz *et al.*, 1992). We have recently reported differences in the splicing patterns of mRNA transcripts encoding the Tat proteins of these two virus strains (see below)(Setiyaningsih *et al.*, 2008).

Atypical responses to infection associated with lower peak VL and reduced or absent clinical signs are seen in 15% experimentally infected cattle irrespective of the strain used for inoculation. It has been postulated that, similar to EIAV infection in horses (Cook *et al.*, 2003), a threshold level of JDV is required for induction of acute disease in Bali cattle. Similarly, high levels of $SIV_{smmPBj14}$ in the plasma of pig-tailed macaques are associated with a virus threshold beyond which a synergistic cycle of lymphoid activation and viral replication accelerates uncontrollably resulting in an acutely lethal syndrome (Schwiebert and Fultz, 1994). The comparison of plasma VL on the first day of elevated temperatures in cattle infected with different strains of JDV clearly shows that $JDV_{PUL/01}$ infected cattle develop fever at a lower plasma VL compared to cattle infected with $JDV_{TAB/87}$ (Table 11.1)(Desport *et al.*, 2009a).

## Persistence and diagnosis

BIV and JDV both persist within their hosts and correlates of infection include seroconversion to p26 antigens and detection of proviral DNA in PBMC. The cross-reactivity between the responses to p26 after infection with BIV or JDV means that there is no serological test that can differentiate between the two infections (Kertayadnya *et al.*, 1993; Desport *et al.*, 2005). There are exceedingly few BIV-infected cells in the peripheral blood of chronically infected calves and none were detected in lymphoid follicles by *in situ* hybridization (Carpenter *et al.*, 1992). Baron *et al.* found that DNA extracted from at least $2.5 \times 10^5$ PBMC taken 4–8 weeks post infection is required to obtain a PCR product using BIV pol primers. No PCR products could be detected at 12 months post infection and capsid antibody responses which had developed after 3 weeks had also become undetectable (Baron *et al.*, 1998).

In calves, sheep and goats infected with BIV the first antibodies detected against BIV are specific for p26 and remain detectable for more than 2 years in experimentally infected cattle. After several months the predominant antibodies shift to BIV SU whereas p26 antibodies diminish (Whetstone *et al.*, 1991). Between 33% and 50% of experimentally infected sheep and cattle eventually become seronegative raising the concern of naturally occurring latently infected animals. A cyclic peptide from an immunodominant region in the N-terminal portion of TM is now used for serodiagnosis after the observation that antibodies to it could be detected by 4 weeks and persisted to at least 50 weeks after experimental BIV infection (Scobie *et al.*, 1999).

Proviral DNA can be detected in PBMC of seronegative bulls indicating that false negatives could be a problem if only serological tests are used to detect infections with BIV (Jacobs *et al.*, 1998). Others have reported a close correlation between PCR and serology results from field samples (Zhang *et al.*, 1997c). The sensitivity of PCR assays for detection of BIV proviral DNA has been improved by the development of nested PCR methods and 5′ *Taq* nuclease assays (Suarez *et al.*, 1995; Lew *et al.*, 2004). Size variation in the nested PCR products generated using *env* primers provides some preliminary evidence for the presence of quasispecies during BIV infection.

A low transient plasma viraemia was detected after experimental infection of Bali cattle with BIV R29, although no clinical signs were observed. BIV RNA was detected at titres of up to $10^4$ genome copies/ml in the plasma of 8/13 infected cattle and proviral DNA could be detected in PBMC by d8 in some cattle, and d20 in all infected cattle. However, proviral DNA levels became undetectable quickly in PBMC with only 2/13 cattle remaining positive at 40 days post infection. All cattle seroconverted to the Tm peptide antigen at around 20 days post infection but, surprisingly, responses to p26 were poor. This is not a host-specific phenomenon as strong responses to p26 were detected after 7 days in an earlier pilot study in which three Bali cattle were infected with BIV. However, seroconversion occurred before the detection of proviral DNA in this study suggesting that levels of p26 present in the inoculum may account for seroconversion and that this is not necessarily a correlate of infection, at least in the early stages (McNab *et al.*, 2009).

A Bayesian comparison of antibody detection using immunofluorescent staining and nested PCR for the detection of BIV proviral DNA concludes that although the PCR was more sensitive, substantial misclassification of infection would be expected in epidemiological studies irrespective of the assay used (Orr *et al.*, 2003).

Infectious virus titres drop to 10 $ID_{50}$/ml within 42 days of infection (Fig. 11.1) (Soeharsono *et al.*, 1995b) and proviral DNA cannot be detected reliably in PBMC from cattle that have recovered from natural infection with JDV (Lewis *et al.*, 2009) or BIV, particularly 12 months after experimental infection (Baron *et al.*, 1998; Saman *et al.*, 1999). JDV proviral DNA could not be amplified from 79% of seropositives indicating that the proviral load in PBMC of cattle that have recovered from infection with JDV is very low to undetectable. Given the genetic stability of JDV strains within Bali this is unlikely to be due to sequence variation and it is more likely to be due to the limitations in sensitivity of the assay (Desport *et al.*, 2007). Additional factors could be related to the tropism for B-lineage cells which make up only 11% of the peripheral blood lymphocyte population in normal cattle (Paul *et al.*, 1979) or to the low levels of virus replication that occur in infections with BIV (McNab *et al.*, 2009).

Perturbations of the immune system are a consequence of many lentiviral infections and variable Gag responses are observed after experimental infection with BIV and JDV. BIV CA antibodies were detectable from 2 weeks until at least 2.5 years after infection (Whetstone *et al.*, 1990) whilst Gag responses were found to have declined by 40 weeks after infection and in a second study to remain low or undetectable until 190 weeks after infection (Isaacson *et al.*, 1995). We have observed weak to undetectable CA antibody responses in 15% of JDV experimentally infected cattle which, despite having detectable plasma viraemia, also failed to develop the classical febrile response to infection. Interestingly, all except one of these animals developed strong TM responses detected using the JDV $TM^c$ peptide ELISA (Lewis *et al.*, 2009).

Capsid antibody responses are used as diagnostic indicators of bovine lentiviral infections (Burkala *et al.*, 1999; Meas *et al.*, 2000; Barboni *et al.*, 2001) but the absence of detectable responses in some animals suggests that these diagnostic assays may be underestimating the prevalence of infections.

Both of these viruses have been shown to persist in their hosts without progression of disease. Infectious virus was found to persist for at least 25 months after clinical recovery in Bali cattle and when recovered cattle were re-challenged with JDV up to 22 months after the initial infection, no recurrence of disease was detected in 94% of cattle (Soeharsono *et al.*, 1990).

## Role of accessory proteins and LTRs

The bovine lentiviruses differ from other members of the lentivirus family in the accessory proteins that they do or do not encode. Different types of retroviral Gag proteins may use distinct mechanisms for assembly which may be reflected by different core structures within the virus particles that are generated (Guo *et al.*, 2004). Uniquely among the lentiviruses, the BIV Gag precursor protein is cleaved into 6 major proteins in the order $NH_2$-MA-p2L-CA-p3-NC-p2-COOH (Tobin *et al.*, 1994). The first 8 residues of the p3 spacer sequence between CA and NC are important in the assembly of BIV Gag. This includes a pentapeptide motif (LVAAM) which appears to be conserved in EIAV (LAKAL), FIV (LAEAL), HIV-1 (LAEAM) and VMV (LAQAL). Cleavage sites for the Gag proteins of JDV are not yet determined but a pentapeptide motif consisting of LAEAF is also present in this spacer region indicating a similar mode of assembly to BIV.

JDV encodes the most potent Tat of any of the lentiviruses which can functionally substitute for HIV and other lentivirus Tat proteins (Chen *et al.*, 1999). Differences in lentiviral replication dynamics have also been attributed, at least to some extent, to the occurrence of variable splicing events at different stages of infection. BIV Tat is encoded by a multiply spliced mRNA transcript but only exon 2 encodes a 103 amino acid product which can transactivate not only the BIV LTR but also the HIV LTR (Liu *et al.*, 1992). BIV *tat* mRNA splice variants, encoding Tat proteins of 103 and 108 amino acids, are identical for the first 98 residues but have different transactivation abilities *in vitro* which are attributed to the differences in their C-termini (Fong *et al.*, 1997). Another variant BIV Tat protein which contains the first 98 amino acids of Tat 103 followed by 138 amino acids from the 3′ end of Rev has an increased transactivation activity *in vitro* compared to the original Tat103 and this is predicted to be due to the basic domain of Rev (St-Louis *et al.*, 2005). Several different *tat* splice variants are found in PBMC during acute Jembrana disease and Tat is detected early in infection in PBMC lysates taken before the onset of fever. In $JDV_{PUL/01}$, an additional non-coding exon is present in some transcripts which may be a factor in the different replication dynamics observed with this strain compared to $JDV_{TAB/87}$ (Setiyaningsih *et al.*, 2008). Non-coding exons have not been identified in BIV transcripts but they have been described in HIV and are associated with changes in *in vitro* gene expression levels or related to the cell type from which the transcript was derived (Robert-Guroff *et al.*, 1990).

Lentiviral Tat proteins trigger viral gene expression by the recruitment of the cellular transcription factor P-TEFb to the viral TAR RNA target which is mediated by the cooperative interaction of Tat and cyclin T1 (an essential component of P-TEFb).

The JDV Tat RNA-binding domain is unusual in that it can bind to HIV-1 TAR using cyclin T1 for high affinity binding and it can also bind BIV TAR with high affinity on its own by adopting a β-hairpin conformation when complexed to the RNA (Xie *et al.*, 2003). BIV Tat is able to bind to its TAR in the absence of any cellular cofactor but it can also specifically recruit cyclin T1 and this is essential for BIV Tat function. Because the cyclin T1 protein does not contribute to TAR binding, BIV Tat is able to function in a wide range of host cells including those that do not support HIV-1 Tat function (Barboric *et al.*, 2000). Therefore although BIV Tat is apparently dependant on the same cofactors as other lentiviral Tat proteins, it is unique in terms of the mechanism used to recruit the BIV Tat-cyclin T1 complex to the viral LTR promoter (Bogerd *et al.*, 2000). CDK9/Cyclin T1 expression is related to particular stages of lymphoid differentiation/activation and expression is found in human precursor B and T-cells, peripheral lymphoid tissues, germinal centre cells and scattered B- and T-cell blasts in interfollicular areas. Mantle cells, plasma cells, and small resting T-lymphocytes do not express either molecule (Bellan *et al.*, 2004). In human monocytes isolated from healthy blood donors, cyclin T1 protein expression is low and is induced to high levels within the first few days of differentiation by a post-transcriptional mechanism. After 1–2 weeks of macrophage differentiation, however, cyclin T1 expression is shut off. Treatment of macrophages with lipopolysaccharide can re-induce cyclin T1, indicating that the activation status of macrophages can regulate cyclin T1 expression. Recent results indicate that HIV-1 infection is able to induce cyclin T1 expression in macrophages. Regulation of cyclin T1 expression may be involved in establishing latent infection in macrophages and opportunistic or super-infection may reactivate virus by inducing cyclin T1 expression (Liou *et al.*, 2002, 2004b). HIV-1 infection early in macrophage differentiation results in sustained cyclin T1 expression, while infection at late times in differentiation results in the reinduction of cyclin T1. Expression of the viral Nef protein from an adenovirus vector suggests that Nef contributes to the HIV-1 induction of cyclin T1. These findings indicate that HIV-1 infection hijacks a component of the innate immune response in macrophages that results in enhancement rather than inhibition of viral replication (Liou *et al.*, 2004a). A striking feature of both BIV and JDV genomes is the absence of a *nef* ORF which is located in the 3′ region of *env* of the primate lentiviruses. However, both BIV and JDV encode an accessory protein, Tmx, from this region of the genome and although it is genetically unrelated to Nef, it is tempting to speculate that it may fulfil similar functions. A putative polyclonal B-cell stimulatory epitope is present in the carboxyl end of the envelope glycoprotein of HIV-1 which is specifically associated with Nef (Chirmule *et al.*, 1990; Chirmule *et al.*, 1994). Although the function of Tmx is unknown, it is possibly involved in the proliferation of B-cells that occurs during *in vivo* infections with these viruses (Garvey *et al.*, 1990; Chadwick *et al.*, 1995). Alternatively, or in addition, the lymphocytosis could be related to expression of exogenous Tat early in infection as HIV-1 Tat has been shown to cause B-cell hyperactivation *in vitro* (Huang *et al.*, 1997).

Both exogenous and endogenous Tat induce apoptotic cell death *in vitro* and this is proposed to be mediated through the interaction of the cysteine rich region with both tubulin and microtubule dynamics. Thus, BIV Tat may affect cellular signalling pathways in addition to regulating viral genome replication (Xuan *et al.*, 2008). BIV Tat can translocate through the plasma membrane in an endocytosis-independent way and the arginine residues in the arginine-rich motif (RRRGTRGKGRRIRR) are responsible for the cellular uptake of the protein. BIV Tat is a pleiotropic factor in both infected and uninfected cells, localizing predominantly in the nucleus and resulting in LTR activation and NF-κB induction. Extracellular BIV Tat was found to affect both viral and host gene expression of neighbouring uninfected cells making the cellular environment potentially amenable to virus invasion or reactivation of latent viruses (Deng *et al.*, 2008).

JDV Tat is produced from a single exon *tat-1* transcript and has an arginine-rich translocation motif which is highly conserved between even the most divergent strains of JDV (Setiyaningsih *et al.*, 2008) and was found to be responsible for the internalization of extracellular Tat with both a nuclear and cytoplasmic distribution. This is proposed to be a mechanism whereby JDV, like BIV, can influence cellular gene expression in infected and neighbouring uninfected cells and is potentially a key virulence determinant (Deng *et al.*, 2006). Apoptotic death induced by some viruses results in the formation of small membrane bound apoptotic bodies which can pinch off from the dying cell and be consumed by the phagocytic action of neighbouring cells. This is thought to facilitate the spread of virus without initiating a host response or the requirement of specific viral receptors on the neighbouring cell (Teodoro and Branton, 1997). The observation that JDV infected cells in the spleen of infected cattle occur in groups or clusters (Fig. 11.3) certainly supports the possibility that neighbouring cells become infected via one or both of these mechanisms at this stage of the disease.

A major deletion of 157bp is observed in the JDV LTR compared to the BIV LTR which is 192bp longer in total. This results in the absence of motifs for several enhancer elements in the JDV LTR which are present in BIV including the CCAAT/enhancer binding proteins (C/EBP) motif which is found in most eukaryotic and retroviral promoters. C/EBP proteins were found to be limiting for HIV-1 transcription but not infection and primary macrophages did not support the replication of HIV-1 harbouring mutant C/EBP binding sites in the LTR (Henderson *et al.*, 1996). However, primary $CD4^+$ T-cells supported replication of wild-type and mutant HIV-1 equally well indicating that this is a macrophage-specific regulatory mechanism for HIV-1 replication (Henderson and Calame, 1997). Whilst it is tempting to link the unusual absence of this motif in the JDV LTR with an inability to replicate in macrophages, there are other transcription factor binding motifs which are present in the JDV LTR that have also been associated with lentivirus replication in macrophages. There are 3 potential PU.1 sites (consensus sequence GGAA) in the JDV LTR and these have been identified as being important for EIAV expression in macrophages although other elements within the enhancer region of the LTR are also required (Hines *et al.*, 2004). Ultimately a JDV infectious clone and cell culture system are required in order to determine the significance of the motifs that are present in the JDV LTR and to confirm whether macrophages are productively infected by JDV.

## Immunological control – neutralizing antibodies and vaccination

Infections with most lentiviruses lead to the development of neutralizing antibodies over a number of months which can be quite specific to the inoculation strain. BIV infection does not generally elicit high titres of cross-neutralizing antibodies in studies with both R29 and Florida isolates although VN antibodies were detected in animals from which virus was most easily and consistently recovered (Carpenter *et al.*, 2000). The spread of Jembrana disease is currently controlled by ring vaccination of susceptible cattle in outbreak areas with two doses of JDVacc given with a 4 week interval. JDVacc is an inactivated tissue derived vaccine which is prepared by mixing spleen tissue from an acutely infected animal with a mineral oil adjuvant. Initial vaccine efficacy trials indicated that the vaccine did not prevent Jembrana disease but was able to ameliorate the clinical signs of disease (Hartaningsih *et al.*, 2001). A more recent small trial with JDVacc in four cattle revealed that there was a reduction in total viral load compared to the controls (Ditcham *et al.*, 2009). By estimating a threshold of virus transmission at $\geq 10^6$ genome copies/ml, it was calculated that the vaccine was able to reduce the number of infectious days by 33% (Table 11.1). Whilst this threshold was only an estimate, >99% of the total viral load during the acute stage of the disease is above 106 genome copies/ml plasma. Vaccinated animals exhibited lower peak VLs compared to the controls (Table 11.1), indicating that a low level of virus neutralization or reduction in early spread of virus may be due to the humoral immune responses generated after vaccination. The most significant effects of JDVacc appear after the peak VL when vaccinates cleared circulating virus much faster than the control animals. The elevated rectal temperatures at lower viral titres in the vaccinated group indicate that effective viral clearance is due to a primed CMI response and highlights the balance between effective immune activation and potentiation of viral replication. Neutralizing antibodies were not detected in pooled sera from JDVacc vaccinated animals that resisted challenge. In contrast, sera from recovered cattle was found to contain neutralizing activity but only if it was collected >4 months after recovery (Hartaningsih *et al.*, 2001).

## Concluding remarks

The bovine lentiviruses exhibit some characteristics that make them unique in the lentivirus family whilst their genome organization, morphology and cross reactivity with other members of the group place them firmly in this virus genus. Lentiviruses have a common preference for monocyte/macrophage-lineage cells, tending towards establishing persistent or latent infections in them. It is clear that these cells together with B-cells are both infected with and affected by infection with BIV whilst infection of T-cells is variable but they are likely to be involved with the immune dysfunctions that develop over time.

The viruses exhibit different core structures which may be related to differences in the processing and cleavage of their respective Gag precursors although the p3 spacer which is important for BIV assembly appears to be present in JDV.

Reduced plasma viraemia during the acute stage of infection with either virus appears to correlate with low or variable responses to p26 and after the initial plasma viraemia has subsided, the levels of circulating virus and proviral DNA are uniformly low. These factors affect the diagnostic sensitivity of assays that are currently used and it is likely that the seroprevalence of JDV and BIV infections is being underestimated.

The BIV genome appears to be genetically stable within an infected animal but isolates from different geographical locations show diversity more typical of this virus family. JDV strains in Bali are highly conserved but a more divergent strain from Kalimantan (Indonesian Borneo) suggests that the Bali strains may be a monophyletic cluster of geographically nearby isolates, similar to the phenomenon which has been observed for FIV (Biek *et al.*, 2003).

JDV is often described as an atypical lentivirus because of the acute nature of the disease, the low level of viral variation, the delayed antibody responses and the ensuing immunological control that develops post infection that prevents heterologous infection and relapses. The apparent tropism of JDV for B-lineage cells provides the basis for understanding these observations and indicates a fundamental difference from the other members of the lentivirus family. The original observation that IgG containing cells were depleted during acute Jembrana disease is almost certainly due to virus-induced changes in morphology and their failure to survive being hijacked. The role of macrophages in JDV infection of Bali cattle remains unclear. JDV is certainly present in the circulating mononuclear cell population and yet monocytes are not infected. Tissue macrophages do not share the same tissue location as capsid-containing cells but not all JDV capsid positive cells are identified using the B-lineage MAbs suggesting the involvement of another cell type. The observation that clusters of infected cells occur within lymphoid tissues indicates that infection of neighbouring cells without the necessary viral receptors may occur. All of the other lentiviruses, including BIV, are able to infect macrophages and JDV would indeed be an exceptional lentivirus if it were found to be lacking this tropism.

## References

Ackermann, M.R., DeBey, B.M., Stabel, T.J., Gold, J.H., Register, K.B., and Meehan, J.T. (1994). Distribution of anti-CD68 (EBM11) immunoreactivity in formalin-fixed, paraffin-embedded bovine tissues. Vet Pathol *31*, 340–348.

Andresdottir, V., Tang, X., Agnarsdottir, G., Andresson, O.S., Georgsson, G., Skraban, R., Torsteinsdottir, S., Rafnar, B., Benediktsdottir, E., Matthiasdottir, S., *et al.* (1998). Biological and genetic differences between lung- and brain-derived isolates of maedi-visna virus. Virus Genes *16*, 281–293.

Avidan, O., and Hizi, A. (2008). Expression and characterization of the integrase of bovine immunodeficiency virus. Virology *371*, 309–321.

Barboni, P., Thompson, I., Brownlie, J., Hartaningsih, N., and Collins, M.E. (2001). Evidence for the presence of two bovine lentiviruses in the cattle population of Bali. Vet. Microbiol. *80*, 313–327.

Barboric, M., Taube, R., Nekrep, N., Fujinaga, K., and Peterlin, B.M. (2000). Binding of Tat to TAR and recruitment of positive transcription elongation factor b occur independently in bovine immunodeficiency virus. J. Virol. *74*, 6039–6044.

Baron, T., Betemps, D., Mallet, F., Cheynet, V., Levy, D., and Belli, P. (1998). Detection of bovine immunodeficiency-like virus infection in experimentally infected calves. Arch. Virol. *143*, 181–189.

Bellan, C., De Falco, G., Lazzi, S., Micheli, P., Vicidomini, S., Schurfeld, K., Amato, T., Palumbo, A., Bagella, L., Sabattini, E., *et al.* (2004). CDK9/CYCLIN T1 expression during normal lymphoid differentiation and malignant transformation. J. Pathol. *203*, 946–952.

Biek, R., Rodrigo, A.G., Holley, D., Drummond, A., Anderson, C.R., Jr, Ross, H.A., and Poss, M. (2003). Epidemiology, genetic diversity, and evolution of endemic feline immunodeficiency virus in a population of wild cougars. J. Virol. *77*, 9578–9589.

Bielefeldt-Ohmann, H., Sabara, M., Lawman, M.J., Griebel, P., and Babiuk, L.A. (1988). A monoclonal antibody detects macrophage maturation antigen which appears independently of class II antigen expression. Reactivity of monoclonal EBM11 with bovine macrophages. J. Immunol. *140*, 2201–2209.

Bogerd, H.P., Wiegand, H.L., Bieniasz, P.D., and Cullen, B.R. (2000). Functional differences between human and bovine immunodeficiency virus Tat transcription factors. J. Virol. *74*, 4666–4671.

Boothe, A.D., and Van der Maaten, A.J. (1974). Ultrastructural studies of a visna-like syncytia-producing virus from cattle with lymphocytosis. J. Virol. *13*, 197–204.

Brandtzaeg, P. (1988). The new monoclonal antibody (Mac 387) that reacts with macrophages on paraffin sections detects the well-known leukocyte L1 antigen. J. Histochem. Cytochem. *36*, 1203–1206.

Burkala, E.J., Ellis, T.M., Voigt, V., and Wilcox, G.E. (1999). Serological evidence of an Australian bovine lentivirus. Vet. Microbiol. *68*, 171–177.

Carpenter, S., Miller, L.D., Alexandersen, S., Whetstone, C.A., VanDerMaaten, M.J., Viuff, B., Wannemuehler, Y., Miller, J.M., and Roth, J.A. (1992). Characterization of early pathogenic effects after experimental infection of calves with bovine immunodeficiency-like virus. J. Virol. *66*, 1074–1083.

Carpenter, S., Vaughn, E.M., Yang, J., Baccam, P., Roth, J.A., and Wannemuehler, Y. (2000). Antigenic and genetic stability of bovine immunodeficiency virus during long-term persistence in cattle experimentally infected with the BIV(R29) isolate. J. Gen. Virol. *81*, 1463–1472.

Chadwick, B.J., Coelen, R.J., Wilcox, G.E., Sammels, L.M., and Kertayadnya, G. (1995). Nucleotide sequence analysis of Jembrana disease virus: a bovine lentivirus associated with an acute disease syndrome. J. Gen. Virol. *76 (Pt 7)*, 1637–1650.

Chadwick, B.J., Desport, M., Brownlie, J., Wilcox, G.E., and Dharma, D.M. (1998). Detection of Jembrana disease virus in spleen, lymph nodes, bone marrow and other tissues by *in situ* hybridization of paraffin-embedded sections. J. Gen. Virol. *79 (Pt 1)*, 101–106.

Chakrabarti, L., Hurtrel, M., Maire, M.A., Vazeux, R., Dormont, D., Montagnier, L., and Hurtrel, B. (1991). Early viral replication in the brain of SIV-infected rhesus monkeys. Am. J. Pathol. *139*, 1273–1280.

Chen, H., Wilcox, G., Kertayadnya, G., and Wood, C. (1999). Characterization of the Jembrana disease virus tat gene and the cis- and trans-regulatory elements in its long terminal repeats. J. Virol. *73*, 658–666.

Chen, R., Wang, H., and Mansky, L.M. (2002). Roles of uracil-DNA glycosylase and dUTPase in virus replication. J. Gen. Virol. *83*, 2339–2345.

Chirmule, N., Kalyanaraman, V.S., Saxinger, C., Wong-Staal, F., Ghrayeb, J., and Pahwa, S. (1990). Localization of B-cell stimulatory activity of HIV-1 to the carboxyl terminus of gp41. AIDS Res. Hum. Retroviruses *6*, 299–305.

Chirmule, N., Oyaizu, N., Saxinger, C., and Pahwa, S. (1994). Nef protein of HIV-1 has B-cell stimulatory activity. Aids *8*, 733–734.

Clements, J.E., and Zink, M.C. (1996). Molecular biology and pathogenesis of animal lentivirus infections. Clin. Microbiol. Rev. *9*, 100–117.

Cook, R.F., Cook, S.J., Berger, S.L., Leroux, C., Ghabrial, N.N., Gantz, M., Bolin, P.S., Mousel, M.R., Montelaro, R.C., and Issel, C.J. (2003). Enhancement of equine infectious anemia virus virulence by identification and removal of suboptimal nucleotides. Virology *313*, 588–603.

Cooper, C.R., Hanson, L.A., Diehl, W.J., Pharr, G.T., and Coats, K.S. (1999). Natural selection of the Pol gene of bovine immunodeficiency virus. Virology *255*, 294–301.

Deng, G., Qiao, W., Su, Y., Sha, R., Geng, Y., and Chen, Q. (2006). Internalization of Jembrana disease virus Tat: possible pathway and implication. Virus Res. *121*, 122–133.

Deng, G., Su, Y., Mu, J., Sha, R., Geng, Y., Qiao, W., and Chen, Q. (2008). Molecular basis of the internalization of bovine immunodeficiency virus Tat protein. Virus Genes *36*, 85–94.

Desport, M., Stewart, M.E., Sheridan, C.A., Ditcham, W.G., Setiyaningsih, S., Tenaya, W.M., Hartaningsih, N., and Wilcox, G.E. (2005). Recombinant Jembrana disease virus gag proteins identify several different antigenic domains but do not facilitate serological differentiation of JDV and nonpathogenic bovine lentiviruses. J. Virol. Methods *124*, 135–142.

Desport, M., Stewart, M.E., Mikosza, A.S., Sheridan, C.A., Peterson, S.E., Chavand, O., Hartaningsih, N., and Wilcox, G.E. (2007). Sequence analysis of Jembrana disease virus strains reveals a genetically stable lentivirus. Virus Res. *126*, 233–244.

Desport, M., Ditcham, W.G., Lewis, J.R., McNab, T.J., Stewart, M.E., Hartaningsih, N., and Wilcox, G.E. (2009a). Analysis of Jembrana disease virus replication dynamics *in vivo* reveals strain variation and atypical responses to infection. Virology *386*, 310–316.

Desport, M., Tenaya, I.W.M., McLachlan, A., McNab, T.J., Rachmat, J., Hartaningsih, N., and Wilcox, G.E. (2009b). *In vivo* infection of IgG containing cells by Jembrana disease virus during acute infection. Virology *393*, 221–227.

Dharma, D.M., Budiantono, A., Campbell, R.S., and Ladds, P.W. (1991). Studies on experimental

Jembrana disease in Bali cattle. III. Pathology. J. Comp. Pathol. *105*, 397–414.

Dharma, D.M., Ladds, P.W., Wilcox, G.E., and Campbell, R.S. (1994). Immunopathology of experimental Jembrana disease in Bali cattle. Vet. Immunol. Immunopathol. *44*, 31–44.

Ditcham, W.G., Lewis, J.R., Dobson, R.J., Hartaningsih, N., Wilcox, G.E., and Desport, M. (2009). Vaccination reduces the viral load and the risk of transmission of Jembrana disease virus in Bali cattle. Virology *386*, 317–324.

English, R.V., Johnson, C.M., Gebhard, D.H., and Tompkins, M.B. (1993). *In vivo* lymphocyte tropism of feline immunodeficiency virus. J. Virol. *67*, 5175–5186.

Fischer-Smith, T., Croul, S., Adeniyi, A., Rybicka, K., Morgello, S., Khalili, K., and Rappaport, J. (2004). Macrophage/microglial accumulation and proliferating cell nuclear antigen expression in the central nervous system in human immunodeficiency virus encephalopathy. Am. J. Pathol. *164*, 2089–2099.

Flaming, K., van der Maaten, M., Whetstone, C., Carpenter, S., Frank, D., and Roth, J. (1993). Effect of bovine immunodeficiency-like virus infection on immune function in experimentally infected cattle. Vet. Immunol. Immunopathol. *36*, 91–105.

Flaming, K.P., Frank, D.E., Carpenter, S., and Roth, J.A. (1997). Longitudinal studies of immune function in cattle experimentally infected with bovine immunodeficiency-like virus and/or bovine leukemia virus. Vet. Immunol. Immunopathol. *56*, 27–38.

Foley, B. (2000). An overview of the molecular phylogeny of lentiviruses.: Theoretical Biology and Biophysics Group, Los Alamos National Laboratory, Los Alamos, NM.).

Fong, S.E., Greenwood, J.D., Williamson, J.C., Derse, D., Pallansch, L.A., Copeland, T., Rasmussen, L., Mentzer, A., Nagashima, K., Tobin, G., and Gonda, M.A. (1997). Bovine immunodeficiency virus tat gene: cloning of two distinct cDNAs and identification, characterization, and immunolocalization of the tat gene products. Virology *233*, 339–357.

Garvey, K.J., Oberste, M.S., Elser, J.E., Braun, M.J., and Gonda, M.A. (1990). Nucleotide sequence and genome organization of biologically active proviruses of the bovine immunodeficiency-like virus. Virology *175*, 391–409.

Geng, Y., Kashanchi, F., and Wood, C. (1992). Activation of bovine immunodeficiency-like virus expression by bovine herpesvirus type 1. Virology *187*, 832–836.

Gonda, M.A., Braun, M.J., Carter, S.G., Kost, T.A., Bess, J.W., Jr, Arthur, L.O., and Van der Maaten, M.J. (1987). Characterization and molecular cloning of a bovine lentivirus related to human immunodeficiency virus. Nature *330*, 388–391.

Greywoode, G.I., McCarthy, S.P., and McGee, J.O. (1990). Labelling of cells of the mononuclear phagocyte system in routinely processed archival biopsy specimens with monoclonal antibody EBM/11. J. Clin. Pathol. *43*, 992–996.

Guo, X., Hu, J., Whitney, J.B., Russell, R.S., and Liang, C. (2004). Important role for the CA-NC spacer region in the assembly of bovine immunodeficiency virus Gag protein. J. Virol. *78*, 551–560.

Hartaningsih, N., Wilcox, G.E., Kertayadnya, G., and Astawa, M. (1994). Antibody response to Jembrana disease virus in Bali cattle. Vet. Microbiol. *39*, 15–23.

Hartaningsih, N., Dharma, D.M., Soeharsono, S., and Wilcox, G.E. (2001). The induction of a protective immunity against Jembrana disease in cattle by vaccination with inactivated tissue-derived virus antigens. Vet. Immunol. Immunopathol. *78*, 163–176.

Heaton, P.R., Johnstone, P., and Brownlie, J. (1998). Investigation of the cellular tropism of bovine immunodeficiency-like virus. Res. Vet. Sci. *65*, 33–40.

Henderson, A.J., Connor, R.I., and Calame, K.L. (1996). C/EBP activators are required for HIV-1 replication and proviral induction in monocytic cell lines. Immunity *5*, 91–101.

Henderson, A.J., and Calame, K.L. (1997). CCAAT/enhancer binding protein (C/EBP) sites are required for HIV-1 replication in primary macrophages but not CD4(+) T-cells. Proc. Natl. Acad. Sci. U.S.A. *94*, 8714–8719.

Hines, R., Sorensen, B.R., Shea, M.A., and Maury, W. (2004). PU.1 binding to ets motifs within the equine infectious anemia virus long terminal repeat (LTR) enhancer: regulation of LTR activity and virus replication in macrophages. J. Virol. *78*, 3407–3418.

Huang, L., Li, C.J., and Pardee, A.B. (1997). Human immunodeficiency virus type 1 TAT protein activates B-lymphocytes. Biochem. Biophys. Res. Commun. *237*, 461–464.

Isaacson, J.A., Roth, J.A., Wood, C., and Carpenter, S. (1995). Loss of Gag-specific antibody reactivity in cattle experimentally infected with bovine immunodeficiency-like virus. Viral Immunol. *8*, 27–36.

Jacobs, R.M., Smith, H.E., Whetstone, C.A., Suarez, D.L., Jefferson, B., and Valli, V.E. (1994). Haematological and lymphocyte subset analyses in sheep inoculated with bovine immunodeficiency-like virus. Vet. Res. Commun. *18*, 471–482.

Jacobs, R.M., McCutcheon, L.J., Valli, V.E., and Smith, H.E. (1996). Histopathological changes in the lymphoid tissues of sheep exposed to the bovine immunodeficiency-like virus. J. Comp. Pathol. *114*, 23–30.

Jacobs, R.M., Jefferson, B.J., and Suarez, D.L. (1998). Prevalence of bovine immunodeficiency-like virus in bulls as determined by serology and proviral detection. Can. J. Vet. Res. *62*, 231–233.

Kertayadnya, G., Wilcox, G.E., Soeharsono, S., Hartaningsih, N., Coelen, R.J., Cook, R.D., Collins, M.E., and Brownlie, J. (1993). Characteristics of a retrovirus associated with Jembrana disease in Bali cattle. J. Gen. Virol. *74 (Pt 9)*, 1765–1778.

Lew, A.E., Bock, R.E., Miles, J., Cuttell, L.B., Steer, P., and Nadin-Davis, S.A. (2004). Sensitive and specific detection of bovine immunodeficiency virus and bovine syncytial virus by 5′ Taq nuclease assays with fluorescent 3′ minor groove binder-DNA probes. J. Virol. Methods *116*, 1–9.

Lewis, J.R., McNab, T.J., Tenaya, M., Hartaningsih, N., Wilcox, G.E., and Desport, M. (2009). Comparison of Immunoassay and Real-Time PCR Methods for the Detection of Jembrana disease virus Infection in Bali Cattle J. Virol. Methods *159*, 81–86.

Li, Y., and Carpenter, S. (2001). Cis-acting sequences may contribute to size variation in the surface glycoprotein of bovine immunodeficiency virus. J. Gen. Virol. *82*, 2989–2998.

Liou, L.Y., Herrmann, C.H., and Rice, A.P. (2002). Transient induction of cyclin T1 during human macrophage differentiation regulates human immunodeficiency virus type 1 Tat transactivation function. J. Virol. *76*, 10579–10587.

Liou, L.Y., Herrmann, C.H., and Rice, A.P. (2004a). Human immunodeficiency virus type 1 infection induces cyclin T1 expression in macrophages. J. Virol. *78*, 8114–8119.

Liou, L.Y., Herrmann, C.H., and Rice, A.P. (2004b). HIV-1 infection and regulation of Tat function in macrophages. Int. J. Biochem. Cell Biol. *36*, 1767–1775.

Liu, Z.Q., Sheridan, D., and Wood, C. (1992). Identification and characterization of the bovine immunodeficiency-like virus tat gene. J. Virol. *66*, 5137–5140.

Madden, C.R., and Shih, D.S. (1996). Analysis of the long terminal repeat from a cytopathic strain of equine infectious anemia virus. Virology *225*, 395–399.

McGeoch, D.J. (1990). Protein sequence comparisons show that the 'pseudoproteases' encoded by poxviruses and certain retroviruses belong to the deoxyuridine triphosphatase family. Nucleic Acids Res. *18*, 4105–4110.

McNab, T., Desport, M., Tenaya W.M., Hartaningsih, N., and Wilcox, G. (2009). Bovine immunodeficiency virus produces a transient viraemic phase soon after infection in *Bos javanicus*. Vet. Microbiol. doi:10.1016/j.vetmic.2009.09.012

Meas, S., Ohashi, K., Tum, S., Chhin, M., Te, K., Miura, K., Sugimoto, C., and Onuma, M. (2000). Seroprevalence of bovine immunodeficiency virus and bovine leukemia virus in draught animals in Cambodia. J. Vet. Med. Sci. *62*, 779–781.

Meas, S., Ohashi, K., Sugimoto, C., and Onuma, M. (2001). Phylogenetic relationships of bovine immunodeficiency virus in cattle and buffaloes based on surface envelope gene sequences. Brief report. Arch. Virol. *146*, 1037–1045.

Meas, S., Usui, T., Ohashi, K., Sugimoto, C., and Onuma, M. (2002). Vertical transmission of bovine leukemia virus and bovine immunodeficiency virus in dairy cattle herds. Vet. Microbiol. *84*, 275–282.

Morin, T., Guiguen, F., Bouzar, B.A., Villet, S., Greenland, T., Grezel, D., Gounel, F., Gallay, K., Garnier, C., Durand, J., *et al.* (2003). Clearance of a productive lentivirus infection in calves experimentally inoculated with caprine arthritis-encephalitis virus. J. Virol. *77*, 6430–6437.

Munro, R., Lysons, R., Venables, C., Horigan, M., Jeffrey, M., and Dawson, M. (1998). Lymphadenopathy and non-suppurative meningo-encephalitis in calves experimentally infected with bovine immunodeficiency-like virus (FL112). J. Comp. Pathol. *119*, 121–134.

Nash, J.W., Hanson, L.A., and St Cyr Coats, K. (1995a). Bovine immunodeficiency virus in stud bull semen. Am. J. Vet. Res. *56*, 760–763.

Nash, J.W., Hanson, L.A., and St Cyr Coats, K. (1995b). Detection of bovine immunodeficiency virus in blood and milk-derived leukocytes by use of polymerase chain reaction. Am. J. Vet. Res. *56*, 445–449.

O'Neil, S.P., Mossman, S.P., Maul, D.H., and Hoover, E.A. (1999). *In vivo* cell and tissue tropism of SIVsmmPBj14-bcl.3. AIDS Res. Hum. Retroviruses *15*, 203–215.

Onuma, M., Koomoto, E., Furuyama, H., Yasutomi, Y., Taniyama, H., Iwai, H., and Kawakami, Y. (1992). Infection and dysfunction of monocytes induced by experimental inoculation of calves with bovine immunodeficiency-like virus. J. Acquir. Immune Defic. Syndr. *5*, 1009–1015.

Orr, K.A., O'Reilly, K.L., and Scholl, D.T. (2003). Estimation of sensitivity and specificity of two diagnostics tests for bovine immunodeficiency virus using Bayesian techniques. Prev. Vet. Med. *61*, 79–89.

Palmer, E., Sporborg, C., Harrison, A., Martin, M.L., and Feorino, P. (1985). Morphology and immunoelectron microscopy of AIDS virus. Arch. Virol. *85*, 189–196.

Paul, P.S., Senogles, D.R., Muscoplat, C.C., and Johnson, D.W. (1979). Enumeration of T-cells, B-cells and monocytes in the peripheral blood of normal and lymphocytotic cattle. Clin. Exp. Immunol. *35*, 306–316.

Pifat, D.Y., Ennis, W.H., Ward, J.M., Oberste, M.S., and Gonda, M.A. (1992). Persistent infection of rabbits with bovine immunodeficiency-like virus. J. Virol. *66*, 4518–4524.

Poston, R.N., and Hussain, I.F. (1993). The immunohistochemical heterogeneity of atheroma macrophages: comparison with lymphoid tissues suggests that recently blood-derived macrophages can be distinguished from longer-resident cells. J. Histochem. Cytochem. *41*, 1503–1512.

Robert-Guroff, M., Popovic, M., Gartner, S., Markham, P., Gallo, R.C., and Reitz, M.S. (1990). Structure and expression of tat-, rev-, and nef-specific transcripts of human immunodeficiency virus type 1 in infected lymphocytes and macrophages. J. Virol. *64*, 3391–3398.

Rovid, A.H., Carpenter, S., and Roth, J.A. (1995). Monocyte function in cattle experimentally infected with bovine immunodeficiency-like virus. Vet. Immunol. Immunopathol. *45*, 31–43.

Rovid, A.H., and Roth, J.A. (1997). Enhancement of monocyte migration and phagocytosis by the bovine immunodeficiency-like virus Gag proteins. J. Acquir. Immune Defic. Syndr. Hum. Retrovirol. *14*, 18–25.

Saman, E., Van Eynde, G., Lujan, L., Extramiana, B., Harkiss, G., Tolari, F., Gonzalez, L., Amorena, B., Watt, N., and Badiola, J. (1999). A new sensitive serological assay for detection of lentivirus infections in small ruminants. Clin Diagn Lab Immunol *6*, 734–740.

Schiltz, R.L., Shih, D.S., Rasty, S., Montelaro, R.C., and Rushlow, K.E. (1992). Equine infectious anemia virus gene expression: characterization of the RNA splicing pattern and the protein products encoded by open reading frames S1 and S2. J. Virol. *66*, 3455–3465.

Schwiebert, R., and Fultz, P.N. (1994). Immune activation and viral burden in acute disease induced by simian immunodeficiency virus SIVsmmPBj14: correlation between *in vitro* and *in vivo* events. J. Virol. *68*, 5538–5547.

Scobie, L., Venables, C., Hughes, K., Dawson, M., and Jarrett, O. (1999). The antibody response of cattle infected with bovine immunodeficiency virus to peptides of the viral transmembrane protein. J. Gen. Virol. *80 (Pt 1)*, 237–243.

Sellon, D.C., Perry, S.T., Coggins, L., and Fuller, F.J. (1992). Wild-type equine infectious anemia virus replicates *in vivo* predominantly in tissue macrophages, not in peripheral blood monocytes. J. Virol. *66*, 5906–5913.

Setiyaningsih, S., Desport, M., Stewart, M.E., Hartaningsih, N., and Wilcox, G.E. (2008). Sequence analysis of mRNA transcripts encoding Jembrana disease virus Tat-1 *in vivo*. Virus Res. *132*, 220–225.

Smit, T.K., Wang, B., Ng, T., Osborne, R., Brew, B., and Saksena, N.K. (2001). Varied tropism of HIV-1 isolates derived from different regions of adult brain cortex discriminate between patients with and without AIDS dementia complex (ADC): evidence for neurotropic HIV variants. Virology *279*, 509–526.

Snider, T.G., 3rd, Luther, D.G., Jenny, B.F., Hoyt, P.G., Battles, J.K., Ennis, W.H., Balady, J., Blas-Machado, U., Lemarchand, T.X., and Gonda, M.A. (1996). Encephalitis, lymphoid tissue depletion and secondary diseases associated with bovine immunodeficiency virus in a dairy herd. Comp Immunol Microbiol Infect Dis *19*, 117–131.

Snider, T.G., 3rd, Hoyt, P.G., Coats, K.S., Graves, K.F., Cooper, C.R., Storts, R.W., Luther, D.G., and Jenny, B.F. (2003). Natural bovine lentiviral type 1 infection in Holstein dairy cattle. I. Clinical, serological, and pathological observations. Comp Immunol Microbiol Infect Dis *26*, 89–101.

Soeharsono, S., Hartaningsih, N., Soetrisno, M., Kertayadnya, G., and Wilcox, G.E. (1990). Studies of experimental Jembrana disease in Bali cattle. I. Transmission and persistence of the infectious agent in ruminants and pigs, and resistance of recovered cattle to re-infection. J. Comp. Pathol. *103*, 49–59.

Soeharsono, S., Wilcox, G.E., Putra, A.A., Hartaningsih, N., Sulistyana, K., and Tenaya, M. (1995a). The transmission of Jembrana disease, a lentivirus disease of Bos javanicus cattle. Epidemiol Infect *115*, 367–374.

Soeharsono, S., Wilcox, G.E., Dharma, D.M., Hartaningsih, N., Kertayadnya, G., and Budiantono, A. (1995b). Species differences in the reaction of cattle to Jembrana disease virus infection. J. Comp. Pathol. *112*, 391–402.

Soesanto, M., Soeharsono, S., Budiantono, A., Sulistyana, K., Tenaya, M., and Wilcox, G.E. (1990). Studies on experimental Jembrana disease in Bali cattle. II. Clinical signs and haematological changes. J. Comp. Pathol. *103*, 61–71.

St-Louis, M.C., Abed, Y., and Archambault, D. (2005). The bovine immunodeficiency virus: cloning of a tat/rev cDNA encoding a novel Tat protein with enhanced transactivation activity. Arch. Virol. *150*, 1529–1547.

StCyr Coats, K., Pruett, S.B., Nash, J.W., and Cooper, C.R. (1994). Bovine immunodeficiency virus: incidence of infection in Mississippi dairy cattle. Vet. Microbiol. *42*, 181–189.

Stewart, M., Desport, M., Hartaningsih, N., and Wilcox, G. (2005). TaqMan real-time reverse transcription-PCR and JDVp26 antigen capture enzyme-linked immunosorbent assay to quantify Jembrana disease virus load during the acute phase of *in vivo* infection. J. Clin. Microbiol. *43*, 5574–5580.

Suarez, D.L., VanDerMaaten, M.J., Wood, C., and Whetstone, C.A. (1993). Isolation and characterization of new wild-type isolates of bovine lentivirus. J. Virol. *67*, 5051–5055.

Suarez, D.L., Van der Maaten, M.J., and Whetstone, C.A. (1995). Improved early and long-term detection of bovine lentivirus by a nested polymerase chain reaction test in experimentally infected calves. Am. J. Vet. Res. *56*, 579–586.

Suarez, D.L., and Whetstone, C.A. (1997). Size variation within the second hypervariable region of the surface envelope gene of the bovine lentivirus BIV in experimentally and naturally infected cattle. J. Virol. *71*, 2482–2486.

Teodoro, J.G., and Branton, P.E. (1997). Regulation of apoptosis by viral gene products. J. Virol. *71*, 1739–1746.

Terai, C., and Carson, D.A. (1991). Pyrimidine nucleotide and nucleic acid synthesis in human monocytes and macrophages. Exp. Cell. Res. *193*, 375–381.

Tobin, G.J., Sowder, R.C., 2nd, Fabris, D., Hu, M.Y., Battles, J.K., Fenselau, C., Henderson, L.E., and Gonda, M.A. (1994). Amino acid sequence analysis of the proteolytic cleavage products of the bovine immunodeficiency virus Gag precursor polypeptide. J. Virol. *68*, 7620–7627.

Turelli, P., Guiguen, F., Mornex, J.F., Vigne, R., and Querat, G. (1997). dUTPase-minus caprine arthritis-encephalitis virus is attenuated for pathogenesis and accumulates G-to-A substitutions. J. Virol. *71*, 4522–4530.

Unger, R.E., Stout, M.W., and Luciw, P.A. (1991). Simian immunodeficiency virus (SIVmac) exhibits complex splicing for tat, rev, and env mRNA. Virology *182*, 177–185.

Van der Maaten, M.J., Boothe, A.D., and Seger, C.L. (1972). Isolation of a virus from cattle with persistent lymphocytosis. J. Natl. Cancer Inst. *49*, 1649–1657.

Walder, R., Kalvatchev, L., Perez, F., Garzaro, D., and Barrios, M. (2001). Bovine immunodeficiency virus in experimentally infected rabbit: tropism for lymphoid and nonlymphoid tissues. Comp Immunol Microbiol Infect Dis *24*, 1–20.

Whetstone, C.A., VanDerMaaten, M.J., and Black, J.W. (1990). Humoral immune response to the bovine immunodeficiency-like virus in experimentally and naturally infected cattle. J. Virol. *64*, 3557–3561.

Whetstone, C.A., VanDerMaaten, M.J., and Miller, J.M. (1991). A western blot assay for the detection of antibodies to bovine immunodeficiency-like virus in experimentally inoculated cattle, sheep, and goats. Arch. Virol. *116*, 119–131.

Whetstone, C.A., Suarez, D.L., Miller, J.M., Pesch, B.A., and Harp, J.A. (1997). Bovine lentivirus induces early transient B-cell proliferation in experimentally inoculated cattle and appears to be pantropic. J. Virol. *71*, 640–644.

Wilcox, G.E., Kertayadnya, G., Hartaningsih, N., Dharma, D.M., Soeharsono, S., and Robertson, T. (1992). Evidence for a viral aetiology of Jembrana disease in Bali cattle. Vet. Microbiol. *33*, 367–374.

Wright, S.M., Mleczko, A., and Coats, K.S. (2002). Bovine immunodeficiency virus expression *in vitro* is reduced in the presence of beta-chemokines, MIP-1alpha, MIP-1beta and RANTES. Vet. Res. Commun. *26*, 239–250.

Wu, D., Murakami, K., Morooka, A., Jin, H., Inoshima, Y., and Sentsui, H. (2003). *In vivo* transcription of bovine leukemia virus and bovine immunodeficiency-like virus. Virus Res. *97*, 81–87.

Xie, B., Wainberg, M.A., and Frankel, A.D. (2003). Replication of human immunodeficiency viruses engineered with heterologous Tat-transactivation response element interactions. J. Virol. *77*, 1984–1991.

Xuan, C., Qiao, W., Li, J., Peng, G., Liu, M., Chen, Q., Zhou, J., and Geng, Y. (2008). BTat, a trans-acting regulatory protein, contributes to bovine immunodeficiency virus-induced apoptosis. Cell. Microbiol. *10*, 31–40.

Zhang, S., Troyer, D.L., Kapil, S., Zheng, L., Kennedy, G., Weiss, M., Xue, W., Wood, C., and Minocha, H.C. (1997a). Detection of proviral DNA of bovine immunodeficiency virus in bovine tissues by polymerase chain reaction (PCR) and PCR *in situ* hybridization. Virology *236*, 249–257.

Zhang, S., Wood, C., Xue, W., Krukenberg, S.M., Chen, Q., and Minocha, H.C. (1997b). Immune suppression in calves with bovine immunodeficiency virus. Clin Diagn Lab Immunol *4*, 232–235.

Zhang, S., Xue, W., Wood, C., Chen, Q., Kapil, S., and Minocha, H.C. (1997c). Detection of bovine immunodeficiency virus antibodies in cattle by western blot assay with recombinant gag protein. J. Vet. Diagn. Invest. *9*, 347–351.

# Lentivirus Coinfections and Superinfections

12

Sue VandeWoude and Mary Poss

Abstract

The phenomenon of lentiviral superinfection – that is, infection with a second unique strain of virus after a primary viraemia has already been established – is an important aspect of lentivirus biology, and is being increasingly recognized as a method for rapid emergence of new strains via the process of recombination. Though there are both host and viral barriers to simultaneous coinfection following exposure with two unique lentiviruses, it is well established in both natural and experimental systems that infection with one strain of lentivirus does not impart resistance to superinfection. Viral kinetics and dynamics of infection are altered in coinfections relative to single infections. Superinfection may result in either diminished or enhanced virulence of the primary or secondary strain and may occur with high prevalence in individuals with repeated exposures. Evaluation of the incidence of naturally occurring superinfections, and biological behaviour of experimental superinfections, either with host-adapted or non-host-adapted strains, provides insights into potential pitfalls and additional strategies to consider during lentiviral vaccine development. Analysis of lentiviral genome recombination events as a sequelae of superinfection enhances understanding of mechanisms of lentiviral evolution and virus-host adaptations. This chapter provides a review and comparison of lentiviral superinfections and resultant recombination outcomes in different hosts, including primate, Felidae, and ruminant species, highlighting what these studies have revealed about lentiviral pathogenesis, host immune response to lentiviral infection, and viral genomic responses to superinfections.

## Viral and cellular restrictions to coinfection

Coinfection of a single host with two or more unique lentiviral strains as a result of multiple exposures would be likely to occur if primary infection does not result in sterilizing immunity, and sufficient numbers of susceptible host target cells are available for secondary viral replication. One possible outcome of superinfection would be that primary and secondary virus replication occurs in different target cells, i.e. two simultaneous and more or less independent infections. If two viruses propagate to sufficient levels, the probability that a single cell could be coinfected with two strains increases. In general, there are several mechanisms by which viral coinfections might be limited (Nethe *et al.*, 2005). In chronological order from the point of viral entry into the host, these include: (i) adaptive immunity: circulating neutralizing antibodies cross-reactive against a coinfecting strain could bind virus penetrating mucosal barriers and eliminate virions from circulation, or cytotoxic T-cells primed by the primary infection may recognize and eliminate cells infected by the secondary virus; (ii) innate immunity: a general antiviral state due to release of innate factors during primary infection could work *in trans* to disrupt secondary infection; (iii) resource limitation: elimination or diminution of susceptible target cells by primary infection or down-regulation of receptors on target cells infected with primary virions, may limit infection of a second unique viral infection; (iv) intracellular competition: exhaustion of cellular enzymes required for successful viral integration, replication, and assembly; activation of intracellular restriction factors; and (v) interference of primary virion constituents with replication or assembly of a second strain to prevent successful replication of a second virus (Fig. 12.1). These blocks to coinfection assume that the primary and secondary viruses utilize identical or similar receptor repertoires, and have similar cellular tropisms, which may not be the case for highly divergent clades, or for coinfections involving lentiviruses that have originated from different host species. In this latter situation, it would be possible for two infecting strains to replicate simultaneously within different target cells, with a lower probability that single cell coinfection would occur. Successful coinfection could result in discordant replication of the primary or secondary strain, or in chimeric progeny virions, which may have decreased fitness relative to parental strains (resulting in loss of chimeric phenotype), or with equivalent or increased fitness relative to parental strains, resulting in enhanced replication and emergence of new viral strains.

## Viral genomic recombination during superinfection

Lentivirus genomes consist of two identical single strands of genomic RNA. It is normal for the virus polymerase to switch strands during minus strand DNA synthesis (Coffin, 1979; Hu and Temin, 1990a,b). Recombination can also occur during synthesis of the plus sense DNA strand by a strand displacement-assimilation method (Junghans *et al.*, 1982). Both HIV-1 and FIV reverse transcriptases (RT) can perform strand displacement DNA synthesis (Amacker *et al.*, 1995). The structure of the lentiviral genome and intrinsic mechanism of reverse transcription (RT) facilitates the production of mosaic provirus. If a cell is infected with a single virus particle with two identical copies of genomic RNA, there is no genetic consequence of RT movement between the two single strands of RNA. However, if more than one lentivirus infects a single cell, genomic strands will not be identical and recombinant viruses can be produced. Several low probability events must occur if the progeny of viral recombinants are to persist. First, an individual cell must be infected with more than one genetically distinct virus, overcoming all barriers to superinfection outlined above. Both viruses must reverse transcribe their genomes and successfully integrate. A copy of each viral genome must be transcribed and packaged in the same viral genome. This virus must infect a new cell and reverse transcription must yield a provirus with a full complement of functional genes. Finally, the recombination progeny must be at least as fit as parental viruses infecting the same host and be able to establish infection in a new host. Despite the low probability of many of these events, viral recombinants comprise measurable proportions of the population in most natural lentivirus infections as reviewed in later sections of this chapter. With methodology currently employed by most investigators, it is difficult to distinguish dual infection from infection with a recombinant virus. Detecting coinfection is far more straightforward using conventional methodologies, i.e. sequence analysis of a single PCR-amplification product representing a small region of the viral genome. With the advent of complete genome sequencing, the incidence of circulating recombinants and generation of recombinants in coinfected animals will be more accurately demonstrated.

## Naturally occurring lentiviral superinfections

### Naturally occurring HIV coinfections

The premise that retroviruses are highly host specific is supported by congruence of retrovirus and host phylogeny. However, both of the contemporary human lentivirus infections, HIV-1 and HIV-2, derive from cross-species infection events (Apetrei *et al.*, 2004). Thus, although cross-species transmission events are rare, they do occur. In fact, the high genetic diversity within circulating strains of HIV-1, and to a lesser extent HIV-2, today is due, in part, to several cross-species transmission events from the respective primate hosts; chimpanzee (*Pan troglodytes*) and sooty mangebey (*Cercocebus atys*) (Dykhuizen *et al.*, 1998; Blackard *et al.*, 2002). HIV-1 strains have been grouped into three major classes based on phylogenetic relatedness. Group M comprises subgroups A–K and recombinants, which are

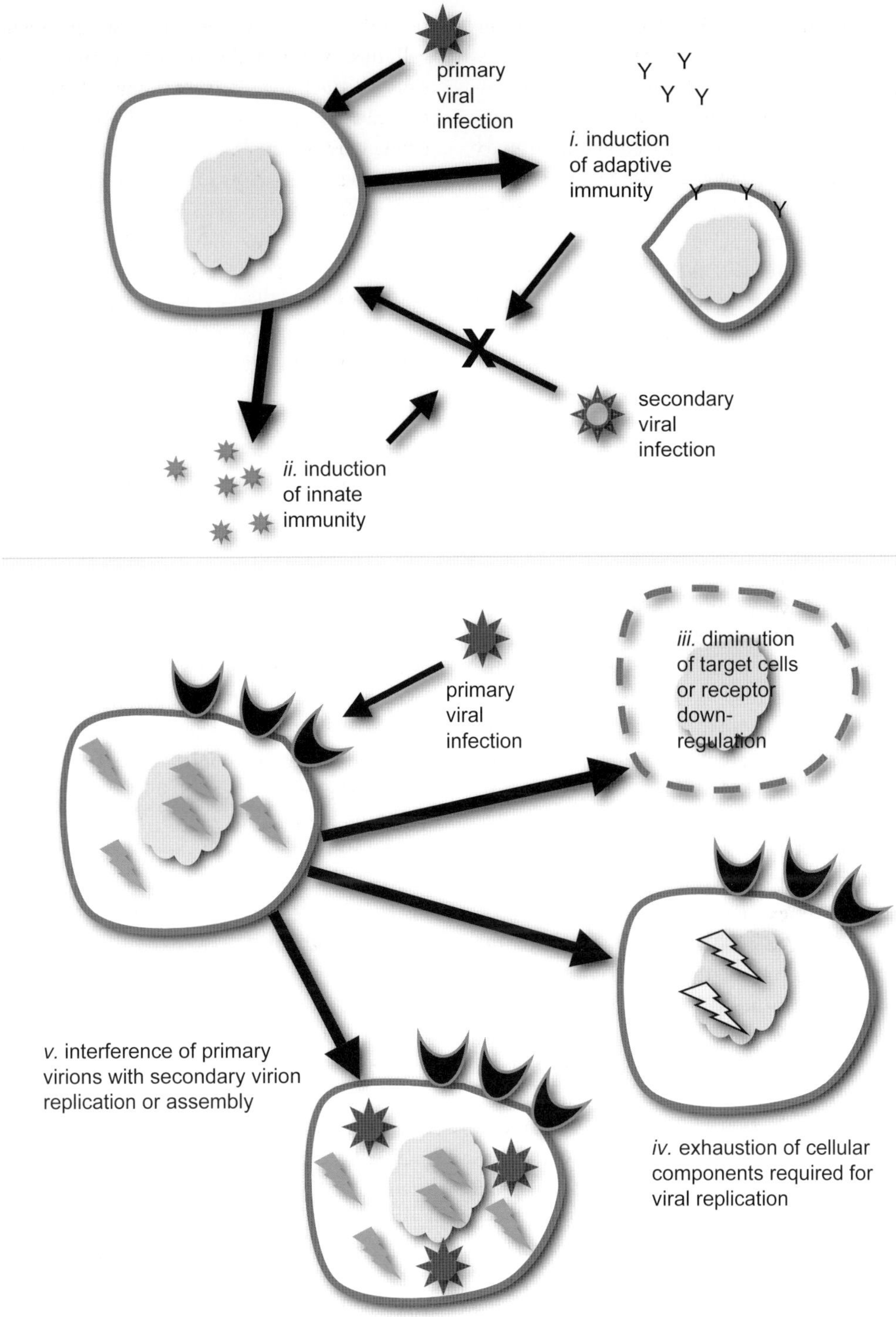


**Figure 12.1** Mechanisms for inhibition of viral coinfection. Primary viral infection may induce (i) adaptive or (ii) innate immunity that interferes with secondary infection. Primary viral infection can also impair the host target cell by (iii) killing the cell, or down-regulating cellular receptors; (iv) exhausting cellular resources (depicted as thunderbolts) that are necessary to support secondary infection; or (v) accumulation of primary virion components (depicted as starburst) may inhibit successful complete viral replication of a second virion.

associated with the current HIV-1 pandemic. Two other groups, N and O, represent introductions of different $SIV_{cpz}$ strains into the human population. Viruses within these groups are not as transmissible or as virulent as those in group M and consequently, they have lower genetic diversity than group M viruses. HIV-2 isolates from the majority of infected individuals form two main groups, A and B, which represent expansion from introductions of two different $SIV_{sm}$ strains to humans (Damond *et al.*, 2001). At least six other cross-species transmission events of $SIV_{sm}$ have resulted in infection of individuals with no onward transmission (Chen *et al.*, 1996; Yamaguchi *et al.*, 2000; Damond *et al.*, 2001). Thus, in the absence of any further cross-species transmission events, humans have the potential to become superinfected with two or more human lentiviruses that are from the same subgroup (both subgroup A viruses) from different subgroups (a subgroup A and B virus) from different groups (a group N and O virus) or from different species (HIV-1 and HIV-2).

Superinfection with more than one strain of HIV should be limited if the host immune response to the primary virus provides any protection against infection by the second lentivirus. The first report of the ability to infect an already infected host came from experimental studies in chimpanzees (Schwiebert and Fultz, 1994). Subsequently, superinfection has been reported in many individuals, including those from longitudinal cohorts where more precise documentation of primary infectious status is available (Levy, 2003; Smith *et al.*, 2005). Virtually all combinations of human lentiviruses have been documented in superinfections. Significantly, superinfection with two viruses of the same clade has been reported in individuals who have antibodies capable of neutralizing the secondary virus (Blish *et al.*, 2008). These data cast doubt on the general efficacy of host immunity to prevent subsequent infection. It is likely that features predisposing to superinfection will be complex, including contact frequency among infected individuals, genetic divergence of primary and secondary viruses and the dynamic immunological responses of the host.

The consequences of coinfection are not clear; there are reports of both protection and synergism occurring in coinfected individuals. Of particular importance was the observation that HIV-2 infected women had apparent resistance to superinfection with HIV-1, despite having a higher incidence of STDs (Travers *et al.*, 1995). Other reported consequences may be a bias towards specific strains that can superinfect (Sarr *et al.*, 2000), which would affect the makeup of recombinant strains. The most serious effect of coinfection – enhanced disease – has also been reported. The majority of superinfections resulting in more rapid onset of disease involved different strains of subtype B virus (Afkhami-Goli *et al.*, 2009). The above data may indicate that the genetic distance between the primary and secondary virus strains may dictate whether a superinfection event will result in modulation or enhancement of disease.

In addition to the nine HIV-1 parental subgroups, there are multiple circulating recombinant forms (CRF). As discussed above, the production of viral recombinants with a genotype distinct from the parental strain requires that a cell be simultaneously infected with two genetically distinct viral strains. This distinction is important because strand switching between the two single stranded RNA genomes of a lentivirus is a normal event. There are both viral and cellular mechanisms that prevent infection of an already infected cell by another retrovirus as discussed above (Nethe *et al.*, 2005). If two distinct viruses breach the molecular restriction factors to dual infection and produce a recombinant virus, the resultant mosaic progeny must successfully infect another cell, replicate and be transmitted. To be classified as a CRF, the virus must be detected in at least 3 individuals.

There are currently 43 documented CRF involving HIV-1 subtypes; many more recombinant viruses have been identified from individuals. CRF AE and AG are among the most prevalent strains of HIV-1 detected (McCutchan, 2006), suggesting that these subtype A recombinants are as fit as the parental viruses. There are no reports of HIV-1–HIV-2 recombinants despite reports of superinfection with these two viruses. However, a replication competent recombinant between the distantly related group M and O viruses has been detected in an individual, indicating that genetic distance alone does not limit recombination potential.

HIV superinfection has the potential to increase the complexity of managing an extremely difficult pandemic. Although the consequences of superinfection can be variable, production of new recombinants increases the potential for the emergence of phenotypic variants, including drug resistant variants, viruses with enhanced replicative capacity, enhanced virulence or transmissibility, or capacity to replicate in new cell types. These considerations illustrate the need to educate individuals undergoing treatment about the risk of becoming simultaneously infected with a different strain of HIV. It is, of course, also possible that coinfection with HIV will result in emergence of strains with a more symbiotic host–virus relationship, as seems to exist in African primates or non-domestic Felidae infected with species-specific SIVs or FIVs. It is for this reason that continued characterization of emergent HIVs and lentivirus evolutionary biology is essential for relating viral genetic fingerprints to disease phenotypes.

## Naturally occurring non-human primate lentiviral superinfections: emergence via recombination

There are no reports of SIV superinfection occurring in non-human primates naturally infected with SIV. However, multiple lines of evidence point to this phenomenon as an important mechanism leading to generation of new SIV strains capable of an expanded host range. Six 'classical' SIV lineages arising from chimpanzee ($SIV_{cpz}$), sooty mangabey, African green monkey, L'Hoest's monkey, Syke's monkey, and colobus monkey have been described. As the number of complete genome sequences from the more than 40 African primate species with naturally occurring SIV infection has accumulated, many of these lineages have been shown to be recombinant forms of other existing strains. For example, the 5′ region of $SIV_{cpz}$ is most closely related to $SIV_{rcm}$ (red capped mangabey), whereas the 3′ half of the genome is more closely related to $SIV_{gsn}$ (greater spot-nosed). $SIV_{cpz}$ and $SIV_{gsn}$ also contain the vpu accessory gene, further supporting the case for recombination between two SIV genomes. This may have arisen following chimpanzee simultaneous exposure to divergent SIV lineages ($SIV_{rcm}$ and $SIV_{gsn}$, for example), or to a recombinant that arose in a primary host with coinfection (i.e. red capped mangabey exposed to $SIV_{gsn}$), which ultimately lead to infection/selection of a successful recombinant in chimpanzees (VandeWoude and Apetrei, 2006). Further, the two primary $SIV_{cpz}$ strains (isolated from *troglodytes* and *schweinfurthii* subspecies of *P. troglodytes*), appear to have different evolutionary histories, suggesting the schweinfurthii subtype underwent additional recombination events following SIV superinfection (Leitner *et al.*, 2007). While this specific example appears exceedingly complex, full genome phylogenetic analysis suggests at least eight SIVs have arisen following recombination events between ancestral SIVs in other host species (Apetrei *et al.*, 2004). Further, several species of primates (i.e. moustached monkeys, *Cercophithecus cebus*; wild mandrillus sphinx) reveal multiple circulating subtypes of viruses in wild populations, suggesting a mechanism for frequent superinfection/recombination events (Aghokeng *et al.*, 2007). Sampling and targeted analysis of free-ranging African primates infected with SIV would probably reveal examples of naturally occurring SIV superinfections as has been described in humans, non-domestic and domestic Felidae and ruminants.

## Naturally occurring superinfection in the domestic cat

Several large scale studies have evaluated naturally occurring FIVs in domestic cats in Australia, North America (Canada and USA), Europe, and New Zealand, revealing that superinfection, probably as a result of multiple exposures resulting in successful FIV transmission, occurs with relatively high frequency (Kyaw-Tanner and Robinson, 1996; Bachmann *et al.*, 1997; Hayward and Rodrigo, 2008). Recombination between different substrains characterized by env phylogeny occurs with a higher frequency than dual-infection, suggesting multiple virus types can infect a single cell, and FIV infection does not confer protection against subsequent exposures as noted above for HIV (Kyaw-Tanner and Robinson, 1996; Bachmann *et al.*, 1997; Carpenter *et al.*, 1998; Hayward *et al.*, 2007; Hayward and Rodrigo, 2008). In total, five studies analysed over 380 FIV-positive individuals; coinfection was documented in three animals and 7% of the isolates demonstrated recombination between defined parental clades (Kyaw-Tanner and Robinson, 1996; Carpenter *et al.*, 1998; Reggeti and Bienzle, 2004). A recombinant virus was recovered from a cat simultaneously coinfected with two parental strains, providing evidence for the mechanism of recombinant viral emergence (Hayward and Rodrigo, 2008). Related recombinant viruses successfully propagate in multiple individuals, demonstrating persistence of the chimeric virus in local geographic regions and emergence of new subtypes (Reggeti and Bienzle, 2004; Hayward and Rodrigo, 2008), similar to multiple CRF described in HIV infection.

## Lentiviral superinfection and recombination in non-domestic cats

Over 20 species of non-domesticated cats have been determined to be seroreactive against FIV antigens. Certain species, namely pumas (*Puma concolor*, also called cougars or Florida panthers), and African lions (*Panthera leo*), may have sero-prevalence approaching 100% in some subpopulations (Troyer *et al.*, 2004). Genetic analysis of FIVs of pumas ($FIV_{pco}$) and lions ($FIV_{ple}$) had determined that superinfection does occur and may result in recombination. It seems probable that the circumstances of FIV are similar to that of natural SIV infection in non-human primates, namely that recombination of coinfections gives rise to new strains capable of infecting new hosts. (VandeWoude and Apetrei, 2006; Pecon-Slattery *et al.*, 2008; Troyer *et al.*, 2008).

Despite high seroprevalence of $FIV_{pco}$ in most populations examined, pumas are solitary animals, so coinfection opportunities are limited in this species. This is supported by phylogenetic congruity between sequences from the viral *pol* and *env* genes, and little evidence of dual infection in $FIV_{pco}$ sequences from over 150 infected cougars in the Northern Rockies ecosystem. However, samples from seven animals showed discordant phylogenies (Biek *et al.*, 2006) and coinfection was confirmed by using heteroduplex mobility assays on additional amplicons obtained from the animals (M.L. Poss, unpublished).

Full length $FIV_{pco}$ genomes were obtained from animals with sequences representative of the phylogenetic diversity within the region and from several animals with evidence of dual infection. There was no evidence of recombination in the 5′ end of the genome; however, six statistically significant regions of recombination clustered in the 3′ half of the genome near the 5′ end of accessory genes, and three recombinant viruses were identified by phylogenetic incongruence (Bruen and Poss, 2007). The recombination event essentially transferred the vif and orfA genes from one virus into the structural gene backbone of a second virus. The virus donating the accessory genes was dominant in the early 1990's in the Greater Yellowstone region but was not detected 10 years later. Thus, there is evidence for virus recombination and parental virus extinction in the Northern Rockies Ecosystem, though of eight viral subtypes identified in this population, only one had a recombinant history.

African lions harbour six phylogenetically distinct clades of FIV, based upon ~600 basepair region of *pol* (Antunes *et al.*, 2008). Superinfection with multiple clades has been documented in 43% of the infections of certain lion prides (Troyer *et al.*, 2004), suggesting that antiviral immunity and cellular restrictions to superinfection are limited or non-existent in this species. Further, intra-subtype strains cluster geographically (Antunes *et al.*, 2008). Evaluation of full genome sequences of two parental clades (B and E) demonstrated high levels of genetic diversity within and between FIVs isolated from other species. Lion subtypes B and E are monophyletic across all gene regions except *env*, where the SU and TM regions vary substantially. Subtype E *env* is more related to domestic cat sequences than to subtype B, probably arising from a recombination event during superinfection with two FIV strains of highly divergent origin resulting in a mosaic virus able to propagate successfully in lion cells. This evidence suggests that FIV infection was introduced into the African lion population several times, possibly via multiple other primary hosts, similar to evolutionary events determined to occur in African non-human primate species (Pecon-Slattery *et al.*, 2008). The recombination event leading to subtype heterogeneity has resulted in divergent receptor usage between clades; clade E subtype uses a CD134/CXCR4 receptor repertoire, similar to domestic cat FIV, whereas subtype B uses an as yet undefined cellular entry mechanism (McEwan *et al.*, 2008).

### Lentiviral superinfection and recombination in ruminants

Sheep and goats are each infected with lentiviruses, visna–maedi virus (VMV), and caprine arthritis encephalitis virus (CAEV), which were originally presumed to be species specific. However, there is substantial evidence for natural transmission of these small ruminant lentiviruses among sheep and goats based on both serological and genetic methods (Zanoni *et al.*, 1991; Ravazzolo *et al.*, 2001; Rolland *et al.*, 2002; Pisoni *et al.*, 2007). Although goats can be coinfected with both VMV and CAEV experimentally (Jolly and Narayan, 1989) there are few reports of coinfection with small ruminant lentiviruses (SRLV) under natural conditions. However, dual infection of some goats with both VMV and CAEV has been reported and an interspecies recombinant was identified (Pisoni *et al.*, 2007). This recombination event occurred within the *env* gene. The detection of an intragenic recombinant in *env* is of interest because it is a principal determinant of the pathogenic potentials of VMV and CAEV (Andresdottir *et al.*, 2002). Further, the recombinant virus was detected by PCR in host tissue and recovered from culture, suggesting that it was replication competent in the host (Pisoni *et al.*, 2007). It is noteworthy that animal management conditions may influence the occurence of coinfection events because the animals evaluated in this report were from a large commercial herd that had not implemented lentivirus control methods. The high population density in some herds and management practices of herding mixed flocks of sheep and goats may facilitate cross-species infection.

There is serological evidence in Bali cattle in Indonesia for the occurrence of coinfections with the bovine lentiviruses, bovine immunodeficiency virus (BIV) and Jembrana disease virus (JDV) (Barboni *et al.*, 2001). JDV is widely believed to represent a cross-species transmission event originating from BIV. Though this has never been substantiated, the close genetic relationship between these viruses makes this a plausible scenario. Serological differentiation between these infections is impossible with the methodologies that are currently available (Desport *et al.*, 2005). Some Bali cattle experimentally infected with BIV developed a mild transient plasma viraemia within 4–9 days of infection and seroconvert to Env antigens by 12 days post infection. Superinfection with JDV 42 days after BIV infection failed to reduce the JDV viral load in superinfected cattle compared to control JDV only cattle, despite strong responses to Env antigens prior to superinfection. These experiments indicate that prior infection with BIV did not offer any real protection against subsequent exposure to JDV, although it is important to note that BIV infection did not exacerbate the effects of JDV. Some reactivation of BIV was observed in a proportion of superinfected cattle during the peak period of JDV replication (T. McNab, unpublished).

## Experimental coinfection studies

### SIV vaccine models

In contrast to the paucity of publications describing naturally occurring SIV superinfections, many experimental studies have evaluated the capacity of one lentivirus to interfere with the replication during challenge infection – primarily as a means to establish an effective HIV vaccine. While this primary goal has not been achieved, these studies offer the opportunity to evaluate mechanisms that underlie vaccine failure and/or prevent superinfection. Experimental infections with multiple SIVs have occurred using viruses of the same subtype, SIVs of different types, or with chimeric or attenuated HIV infection as a primary infection, followed by challenge with a virulent SIV. These studies have nearly always occurred in Asian macaques, which to date have not been shown to be infected with species-specific SIV in the wild. African SIVs, typically originating from African green or sootey mangabey monkeys, can induce pathogenic immunodeficiency in several species of macaque; however, recent studies of the emergence of pathogenic isolates demonstrate that several inadvertent serial passages of tissues from infected African species occurred before pathogenic disease emerged (Apetrei *et al.*, 2005; Apetrei *et al.*, 2006).

Vaccination with live attenuated SIV may confer protection against homologous wild-type infection. When a heterologous *ex vivo* derivative of $SIV_{mac}251$ was used as challenge virus following vaccination with live attenuated

$SIV_{mac}C8$, lower viral loads of $SIV_{mac}251$ were detected 20 weeks PI that was not correlated with neutralizing antibody or anti-Nef CD8 cytotoxic T-cells. Further, at 3 weeks post superinfection, elevated levels of vaccine virus replication were observed relative to singly infected animals, suggesting that dual-infection enhanced replication of the primary virus (Berry *et al.*, 2008). Others have concluded that adaptive anamnestic T-cell responses were not the mechanism of protection against superinfection (Stebbings *et al.*, 2002). Other vaccine trials using SIV deletion mutants, notably SIVΔnef, have demonstrated complete protection against subsequent superinfection, though this often occurred in the absence of a measurable adaptive immune response and both challenge and vaccine viruses persist for years post-challenge which often resulted in recombination and reversion to virulent SIVs (Cranage *et al.*, 1997; Titti *et al.*, 1997; Sernicola *et al.*, 1999; Sharpe *et al.*, 2004). Studies evaluating the capacity of SIV deletion mutants to recombine and regain replicative competency following mucosal infection demonstrated that recombination eventually occurs in a majority of animals. Rescue of viral competency following recombination was rapid (within 2 weeks) when complementary deletion mutants were inoculated simultaneously and infection with the more attenuated mutant ($SIV_{mac}239\Delta vpx\Delta vpr$) followed by the more virulent isolate ($SIV_{mac}239\Delta nef$) was more likely to result in rescue recombination than the converse infection scheme (Kim *et al.*, 2005). Additional studies document spontaneous reversion of SIVΔnef to wild type, possibly by an *in vivo* duplication event, (Whatmore *et al.*, 1995; Stahl-Hennig *et al.*, 1996), and further describe that protection afforded by SIVΔnef is dose dependent and influenced by early infection dynamics (Cranage *et al.*, 1998). These findings underscore the difficulty of designing a modified live vaccination scheme that does not pose significant risk of rapid reversion to wild-type phenotype following superinfection via duplication or recombination.

In addition to apathogenic host-adapted SIV modified live experiments, SIV vaccination studies have also been conducted using avirulent HIV-2 isolates, or an HIV-SIV hybrid virus (SHIV, see Chapter 6). These studies have provided conflicting results, but typically, apathogenic infections with SHIV or SIV resulted in partial to complete protection against virulent disease. However, if initial infections were established with pathogenic isolates, widespread coinfection with a second strain was not prevented (Stephens *et al.*, 1997), though viral loads could be contained over time (Dunn *et al.*, 1997). HIV-2 infection of macaques typically results in productive yet apathogenic disease as viral replication in monkeys is restricted by intracellular restriction factors (see Chapter 4). Subsequent exposure of HIV-2 infected animals affords partial protection against virulent SIV disease (Putkonen *et al.*, 1995) that may eventually result in HIV-2 activation (Petry *et al.*, 1995). The as yet undescribed mechanisms may be related to activation of innate responses induced by infection with a non-adapted but related virus. A recent set of similar experiments have demonstrated that $SIV_{rcm}$, which has limited pathogenicity in rhesus macaques, affords partial protection against superinfection with pathogenic $SIV_{mac}251$ by undefined mechanisms (Ling *et al.*, 2008). Non-human primate coinfection studies that utilize a non-adapted but related lentivirus to interfere with virulent disease induced by a species-adapted pathogenic virus are similar to studies performed in domestic cats infected with puma derived FIV as described below. Collectively these experiments provide important model systems for the study of innate immune mechanisms that restrict lentiviral infections.

## FIV lentiviral pathogenesis models

Several studies have demonstrated that intentional exposure of domestic cats to two strains of domestic-cat derived FIVs (typically with varying pathogenicity) results in coinfection. Dynamics of infection have suggested that the primary infection tends to become the dominant isolate after weeks to years of coinfection, even if the second virus would have typically been considered the more virulent, replicatively competent virus (Okada *et al.*, 1994; Pistello *et al.*, 1999). Recombination has also been demonstrated as a consequence of experimental FIV coinfection (Kann *et al.*, 2007). One study reported delayed infection of bone marrow and differential neutralizing antibody titres during coinfections that occurred 30 weeks after primary exposure (Okada *et al.*, 1994). A second report determined that exposure to a virulent FIV following a 7 month infection with an attenuated strain resulted in an eventual decrease in viral loads of the virulent strain, improved clinical outcomes, and altered viral distribution profiles in single vs. coinfected animals. These significant differences were not noted until two years after coinfections were established (Pistello *et al.*, 1999).

An FIV derived from pumas (PLV1695) has been shown to establish a persistent but asymptomatic infection in domestic cats, despite the fact that it is genetically highly divergent from domestic cat strains (VandeWoude *et al.*, 2002; Terwee *et al.*, 2005). Virus titres peaked at approximately 3 weeks post infection and declined slowly over an 8-month period. Adaptive immune responses were not detected until after peak viral load, and immunosuppressive doses of corticosteroids did not result in rebound viraemia (Terwee *et al.*, 2005). Analysis of full-length proviral PLV genome sequences recovered from circulating PBMC at various times post infection revealed an increased frequency of G to A substitutions and numerous defective genomes following replication in a domestic cat host. These findings are consistent with editing of the viral genome by a host cytidine deaminase (Poss *et al.*, 2006). Five sites in $FIV_{pco}$ RT-*pol* were shown to be under positive selection; also significantly, virus recovered from cats that contained one of these selected sites in RT had measurable recombination rates and phylogenetic evidence of recombination (Poss *et al.*, 2007). When cats infected with PLV-1695 were challenged with a virulent strain of FIV ($FIV_{C\text{-}36}$), coinfection resulted; however, CD4 cell depletion was prevented, and significant differences were noted in cytokine profiles and immunocytes over time (Terwee *et al.*, 2008). By 24 days after $FIV_{C\text{-}36}$ infection, the dual infected cats had a unique immunological profile compared to the other three groups, which was characterized by elevated levels of CD8, CD25, and FAS expressing cells, increased numbers of lymphocytes and neutrophils and elevated IL4 and IFN$\gamma$ (Roy *et al.*, 2009).

These reports demonstrate that innate host immune defences such as cytidine deaminase can interfere with cross-species transmission of host-adapted lentiviruses. Concurrently, viral recombination rates can be enhanced during cross-species infection, potentially increasing the emergence of virus with increased reproductive fitness, particularly if the host is already infected with endemic lentiviruses. Finally, coinfection with distinct strains of lentiviruses results in dynamic differences in innate immune responses, resulting in distinct disease phenotypes and immunological profiles during dual vs single infection.

## Summary and discussion

Despite the fact that many mechanisms exist to prevent lentiviral coinfections, it is clear that both closely and distantly related lentiviral strains can simultaneously infect a single host. This is demonstrated by naturally occurring superinfections as well as experimental coinfection studies in primates, felids, and ruminant species. Lentiviral coinfection can clearly alter dynamics of primary infection, either to the benefit or detriment of the host; adaptive immune responses do not appear to limit lentiviral coinfections, though infection with HIV-2 has demonstrated some epidemiological evidence of protection against HIV-1 coinfections, and a puma-derived FIV provided partial protection against virulent FIV. Multiple viral transmissions are enhanced by situations that result in close association of infected individuals, resulting in opportunities for emergence of new strains capable of infecting primary hosts or a new species.

Due to the nature of lentiviral replication, viral recombination is a common outcome in natural and experimental coinfections, and may be enhanced when a lentivirus infects a new host. Lentiviral evolution and expansion into new species seems to be at least partially attributable to the fact that coinfection can occur and result in successful propagation of new viruses via recombination events. Lentivirus populations are therefore dynamic, and viral extinction and recolonization events are probably the rule, not the exception.

### References

Afkhami-Goli, A., Liu, S.H., Zhu, Y., Antony, J.M., Arab, H., and Power, C. (2009). Dual lentivirus infection potentiates neuroinflammation and neurodegeneration: viral copassage enhances neurovirulence. J. Neurovirol. *15*, 139–152.

Aghokeng, A.F., Bailes, E., Loul, S., Courgnaud, V., Mpoudi-Ngolle, E., Sharp, P.M., Delaporte, E., and

Peeters, M. (2007). Full-length sequence analysis of SIVmus in wild populations of mustached monkeys (*Cercopithecus cephus*) from Cameroon provides evidence for two co-circulating SIVmus lineages. Virology *360*, 407–418.

Amacker, M., Hottiger, M., and Hubscher, U. (1995). Feline immunodeficiency virus reverse transcriptase: expression, functional characterization, and reconstitution of the 66- and 51-kilodalton subunits. J. Virol. *69*, 6273–6279.

Andresdottir, V., Skraban, R., Matthiasdottir, S., Lutley, R., Agnarsdottir, G., and Thorsteinsdottir, H. (2002). Selection of antigenic variants in maedi-visna virus infection. J. Gen. Virol. *83*, 2543–2551.

Antunes, A., Troyer, J.L., Roelke, M.E., Pecon-Slattery, J., Packer, C., Winterbach, C., Winterbach, H., Hemson, G., Frank, L., Stander, P., *et al.* (2008). The evolutionary dynamics of the lion Panthera leo revealed by host and viral population genomics. PLoS Genet *4*, e1000251.

Apetrei, C., Robertson, D.L., and Marx, P.A. (2004). The history of SIVS and AIDS: epidemiology, phylogeny and biology of isolates from naturally SIV infected non-human primates (NHP) in Africa. Front Biosci *9*, 225–254.

Apetrei, C., Kaur, A., Lerche, N.W., Metzger, M., Pandrea, I., Hardcastle, J., Falkenstein, S., Bohm, R., Koehler, J., Traina-Dorge, V., *et al.* (2005). Molecular epidemiology of simian immunodeficiency virus SIVsm in U.S. primate centers unravels the origin of SIVmac and SIVstm. J. Virol. *79*, 8991–9005.

Apetrei, C., Lerche, N.W., Pandrea, I., Gormus, B., Silvestri, G., Kaur, A., Robertson, D.L., Hardcastle, J., Lackner, A.A., and Marx, P.A. (2006). Kuru experiments triggered the emergence of pathogenic SIVmac. Aids *20*, 317–321.

Bachmann, M.H., Mathiason-Dubard, C., Learn, G.H., Rodrigo, A.G., Sodora, D.L., Mazzetti, P., Hoover, E.A., and Mullins, J.I. (1997). Genetic diversity of feline immunodeficiency virus: dual infection, recombination, and distinct evolutionary rates among envelope sequence clades. J. Virol. *71*, 4241–4253.

Barboni, P., Thompson, I., Brownlie, J., Hartaningsih, N., and Collins, M.E. (2001). Evidence for the presence of two bovine lentiviruses in the cattle population of Bali. Vet. Microbiol. *80*, 313–327.

Berry, N., Stebbings, R., Ferguson, D., Ham, C., Alden, J., Brown, S., Jenkins, A., Lines, J., Duffy, L., Davis, L., *et al.* (2008). Resistance to superinfection by a vigorously replicating, uncloned stock of simian immunodeficiency virus (SIVmac251) stimulates replication of a live attenuated virus vaccine (SIVmacC8). J. Gen. Virol. *89*, 2240–2251.

Biek, R., Rodrigo, A.G., Holley, D., Drummond, A., Anderson, C.R., Jr, Ross, H.A., and Poss, M. (2003). Epidemiology, genetic diversity, and evolution of endemic feline immunodeficiency virus in a population of wild cougars. J. Virol. *77*, 9578–9589.

Biek, R., Drummond, A.J., and Poss, M. (2006). A virus reveals population structure and recent demographic history of its carnivore host. Science *311*, 538–541.

Blackard, J.T., Cohen, D.E., and Mayer, K.H. (2002). Human immunodeficiency virus superinfection and recombination: current state of knowledge and potential clinical consequences. Clin. Infect. Dis. *34*, 1108–1114.

Blish, C.A., Dogan, O.C., Derby, N.R., Nguyen, M.A., Chohan, B., Richardson, B.A., and Overbaugh, J. (2008). Human immunodeficiency virus type 1 superinfection occurs despite relatively robust neutralizing antibody responses. J. Virol. *82*, 12094–12103.

Bruen, T.C., and Poss, M. (2007). Recombination in feline immunodeficiency virus genomes from naturally infected cougars. Virology *364*, 362–370.

Carpenter, M.A., Brown, E.W., MacDonald, D.W., and O'Brien, S.J. (1998). Phylogeographic patterns of feline immunodeficiency virus genetic diversity in the domestic cat. Virology *251*, 234–243.

Chen, Z., Telfier, P., Gettie, A., Reed, P., Zhang, L., Ho, D.D., and Marx, P.A. (1996). Genetic characterization of new West African simian immunodeficiency virus SIVsm: geographic clustering of household-derived SIV strains with human immunodeficiency virus type 2 subtypes and genetically diverse viruses from a single feral sooty mangabey troop. J. Virol. *70*, 3617–3627.

Coffin, J.M. (1979). Structure, replication, and recombination of retrovirus genomes: some unifying hypotheses. J. Gen. Virol. *42*, 1–26.

Cranage, M.P., Whatmore, A.M., Sharpe, S.A., Cook, N., Polyanskaya, N., Leech, S., Smith, J.D., Rud, E.W., Dennis, M.J., and Hall, G.A. (1997). Macaques infected with live attenuated SIVmac are protected against superinfection via the rectal mucosa. Virology *229*, 143–154.

Cranage, M.P., Sharpe, S.A., Whatmore, A.M., Polyanskaya, N., Norley, S., Cook, N., Leech, S., Dennis, M.J., and Hall, G.A. (1998). *In vivo* resistance to simian immunodeficiency virus superinfection depends on attenuated virus dose. J. Gen. Virol. *79*, 1935–1944.

Damond, F., Apetrei, C., Robertson, D.L., Souquiere, S., Lepretre, A., Matheron, S., Plantier, J.C., Brun-Vezinet, F., and Simon, F. (2001). Variability of human immunodeficiency virus type 2 (hiv-2) infecting patients living in france. Virology *280*, 19–30.

Desport, M., Stewart, M.E., Sheridan, C.A., Ditcham, W.G., Setiyaningsih, S., Tenaya, W.M., Hartaningsih, N., and Wilcox, G.E. (2005). Recombinant Jembrana disease virus gag proteins identify several different antigenic domains but do not facilitate serological differentiation of JDV and nonpathogenic bovine lentiviruses. J. Virol. Methods *124*, 135–142.

Dunn, C.S., Hurtrel, B., Beyer, C., Gloeckler, L., Ledger, T.N., Moog, C., Kieny, M.P., Mehtali, M., Schmitt, D., Gut, J.P., *et al.* (1997). Protection of SIVmac-infected macaque monkeys against superinfection by a simian immunodeficiency virus expressing envelope glycoproteins of HIV type 1. AIDS Res. Hum. Retroviruses *13*, 913–922.

Dykhuizen, M., Mitchen, J.L., Montefiori, D.C., Thomson, J., Acker, L., Lardy, H., and Pauza, C.D. (1998). Determinants of disease in the simian immunodeficiency virus-infected rhesus macaque: characterizing animals with low antibody responses and rapid progression. J. Gen. Virol. 79, 2461–2467.

Hayward, J.J., Taylor, J., and Rodrigo, A.G. (2007). Phylogenetic analysis of feline immunodeficiency virus in feral and companion domestic cats of New Zealand. J. Virol. *81*, 2999–3004.

Hayward, J.J., and Rodrigo, A.G. (2008). Recombination in feline immunodeficiency virus from feral and companion domestic cats. Virol J 5, 76.

Hu, W.S., and Temin, H.M. (1990a). Retroviral recombination and reverse transcription. Science *250*, 1227–1233.

Hu, W.S., and Temin, H.M. (1990b). Genetic consequences of packaging two RNA genomes in one retroviral particle: pseudodiploidy and high rate of genetic recombination. Proc. Natl. Acad. Sci. U.S.A. *87*, 1556–1560.

Jolly, P.E., and Narayan, O. (1989). Evidence for interference, coinfections, and intertypic virus enhancement of infection by ovine-caprine lentiviruses. J. Virol. *63*, 4682–4688.

Junghans, R.P., Boone, L.R., and Skalka, A.M. (1982). Retroviral DNA H structures: displacement-assimilation model of recombination. Cell *30*, 53–62.

Kann, R., Seddon, J., Kyaw-Tanner, M., and Meers, J. (2007). Co-infection with different subtypes of feline immunodeficiency virus can complicate subtype assignment by phylogenetic analysis. Arch. Virol. *152*, 1187–1193.

Kim, E.Y., Busch, M., Abel, K., Fritts, L., Bustamante, P., Stanton, J., Lu, D., Wu, S., Glowczwskie, J., Rourke, T., *et al.* (2005). Retroviral recombination *in vivo*: viral replication patterns and genetic structure of simian immunodeficiency virus (SIV) populations in rhesus macaques after simultaneous or sequential intravaginal inoculation with SIVmac239Deltavpx/Deltavpr and SIVmac239Deltanef. J. Virol. *79*, 4886–4895.

Kyaw-Tanner, M.T., and Robinson, W.F. (1996). Quasispecies and naturally occurring superinfection in feline immunodeficiency virus infection. Arch. Virol. *141*, 1703–1713.

Leitner, T., Dazza, M.C., Ekwalanga, M., Apetrei, C., and Saragosti, S. (2007). Sequence diversity among chimpanzee simian immunodeficiency viruses (SIVcpz) suggests that SIVcpzPts was derived from SIVcpzPtt through additional recombination events. AIDS Res. Hum. Retroviruses *23*, 1114–1118.

Levy, J.A. (2003). Is HIV superinfection worrisome? Lancet *361*, 98–99.

Ling, B., Veazey, R.S., and Marx, P.A. (2008). Nonpathogenic CCR2-tropic SIVrcm after serial passage and its effect on SIVmac infection of Indian rhesus macaques. Virology *379*, 38–44.

McCutchan, F.E. (2006). Global epidemiology of HIV. J. Med. Virol. *78* Suppl. *1*, S7–S12.

McEwan, W.A., McMonagle, E.L., Logan, N., Serra, R.C., Kat, P., Vandewoude, S., Hosie, M.J., and Willett, B.J. (2008). Genetically divergent strains of feline immunodeficiency virus from the domestic cat (Felis catus) and the African lion (Panthera leo) share usage of CD134 and CXCR4 as entry receptors. J. Virol. *82*, 10953–10958.

Nethe, M., Berkhout, B., and van der Kuyl, A.C. (2005). Retroviral superinfection resistance. Retrovirology *2*, 52.

Okada, S., Pu, R., Young, E., Stoffs, W.V., and Yamamoto, J.K. (1994). Superinfection of cats with feline immunodeficiency virus subtypes A and B. AIDS Res. Hum. Retroviruses *10*, 1739–1746.

Pecon-Slattery, J., McCracken, C.L., Troyer, J.L., VandeWoude, S., Roelke, M., Sondgeroth, K., Winterbach, C., Winterbach, H., and O'Brien, S.J. (2008). Genomic organization, sequence divergence, and recombination of feline immunodeficiency virus from lions in the wild. BMC Genomics *9*, 66.

Petry, H., Dittmer, U., Stahl-Hennig, C., Coulibaly, C., Makoschey, B., Fuchs, D., Wachter, H., Tolle, T., Morys-Wortmann, C., Kaup, F.J., and *et al.* (1995). Reactivation of human immunodeficiency virus type 2 in macaques after simian immunodeficiency virus SIVmac superinfection. J. Virol. *69*, 1564–1574.

Pisoni, G., Bertoni, G., Puricelli, M., Maccalli, M., and Moroni, P. (2007). Demonstration of coinfection with and recombination by caprine arthritis-encephalitis virus and maedi-visna virus in naturally infected goats. J. Virol. *81*, 4948–4955.

Pistello, M., Matteucci, D., Cammarota, G., Mazzetti, P., Giannecchini, S., Del Mauro, D., Macchi, S., Zaccaro, L., and Bendinelli, M. (1999). Kinetics of replication of a partially attenuated virus and of the challenge virus during a three-year intersubtype feline immunodeficiency virus superinfection experiment in cats. J. Virol. *73*, 1518–1527.

Poss, M., Idoine, A., Ross, H.A., Terwee, J.A., Vandewoude, S., and Rodrigo, A. (2007). Recombination in feline lentiviral genomes during experimental cross-species infection. Virology *359*, 146–151.

Putkonen, P., Walther, L., Zhang, Y.J., Li, S.L., Nilsson, C., Albert, J., Biberfeld, P., Thorstensson, R., and Biberfeld, G. (1995). Long-term protection against SIV-induced disease in macaques vaccinated with a live attenuated HIV-2 vaccine. Nat. Med. *1*, 914–918.

Ravazzolo, A.P., Reischak, D., Peterhans, E., and Zanoni, R. (2001). Phylogenetic analysis of small ruminant lentiviruses from Southern Brazil. Virus Res. *79*, 117–123.

Reggeti, F., and Bienzle, D. (2004). Feline immunodeficiency virus subtypes A, B and C and intersubtype recombinants in Ontario, Canada. J. Gen. Virol. *85*, 1843–1852.

Rolland, M., Mooney, J., Valas, S., Perrin, G., and Mamoun, R.Z. (2002). Characterisation of an Irish caprine lentivirus strain--SRLV phylogeny revisited. Virus Res. *85*, 29–39.

Roy, S., Lavine, J., Chiaromonte, F., Terwee, J., VandeWoude, S., Bjornstad, O., and Poss, M. (2009). Multivariate statistical analyses demonstrate unique host immune responses to single and dual lentiviral infection. PLoS ONE *4*, e7359.

Sarr, A.D., Sankale, J.L., Hamel, D.J., Travers, K.U., Gueye-Ndiaye, A., Essex, M., Mboup, S., and Kanki, P.J. (2000). Interaction with human immunodeficiency virus (HIV) type 2 predicts HIV type 1 genotype. Virology *268*, 402–410.

Schwiebert, R., and Fultz, P.N. (1994). Immune activation and viral burden in acute disease induced by simian immunodeficiency virus SIVsmmPBj14: correlation between *in vitro* and *in vivo* events. J. Virol. *68*, 5538–5547.

Sernicola, L., Corrias, F., Koanga-Mogtomo, M.L., Baroncelli, S., Di Fabio, S., Maggiorella, M.T., Belli, R., Michelini, Z., Macchia, I., Cesolini, A., *et al.* (1999). Long-lasting protection by live attenuated simian immunodeficiency virus in cynomolgus monkeys: no detection of reactivation after stimulation with a recall antigen. Virology *256*, 291–302.

Sharpe, S.A., Cope, A., Dowall, S., Berry, N., Ham, C., Heeney, J.L., Hopkins, D., Easterbrook, L., Dennis, M., Almond, N., and Cranage, M. (2004). Macaques infected long-term with attenuated simian immunodeficiency virus (SIVmac) remain resistant to wild-type challenge, despite declining cytotoxic T-lymphocyte responses to an immunodominant epitope. J. Gen. Virol. *85*, 2591–2602.

Smith, D.M., Richman, D.D., and Little, S.J. (2005). HIV superinfection. J. Infect. Dis. *192*, 438–444.

Stahl-Hennig, C., Dittmer, U., Nisslein, T., Pekrun, K., Petry, H., Jurkiewicz, E., Fuchs, D., Wachter, H., Rud, E.W., and Hunsmann, G. (1996). Attenuated SIV imparts immunity to challenge with pathogenic spleen-derived SIV but cannot prevent repair of the nef deletion. Immunol. Lett. *51*, 129–135.

Stebbings, R.J., Almond, N.M., Stott, E.J., Berry, N., Wade-Evans, A.M., Hull, R., Lines, J., Silvera, P., Sangster, R., Corcoran, T., *et al.* (2002). Mechanisms of protection induced by attenuated simian immunodeficiency virus. Virology *296*, 338–353.

Stephens, E.B., Joag, S.V., Atkinson, B., Sahni, M., Li, Z., Foresman, L., Adany, I., and Narayan, O. (1997). Infected macaques that controlled replication of SIVmac or nonpathogenic SHIV developed sterilizing resistance against pathogenic SHIV(KU-1). Virology *234*, 328–339.

Terwee, J.A., Yactor, J.K., Sondgeroth, K.S., and Vandewoude, S. (2005). Puma lentivirus is controlled in domestic cats after mucosal exposure in the absence of conventional indicators of immunity. J. Virol. *79*, 2797–2806.

Terwee, J.A., Carlson, J.K., Sprague, W.S., Sondgeroth, K.S., Shropshire, S.B., Troyer, J.L., and VandeWoude, S. (2008). Prevention of immunodeficiency virus induced CD4+ T-cell depletion by prior infection with a non-pathogenic virus. Virology *377*, 63–70.

Titti, F., Sernicola, L., Geraci, A., Panzini, G., Di Fabio, S., Belli, R., Monardo, F., Borsetti, A., Maggiorella, M.T., Koanga-Mogtomo, M., *et al.* (1997). Live attenuated simian immunodeficiency virus prevents super-infection by cloned SIVmac251 in cynomolgus monkeys. J. Gen. Virol. *78 (Pt 10)*, 2529–2539.

Travers, K., Mboup, S., Marlink, R., Gueye-Nidaye, A., Siby, T., Thior, I., Traore, I., Dieng-Sarr, A., Sankale, J.L., Mullins, C., and *et al.* (1995). Natural protection against HIV-1 infection provided by HIV-2. Science *268*, 1612–1615.

Troyer, J.L., Pecon-Slattery, J., Roelke, M.E., Black, L., Packer, C., and O'Brien, S.J. (2004). Patterns of feline immunodeficiency virus multiple infection and genome divergence in a free-ranging population of African lions. J. Virol. *78*, 3777–3791.

Troyer, J.L., Vandewoude, S., Pecon-Slattery, J., McIntosh, C., Franklin, S., Antunes, A., Johnson, W., and O'Brien, S.J. (2008). FIV cross-species transmission: an evolutionary prospective. Vet. Immunol. Immunopathol. *123*, 159–166.

VandeWoude, S., Hageman, C.A., O'Brien, S.J., and Hoover, E.A. (2002). Nonpathogenic lion and puma lentiviruses impart resistance to superinfection by virulent feline immunodeficiency virus. J. Acquir. Immune Defic. Syndr. *29*, 1–10.

VandeWoude, S., and Apetrei, C. (2006). Going wild: lessons from naturally occurring T-lymphotropic lentiviruses. Clin. Microbiol. Rev. *19*, 728–762.

Whatmore, A.M., Cook, N., Hall, G.A., Sharpe, S., Rud, E.W., and Cranage, M.P. (1995). Repair and evolution of nef *in vivo* modulates simian immunodeficiency virus virulence. J. Virol. *69*, 5117–5123.

Yamaguchi, J., Devare, S.G., and Brennan, C.A. (2000). Identification of a new HIV-2 subtype based on phylogenetic analysis of full-length genomic sequence. AIDS Res. Hum. Retroviruses *16*, 925–930.

Zanoni, R.G., Nauta, I.M., Pauli, U., and Peterhans, E. (1991). Expression in Escherichia coli and sequencing of the coding region for the capsid protein of Dutch maedi-visna virus strain ZZV 1050: application of recombinant protein in enzyme-linked immunosorbent assay for the detection of caprine and ovine lentiviruses. J. Clin. Microbiol. *29*, 1290–1294.

# Index

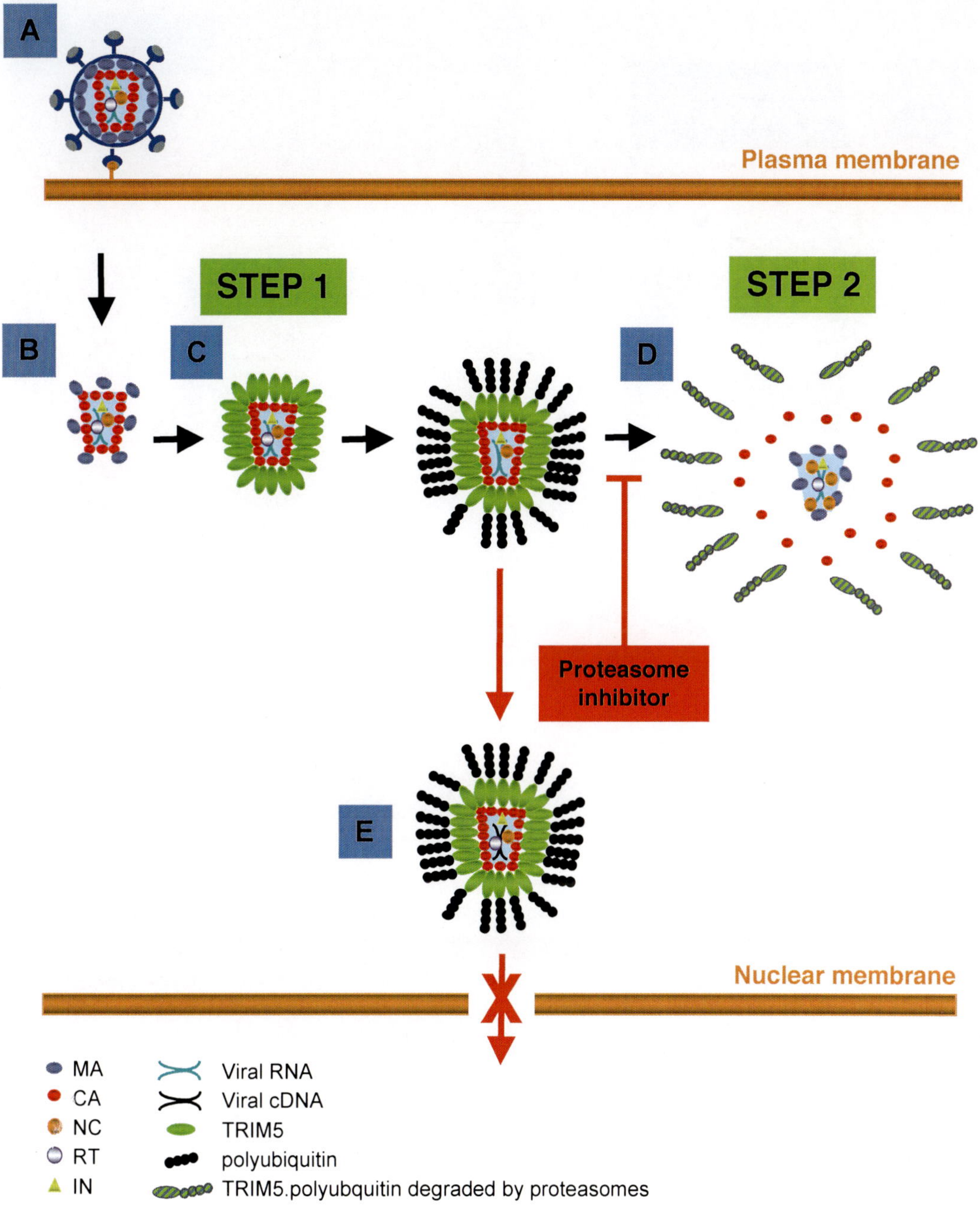


**Plate 4.2** Potential model of TRIM5 restriction of retroviruses. For details, see page 120.

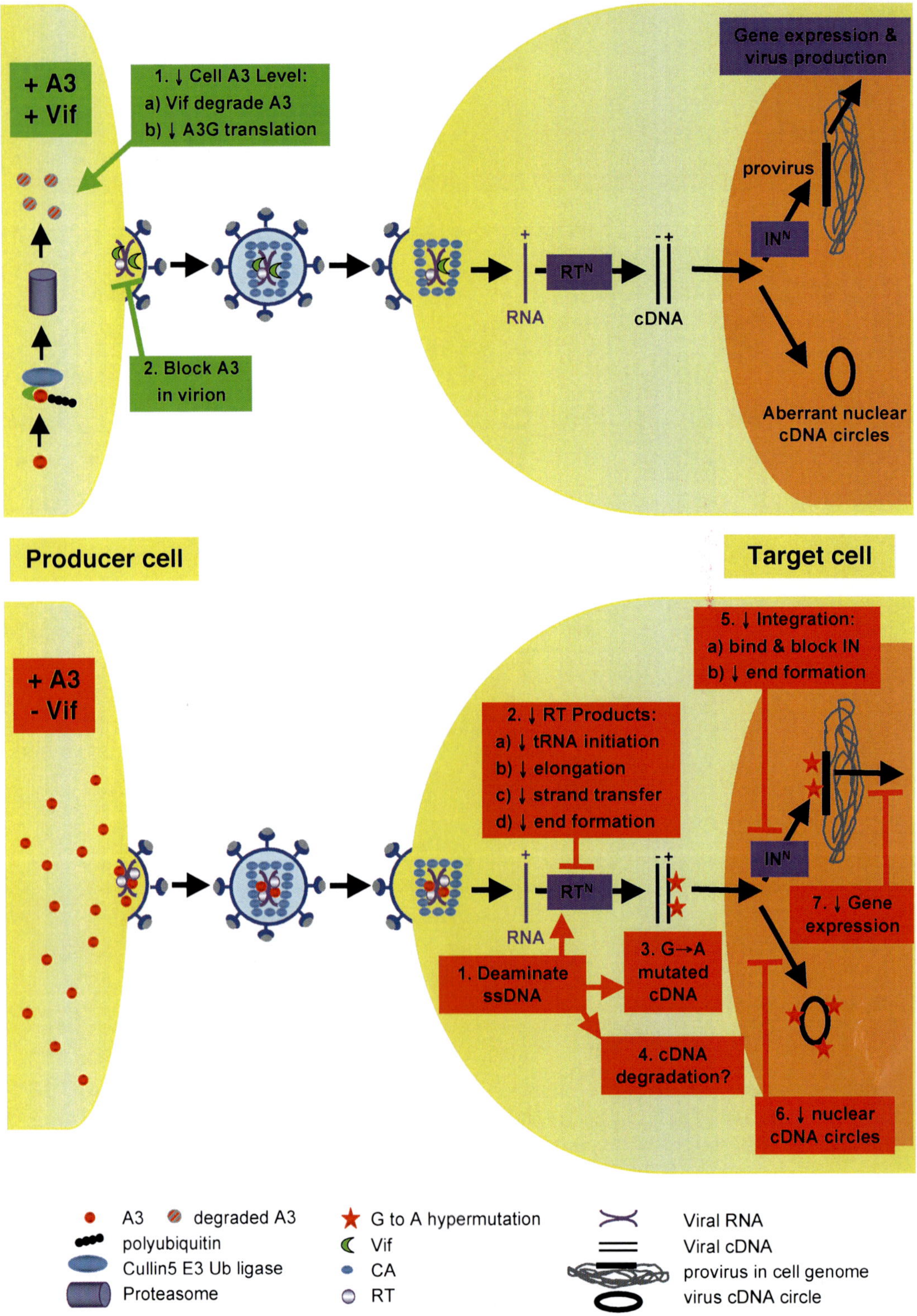


**Plate 4.3** Virion-associated huA3G restricts HIV-1 and HIV-1 avoids huA3G via multiple mechanisms. For details, see page 128.